101 839 409 5

AF616100

Handbook of Frozen Food Processing and Packaging

FOOD SCIENCE AND TECHNOLOGY

A Series of Monographs, Textbooks, and Reference Books

76. Food Chemistry: Third Edition, *edited by Owen R. Fennema*
77. Handbook of Food Analysis: Volumes 1 and 2, *edited by Leo M. L. Nollet*
78. Computerized Control Systems in the Food Industry, *edited by Gauri S. Mittal*
79. Techniques for Analyzing Food Aroma, *edited by Ray Marsili*
80. Food Proteins and Their Applications, *edited by Srinivasan Damodaran and Alain Paraf*
81. Food Emulsions: Third Edition, Revised and Expanded, *edited by Stig E. Friberg and Kåre Larsson*
82. Nonthermal Preservation of Foods, *Gustavo V. Barbosa-Cánovas, Usha R. Pothakamury, Enrique Palou, and Barry G. Swanson*
83. Milk and Dairy Product Technology, *Edgar Spreer*
84. Applied Dairy Microbiology, *edited by Elmer H. Marth and James L. Steele*
85. Lactic Acid Bacteria: Microbiology and Functional Aspects, Second Edition, Revised and Expanded, *edited by Seppo Salminen and Atte von Wright*
86. Handbook of Vegetable Science and Technology: Production, Composition, Storage, and Processing, *edited by D. K. Salunkhe and S. S. Kadam*
87. Polysaccharide Association Structures in Food, *edited by Reginald H. Walter*
88. Food Lipids: Chemistry, Nutrition, and Biotechnology, *edited by Casimir C. Akoh and David B. Min*
89. Spice Science and Technology, *Kenji Hirasa and Mitsuo Takemasa*
90. Dairy Technology: Principles of Milk Properties and Processes, *P. Walstra, T. J. Geurts, A. Noomen, A. Jellema, and M. A. J. S. van Boekel*
91. Coloring of Food, Drugs, and Cosmetics, *Gisbert Otterstätter*
92. *Listeria*, Listeriosis, and Food Safety: Second Edition, Revised and Expanded, *edited by Elliot T. Ryser and Elmer H. Marth*
93. Complex Carbohydrates in Foods, *edited by Susan Sungsoo Cho, Leon Prosky, and Mark Dreher*
94. Handbook of Food Preservation, *edited by M. Shafiur Rahman*
95. International Food Safety Handbook: Science, International Regulation, and Control, *edited by Kees van der Heijden, Maged Younes, Lawrence Fishbein, and Sanford Miller*
96. Fatty Acids in Foods and Their Health Implications: Second Edition, Revised and Expanded, *edited by Ching Kuang Chow*

97. Seafood Enzymes: Utilization and Influence on Postharvest Seafood Quality, *edited by Norman F. Haard and Benjamin K. Simpson*
98. Safe Handling of Foods, *edited by Jeffrey M. Farber and Ewen C. D. Todd*
99. Handbook of Cereal Science and Technology: Second Edition, Revised and Expanded, *edited by Karel Kulp and Joseph G. Ponte, Jr.*
100. Food Analysis by HPLC: Second Edition, Revised and Expanded, *edited by Leo M. L. Nollet*
101. Surimi and Surimi Seafood, *edited by Jae W. Park*
102. Drug Residues in Foods: Pharmacology, Food Safety, and Analysis, *Nickos A. Botsoglou and Dimitrios J. Fletouris*
103. Seafood and Freshwater Toxins: Pharmacology, Physiology, and Detection, *edited by Luis M. Botana*
104. Handbook of Nutrition and Diet, *Babasaheb B. Desai*
105. Nondestructive Food Evaluation: Techniques to Analyze Properties and Quality, *edited by Sundaram Gunasekaran*
106. Green Tea: Health Benefits and Applications, *Yukihiko Hara*
107. Food Processing Operations Modeling: Design and Analysis, *edited by Joseph Irudayaraj*
108. Wine Microbiology: Science and Technology, *Claudio Delfini and Joseph V. Formica*
109. Handbook of Microwave Technology for Food Applications, *edited by Ashim K. Datta and Ramaswamy C. Anantheswaran*
110. Applied Dairy Microbiology: Second Edition, Revised and Expanded, *edited by Elmer H. Marth and James L. Steele*
111. Transport Properties of Foods, *George D. Saravacos and Zacharias B. Maroulis*
112. Alternative Sweeteners: Third Edition, Revised and Expanded, *edited by Lyn O'Brien Nabors*
113. Handbook of Dietary Fiber, *edited by Susan Sungsoo Cho and Mark L. Dreher*
114. Control of Foodborne Microorganisms, *edited by Vijay K. Juneja and John N. Sofos*
115. Flavor, Fragrance, and Odor Analysis, *edited by Ray Marsili*
116. Food Additives: Second Edition, Revised and Expanded, *edited by A. Larry Branen, P. Michael Davidson, Seppo Salminen, and John H. Thorngate, III*
117. Food Lipids: Chemistry, Nutrition, and Biotechnology: Second Edition, Revised and Expanded, *edited by Casimir C. Akoh and David B. Min*
118. Food Protein Analysis: Quantitative Effects on Processing, *R. K. Owusu-Apenten*
119. Handbook of Food Toxicology, *S. S. Deshpande*
120. Food Plant Sanitation, *edited by Y. H. Hui, Bernard L. Bruinsma, J. Richard Gorham, Wai-Kit Nip, Phillip S. Tong, and Phil Ventresca*
121. Physical Chemistry of Foods, *Pieter Walstra*
122. Handbook of Food Enzymology, *edited by John R. Whitaker, Alphons G. J. Voragen, and Dominic W. S. Wong*
123. Postharvest Physiology and Pathology of Vegetables: Second Edition, Revised and Expanded, *edited by Jerry A. Bartz and Jeffrey K. Brecht*
124. Characterization of Cereals and Flours: Properties, Analysis, and Applications, *edited by Gönül Kaletunç and Kenneth J. Breslauer*
125. International Handbook of Foodborne Pathogens, *edited by Marianne D. Miliotis and Jeffrey W. Bier*
126. Food Process Design, *Zacharias B. Maroulis and George D. Saravacos*
127. Handbook of Dough Fermentations, *edited by Karel Kulp and Klaus Lorenz*
128. Extraction Optimization in Food Engineering, *edited by Constantina Tzia and George Liadakis*
129. Physical Properties of Food Preservation: Second Edition, Revised and Expanded, *Marcus Karel and Daryl B. Lund*
130. Handbook of Vegetable Preservation and Processing, *edited by Y. H. Hui, Sue Ghazala, Dee M. Graham, K. D. Murrell, and Wai-Kit Nip*
131. Handbook of Flavor Characterization: Sensory Analysis, Chemistry, and Physiology, *edited by Kathryn Deibler and Jeannine Delwiche*
132. Food Emulsions: Fourth Edition, Revised and Expanded, *edited by Stig E. Friberg, Kare Larsson, and Johan Sjoblom*
133. Handbook of Frozen Foods, *edited by Y. H. Hui, Paul Cornillon, Isabel Guerrero Legarret, Miang H. Lim, K. D. Murrell, and Wai-Kit Nip*
134. Handbook of Food and Beverage Fermentation Technology, *edited by Y. H. Hui, Lisbeth Meunier-Goddik, Ase Solvejg Hansen, Jytte Josephsen, Wai-Kit Nip, Peggy S. Stanfield, and Fidel Toldrá*

135. Genetic Variation in Taste Sensitivity, *edited by John Prescott and Beverly J. Tepper*
136. Industrialization of Indigenous Fermented Foods: Second Edition, Revised and Expanded, *edited by Keith H. Steinkraus*
137. Vitamin E: Food Chemistry, Composition, and Analysis, *Ronald Eitenmiller and Junsoo Lee*
138. Handbook of Food Analysis: Second Edition, Revised and Expanded, Volumes 1, 2, and 3, *edited by Leo M. L. Nollet*
139. Lactic Acid Bacteria: Microbiological and Functional Aspects: Third Edition, Revised and Expanded, *edited by Seppo Salminen, Atte von Wright, and Arthur Ouwehand*
140. Fat Crystal Networks, *Alejandro G. Marangoni*
141. Novel Food Processing Technologies, *edited by Gustavo V. Barbosa-Cánovas, M. Soledad Tapia, and M. Pilar Cano*
142. Surimi and Surimi Seafood: Second Edition, *edited by Jae W. Park*
143. Food Plant Design, *Antonio Lopez-Gomez; Gustavo V. Barbosa-Cánovas*
144. Engineering Properties of Foods: Third Edition, *edited by M. A. Rao, Syed S.H. Rizvi, and Ashim K. Datta*
145. Antimicrobials in Food: Third Edition, *edited by P. Michael Davidson, John N. Sofos, and A. L. Branen*
146. Encapsulated and Powdered Foods, *edited by Charles Onwulata*
147. Dairy Science and Technology: Second Edition, *Pieter Walstra, Jan T. M. Wouters and Tom J. Geurts*
148. Food Biotechnology, Second Edition, *edited by Kalidas Shetty, Gopinadhan Paliyath, Anthony Pometto and Robert E. Levin*
149. Handbook of Food Science, Technology, and Engineering - 4 Volume Set, *edited by Y. H. Hui*
150. Thermal Food Processing: New Technologies and Quality Issues, *edited by Da-Wen Sun*
151. Aflatoxin and Food Safety, *edited by Hamed K. Abbas*
152. Food Packaging: Principles and Practice, Second Edition, *Gordon L. Robertson*
153. Seafood Processing: Adding Value Through Quick Freezing, Retortable Packaging, and Cook-Chilling, *V. Venugopal*
154. Ingredient Interactions: Effects on Food Quality, Second Edition, *edited by Anilkumar Gaonkar and Andrew McPherson*
155. Handbook of Frozen Food Processing and Packaging, *edited by Da-Wen Sun*
156. Vitamins In Foods: Analysis, Bioavailability, and Stability, *George F. M. Ball*

Handbook of Frozen Food Processing and Packaging

edited by
Da-Wen Sun

Taylor & Francis
Taylor & Francis Group
Boca Raton London New York

A CRC title, part of the Taylor & Francis imprint, a member of the Taylor & Francis Group, the academic division of T&F Informa plc.

Published in 2006 by
CRC Press
Taylor & Francis Group
6000 Broken Sound Parkway NW, Suite 300
Boca Raton, FL 33487-2742

Printed in the United States of America on acid-free paper
10 9 8 7 6 5 4 3 2 1

International Standard Book Number-10: 1-57444-607-X (Hardcover)
International Standard Book Number-13: 978-1-57444-607-4 (Hardcover)
Library of Congress Card Number 2005049925

Library of Congress Cataloging-in-Publication Data

Handbook of frozen food packaging and processing / edited by Da-Wen Sun.
p. cm. -- (Food science and technology ; 155)
Includes bibliographical references and index.
ISBN 1-57444-607-X (alk. paper)
1. Frozen foods. 2. Frozen foods--Packaging. I. Sun, Da-Wen. II. Food science and technology (CRC Press) ; 155.

TP372.3.H36 2005
664'.02853--dc22 2005049925

Visit the Taylor & Francis Web site at
http://www.taylorandfrancis.com

and the CRC Press Web site at
http://www.crcpress.com

Preface

The frozen food industry is one of the biggest sectors in the food industry. With the development of freezing technology, there has been a steady increase in the number and variety of frozen foods. Besides convenience to consumers, the popularity of frozen food is also due to the fact that it continues to demonstrate a good food safety record, because the freezing of food can effectively reduce the activity of microorganisms and enzymes, thus retarding deterioration. In addition, crystallization of water reduces the amount of water in food items and inhibits microbial growth. Nowadays, freezing is involved in almost every food process. Therefore, the purpose of this handbook is to provide engineers and technologists working in research, development, and operations in the food industry with critical, comprehensive, and readily accessible information on the art and science of frozen foods. This book is also intended to assemble essential, authoritative, and complete references and data that can be used by researchers in universities and research institutions. Each chapter is written by an author who has both academic and professional credentials.

The contents of the book are divided into five parts: fundamentals of freezing, facilities for the cold chain, quality and safety of frozen foods, monitoring and measuring techniques for quality and safety, and packaging of frozen foods.

Part I deals with fundamental topics relating to freezing, which include physical–chemical principles during freezing process (Chapter 1), glass transition phenomena in frozen foods (Chapter 2), principles of refrigeration (Chapter 3), food microbiology (Chapter 4), thermalo–physical properties of frozen foods (Chapter 5), calculation of freezing time and freezing load (Chapter 6), mathematical modeling of the freezing process (Chapter 7), and some recent developments in the freezing process (Chapter 8).

Part II focuses on freezing-related equipment and facilities. This part begins with a chapter on industrial freezing methods and equipment (Chapter 9), then discusses storage facility and design for frozen food (Chapter 10), methods and facilities used for frozen food transportation (Chapter 11), descriptions on retail display units (Chapter 12), and those used in household refrigerators and freezers (Chapter 13). This part ends with a chapter on techniques for monitoring and control of the cold chain (Chapter 14).

Part III stresses the importance of quality, safety, and nutritional values of frozen foods. These issues for each type of frozen foods are discussed in detail in a separate chapter. The frozen foods covered in this part are meat and meat products (Chapter 15), poultry and poultry products (Chapter 16), fish, shellfish, and related products (Chapter 17), vegetables (Chapter 18), fruits (Chapter 19), dairy products (Chapter 20), ready meals (Chapter 21), bakery products (Chapter 22), and eggs and egg products (Chapter 23).

Part IV describes methods and techniques used to measure and maintain the quality and safety of frozen foods. These are objective methods such as physical measurements (Chapter 24), chemical measurements (Chapter 25), and sensory methods (Chapter 26). Testing of microorganisms causing foodborne illnesses and spoilage is presented in Chapter 27, and Chapter 28 deals with shelf-life testing and prediction for frozen foods.

Part V discusses topics related to packaging of frozen foods. These are: introduction to frozen food packaging (Chapter 29), various packaging materials for frozen foods such as polymer (Chapter 30), paper and cardboard (Chapter 31), and other packaging materials (Chapter 32). Description of packaging machinery then follows in Chapter 33, and finally the book concludes with a chapter on future developments in frozen food packaging (Chapter 34).

Preface

Editor

Born in Southern China, Professor Da-Wen Sun is an internationally recognized figure for his leadership in food engineering research and education. His main research activities include cooling, drying, and refrigeration processes and systems, quality and safety of food products, bioprocess simulation and optimization, and computer vision technology. Especially, his innovative studies on vacuum cooling of cooked meats, pizza quality inspection by computer vision, and edible films for shelf-life extension of fruit and vegetables have been widely reported in national and international media. The results of his work have been published in over 150 peer-reviewed journal papers and more than 200 conference papers.

He received a first class BSc Honours and MSc in mechanical engineering, and a PhD in chemical engineering in China before working in various universities in Europe. He became the first Chinese national to be permanently employed in an Irish University when he was appointed College Lecturer at National University of Ireland, Dublin (University College Dublin) in 1995, and was then promoted to Senior Lecturer. He is now the Professor and Director of the Food Refrigeration and Computerized Food Technology Research Group at Department of Biosystems Engineering, University College Dublin.

As a leading academic in food engineering, Dr. Sun has significantly contributed to the field of food engineering. He has trained many PhD students, who have made their own contributions to the industry and academia. He has also given lectures on advances in food engineering on a regular basis in academic institutions internationally and delivered keynote speeches at international conferences. As a recognized authority on food engineering, he has been conferred adjunct/visiting/consulting professorships from ten top universities in China including Shanghai Jiao Tong University, Zhejiang University, Harbin Institute of Technology, China Agricultural University, South China University of Technology, Southern Yangtze University, and so on. In recognition of his significant contribution to food engineering worldwide, the International Commission of Agricultural Engineering (CIGR) awarded him the CIGR Merit Award in 2000, and the Institution of Mechanical Engineers (IMechE) based in the UK named him "Food Engineer of the Year 2004."

He is a fellow of the Institution of Agricultural Engineers. He has also received numerous awards for teaching and research excellence, including the President's Research Award of University College Dublin twice. He is the Chair of CIGR Section VI on Postharvest Technology and Process Engineering, Guest Editor of *Journal of Food Engineering* and *Computers and Electronics in Agriculture*, and editorial board member for *Journal of Food Engineering*, *Journal of Food Process Engineering*, and *Czech Journal of Food Sciences*. He is also a Chartered Engineer registered in the UK Engineering Council.

Contributors

Nevin Amos
ZESPRI International Limited
Mt Maunganui South
New Zealand

Edgar Chambers IV
The Sensory Analysis Center
Kansas State University
Manhattan, Kansas

Giovanni Cortella
Dipartimento di Energetica e Macchine
Università di Udine
Udine, Italy

Philip G. Creed
Bournemouth University
Poole, Dorset, UK

Paola D'Agaro
Dipartimento di Energetica e Macchine
Università di Udine
Udine, Italy

Marilyn C. Erickson
Department of Food Science and Technology
Center for Food Safety
University of Georgia
Griffin, Georgia

Silvia Estrada-Flores
Food Science Australia
North Ryde, NSW, Australia

Martin George
Campden and Chorleywood
Food Research Association
Chipping Campden
Gloucestershire, UK

Maria C. Giannakourou
Laboratory of Food Chemistry
and Technology
School of Chemical Engineering
National Technical University
of Athens
Athens, Greece

Virginia Giannou
Laboratory of Food Chemistry and
Technology
School of Chemical Engineering
National Technical University
of Athens
Athens, Greece

C.O. Gill
Agriculture and Agri-Food Canada
Lacombe Research Centre
Lacombe, Alberta

H. Douglas Goff
Department of Food Science
University of Guelph
Guelph, Ontario

Gerrit Hasselmann
Laboratory for Packaging
Test and Research
Fraunhofer Institute Material
Flow und Logistics
Dortmund, Germany

Angela Hunt
Seafood Laboratory
Oregon State University
Astoria, Oregon

Jacek Jaczynski
Animal and Veterinary Sciences
West Virginia University
Morgantown
West Virginia

Stephen James
Food Refrigeration and Process
Engineering Research Centre (FRPERC)
University of Bristol
Langford, Bristol, UK

Stefan Kasapis
Department of Chemistry
National University of Singapore
Singapore

Laurence Ketteringham
Food Refrigeration and Process Engineering Research Centre (FRPERC)
University of Bristol
Langford, Bristol, UK

Nahed Kotrola, Ph.D.
Food and Beverage Division
Ecolab
Auburn, Alabama

John M. Krochta
Department of Food Science and Technology
University of California
Davis, California

Lih-Shiuh Lai
Department of Food Science and Biotechnology
National Chung Hsing University
Taichung, Taiwan

Alain Le Bail
UMR CNRS GEPEA (6144 SPI)
ENITIAA-Ecole des Mines de Nantes
University of Nantes
Nantes, France

Kwang Ho Lee
Korea Food and Drug Administration
Eunpyunggu
Seoul, Korea

Simon J. Lovatt
AgResearch Ltd.
Hamilton, New Zealand

Andrea Maestrelli
Istituto Sperimentale per la Valorizzazione Tecnologica dei Prodotti Agricoli (IVTPA)
Milano, Italy

Parameswarakumar Mallikarjunan
Biological Systems Engineering Department
Virginia Polytechnic Institute and State University
Blacksburg, Virginia

Rodolfo H. Mascheroni
CIDCA (CONICET–UNLP)
La Plata, Buenos Aires,
Argentina

Rajeshwar S. Matche
Food Packaging Technology Department
Central Food Technological Research Institute
Mysore, Karnataka, India

Sherry McGraw
The Sensory Analysis Center
Kansas State University
Manhattan, Kansas

Brian McKenna
Food Science Department
University College Dublin (National University of Ireland, Dublin)
Dublin, Ireland

Gauri S. Mittal
School of Engineering
University of Guelph
Guelph, Ontario

Sandra Moorhead
Food Science Department
University of Guelph
Guelph, Ontario

Mike F. North
AgResearch Ltd.
Hamilton, New Zealand

George-John E. Nychas
Department of Food Science and Technology
Laboratory of Microbiology and Biotechnology of Foods
Agricultural University of Athens
Athens, Greece

Jae W. Park
Seafood Laboratory and Department of Food Science and Technology
Oregon State University
Astoria, Oregon

Wenceslao Canet Parreño
Instituto del Frío, CSIC
Science and Technology of Vegetable Products Department
Ciudad Universitaria
Madrid, Spain

Q. Tuan Pham
School of Chemical Engineering and Industrial Chemistry
University of New South Wales
Sydney, Australia

Viviana O. Salvadori
CIDCA (CONICET–UNLP)
La Plata, Buenos Aires,
Argentina

Amalia Scannell
Department of Food Science
National University of Ireland
University College Dublin
Dublin, Ireland

Kathleen A. Smiley
The Sensory Analysis Center
Kansas State University
Manhattan, Kansas

Da-Wen Sun
Food Refrigeration and Computerised Food Technology (FRCFT) Research Group
National University of Ireland
Dublin, Ireland

David Tanner
Food Science Australia
North Ryde, NSW,
Australia

Petros S. Taoukis
Laboratory of Food Chemistry and Technology
School of Chemical Engineering
National Technical University of Athens
Athens, Greece

Danila Torreggiani
Istituto Sperimentale per la Valorizzazione Tecnologica dei Prodotti Agricoli (IVTPA)
Milano, Italy

María Dolores Alvarez Torres
Instituto del Frío, CSIC
Science and Technology of Vegetable Products Department
Ciudad Universitaria
Madrid, Spain

Constantina Tzia
Laboratory of Food Chemistry and Technology
School of Chemical Engineering
National Technical University of Athens
Athens, Greece

Lijun Wang
Department of Biological Systems Engineering
University of Nebraska-Lincoln
Lincoln, Nebraska

Curtis L. Weller
Department of Biological Systems Engineering
University of Nebraska-Lincoln
Lincoln, Nebraska

André Wötzel
Packaging Test and Research Laboratory
Fraunhofer Institute Material Flow und Logistics
Dortmund, Germany

Noemi Zaritzky
Centro de Investigación y Desarrollo en Criotecnología de Alimentos (CIDCA)
UNLP-CONICET
Departamento Ingeniería Química
Universidad Nacional de La Plata
Buenos Aires,
Argentina

Liyun Zheng
Food Refrigeration and Computerised Food Technology (FRCFT) Research Group
National University of Ireland
Dublin, Ireland

Table of Contents

Part I

Fundamentals of Freezing

1 Physical–Chemical Principles in Freezing

Noemi Zaritzky
Centro de Investigación y Desarrollo en Criotecnología de Alimentos (CIDCA)
UNLP-CONICET and Departamento Ingeniería Química, Facultad Ingeniería,
Universidad Nacional de La Plata, Argentina

CONTENTS

I. INTRODUCTION

Water is the most abundant substance on the Earth, and the major component of most foods and biological specimens Water is essential for life. In almost all living cells water is the most abundant molecule accounting for 60–90% of the mass of the cell. Water is also a very important component in foods, affecting quality attributes and shelf life stability [1]. Freezing is regarded as one of the best methods for long-term food preservation. During freezing, water is converted to ice, thus chemical reactions and microbial growth are reduced at low temperatures; this apart,, the formation of ice removes water from food systems, lowering the water activity. In this Chapter, water and ice structures, ice formation (nucleation and crystal growth), state diagrams, vitrification, freezing mechanisms in plant and animal tissues, and the physical and chemical effects of freezing will be discussed.

II. THE STRUCTURE OF WATER AND ICE

Water is a V-shaped molecule composed of two light hydrogen atoms and one relatively heavy oxygen atom. The approximately 16-fold difference in mass between the two atoms leads to its ease of rotation and allows the significant relative movements of the hydrogen nuclei. Natural water is actually a mixture of several species differing in molecular weight. There are three stable isotopes of hydrogen: ^{1}H, ^{2}H (deuterium), ^{3}H (tritium), and six of oxygen (^{14}O, ^{15}O, ^{16}O, ^{17}O, ^{18}O, ^{19}O) [2].

Oxygen has six valence electrons and each hydrogen atom has one electron [1]. Two orbitals of the oxygen atom participate in covalent σ bonding (40% partial ionic character) with the two hydrogen atoms; an electron pair is shared between each hydrogen atom that is covalently bonded to the oxygen, leaving two lone pairs of electrons on the oxygen atom (Figure 1.1).

Oxygen is more electronegative than hydrogen and attracts electron more strongly than hydrogen. As a result, an uneven geometrical distribution of charge occurs within each O—H bond of the water molecule, with oxygen bearing a partial negative charge and hydrogen a partial positive charge (Figure 1.1). The uneven distribution of charge within a polar bond leads to a dipole moment. The polar property allows water to separate polar solute molecules and to dissolve many substances. As the positive charges are constrained within the respective nuclei, the electronic environment of water is determined by the location of the outer electrons that may be found in various locations with characteristic probabilities. Figure 1.1 shows a schematic plot of the electron density around the atoms, and the water molecule structure. The average electron density around the oxygen atom is about 10 times that around the hydrogen atoms. The two zones behind the oxygen atom in Figure 1.1 represents a higher electron density with a slightly negative average charge density, leaving the two protons with a lower electron density and a slightly positive charge. As there is a natural repulsion between the unshared electron pairs and the electrons in the covalent bonds with hydrogen atoms, a three-dimensional plot of these regions with uneven distribution of charge shows that they nearly form the vertices of a regular tetrahedron around the central oxygen atom.

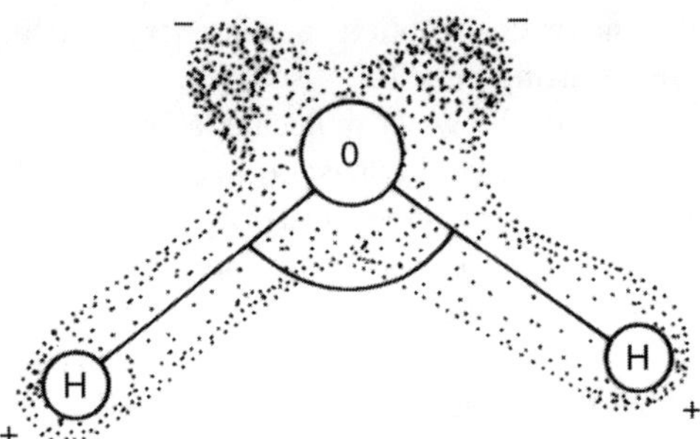

FIGURE 1.1 Schematic plot of the electron density around the atoms in the water molecule.

The water molecule is often described as having four, approximately tetrahedrally arranged, sp^3-hybridized electron pairs, two of which are associated with hydrogen atoms with two remaining lone pairs. Water molecules (H_2O) are symmetric with two mirror planes of symmetry and a twofold rotation axis. For an isolated water molecule (vapor state), the experimental values of the O—H intermolecular distance is 0.95718 Å and the H—O—H angle is 104.474° [3–5], very close to the angle between the vertices of a regular tetrahedron (109.47°).

A. Hydrogen Bonds

Water can interact strongly with other polar molecules. An important consequence of the water molecular polarity is the attraction of water molecules for one another. The attraction between one of the slightly positive hydrogen atoms of one water molecule and the slightly negative oxygen atom of another molecule produces a hydrogen bond (Figure 1.2). A water molecule can form up to four hydrogen bonds: the oxygen atom of a water molecule is the hydrogen acceptor for two hydrogen atoms and each OH group serves as hydrogen donor [2]; each water molecule is a hydrogen donor in two of these bonds and a hydrogen acceptor in the other two. In the hydrogen bond between two water molecules, the hydrogen atom remains covalently bonded to its oxygen atom (with a dissociation energy of about 492 kJ mol^{-1}). The distance between this hydrogen atom and the oxygen atom of the other molecule (1.86 Å) is about twice the length of the covalent bond (0.957 Å) (Figure 1.2). Hydrogen bonds are about 20 times weaker than covalent bonds and

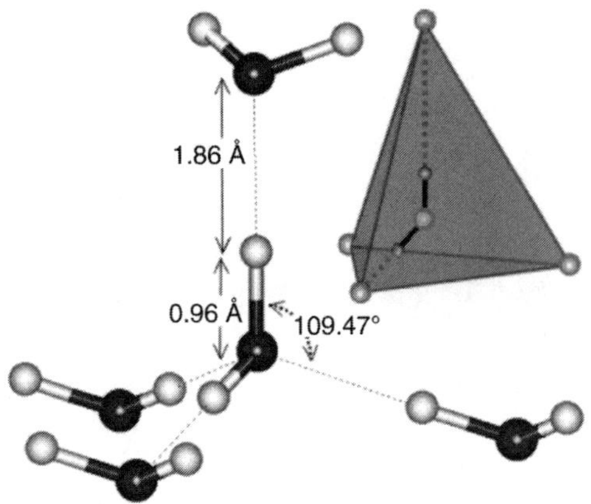

FIGURE 1.2 Hydrogen bonding of water molecules in the tetrahedral configuration of hexagonal ice. Dark spheres correspond to oxygen atoms and light spheres to hydrogen atoms.

electrostatic forces make the major contribution to this energy. In liquid water these hydrogen bonds tend to be very transient (lasting only 10^{-11} sec) [3].

With the tetrahedral geometry of the water molecule, a large array of such hydrogen bonding can be established. Water molecules in liquid phase form an infinite three-dimensional hydrogen-bonded network with localized and structured clustering. The very numerous hydrogen bonds offsets their weak and transient nature. Liquid water has a partially ordered structure in which hydrogen bonds are constantly formed and broken up. The bond angle of the isolated water molecule in vapor phase (104.5°) and the O—H intermolecular distance of 0.95718 Å are not maintained in liquid water, where O—H length is 0.991 Å, and H—O—H angle 105.5° [6]; these slightly higher values are caused by the hydrogen bonds that produces a weakening of the covalent bonding. These bond lengths and angles are likely to change, because of polarization shifts, in different hydrogen-bonded environments and when the water molecules are bound to solutes and ions.

B. Hexagonal Ice (Ice I_h)

There are 11 different forms of crystalline ice. The hexagonal form known as ice Ih is the normal form of ice in frozen food. Hexagonal ice has triple points with liquid and gaseous water (0.01°C, 612 Pa), with liquid water and ice-three (−22.0°C, 207.5 MPa) and with ice-two and ice-three (−34.7°C, 212.9 MPa) [3].

In hexagonal ice, each molecule of water participates in four hydrogen bonds although in the liquid phase, some of the weaker hydrogen bonds must be broken to allow the molecules to move around [3]; hydrogen bonding is more regular in ice than in liquid water.

The hydrogen bonds in hexagonal ice are arranged tetrahedrally around the oxygen atom of each water molecule and are held relatively static. In ice Ih, the distance to the nearest oxygen neighbor is 2.82 Å and the formed angles between the bond–bond, bond–lone pair and lone pair–lone pair electrons are all of 109.47°, typical of a tetrahedrally coordinated lattice structure (Figure 1.2). The average energy required to break each hydrogen bond in ice is estimated to be 23 kJ mol^{-1} while that to break each hydrogen bond in liquid water is less than 20 kJ mol^{-1} [2]. Water molecules in ice can only vibrate back and forth while liquid molecules can move fast enough, however, they are still attached to each other. The ability of water molecules in ice to form four hydrogen bonds and the strength of these bonds give the ice a high melting point. The geometrical regularity of the bonds contributes to the strength of the ice crystals. Hexagonal ice has an open low-density structure with a rigid lattice within empty spaces (Figure 1.3). The density of ice Ih at 0°C is 916.8 kg m^{-3} lower than the density of liquid water (999.8 kg m^{-3}) at the same temperature.

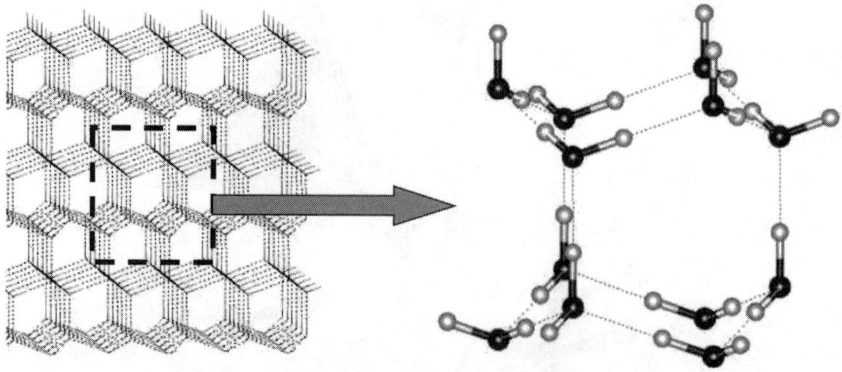

FIGURE 1.3 Structure of ice Ih: hexagonal lattice. Dark spheres correspond to oxygen atoms in the water molecules. Every water molecule is hydrogen bonded to four others.

When water freezes into ice, the macroscopic crystalline structure takes the form of a hexagonal prism (Figure 1.4). The hexagonal prism, includes two hexagonal "basal" faces and six rectangular "prism" faces; the crystallographic *c*-axis is in the vertical direction.

The water molecules stack together to form a regular crystalline lattice that has sixfold symmetry. The crystals may be considered as formed by parallel sheets lying on top of each other. The unit cell is formed by a group of four molecules (two above and two below); two and two-halves of which constitute the hexameric box where planes consist of chair-form hexamers (the two horizontal planes, opposite, forming the basal plane) or boat-form hexamers (the three vertical planes, opposite, forming prism faces) (Figure 1.3 and Figure 1.4).

The hexagonal ice crystal has unit cell dimensions of 4.511 Å along the *a*-axis and 7.351 Å along the *c*-axis. Hexagonal ice crystals may form by growing in the direction of the *c*-axis or by growing more rapidly perpendicular to the *c*-axis. Molecular forces, which operate at the molecular scale to produce the crystal lattice, can control the shape of the crystals; facet planes appear on many growing crystals because some surfaces grow more slowly than others. Water molecules can readily attach to the rough surfaces of a crystal, which grow relatively quickly. The facet planes tend to be smoother on a molecular scale; water molecules cannot easily attach to the smooth surfaces, and the facet surfaces advance more slowly [4]. After all the rough surfaces have grown out, what remains are the slow-moving facet surfaces given the final form of the crystals. The hexagonal prism can be "plate-like" or "column-like", depending on which facet surfaces grow most quickly.

C. Properties of Water and Ice

Water is a rather unusual substance having high boiling and freezing points, high specific heat, high latent heats of fusion and vaporization, high surface tension, high polarity, and unusual density changes. The ability of water to form three-dimensional hydrogen bonding explains many of these anomalous properties [3,4].

- Water is a liquid rather than a gas at ambient temperature. By molecular weight (MW), it should be a gas, considering that CO_2 (MW = 44), O_2 (MW = 32), CO (MW = 28), N_2 (MW = 28), CH_4 (MW = 18), and H_2 (MW = 2) are all gases at room temperature.

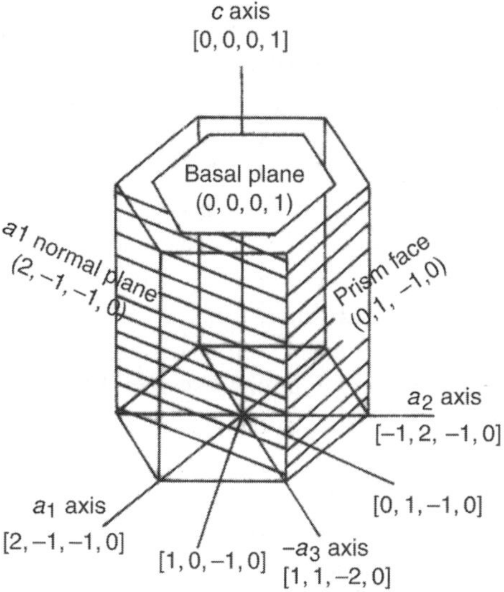

FIGURE 1.4 Ih ice crystal: the hexagonal prism. Crystallographic planes are represented between parentheses and crystallographic axis between brackets.

- Water has a relatively high melting point (0°C at 1 atm); it is over 100°C higher than expected by extrapolation of the melting points of other dihydrides of similar structure formed with atoms close to oxygen in Group VIA of the periodic chart. These dihydrides listed in order of increasing molecular weight are: H_2O, H_2S, H_2Se, H_2Te. The explanation is that in ice (Ih), all water molecules participate in hydrogen bonds that are held relatively static. In liquid water, some of the weaker hydrogen bonds must be broken to allow the molecules move around. The large energy required for breaking these bonds must be supplied during the melting process.
- Water has a high boiling point (100°C at 1 atm); it is over 150°C higher than expected by extrapolation of the boiling points of other hydrides. There is considerable hydrogen-bonding in liquid water, which prevents water molecules from being easily released from the surface of water reducing the vapor pressure. As boiling cannot occur until this vapor pressure equals the external pressure then the change of phase is produced at a high temperature. Besides the pressure/temperature range for the liquid phase in water is much larger than for most other materials (e.g., under ambient pressure the liquid range of water is 100°C, whereas for both H_2S and H_2Se it is about 25°C).
- Water has a high heat of vaporization (40.7 kJ mol^{-1}, at 1 atm). In water at 100°C there is still considerable hydrogen bonding (~75%); as these bonds need to be broken, a high amount of energy is required to convert liquid water into vapor.
- The high specific heat of liquid water is attributed to the cohesive properties; water molecules resist the net breaking of hydrogen bonds during heating.
- The critical point of water (374°C) is over 250°C higher than expected by extrapolation of the critical points of other Group VIA dihydrides. The critical point can only be reached when the interactions between the water molecules fall below a certain threshold level. Owing to the strength and extent of the hydrogen bonding, much energy is needed to cause this reduction in molecular interaction and this requires higher temperatures. Even close to the critical point, a considerable number of hydrogen bonds remain, but no longer tetrahedrally arranged.
- Water has a high surface tension (72.75 dyn cm^{-1} at 20°C). Water molecules at the liquid–gas surface are pulled towards the bulk liquid phase by the hydrogen bonds. To increase the surface area, a relatively large energy is required to remove a molecule from the interior bulk water to the surface where there are less hydrogen bonds. The surface tension of water is large compared to other molecules. Lowering the temperature greatly increases the hydrogen bonding causing increased surface tension.
- One of the anomalies of water is the contraction of liquid on melting. When water freezes at 0°C, at atmospheric pressure, its volume increases by about 9%. The structure of ice Ih is open with a low packing efficiency where all the water molecules are involved in four straight tetrahedrally oriented hydrogen bonds. On water melting, some of these bonds break, others bend and the network structure undergoes a partial collapse allowing unbounded molecules to approach more closely increasing the number of nearest neighbors. This is in contrast to normal liquids, that usually contract on freezing and expand on melting. This is because within the solid phase, molecules are in fixed positions but require more space to move around in liquid phase. In the case of water the volume expansion when going from liquid to solid, under ambient pressure, causes tissue damage in biological organisms on freezing [3].
- Water has a high latent heat of fusion (335 kJ kg^{-1} at 0°C) owing to the energy that goes into hydrogen bond formation in the ice crystal.

  Other interesting water properties are:
- High pressures tend to inhibit the solidification of water rather than enhance it.
- Water as a hydrophilic solvent tends to dissolve hydrophilic substances very efficiently (both polar molecules and those which ionize upon dissolving).

- The thermal conductivity of water is large compared to those of other liquids and the thermal conductivity of ice is moderately large compared to other nonmetallic solids. Besides, at 0°C the thermal conductivity of ice is four times that of water at the same temperature and the thermal diffusivity of ice is nine times greater than that of water [1].

III. THE FREEZING PROCESS

Freezing involves different factors in the conversion of water to ice: thermodynamic factors that define the characteristics of the system under equilibrium conditions, and kinetic factors that describe the rates at which equilibrium might be attained.

The freezing process includes two successive processes: the formation of ice crystals (nucleation), and the subsequent increase in crystal size (growth) [4,7].

A. Homogeneous and Heterogeneous Nucleation

When ice and water coexist at atmospheric pressure, the temperature of the system reaches the freezing point of pure water ($T_f = 0°C$) as long as both liquid and solid are present; the amount of ice remains constant and no energy is either added or removed from the mixture. The freezing point of water (or the melting point for ice, T_m) is an equilibrium point. However, if water is cooled to 0°C it will not freeze; it is necessary to get temperatures (T) substantially below the freezing point (T_f) before ice begins to form. Supercooling (or undercooling), defined as $\Delta T_s = T_f - T$, of pure water is necessary for nucleation to occur.

Nucleation refers to the process by which a minimum crystal is formed with a critical radius which can then expand and grow. During nucleation the latent heat of solidification is released; molecules aggregate into an ordered particle of a sufficient size to survive and serve as a site for further crystal growth [4]. At the surface of the crystals, there is a constant interchange of water molecules between the solid and liquid phases. If the crystal surface is planar, then the number of molecules which leave the crystal is equal to the number of molecules which adhere to it. If there is a corner on the crystal, with a given curvature, this number of molecules will not be equal. The molecules that are part of the crystal at the corner will be less strongly bonded to the crystal because they do not have as many neighbors to bond with and they are more easily removed from the crystal. In contrast, molecules from the liquid are less likely to join the crystal at the corner. This leads to a net loss of molecules from the corner, producing melting at a temperature for which there is equilibrium on a planar crystal surface. The melting point of a crystal is then a function of its radius of curvature; for a given temperature there will be a critical radius which defines the minimum size that a crystal can have and still be stable.

Nucleation may be homogeneous or heterogeneous: homogeneous ice nucleation is produced in water free from all impurities, and heterogeneous nucleation (catalytic nucleation) takes place when water molecules aggregate in a crystalline arrangement on nucleating agents such as active surfaces; this type of nucleation predominates in food systems. As can be observed in Figure 1.5 homogeneous nucleation requires a higher supercooling than heterogeneous nucleation.

The analysis of the thermodynamic aspects involved in nuclei formation in pure systems (homogeneous nucleation) allows the estimation of the critical radius. If a solid phase and a liquid phase such as ice and liquid water are in equilibrium, their molar free energies are equal and there will be no temperature difference across a planar interface between the two phases. The equilibrium temperature is T_f which is defined as the freezing point if solid and liquid phases are in equilibrium. However, if the interface is curved, then a pressure difference (ΔP) must exist to account for the difference in surface energy. Considering the ice nucleus as a spherical particle of radius r, the work done by an increase in pressure (ΔP) on a sphere of radius r when the

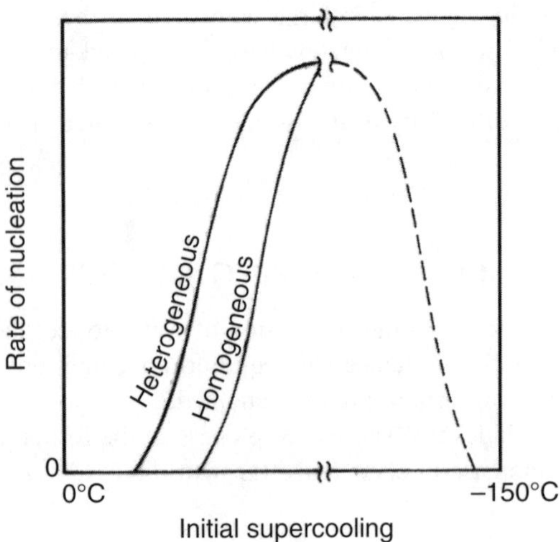

FIGURE 1.5 Effect of the initial supercooling on homogeneous and heterogeneous rates of ice nucleation.

radius is increased to $r + \mathrm{d}r$ is given by:

$$\mathrm{d}W = \Delta P\,\mathrm{d}V = \Delta P\,4\pi r^2\,\mathrm{d}r \tag{1.1}$$

where $V = (4/3)\pi r^3$ is the volume of the sphere and $\mathrm{d}V = 4\pi r^2\,\mathrm{d}r$.

Besides, the work done by expanding the area of the same sphere ($A = 4\pi\, r^2$) when the radius is increased to $r + \mathrm{d}r$ is given by:

$$\mathrm{d}W = \sigma\,\mathrm{d}A = \sigma\,8\pi r\,\mathrm{d}r \tag{1.2}$$

where σ is the surface tension (free energy per unit area of the interface).

At equilibrium, both energies will be equal, so combining Equation (1.1) and Equation (1.2) the following is obtained:

$$\Delta P = \frac{2\sigma}{r} \tag{1.3}$$

The pressure difference ΔP just compensates for the departure ΔT_s of the temperature from that existing across a planar interface.

If the solid phase is incompressible, then ΔP is equal to the volume free energy difference (ΔG_p) that would exist between solid and liquid at $T - \Delta T_s$, if the solid and liquid phases were at the same pressure. Substituting ΔP in Equation (1.3) by the difference in volume free energy ΔG_p the following is obtained:

$$\Delta G_p = \frac{2\sigma}{r^*} \tag{1.4}$$

where r^* is the critical radius at which equilibrium is established. The free energy at a constant pressure (ΔG_p) is given by:

$$\Delta G_P = \Delta H - T_K \Delta S \tag{1.5}$$

where T_K is the absolute temperature. As the change in free energy is zero at the equilibrium freezing temperature T_{Kf}, then:

$$\Delta S = \frac{\Delta H}{T_{Kf}} \tag{1.6}$$

Substituting Equation (1.6) in Equation (1.5) and introducing the volume latent heat of fusion (L_f) and the supercooling $\Delta T_s = T_{Kf} - T_K = T_f - T$, the following is obtained:

$$\Delta G_P = \Delta H - \frac{T_K \Delta H}{T_{Kf}} = \Delta H \frac{(T_f - T)}{T_{Kf}} = L_f \frac{\Delta T_s}{T_{Kf}} \tag{1.7}$$

Replacing ΔG_p in Equation (1.4) the expression for the critical radius is

$$r^* = \frac{2\sigma T_{Kf}}{L_f \Delta T_s} = \frac{2\sigma T_{Kf} v}{\lambda_A \Delta T_s} \tag{1.8}$$

where T_{Kf} is the absolute freezing temperature of pure water (273 K), λ_A the molar latent heat of fusion for pure water, and v the molar volume of water.

Liquid water consists of clusters of molecules which are undergoing constant collisions with other molecules and clusters, sometimes breaking apart and sometimes forming larger clusters. If pure water is cooled, it is necessary to obtain a cluster of a size sufficient to match the critical radius or that particular cluster will not be stable.

From the analysis of the dependence of the critical radius as a function of temperature [Equation (1.8)] it can be observed that at 0°C ($\Delta T_s = 0$, without supercooling) a cluster of infinite radius would be required to be stable; furthermore, Equation (1.8) shows that there is a minimum temperature at which the critical radius must exist.

The homogeneous nucleation temperature in pure water is about −45°C, the minimum temperature to which pure water can be cooled to, before freezing occurs spontaneously.

One of the general equations that describe the rate of nucleation J, expressed as the number of nuclei formed per unit volume and time, is given by [8]:

$$J = CT \exp\left(\frac{-BT_f^2}{\Delta T_s^2 T}\right) \tag{1.9}$$

where B and C are coefficients depending on the type of product in which nucleation occurs, T the system temperature, ΔT_s the supercooling, and T_f the initial freezing temperature of the system. Nucleation is a statistical phenomenon and Equation (1.9) shows that the greater the supercooling the higher the rate of nuclei formed per unit of volume sample is. The probability of nucleation depends on the volume of the sample. When the sample volume is small the probability of nucleation is low, and very low freezing temperatures are required. Nucleation rate is highly dependent on the temperature of the freezing medium, on supercooling, and on the viscosity of the liquid. High initial supercooling increases nucleation rates (Figure 1.5).

Heterogeneous nucleation is produced in water containing impurities or large particle in the solution that act as active surfaces (Figure 1.5) for the formation of ice nuclei [4]. When the water molecules wet the surface of an impurity (that is large compared to water molecules) with a certain contact angle, then a portion of a sphere can form which has the critical radius that will therefore be stable; the supercooling necessary for heterogeneous nucleation is then lower than that for homogeneous nucleation (Figure 1.5).

For viscous systems, volume diffusion can influence the rate of nucleation. At high solute concentrations or high viscosities the viscous barrier to nucleation becomes more important than the supercooling and nucleation is inhibited. At very low temperatures, below the glass transition temperature of pure water which is approximately −135°C, nucleation no longer is produced (Figure 1.5). The glassy state is described elsewhere in the book.

B. Crystal Growth

As long as a stable ice nucleus is formed, further growth is possible by addition of molecules to the solid–liquid interphase. Growth is not instantaneous, and is controled by the rate of removal of the latent heat released during the phase change, as well as by the rate of mass transfer in the case of solutions (diffusion of water molecules from the surrounding solution to the surface of the ice crystals and counter-diffusion of solutes away from the growing crystal surface). The rate of crystalline growth (G) is also a function of the supercooling reached by the specimen according to the following phenomenological expression [7,8]:

$$G = \beta(\Delta T_s)^n \tag{1.10}$$

However, heat transfer is not the only factor that governs crystal growth or the rate of ice propagation. If ice is crystallizing from a solution, the solutes must be rejected from the region occupied by the pure ice crystals. Ice growth is also governed by mass transfer as water molecules must diffuse and add into the growing ice crystal and at the same time solutes have to diffuse away from the crystal.

Crystal size varies inversely with the number of nuclei formed. At high freezing rates, a high number of nuclei are formed and the mass of ice is distributed in a large number of small crystals. At low freezing rates fewer nuclei are formed leading to large crystal sizes.

C. Freezing Curves

Figure 1.6 shows typical time–temperature relationships during the freezing of small samples (without thermal gradients) of pure water (Figure 1.6a) and of an aqueous solution (Figure 1.6b). Cooling of pure water (Figure 1.6a) involves in the first stage, the removal of sensible heat (4.18 kJ/kg °C). Nucleation is necessary for freezing to initiate, and the temperature can fall below 0°C without the formation of ice crystals. Point S indicates the supercooling of the liquid before crystallization begins [4, 9]. Once the critical mass of nuclei is reached, the system nucleates

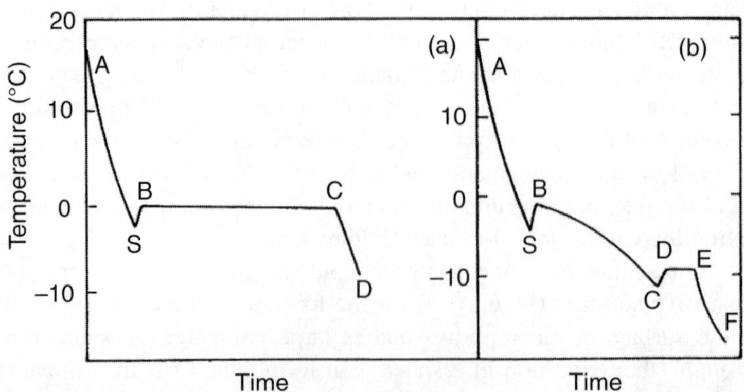

FIGURE 1.6 Typical time–temperature relationships during the freezing of: (a) pure water; (b) an aqueous solution.

at point S (Figure 1.6a) and releases its latent heat faster than heat removed from the system. The abrupt increase in temperature (point S to point B), because of the liberation of the heat of solidification after initial supercooling, represents the onset of ice crystallization. Once crystallization begins, the temperature reaches point B, the freezing point of pure water (0°C). While the solid and liquid are in equilibrium, the temperature remains at the freezing point until all of the water has been converted to ice (point C). In pure water, the "plateau" from B to C represents the time during crystal growth is occurring. The freezing time for the point where the thermocouple is inserted is usually considered as the time from the onset of nucleation to the end of the crystal growth phase. On completion of solidification, further removal of heat results in a decrease in temperature towards D.

The freezing of food systems is more complex than the freezing of pure water [4]. Food systems contain water and solutes, and the behavior is similar to that of an aqueous solution. When freezing aqueous solutions (Figure 1.6b), the cooling curve also shows supercooling (point S). Nucleation occurs at point S and the released heat of crystallization raises the temperature from S to B. The point B represents the freezing point of the solution, which is lower than the freezing point of pure water. The freezing point depression is determined by the number of dissolved solute molecules. Besides, in aqueous solutions, supercooling is generally lower than in pure water as the added solute promotes heterogeneous nucleation, accelerating the nucleation process. In very concentrated solutions, it is sometimes difficult to induce supercooling [9]. Further cooling from B to C (Figure 1.6b) results in the growth of ice crystals and a substantial ice formation. A gradual increase in solute concentration is produced as water is separated in the form of relatively pure ice crystals, and the declining freezing point (negative slope of B–C in Figure 1.6b) reflects the change in concentration. Solute concentration increases during the freezing process and eventually reaches its eutectic temperature. Supersaturation, indicated by point C in Figure 1.6b, can be observed before the crystallization of the solute. Latent heat of solute crystallization is released in C, causing a slight increase of temperature from C to D (Figure 1.6b). At temperature D, the solution assumes the eutectic equilibrium composition that remains constant during eutectic solidification and constant temperature (D–E). Cooling below E is produced after the solution is solidified completely.

1. Phase and State Diagrams

Phase diagrams are used to describe equilibrium situations in which two or more phases of matter exist together in pure substances or in solutions. Phase diagram shows the preferred physical states of matter at different temperatures and pressures. Each line gives the conditions when two phases coexist but a change in temperature or pressure may cause the phases to change abruptly from one to the other. Working at atmospheric pressure the freezing process in solutions can be analyzed, using temperature versus solute concentration diagrams. However, these phase diagrams only indicate the conditions in which equilibrium phase transformation can occur. In contrast, state diagrams provide more information because they contain equilibrium as well as information on conditions of nonequilibrium and metastable equilibrium states [10,11] such as the glass transition conditions. A schematic temperature–composition, state diagram for an aqueous system with a single solute is shown in Figure 1.7. When freezing solutions, the equilibrium thermodynamic process can be represented as an equilibrium freezing (liquid) curve (Figure 1.7), which extends from the melting temperature (T_m) of pure water (0°C) to the eutectic temperature (T_{eu}) of the solute, which is the point at which the solute has been freeze-concentrated to its saturation concentration. As temperature decreases, water is removed in the form of ice, and the solute in the unfrozen phase is freeze-concentrated. An equilibrium freezing temperature exists for each ice/unfrozen phase ratio, which is a function of the solute concentration. As the solution is progressively frozen, more water is turned into ice and the residual solution becomes more concentrated. Solutions and food systems do not have a sharp freezing point like water; latent heat is released

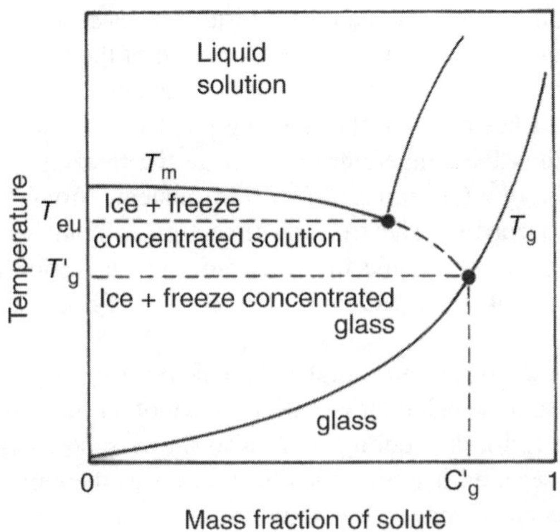

FIGURE 1.7 Schematic state diagram at constant pressure of an aqueous binary solution showing the equilibrium freezing curve, the solubility line and the glass transition temperatures.

gradually over a large range of temperature. The mass of ice in equilibrium with the unfrozen solution is a function of temperature, which is characteristic of the type of foodstuff and depends on water content. For example, in the case of beef tissue, with a total water content of 74 g water/g tissue, 80% of the water is converted to ice at $-7°C$ [12].

In the case of complex systems with multiple solutes and in foods many different eutectic points might be expected, but each plateau would be quite short if small quantities of solutes were involved. Solute crystallization at the eutectic point is unlikely due to the very low temperatures, extremely high viscosities, and resulting low diffusion rates and limited solute mobility [13].

2. Freezing Point Depression

Freezing point depression in solutions is a colligative property that depends on the concentration of solute particles, which lowers the effective number of solvent molecules that can produce the phase transition from liquid to solid. Freezing point depression is directly proportional to the molal concentration of solute. Using basic thermodynamic principles, it is possible to predict the melting point of different foodstuffs (T_{fs}). The following equation allows to calculate the freezing point depression ($\Delta T_f = T_f - T_{fs}$) with reference to the freezing point (T_f) of the pure solvent (water), as a function of a nonvolatile nonelectrolyte solute concentration [14]:

$$\Delta T_f = \frac{R T_{Kf}^2 M_A m}{1000 \lambda_A} = 1.86\,\text{m} \tag{1.11}$$

where λ_A is the latent heat of fusion for pure water (6003 kJ mol^{-1}); M_A the molecular weight of water (18 g mol^{-1}); m the molality of the solution representing the food system (number of moles of solute/1000 g of solvent); T_{Kf} the freezing point of pure water (273 K). Foodstuffs with higher solute content show a lower melting point; typical values of initial freezing points are: $-1.1°C$ for beef, -0.9 to $-2.7°C$ for fruits, -0.8 to $-2.8°C$ for vegetables, and about $-0.5°C$ for eggs and milk [4].

D. Freezing under Thermal Gradients

1. Freezing Rate Definitions

The rate of freezing determines the ice crystal size; the faster the rate, the more the nucleation, and a greater number of crystals of smaller size will result. The first simple definition of freezing rate is the rate of temperature change. In very small specimens, temperature gradients can be practically neglected and all the points in the sample will have similar freezing rates. However, in large-sized systems, temperature gradients along the sample are established; then freezing rate is position dependent and varies along the frozen sample. High freezing rates are observed on the surface in contact with the refrigerant decreasing towards the thermal center [15].

For a given point in the sample, the freezing rate can be represented by the characteristic local freezing time (t_c) that is the time needed to change the temperature from the initial freezing point to a temperature for which, for example, the 80% of the total water content is converted to ice. Each point along the sample will have a different t_c value [15].

A better definition of freezing rate might be to consider the average rate of ice formation, or the rate of advance of the freezing interface, which is related to the rate of heat removal. The freezing process is for practical purposes complete when most of the freezable water at the thermal center of the product has been converted to ice. According to the International Institute of Refrigeration [16], the freezing rate of a food is defined as the ratio between the minimum distance from the surface to the thermal center, and the time elapsed between the surface reaching 0°C and the thermal center 10°C colder than the temperature of initial ice formation. In commercial practice [16], freezing rates vary between 0.2–100 cm/h; 0.2–0.5 cm/h correspond to slow freezing (bulk freezing in cold chambers), 0.5–3 cm/h to quick freezing (air blast and contact plate freezers), 5–10 cm/h to rapid freezing (individual quick freezing of small sized products in fluidized beds), and 10–100 cm/h to ultra rapid freezing by spraying or immersion in cryogenic fluids (liquid nitrogen, carbon dioxide).

2. Nucleation and Ice Crystal Growth in Water and Aqueous Solutions

It is important to recognize that the presence of a uniform high number of small crystals at high freezing rates and a few large ice crystals at slow freezing rates is only valid for small specimens without thermal gradients. When freezing large samples, ice phase nucleates only in a supercooled region that is in contact with the cooling medium (heat sink). The heat released at the crystal surface (ice–liquid interphase) and the sensible heat from the unfrozen liquid phase are both transferred through the frozen phase by conduction. A temperature profile is established in the system being the temperatures in front of the interphase higher than in the frozen phase, which is a situation that suppresses thermal supercooling ahead of the ice–liquid interphase. When pure water is being frozen under these thermal conditions the growing crystal surface will remain essentially smooth [4,8,15]. If for any reason a protuberance appears at the ice–liquid interphase, the extreme of this irregular surface would be in a zone where supercooling is absent and then the protuberance will tend to disappear.

Similarly, when freezing solutions, the heat released during the phase change is also transferred across the frozen phase, and the temperature increases when moving away from the interphase; the heat flux direction during freezing is indicated in Figure 1.8. Thermal supercooling takes place only in the external surface layers and nucleation is only produced at the border that is in contact with the refrigerating medium. The temperature rise caused by the crystallization of ice impedes any subsequent nucleation in the system. However, in the case of solutions as an ice nucleus begins to grow, solutes are rejected from the ice phase and accumulate at the solid–liquid interphase. This situation leads to a solute concentration gradient in the liquid which surrounds the ice front giving rise to a modification in the solid–liquid equilibrium temperatures (Figure 1.8). Equilibrium temperatures decrease with increasing solute concentration, thus, a zone where supercooling

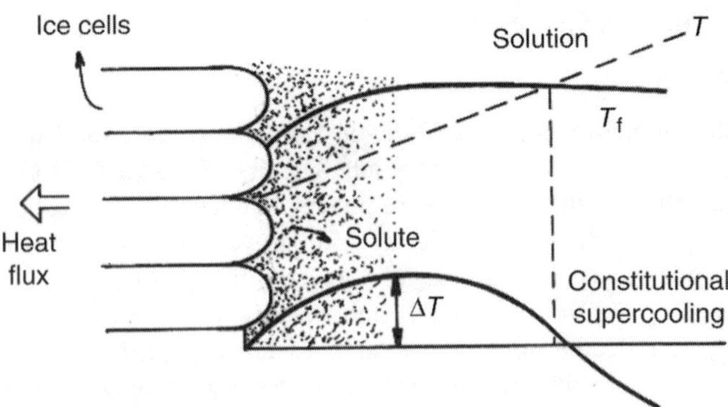

FIGURE 1.8 Constitutional supercooling during freezing of solutions and the formation of an irregular ice–liquid interphase. The accumulation of solutes in the unfrozen phase near the interphase produces a concentration gradient that gives rise to a modification in the solid–liquid equilibrium temperatures. Supercooling increases in front of the interphase.

($\Delta T_s = T_f - T$) increases in front of the interphase can be generated and it is denominated constitutional supercooling. The existence of this constitutional supercooling a growing ice crystal results in an unstable condition; as there is an increasing supercooling which is maximal just before the interface, a planar ice front will be susceptible to small perturbations [8]. If a ice protrusion of the interface advances just slightly ahead of the plane, then its growth rate will increase in the supercooled zone. Such an instability will grow through the supercooled region; ice cells will grow adjacent to each other with segregation of solute between them. The cells exclude solute to the sides as well as in front and the regions between cells will contain concentrated solute.

Growth of ice crystals is produced from the border towards the interior forming columns, where those oriented closest to the direction of the thermal gradient predominate, increasing the diameter of these columns from the refrigerated border to the center of the food system [15]. If the conditions leading to cellular growth are particularly pronounced, then the cells may turn to dendrites, which are protuberances with side branches.

As the ice crystal is built on a hexagonal symmetry, these side branches will follow that symmetry. Once dendritic breakdown (the formation of growing dendrites from a planar interface) occurs, the excluded solute will be confined in regions that are surrounded between the growing dendrites.

The final crystal shape (habit or morphology) [7] with individual faces depends not only on the crystal system classification but also on the conditions under which the crystal was formed and grown. When crystals are grown at very high freezing rates the crystals have dendritic shapes. The tree-like formations involve growth of a main trunk from which branches extend to the solution.

IV. VITRIFICATION

In this section, the fundamental concepts of vitrification in pure water and aqueous solutions will be discussed. Ice crystallization occurs by two successive processes of nucleation and growth. During freezing of pure water, the viscosity of the liquid phase rises. If the liquid is cooled very quickly the viscosity may reach very high values that molecular rearrangements in the liquid become extremely slow avoiding ice crystallization. The liquid is in a metastable state until it gets below the glass transition temperature (T_g) where the system is an amorphous solid or glass. A glass is defined as a nonequilibrium, metastable, amorphous, disordered solid of very high viscosity [17]. Glass transition (or glass–liquid transition) is produced when a supercooled melt is converted into a glass during cooling or when the reverse transformation occurs upon heating. Both the supercooled melt and the glass are noncrystalline states; the glass is an out-of-equilibrium state where the liquid-like structure of the melt is maintained as a rigid solid, and the supercooled melt, observed between the glass–liquid transition and the melting point, can be a viscoelastic "rubber" in the case of a polymeric material, or a viscous liquid for low molecular weight materials. The glass–liquid transition is a kinetic and relaxation process associated with the relaxation of the material. At temperatures above the T_g, the material, if submitted to a perturbation, can recover after a characteristic relaxation time; the supercooled melt is a viscoelastic material having a relaxation time that is similar to the experimental timescale. The process of glass formation is called vitrification and the system is considered to be vitrified if its viscosity is extremely high (10^{10} to 10^{14} Pa sec) [1,9,18]. The T_g of pure water is close to $-135°C$; achieving vitrification with pure water requires very small specimens and extremely high cooling rates ($>10^7$ K sec^{-1}) [19]. Vitrification can also be achieved by adding solutes that impede the process of crystal growth. The schematic state diagram of Figure 1.7 is useful to analyze vitrification in aqueous solutions. As was described previously the equilibrium thermodynamic freezing process can be represented by the equilibrium liquid–solid curve (Figure 1.7), which gives the melting temperature as a function of solute concentration. The curve extends from the melting temperature (T_m) of pure water (0°C) to the eutectic temperature (T_{eu}). Along the freezing process, the solution becomes more concentrated. Co-crystallization of solute at T_{eu} is unlikely because of the high viscosity of the system produced by solute concentration and low temperature [20]. Freeze concentration continues beyond T_{eu} into a non-equilibrium state through a viscoelastic liquid–solid glass state transition because of the reduction in molecular motion and diffusion kinetics [21,22]. The glass transition curve extends from the glass transition temperature (T_g) of pure water ($-135°C$) to the T_g of pure solute. Rapidly cooled solutions exhibit less ice formation and the T_g of the unfrozen portion is usually low. At subzero temperatures, the formation of an amorphous state is time-dependent, as the limiting factor of the process (water removal in the form of ice) becomes more difficult as concentration increases. The marked effect of viscosity on mass transfer properties acts as the limiting factor for ice growth. In addition, under conditions where heat removal is rapid, a high level of supercooling at the interface decreases the propagation rate and freezing becomes progressively slower as ice crystallization is hindered, consequently more time is required for crystal growth at each temperature. In aqueous solutions, as the concentration of solutes increases, the temperature T_g at which vitrification occurs also increases, and the cooling rate necessary to achieve vitrification is reduced with respect to pure water. Vitrification in solutions is then easier to be produced than in pure water, because the addition of solutes decreases the probability of ice nucleation and growth. The reasons are that solutes are incompatible with the ice structure and that the viscosity at any temperature is usually larger when solutes are present, making it difficult for the motion and reorientation of the water molecules into the ice structure; the higher viscosity hinders both nucleation and growth.

The intersection of the nonequilibrium extension of the liquid curve beyond T_{eu} (Figure 1.7) and the glass transition curve is given by T'_g, defined as the maximally freeze-concentrated glass transition temperature of the frozen system where the unfrozen water in the matrix is unable to freeze and then ice formation ceases within the timescale of normal measurement [9,17,20,21,23–25]. Below T'_g the unfrozen matrix takes on solid properties because of reduced molecular motion, which is responsible for the marked reduction in translational, not rotational mobility [18,22]. At this temperature the concentration of solute within the glass is C'_g (Figure 1.7). If a product is stored at a temperature below T'_g it may be expected to be composed of ice and a freeze-concentrated phase in the glassy state and long-term stability may be predicted. If the storage temperature is between T'_g and T_m, the freeze-concentrated phase is not in the glassy state, it is more diluted and processes governed by diffusion are not inhibited. These processes can lead to deterioration during food storage [1]. In rapidly cooled systems in the glassy state, ice formation can occur during rewarming (exothermic devitrification) at a temperature above T'_g

[10,22] and ice is produced by crystallization of the immobilized water, before the onset of ice melting. Recrystallization and crystal growth may cause cell injury and loss of biological activity of cells.

Vitrification can occur in biological systems at ambient temperatures (desiccation) or subzero temperatures (cooling), and has been suggested as a mechanism for membrane protection during dehydration [19]. In both cases, if the viscosity rises to $\sim 10^{14}$ Pa sec (caused by either higher concentrations or lower temperatures) then the solution is vitrified. In cells or lamellar phases at low hydration, the vitrification will occur where the sugars are located. Membranes can be protected from dehydration by vitrification by the following mechanisms: (i) Once a glass has formed, further dehydration will be limited, then decreasing the subzero temperature will have little effect on the intermembrane separation. In this case, the membranes will have an effective hydration higher than at equilibrium. (ii) A glass may allow the membranes to remain in the fluid lamellar phase at hydrations and temperatures that normally would lead to deleterious phase transitions.

Besides, vitrification is considered one of the most promising approaches to the cryopreservation of biological materials.

V. MECHANISMS OF ICE FORMATION IN CELLS AND TISSUES

A. Intracellular and Extracellular Ice Crystals in Frozen Cells and Tissues

In cell suspensions and tissues the existence of barriers to the water movement such as cell membranes, introduces complexity to the mechanisms of freezing because the inside and the outside environments have to be considered [26,27].

Freezing of food tissues can lead to extracellular ice and also under determined conditions to intracellular ice. The membrane permeability and the internal properties of the cell are important factors that affect ice formation. The location of ice crystals in tissues and cellular suspensions is a function of the nature of the cells, the freezing rate, and the specimen temperature. It is generally accepted that crystallization, regardless of freezing rate, is initiated in the extracellular fluid [4,28].

Different theories have been proposed to explain that further crystallization can continue exclusively in extracellular regions, or can be produced also in intracellular regions. Slow freezing (lower than 1°C/min) of plant tissue, animal tissue, or cellular suspensions generally causes ice crystals to form exclusively in extracellular areas [4]. When ice starts to form in the extracellular space, solute concentration increases and water activity decreases in the unfrozen external region. As cells contain a higher concentration of nondiffusible ions than the surrounding fluid, the total concentration of ionic particles will be greater inside the cell than in the extracellular space, and a lower freezing point would be expected for the intracellular space. At relatively high subfreezing temperatures, ice crystals cannot penetrate cellular membranes, and the intracellular fluid remains in a supercooled condition without intracellular ice crystals. As water activity of the intracellular fluid at any given temperature is higher than that of the extracellular fluid, water diffuses from the cells and is deposited on the extracellular ice crystals in order to equilibrate the chemical potential in both fluids; supercooling in the intracellular spaces is then minimized decreasing the probability of intracellular nucleation. Slow freezing results in considerable shrinkage of the cells and formation of exclusively large extracellular ice crystals. In contrast, tissues and cellular suspensions that are frozen rapidly at very low temperature show both intra- and extracellular ice crystals with a uniform distribution. Rapid freezing produces intracellular crystallization and results in numerous small ice crystals, minimum dislocation of water, and in a case of food systems an appearance which is similar to the original unfrozen system. The formation of intracellular ice is affected by several factors [27]. One factor is the cell permeability that controls the loss of water through the membrane to the external environment when the osmotic gradient is established. The migration of water from the intracellular space increases the internal solute concentration, reducing the internal freezing point and the degree of intracellular supercooling. A high membrane permeability helps to prevent intracellular freezing and to sustain a large supercooling. At low freezing rates with either low or high water cell permeability, the rate of change of the external unfrozen matrix concentration is slow, and water can migrate from the interior fast enough to minimize internal supercooling. Under these conditions the cell dehydrates, and the water is deposited on the external ice crystals. When the freezing rate is high and the water cell permeability is low [27], the extracellular unfrozen matrix increases its solute concentration rapidly. However, as water cannot be transferred rapidly, the intracellular region becomes increasingly supercooled. At some critical supercooling, the internal contents will freeze; intracellular freezing will cause structure damage, but there will be little water transfer from the intracellular space. When there is a fast freezing and a high water permeability, as the unfrozen matrix concentration increases, water migrates from the intra to the extracellular spaces and maintains minimal intracellular supercooling. In this case, the cell dehydrates but does not freeze [26]. Intracellular freezing is favored by rapid cooling to a low temperature so that the opportunity for cellular dehydration is minimized. Under these conditions, there is a high probability for intracellular ice nucleation or for the growth of extracellular ice crystals through the cell membrane. Cell membranes act as effective barriers to crystal growth at high subfreezing temperatures, such as those encountered during slow freezing, whereas during rapid cooling to some critical low temperature (in the neighborhood of $-10°C$), the barrier properties of membranes tend to disappear. Mazur [28,29] suggested two possible ways in which a membrane might lose its barrier properties as the temperature decreases: (i) low temperatures may damage the membranes either directly or indirectly associated with the concentration of solutes during freezing, or (ii) the membrane may remain unaltered, but as the temperature decreases ice crystals are able to exist with smaller radii of curvature, and they can grow through the tiny water-filled pores of the membrane.

There are different hypotheses which describe intracellular freezing. One of the theories of intracellular freezing [30] holds that critical supercooling of the protoplasm leads to spontaneous nucleation. The second theory asserted that when the minimum radius of growing ice crystals in the extracellular space matched the radius of aqueous pores in the cell membrane, then these growing crystals would move through the pores and nucleate the protoplasm. The third hypothesis postulated that electrical transients at the ice interface could cause the plasma membrane to rupture, thereby allowing ice from the extracellular compartment to nucleate the intracellular compartment. A more recent hypothesis consistent with experimental observations was proposed in which the plasma membrane is ruptured when a critical gradient in osmotic pressure across the membrane is exceeded and the protoplasm is nucleated by extracellular ice [30]. The osmotically driven water flux occurring in cells during freezing is viewed as the agent responsible for producing a rupture of the plasma membrane, thus allowing extracellular ice to propagate into the cytoplasm. This theory gives an accurate description of the phenomenology of intracellular ice formation [30].

The freezing of large tissue pieces is commonly produced under thermal gradients with freezing rates that are high in the external regions in contact with the refrigerant medium and decrease towards the thermal center of the sample. Histological analysis in frozen meat tissues in conditions where the heat flow was parallel or perpendicular to the muscle fibers allowed the observation of ice crystal sizes and distribution as a function of the local freezing rate [31,32]. The formation of intracellular ice was only reported in a narrow zone adjacent to the area in contact with the cooling medium, submitted to high freezing rates. The authors [31,32] expressed the freezing rate by means of the characteristic local freezing time (t_c), which that was defined as the time necessary to change the temperature from $-1°C$ (initial freezing time for beef tissue) to $-7°C$ (80% of total water is converted to ice) in a given point of the system. The existence of intracellular ice constitutes an index of high freezing rates. Intracellular ice was observed for t_c values lower than 0.5 min; ice crystals nucleated in the refrigerated surface grew towards the thermal center of the meat piece in the form of columns (cell growth). As freezing rate decreased, intracellular ice disappeared, and only the growth of extracellular columns was observed at the expense of the water from the meat fiber; because of this dehydration process, the shape of the fibers becomes irregular and distorted. Measurements of the average equivalent diameter of the ice crystals showed that their sizes increase with the local characteristic freezing time in the frozen tissue [15].

B. Freezing Injury in Living Cells

To preserve living cells the challenge is to determine how they can survive both the freezing process and the subsequent return to physiological conditions. Two distinct mechanisms of cell injury during freezing and thawing were proposed [33], one occurring at low cooling rates where the cell remains close to osmotic equilibrium (solution injury) and the other at high freezing rates in which there is supercooled water within the cell and intracellular ice formation (intracellular ice injury).

Solution injury is produced when cells are cooled too slowly, then the outside environment of the cell freezes first and extracellular ice forms. Extracellular ice creates a chemical potential difference across the membrane of the cell producing a flux of water which dehydrates and shrinks the cell. The slower the cells are cooled, the longer the dehydration occurs, causing irreversible damage. The recovery of the cells is high when the cells have only been exposed to the freeze-concentrated solution for a short period of time. However, as the temperature drops, the cells are exposed to even more concentrated solutions and the total time of exposure to the freeze-concentrated solution also increases; in this case the survival of the cells decreases markedly.

On the other hand, intracellular ice injury appears when cells are cooled too quickly and the cell retains water; this water expands during freezing and intracellular ice crystals damage the cell itself [33]. Although the avoidance of intracellular freezing is usually necessary for cell survival, it is not sufficient. Slow freezing itself can be injurious; as ice forms outside the cell, the residual unfrozen medium forms channels of decreasing size and increasing solute concentration. The cells shrink in osmotic response to the rising solute concentration. Prior theories have ascribed slow freezing injury to the concentration of solutes or the cell shrinkage. More recent experiments, however, indicate that the damage is more because of the decrease in the size of the unfrozen channels. This new view of the mechanism of slow freezing injury ought to facilitate the development of procedures for the preservation of complex assemblages of cells of biological, medical, and agricultural significance [33]. Cryoprotectants can help reduce the damage caused by both solution injury and intracellular ice injury [34].

In the case of food systems, conditions which produce intracellular crystallization (rapid freezing) result in numerous small ice crystals, minimum dislocation of water, and an appearance which is similar to the original unfrozen system. Food quality is usually superior to that obtained by slow freezing. In contrast, in living matter, intracellular freezing is usually associated with lethality, especially if the intracellular crystals are abundant and large in relation to cell size [28].

The location of ice crystals have more influence on the retention of viability in frozen biological specimens than on the quality of frozen foods [34].

VI. PHYSICAL AND CHEMICAL CHANGES DURING FREEZING AND FROZEN STORAGE IN PLANT AND ANIMAL TISSUES

A. Structure Characteristics of Plant and Muscle Tissues

The freezing process is often associated with damage; ice formation involves a series of physicochemical modifications that decrease food quality. Freezing damage in tissues refers to irreversible changes due to the freezing process that become apparent after thawing; it is important to know the structural characteristics of the tissues to understand the damage associated with the freezing process.

Plant tissues, consist of an outer epidermis, parenchymateous cells, and supportive tissue [35]. The epidermis, which is structurally adapted to provide protection against biological or physical stress, consists of tightly packed cells containing waxy material. The parenchymatous tissue performs much of the metabolic activity of the plant and is constituted by semirigid, polyhedral cells containing cellulosic cell walls bounded by pectinaceous middle lamella and often including an extensive network of air spaces. Mature plant cells contain a number of organelles not found in animal cells such as chloroplasts, chromoplasts, large vacuols, protein bodies, amyloplasts, and starch granules. The vacuole, which may comprise most of the mature plant cells, contains organic acids, phenols, and hydrolytic enzymes that can be released when the fragile membranes are disrupted by freezing. Firmness and crispiness (textural properties associated with fruits and vegetables) are attributed to the osmotic pressure developed within the cell when pressure is exerted on the rigid cell walls. Exposure of cell wall to hydrolytic enzymes that attack pectins, hemicelluloses, and noncellulosic carbohydrate material constituents would dissipate the osmotic pressure. Another organella in plant tissues is the choloroplast containing chlorophyll that affects color quality of many plant foods during storage [35].

In contrast with plant tissues, in muscle tissues, the presence of myofibrils and the sarcoplasmic reticulum is important. Muscle cells (myofibrils) are long parallel bundles of contractile proteins (myosin and actin); these flexible and elongated fibers are aligned with a parallel arrangement, with minimal air spaces and separated by an extracellular matrix rich in glycoprotein. A large portion of hydrolytic enzymes is located in the lysosome (an organelle similar to that of the vacuole in the plant cell). After animal death, meat is left in a contraction state until hydrolytic enzymes present in the cytoplasm can disrupt the proteins and tenderize the meat. Although the muscle sarcolemma tends to have a greater hydraulic permeability than the plant cell wall and membrane, internal cell freezing is more common in animal cells than in plant cells.

B. Modifications Produced by Freezing and Frozen Storage

In cellular foods the growth of ice crystals disrupts structure by both physical fracture and the osmotic pressures exerted by the extracellular concentration of solutes as ice is formed. During freezing of cell systems the most important physical changes are modifications in cell volume, dislocation of water, mechanical damage, and freeze-cracking. Physical changes that can be produced during frozen storage are moisture migration, freezer burn, and ice recrystallization. Apart from this, chemical modifications are also produced during freezing and frozen storage of tissues, such as enzymatic reactions, lipid oxidation, and protein denaturation. The increase of solute concentration during freezing and the decompartmentation of cell organelles can affect significantly the rate of these chemical reactions.

C. Physical Modifications Induced by Freezing

1. Changes in Cell Volume, Water Dislocation during Freezing, and Mechanical Damage

During freezing, cell volume changes; pure water expands approximately 9% when it is transformed into ice. Most foods and living specimens also expand on freezing, but to a lesser extent than pure water. As most other constituents contract as the temperature is lowered, it is apparent that the volume change will not be uniform throughout the system. Areas containing ice crystals will expand and others will contract leading to mechanical damage. The presence of intercellular air spaces, which are common in plant tissue, can accommodate growing crystals and minimize changes in the external dimensions of the specimen. During extracellular freezing, dehydration and shrinkage of the cells may cause rupture or folding of cell membranes [27]. Mechanical damage from ice crystals to the tissue structures occurs when flexible cell components are stressed in areas where ice is present. Ice crystals continue to grow in size and exert additional stress on fragile cellular structures. As flexing of cellular tissues occurs, ice can grow into this newly created volume and prevent the structure from relaxing back into its original shape [4]. Mechanical damage to the texture of food tissues during freezing is more likely in plant tissue than in muscle. The texture damage in frozen–thawed plant tissues is attributed to the semirigid nature of the cells. Muscle cells are less likely to be damaged as a consequence of freezing and thawing and structural change is evidenced by cell separation.

Freezing rates influence the size of the ice crystals and then can also affect the surface color of frozen systems. High freezing rates lead generally to pale colors, because the small ice crystals produce scattering of the incident light [36].

Slow freezing produces extracellular ice and leads to moisture movement through osmotic mechanism producing water dislocation. Freezing can be considered as a dehydration process in which frozen water is removed from the original location in the product to form ice crystals. During thawing, water may not be reabsorbed in the original regions, leading to the release of drip. Factors that affect drip losses are size and location of ice crystals, rate of thawing, the extent of water reabsorption, the status of the tissue before freezing, and the water-holding capacity of the tissue. In vegetable tissues water does not reabsorb into the cells, however in animal tissues reabsorption of water may occur.

2. Freeze-Cracking

High freezing rates lead to small ice crystal sizes and to better quality in food systems. The formation of small ice crystals contributes to a homogeneous structure; little damage to the tissue can be detected and drip losses are minimal. However, some products may crack when they are submitted to very high freezing rates, or very low temperatures as in cryogenic fluids. Freeze-cracking has been reported in the literature for different food products, and was reviewed by Hung [37]. Kim and Huang [38] suggested that the crust formed during freezing on the surface of a product serves as a shell that prevents further volume expansion when the internal portion of the unfrozen material undergoes phase transition. If the internal stress is higher than the frozen material strength, the product will crack during freezing. Systems with high void spaces show a higher probability that internal stress will dissipate, instead of accumulating, reducing the possibility of freeze-cracking. Precooling prevents freeze-cracking because it reduces the differences in temperature between the product and the freezing medium. Precooling also reduces the time delay between the freezing of the border and the center of the system; thus the center of the food expands during ice formation at an earlier stage. When the phase change of the core region occurs before the surface becomes brittle, food products can support the internal pressure and freeze-cracking is not produced. Rapid freezing coupled with low final temperatures will nearly always result in severe cracking of specimens containing large percentages of water and that cracking was probably the result of nonuniform contraction following solidification.

3. Moisture Migration

During frozen storage, the existence of temperature gradients within a product may result in moisture migration, relocating the water within the product. This is a consequence of the temperature dependence of water vapor pressure. Water vapor will tend to transfer to regions of low vapor pressure. There is an overall tendency for moisture to move into the void spaces around the product and to accumulate on the product surface and on the internal package surface. In packaged frozen food, moisture migration [39] leads to ice formation inside the package. Temperature fluctuations (cooling–warming cycles) lead to a net migration of moisture from the interior towards the surface of the foodstuff, or to the wrap. The temperature of the packaging material follows the temperature fluctuations in the room faster than the product itself. As the surrounding temperature decreases, moisture inside the pores sublimes and diffuses to the packaging film; when ambient temperature increases, the ice on the wrap tends to diffuse back to the surface of the food, however, reabsorption of water in the original location is impossible, and the process can be considered irreversible. Moisture migration can be minimized by maintaining small temperature fluctuations and small internal temperature gradients and by the inclusion of internal barriers within the product and within the packaging.

4. Freezer Burn

Freezer burn is a surface desiccation defect that can occur when frozen tissues are stored without an adequate moisture barrier packaging. It manifests as an opaque dehydrated surface, produced by moisture losses in frozen foods. Freezer burn increases oxygen contact with the food surface area and raises oxidative reactions, which irreversibly alter color, texture, and flavor. It is caused by the sublimation of ice on the surface region of the tissue where the water pressure of the ice is higher than the vapor pressure in the environment. In cold storage rooms, the temperature of the freezing coil is always lower than the surrounding air therefore ice will form and accumulate on the coil. As moisture is removed, the relative humidity of the air in the cold room drops. As the water vapor pressure over the surface of the frozen product is higher than that of the air a constant loss of water in the form of vapor is produced from unprotected materials. Because it is difficult for moisture to transfer back to the initial location of the void, sublimation continues as long as this vapor pressure difference continues. Glazing, dipping, or spraying a thin layer of ice on the surface of a unwrapped frozen product helps to prevent drying. Freeze burn is prevented if a product is packed in tight-fitting, water- and vapor-proof material, because evaporation cannot take place.

5. Recrystallization of Ice

During frozen storage, ice crystals undergo metamorphic changes. In frozen aqueous solutions, recrystallization is the process by which the average ice crystal size increases with time. Small ice crystals are thermodynamically unstable, having a high surface-to-volume ratio and therefore a high excess of surface free energy. To minimize free energy, the net result is that the number of crystals decreases at constant ice phase volume but their mean size increases [39,40]. Recrystallization reduces the advantages of fast freezing and includes any change in the number, size, shape, orientation, or perfection of crystals following completion of initial solidification [4]. Recrystallization basically involves small crystals disappearing, large crystals growing, and crystals fusing together and affects the quality of the products because small ice crystals promote a better quality, meanwhile large crystals often produces damage during freezing. As the temperature of an aqueous specimen increases within the subfreezing range, the rate of recrystallization also increases. Recrystallization in frozen systems has been studied in detail by Luyet and coworkers [41,42]. There are different types of recrystallization processes described in literature [4,7]: (a) isomass, (b) migratory, (c) accretive, (d) pressure-induced, and (e) irruptive.

a. *Surface Isomass Recrystallization*

This includes changes in the shape or internal structure of a crystal and reduction of the defects as the crystal tends to a lower energy level maintaining a constant mass of ice. This "rounding off" process may be produced by surface diffusion of the water molecules. Ice crystals of irregular shape and large surface-to-volume ratio (dendritic crystals) adopt a more compact configuration with a smaller surface-to-volume ratio and a lower surface energy. Sharper surfaces are less stable than flatter ones and will show a tendency to become smoother over time.

b. *Migratory Recrystallization or Grain Growth*

This refers to the tendency of large crystals in a polycrystal system to grow at the expense of the smaller ones. Ostwald ripening refers to migratory recrystallization that occurs at constant temperature and pressure due to differences in surface energy between crystals.

Melting–diffusion–refreezing or sublimation–diffusion–condensation are possible mechanisms leading to an increase in average crystal size, a decrease in the number of crystals, and a decrease in surface energy of the entire crystalline phase. At constant temperature and pressure, migratory recrystallization is the result of differences in the surface energies of large and small crystals. The small crystals, with a very small radii of curvature, cannot bind their surface molecules as firmly as larger crystals, thus, small crystals exhibit lower melting points than large ones. Migratory recrystallization is enhanced by temperature fluctuation inducing a melt–refreeze behavior due to ice content fluctuations. Melt–refreeze behavior can lead to complete disappearance of smaller crystals during warming and growth of larger crystals during cooling, or to a decrease in size of crystals during partial melting and regrowth of existing crystals during cooling. Melt–refreeze should occur to a greater extent at higher temperatures and more rapidly for smaller crystals.

c. Accretive Recrystallization

This is produced when contacting crystals join together increasing crystal size and decreasing the number of crystals and surface energy of the crystalline phase. The proposed mechanism of crystal aggregation is surface diffusion. Accretion refers to a natural tendency of crystals in close proximity to fuse together; the concentration gradients in the areas between them are high, thus, material is transported to the point of contact between crystals and a neck is formed. Further "rounding off" will occur because a high curvature surface like this has a natural tendency to become planar.

The number of molecules leaving a curved surface is larger than the number of molecules arriving on that surface. The continuous exchange of molecules at the interface serves to reduce the curvature of a single crystal (forming a sphere) or to reduce the number of small crystals by adding to the larger crystals.

d. Pressure-Induced Recrystallization

If force is applied to a group of crystals, those crystals with their basal planes aligned with the direction of force will grow at the expense of those in other orientations. This type of recrystallization can result in an increase in crystal size, a decrease in the number of crystals, and a reorientation so that more crystals will have their *c*-axis normal to the direction of the force. This type of recrystallization is uncommon in foods or living matter.

e. Irruptive Recrystallization

Under conditions of very fast freezing, aqueous specimens will solidify in a partially noncrystalline state and not all the freezable water is converted to ice. Upon warming to some critical temperature, crystallization of ice will be produced abruptly. This phenomenon is described as "irruptive recrystallization", however "devitrification" is also used when the frozen specimen is totally noncrystalline after initial solidification.

Rates of ice recrystallization in frozen solutions and in frozen muscle tissue were reported by Zaritzky and coworkers [43–46] proposing that the driving force for recrystallization of ice is the difference in the surface energy of two adjacent crystals, with this energy being proportional to the crystal curvature. Ice crystal size distributions were measured from the micrographs and a direct relationship between crystal size and the number of sides in the crystal was established; small crystals with three or four sides show concave surfaces and tend to disappear because the crystal boundaries move toward the center of curvature. Six-sided crystals have planar surfaces and are stable, and those with a higher number of sides tend to grow. Histograms of the relative frequencies of crystal diameters as a function of equivalent diameter were obtained for different freezing rates and storage conditions. In meat tissues, it was demonstrated that ice crystal size reaches a limiting value D_l, which is related to the tissue matrix characteristics. The following equation was proposed considering that the driving force of this phenomenon is the difference between the instantaneous curvature of the system and the limit curvature [46]:

$$\frac{\mathrm{d}D}{\mathrm{d}t} = k\left(\frac{1}{D} - \frac{1}{D_l}\right) \tag{1.12}$$

where D is the mean equivalent ice crystal diameter at time t, D_l the limit equivalent diameter, and k the kinetic constant. Integration of Equation (1.12) leads to the following expression:

$$\ln\left(\frac{D_l - D_0}{D_l - D}\right) + \frac{1}{D_l}(D_0 - D) = \frac{k}{D_l^2}t \tag{1.13}$$

where D_0 is the mean initial equivalent diameter. This model satisfactorily fitted experimental data at short and long storage times [44].

Further, the Ostwald ripening principles were applied to ice recrystallization in food systems; this theory predicts that the recrystallization process can be described by [47]:

$$D = D_0 + kt^{1/n} \tag{1.14}$$

where D is the mean crystalline diameter, D_0 the initial diameter, k the recrystallization rate, and n the power law exponent. Recrystallization was studied in either model sugar systems or in ice cream [40,48,49] and results showed that ice crystals increased in size as a function of time to a power $(1/n)$ between 0.33 and 0.5 [47].

Hydrocolloid stabilizers (locust bean gum, guar gum, carrageenan, xanthan gum) are often added to foods to control ice recrystallization. The addition of hydrocolloids is important in the case of ice cream [48–50] and also in frozen gelatinized starch-based systems [51–55]. However, the action mechanisms of the different stabilizers on ice recrystallization is still not clear.

D. Chemical Changes Produced by Freezing

1. Concentration of Nonaqueous Constituents During Freezing

During the freezing of aqueous solutions, cellular suspensions or tissues, water is transferred into ice crystals and the nonaqueous constituents concentrate in the unfrozen solution [4]. When a solution is frozen slowly over a range of temperatures wherein eutectics do not form, the ice crystals have no impurities, equilibrium conditions can be reached, and the concentration of the unfrozen solution depends only on temperature. Slow freezing results in a maximum ice crystal purity and maximum concentration of solutes in the unfrozen phase. In contrast, rapid freezing results in a considerable entrapment of solutes by growing crystals and a lower concentration of solutes in the unfrozen phase. After freezing, many solutes may be supersaturated in the unfrozen phase; they may crystallize or precipitate and eutectic mixtures may be formed changing the relative concentration of solutes. The increasing concentration of solutes in the unfrozen matrix increases the ionic strength and can produce pH changes affecting the biopolymer structures in the matrix. Charged molecules may react differently because of the increasing ionic strength; time of exposure to high solute concentrations during the freezing or thawing process can lead to significant modifications. Besides the unfrozen phase may change other properties during freezing, such as titratable acidity, viscosity, surface and interfacial tension, and oxidation–reduction potential; water structure and water–solute interactions may be altered and interactions between macromolecules such as proteins increase. Solute concentration affects protein aggregation and precipitation, with many of these reactions being irreversible. Changes in pH during freezing of buffer systems were caused either by ice formation alone (concentration of solutes in the unfrozen phase) or by crystallization of buffer salts in conjunction with ice formation (eutectic formation) Changes of 0.3–2.0 pH units were reported in some tissues during frozen storage [4].

2. Effect of Freezing on Chemical Reactions

Freezing can give unusual effects on chemical reactions. Temperature and concentration of the reactants in the unfrozen phase (freeze concentration effects) are the main factors responsible for changes in the kinetics of enzymatic and nonenzymatic reactions during freezing. Furthermore, decompartmentation of cell organelles in tissues during freezing causes mixing of cell components affecting reaction rates; thus reactions that normally do not occur in intact cells may occur as a consequence of the freezing process. In food tissues the formation of ice crystals can release enzymes and chemical substances from enclosed contents affecting the product during freezing and storage, leading to quality deterioration. Most enzymes exhibit substantial activity after freezing and thawing and many enzymes show significant activity in partially frozen systems. Freeze-induced rate enhancements are also common to many kinds of nonenzymatic reactions.

In many frozen systems, reaction rates as a function of temperature go through a maximum at some temperature below the initial freezing point. This is a consequence of opposing factors: low temperatures that decrease reaction rates, and increasing solute concentration in the unfrozen phase that may increase rates [4]. Freezing or thawing processes in which the system stays around $-5^{\circ}C$ for an extended period often show significant solute-induced freezing damage. For example oxidation of myoglobin (meat pigment) was accelerated at temperatures close to $-5^{\circ}C$ [56,57]. Although many freeze-induced rate enhancements can be explained by the freeze-concentration effect, one or more of the following factors may also be involved: a possible catalytic effect of ice crystals, a greater proton mobility in ice than in water, a favorable substrate–catalyst orientation caused by freezing or a greater dielectric constant for water than ice [4].

Important chemical changes that can occur during freezing and frozen storage are enzymatic reactions, protein denaturation, lipid oxidation, degradation of pigments and vitamins, and flavor deterioration.

a. *Enzyme Activity*

Storage at low temperatures can decrease the activity of enzymes in tissues but not inactivate them. In raw products, hydrolytic enzymes (hydrolases such as lipases, phospholipases, proteases, and so on, which catalyze the transfer of groups to water) may remain active during frozen storage. Hydrolytic enzymes can produce quality deterioration in products that are not submitted to thermal treatments before freezing however blanching of vegetables or cooking of meat inactivate these enzymes [35].

Lipolytic enzymes such as lipases and phospholipases, hydrolyze ester linkages of triacylglycerols and phospholipids, respectively. If they are not controlled during storage, the hydrolysis of lipids can lead to undesirable flavor and textural changes. Certain lipases can remain active in frozen food systems stored even at $-29^{\circ}C$. Lipase activity is evident in the accumulation of free fatty acids. Freezing may accentuate lipolysis by disrupting the lysosomal membrane that releases hydrolytic enzymes, especially at low freezing rates and under fluctuating temperatures. The increase of salt concentration during freezing may accelerate lysosomal release of lipases. During storage, lipolytic activity has detrimental consequences; the release of short-chain free fatty acids can lead to hydrolytic rancidity, producing off flavors and may interact with proteins, forming complexes that affect texture. Proteases catalyzes the hydrolysis of proteins to peptides and aminoacids; in meat this endogenous enzymes are considered beneficial, providing tenderization of the muscle during rigor mortis [35]. Conditioned meat on freezing not only retained the texture quality, but also has a smaller tendency to drip on thawing.

The browning of plant tissue is caused by enzymatic oxidation of phenolic compounds in the presence of oxygen. Disruption of cells by ice crystals can start enzymatic browning by facilitating contact between *o*-diphenol oxidase and its substrate. The oxido-reductases are of primary importance because their action leads to off flavor and pigment bleaching in vegetables, and to browning in some fruits.

In vegetable and fruit tissues, endogenous pectin methyl estearases catalyzes the removal of the methoxyl groups from pectins. Hydrolytic enzymes, like chlorophylases and anthocyanases present in plants, may catalyze destruction of pigments in frozen tissues affecting the color, if they are not inactivated by blanching.

Hydrolytic rancidity, textural softening, and color loss are direct consequences of hydrolytic enzyme activities, although textural toughening and acceleration of lipid oxidation may be secondary consequences.

b. Protein Denaturation

The main causes of freeze-induced damage to proteins are ice formation and recrystallization, dehydration, salt concentration, oxidation, changes in lipid groups, and the release of certain cellular metabolites. During freezing, proteins are exposed to an increased concentration of salts in the unfrozen phase; the high ionic strength can produce competition with existing electrostatic bonds, modifying the native protein structure. Losses in functional properties of proteins are commonly analyzed by comparing water-holding capacity, viscosity, gelation, emulsification, foaming, and whipping properties. Freezing has an important effect in decreasing water-holding capacity of muscle systems on thawing. This decrease occurs during freezing because water–protein associations are replaced by protein–protein associations or other interactions [58].

Dehydration of the cells caused by ice formation is an important factor that leads to protein denaturation. Proteins exposed to the aqueous medium of the biological tissues have a hydrophobic interior and charged (or polar) side chains in the surface. The migration of water molecules from the interior of the cells during extracellular freezing leads to a more dehydrated state disrupting protein–solvent interactions; protein molecules exposed to a less polar medium increase the exposure of hydrophobic chains modifying protein conformation; protein–protein interactions are produced to maintain the minimum free energy, resulting in protein denaturation and formation of aggregates.

c. Lipid Oxidation

It is produced in frozen foods leading to loss of quality (flavor, appearance, nutritional value, and protein functionality). Lipid oxidation is a complex process that proceeds upon a free radical process [59]. During the initiation stage, a hydrogen atom is removed from a fatty acid, leaving a fatty acid alkyl radical that is converted in the presence of oxygen to a fatty acid peroxyl radical. In the next step, the peroxyl radical abstracts a hydrogen from an adjacent fatty acid forming a hydroperoxide molecule and a new fatty acid alkyl radical. Breakdown of the hydroperoxide is responsible for further propagation of the free radical process. Decomposition of hydroperoxides of fatty acids to aldehydes and ketones is responsible for the characteristic flavors and aromas. Enzymatic and nonenzymatic pathways can initiate lipid oxidation. One of the enzymes that is considered important in lipid oxidation is lipoxygenase that is present in many plants and animals. Lipoxygenase is the main enzyme responsible for pigment bleaching and off odors in frozen vegetables. If the enzymes are not inactivated by blanching, they can generate offensive flavors and also loss of pigment colors.

VII. CONCLUSIONS

Freezing is one of the best methods for food preservation. Water is removed and converted into ice crystals causing complex modifications. The analysis of the physicochemical aspects involved in ice formation allows a better understanding of the different phenomena occurring during freezing.

Life of animals and plants is a water-based phenomenon. The unusual properties of water and ice arise from their angled shape and the hydrogen intermolecular bonds that they can form. Supercooling is necessary to overcome the free energy that accompanies the formation of a new phase (an ordered solid particle) from the melted phase. At a given temperature, which depends mainly on the rate of cooling and the sample volume, nucleation will occur. Small clusters of molecules with an ice-like structure are continuously forming and breaking up; if one of these nuclei reaches a critical size, then it becomes energetically favorable for more water molecules to grow on this nucleus, and the ice will propagate rapidly through the entire sample.

Although in small samples high freezing rates produced a large number of ice crystals, in large samples, nucleation is only produced in the zone that is in contact with the refrigerant. The size of the ice crystals depends on freezing rate. Constitutional supercooling explains the formation of an irregular ice–liquid interphase and ice columns growing from the border to the thermal center.

In the case of cells and tissues the presence of intra and extracellular ice has influence on the damage produced by freezing. Intracellular ice is only formed at high freezing rates; slow freezing produces water dislocation, cellular dehydration, and extracellular ice. A cell wall, or membrane, that is not a good barrier to water molecules will favor intracellular dehydration and growth of extracellular ice. In an attempt to balance the chemical potential, intracellular water migrates outward, leading to cell dehydration, and to an increase in the ionic strength of the cell. Ice crystal growth can cause membrane distortions and stress on rigid structures, producing mechanical damage.

If a system is cooled sufficiently quickly so that nucleation cannot occur then it is possible to avoid ice formation. This process is called vitrification and results in an amorphous solid or glass. Achieving vitrification with pure water requires very small amounts of water and very high cooling rates, although with high concentrations of solutes, solutions can be vitrified relatively easily.

Freezing damage is associated with ice formation, either directly through the mechanical effects produced by ice crystals or indirectly by an increase in solute concentration in the unfrozen phase. Changes in the ionic strength of the unfrozen solution affect functional properties and stability of biomolecules, chemical reaction kinetics, and water-holding capacity.

Physical changes in frozen foods include drip losses, moisture migration, freeze-cracking, and ice recrystallization. The growth in size of ice crystals can influence the damage during frozen storage and therefore lead to loss in quality. Recrystallization at constant or fluctuating temperatures occurs because systems tend to move toward a state of equilibrium where free energy is minimized.

Chemical changes that can be detected during freezing and frozen storage are protein denaturation, lipid oxidation, enzymatic browning, flavor deterioration, and degradation of pigments and vitamins. Formation of ice crystals can cause disruption in the frozen tissues, leading to the release of enzymes and chemical substances that affect food quality. All these physicochemical changes contribute to affecting the quality of frozen food.

NOMENCLATURE

A	area (m^2)
B, C	coefficients in Equation (1.9)
C'_g	concentration of solute within the glass (mass fraction of solute)
D	mean equivalent ice crystal diameter at time t (μm)
D_l	limit equivalent diameter of the crystal (μm)
D_0	mean initial equivalent diameter of the crystal (μm)
G	rate of ice crystal growth (μm sec^{-1})
J	rate of nucleation (number of nuclei formed per unit volume and time)
k	recrystallization kinetic constant
L_f	volume latent heat of fusion (J m^{-3})
m	molality of the solution representing the food system (number of moles of solute per 1000 g of solvent)

M_A	molecular weight of water (18 g mol^{-1})
MW	molecular weight (g mol^{-1})
n	power law exponent.
r	radius of the ice nucleus (μm)
r^*	critical radius of the nucleus at which equilibrium is established (μm)
R	universal gas constant (8314 Pa m^3 kg mol^{-1} K^{-1})
t	time (min)
t_c	characteristic local freezing time (min)
T	temperature (°C)
T_{eu}	eutectic temperature (°C)
T_f	freezing temperature of pure water (0°C)
T_{fs}	initial freezing point of foodstuffs (systems containing solutes) (°C)
T_g	glass transition temperature (°C)
T'_g	maximally freeze concentrated glass transition temperature of the frozen system (°C)
T_K	absolute temperature (K)
T_{Kf}	absolute freezing temperature of pure water (273 K)
T_m	melting point for ice (°C)
v	molar volume of water (m^3 mol^{-1})
V	volume of the sphere (m^3)
W	work to increase the surface (J)

Greek letters

β	coefficient in Equation (1.10)
ΔG_p	volume free energy difference (J m^{-3})
ΔH	volume enthalpy change (J m^{-3})
ΔP	pressure difference (Pa)
ΔT_f	freezing point depression ($\Delta T_f = T_f - T_{fs}$) (°C)
ΔT_s	Supercooling or undercooling ($\Delta T_s = T_f - T$ for pure water or $\Delta T_s = T_{fs} - T$ for systems containing solutes) (°C)
σ	surface tension or free energy per unit area of the interface (J m^{-2}).
ΔS	entropy change (J m^{-3} K^{-1})
λ_A	molar latent heat of fusion for pure water (6003 kJ mol^{-1})

REFERENCES

1. O Fennema. Water and ice. In: O Fennema, Ed., *Food Chemistry*, 3rd ed., New York: Marcel Dekker, 1996, pp. 17–94.
2. R Ruan, PL Chen. *Water in Foods and Biological Materials. A Nuclear Magnetic Resonance Approach*, New York: Technomic Publishing Co., 1998, pp. 51–71.
3. M Chaplin. Water structure and behavior, Water and aqueous systems research, London, South Bank University, London, UK, 2004.
4. OR Fennema, WD Powrie, EH Marth. *Low Temperature Preservation of Foods and Living Matter*, New York: Marcel Dekker, 1973, pp. 3–207.
5. CW Kern, M Karplus. The water molecule. In: F Franks, Ed., *Water. A Comprehensive Treatise*, Vol. 1, New York: Plenum Press, 1972, pp. 21–91.
6. PL Silvestrelli, M Parrinello. Structural, electronic, and bonding properties of liquid water from first principles. *Journal of Chemical Physics* 111:3572–3580, 1999.
7. R Hartel. *Crystallization in Foods*. Maryland: Aspen Publishers Inc., 2001, pp. 192–231.
8. A Calvelo. Recent studies on meat freezing, Chapter 5. In: R Lawrie, Ed., *Developments in Meat Science*. Vol. 2. Barking, UK: Applied Science Publishers, 1981, pp. 125–158.

9. F Franks. *Biophysics and Biochemistry at Low Temperatures*. Cambridge University Press, Cambridge, 1985, pp. 39–52.
10. Y Roos, M Karel. Applying state diagrams to food processing and development. *Food Technology*, 45:66–71, 107, 1991.
11. Y Roos. *Phase Transitions in Foods*, Food Science and Technology International Series. San Diego: Academic Press, 1995.
12. L Riedel. Calorimetric investigations of the meat freezing process. *Kaltetechnik*, 9:38–40, 1957.
13. L Slade, H Levine. Beyond water activity: recent advances based on an alternative approach to the assessment of food quality and safety. *Critical Reviews in Food Science and Nutrition.*, 30:115/360, 1991.
14. DR Heldman. Predicting the relationship between unfrozen water fraction and temperature during food freezing using freezing point depression, *ASHRAE Transactions*, 91 (part 2B), 371–384, 1974.
15. N Zaritzky. Factors affecting the stability of frozen foods. In: CJ Kennedy, Ed., *Managing Frozen Foods*. Cambridge, England: CRC, Woodhead Publishing Limited, 2000, pp. 111–133.
16. Anonymous, Recommendations for the processing and handling of frozen foods. International Institute of Refrigeration, Paris, 1972, pp. 82.
17. M Le Meste, D Champion, G Roudaut, G Blond, D Simatos. Glass transition and food technology. A critical appraisal. *Journal of Food Science* 67(7):2444–2458, 2002.
18. L Slade, H Levine. Glass transitions and water-food structure interactions. In: JE Kinsella, SL Taylor, Eds., *Advances in Food Nutrition Research*. Vol. 38, San Diego: Academic Press, 1995, pp. 103–269.
19. J Wolfe, G Bryant. Freezing, drying and/or vitrification of membrane-solute-water systems. *Cryobiology*, 39:103–129, 1999.
20. HD Goff. Measurement and interpretation of the glass transition in frozen foods. In: MC Erickson, YC Hung, Eds., *Quality in Frozen Food*. New York: Chapman & Hall, 1997, pp. 29–50.
21. HD Goff. Measuring and interpreting the glass transition in frozen foods and model systems. *Food Research International*, 27:187–189, 1994.
22. Y Roos, M Karel. Non-equilibrium ice formation in carbohydrate solutions. *Cryo Letters* 12:367–376, 1991.
23. H Levine, L Slade. Principles of cryostabilization technology from structure/property relationships of carbohydrate/water systems – A review. *Cryo Letters* 9:21–60, 1988.
24. H Levine, L Slade. Eds., *Water Relationship in Foods*. New York: Plenum Press, 1991, pp. 251–273.
25. Y Roos, M Karel. Amorphus and state and delayed ice formation in sucrose solutions. *International Journal of Food Science and Technology* 26:553–566, 1991.
26. DS Reid. Overview of physical/chemical aspects of freezing. In: MC Erickson, YC Hung. Eds., *Quality in Frozen Food*. New York: Chapman & Hall, 1997, pp. 10–28.
27. DS Reid. Basic physical phenomena in the freezing and thawing of plant and animal tissue, In: CP Mallet, Ed., *Frozen Food Technology*. Glaslow: Blackie Academic and Professional,1994, pp. 1–19.
28. P Mazur. Physical and chemical basis of injury in single celled microorganisms subjected to freezing and thawing, In: HT Meryman, Ed., *Cryobiology*. New York: Academic Press, 1966, pp. 213–315.
29. P Mazur. *Cryobiology*, The freezing of biological systems, *Science* 168:939–949, 1970.
30. K Muldrew, LE McGann. The osmotic rupture hypothesis of intracellular freezing injury. *Biophysical Journal* 66:532–541, 1994.
31. A Bevilacqua, N Zaritzky, A Calvelo. Histological measurements of ice in frozen beef. *Journal of Food Technology* 14:237–251, 1979.
32. A Bevilacqua, N Zaritzky. Ice morphology in frozen beef, *Journal of Food Technology* 15:589–597, 1980.
33. P Mazur. Freezing of living cells: mechanisms and implications. *American Journal of Physiology*, 143:C125–C142, 1984.
34. GA Mac Donald, TC Lanier. Cryoprotectants for improving frozen food quality. In: MC Erickson, YC Hung, Eds., *Quality in Frozen Food*. New York: Chapman & Hall, 1997, pp. 197–232.
35. RV Sista, MC Erickson, RL Shewfelt. Quality deterioration in frozen foods associated with hydrolitic enzyme activities. In: MC Erickson, YC Hung, Eds., *Quality in Frozen Food*. New York: Chapman & Hall, 1997, pp. 101–110.

36. N Zaritzky, M Añón, A Calvelo. Rate of freezing effect on the colour of frozen beef liver. *Meat Science* 7:299–312, 1982.
37. YC Hung. Freeze cracking. In: MC Erickson, YC Hung, Eds., *Quality in Frozen Food.* New York: Chapman & Hall, 1997, pp. 92–99.
38. NK Kim, YC Huang. Freeze cracking in foods as affected by physical properties. *Journal of Food Science* 59:669/664, 1994.
39. QT Pham, RF Mawson. Moisture migration and ice recrystallization in frozen food. In: MC Erickson, YC Hung, Eds., *Quality in Frozen Food.* New York: Chapman & Hall, 1997, pp. 67–91.
40. RL Sutton, A Lips, G Piccirillo, A Sztehlo. Kinetics of ice recrystallization in aqueous fructose solutions. *Journal of Food Science* 61 (4):741–745, 1996.
41. G Rapatz, BJ Luyet. Recrystallization at high subzero temperatures in gelatin gels subjected to various subcooling treatments. *Biodynamica* 8:85–105, 1959.
42. AP Mac Kenzie, BJ Luyet. Electron microscope study of recrystallization in rapidly frozen gelatin gels. *Biodynamica* 10:95–122, 1967.
43. AE Bevilacqua, NE Zaritzky. Ice recrystallization in frozen beef. *Journal of Food Science*, 47:1410–1414, 1982.
44. MN Martino, NE Zaritzky. Ice recrystallization in a model system and in frozen muscle tissue, *Cryobiology* 26:138–148, 1989.
45. M Martino, N Zaritzky. Effects of temperature on recrystallization of polycrystalline ice. *Sciences des Aliments*, 7:147–166, 1987.
46. MN Martino, NE Zaritzky. Ice crystal size modifications during frozen beef storage. *Journal of Food Science* 53:1631–1637, 1649, 1988.
47. DP Donhowe, R Hartel. Recrystallization of ice in ice cream during controlled accelerated storage. *International Dairy Journal* 6:1191–1208, 1996.
48. EK Harper, CF Shoemaker. Effect of locust beam gum and selected sweetening agents on ice recrystallization rates. *Jounal of Food Science* 48:1801–1803, 1983.
49. RL Sutton, A Lips, G Piccirillo. Recrystallization in aqueous fructose solutions as affected by locust bean gum. *Jounal of Food Science* 61 (4):746–748, 1996.
50. A Regand, HD Goff. Structure and ice recrystallization in frozen stabilized ice cream model solutions, *Food Hydrocolloids* 17:95–102, 2003.
51. C Ferrero, M Martino, N Zaritzky. Stability in frozen starch pastes. Effect of freezing storage and xanthan gum addition. *Journal of Food Processing and Preservation* 17 (3):191–211, 1993.
52. C Ferrero, M Martino, N Zaritzky. Corn starch, xanthan gum interaction and its effect on the stability during storage of frozen gelatinized suspensions. *Starke* 46:300–308, 1994.
53. C Ferrero, M Martino, N Zaritzky. Effect of freezing rate and xanthan gum on the properties of corn starch and wheat flour pastes. *International Journal of Food Science and Technology* 28:481–498, 1993.
54. C Ferrero, M Martino, N Zaritzky. Effect of hydrocolloids on starch thermal transitions, as measured by DSC. *Journal of Thermal Analysis* 47:1247–1266, 1996.
55. C Ferrero, N Zaritzky. Effect of freezing rate and frozen storage on starch-sucrose-hydrocolloid systems. *Journal of the Science of Food and Agriculture* 80:2149–2158, 2000.
56. MC Lanari, AE Bevilacqua, NE Zaritzky. Pigments modifications during freezing and frozen storage of packaged beef. *Journal of Food Process Engineering* 12:49–66, 1990.
57. MC Lanari, NE Zaritzky. Effect of packaging and frozen storage temperatures on beef pigments. *International Journal of Food Science and Technology* 26:467–478, 1991.
58. YL Xiong. Protein denaturation and functionality losses. In: MC Erickson, YC Hung, Eds., *Quality in Frozen Food.* New York: Chapman & Hall, 1997, pp. 111–140.
59. MC Erickson. Lipid oxidation: flavor and nutritional quality deterioration in frozen food. In: MC Erickson, YC Hung, Eds., *Quality in Frozen Food.* New York: Chapman & Hall, 1997, pp. 141–173.

2 Glass Transitions in Frozen Foods and Biomaterials

Stefan Kasapis
National University of Singapore, Singapore

CONTENTS

I. INTRODUCTION

High solid systems refer mainly to mixtures of biopolymers and co-solutes and as such are increasingly popular in the industrial world [1]. The mixtures are used as a base to formulate products with a variety of textures and sensory stimuli but mechanistic knowledge behind these properties has been lacking. In recent times, the importance of the rubber to glass transition and the development of the glassy state became widely appreciated in the understanding and controlling the quality of materials [2]. The emphasis now is on mapping out the relationship between the kinetics of vitrification and the metastability of systems to produce innovative methods of processing and product formulations [3].

Popular science dictionaries define glass as a liquid, which is unable to flow during the timescale of practical observation. Molecules in the liquid remain in a random orientation due to the viscosity increase that prevents them from arranging into regular patterns. Therefore, the essential requirement for glass formation is a high cooling rate to inhibit preliminary nucleation and crystal growth. The temperature at which the sample acquires glassy consistency is known as T_g but it is not that well defined as, for example, the melting point (T_m), because the process of vitrification may take place over a wide range of temperatures. The resulting glassy system is thermodynamically unstable, but derives kinetic stability from its high viscosity [4].

There has been an extensive work in the literature about the vitrification of pure compounds. For instance, values of $T_{g(anh)}$ of some members of the series of glucose carbohydrates are: 38.5°C for glucose, 95°C for maltose, 130°C for maltotriose, 175°C for maltohexose, and about 185°C for starch [5]. Examples of partial and total glassy behavior include hair, dry cotton shirts, biscuits, coffee granules, pasta, spaghetti, ice cream, as well as inorganic oxide systems, organic and inorganic polymers, and carbohydrate or protein matrices in aqueous environment or in mixture with high levels of sugars. An important consideration in the discussion of the behavior of these foodstuffs is the concept of plasticization and its effect on the glass transition temperature [6]. A plasticizer is defined as a substance incorporated in a material to increase the material's workability, flexibility, or extensibility. For example, proteins or polysaccharides are plasticized by low-molecular-weight diluents. Water is the most effective diluent-plasticizer and increasing concentrations dramatically reduce the glass transition temperature.

Although the glassy consistency is widely observed, a theoretical treatment is far from trivial. Various ideas have been put forward to rationalize the discontinuities in molecular processes observed in the vicinity of T_g, but a simple unified theory of the phenomenon is yet to be achieved. The prevailing theories focus on thermodynamic, kinetic, or free volume aspects and use a single property or parameter to characterize the glass [7]. These are described as:

(1) The process is considered to be a second-order thermodynamic transition in which the material undergoes a change in state but not in phase. A first-order transition exhibits a discontinuity in the primary thermodynamical variables of volume, enthalpy, and free energy. Instead, the glass transition region records marked changes in the first derivative variables of the coefficient of expansion (α_p), heat capacity (C_p), and so on [8]. Furthermore, the spike in α_p and C_p observed at the crystallization temperature (first-order transition) has no counterpart during vitrification. The theory argues that if measurements could be made infinitely slow, the true underlying transition temperature, T_2, would be attained, at which the configurational entropy of the system becomes zero. Using the quasi-lattice model of Flory [9], the energy barrier to intramolecular rotation was identified as the most critical variable and the T_2 was calculated to lie approximately 50 K below the experimental T_g. The theory was successful in predicting the effects of molecular weight, copolymerization, plasticization, and crosslinking on T_g but the validity of describing a kinetically determined transition as a system at equilibrium is questionable. Furthermore, T_2 cannot be measured experimentally and thus its existence cannot be proved.

(2) The experimental measurement of the glass transition temperature is kinetically determined because it depends on the applied frequency of oscillation, cooling or heating rate, and sample history [10]. Work has been carried out calorimetrically and experiments involved annealing the sample to a temperature above the experimental T_g until equilibrium was established and then cooling rapidly to the temperature of interest. The temperature jumps demonstrated considerable volume relaxation and hysteresis effects in materials. A measure of the time-dependent relaxation modes in the glassy state could be given by pinpointing a temperature at which the value of a property would approximate the equilibrium value [11]. Thus kinetic postulates do not attempt a molecular understanding of the glassy state, but rather model the observed rate-dependent behavior in terms of two or more relaxation timescales.

(3) The approach used extensively by material scientists to develop a mechanistic understanding of the rubber to glass transition is based on the concept of macromolecular free volume. According to Ferry [12], holes between the packing irregularities of long-chain segments or the space required for their string-like movements accounts for free volume (u_f). Adding to that the space occupied by the van der Waals radii of polymeric contours and the thermal vibrations of individual residues, that is, the occupied volume (u_0), we come up with the total volume per unit mass (u) of a macromolecule. In polymer melts, the proportion of free volume is usually 30% of the total volume and the theory predicts that it collapses to about 3% at the glass transition temperature [13]. At this point, the thermal expansion coefficient of free volume (α_f) undergoes a discontinuity, which reflects a change in slope in the graph of the linear dependence of total volume with temperature. A schematic representation of the concept of free volume is given in Figure 2.1.

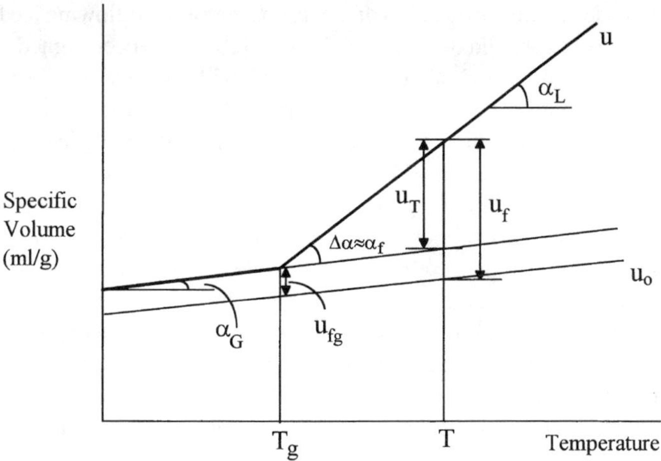

FIGURE 2.1 (1) If the occupied volume u_0 is a constant fraction of the total volume below T_g, then a line can be drawn nearly parallel to the total specific volume (u) below T_g, with the difference a small constant fraction of u. (2) Above the "knee" temperature T_g, the expansion of u_0 does not match the overall expansion, leaving an increasing volume difference that is termed u_f (free volume). (3) At and below T_g there is a certain small fraction of free volume u_{fg} that is assumed to be constant. (4) The difference between the expansivity below T_g (α_G) and that above T_g (α_L) leads to a free volume component increasing with temperature (u_T) according to the relation $u_T = (\alpha_L - \alpha_G)(T - T_g)$ if T is the temperature of observation. (5) The free volume total is $u_f = u_{fg} + u_T$, or $u_f = u_{fg} + (\alpha_L - \alpha_G)(T - T_g)$ as a function of temperature. The difference ($\Delta\alpha$) between α_L and α_G is written as α_f, the thermal expansion coefficient of the free volume.

The free volume concept is popular partly due to it being intuitively appealing. Often (but not invariably), it is able to explain the observed trends correctly in synthetic polymers, low-molecular weight organic liquids, and inorganic compounds, and is easy for researchers in materials science coming from many different backgrounds [14,15]. This has prompted calls for the universality of the approach in glass-forming systems where changes in the free volume appear to be independent of chemical features. Nevertheless, there is a tendency to apply the approach to a number of processes in frozen foods without a direct mechanistic justification, which shall be critically evaluated in this chapter.

II. UNFREEZABLE WATER

Although there is a great debate as to the physics of free water, which is related to water activity and the concept of bound water, headway in product development and preservation can be made by considering a "dynamic" portion of the water content as being unavailable for chemical or microbial processes. This is well known to a food technologist as the unfreezable water [16,17]. Cooling of an aqueous solution of protein or carbohydrate forms an unfrozen matrix suspending a discontinuous phase of ice crystals. High cooling rates, as compared with those of water diffusion and crystallization, result in matrices of low solute content and glass transition temperatures. Solute is amorphous and so is water, but the density and "nature" of the glass should be distinct from that of pure water occurring at $-134°C$ [18]. Slow cooling, on the contrary, creates conditions closer to equilibrium in the sense that ice formation is enhanced thus concentrating up the solute phase. Credit should be given to Franks [19] for pointing out that the water "kept" in the amorphous phase is not really bound in an energetic sense but it renders unfreezable due to the slow down of diffusional mobility at conditions close to T_g. In many respects, the physicochemical properties of these water molecules are closer to liquid water than ice.

There is an ongoing debate on the question of solute concentration following ice formation in these supercooled solutions when expressed on a dry weight basis [20]. It has been argued that the percentage of unfreezable water for different foods should be constant at 22 $\pm$ 2% or 0.28 $\pm$ 0.03 g water/g dry matter calculated by the latent heat of melting of ice (ΔH_m) [21,22]. However, work on commercial glucose (corn) syrups showed that the composition of unfrozen water can be bracketed within the range of 0.5–0.9 g per g of dry solute [23], i.e., much higher than the earlier suggestion. It was further established that for this homologous family, increasing molecular weight reduces the amount of bound water in the amorphous phase. A similar relationship did not hold for the non-homologous sugars, and polyols also used widely as water binders in frozen foods. Hatley et al. [24] suggested that, in view of the difficulties inherent to experimental methodology for estimating the amount of unfrozen water, measurements should be made from concentrated systems (0–20% water) in conjunction with a direct measurement on a freeze-concentrated sample. It remains to be seen if there is a valid universal observation for protein and carbohydrates in model systems and foodstuffs because the extent of interaction between water molecules and various materials is quite distinct.

III. THE CONCEPT OF "STATE DIAGRAM" IN FOOD SYSTEMS

Lately, technology transfer from the materials science of aqueous solid solutions to the functional attributes of ingredients used in the food processing industry occurred by developing the so-called "state diagram" [25]. In its simplest form, a state diagram represents the pattern of change in the physical state of a material as a function of increasing levels of solids [26]. The basic understanding of physical properties in foodstuffs can be extended by bringing together a combination of equilibrium and metastable processes as a function of time, pressure, and temperature of processing and preservation [27]. Figure 2.2 reproduces a schematic state diagram that shows the physical state of materials in relation to temperature and concentration.

Most foods are complex systems and on cooling various components would reach their saturation concentration at a specified temperature. Line AB represents equilibria between the dissolved and the crystalline phase of a given solute, and the positive slope is an indication of a more soluble state with increasing temperature [28]. AB is also known as the solubility or eutectic curve. In most cases, however, solute crystallization is avoided at the eutectic point owing to high viscosity at low

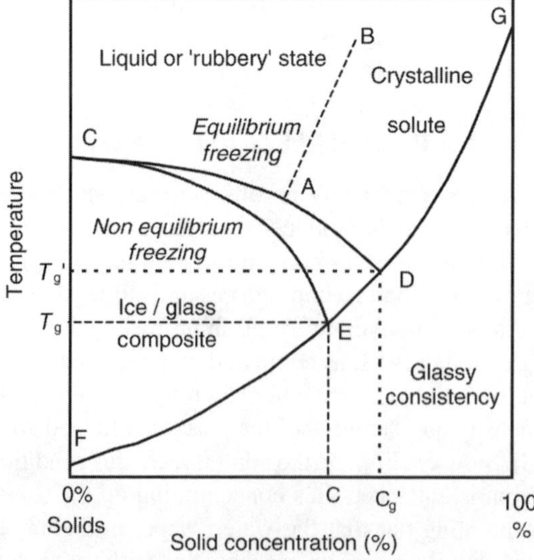

FIGURE 2.2 A schematic representation of the effect of concentration, temperature, and thermal rate on the stability of solutions and rubber-like materials, partially crystalline systems, and glassy products.

temperatures that prevents diffusional mobility of molecules. Instead, slow cooling of the preparation (say, from ambient temperature) will result in water crystallization, and equilibria between unfrozen water and ice formation are provided by the freezing curve (CD). This is also known as the liquidus curve and it develops a negative slope with increasing solute concentration due to depression of the freezing point [29].

To obtain points on the freezing curve, one has to place a stainless steel cylinder filled with sample in a programmable freezer for cooling. During experimentation, the scan rate is determined from the initial rate of cooling and it can be between 1 and 2°C/min. The temperature change as a function of time is logged by an automatic thermocouple positioned deep into the sample [30]. A typical cooling curve is shown in Figure 2.3. Cooling below the initial freezing point of a sample without formation of ice results in a supercooled state. Following initial supercooling, the critical mass of nuclei is reached with the system nucleating at point "a." At this stage, the sample releases its latent heat of fusion faster than the amount of heat removed from the system causing an instantaneous increase in temperature to the equilibrium freezing point "b" [31]. Temperatures at point "b" are usually considered for plotting the freezing curve in Figure 2.2 [32].

Several empirical and theoretical models have been used to predict the freezing behavior of foods in the state diagram. The theoretical Clausius–Clapeyron equation is well known in conjunction with implementing a nonlinear regression analysis, but there is a drawback in that it underestimates the "effective molecular weight" (EMW) of polymeric solids in the sample. The equation was designed to address freezing in ideal conditions, which approximate a very dilute solution. It can be improved by introducing a parameter for nonideal behavior due to unfreezable water (B = water unavailable for freezing/total solids), as follows [33]:

$$\delta = -\frac{\beta}{\lambda_w}\ln\left[\frac{1 - X_s - BX_s}{1 - X_s - BX_s + EX_s}\right] \tag{2.1}$$

where δ is the freezing point depression ($T_w - T_f$), T_f the freezing point of food (°C), T_w the freezing point of water (°C), β the molar freezing point constant of water (1860 kg K/kg mol), λ_w the molecular weight of water, X_s the mass fraction of solids, and E the molecular weight ratio of water and solids (λ_w/λ_s). Equation (2.1) deals successfully with the freezing properties of date pastes at which sugars are the main components and returns an EMW value of about 200 [34].

This prediction is slightly higher than the molecular weight of glucose or fructose and thus reflecting the contribution of small amounts of biomacromolecules to the cryohydric properties of the paste. Nevertheless, predicted values of EMW for the protein of selected seafood

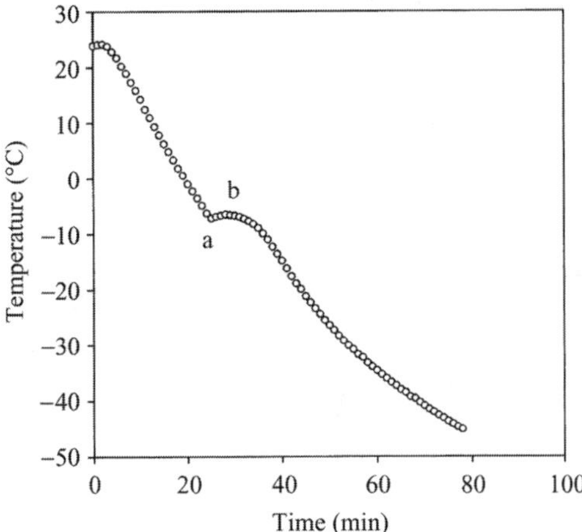

FIGURE 2.3 Cooling curve of tuna meat at 54.6% solids showing the onset of ice crystallization (a) and the equilibrium freezing point (b).

invertebrates and fish (e.g., shrimp, abalone, tuna, etc.) fall far short of the expected macromolecular dimensions published in the literature [35]. Furthermore, the fit of Equation (2.1) increasingly deviates from the experimental freezing points at high levels of solute. A possible explanation that merits further investigation argues that adding the concept of bound water is not sufficient to account for the complex process of a phase or state transition. This brings in context the macromolecular effects and the formation of a three-dimensional network that further enhances the immobility of water molecules, a theme which today is addressed as part of the process of vitrification [36].

Indeed, quenching of foodstuffs from the melt or the rubbery state eliminates crystallization and yields a glassy consistency at which there is a significant arrest of translational motions, with materials becoming extremely rigid. Curve FEDG reproduces this barrier below which compounds involved in deterioration reactions take many months or even years to diffuse over molecular distances and approach each other to react [37]. Once a few glass transition temperatures, including that of vitrified water, are obtained the state diagram can be completed using empirical modeling. In 1952, Gordon and Taylor [38] proposed an empirical equation to predict the glass transition temperature of mixtures comprising amorphous synthetic polymers. Today, this is commonly used to predict the vitrification properties of multicomponent mixtures, such as foodstuffs and biological materials [39]:

$$T_{gm} = \frac{X_s T_{gs} + k X_w T_{gw}}{X_s + k X_w} \tag{2.2}$$

where T_{gm}, T_{gs}, and T_{gw} are the glass transition temperatures of the mixture, solids, and water, respectively, X_s and X_w are the mass fraction of solids and water, and k is the Gordon–Taylor parameter, which from the thermodynamic standpoint is equivalent to the ratio of specific- heat change of components at their T_g [40]. The Gordon–Taylor equation can be recast in a linear form assuming that the weight fraction of the water molecules is negligible in samples subjected to prolonged drying. Thus the intercept and the gradient of the linearized form of the equation afford estimation of the T_{gs} and k, respectively. Equation (2.2) is able to follow the concentration dependence of vitrification in a wide range of foodstuffs and in the case of date pastes it produced $T_{gs} = 57.4°C$ and $k = 3.2$. The T_{gs} corresponds to point G in Figure 2.2. This appears to be a reasonable estimate, taking into account that the T_{gs} values of fructose and glucose are 5 and 31°C, respectively, and those of pectin and starch, that is, date components, are in excess of 100°C [41]. The predicted value of k is congruent with those in the literature for strawberries and horseradish (4.7 and 5.3, respectively [42]), but the parameter remains an index of reference without physical significance.

At intermediate rates of cooling, which are relevant to the lowering of temperature in a freezing food process, solute crystallization is avoided thus reaching a supersaturated state. Eventually, partial ice formation will take place and the remaining water with the solute will form an amorphous phase suspending the ice inclusions [43]. Regardless of the initial composition of the material, vitrification of the freeze-concentrated phase should yield two parameters, C_g and T_g, which reflect the physical state of the solute (point E in Figure 2.2 [44]). Lowering the cooling rate or annealing the sample at temperatures around T_g induces additional ice formation thus shifting point E to the right along the glass curve. Eventually points E and D will become coincident thus defining two equilibrium parameters, T_g' and C_g'. The former is the particular T_g of the maximally freeze-concentrated solute and water matrix surrounding the ice crystals in a frozen system [45]. The latter is the composition of solute at T_g'. The previous section discussed ideas regarding the levels of unfreezable water at equilibrium, which determine the range of C_g'. It has been further noted that as the molecular weight of soluble solids increases, the values of T_g' and C_g' move up the temperature and composition axes (toward 0°C and 100% solids, respectively [46,47]). Product manipulation in frozen food applications of this observation will be discussed in later sections.

IV. MEASUREMENT OF THE GLASS TRANSITION

A. Sample Preparation and Moisture Determination

Vitrification phenomena are determined on fresh and freeze-dried foods rehydrated to different moisture contents in preparation for freezing and subsequent analysis. In general, materials are brought chilled to the laboratory, and the mass and size of the sample are recorded. The water content and total solids of the fresh material are measured gravimetrically by drying in an air convection drier at 105°C at least for 2 h. The remaining samples are washed and frozen in an automatically controlled freeze-drier. The plate temperature and vacuum in the chamber and the condensing plate temperature are set for drying that takes a few days. Drying is designed to achieve a moisture content below 4% on a wet basis. The freeze-dried preparation is then homogenized in a laboratory-scale grinder to form powder, and samples are stored in an air-sealed container at refrigeration temperature for further use.

Composition analysis on several samples involves averaging measurements of protein, fat, and ash according to the Association of Official Analytical Chemists (AOAC). Crude carbohydrates are estimated by difference. Humidification of the freeze-dried samples is achieved by placing them in open weighing bottles and storing in air-sealed glass jars while maintaining equilibrium relative humidity with saturated salt solutions. Salts normally used achieve water activity values that range from 0.12 (LiCl) and 0.44 (K_2CO_3) to 0.94 (KNO_3) at 25°C [48]. A test tube containing thymol to prevent mold growth during storage is also added. Twice per week, samples should be removed and weighed until the mass loss or gain reached $\leq$0.001 g for successive weighing. Equilibrium is reached within 3–8 weeks depending on the water activity of the sample.

B. Conventional Differential Scanning Calorimetry

For almost half a century, differential scanning calorimetry (DSC) has been used to measure as a function of temperature the difference in energy inputs into a substance and its reference, with both materials being subjected to a temperature control program [49]. The most common instrument design for making DSC measurements is the heat flux design shown in Figure 2.4. In this design, a metallic disk (made of constantan alloy) is the primary means of heat transfer to and from the sample and reference. The sample, contained in a metal pan, and the reference (an empty pan) sit on raised platforms formed in the constantan disc. As heat is transferred through the disc, the differential heat flow to the sample and the reference is measured by area thermocouples formed by the junction of the constantan disc and chromel wafers, which cover the underside of the platforms [50].

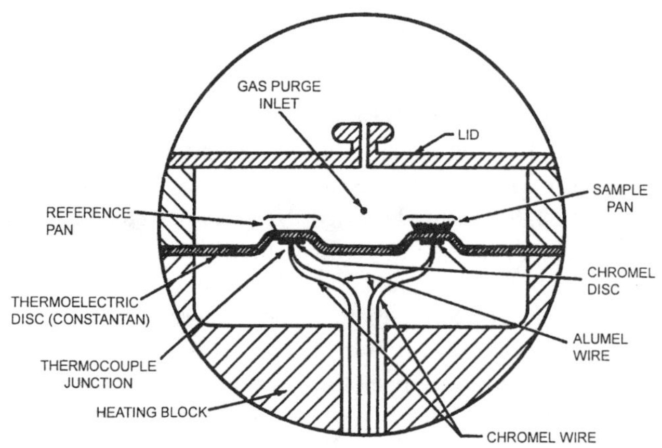

FIGURE 2.4 Heat flux schematic of conventional differential scanning calorimetry.

The thermocouples are connected in series and measure the differential heat flow using the thermal equivalent of Ohm's law: $dQ/dt = \Delta T/R_D$, where dQ/dt is the heat flow, ΔT the temperature difference between reference and sample, and R_D the thermal resistance of the constantan disc. Chromel and alumel wires attached to the chromel wafers form thermocouples which directly measure sample temperature. Purge gas is admitted to the sample chamber through an orifice in the heating block before entering the sample chamber. The result is a uniform, stable thermal environment which assures good baseline flatness and sensitivity (low signal-to-noise ratio [51]). A common DSC application is the precise measurement of a transition temperature whether melting of a crystal or the polymorphic process of a polymer [52]. Drawing a baseline underneath the exothermic or endothermic peak and then subtracting this baseline from the experimental trace allows accurate estimation of the enthalpy (ΔH) and the midpoint temperature (T_m) of the molecular process. Thus the enthalpy of the ice melting peak can be combined with a calibration for pure water to yield the amount of ice in the frozen sample and, hence, by difference from the known weight of the total water in the initial preparation, the moisture content of the unfrozen phase at T_g'.

DSC can also trace vitrification processes by providing a direct, continuous measurement of a sample's heat capacity. In doing so, the apparatus is calibrated for heat flow using a traceable indium standard ($\Delta H_f = 28.3$ J g^{-1}) and for the heat capacity response using a sapphire standard. These days, refrigerated cooling systems can achieve temperatures down to -100°C. Initial cooling to well below or repeated annealing around T_g' ensures maximum freeze concentration in samples. The upper temperature bound can exceed 300°C. A nitrogen gas flow of about 25 ml/min is required to avoid water condensing in the measuring cell. A blank aluminum DSC pan is used as the reference sample and samples of 5–15 mg are placed in hermetically sealed pans. Scan rates employed in the literature range from 1 to 20°C/min. Results are given as average of at least three replicates $\pm$ standard deviation and T_g' values should have a reproducibility of ± 1.0°C [53]. Improved identification of the several and some times small endothermic and exothermic peaks in heat flow occurring typically at subzero temperatures is achieved by plotting the first derivative thermograms [54,55].

It has been argued that from a fundamental viewpoint, derivation of a mechanical glass transition temperature is more reliable than the values obtained from calorimetric measurements. It is true that there is no clear-cut relationship between molecular mobility and thermal event in calorimetric experiments which forces researchers to resort to limiting factors in the form of T_{g1}, T_{g2}, and T_{g3} for the onset, middle, and completion of a particular case [56]. Furthermore, calorimetrically determined glass transition temperatures are affected by the heating rate, which should be reported [57]. Nevertheless, glass formation is in the nature of a second-order thermodynamic transition, which is accompanied by a heat capacity change and detected readily by calorimetry.

C. Modulated Differential Scanning Calorimetry

Despite its utility, DSC does have some important limitations. In pure systems, different types of transitions such as melting and recrystallization in a semicrystalline material may overlap. In multicomponent systems, transitions of the different compounds may partially overlap. To increase the sensitivity and resolution of thermal analysis, provide the heat capacity and heat flow in a single experiment and measure the thermal conductivity, 10 years ago, modulated DSC (MDSC) was developed and commercialized. As a result, complex transitions can be separated into molecular processes with examples including the enthalpic relaxation that occurs at the glass transition region and changes in heat capacity during the exothermic cure reaction of a thermoset [58].

MDSC is a technique which also measures the difference in heat flow between a sample and an inert reference as a function of time and temperature. In addition, the same "heat flux" cell design is used. However, in MDSC a different heating profile is applied to the sample and reference. Specifically, a sinusoidal modulation (oscillation) is overlaid on the conventional linear heating or cooling ramp to yield a profile in which the average sample temperature continuously changes with time

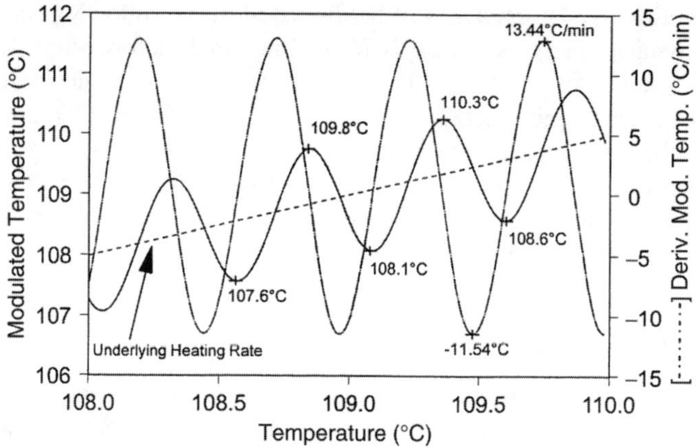

FIGURE 2.5 Typical modulated differential scanning calorimetry heating profile.

but not in a linear fashion [59]. The solid line in Figure 2.5 shows the overall profile for an MDSC heating experiment. This is the net effect of imposing a complex heating profile on the sample and it can be analyzed into two simultaneously running experiments: one experiment at the traditional linear (average) heating rate (dashed line in Figure 2.5) and one at a sinusoidal (instantaneous) heating range (dashed-dot line in Figure 2.5).

Due to the modulated heating rate, there is a resultant modulated heat flow curve and the total heat flow recorded as the final quantitative result is continuously calculated as the moving average of the raw modulated heat flow signal. The reversing component of the total heat flow is calculated by multiplying the measured heat capacity with the average (underlying) heating rate used in the experiment [60]. The kinetic (nonreversing) component of the total heat flow is determined as the arithmetic difference between the total heat flow and the heat capacity component. Phenomena such as glass transitions and melting are reversing or heat capacity events. Nonreversing signals contain kinetic events such as crystallization, crystal perfection and reorganization, cure, and decomposition [61].

D. Rheological Analysis

1. Viscosity (η)

This is the simplest rheological parameter to measure and it was considered as a factor in determining whether a liquid will crystallize or form a glass during freezing. It has been widely held that there is a unique relationship between viscosity and glass transition temperature with the former being about 10^{12} Pa s. Thus "as T_g falls below the ambient temperature due to plasticization by water, the viscosity falls below the characteristic η_g at T_g" [62]. In reality, this is only a myth, albeit a difficult one to beat, because a specific relationship between η and T_g is rarely the case. For example, viscosity increases with increasing molecular weight at the glass transition temperature and in the case of polystyrene it varies from $10^{11.8}$ to 10^{16} Pa s within the M_w range of 16.4–600 kDa [63].

There are also practical issues pertaining to the difficulty of experimenting with unidirectional viscosity on shear at conditions of extreme sample rigidity [64–66]. Undercooled glucose, sucrose, maltodextrin, or maltose–water mixtures assume all the superficial aspects of a solid and at subzero temperatures it is found impractical to carry on viscosity determinations by the standard method of concentric cylinders. Readings do not exceed $10^{6.5}$ Pa s and a long extrapolation to 10^{12} Pa s is implemented in an attempt to predict the value of T_g. However, without concrete evidence of an exponential (Arrhenius) temperature dependence of viscosity or a Williams, Landel and Ferry (WLF) function of molecular processes, the arbitrary treatment of results is fundamentally flawed (the Arrhenius and WLF models will be discussed in the following section in connection with dynamic mechanical analyses). Kasapis [67] using small-deformation dynamic oscillation at $-55°C$ recorded values of about 10^{11} Pa s for the complex dynamic viscosity (η^*) of acid pigskin gelatin with sucrose and glucose syrup (85% solids). Further difficulties in developing a viscosity-related T_g became apparent, as η^* descended steeply from about 10^{11} to 10^8 Pa s with the increasing experimental frequency range from 0.1 to 100 rad/s at $-55°C$. Thus the absence of a "plateau" in the frequency or shear rate dependence of viscosity for biomaterials makes predictions of T_g from viscosity readings rather tenuous.

2. Dynamic Mechanical Analysis

This analysis constitutes a sophisticated approach to assess the viscoelastic nature of materials and owing to the advent of microcomputing in recent years is becoming commonplace in research and development laboratories. It provides readings of the storage modulus (G'), which is the elastic component of the network, loss modulus (G''; viscous component), and dynamic viscosity (η^*) mentioned earlier. Variations with time and temperature can further be assessed as a measure of the "phase lag" δ ($\tan\delta = G''/G'$) of the relative liquid-like and solid-like texture of a food product [68]. Samples are loaded onto the temperature-preset platen of the rheometer and further cooled or heated at a scan rate of up to 5°C/min (Figure 2.6). Thermal runs may be interrupted at constant temperature intervals of 3–5°C to record frequency sweeps from 0.1 to 100 rad/s. It is advisable to implement a strain sweep at the end of the experimental routine to confirm that the small deformation analysis was carried out within the linear viscoelastic region [69].

Today, it is recognized that the phase transitions of food materials can be treated with the "synthetic polymer approach." To a large extent, the approach focuses on the WLF work although the recently introduced coupling theory, built on the physics of intermolecular interactions and cooperativity of polymeric segments, appears to be promising for future studies [70]. The quantitative form of this methodology on the mechanical properties of materials can be summarized as follows (WLF equation [71]):

$$\log a_T = -\frac{C_1^0(T - T_0)}{C_2^0 + T - T_0} \tag{2.3}$$

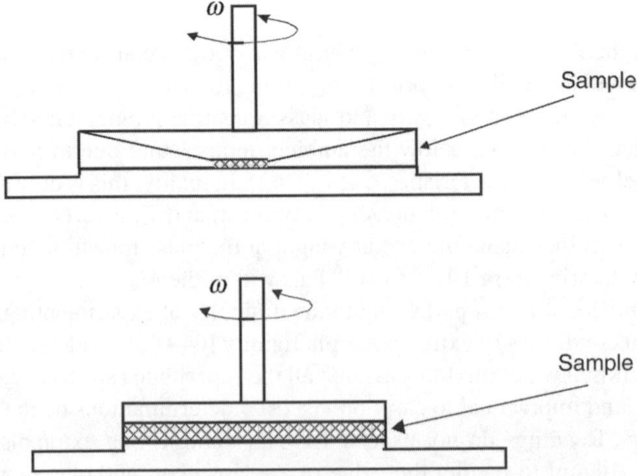

FIGURE 2.6 Measuring geometries of cone-and-plate and parallel plate used in mechanical analysis of frozen foods.

Given current technology, it is still difficult to measure a change that occurs in less than a tenth of a second. Conversely, measuring a change that occurs over a period of a week is prohibitive in terms of laboratory time. To extrapolate over long or short times, the WLF equation builds on a reference temperature (T_0) arbitrarily chosen within the glass transition region. Using the method of thermorheological simplicity, data of $\log G'$ and $\log G''$ obtained from frequency sweeps at a series of experimental temperatures (T), are shifted along the log time axis of T_0 until they fall into a single smooth curve [72]. Thus, the shift factor, a_T, determines how much the timescale of measurement shifts with temperature, with C_1^0 and C_2^0 being the WLF constants. The WLF equation acquires physical significance when examined in the light of the theory of free volume discussed in Figure 2.1. The constants then become

$$C_1^0 = \frac{B}{2.303 f_0} \quad \text{and} \quad C_2^0 = \frac{f_0}{\alpha_f} \tag{2.4}$$

where the fractional free volume, f_0, is the ratio of free to total volume of the molecule, α_f the thermal expansion coefficient, and B is usually set to 1 [73].

It cannot be emphasized enough that application of the WLF equation to the vitrification of the amorphous phase in frozen foods is only justified when the kinetics of the molecular process are rationalized within the framework of the free volume theory (discussed earlier). If not, the equation is reduced to nothing more than an exponential fit to a set of data. Furthermore, research in frozen foods had been carried out using mainly calorimetry, a technique that is not suited to the derivation of the WLF constants [74]. As a last resort without specific data, C_1^0 and C_2^0 were obtained from work on synthetic polymers where estimates at the glass transition region were 17.44° and 51.6°, respectively. However, these are average values and according to Ferry "it is evident that the actual variation from one polymer to another is too great to permit use of these 'universal' values" [12]. Clearly, the problem is exacerbated when that type of numerical approximations are used without critical evaluation in frozen foodstuffs.

The WLF theory becomes inappropriate at temperatures below T_g or higher than $T_g + 100°C$ when the temperature dependence of relaxation processes is heavily controlled by specific features, for example, the chemical structure of molecules in the melt [75]. Thus, for a number of chemical and physical reactions in the glassy state, the extent of temperature dependence was found to conform with the Arrhenius rate law in a modified form that includes a reference temperature, T_0 [76]:

$$\log a_T = \frac{E_a}{2.303R}\left(\frac{1}{T} - \frac{1}{T_0}\right) \tag{2.5}$$

where R is the gas constant. If the dependence of the relaxation times, and hence the material properties, on temperature follows the Arrhenius equation, we will obtain a straight line with the gradient reflecting the activation energy (E_a) of the particular process.

Besides the fundamental WLF/Arrhenius approach, several empirical indicators of the glass transition temperature in frozen foods can be found in the literature. The empirical nature of these indicators does not allow for a physical interpretation of the predictions of T_g', which can be entirely different values for the same preparation. Indeed, there is a debate in the literature as to which indicator represents the best estimate of T_g' but, in our view, there is little to be gained from a discussion of that nature [77]. The indices are determined as the thermal profile of the storage and loss moduli at the point where the G' and G'' traces fall rapidly with increasing temperature, the point where the tan δ trace reaches a maximum in the glass transition region, and so on [78]. For example, mechanical work on 80% aqueous solutions of fructose, glucose, and sucrose considered the onset values of G'' increase as a convenient indicator of T_g' for the frozen

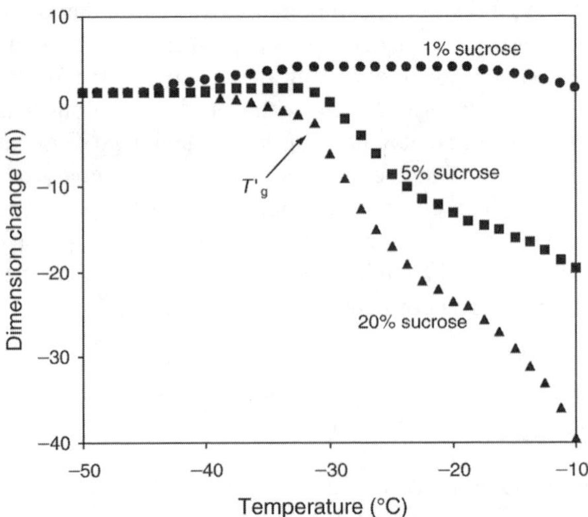

FIGURE 2.7 Temperature dependence of frozen sucrose preparations and derivation of the glass transition temperature using thermal mechanical analysis.

systems. This was followed by decreasing values of G'' due to devitrification at $T > T_g'$ thus creating a peak in the thermal profile. The respective G'' values were −48, −44, and −43°C, which were found to correlate well with the endpoint values of the glass transition temperature obtained by DSC.

Figure 2.7 reproduces a typical thermomechanical profile recorded regularly in R&D units especially in relation to frozen food projects. In doing so, a temperature-programmed penetrometer affords a simple means of measuring the softening of ingredients and products within the temperature range of processing and storage [79]. When experimentally feasible, the thermomechanical analyzer (TMA) is fitted with a dilatometer by which a record of volume change of the sample as it is heated is made. There is a dramatic change in the expansion rate (μm/°C) of the frozen sample during heating from the glassy state to the melt with increasingly concentrated sucrose solutions. The heating curve at 20% sucrose was used to extract an estimate for the T_g' of about −32°C. Comparison with the values reported in the previous paragraph unveils the difficulty of pinpointing the "true" T_g' from different techniques and without an interlinking fundamental reasoning [80]. Adding to the difficulty is that some of the reported values vary widely due to the applied heating rate. As shown in Figure 2.8, there is a considerable increase in T_g' values from −45 to −40°C with higher heating rates (0.5–15°C/min) in frozen solutions of 30% glucose [81,82]. This is due to the thermal lag caused by the conventional type of heating used in TMA, which makes it increasingly difficult to uniformly heat the measuring compartment and sample at rapid heating rates.

V. GLASS TRANSITIONS IN FROZEN FRUITS, FRUIT JUICES, AND MODEL CARBOHYDRATE SOLUTIONS

In these systems, carbohydrates are the main solutes and their thermal behavior during freeze-drying (sublimation) relates to the overall stability and quality control as seen in the loss of physical structure (collapse) or the retention of aroma in the final product. Work on fresh strawberries and the rehydrated powder of the fruit following freeze-drying focused on the effect of moisture on thermal transitions, as a typical DSC methodology in aqueous carbohydrate systems [83].

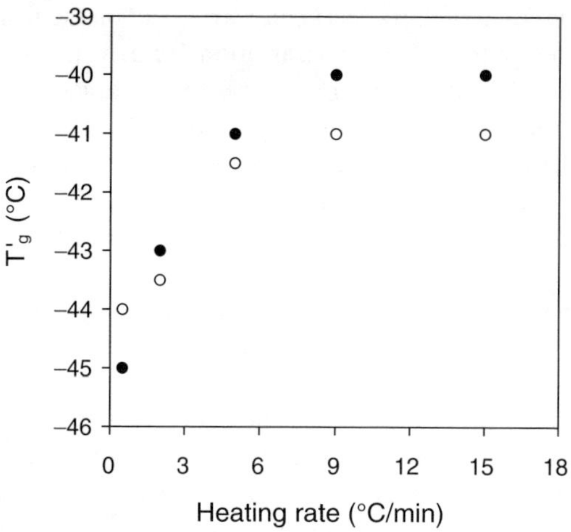

FIGURE 2.8 Duplicate measurements of the effect of heating rate on the mechanical glass transition temperature obtained for 30% glucose solutions.

Nevertheless, it is rather difficult to pinpoint the glass transition temperature of dried strawberry powder because the onset (T_{g1}) and completion (T_{g2}) of changes in the heat flow trace can be as apart as 16.5°C. Furthermore, the thermal stability of the amorphous network is enhanced with increasing temperature of freeze-drying but no justification is offered as to why. A freeze-drying temperature of 20°C was chosen arbitrarily for humidification studies of the dried preparation.

As expected, increasing the moisture content dramatically reduced the glass transition temperature of the strawberry–water preparation. A linear relationship was constructed between T_{g1} and water activity (a_w) but the plasticized profile of T_{g2} remains uncertain. Samples with water activity higher than 0.75 (i.e., >23.3% equilibrium water content) are dominated by ice melting, which masks possible glass transition phenomena. The spectrum is rather "spiky" and several parameters have been introduced to accommodate the various endo- and exothermic peaks. However, the physical significance or correlation of "ante-melting" (T_{am}), "incipient melting" (T_{im}), and "incipient intensive melting" (T'_{im}) that may relate to the viscosity changes in the concentrated amorphous phase are not clearly defined. Levine and Slade [84] observed that T_{am} and T_{im} coincide with the onset and completion of the glass transition of the maximally freeze-concentrated solute/water matrix (T_g') thus being of no additional benefit to interpretation of thermal events. Similar results regarding the moisture dependence of ice melting and the glass transition were found for several sugar preparations (fructose, glucose, sucrose, etc.), fruit juices and naturally occurring food materials with a relatively high carbohydrate content (up to 10% in white cabbage, apple, etc.) [85–89].

The aforementioned work was happening at around the same time when Levine and Slade [90] embraced a physicochemical method for the interpretation of vitrification phenomena inspired largely by the "sophisticated synthetic polymer approach" [91]. Using DSC, they collected hundreds of T_g values for commercial starch hydrolysis products (SHP) and polyhydroxy compounds with a view to emphasizing the utility of the glassy state in the quality control of natural and processed foods. For maximally frozen 20% (w/w) SHP solutions, a linear relationship was constructed between the increasing dextrose equivalent (DE) of the materials and their decreasing T_g'. Regarding the polyhydroxy compounds (sugars, glycosides, polyols), a similar linearity was established between increasing T_g' and decreasing 1/molecular weight (MW) of the materials. Both results proved to be of considerable utility in frozen produce, and in preventing structural collapse during freeze-drying and storage [92,93].

Clearly, considerable amount of work has been carried out on the frozen carbohydrate systems, but the exact nature of the complex subzero transitions remain the subject of controversy [94]. Figure 2.9 reproduces the DSC thermograms of 20% (w/w) glucose and maltodextrin solutions that have been cooled rapidly to about −80°C. Subsequent heating was implemented at a rate of 5°C/min thus unveiling transitions T_A, T_B, and T_C with increasing temperature in the glucose preparation. This is a generic type of behavior in frozen sugar solutions [95], and one school of thought identifies the endothermic T_A with the T_g of a partially vitrified solution due to rapid cooling as compared with ice formation. Slow heating should allow exothermic devitrification (i.e., crystallization) of some of the previously unfrozen water (T_B) thus creating thermal event T_C which is the T_g' of a system close to a maximally freeze-concentrated state [96,97]. The final transition is the equilibrium melting of ice (T_M).

Maltodextrins, on the contrary, exhibit only one endothermic event (T_g') occurring before the ice melting presumably due to comparable cooling and freezing rates in this system. The approach implies that repeated annealing would eliminate T_A but this is not the experimental observation, with both thermal events (T_A and T_C) remaining distinct. An alternative interpretation argues that complete freeze concentration occurs at T_A, which is now equivalent to T_g', whereas the temperature range from T_C and beyond is associated with ice melting. In support of this view, annealing of 66% sucrose solution showed a clear T_g' followed by a single melting endotherm of ice with onset at −34°C, a result that contradicted the earlier view of T_g' occurring at −32°C [98–100]. The temperature gap between the two events is owing to increasing thermal stability of the first-order transition of ice as compared to glass (second-order transition).

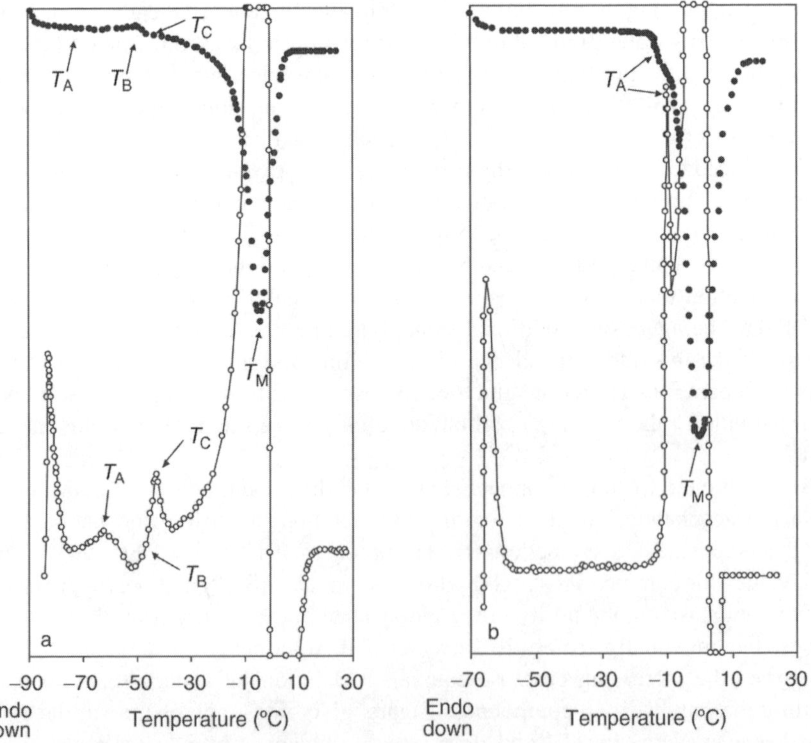

FIGURE 2.9 DSC thermograms for 20% solutions of (a) glucose and (b) maltodextrin with dextrose equivalent 10 showing the heat flow traces and their first derivative (top and bottom, respectively).

VI. GLASS TRANSITIONS IN ICE CREAM AND OTHER FABRICATED PRODUCTS

Proteins and polysaccharides, often in the form of complex multicomponent mixtures, play a fundamental role in developing functionality and in frozen foods are used widely to create a structured body and to maintain the rubbery or glassy texture required by the consumer [101]. Furthermore, increasing consumer awareness of the health implications of a high-calorie, low-fiber diet encouraged the use of polysaccharides in fabricated products, in addition to their properties of binding water and generating acceptable "mouthfeel." Increasing length of the individual biopolymer coils in solution will result in the formation of an entangled network at which the relaxation time of topological interactions will be heavily governed by the polymer molecular weight (and branching). Flow of the solution requires chains to move through the entangled network of neighboring coils, the restriction of mobility increases steeply with increasing network density, giving rise to a high concentration dependence of viscosity [102].

Stated it may be the effect of molecular weight on rheological properties, calorimetric studies on synthetic polymers reveal that there is a minimal effect on T_g at conditions above the critical molecular weight of coil overlap and entanglement (MW* [103]). For maltodextrins in ice cream, the linear T_g'/DE function, mentioned in the previous section, was recast to a linearity between T_g' and $1/\mathrm{MW}$ only at coil dimensions below MW* (DE $\sim$ 6), whereas above the entanglement point T_g values fell on a plateau region [104,105]. Besides maltodextrin, various plant polysaccharides (locust bean gum, carrageenan, etc.) have been incorporated in the ice cream formulation to prevent gradual ice formation occurring during storage. The stabilizers have been partially successful to reduce initial crystal size and subsequent growth [106–108]. Increasing viscosity slows down the diffusional mobility of water molecules in the freeze-concentrated serum phase, as compared with an unstabilized ice cream, but the precise relationship between this process and T_g' is uncertain [109]. Phase separation between the ice cream ingredients has been postulated as a critical mechanism for ice recrystallization, and transmission electron microscopy for model systems of sugars, proteins, and polysaccharides under temperature cycling conditions showed phase-separated domains. This is currently the subject of much research leading to a pattern of T_g' manipulation in relation to ice cream stabilization with cryostabilizing agents such as gelatin, dextran, and so on [110–113].

Ice cream formulations incorporate a few percent of biopolymers and a contrary view argues that their effect on raising T_g' through increasing viscosity is minimal [114,115]. Thus incorporation of dextran, guar gum, or xanthan gum in a sugar preparation showed no variation in the glass transition curve of the state diagram using data from DSC thermograms. However, modest enhancement of the thermal stability of the amorphous phase with dextran was recorded in thermal mechanical analysis, with the T_g' being about 5°C higher than the unstabilized serum. The marginal effect on T_g' may be attributable to the use of non/weak gelling polysaccharides. Calorimetry provides information primarily on the mobility of the sugar phase and the small addition of biopolymer is a mere cross-contamination. It does appear, however, that the increase in the glass transition temperature is related to the ability of the biopolymer to form a network, a process which rheology is extremely well qualified to follow [116].

Thus the glass transition temperature measured by calorimetry remains unaltered by the presence of low levels of polysaccharide, suggesting that the mobility of the sugar is unaffected by the presence of the polysaccharide. However, the mechanical profile of the rubber-to-glass transition is strongly influenced by the polysaccharide particularly if it is network forming. It has been proposed that the magnitude of this polysaccharide contribution to rheology should be represented by a "network T_g," the greater the extent to which this differs from the calorimetric T_g, the larger the influence of the macromolecule on the rheology [117]. Reduction in the diffusion kinetics and increase in the relaxation time of the unfrozen phase in the presence of polysaccharides, as monitored by stress relaxation studies, further confirmed the utility of a three-dimensional network in the stabilization of systems at subzero temperatures [118–120]. The understanding was used in the controlled incorporation of air bubbles surrounded by partially coalesced fat globules and an amorphous biopolymer–water matrix in whipped cream and ice cream. Favorable textural properties and greater resistance of the product to drainage are the outcomes to such an undertaking [121].

Besides foamed dairy emulsions, there is some work on the texture of the vitrified fish muscle. Thus, differences between the temperature of the brittle-ductile transition and the conventional DSC T_g were found for traditional Japanese fish dishes presumably due to the contribution of the fish protein to structure formation [122]. In such complex products, many transitions have been recorded by DSC, which are highly dependent on the annealing temperature [123]. Some of them can be related to T_g' values reported for protein muscle, but the remaining thermal events are quite distinct and may play an important role in the preservation of texture in cod and tuna dishes. In bakery products, loss of crispness, changes in crumb firmness, and crumbliness are major factors affecting loss of product freshness (staling). Freezing slows down the staling rate but, as before, the storage of frozen bakery products is far too complex to rationalize on the basis of a single glass transition temperature [124]. Difficulties due to the heterogeneity of frozen bakery products were also highlighted when a combined temperature–pressure protocol was used to induce phase transitions. High pressure is a technology of the future with industrial potential on the subzero domain for quality control of the food matrix [125].

VII. T_g PERSPECTIVE OF COLLAPSE PHENOMENA, CHEMICAL REACTIONS, AND ENZYMIC ACTIVITY

In their classical paper, White and Cakebread [126] discussed the importance of the physicochemical state of ingredients during various processes (e.g., freeze-drying), and aging of products (frozen dairy, confectioneries, etc.). Among others, critical issues for quality control include the mechanical collapse of the amorphous phase surrounding the ice crystals, thus leading to shrinkage of the product, possible enzymatic activity below T_g, loss of encapsulated volatiles, flavor retention at subzero storage, and preservation of bioactive molecules in glassy matrices [127,128]. Clearly, material collapse is prevented at temperatures below T_g and when it happens at higher temperatures is an irreversible process leading to loss of porosity and a dense glass in the matrix [129]. This glass can then be temperature-cycled reversibly to assess its physical properties.

The WLF theory has been extensively used in the past to provide a quantitative mechanism for collapse in frozen and low-moisture foods [130]. It has not been appreciated that the theory is not intended to describe "any" diffusion-controlled relaxation. For example, it is not suited to the diffusional mobility of linear or branched polysaccharides in the flow region at which the "reptation" or "virtual tube" theory of macromolecular disentanglement takes over [131]. The merit of the WLF theory lies on its partnership with the concept of free volume, which is only applicable within the glass transition region. Under these conditions, the time–temperature profiles of viscoelastic functions are quite similar despite the wide differences in chemical composition and even in polymer concentration or the presence of diluent.

This, of course, is not the case with collapse, which is governed by the conformational and network properties of materials to support a porous matrix at the macromolecular level [132]. Once the matrix has collapsed leading to the formation of a dense glass, Equation (2.3) can be used in association with free volume that may be present as holes of the order of molecular (monomeric) dimensions or smaller voids due to packing irregularities. The equation may be able to follow the exponential temperature dependence of mechanical collapse but without physical science, the fit remains a mere polynomial function. Similar sentiments should be echoed for the application of the equation to the spontaneous agglomeration (caking) of solid powder particles occurring during storage or processing [133].

In the case of chemical and enzymatic processes, it is believed that these slow down considerably if not inhibited entirely at temperatures below T_g' in frozen foods [134]. In this context, enzymatic activity relates to the diffusion-controlled substrate/enzyme interaction whereas chemistry is mainly interested in the prevention of flavor/color degradation and oxidative reactions such as fat rancidity. These changes may impart further on texture, thus leading to the crystallization of soluble solids, grain growth of ice, and so on. The technology of cryostabilization aims to prevent deleterious effects on the overall quality of the product and uses the glass transition temperature as an effective means of retarding them [135]. It appears that, although the enzymic activity is curtailed at $T < T_g'$, the enzyme itself is preserved thus being able to resume work on the substrate once the sample is removed from storage ($T > T_g'$). For example, that type of result was confirmed for polyphenoloxidase and peroxidase activity in sucrose, fructose, and glycerol–water media when it was assayed spectrophotometrically at subzero temperatures [136].

Kerr et al. [137] demonstrated that the rate of hydrolysis of disodium-*p*-nitrophenyl phosphate (catalyzed by alkaline phosphatase) at $T < T_g'$ reduced to near zero in frozen sugar and maltodextrin solutions. Neither the Arrhenius nor the WLF frameworks were able to follow convincingly the reaction rates of the molecular process. Regardless of the choice of the best model to follow various chemical processes during vitrification, it has been confirmed that the rates of "all important" nonenzymatic browning are strongly related to moisture content and the glass transition temperature. This imparts to the quality and palatability of various frozen preparations [138].

Other studies focused on the chemistry of maintaining a natural and "agreeable" color in frozen postharvest without adding sulfur dioxide. It was found that incorporation of different sugars into apricot cubes modified the amount of unfreezable water and hence their subzero phase transition. Color stability was enhanced with this manipulation of T_g', with maltose showing the highest protective effect during frozen storage at $T < T_g'$ [139]. However, frozen storage of products at $T > T_g'$, resulted in color deterioration with stabilizers, with no clear relationship between the loss in stability of anthocyanin pigment and the amplitude of the difference between T_g' and the storage temperature [140,141]. Finally, loss of vitamin C in frozen green vegetables (spinach, peas, green beans, and okra) was studied to establish a relationship between the time and temperature of processing and storage. It was stated that the vitrification properties of plant tissue affect the rate of vitamin C loss significantly, with okra being the best retention medium [142].

VIII. CONCLUSIONS

A cursory exploration of the recent literature on frozen edible materials using a scientific search engine downloads a vast multitude of documents. There is no question that the application of the glass transition temperature to partially frozen biomaterials and the outstanding conception of the state diagram enhanced the understanding of phase transitions in these model systems. Thus "phase I" of the scientific quest for building up a database of functional properties in frozen solutions and gels has largely been accomplished. The future lies in the use of the fundamental knowledge in real food products, which are complex mixtures of various ingredients. At the moment, it seems that there is a gap between the voluminous literature on basic studies and a clear pathway for processing, preservation, and innovation in frozen food produce. State diagrams have been effective tools in mapping out the physical behavior of pure ingredients but it is high time to be tested in heterogeneous bakery, meat, and fish embodiments. In real foods, rationalization of physicochemical stability and mouthfeel on the basis of a single glass transition curve as a function of a total (agglomerate) level of solids is questionable. Furthermore, one feels compelled to note that investigations on structure have been carried out mainly using thermal analysis, which is not the technique of choice in synthetic polymer research. A new concept of network T_g has been introduced to the literature and mechanical analysis in combination with valid application of the free volume theory should be utilized to complement DSC results. In this context, fundamental understanding of the morphology of biopolymer networks and the implication of their interactions in a binary system (in the form, e.g., of phase separation) is needed in frozen foodstuffs like ice cream.

NOMENCLATURE

α_f	thermal expansion coefficient of free volume (1/°C)
α_p	thermal expansion coefficient (1/°C)
a_T	shift factor in Equation (2.3)
a_w	water activity (fraction)
β	molar freezing point constant of water in Equation (2.1) (1860 kg K/kg mol)
B	ratio of water unavailable for freezing to total solids in Equation (2.1)
C_1^0	WLF constant in Equation (2.3)
C_2^0	WLF constant in Equation (2.3)
$C_g{}'$	composition of solute at $T_g{}'$ (%)
C_p	heat capacity
δ	freezing point depression in Equation (2.1)
ΔH_f	latent heat of fusion of a solid (J/g)
ΔH_m	latent heat of melting of ice (J/g)
E	molecular weight ratio of water and solids in Equation (2.1)
E_a	activation energy in Equation (2.5) (J/mol)
f_0	ratio of free to total volume of a molecule in Equation (2.4)
η	viscosity (Pa s)
η^*	complex dynamic viscosity (Pa s)
η_g	viscosity at the glass transition temperature (Pa s)
G'	storage modulus (Pa)
G''	loss modulus (Pa)
k	Gordon–Taylor parameter in Equation (2.2)
λ_s	molecular weight of solids in Equation (2.1)
λ_w	molecular weight of water in Equation (2.1)
R	gas constant in Equation (2.5) (8.314 J/mol K)
T	experimental temperature in Equation (2.3) (K)
T_0	reference temperature in Equation (2.3) (K)
T_f	freezing point of food in Equation (2.1) (°C)
T_g	glass transition temperature (°C)
$T_g{}'$	glass transition of the maximally freeze-concentrated solute/water matrix surrounding the ice crystals in a frozen system (°C)
$T_{g(anh)}$	glass transition temperature of anhydrous pure materials (°C)
T_{gm}	glass transition temperature of the mixture in Equation (2.2) (°C)
T_{gs}	glass transition temperature of solids in Equation (2.2) (°C)
T_{gw}	glass transition temperature of water in Equation (2.2) (°C)
T_m	melting temperature (°C)
T_w	freezing point of water in Equation (2.1) (°C)
$\tan\delta$	ratio of loss to storage modulus
u	total volume per unit mass
u_f	free volume
u_0	occupied volume
X_s	mass fraction of solids in Equation (2.1)
X_w	mass fraction of water in Equation (2.2)

REFERENCES

1. M Padmanabhan. The application of rheological thermal analysis to foods. In: P Fischer, I Marti, EJ Windhab, Eds., *Proceedings of the 3rd International Symposium on Food Rheology and Structure*. ETH Zürich: Laboratory of Food Process Engineering, 2003, pp. 57–63.
2. E Shalaev, F Franks. Solid–liquid state diagrams in pharmaceutical lyophilisation: crystallisation of solutes. In: H Levine, Ed., *Amorphous Food and Pharmaceutical Systems*. Cambridge: The Royal Society of Chemistry, 2002, pp. 145–157.
3. K Binder, J Baschnagel, W Paul. Glass transition of polymer melts: test of theoretical concepts by computer simulation. *Progress in Polymer Science* 28:115–172, 2003.
4. G Allen. A history of the glassy state. In: JMV Blanshard, PJ Lillford, Eds., *The Glassy State in Foods*. Nottingham: Nottingham University Press, 1993, pp. 1–12.
5. H Levine, L Slade. Principles of 'cryostabilization' technology from structure/property relationships of carbohydrate/water systems — a review. *Cryo-Letters* 9:21–63, 1988.
6. PA Perry, AM Donald. The effect of sugars on the gelatinisation of starch. *Carbohydrate Polymers* 49:155–165, 2002.
7. MS Rahman. Glass transition and other structural changes in foods. In: *Handbook of Food Preservation*. New York: Marcel Dekker, 1999, pp. 75–93.
8. B Frick, D Richter. The microscopic basis of the glass transition in polymers from neutron scattering studies. *Science* 267:1939–1947, 1995.
9. PJ Flory. *Principles of Polymer Chemistry*. Ithaca, NY: Cornell University Press, 1953.
10. RGC Arridge. The glass transition. In: *Mechanics of Polymers*. Oxford: Clarendon Press, 1975, pp. 24–50.
11. ML Mansfield. An overview of theories of the glass transition. In: JMV Blanshard, PJ Lillford, Eds., *The Glassy State in Foods*. Nottingham: Nottingham University Press, 1993, pp. 103–122.
12. JD Ferry. Dependence of viscoelastic behavior on temperature and pressure. In: *Viscoelastic Properties of Polymers*. New York: John Wiley, 1980, pp. 264–320.
13. D Cangialosi, H Schut, A van Veen, SJ Picken. Positron annihilation lifetime spectroscopy for measuring free volume during physical aging of polycarbonate. *Macromolecules* 36:142–147, 2003.
14. B Wang, W Gong, WH Liu, ZF Wang, N Qi, XW Li, MJ Liu, SJ Li. Influence of physical aging and side group on the free volume of epoxy resins probed by positron. *Polymer* 44:4047–4052, 2003.
15. G Dlubek, V Bondarenko, J Pionteck, M Supej, A Wutzler, R Krause-Rehberg. Free volume in two differently plasticized poly(vinyl chloride)s: a positron lifetime and PVT study. *Polymer* 44:1921–1926, 2003.
16. CG Biliaderis. Differential scanning calorimetry in food research — a review. *Food Chemistry* 10:239–265, 1983.
17. TW Schenz. Glass transitions and product stability — an overview. *Food Hydrocolloids* 9:307–315, 1995.
18. GP Johari, A Hallbrucker, E Mayer. The glass–liquid transition of hyperquenched water. *Nature* 330:552–553, 1987.
19. F Franks. The amorphous aqueous state — some personal reminiscences. In: H Levine, Ed., *Amorphous Food and Pharmaceutical Systems*. Cambridge: The Royal Society of Chemistry, 2002, pp. v–ix.
20. K-I Izutsu, S Kojima. Miscibility of components in frozen solutions and amorphous freeze-dried protein formulations. In: H Levine, Ed., *Amorphous Food and Pharmaceutical Systems*. Cambridge: The Royal Society of Chemistry, 2002, pp. 216–219.
21. YH Roos. Phase transitions and unfreezable water content of carrots, reindeer meat and white bread studied using differential scanning calorimetry. *Journal of Food Science* 51:684–689, 1986.
22. YH Roos. Melting and glass transitions of low molecular weight carbohydrates. *Carbohydrate Research* 238:39–48, 1993.
23. L Slade, H Levine. Beyond water activity: recent advances based on an alternative approach to the assessment of food quality and safety. In: FM Clydesdale, Ed., *Critical Reviews in Food Science and Nutrition*. Boca Raton: CRC Press, 1991, pp. 115–360.
24. RHM Hatley, C van den Berg, F Franks. The unfrozen water content of maximally freeze concentrated carbohydrate solutions: validity of the methods used for its determination. *Cryo-Letters* 12:113–124, 1991.

25. M Karel, MP Buera, Y Roos. Effects of glass transitions on processing and storage. In: JMV Blanshard, PJ Lillford, Eds., *The Glassy State in Foods*. Nottingham: Nottingham University Press, 1993, pp. 13–34.
26. Y Bai, MS Rahman, CO Perera, B Smith, LD Melton. State diagram of apple slices: glass transition and freezing curves. *Food Research International* 34:89–95, 2001.
27. S Ablett, AH Clark, MJ Izzard, PJ Lillford. Modelling of heat capacity–temperature data for sucrose–water systems. *Journal of Chemical Society — Faraday Transactions* 88:795–802, 1992.
28. YH Roos, M Karel. Applying state diagrams to food processing and development. *Food Technology* 45:66–71, 107, 1991.
29. MS Rahman. Phase transitions in foods. In: *Food Properties Handbook*. Boca Raton: CRC Press, 1995, pp. 87–177.
30. MS Rahman, RH Driscoll. Freezing points of selected seafoods (invertebrates). *International Journal of Food Science and Technology* 29:51–61, 1994.
31. MS Rahman, N Guizani, M Al-Khaseibi, SA Al-Hinai, SS Al-Maskri, K Al-Hamhami. Analysis of cooling curve to determine the end point of freezing. *Food Hydrocolloids* 16:653–659, 2002.
32. MS Rahman. The accuracy of prediction of the freezing point of meat from general models. *Journal of Food Engineering* 21:127–136, 1994.
33. CS Chen. Effective molecular weight of aqueous solutions and liquid foods calculated from the freezing point depression. *Journal of Food Science* 51:1537–1553, 1986.
34. S Kasapis, MS Rahman, N Guizani, M Al-Aamri. State diagram of temperature vs. date solids obtained from the mature fruit. *Journal of Agricultural and Food Chemistry* 48:3779–3784, 2000.
35. MS Rahman, S Kasapis, N Guizani, OS Al-Amri. State diagram of tuna meat: freezing curve and glass transition. *Journal of Food Engineering* 57:321–326, 2003.
36. JR Mitchell. Water and food macromolecules. In: SE Hill, DA Ledward, JR Mitchell, Eds., *Functional Properties of Food Macromolecules*. Gaithersburg: Aspen Publishers, 1998, pp. 50–76.
37. TW Schenz. Relevance of the glass transitions on product functionality. In: GO Phillips, PA Williams, DJ Wedlock, Eds., *Gums and Stabilisers for the Food Industry 8*. Oxford: IRL Press, 1996, pp. 331–340.
38. M Gordon, JS Taylor. Ideal copolymers and the second-order transitions of synthetic rubbers. I. Non-crystalline copolymers. *Journal of Applied Chemistry* 2:493–500, 1952.
39. JL Kokini, AM Cocero, H Madeka, E de Graaf. The development of state diagrams for cereal proteins. *Trends in Food Science and Technology* 5:281–288, 1994.
40. PR Couchman, FE Karasz. A classical thermodynamic discussion of the effect of composition on glass-transition temperatures. *Macromolecules* 11:117–119, 1978.
41. YH Roos, M Karel. Effects of glass transitions on dynamic phenomena in sugar containing food systems. In: JMV Blanshard, PJ Lillford, Eds., *The Glassy State in Foods*. Nottingham: Nottingham University Press, 1993, pp. 207–222.
42. YH Roos. Water activity and physical state effects on amorphous food stability. *Journal of Food Processing and Preservation* 16:433–447, 1993.
43. PJA Sobral, VRN Telis, AMQB Habitante, A Sereno. Phase diagram for freeze-dried persimmon. *Thermochimica Acta* 376:83–89, 2001.
44. YH Roos, M Karel. Nonequilibrium ice formation in carbohydrate solutions. *Cryo-Letters* 12:367–376, 1991.
45. TW Schenz, B Israel, MA Rosolen, Thermal analysis of water-containing systems. In: H Levine, L Slade, Eds., *Water Relationships in Food*. New York: Plenum Press, 1991, pp. 199–214.
46. S Ablett, MJ Izzard, PJ Lillford. Differential scanning calorimetric study of frozen sucrose and glycerol solutions. *Journal of Chemical Society — Faraday Transactions* 88:789–794, 1992.
47. YH Roos, M Karel. Water and molecular weight effects on glass transitions in amorphous carbohydrates and carbohydrate solutions. *Journal of Food Science* 56:1676–1681, 1991.
48. SS Sablani, RM Myhara, OG Mahgoub, Z Al-Attabi, M Al-Mugheiry. Water sorption isotherms of freeze-dried fish sardines. *Drying Technology* 19:671–678, 2001.
49. DJ Wright. Thermoanalytical methods in food research. In: HW-S Chan, Ed., *Biophysical Methods in Food Research*. Oxford: Blackwell Scientific, 1984, pp. 1–35.
50. E Verdonck, K Schaap, LC Thomas. A discussion of the principles and applications of modulated temperature DSC (MTDSC). *International Journal of Pharmaceuticals* 192:3–20, 1999.

51. TR Noel, R Parker, SG Ring. Effect of molecular structure on the conductivity of amorphous carbohydrate–water–KCl mixtures in the supercooled liquid state. *Carbohydrate Research* 338:433–438, 2003.
52. S Dierckx, A Huyghebaert. Effects of sucrose and sorbitol on the gel formation of a whey protein isolate. *Food Hydrocolloids* 16:489–497, 2002.
53. A Boutebba, M Milas, M Rinaudo. Order–disorder conformational transition in succinoglycan: calorimetric measurements. *Biopolymers* 42:811–819, 1997.
54. F Franks. Complex aqueous systems at subzero temperatures. In: D Simatos, JL Multon, Eds., *Properties of Water in Foods*. Dordrecht: Martinus Nijhoff, 1985, pp. 497–509.
55. MR Ollivon. Calorimetric and thermodielectrical measurements of water interactions with some food materials. In: H Levine, L Slade, Eds., *Water Relationships in Food*. New York: Plenum Press, 1991, pp. 175–189.
56. MG Prolongo, C Salom, RM Masegosa. Glass transitions and interactions in polymer blends containing poly(4-hydroxystyrene) brominated. *Polymer* 43:93–102, 2002.
57. MF Mazzobre, JM Aguilera, MP Buera. Microscopy and calorimetry as complementary techniques to analyze sugar crystallisation from amorphous systems. *Carbohydrate Research* 338:541–548, 2003.
58. V Truong, BR Bhandari, T Howes, B Adhikari. Analytical models for the prediction of glass transition temperature of food systems. In: H Levine, Ed., *Amorphous Food and Pharmaceutical Systems*. Cambridge: The Royal Society of Chemistry, 2002, pp. 31–58.
59. JEK Schawe. Principles for the interpretation of modulated temperature DSC measurements. Part 1. Glass transition. *Thermochimica Acta* 261:183–194, 1996.
60. A Boller, C Schick, B Wunderlich. Modulated differential scanning calorimetry in the glass transition region. *Thermochimica Acta* 266:97–111, 1995.
61. PA Sopade, B Bhandari, B D'Arcy, P Halley, N Caffin. A study of vitrification of Australian honeys at different moisture contents, In: H Levine, Ed., *Amorphous Food and Pharmaceutical Systems*. Cambridge: The Royal Society of Chemistry, 2002, pp. 169–183.
62. H Levine, L Slade. 'Collapse' phenomena — a unifying concept for interpreting the behaviour of low moisture foods. In: JMV Blanshard, JR Mitchell, Eds., *Food Structure — Its Creation and Evaluation*. London: Butterworths, 1988, pp. 149–180.
63. DJ Plazek. A myopic review of the viscoelastic behavior of polymers. *Journal of Non-Crystalline Solids* 131–133:836–851, 1991.
64. TR Noel, SG Ring, MA Whittam. Kinetic aspects of the glass-transition behaviour of maltose–water mixtures. *Carbohydrate Research* 212:109–117, 1991.
65. WL Kerr, DS Reid. Temperature dependence of the viscosity of sugar and maltodextrin solutions in coexistence with ice. *Lebensmittel-Wissenschaft und Technologie* 27:225–231, 1994.
66. E Maltini, M Anese. Evaluation of viscosities of amorphous phases in partially frozen systems by WLF kinetics and glass transition temperatures. *Food Research International* 28:367–372, 1995.
67. S Kasapis. Critical assessment of the application of the WLF/free volume theory to the structural properties of high solids systems: a review. *International Journal of Food Properties* 4:59–79, 2001.
68. A Walton. Modern rheometry in characterising the behaviour of foods. *Food Science and Technology Today* 14:144–146, 2000.
69. S Kasapis, IM Al-Marhoobi, JR Mitchell. Molecular weight effects on the glass transition of gelatin/co-solute mixtures. *Biopolymers* 70:169–185, 2003.
70. KL Ngai, DJ Plazek. Identification of different modes of molecular motion in polymers that cause thermorheological complexity. *Rubber Chemistry and Technology* 68:376–434, 1995.
71. JD Ferry. The transition zone from rubber-like to glass-like consistency. In: *Viscoelastic Properties of Polymers*. New York: John Wiley, 1980, pp. 321–365.
72. S Kasapis, IM Al-Marhoobi, M Deszczynski, JR Mitchell, R Abeysekera. Gelatin vs. polysaccharide in mixture with sugar. *Biomacromolecules* 4:1142–1149, 2003.
73. B Neway, MS Hedenqvist, UW Gedde. Effect of thermal history on free volume and transport properties of high molar mass polyethylene. *Polymer* 44:4003–4009, 2003.
74. YH Roos. Characterisation of food polymers using state diagrams. *Journal of Food Engineering* 24:339–360, 1995.
75. S Kasapis. Structural properties of high solids biopolymer systems. In: SE Hill, DA Ledward, JR Mitchell, Eds., *Functional Properties of Food Macromolecules*. Gaithersburg: Aspen, 1998, pp. 227–251.

76. S Kasapis. The use of Arrhenius and WLF kinetics to rationalise the rubber-to-glass transition in high sugar/κ-carrageenan systems. *Food Hydrocolloids* 15:239–245, 2001.
77. M Peleg. A note on the tan δ (T) peak as a glass transition indicator in biosolids. *Rheological Acta* 34:215–220, 1995.
78. JR Mitchell. Hydrocolloids in low water and high sugar environments. In: PA Williams, GO Phillips, Eds., *Gums and Stabilisers for the Food Industry 10*. Cambridge: The Royal Society of Chemistry, 2000, pp. 243–254.
79. J Rieger. The glass transition temperature T_g of polymers — comparison of the values from differential thermal analysis (DTA, DSC) and dynamic mechanical measurements (torsion pendulum). *Polymer Testing* 20:199–204, 2001.
80. YH Roos, M Karel. Amorphous state and delayed ice formation in sucrose solutions. *International Journal of Food Science and Technology* 26:553–566, 1991.
81. TJ Maurice, YJ Asher, S Thomson. Thermomechanical analysis of frozen aqueous systems. In: H Levine, L Slade, Eds., *Water Relationships in Food*. New York: Plenum Press, 1991, pp. 215–223.
82. S Kasapis, IM Al-Marhoobi, JR Mitchell. Testing the validity of comparisons between the rheological and the calorimetric glass transition temperatures. *Carbohydrate Research* 338:787–794, 2003.
83. YH Roos, Effect of moisture on the thermal behaviour of strawberries studied using differential scanning calorimetry. *Journal of Food Science* 52:146–149, 1987.
84. H Levine, L Slade. Interpreting the behavior of low-moisture foods. In: TM Hardman, Ed., *Water and Food Quality*. London: Elsevier, 1989, pp. 71–134.
85. YH Roos, M Karel, JL Kokini. Glass transitions in low moisture and frozen foods: effects on shelf life and quality. *Food Technology* November:95–108, 1996.
86. K Paakkonen, L Plit. Equilibrium water content and the state of water in dehydrated white cabbage. *Journal of Food Science* 56:1597–1599, 1991.
87. S Ablett, MJ Izzard, PJ Lillford, I Arvanitoyannis, JMV Blanshard. Calorimetric study of the glass transition occurring in fructose solutions. *Carbohydrate Research* 246:13–22, 1993.
88. ME Sahagian, HD Goff. Effect of freezing rate on the thermal, mechanical and physical aging properties of the glassy state in frozen sucrose solutions. *Thermochimica Acta* 246:271–283, 1994.
89. MM Sa, AM Figueiredo, AM Sereno. Glass transitions and state diagrams for fresh and processed apple. *Thermochimica Acta* 329:31–38, 1999.
90. H Levine, L Slade. Thermomechanical properties of small carbohydrate–water glasses and 'rubbers.' *Journal of Chemical Society — Faraday Transactions* 84:2619–2633, 1988.
91. JD Ferry. Some reflections on the early development of polymer dynamics: viscoelasticity, dielectric dispersion, and self-diffusion. *Macromolecules* 24:5237–5245, 1991.
92. H Levine, L Slade. A polymer physico-chemical approach to the study of commercial starch hydrolysis products. *Carbohydrate Polymers* 6:213–244, 1986.
93. D Courtehoux, Y Le Bot, Ph. Lefevre, G Ribadeau Dumas. *Polyols: The Key to "Lite" Confectionery. Europe: Food Technology International*, 1996, pp. 47–51.
94. HD Goff. Measurement and interpretation of the glass transition in frozen foods. In: MC Erickson, Y-C Hung, Eds., *Quality in Frozen Food*. New York: Chapman & Hall, 1997, pp. 29–50.
95. SR Aubuchon, LC Thomas, W Theuerl, H Renner. Investigations of the sub-ambient transitions in frozen sucrose by modulated differential scanning calorimetry (MDSC®). *Journal of Thermal Analysis* 52:53–64, 1998.
96. H Levine, L Slade. Cryostabilization technology: thermoanalytical evaluation of food ingredients and systems. In: VR Harwalker, CY Ma, Eds., *Thermal Analysis of Foods*. New York: Elsevier, 1990, pp. 221–305.
97. I Arvanitoyannis, JMV Blanshard, S Ablett, MJ Izzard, PJ Lillford. Calorimetric study of the glass transition occurring in aqueous glucose: fructose solutions. *Journal of the Science of Food and Agriculture* 63:177–188, 1993.
98. MJ Izzard, S Ablett, PJ Lillford. A calorimetric study of the glass transition occurring in sucrose solutions. In: E Dickinson, Ed., *Food Polymers, Gels, and Colloids*. Cambridge: The Royal Society of Chemistry, 1991, pp. 289–300.
99. YH Roos. Phase transitions and transformations in food systems. In: DR Heldman, DB Lund, Eds., *Handbook of Food Engineering*. New York: Marcel Dekker, 1992, pp. 145–197.

100. YH Roos, M Karel. Phase transitions of amorphous sucrose and frozen sucrose solutions. *Journal of Food Science* 56:266–267, 1991.
101. RK Richardson, S Kasapis. Rheological methods in the characterisation of food biopolymers. In: DLB Wetzel, G Charalambous, Eds., *Instrumental Methods in Food and Beverage Analysis.* Amsterdam: Elsevier, 1998, pp. 1–48.
102. T McLeish. On the trail of topological fluids. *Physics World* (March):32–37, 1995.
103. S Montserrat, F Roman, P Colomer. Vitrification and dielectric relaxation during the isothermal curing of an epoxy-amine resin. *Polymer* 44:101–114, 2003.
104. H Levine, L Slade. Water as a plasticizer: physico-chemical aspects of low-moisture polymeric systems. In: F Franks, Ed., *Water Science Reviews 3 — Water Dynamics*. Cambridge: Cambridge University Press, 1988, pp. 79–185.
105. L Slade, H Levine. Glass transitions and water–food structure interactions. In: JE Kinsella, SL Taylor, Eds., *Advances in Food and Nutrition Research*. San Diego: Academic Press, 1995, pp. 103–269.
106. HD Goff. Low-temperature stability and the glassy state in frozen foods. *Food Research International* 25:317–325, 1992.
107. HD Goff, KB Caldwell, DW Stanley, TJ Maurice. The influence of polysaccharides on the glass transition in frozen sucrose solutions and ice cream. *Journal of Dairy Science* 76:1268–1277, 1993.
108. JV Patmore, HD Goff, S Fernandes. Cryo-gelation of galactomannans in ice cream model systems. *Food Hydrocolloids* 17:161–169, 2003.
109. RL Sutton, ID Evans, JF Crilly. Modelling ice crystal coarsening in concentrated disperse food systems. *Journal of Food Science* 59:1227–1233, 1994.
110. HD Goff. Measuring and interpreting the glass transition in frozen foods and model systems. *Food Research International* 27:187–189, 1994.
111. HD Goff. The use of thermal analysis in the development of a better understanding of frozen food stability. *Pure and Applied Chemistry* 67:1801–1808, 1995.
112. DS Reid, W Kerr, J Hsu. The glass transition in the freezing process. *Journal of Food Engineering* 22:483–494, 1994.
113. A Regand, HD Goff. Structure and ice recrystallisation in frozen stabilized ice cream model systems. *Food Hydrocolloids* 17:95–102, 2003.
114. G Blond. Mechanical properties of frozen model solutions. *Journal of Food Engineering* 22:253–269, 1994.
115. D Simatos, G Blond, F Martin. Influence of macromolecules on the glass transition in frozen systems. In: E Dickinson, D Lorient, Eds., *Food Macromolecules and Colloids*. Cambridge: The Royal Society of Chemistry, 1995, pp. 519–533.
116. S Kasapis, JR Mitchell. Definition of the rheological glass transition temperature in association with the concept of iso-free-volume. *International Journal of Biological Macromolecules*, 29:315–321, 2001.
117. S Kasapis, JR Mitchell, R Abeysekera, W MacNaughtan. Rubber-to-glass transitions in high sugar/biopolymer mixtures. *Trends in Food Science and Technology*, 15:298-304, 2004.
118. ME Sahagian, HD Goff. Influence of stabilizers and freezing rate on the stress relaxation behaviour of freeze-concentrated sucrose solutions at different temperatures. *Food Hydrocolloids* 9:181–188, 1995.
119. ME Sahagian, HD Goff. Thermal, mechanical and molecular relaxation properties of stabilized sucrose solutions at sub-zero temperatures. *Food Research International* 28:1–8, 1995.
120. HD Goff, E Verespej, D Jermann. Glass transitions in frozen sucrose solutions are influenced by solute inclusions within ice crystals. *Thermochimica Acta* 399:43–55, 2003.
121. DW Stanley, HD Goff, AK Smith. Texture–structure relationships in foamed dairy emulsions. *Food Research International* 29:1–13, 1996.
122. H Watanabe, CQ Tang, T Suzuki, T Mihori. Fracture stress of fish meat and the glass transition. *Journal of Food Engineering* 29:317–327, 1996.
123. KN Jensen, BM Jorgensen, J Nielsen. Low-temperature transitions in cod and tuna determined by differential scanning calorimetry. *Lebensmittel-Wissenschaft und Technologie* 36:369–374, 2003.
124. SP Cauvain. Improving the control of staling in frozen bakery products. *Trends in Food Science and Technology* 9:56–61, 1998.

125. A LeBail, L Boillereaux, A Davenel, M Hayert, T Lucas, JY Monteau. Phase transition in foods: effect of pressure and methods to assess or control phase transition. *Innovative Food Science and Emerging Technologies* 4:15–24, 2003.
126. GW White, SH Cakebread. The glassy state in certain sugar-containing food products. *Journal of Food Technology* 1:73–82, 1966.
127. M Karel, JM Flink. Some recent developments in food dehydration research. In: AS Mujumdar, Ed., *Advances in Drying*. Washington: Hemisphere Publishing, 1983, pp. 103–153.
128. VN Morozov, SG Gevorkian. Low-temperature glass transition in proteins. *Biopolymers* 24:1785–1799, 1985.
129. YH Roos. Glass transition-related physicochemical changes in foods. *Food Technology* (October): 97–102, 1995.
130. YH Roos. Reaction kinetics. In: *Phase Transitions in Foods*. San Diego: Academic Press, 1995, pp. 271–312.
131. S Kasapis, ER Morris, M Gross, K Rudolph. Solution properties of levan polysaccharide from *Pseudomonas syringae* pv. phaseolicola, and its possible primary role as a blocker of recognition during pathogenesis. *Carbohydrate Polymers* 23:55–64, 1994.
132. MS Rahman. A theoretical model to predict the formation of pores in foods during drying. *International Journal of Food Properties* 6:61–72, 2003.
133. G Tardos, D Mazzone, R Pfeffer. Measurement of surface viscosities using a dilatometer. *Canadian Journal of Chemical Engineering* 62:884–887, 1984.
134. M Karel. Effects of water activity and water content on mobility of food components, and their effects on phase transitions in food systems. In: D Simatos, JL Multon, Eds., *Properties of Water in Foods*. Dordrecht: Martinus Nijhoff, 1985, pp. 153–169.
135. C Van Den Berg, F Franks, P Echlin. The ultrastructure and stability of amorphous sugars. In: JMV Blanshard, PJ Lillford, Eds., *The Glassy State in Foods*. Nottingham: Nottingham University Press, 1993, pp. 249–267.
136. L Manzocco, MC Nicoli, M Anese, A Pitotti, E Maltini. Polyphenoloxidase and peroxidase activity in partially frozen systems with different physical properties. *Food Research International* 31:363–370, 1999.
137. WL Kerr, MH Lim, DS Reid, H Chen. Chemical reaction kinetics in relation to glass transition temperatures in frozen food polymer solutions. *Journal of the Science of Food and Agriculture* 61:51–56, 1993.
138. MS Rahman. Food preservation by freezing. In: *Handbook of Food Preservation*. New York: Marcel Dekker, 1999, pp. 259–284.
139. E Forni, A Sormani, S Scalice, D Torreggiani. The influence of sugar composition on the colour stability of osmodehydrofrozen intermediate moisture apricots. *Food Research International* 30:87–94, 1997.
140. D Torreggiani, E Forni, I Guercilena, A Maestrelli, G Bertolo, GP Archer, CJ Kennedy, S Bone, G Blond, E Contreras-Lopez, D Champion. Modification of glass transition temperature through carbohydrates additions: effect upon colour and anthocyanin pigment stability in frozen strawberry juices. *Food Research International* 32:441–446, 1999.
141. A Rizzolo, RC Nani, D Viscardi, G Bertolo, D Torreggiani. Modification of glass transition temperature through carbohydrates addition and anthocyanin and soluble phenol stability of frozen blueberry juices. *Journal of Food Engineering* 56:229–321, 2003.
142. MC Giannakourou, PS Taoukis. Kinetic modelling of vitamin C loss in frozen green vegetables under variable storage conditions. *Food Chemistry* 83:33–41, 2003.

3 An Overview of Refrigeration Cycles

Da-Wen Sun
National University of Ireland, Dublin, Ireland

CONTENTS

I. INTRODUCTION

Freezing is the process of removing heat for producing and maintaining temperatures below initial freezing point [1]. In the food industry, freezing is the most popular long-term preservation method for food products. Besides convenience to consumers, the popularity of frozen foods is also due to the fact that frozen foods continue to demonstrate a good food safety record, as freezing can effectively reduce the activity of microorganisms and enzymes, thus preventing deterioration [1,2]. In addition, crystallization of water reduces the amount of liquid water in food items and inhibits microbial growth [3].

Food freezing consists of three parts: (a) cooling to remove sensible heat and to reduce the product temperature to the freezing point; (b) removal of the product's latent heat of fusion and changing the water to ice crystals; and (c) continued cooling below the freezing point and thus reducing the product temperature to the desired frozen storage temperature [4,5]. The longest part of the process is the removal of the latent heat of fusion as water turns to ice crystals, which determines the freezing rate [1,2].

The freezing systems used in the food industry are generally operated on the basis of some refrigeration cycles [6]. At present, the refrigeration market is dominated by electricity-powered mechanical vapor-compression units. As electricity generation produces large amounts of CO_2, contributing to global warming, many governments in the world have committed themselves to reduce the emission of greenhouse gases. A solution to the global warming problem is to develop refrigeration systems powered by waste thermal energy [7–12]. Utilization of low-grade thermal energy has been a research topic for many decades as the energy is widely available from sources, such as industrial processes, flat plate solar collectors, and exhausts from automobiles. As refrigeration is one of the economically feasible and environmentally friendly applications for harnessing low-grade thermal energy, alternative or novel refrigeration cycles powered by thermal energies have been developed, aimed at significant savings in electrical energy consumption [13–15].

II. FUNDAMENTALS

The description and analysis of refrigeration cycles require knowledge of thermodynamics and heat transfer. Therefore, it is important to review some relevant fundamental principles that are important for the calculation of refrigeration cycles [16–19].

A. Temperature

Temperature is the most basic and common term used in food freezing. The temperature of a substance is an indication of the ability of the substance to exchange energy with another substance that is in contact with it. The temperature scale used in the food industry is normally the Celsius scale, which has two reference points: are freezing point of water (0°C) and boiling point of water (100°C) at atmospheric pressure. In the Celsius scale, t (°C) is normally used as the symbol for temperatures. However, in refrigeration cycle calculation, the Kelvin scale or the absolute temperature scale is also used. The Kelvin scale uses the absolute zero (0 K or −273.15°C) as its reference point, and it has the same degree intervals as the Celsius scale. In the Kelvin scale, T (K) is normally used as the symbol for temperatures. Therefore, temperature differences in the Celsius scale can also be stated in the Kelvin scale. In the Kelvin scale, the freezing and boiling points of water at atmospheric pressure are +273.15 and +373.15°C, respectively.

B. Heat

In the food industry, the term "heat" is normally used to refer to the thermal energy. A change in this energy may result in a change in temperature or a change between the solid, liquid, and gaseous states. Freezing is a process of removing heat and thus showing reduction in temperature.

Heat is a form of the internal energy of a substance and it is related to the molecular structure and the degree of molecular activity. Therefore, heat consists of the kinetic and potential energies of the molecules. Depending on the degree of molecular activity, heat can be divided into sensible heat and latent heat. The sensible heat is the sum of the kinetic energies of the molecules, and therefore, the change of sensible heat will cause a change of temperature. In contrast, latent heat is associated with overcoming of molecular forces that bind the molecules to each other and the breaking away of the molecules, and therefore, the change of latent heat does not cause any change in temperature but a change in phase (solid to liquid, liquid to gas, or vice versa).

C. Enthalpy and Specific Heat

Enthalpy h is often used to calculate the change of heat in refrigeration cycles. Enthalpy is defined as the sum of internal energy and flow work and it has a unit of kJ/kg. If only steady flow is involved in a process, the flow work will not change significantly, and the difference in enthalpy will be the quantity of heat gained or lost. Therefore, enthalpy covers both sensible heat and latent heat. The value of enthalpy is always based on some arbitrarily chosen datum plane, so enthalpy difference is normally used in calculation.

If a change of enthalpy can be sensed by a change in temperature, this change is expressed as the specific heat, that is, the specific heat is the change in enthalpy per degree of temperature. In other words, the specific heat of a substance is the quantity of energy required to raise the temperature of a unit mass by one degree and hence it has the unit of kJ/(kg K). There are two common specific heats, one for constant volume process c_v and the other for constant pressure process $c_{p.}$ As the refrigeration process normally occurs at a constant pressure, c_p is used. If a change of enthalpy does not cause any change in temperature, but a change in phase, this change is the latent heat [18,19].

D. Entropy

The entropy s of a unit mass of a substance at any given condition is an expression of the total energy transferred to the material per degree to bring the substance to that condition from some arbitrarily chosen datum plane [18], therefore, it has the unit of kJ/(kg K).

In a compression or expansion process, if a gas or vapor undergoes such a process without friction and without adding or removing heat, the entropy of the substance remains constant, which is defined as the isentropic process. If a compression or expansion process is isentropic, the change in enthalpy represents the amount of work per unit mass involved in the process [17].

E. Energy Conservation for Steady Flow

The principle of energy conservation applies to every process. The principle simply states that energy cannot be created or destroyed, and the total amount of energy remains constant. Therefore, for a system in a refrigeration cycle, the energy conservation can be expressed as the rate of energy with the flow stream entering the system plus the rate of heat added minus the rate of work performed and minus the rate of energy with the flow stream leaving the system equals the rate of energy change in the system [16]. This expression can be described mathematically as:

$$m\left(h_{\text{in}} + \frac{V_{\text{in}}^2}{2} + gz_{\text{in}}\right) + Q - W - m\left(h_{\text{out}} + \frac{V_{\text{out}}^2}{2} + gz_{\text{out}}\right) = \frac{dE}{d\theta} \tag{3.1}$$

In most of refrigeration systems, as the mass flow rate remains almost constant, the flow can be assumed as steady flow, therefore, the term describing the rate of energy change in Equation (3.1) can be omitted. Furthermore, the changes in potential and kinetic energies of the flow are very small, when compared with the change of enthalpy, heat transferred or work done, their effects can be neglected. Hence, for a system in a refrigeration cycle, Equation (3.1) can be rewritten as

$$Q - W = m\Delta h = m(h_{\text{out}} - h_{\text{in}}) \tag{3.2}$$

For a system without the involvement of work such as a condenser, evaporator, or heat exchanger, the heat transferred to or from these systems can be calculated by the change of enthalpy multiplied by the mass flow rate:

$$Q = m(h_{\text{out}} - h_{\text{in}}) \tag{3.3}$$

In contrast, for a system on which work is done such as a compressor or a pump, the amount of heat transferred is negligible, therefore, Equation (3.2) can be simplified as

$$W = m(h_{\text{in}} - h_{\text{out}}) = m(P_{\text{in}} - P_{\text{out}})v \tag{3.4}$$

F. Heat Transfer

Heat transfer occurs from a high temperature body to a low-temperature body. The transfer of heat takes place in three different ways: conduction, convection, and radiation. In refrigeration systems, only heat conduction and heat convection generally occur.

Conduction transfers heat through a continuous mass or from one body touching the other. It involves the transfer of energy from the more energetic molecules of a substance to the adjacent less energetic ones due to the interactions between the molecules [20–22]. The Fourier law is used to describe the heat conduction

$$Q = -kA\frac{\mathrm{d}T}{\mathrm{d}x} \tag{3.5}$$

Equation (3.5) indicates that heat conduction takes place in a direction of decreasing temperature, and the rate of heat conduction in a direction is proportional to the temperature gradient in that direction. The ability to conduct heat in a material is related to the characteristics of the material itself. This ability is defined as the thermal conductivity k of the material, and it has the unit of kW/(m K) [23–25].

Convection is the combined effects of heat conduction and fluid flow. It transfers energy between a solid surface and the adjacent liquid or gas, which is in motion. Depending on the flow velocity, convection can be free (or natural) convection or forced convection. In free convection, the fluid flow is caused by buoyancy forces, which are induced by density differences due to the variation of temperature in the fluid. In contrast, in forced convection, the fluid is forced to flow by external means such as a pump or fan. The Newton's law of cooling is used to determine the rate of convection heat transfer,

$$Q = h_t A(T_{\text{sur}} - T_{\text{f}}) \tag{3.6}$$

where h is an experimentally determined parameter defined as the convection heat transfer coefficient with the unit kW/(m^2 K), and its value depends on many relevant factors such as the nature of fluid flow, properties of the fluid, and surface geometry [1,26,27].

G. Phase Change Processes

The refrigerants used in most cooling systems pass between liquid and vapor states in refrigeration cycles. These refrigerants behave similarly during the changing states. Figure 3.1 shows the phase

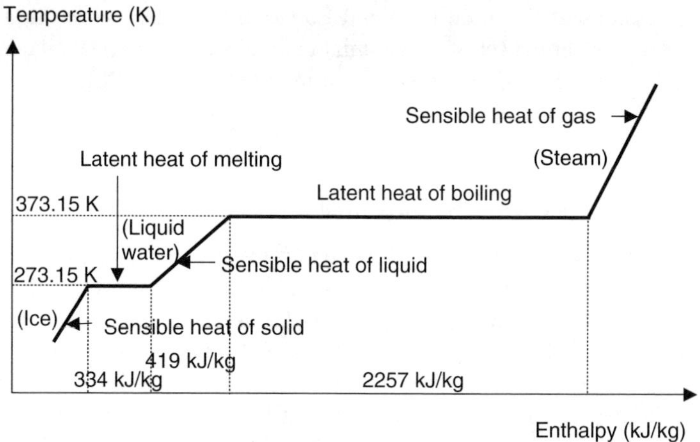

FIGURE 3.1 Schematic representation of the relation between temperature and enthalpy during the phase change process of H_2O.

change process using pure water at atmospheric pressure as an example [16,19]. At temperature below 0°C, water exists in the solid state, that is, ice. If heat (sensible heat) is added, the temperature of the ice begins to rise until it reaches the melting point (or freezing point), which is 0°C. At the melting point, if heat is further added, the ice begins to melt. The heat required to melt the ice is the latent heat, which is 334 kJ/kg. During the melting process, a mixture of ice and liquid water coexists. The melting process continues at the constant melting temperature until the ice is completely changed to the liquid water state, which is also termed as compressed or subcooled liquid. If heat is added continuously, the temperature of the subcooled water starts to rise above the melting point until it reaches the boiling point, which is 100°C (the liquid which is about to vaporize is called saturated liquid). The sensible heat required to raise the temperature of the subcooled water from 0 to 100°C is 419 kJ/kg. If heat is further added, the liquid water begins to boil, and the boiling process occurs at the constant temperature of 100°C and latent heat must be supplied. The latent heat of boiling is 2257 kJ/kg. During boiling, a mixture of saturated liquid and saturated vapor coexists until the saturated liquid water is completely changed to saturated vapor (the vapor which is about to condense is termed as saturated vapor). Then, if further heat is added, the temperature of the vapor will rise above 100°C, and this vapor is defined as superheated vapor as this vapor is not about to condense.

If the pressure in the earlier process is above or below atmospheric pressure, the melting and boiling points will be different. With the increase in pressure, the boiling temperature will be increased, however, the temperature at which liquid water begins to freeze will be decreased.

During the phase change process from liquid to vapor, if the refrigerant is at the saturated states such as saturated liquid or saturated vapor, saturation pressure P_s, enthalpy h_s, or entropy s_s are a function of saturation temperature T_s only, and therefore, if T_s is known, P_s, h_s and s_s can be determined by the following functions

$$P_s = P(T_s), \quad h_s = h(T_s), \quad s_s = s(T_s) \tag{3.7}$$

If the refrigerant is in the other states such as subcooled liquid, mixture of saturated liquid and saturated vapor, or superheated vapor, the enthalpy h or entropy s is a function of temperature T and pressure P and vice versa:

$$T = T(P, h), \quad h = h(T, P), \quad P = P(T, h), \quad s = s(T, P) \tag{3.8}$$

The actual expressions of Equation (3.7) and Equation (3.8) depend on the refrigerant used. These expressions are normally a set of polynomial equations, which are available from literature or can be obtained by curve fitting to published thermodynamic data [28].

H. Equilibrium for Multicomponent Solution

Multicomponent solution is needed for the operation of absorption refrigeration cycle. Depending on the cooling requirement, various solutions are available [29–33], however, the most commonly used ones are the lithium bromide–water (LiBr–H_2O) solution [11] and the water–ammonia (H_2O–NH_3) solution [7,9]. Figure 3.2 shows an equilibrium condition using water–lithium bromide solution as an example [17]. The lithium bromide is a solid salt crystal, which will absorb water vapor to become a liquid solution when it is in contact with the vapor. The pressure of the water vapor in the solution vessel is a function of the H_2O–LiBr solution temperature and concentration. Therefore, many different combinations of temperatures and concentrations of solution can provide the same vapor pressure. If pure water is contained in another vessel, the pressure of the water vapor in this vessel will be a function of temperature of the pure water only. If these two vessels are connected, equilibrium will be established between the two vessels, in other words, the two vessels will have the same water vapor pressure. By changing the temperature in the pure water vessel, the water vapor pressure in the two connected vessels will be changed, and therefore, the solution concentration can be altered depending on the solution temperature.

Obviously, in the multicomponent solution, the solution concentration X is a function of temperature T and vapor pressure P of the solution, therefore, if any of the two variables are known, the third variable can be obtained:

$$P = P(T,X), \quad T = T(P,X), \quad X = X(P,T) \tag{3.9}$$

As long as the state of the solution is determined, the enthalpy h or entropy s of the solution can be calculated, for example, if temperature T and concentration X of the solution are known, its enthalpy or entropy can be determined and vice verse

$$h = h(T,X), \quad s = s(T,X), \quad T = T(X,h), \quad X = X(T,h) \tag{3.10}$$

Again, the actual expressions of Equation (3.9) and Equation (3.10) depend on the solution used. These expressions are available from the literature or can be obtained by curve fitting to published thermodynamic data [7,9,11].

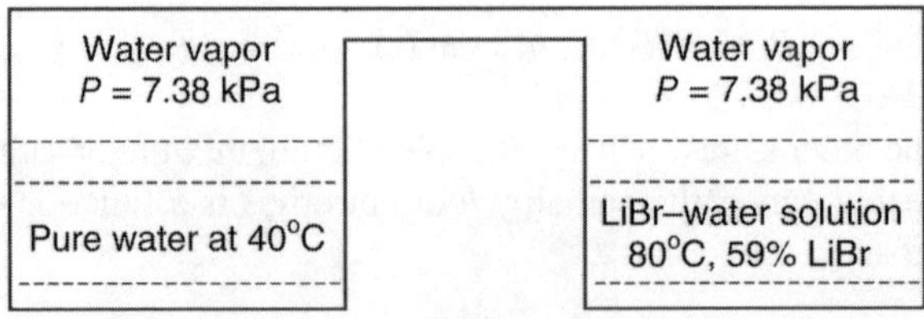

FIGURE 3.2 Equilibrium of multicomponent solution.

I. Coefficient of Performance

The efficiency of a refrigeration cycle is expressed in terms of the coefficient of performance (COP). As the purpose of refrigeration cycles is to remove heat from the refrigerated area, to accomplish this purpose, the cycle requires an energy input as either heat or work, depending on the operation of the actual cycle. Therefore, the COP is defined as

$$\mathrm{COP} = \frac{\text{desired output}}{\text{required input}} = \frac{Q_e}{E_{\text{net,in}}} \tag{3.11}$$

where Q_e is the useful refrigeration and $E_{\text{net,in}}$ is the net energy input in the form of heat or work. These two terms must be in the same units so that COP is dimensionless. The COP value can be greater than unity, which indicates that the amount of heat removed from the refrigerated area can be greater than the amount of energy input.

III. VAPOR COMPRESSION CYCLE

The vapor compression cycle is the most frequently used refrigeration cycle in the food industry. The cycle mainly consists of four main components: a compressor, a condenser, an expansion valve, and an evaporator. The evaporator and condenser can be easily identified in a household refrigerator. The freezer compartment used to store frozen foods serves as the evaporator as the heat from the foods is removed by the refrigerant in the evaporator. The condenser is located behind the refrigerator where the coils dissipate heat to the kitchen.

A. The Carnot Refrigeration Cycle and its Modification

The Carnot refrigeration cycle is the most efficient cycle, and therefore, is an ideal cycle. No other refrigeration cycle can perform better than the Carnot cycle [16]. The cycle $1'-2'-3'-4'$ shown in Figure 3.3 is the Carnot cycle if there is no temperature difference between T_e' and the load and no temperature difference between T_c' and the ambient. Therefore, the Carnot cycle consists of the following thermodynamically reversible processes: adiabatic compression $1'-2'$, isothermal rejection of heat $2'-3'$, adiabatic expansion $3'-4'$, and isothermal addition of heat $4'-1'$ [17]. Among them, process $4'-1'$ is the refrigeration step, which is the ultimate goal of the cycle as it removes heat from the

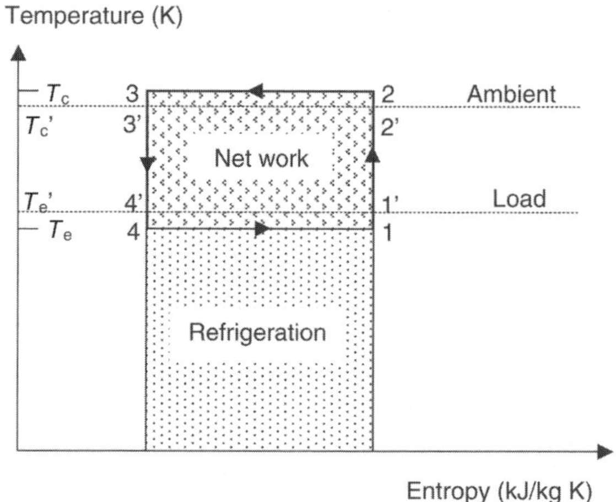

FIGURE 3.3 Carnot refrigeration cycle and its modification.

load, with all the other processes serving to achieving this goal by discharging the removed heat to the ambient.

In the Carnot cycle in Figure 3.3, the desired output of the cycle is the useful refrigeration, which is the area beneath line $4'-1'$, and the desired input to the cycle is the network, which is the area enclosed in rectangle $1'-2'-3'-4'$, therefore, the COP is the ratio of the first area over the second area.

The Carnot cycle assumes no resistance in heat transfer with the load and with the ambient, in reality, no heat transfer without temperature difference can occur as the temperature difference is the driving force for heat flow. Therefore, T_e should be lower than the load temperature, so that the heat from the load can be transferred to the refrigerant and T_c should be higher than the ambient temperature for the heat from the refrigerant to be rejected to the ambient. This modification is illustrated in Figure 3.3, showing the comparison of the ideal Carnot cycle $1'-2'-3'-4'$ with the modified cycle 1–2–3–4. Obviously, the modified cycle has lower COP as the rectangular area enclosed by $2'-2-3-3'$ increases the network and the rectangular area enclosed by $1-1'-4'-4$ not only increases the network but also decreases the useful refrigeration [17]. That is why the Carnot cycle has the highest COP.

B. The Standard Vapor Compression Cycle

The refrigeration cycle in Figure 3.3 is more often shown in the pressure–enthalpy chart; as such a chart can give information on the liquid and vapor states of the refrigerant. Figure 3.4 illustrates the standard vapor compression cycle in the pressure–enthalpy chart [34]. Figure 3.4a is the flow diagram, indicating that the refrigerant vapor 1 from the evaporator is compressed by the compressor to high pressure vapor 2, which is then condensed in the condenser to liquid. The heat of condensation is rejected to the ambient. The liquid 3 exits the condenser, then passes through an expansion valve to undergo an pressure reduction, and then flows into the evaporator to evaporate. The latent heat of evaporation needed is supplied by the cooling load, thus generating refrigeration. The evaporated vapor 4 exits the evaporator and finally flows into the compressor to complete the cycle. The aforementioned process is clearly shown in Figure 3.4b.

The standard vapor compression cycle is different from the Carnot refrigeration cycle. First, the refrigerant 1 entering the compressor is saturated vapor, after compression, the vapor becomes superheated vapor 2. This compression process is assumed to occur at a constant entropy (isentropic). Secondly, the liquid 3 from the condenser passing through the expansion process remains the same value of enthalpy after it exits to become subcooled liquid 4. That is, the expansion process

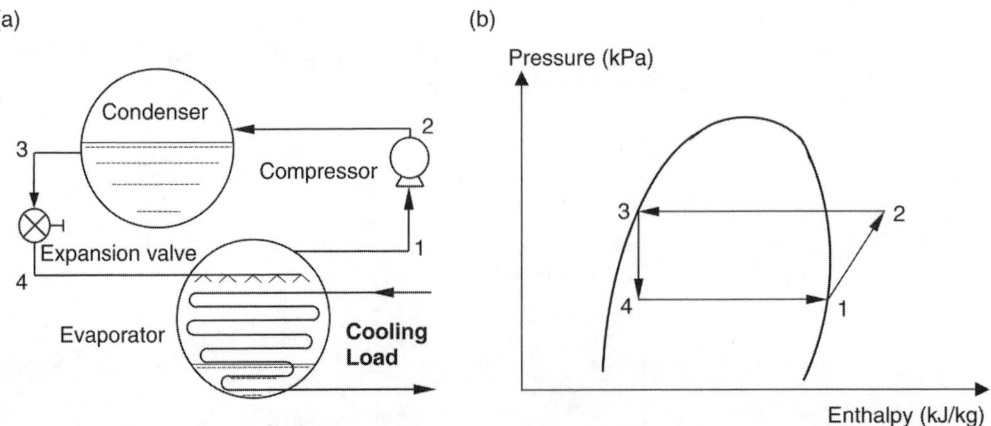

FIGURE 3.4 The standard vapor compression cycle on (a) flow diagram and (b) pressure–enthalpy diagram.

3–4 is a constant-enthalpy (isenthalpic) process and its entropy increases rather than the constant entropy process as required in the Carnot cycle.

Therefore, referring to Figure 3.4b, the standard vapor compression cycle consists of the following processes: reversible and adiabatic compression from saturated vapor to the condenser pressure 1–2, desuperheating and condensation of the refrigerant and reversibly rejecting heat at the condenser pressure 2–3, irreversible expansion at constant enthalpy from saturated liquid to the evaporator pressure 3–4, and reversibly adding heat at the evaporator pressure to cause evaporation to saturated vapor 4–1 [17,34].

C. Cycle Analysis

To calculate the COP for the standard vapor compression cycle, thermodynamic analysis should be performed. Referring to Figure 3.4b, the cycle is characterized by the condenser and evaporator temperatures. Therefore, the refrigerant properties at states 1 and 3 can be specified:

$$T_1 = T_e, \quad P_1 = P(T_e), \quad h_1 = h(T_e), \quad s_1 = s(T_e) \tag{3.12}$$

$$T_3 = T_c, \quad P_3 = P(T_c), \quad h_3 = h(T_c) \tag{3.13}$$

The condensed fluid emerging from the condenser undergoes a pressure reduction via the expansion valve and hence

$$P_4 = P_1, \quad h_4 = h_3, \quad T_4 = T(P_4, h_4) \tag{3.14}$$

The vapor from the evaporator is compressed to the condenser pressure by the compressor before entering the condenser. This compression process is normally assumed to be isentropic, therefore

$$P_2 = P_3, \quad s_2 = s_1, \quad T_2 = T(P_2, s_2), \quad h_2 = h(T_2, P_2) \tag{3.15}$$

Therefore, the energy balance across the evaporator and compressor for COP calculation is performed:

$$Q_e = m(h_1 - h_4), \quad W_c = m(h_2 - h_1) \tag{3.16}$$

Finally, the COP is

$$\text{COP} = \frac{Q_e}{W_c} \tag{3.17}$$

IV. ABSORPTION REFRIGERATION CYCLE

The absorption refrigeration cycle has recently attracted much research attention because of the possibility of using waste thermal energy or renewable energies as the power source, thus reducing the demand for electricity supply. The lithium bromide–water ($LiBr–H_2O$) and water–ammonia ($H_2O–NH_3$) systems are the most common absorption systems, where the components are given as refrigerant–absorbent [7,9,11]. To improve the performance of absorption systems, new refrigerant–absorbent pairs have been developed [29,30]. These pairs include ($H_2O–NH_3$)–LiBr, CH_3OH–(LiBr–$ZnBr_2$) and H_2O–($LiNO_3$–KNO_3–$NaNO_3$) [31], H_2O–glycerol [32], H_2O–LiCi [33], and $LiNO_3$–NH_3 [30].

In the H_2O–LiBr system, water is used as refrigerant, therefore the application of such a system is limited by the freezing point of water and H_2O–LiBr systems are normally used in the air-conditioning industry [11]. In contrast, in the NH_3–H_2O systems, as ammonia is used as the refrigerant, these systems can be applied to food refrigeration or ice making [7,35]. Therefore, only NH_3–H_2O absorption cycle is discussed.

A. Cycle Description

Generally speaking, an absorption cycle consists of a generator, condenser, evaporator, absorber, pump, heat exchanger, and two expansion valves. However, for the NH_3–H_2O absorption cycle, two special components, that is, a rectifier and a dephlegmator are occasionally needed because of the fact that water is volatile. The NH_3–H_2O absorption cycle is shown in Figure 3.5 [11]. As ammonia is evaporated off the generator, it also contains some water vapor, and if the mixture of ammonia and water flows into the evaporator, the water component will elevate the evaporating temperature, in the meantime, the water may also freeze along the pipelines. Therefore, this water must be removed as completely as possible. With the rectifier and dephlegmator as illustrated in Figure 3.5, the vapor driven off at the generator first flows countercurrently to the incoming solution in the rectifier, the solution then passes through the dephlegmator and condenses some water-rich liquid, which drains back to the rectifier. Therefore, only a small amount of water vapor may eventually escape the dephlegmator and flows from the evaporator to the absorber.

The operating principle of the NH_3–H_2O absorption cycle shown in Figure 3.5 is as follows. High-pressure refrigerant vapor 1 generated by the generator condenses into liquid 2 in the condenser, and the heat of condensation is rejected to the environment. The condensed liquid then enters the evaporator 3 to evaporate through a throttling valve, which is used to maintain the pressure difference between the condenser and evaporator. The heat required for evaporation is provided by the cooling load. Vapor 4 evaporated from the evaporator is absorbed by the liquid strong solution 10 coming from the generator in the absorber. The heat of absorption is rejected to the

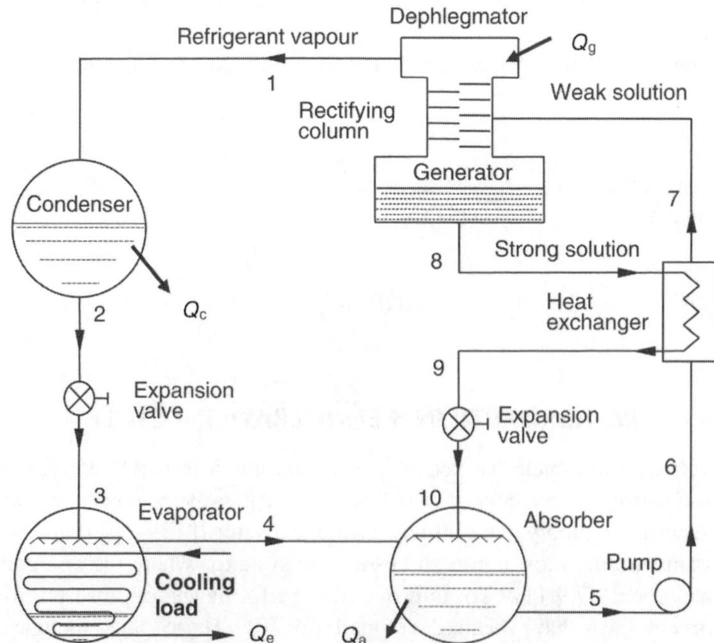

FIGURE 3.5 The schematic of the ammonia–water absorption cycle.

environment. The pump receives low-pressure liquid weak solution 5 from the absorber, elevates the pressure of the weak solution 6, and delivers 7 to the generator. By weak solution (strong solution), it is meant that the ability of the solution to absorb the refrigerant vapor is weak (strong). In the generator, heat from a high-temperature source drives off the refrigerant vapor 1 in the weak solution. The liquid strong solution 8 then returns to the absorber 9 through a throttling valve 10. The function of the throttling valve is to provide a pressure drop to maintain the pressure difference between the generator and the absorber. Therefore, for an absorption cycle, the generator and condenser operate at the same high-pressure level, whereas the evaporator and absorber maintain at the same low-pressure level. To improve cycle performance, a solution heat exchanger is normally added to the cycle as shown in Figure 3.5. This solution heat exchanger is an energy-saving device, which is not an essential item for the successful operation of the cycle.

The heat flow pattern to and from the absorption cycle is that high-temperature heat is required at the generator, whereas low-temperature heat from the substance being refrigerated enters the evaporator. The rejection of heat from the cycle is at the absorber and condenser at temperatures higher than the atmosphere so that the heat can be rejected to the environment.

B. Cycle Analysis

As shown in Figure 3.5, the operation of the absorption cycle is characterized by the temperatures at generator, condenser, absorber, and evaporator and the refrigerant mass flowing through the evaporator or the required refrigerating load. Therefore, the cycle can be thermodynamically analyzed as follows [7,9].

The operating pressures in the aforementioned components are determined by the saturated liquid and vapor concentrations in the condenser and evaporator, respectively, and their temperatures, that is

$$P_c = P_g = P(T_2, X_2), \quad P_e = P_a = P(T_4, Y_4) \tag{3.18}$$

For the strong solution at state 8 and the refrigerant vapour 1 from the generator

$$T_8 = T_g, \quad P_8 = P_g, \quad X_8 = X(T_8, P_8), \quad h_8 = h(T_8, X_8) \tag{3.19}$$

$$T_1 = T_g, \quad P_1 = P_g, \quad h_1 = h(T_1, Y_1) \tag{3.20}$$

For the weak solution at state 5 from the absorber

$$T_5 = T_a, \quad P_5 = P_a, \quad X_5 = X(T_5, P_5), \quad h_5 = h(T_5, X_5) \tag{3.21}$$

Applying the mass conservation principle at the generator yields

$$X_7 = X_5, \quad m_7 X_7 = m_8 X_8 + m_1 Y_1, \quad m_1 = m_7 - m_8 \tag{3.22}$$

Therefore, the mass flow rates of the strong and weak solutions at states 8 and 7, respectively, can be found as

$$m_8 = \frac{X_7 - Y_1}{X_8 - X_7} m_1, \quad m_5 = m_7 = \frac{X_8 - Y_1}{X_8 - X_7} m_1 \tag{3.23}$$

The weak solution from the absorber is pumped to the solution heat exchanger. As a result, the enthalpy at 6 is increased:

$$T_6 = T_5, \quad P_6 = P_g, \quad X_6 = X_5, \quad \upsilon_6 = \upsilon(T_6, X_6), \quad h_6 = h_5 + (P_6 - P_5)\upsilon_6 \tag{3.24}$$

If the effectiveness of the solution heat exchanger is η, the fluid at states 9 and 7 is derived using the earlier values as follows:

$$T_9 = \eta T_6 + (1 - \eta)T_8, \quad X_9 = X_8, \quad h_9 = h(T_9, X_9) \tag{3.25}$$

$$X_7 = X_6, \quad m_6 = m_5, \quad h_7 = h_6 + \frac{m_8}{m_6}(h_8 - h_9), \quad T_7 = T(X_7, h_7) \tag{3.26}$$

Through the expansion valve, the fluid pressure is reduced from high pressure at 9 to low pressure at 10, therefore

$$X_{10} = X_9, \quad h_{10} = h_9, \quad T_{10} = T(X_{10}, h_{10}) \tag{3.27}$$

The condensed liquid refrigerant at 2 from the condenser can be specified:

$$P_2 = P_c, \quad X_2 = Y_1, \quad T_2 = T(P_2, X_2), \quad h_2 = h(T_2, X_2) \tag{3.28}$$

Similarly, the evaporated vapor refrigerant at 4 from the evaporator can be specified:

$$P_4 = P_e, \quad T_4 = T_e, \quad Y_4 = Y_1, \quad h_4 = h(T_4, Y_4) \tag{3.29}$$

The liquid at 2 undergoes a pressure reduction via the expansion valve before entering the evaporator, therefore

$$P_3 = P_e, \quad h_3 = h_2, \quad T_3 = T(h_3, X_3) \tag{3.30}$$

To obtain the COP of the cycle, energy balances at the generator and evaporator are required, thus

$$Q_g = m_1 h_1 + m_8 h_8 - m_7 h_7, \quad Q_e = m_1(h_4 - h_3), \quad W_{me} = (P_6 - P_5)v_6 \tag{3.31}$$

Finally, the performance of the cycle is calculated as

$$\text{COP} = \frac{Q_e}{Q_g + W_{me}} \tag{3.32}$$

The circulation ratio of the cycle is defined as

$$f = \frac{m_5}{m_1} \tag{3.33}$$

The heat rejected by the absorber and condenser can also be calculated by applying the heat conservation principle as

$$Q_a = m_4 h_4 + m_{10} h_{10} - m_5 h_5, \quad Q_c = m_1(h_1 - h_2) \tag{3.34}$$

C. Performance Characteristics

Figure 3.6 shows the typical effect of generator temperatures on COP at various ammonia concentrations [7,9]. The performance characteristics of the NH_3–H_2O absorption refrigeration cycle is different from those of a mechanical vapor compression cycle. The NH_3–H_2O is a binary

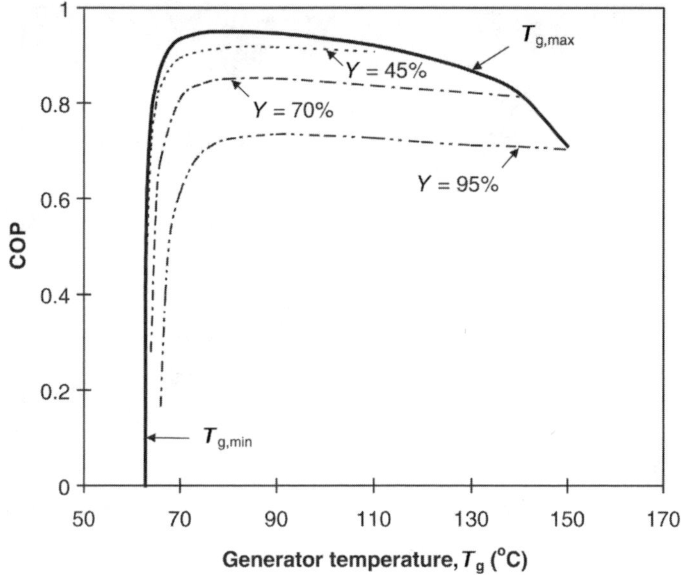

FIGURE 3.6 Effect of generator temperature on COP with various ammonia concentration in the refrigerant.

mixture, with H_2O as the absorbent and NH_3 as the refrigerant. As discussed previously, the refrigerant is not 100% ammonia as some water is contained in it. This is caused by the vapour–liquid behavior of a binary mixture, with one as absorbent and the other as refrigerant. For a fixed ammonia concentration, which is controlled by the rectifying process, the cycle performance varies with the generator temperature and there exists an optimum generator temperature. As the ammonia concentration of the weak solution entering the generator is determined by the absorber conditions, and the condenser and generator are at the same pressure level, the optimum generator temperature determines the optimum concentration of the strong solution leaving the generator, and therefore, the optimum flow rate of the refrigerant vapor emerging from the generator. If the solution temperature at the generator is higher than its optimum value, some thermal energy is wasted, as higher temperature means higher grade of thermal energy. However, if the solution temperature is lower than the optimum value, less refrigerant vapor is produced at the generator. Therefore, the control of generator temperature is an important issue in achieving high performance of the absorption cycle.

Figure 3.6 also shows that if the ammonia concentration in the refrigerant is lowered by the proper control of the rectifying process, COP can be improved due to the increase in refrigerant flow rate for the same amount of the weak solution entering the generator. However, in actual operation, the refrigerant with as high ammonia concentration as possible is preferred as water vapor containing in the refrigerant may freeze along the pipes. Furthermore, if water enters the evaporator, it will elevate the evaporating temperature [7,9].

Furthermore, Figure 3.6 shows the possible operation region for the generator as enclosed by the generator temperature envelope. Any generator temperature for the cycle within the region can produce a refrigerating effect. Figure 3.6 also indicates the minimum and maximum permitted generator temperatures, however, it should be noted that the envelope curve may change if conditions at other components vary [7,9].

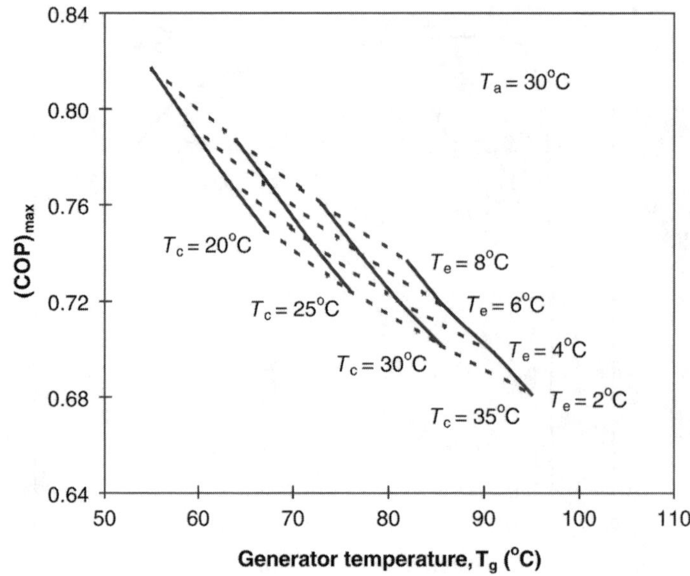

FIGURE 3.7 An example of the optimum design map for NH_3—H_2O absorption cycle.

D. Optimum Design Map

The COP values shown in Figure 3.6 indicate the local maximum for various operating conditions. To provide detailed optimum operating conditions for absorption cycles, optimum design maps can be constructed as the one illustrated in Figure 3.7 [11].

The optimum design map in Figure 3.7 is for NH_3–H_2O absorption cycle with absorber temperature T_a at 30°C. The dashed lines represent constant evaporator temperatures and solid lines for condenser temperatures. As NH_3 is the refrigerant in the absorption cycle in Figure 3.7, optimum design maps for evaporating temperatures lower than 0°C can also be constructed in a similar way. The optimum design map can be used to find the required T_g for achieving the maximum COP value under specified T_e, T_c, and T_a. If there is a shift in one of the conditions, the cycle COP will be lower than the original maximum value, and therefore, relevant conditions need to be re-adjusted according to the maps to establish new optimum conditions for the maximum COP. For example, if $T_e = 4°C$, $T_c = 30°C$, and $T_a = 30°C$, T_g can be found from the map to be 81°C, and the corresponding COP is 0.720, which is the maximum. If T_g shifts to 72°C, the COP value is lowered to 0.695 which is no longer the maximum, indicating that the new conditions are not optimized. To maintain optimum operation, for $T_g = 72°C$, T_c must be decreased to 25°C along the dashed line of $T_e = 4°C$ if the refrigerating temperature remains unchanged, and in this case, the conditions are optimized again and the COP value is 0.744, which is higher than the non-optimum value of 0.695.

Obviously, the optimum design map contains important information for designing new systems and choosing operating conditions for existing systems. If the map is stored in the control system of a cycle, automatic control of absorption systems can be realized for maintaining their optimum performance under various operating conditions [11].

V. EJECTOR REFRIGERATION CYCLE

The ejector refrigeration cycle is normally heat powered and it can be used to harness waste thermal energy or renewable energies. The principle of ejector refrigeration cycle has been known for many decades, however, its potential economic advantages have not always been realized [36]. The ejector refrigeration units experienced their first wave of popularity in the 1930s. However, these units were later supplanted by mechanical vapor compression systems. As an ejector refrigeration cycle has several advantages over conventional vapor compression cycles, which include no moving parts (except the pump), very little wear and susceptibility to breakdown, the use of an easily obtainable and safe refrigerant, and utilization of thermal energy as a power source, recent investigations have shown that it could be an alternative heat-powered refrigeration cycle [37–39].

A. Cycle Description

The layout of the ejector refrigeration cycle is shown in Figure 3.8 [40,41]. The operating principle is as follows. At the generator, the refrigerant is vaporized at high pressure and the refrigerant vapor flows to the ejector. The vapor or primary fluid then enters the primary convergent–divergent ejector nozzle and expands. This expansion causes a low pressure region in the ejector, which induces the vapor or secondary fluid from the evaporator at state point 8. The primary and secondary fluids then mix in the mixing section and enter the constant area section of the ejector, where an aerodynamic transverse shock is usually induced to create a major compression effect. The mixed stream is further compressed to the back pressure of the condenser in the diffuser section of the ejector, which then exits the ejector and flows into the condenser and condenses there. Finally, the condensed liquid is divided into two parts. One part 4 is pumped back to the generator and the other 6 expands through a throttling valve to a low pressure state and enters the evaporator from where it is evaporated to produce the necessary cooling effect. The details of the structure of an ejector can be found in the literature [40,42,43]. Unlike the vapor compression cycle and the absorption cycle, the ejector cycle operates at three distinct pressure levels: the generator at the high pressure, the condenser at the intermediate pressure, and the evaporator at the low pressure.

B. Cycle Analysis

As indicated in Figure 3.8, the operation of an ejector refrigeration cycle is normally characterized by the temperatures at the generator, condenser and evaporator and the total refrigerant mass

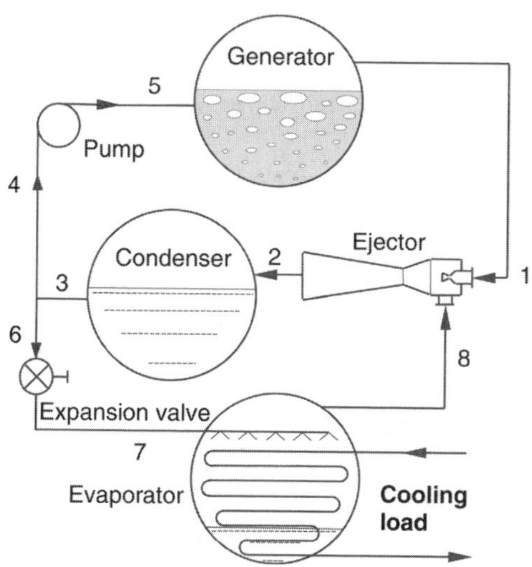

FIGURE 3.8 The layout of an ejector refrigeration cycle.

flowrate through the generator m_1 [40,42]. On the basis of this, starting from the generator, the cycle can be thermodynamically analyzed as follows. The high-pressure vapor at state 1 before entering the ejector is given by the following functional relationships:

$$T_1 = T_g, \quad P_1 = P(T_g), \quad h_1 = h(T_g) \tag{3.35}$$

The low-pressure vapor at state 8 is determined by the evaporator temperature

$$T_8 = T_e, \quad P_8 = P(T_e), \quad h_8 = h(T_e) \tag{3.36}$$

The condensed liquid at state 3 emerging from the condenser can be specified by the following equation:

$$T_3 = T_c, \quad P_3 = P(T_c), \quad h_3 = h(T_c) \tag{3.37}$$

If the properties at states 1, 8, and 3 are known, the entrainment ratio ω of the ejector can be calculated from the following relationships:

$$P_2 = P_3, \quad \omega = f(P_1, T_1, P_8, T_8, P_2, A_r), \quad h_2 = \frac{h_1 + \omega h_8}{1 + \omega}, \quad T_2 = T(P_2, h_2) \tag{3.38}$$

Details of the determination of ω can be found in the literatures [40,42,43]. One part of the condensate is pumped back to the generator, as a result, the enthalpy at 5 is increased:

$$P_5 = P_1, \quad h_5 = h_3 + (P_5 - P_3)v_3, \quad T_5 = T(P_5, h_5) \tag{3.39}$$

The other part of the condensate flows through the expansion valve, and the fluid pressure is reduced from the condenser pressure to the evaporator pressure, therefore

$$P_7 = P_8, \quad h_7 = h_3, \quad T_7 = T(P_7, h_7) \tag{3.40}$$

The mass flow continuity around the cycle yields the following:

$$m_5 = m_1, \quad m_8 = m_7 = \omega m_1, \quad m_2 = m_3 = (1 + \omega)m_1 \tag{3.41}$$

To determine the COP of the cycle, energy conservation principle should be applied at the generator and evaporator, which gives the following energy flow rates

$$Q_g = m_1(h_1 - h_5), \quad Q_e = \omega m_1(h_8 - h_7), \quad W_{me} = (P_5 - P_3)v_3 \tag{3.42}$$

Therefore, COP is determined

$$\text{COP} = \frac{Q_e}{Q_g + W_{me}} \tag{3.43}$$

C. Performance Characteristics

In the ejector refrigeration cycle, the ejector is the key component of the cycle, as the performance of the cycle mainly depends on the performance of the ejector. All the ejectors used have their fixed geometries, and they have several distinct characteristics. Therefore, the ejector cycle demonstrates some interesting phenomena [8,43]. The performance of an ejector is measured by the ratio of the secondary flow (flow coming from the evaporator) m_8 over the primary flow (flow coming from the

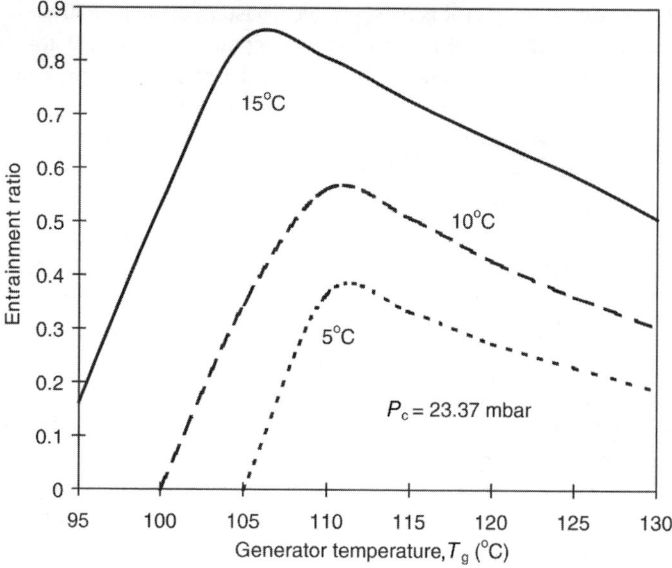

FIGURE 3.9 Effect of the generator temperature on the entrainment ratio of an ejector refrigeration cycle.

generator) m_1. This ratio is termed as the entrainment ratio ω of the ejector. The higher the entrainment ratio, the higher the secondary flow rate, and consequently, the higher the coefficient of performance. Therefore, ejectors should be optimally designed to have the highest possible entrainment ratio.

One of the ejector characteristics is shown in Figure 3.9 [8]. It is noted that there exists an optimum value of generator temperature at which the entrainment ratio reaches the maximum. The optimum generator temperature depends on the evaporator temperature but in a very narrow range. If the generator temperature is higher than the optimum value, the entrainment ratio falls, resulting in a reduction in COP value and cooling capacity of the cycle [8,37].

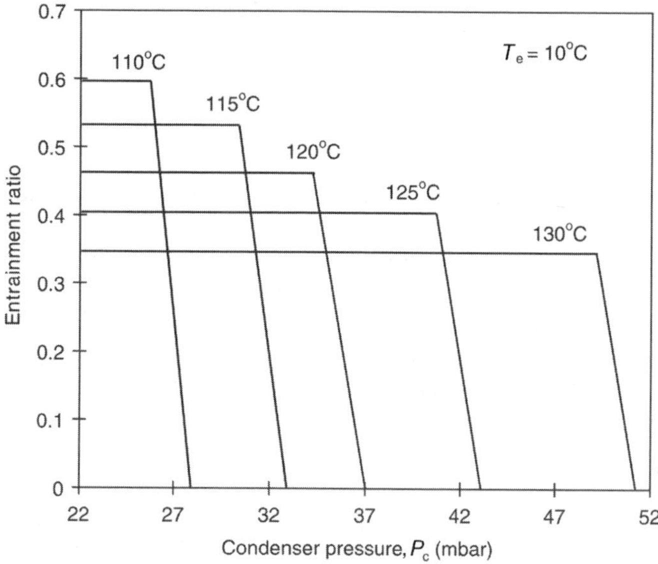

FIGURE 3.10 Effect of the condenser temperature on the entrainment ratio of an ejector refrigeration cycle.

Another characteristic of an ejector is its so-called constant capacity shown in Figure 3.10. The constant capacity means that the entrainment ratio is independent of the ejector back pressure, that is, the condenser pressure P_c. As illustrated in Figure 3.10, if the back pressure is higher than a certain value, the entrainment ratio decreases suddenly and then falls to zero. This pressure at which the entrainment ratio begins to drop is defined as the critical back pressure of the ejector [8,37]. In contrast, if the back pressure is lower than the critical pressure, the entrainment ratio remains constant. This constant capacity phenomenon is caused by the ejector distinct behavior of the choking of the secondary flow. During the operation of an ejector, the walls formed by the expansion of the primary stream m_1 from the nozzle and by the converging mixing section make a hypothetic converging duct termed as "aerodynamic convergent nozzle" [8] for the secondary fluid m_g to pass. The secondary flow accelerates in this "aerodynamic convergent nozzle" to sonic velocity at its exit plane. The sonic velocity is the condition of flow choking [8]. When the secondary flow is choked, that is, the flow reaches its maximum velocity, which is sonic at the exit of the "aerodynamic nozzle," a back pressure lower than the critical pressure cannot be sensed upstream, within the "aerodynamic nozzle", and therefore, the lowering of the back pressure does not affect the flow rate. Under choked conditions, a transverse shock wave normally occurs in the constant area section or diffuser section of the ejector. The lower the back pressure, the further downstream the shock wave will appear. However, if the back pressure is increased, the shock wave tends to move upstream toward the mixing section of the ejector. If back pressure is further increased to above the critical value, it will cause the shock wave to penetrate into the mixing section, leading to the disappearance of choking of the secondary flow, and the falling of the entrainment ratio. This effect is so severe that a little further increase in the back pressure will finally force the shock wave to pass through the mixing section and reach the exit plane of the primary nozzle, causing no secondary flow, and therefore, the entrainment ratio drops to zero. Obviously, fixed geometry ejector refrigeration cycles should operate at these critical conditions and avoid condenser pressures being lower or higher than these critical values [8,37].

The characteristics of the ejector shown in Figure 3.9 and Figure 3.10 indicate that ejector refrigeration cycles cannot perform well unless they work exactly at their design conditions. However, the operating conditions may vary in the actual operation, which will cause the cycles to lose performance or waste energy. Therefore, a variable geometry ejector should be designed to achieve optimum performance over a wide range of operating conditions [43].

Water is normally used in the ejector refrigeration cycle. However, if evaporating temperature lower than 0°C is to be achieved, other refrigerants such as HCFCs and HFCs should be used [41].

D. Optimum Control Map

The operation of an ejector cycle will reach its optimum performance if the ejector operates under critical conditions [8,37]. If these critical pressure data are compiled for an ejector, an optimum control map can then be constructed so that the maximum performance can be obtained [8]. Figure 3.11 shows a typical example of such optimum control maps. These optimum control maps can be stored in the control system of the refrigeration cycle in order to maintain the ejector at its critical operating conditions. An example of using the control map is given subsequently. If the cycle is designed to operate at the critical point O, that is, evaporator temperature of 10°C, generator temperature of 120°C and critical condenser temperature of 26.2°C, the maximum entrainment ratio that the ejector can obtain is 0.43. Any other operating points along the line of OG will have the same performance. Suppose due to the change of environmental conditions, the condenser temperature is lowered to a new point such as M, the cycle is no longer at its critical condition. To control the cycle back to a new critical condition, two possible ways are available, that is, either keeping constant cooling capacity or keeping constant evaporator temperature. If the constant cooling capacity is required, the same entrainment ratio or COP should be maintained.

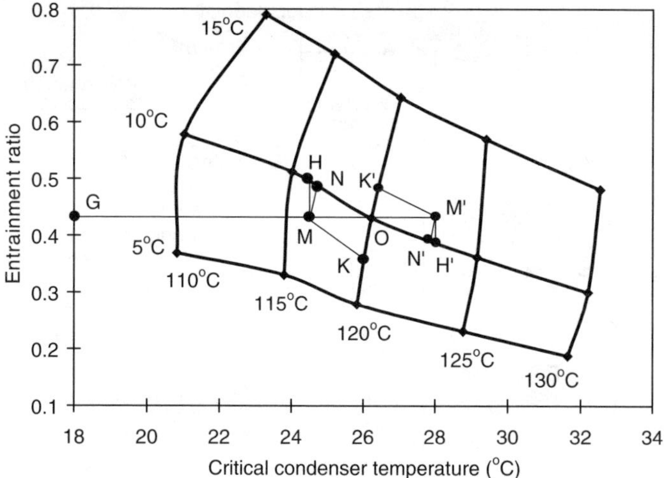

FIGURE 3.11 The optimum control map for an ejector refrigeration cycle to achieve the maximum performance.

This can be achieved by lowering the generator temperature from line OK to line NM, and simultaneously, the evaporator temperature from line ON to line KM, then the new critical condition for the cycle is established at point M. If the priority is to maintain the constant evaporator temperature, the generator temperature should then be lowered from point O to H. In this case, the entrainment ratio will rise from point M to point H, and the new critical condition for the cycle is established at point H. Similarly, suppose in a hot day, the condenser temperature is increased to a higher value at point M′, the generator temperature must be either increased from line OK′ to line N′M′ with M′ as the new critical condition or reduced from point M′ to point H′ with H′ as the new critical condition. Obviously, during the operation of the cycle, the operating condition should be kept at critical points all the time for the maximum performance of the cycle, which can be achieved by automatic control of the temperatures of the generator, condenser, and evaporator temperatures according to the technical data provided in the optimum control maps [8]. Similar optimum control maps may also be constructed for ejector operating on halocarbon compounds to achieve refrigeration below 0°C [41].

VI. NOVEL COMBINED CYCLES

Besides the earlier discussed vapor compression, absorption, and ejector refrigeration cycles, there are other types of refrigeration cycles such as the air cycle and the adsorption cycle [6]. The air cycle operates in a similar way as a vapor compression cycle, but in such a cycle, air is used as refrigerant and therefore no phase changes of the refrigerant are involved [6]. The adsorption refrigeration cycle is either based on the principle of physisorption or chemisorption. An example of the physisorption cycle is the zeolite 13X/water adsorption cycle [6,44,45], and an example of the chemisorption cycle is the metal hydride adsorption cycle [10,46,47]. Among them, the vapor compression, absorption, and ejector refrigeration cycles are the three main types of refrigeration systems. In these cycles, the vapor compression systems are generally electric powered, with the consequence of emission of large amounts of CO_2 and NO_x. Both absorption and ejector systems can be powered by thermal energy, however, their COP values are very low when compared with the vapor compression systems.

To bring together the advantages of the vapor compression, absorption, and ejector refrigeration systems, novel combined cycles have been developed [13–15]. One of them combines an

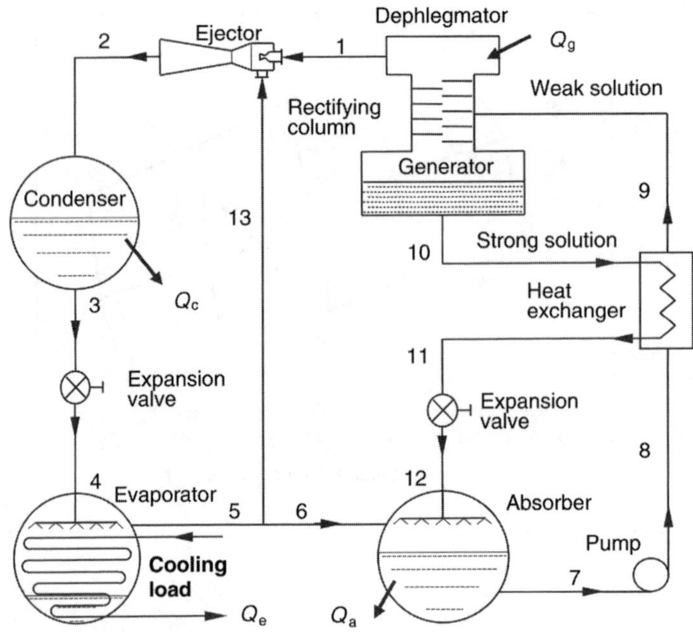

FIGURE 3.12 The combined ejector–absorption cycle.

absorption cycle with an ejector refrigeration cycle, and the other combines the ejector cycle with the vapor compression cycle. As a result, significant improvement to the system COP can be achieved with these novel combined cycles.

A. Combined Ejector–Absorption Refrigeration Cycle

The combined ejector–absorption cycle is shown in Figure 3.12. In this combined cycle, the ejector is integrated into the absorption cycle to increase the refrigerant flow rate from the evaporator and therefore to raise the cooling capacity of the cycle [13].

1. Cycle Analysis

The operating principle of the combined cycle shown in Figure 3.12 is as follows. At the generator, the absorbent–refrigerant solution is heated at the generator by heat source Q_g to produce high-pressure steam refrigerant at 1. This refrigerant (primary fluid) then flows through the primary convergent–divergent nozzle of the ejector and entrains vapor (secondary fluid) evaporated from the evaporator. The primary and secondary fluids are mixed in the ejector, then emerge from it at 2. The combined stream then flows to the condenser and condensed to liquid at 3. The heat of condensation Q_c is rejected to the environment. The condensed liquid 3 expands through a throttling valve to a low-pressure state 4 to enter the evaporator and evaporate there to produce the necessary cooling effect Q_e. Then, some of the evaporated vapor is entrained by the ejector 13 to mix with the primary fluid 1 and the remainder 6 is absorbed by the strong solution to form the weak solution 7. The heat of the absorption Q_a is rejected to the environment. The weak solution 7 is pumped back 8 to the generator 9 via the solution heat exchanger, to gain sensible heat from the strong solution coming from the generator to return to the absorber 12 via the throttling valve 1. Unlike an absorption cycle, the combined cycle operates at three distinct pressure levels, the generator at the high-pressure level, the absorber and condenser at the intermediate-pressure level, and the evaporator at the low-pressure level. This is due to the integration of the ejector in the cycle. If the combined

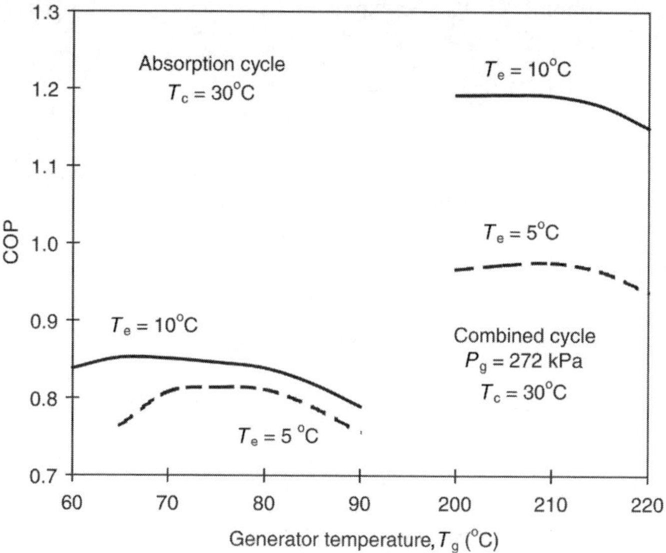

FIGURE 3.13 The comparison of the COP values between a conventional absorption cycle and the combined ejector–absorption cycle.

ejector–absorption cycle is used to generate cooling below 0°C, refrigerants used should be able to evaporate at temperature below 0°C, for example, NH_3 in NH_3–H_2O system.

2. Performance Characteristics

Owing to the additional amount of refrigerant vapor entrained by the ejector, the COP of the combined cycle is higher than that of the absorption cycle [13]. Figure 3.13 shows the predicted COP for the combined cycle when compared with an absorption cycle. It can be seen that for the absorption cycle, the COP varies with generator temperature. The optimum generator temperature for the maximum COP is determined by evaporator temperature. As discussed earlier, the solution concentration entering the generator is determined by the absorber conditions. As the condenser and absorber operate at the same pressure level, the optimum solution concentration leaving the generator is determined by the generator temperature. If the temperature at the generator is higher than its optimum value, some thermal energy is wasted. However, if the temperature is lower than the optimum value, less refrigerant vapor is produced at the generator. Figure 3.13 also shows the COP of the combined cycle. For similar reasons, optimum generator temperature for the combined cycle is also available, however, as the pressures at generator and condenser are at different levels, the pressure at the generator can be chosen as a further controlling parameter over solution concentration.

The results in Figure 3.13 indicate that the COP values of the combined cycle is significantly higher than that of the absorption cycle. For example, for the same evaporator temperatures at 5 and 10°C, the COP values of the combined cycle are about 20 and 40% higher than the absorption cycle, respectively. Thermodynamically, an absorption cycle can be considered to be comprised of a power subcycle (PSC) and a conventional refrigeration subcycle (CRSC). The PSC receives energy in the form of heat Q_g at T_g, delivers some energy W in the form of work to the CRSC, and rejects a quantity of energy Q_a in the form of heat at T_a. The CRSC receives the work W from the PSC to pump heat Q_e at T_e to a sink at T_c, so that Q_c is rejected. Consequently, the

Carnot COP of an absorption cycle can be expressed as

$$(\mathrm{COP})^{\mathrm{C}}_{\mathrm{abs}} = \left(\frac{T_{\mathrm{g}} - T_{\mathrm{a}}}{T_{\mathrm{g}}}\right)_{\mathrm{PSC}} \left(\frac{T_{\mathrm{e}}}{T_{\mathrm{c}} - T_{\mathrm{e}}}\right)_{\mathrm{CRSC}} \tag{3.44}$$

For the combined cycle, the addition of the ejector forms an additional ejector refrigeration subcycle (ERSC) along with the CRSC. Therefore, the work W from the PSC is used to drive both the CRSC and ERSC. In the combined cycle, the Carnot COP can be expressed as

$$(\mathrm{COP})^{\mathrm{C}}_{\mathrm{com}} = \left(\frac{T_{\mathrm{g}} - T_{\mathrm{a}}}{T_{\mathrm{g}}}\right)_{\mathrm{PSC}} \left[\left(\frac{T_{\mathrm{e}}}{T_{\mathrm{c}} - T_{\mathrm{e}}}\right)_{\mathrm{CRSC}} + \left(\frac{T_{\mathrm{e}}}{T_{\mathrm{c}} - T_{\mathrm{e}}}\right)_{\mathrm{ERSC}}\right] \tag{3.45}$$

or

$$(\mathrm{COP})^{\mathrm{C}}_{\mathrm{com}} = 2(\mathrm{COP})^{\mathrm{C}}_{\mathrm{abs}} \tag{3.46}$$

B. Combined Ejector–Vapor Compression Cycle

The combined ejector–vapour compression cycle is shown in Figure 3.14. In this combined cycle, an ejector refrigeration subcycle (ERSC) and a mechanical vapor compression subcycle (VCSC) are integrated into a cycle that operates in a similar way as a vapor compression system [14, 15]. The connection between the two subcycles is the intercooler, which serves as the "evaporator" for the ERSC and as "condenser" for the VCSC. The temperature at the intercooler is set between

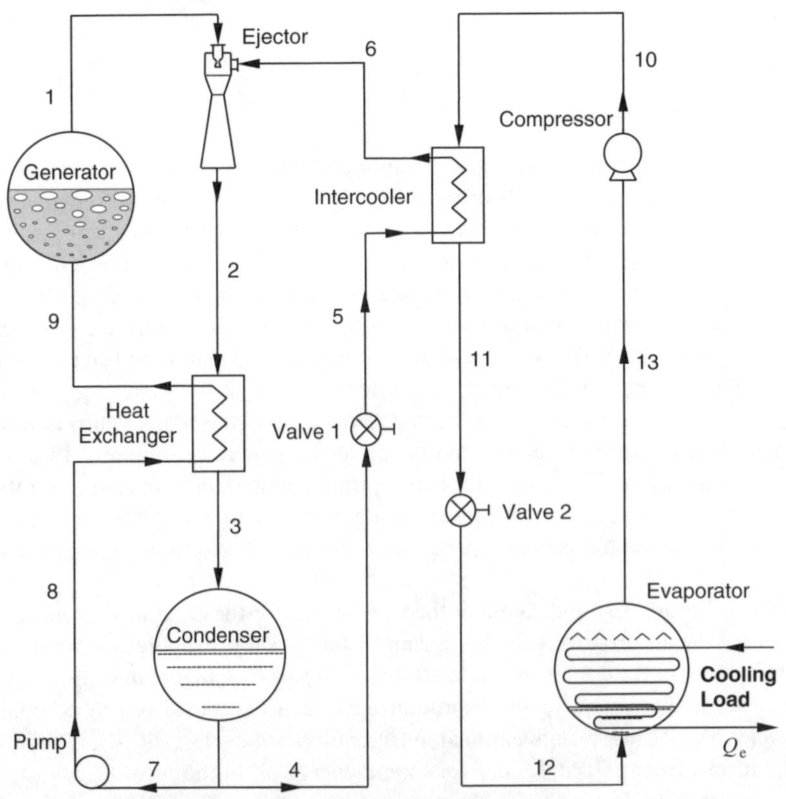

FIGURE 3.14 The combined ejector–vapor compression cycle.

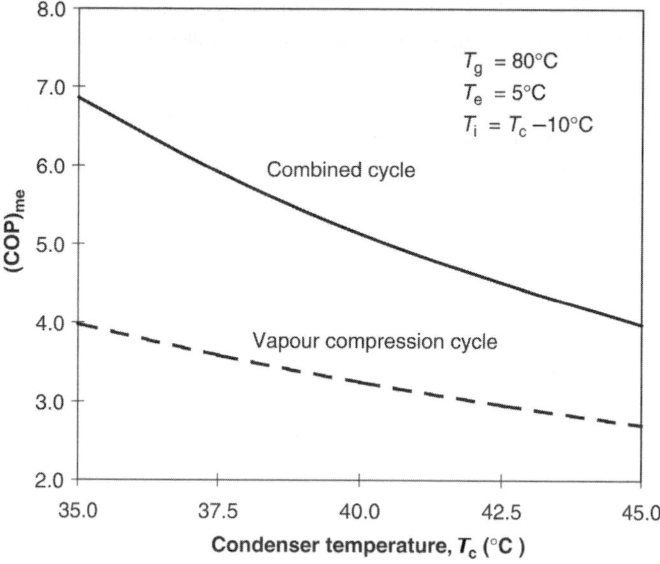

FIGURE 3.15 The comparison of the COP values between a conventional vapor compression cycle and the combined ejector–vapor compression cycle.

the condenser temperature and the evaporator temperature of the combined cycle. This arrangement secures the possibility of using two different refrigerants and makes full use of the advantages of each subcycle. If only one refrigerant is used in the combined cycle, the intercooler and its separating piping can be simplified.

1. Cycle Analysis

The operating principle of the combined cycle shown in Figure 3.14 is as follows. At the generator, the refrigerant is heated to produce high-pressure refrigerant vapor 1, which flows through the ejector and induces vapor 6 from the intercooler. The primary vapor and the entrained vapor then mix in the ejector and combine to a single stream. After exiting the ejector, the combined vapor stream 2 flows via the heat exchanger to the condenser where it condenses into liquid. The condensed liquid is then divided into two parts. One part 7 is pumped back 8 to the generator 9 via the heat exchanger and gains sensible heat from the vapor 2 from the ejector. The other part 4 expands through the throttling valve and enters the intercooler where it is evaporated by the condensation heat from the VCSC, which is generated by condensing the compressed vapor refrigerant 10 from the compressor into liquid 11. The condensed liquid via the throttling valve 11 enters the evaporator and evaporates there to produce the necessary cooling effect Q_e. The vapor 13 is then compressed by the compressor, and finally flows into the intercooler. The heat exchanger in the combined cycle is an energy saving component and it can be omitted. Owing to the existence of two subcycles, the combined cycle operates at four different pressure levels, that is the generator, condenser, and intercooler, and evaporates are all at different pressures.

2. Performance Characteristics

Figure 3.15 shows the comparison between the combined cycle in Figure 3.14 and the conventional vapor compression cycle under the same operating conditions [14]. It is observed that COP of the combined cycle can be more than 50% depending on the condensing temperature. From the energy savings point of view, more than 50% savings in electric energy could be achieved, and as a consequence, more than 50% reduction in NO_x, hydrocarbons, and CO_2 emissions due to electricity generation may be expected. If the COP of the combined cycle is compared with the ejector cycle under same operating conditions, the performance of the ejector in the combined cycle is significantly improved.

Comparison has also been made of the COP values when only one refrigerant is used in both the VCSC and ERSC with those when two different refrigerants are used, respectively, in the subcycles. It is revealed that the combined cycle performs better with dual refrigerants when compared with that if only one refrigerant is used [15].

VII. CONCLUSIONS

In the freezing of food, freezing equipment is needed. The operation of the freezing equipment is normally based on some refrigeration cycles. At present, the most commonly used refrigeration cycle is the mechanical vapor compression cycle powered by electricity. Generally speaking, electricity is generated by burning fossil fuels, which will veritably produce a large amount of CO_2, which is the main contributor to the global warming. In contrast, waste thermal energy is widely available, therefore, utilization of such energy is one of the economically feasible and environmentally friendly options. For this reason, in recent years, thermal powered refrigeration cycles, especially the absorption cycle and the ejector cycle, have been extensively studied. However, comparing with the vapor compression cycle, the COP values of the absorption and ejector cycles are significantly lower. Therefore, two innovative combined cycles, that is, the combined ejector–absorption cycle and the combined ejector–vapor compression cycle have been developed. These combined cycles bring together the advantages of the individual conventional cycles and minimize their disadvantages, and therefore, significant improvement in system COP has been made. If such combined cycles are used in food refrigeration, significant saving in running cost for the food industry could be expected.

NOMENCLATURE

A	area (m^2)
A_r	ejector area ratio
COP	coefficient of performance
c_p	specific heat (kJ/kg K)
E	total energy (kJ)
f	ratio of mass flow rate
g	gravitational acceleration (=9.807 m/s^2)
h	enthalpy (kJ/kg)
h_t	convective heat transfer coefficient (kW/m^2 K)
k	thermal conductivity (kW/m K)
m	mass flow rate (kg/sec)
P	pressure (kPa)
Q	heat transfer rate (kW)
s	entropy (kJ/kg K)
T	temperature (K)
t	temperature (°C)
V	velocity (m/s)
W	power (kW)
x	distance (m)
X	solution concentration, ammonia mass fraction in liquid phase
Y	ammonia mass fraction in vapor phase
z	elevation (m)

Greek letters

Δ	difference
η	effectiveness of heat exchanger
θ	time (s)
υ	specific volume (m^3/kg)
ω	entrainment ratio

Subscripts

a	absorber
abs	conventional absorption cycle
c	condenser, and compressor
com	combined refrigeration cycle
CRSC	conventional refrigeration subcycle
e	evaporator
f	fluid
g	generator
i	intercooler
in	inflow
max	maximum
me	mechanical
min	minimum
net	net
out	outflow
PSC	power subcycle
s	saturation
sur	surface

Superscript

C	carnot refrigeration cycle

REFERENCES

1. AE Delgado, D-W Sun. Heat and mass transfer models for predicting freezing processes—a review. *Journal of Food Engineering* 47 (3):157–174, 2001.
2. B Li, D-W Sun. Novel methods for rapid freezing and thawing of foods—a review. *Journal of Food Engineering* 54 (3):175–182, 2002.
3. K McDonald, D-W Sun. Predictive food microbiology for the meat industry: a review. *International Journal of Food Microbiology* 52:1–27, 1999.
4. B Li, D-W Sun, Effect of power ultrasound on freezing rate during immersion freezing. *Journal of Food Engineering* 55 (3):277–282, 2002.
5. D-W Sun, B Li. Microstructural change of potato tissues frozen by ultrasound-assisted immersion freezing. *Journal of Food Engineering* 57 (4):337–345, 2003.
6. D-W Sun, LJ Wang. Novel refrigeration cycles. In: D-W Sun, Ed., *Advances in Food Refrigeration.* Surrey: Leatherhead Publishing, 2001, pp. 1–69.
7. D-W Sun. The aqua-ammonia absorption system: an alternative option for food refrigeration. *Journal of Food Processing and Preservation* 22:371–386, 1998.
8. D-W Sun. Experiment investigation of the performance characteristics of a steam jet refrigeration system. *Energy Sources* 19:349–367, 1997.

9. D-W Sun. Computer simulation and optimisation of ammonia–water absorption refrigeration systems. *Energy Sources* 19:677–690, 1997.
10. D-W Sun, M Groll, R Werner. Selection of alloys and their influence on the operational characteristics of a two-stage metal hydride heat transformer. *Heat Recovery System and CHP* 12:49–55, 1992.
11. D-W Sun. Thermodynamic design data and optimum design maps for absorption refrigeration systems. *Applied Thermal Engineering* 17 (3):211–221, 1997.
12. D-W Sun. Design of metal hydride reactors. *International Journal of Hydrogen Energy* 17:945–949, 1992.
13. D-W Sun. Evaluation of a novel combined ejector–absorption refrigeration cycle—I: computer simulation. *International Journal of Refrigeration* 19 (3):172–180, 1996.
14. D-W Sun. Solar powered combined ejector–vapour compression cycle for air conditioning and refrigeration. *Energy Conversion and Management* 38 (5):479–491, 1997.
15. D-W Sun. Evaluation of a combined ejector–vapour compression refrigeration system. *International Journal Energy Research* 22:333–342, 1998.
16. YA Cengel, MA Boles. *Thermodynamics an Engineering Approach* 2nd ed., New York: McGraw-Hill, 1994.
17. WS Stoecker, JW Jones. *Refrigeration and Air Conditioning* 2nd ed., Singapore: McGraw-Hill, 1982.
18. RJ Dossat. *Principles of Refrigeration* 3rd ed., New York: Prentice-Hall, 1991.
19. AR Trout, T Welch. *Refrigeration and Air-conditioning* 3rd ed., Oxford: Butterworth-Heinemann, 2000.
20. LJ Wang, D-W Sun. Recent developments in numerical modelling of heating and cooling processes—a review. *Trends in Food Science and Technology* 14 (10):408–423, 2003.
21. LJ Wang, D-W Sun. Evaluation of the performance of slow air, air-blast and water immersion cooling methods in the cooked meat industry by the finite element method. *Journal of Food Engineering* 51 (2):329–340, 2002.
22. LJ Wang, D-W Sun. Numerical analysis of the three dimensional mass and heat transfer with inner moisture evaporation in porous cooked meat joints during vacuum cooling process. *Transactions of the ASAE* 46 (1):107–115, 2003.
23. D-W Sun, X Zhu. Effect of heat transfer direction on the numerical prediction of beef freezing processes. *Journal of Food Engineering* 42 (1):45–50, 1999.
24. X Zhu, D-W Sun. The effects of thermal conductivity calculation on the accuracy of freezing time predicted by numerical methods. *AIRAH Journal* 55 (10):32–34, 2001.
25. K McDonald, D-W Sun, J Lyng. Effect of vacuum cooling on the thermophysical properties of a cooked beef product. *Journal of Food Engineering* 52 (2):167–176, 2002.
26. A Delgado, D-W Sun. Convective heat transfer coefficients. In: D Heldman, Ed., *The Encyclopaedia of Agricultural, Food and Biological Engineering*. New York: Marcel Dekker, 2003.
27. Z Hu, D-W Sun. Predicting local surface heat transfer coefficients during air-blast chilling by different turbulent $k-\varepsilon$ models. *International Journal of Refrigeration* 24 (7):702–717, 2001.
28. ASHRAE. *2001 ASHRAE Handbook Fundamentals*. Atlanta: ASHRAE, 2001.
29. RE Critoph. Evolution of alternative refrigerant–adsorbent pairs for refrigeration cycles. *Applied Thermal Engineering* 16 (11):891–900, 1996.
30. D-W Sun. Comparison of the performances of NH_3–H_2O, NH_3–$LiNO_3$ and NH_3–NaSCN absorption refrigeration systems. *Energy Conversation and Management* 39 (5/6):357–368, 1998.
31. G Grossman, K Gommed. A computer model for simulation of absorption systems in flexible and modular form. *ASHRAE Transactions* 93 (2):2389–2427, 1987.
32. N Bennani, M Prevost, A Coronas. Absorption heat pump cycle: performance analysis of water-glycerol mixture. *Heat Recovery Systems and CHP* 9 (3):257–263, 1989.
33. GS Grover, MA R Eisa, FA Holland. Thermodynamic design data for absorption heat pump systems operating on water–lithium chloride: part I cooling. *Heat Recovery Systems and CHP* 8 (1):33–41, 1988.
34. D-W Sun, Ed., *Advances in Food Refrigeration*. Surrey: Leatherhead Publishing, 2001.
35. RM Lazzarin, A Gasparella, GA Longo. Ammonia–water absorption machines for refrigeration: theoretical and real performances. *International Journal of Refrigeration* 19:239–246, 1996.

36. D-W Sun, IW Eames. Recent developments in the design theories and applications of ejectors—a review. *Journal of the Institute of Energy* 68:65–79, 1995.
37. BJ Huang, CB Juang, FL Hu. Ejector performance characteristics and design analysis of jet refrigeration system. *Journal of Engineering Gas Turbines and Power, Transactions of ASME* 107:792–802, 1985.
38. LO Decker. Consider the cold facts about steam-jet vacuum cooling. *Chemical Engineering Progress* 89 (1):74–77, 1993.
39. IW Eames, S Aphornratana, D-W Sun. The jet pump cycle—a low cost refrigerator option powered by waste heat. *Heat Recovery Systems and CHP* 15:711–721, 1995.
40. D-W Sun, IW Eames. Performance characteristics of HCFC-123 ejector refrigeration cycles. *International Journal of Energy Research* 20:871–885, 1996.
41. D-W Sun. Comparative study of the performance of an ejector refrigeration cycle operating with various refrigerants. *Energy Conversion and Management* 40 (8):873–884, 1999.
42. D-W Sun, IW Eames. Optimisation of ejector geometry and its application in ejector air-conditioning and refrigeration cycles. *Emirates Journal for Engineering Research* 2 (1):16–21, 1997.
43. D-W Sun. Variable geometry ejectors and their applications in ejector refrigeration systems. *Energy* 21 (10):919–929, 1996.
44. RE Critoph. Possible adsorption pairs for use in solar cooling. *International Journal of Ambient Energy* 17 (4):183–190, 1986.
45. LJ Wang, DS Zhu, YK Tan. Heat transfer enhancement on the adsorber of adsorption heat pump. *Adsorption* 5:279–286, 1999.
46. D-W Sun. Thermodynamic analysis of the operation of two-stage metal-hydride heat pumps. *Applied Energy* 54 (1):29–47, 1996.
47. D-W Sun. New methods for evaluating the performance of metal hydride heat pumps. *Journal of the Institute of Energy* 68:121–129, 1995.

4 Microbiology of Frozen Foods

C.O. Gill
Agriculture and Agri-Food Canada, Lacombe Research Centre, Lacombe, Alberta, Canada

CONTENTS

I. INTRODUCTION

Microorganisms that can be present in foods for human consumption may be viruses, bacteria, yeasts, moulds, protozoa, or multicellular parasites. Some foodborne microorganisms in each of these groups can cause disease in humans. Some bacteria, yeasts, and molds can grow in foods to cause spoilage by the production of offensive odors and flavors or by causing undesirable changes in the appearance or texture of a food. Microorganisms that are neither pathogenic nor involved in spoilage processes may also be present in foods, but such organisms generally have been of little interest to food microbiologists.

Viruses, and protozoan or multicellular parasites found in foods multiply or grow and produce eggs, spores, or other infective forms of the organism only when they are within a host [1–4]. Thus, freezing of foods can affect such organisms only by increasing or reducing the rates at which the numbers of infective units decline. The same is true for thermophilic and mesophilic bacteria, yeast, and moulds that cannot grow at chiller temperatures [5]. However, some organisms that are able to grow at sub-zero temperatures may in some circumstances grow rather than simply survive in frozen foods.

II. EFFECTS OF FREEZING ON MICROBIAL ENVIRONMENTS AND MICROBES

A. CHANGES IN THE MICROBIAL ENVIRONMENT DURING FREEZING OF FOODS

Most water in foods will usually contain a more or less complex mixture of solutes. The presence of solutes in water depresses the temperature at which freezing can commence, usually to a temperature between −1 and −3°C [6]. If nuclei from which ice crystals can grow are lacking, the solution may supercool to temperatures below which freezing can start without any ice formations. If nucleation occurs, crystals of pure ice will form in the solution and the concentrations of solutes in the remaining liquid water will increase [7]. Once freezing has occurred, the ice fraction will increase with decreasing temperature until all the water is either frozen or is associated with solutes as water of hydration. Further cooling may then lead to solidification of the remaining saturated solution, or the solution may persist in a liquid, supersaturated state [8].

At any temperature at which ice has formed, the ice will be in equilibrium with the remaining liquid solution. Therefore, the vapor pressure of the solution will be that of ice at the same temperature [9]. The vapor pressure of ice at any freezing temperature is less than that of liquid water at the same temperature. Thus, any partially or wholly frozen food will be arid as compared with the same food when it is not frozen, and the frozen food will become increasingly arid with decreasing temperature. The water available to microorganisms for maintenance of their metabolisms therefore decreases as foods freeze and cool below initial freezing temperatures. The availability of water in a food, frozen or otherwise, can be expressed as its water activity (a_w), which is the ratio of the water vapor pressure of the food to that of pure water at the same temperature [10]. For any frozen food, the a_w will be that of ice at the same temperature (Table 4.1).

As freezing progresses in a food, microorganisms that are free to move in the liquid phase will concentrate in the remaining unfrozen solution [12]. Such planktonic organisms in freezing foods will therefore be exposed not only to low temperatures and low water activities but also to increasing solute concentrations and, perhaps, substantial pH changes [13]. However, in many foods some microorganisms may be attached to or localized within or between immobile, solid elements of the food. Organisms so situated could include bacteria or yeasts attached to particles or surfaces, molds with hyphae ramifying into tissues or between aggregated particles, or larval forms of multicellular parasites in muscle or organ tissues. Immobile organisms may escape exposure to concentrated solutes, but they may be affected by ice crystals that form in their vicinity, or by desiccation of their environment if water sublimes from frozen surfaces to give freeze-dried matter [14]. Foods that develop freeze-dried areas on surfaces are referred to as "freezer burned."

TABLE 4.1
Effect of Freezing Temperatures on the Water Activities of Foods

Temperature (°C)	Water Activity (a_w)
−2	0.981
−5	0.953
−10	0.907
−15	0.864
−20	0.823
−30	0.746

Source: From L Leistner, W Rodel, K Krispien. In: *Water Activity: Influences on Food Quality*, New York: Academic Press, 1981, pp. 885–916.

B. Growth of Microorganisms in Frozen Foods

Bacteria, yeasts, and molds that can grow at chiller temperatures have usually been referred to as psychrophilic or psychrotrophic, depending on the temperature range within which they can grow [15]. The former term is applied to organisms that are intolerant warm temperatures and which have been mainly derived from cold environments. The latter term has been applied to organisms responsible for the spoilage of chilled foods, which usually have maximum temperatures for growth between 30 and 35°C, and to human pathogens such as *Listeria monocytogenes* and *Yersinia enterocolitica* which can grow at temperatures of 40°C or more [16]. Recently, there has been use of the term psychrotolerant instead of psychrophilic, particularly in relation to pathogenic organisms.

The minimum temperature at which an organism can grow is generally regarded as characteristic, although the lowest temperature at which growth occurs is usually higher when an organism is subjected to other stresses in addition to low temperatures [17]. In addition, at least with bacteria, growth may become heterogeneous as temperatures approach the usual minimum for sustained growth, with some cells elongating to form filaments [18]. If temperatures fluctuate periodically from below the usual minimum, which can occur during the cyclic defrosting of refrigeration equipment, growth, and occasional division of filaments may apparently continue indefinitely [19]. Thus, in practice, factors other than the mean temperature may affect the growth of microorganisms in frozen foods.

The minimum temperature for growth in the absence of other stresses has been determined for some microorganisms capable of growth at freezing temperatures by cultivating the organisms in supercooled media. With that technique, growth of vibrios, that is, Gram-negative bacteria at −4°C [20] and growth of bacilli, that is, Gram-positive bacteria at −7°C [21] have been demonstrated. A mold, *Thamnidium elegans*, also grew at −7°C in a supercooled medium, and extrapolation of data for growth rates at freezing temperatures suggested that the minimum temperature for growth could be −10°C [22]. However, in practice, microorganisms in frozen foods will almost inevitably be exposed to osmotic stress or desiccation and, perhaps, to inhibiting concentrations of some solutes as well as to low temperatures. Moreover, even when growth of various organisms is possible, the rate of growth of some may be so slow as to render any increases in their number or size inconsequential for the safety or storage stability of the product. The consequences of inhibition by factors other than temperature and the effects of freezing temperatures on growth rates are evident in descriptions of the aerobic spoilage of red meats and poultry at freezing temperatures [23,24]. At temperatures above −3°C, flora dominated by Gram-negative psychrotrophs develop and the meat is spoiled by putrid odors and flavors within 2 or 3 weeks. At −5°C, although some growth of Gram-positive bacteria may occur, a flora composed largely of yeasts develops. Spoilage is then due to yeast and mold colonies becoming visible after storage of meat for 6 months or more. Yeast and molds may grow at temperatures down to about, but probably not below, −10°C, and without the appearance of visible colonies after 12 months or more. Spoilage of meat by extensive growth of molds is not usually the result of prolonged storage at temperatures near −10°C. Instead such spoilage occurs when temperatures rise above −5°C and growth of bacteria and yeasts is inhibited by desiccation of meat surfaces [25].

Thus, foods stored at temperatures of −10°C or less should be preserved against spoilage as a result of microbial growth. However, temperature gradients commonly exist within refrigerated facilities, and temperatures can rise during periodic defrosting of refrigeration equipment [26]. Consequently, if frozen foods are stored in commercial facilities with refrigeration equipment operating at temperatures close to −10°C, some of the products may be exposed to higher temperatures at which microorganisms may grow to cause spoilage.

C. Injury of Microorganisms during Freezing, Thawing, and Frozen Storage

In principle, microorganisms could be injured, lethally or otherwise during the freezing of foods by low temperatures *per se*, causing damage of cell walls and disruption of membranes by ice formed in the external medium, disruption of cells by the formation of intracellular ice, concentration of damaging solutes in the extracellular water, and dehydration of cells as a consequence of water moving from the cells in response to the increasing osmotic pressure or drying of the extracellular medium. During thawing, ice crystals may enlarge or solutions in a glassy state may melt to expose microorganisms to an environment that is more chemically and physically damaging [27]. During frozen storage, microorganisms may be damaged by physical or chemical reactions between cell components and components of the surrounding medium, or by desiccation of food surfaces. However, injury of microorganisms may be greatly limited if a food contains solutes that protect against freezing injury.

When microorganisms growing at relatively warm temperatures are subjected to a rapid, substantial decrease in temperature, within or beyond the growth temperature range, the cells can be injured with intracellular metabolites and proteins being lost from the cell [28]. At temperatures within the growth temperature range, cold shock proteins are synthesized in response to the sudden stress [29]. However, rates of cooling of foods would generally be too slow to shock bacteria, so injury of bacteria by cold shock is probably of little consequence for the survival of bacteria in frozen foods. Otherwise, cooling of microorganisms below their minimum temperature for growth is not immediately injurious to itself.

Even so, microorganisms held at temperatures below their minima for growth may progressively lose viability [30]. Usually, loss of viability when growth cannot occur is more rapid at higher than at lower temperatures [31]. Thus, organisms that can grow in foods but cannot grow at temperatures below 0°C may be adversely affected by the temperature alone when they experience storage temperatures near the upper end of the range of freezing temperatures for foods. At usual storage temperatures for frozen foods, however, development of damage as a result of the temperatures alone is likely to be of little importance for most microorganisms.

Mechanical damage of cells by ice crystals formed in the extracellular medium does not appear to be a major cause of injury [32]. Microorganisms in freezing media tend to supercool, with the internal water remaining unfrozen at temperatures as low as −15°C [33]. Because of that, the water activity within the cells is above that of the surrounding medium. Consequently, the cells tend to lose water and become dehydrated. When the rate of cooling is relatively slow, water will be lost sufficiently rapidly for the cell contents to remain in osmotic equilibrium with the surrounding medium [34]. Thus, the organism will experience increasing intracellular solute concentrations and, if in contact with the remaining liquid phase, to increasing extracellular solute concentrations as well. However, freezing of cell contents will not occur. Only if the rate of cooling is such that the limit to supercooling of the cell contents is exceeded before the cell is dehydrated will ice crystals form within the cell.

The rates at which cells can dehydrate will depend upon the permeability of the cell membrane to water and the surface area to volume ratio for the cell. That ratio will depend upon the shape and size of the cell (Table 4.2). In general, ice will not form in cells of any size when cooling rates are about or less than 1°C/min [35]. Ice may form in cells of bacteria, yeasts, and molds when cooling rates exceed 10°C/min [36]. For small bacterial cells, freezing unaccompanied by extensive dehydration is probably precluded unless the rate of cooling approaches or exceeds 100°C/min.

If dehydration during freezing was largely responsible for the lethal injury of microorganisms, then the rate of survival could be expected to decline with decreasing temperature. This is observed with some parasites that are also killed by dehydration at ambient temperatures [37]. However, other microorganisms can survive dehydration, although the resistance to desiccation can vary with the physiological state of the organism and the composition of the surrounding medium. Moreover, with yeasts, bacteria, and molds, survival of freezing tends to increase with decreasing temperature, and with increasing rates of cooling up to about 10°C/min [38]. Survival then decreases to a minimum at rates of cooling between 10 and 100°C/min, but increases again with rates of cooling of 1000°C/min and more (Table 4.3). When microorganisms have been

TABLE 4.2
Diameters (Smallest Dimensions) of Foodborne Microorganisms

Type of Organism	Diameter
Viruses	25–75 nm
Bacteria	0.2–2 μm
Yeasts	3–25 μm
Protozoan cysts, oocysts, or spores	10–65 μm
Parasites, larval cysts, or larvae	0.25–20 mm

subjected to the latter, rapid rates of freezing, survival is greater when thawing is rapid than when thawing is slow [39].

Those effects of the rate of cooling are explicable if injury at low rates of freezing is largely due to the increased concentrations of solutes in the extracellular medium [13]. The initial enhancement of survival with increasing rate of cooling is then due to the medium solidifying in a glassy state, without extensive freezing out of pure ice, and so with limited increases of the concentrations of solutes in the liquid phase. At rates of cooling above 10°C/min, increasing amounts of ice are formed within cells, causing injury. However, the sizes of ice crystals decrease with increasing rates of freezing while large intracellular ice crystals cause the greatest damage to cells. Thus, the injury sustained by cells is reduced when small ice crystals are formed intracellularly during very rapid cooling. Despite that, if cells are thawed slowly the sizes of some ice crystals will increase as the cells warm, with greater damage to cells than when thawing is rapid.

In foods, rates of cooling will generally be substantially less than 1°C/min except, perhaps, at surfaces contacted directly by very cold liquids [40]. Therefore, any injury of viruses, bacteria, yeast, and molds during freezing of foods is likely due to high solute concentrations and desiccation rather than damage by intracellular or extracellular ice. However, formation of ice within the organism may well be an important cause of damage to the relatively large protozoan and multicellular parasites.

During frozen storage, the numbers of viable microorganisms in foods can continue to decline, but at rates which may be not only much slower than those that occur during freezing but which also decrease with time [41]. After extended periods of storage, the numbers of some microorganisms in frozen foods may be essentially stable [30], but the numbers of others may continue to decline, to levels at which they cannot be detected (Table 4.4).

TABLE 4.3
Effect of the Rate of Cooling on the Survival of *E. coli* in Water

Cooling Rate (°C/min)	Viable Bacteria (% of Initial Number)
1	30
10	65
100	18
1000	75

Source: From PH Calcott, RA Madeod. *Canadian Journal of Microbiology*, 21:1724–1732, 1975.

TABLE 4.4
Effects on Bacteria in the Spoilage Flora of Freezing and Frozen Storage of Turkey Carcasses

Stage of Processing or Storage	Log Numbers of Bacteria/cm^2			
	Aerobes	Pseudomonads	Coliforms	Enterococci
Before freezing	6.5	5.2	4.8	5.0
After freezing	4.5	2.7	2.5	3.4
Storage, 1 month	4.1	2.1	2.7	1.4
Storage, 2 months	3.1	1.2	2.2	1.1
Storage, 4 months	2.7	1.0	1.4	0.1
Storage, 6 months	2.7	<0.0	0.3	0.1

Source: From AA Kraft, JC Ayres, KF Weiss, WW Marion, SL Balloun, RH Forsythe. *Poultry Science*, 42:128–137, 1963.

III. CRYOPROTECTANTS

Although concentration of some extracellular solutes during freezing may injure cells, the presence of other solutes in the extracellular medium can protect microorganisms against freezing damage. Such cryoprotectants include glycerol and other polyols, glycine, sugars, and other low molecular weight organic compounds such as dimethylsulfoxide or acetamide [43]. However, high molecular weight materials such as starch or soluble proteins and, in some instances, electrolytes can also have cryoprotective effects [44]. As polyols and other low molecular weight compounds readily enter cells and are variously accumulated by xerotolerant organisms in response to osmotic stress [45], it is likely that they similarly protect cell components from the deleterious effects of dehydration during freezing. The cryoprotective effects of electrolytes may also be due to their stabilizing some cell components [13]. Large molecular weight compounds that do not penetrate cells probably act as cryoprotectants by their inhibition of nucleation and growth of ice crystals in the extracellular medium. Cryoprotection by low molecular weight compounds may also be due in part to similar effects on ice formation. In general, the effects of freezing are likely to be less deleterious for microorganisms in complex media, such as those offered by many foods, than in simple media [46,47].

IV. EFFECTS OF FREEZING ON MICROORGANISMS

A. Viruses

Enteric viruses are a major cause of foodborne disease [48]. Those most commonly identified as causes of illness are the hepatitis A picornavirus and the Norwalk-like caliciviruses. Other viruses that may be transmitted in foods are the calicivirus, which is the causative agent of hepatitis E, and various astroviruses and rotaviruses that cause enteric diseases.

The particles of foodborne enteric viruses are at the lower end of the range of viral particle size [49]. The particles have a simple structure, being composed of a nucleic acid core, which is the genome, surrounded by a protein coat. The genomes of most foodborne enteric viruses are single strands of RNA, which can serve directly as messenger RNA for protein synthesis. Replication occurs in the cytoplasm rather than the nucleus of the host cell. Because viruses do not replicate outside the host, food samples must be examined for viruses by molecular techniques. Samples used in such analyses must be small, and foods may contain materials that interfere with assays that identify distinguishing elements of the genome. Therefore, the detection of viruses in foods is generally difficult [50].

In view of the nature of virus particles, their preservation rather than destruction by freezing could be expected. There appear to be no reports on the survival of hepatitis A or E, or Norwalk-like viruses in frozen foods. However, large fractions of polioviruses inoculated into or on foods were found to remain infective after prolonged periods of frozen storage [51,52], while an outbreak of gastroenteritis was reported to be associated with the consumption of ice contaminated with a Norwalk-like virus [53]. Such findings indicate that at least some enteric viruses are likely to remain infective after freezing and frozen storage of foods.

B. Bacteria

Some Gram-positive bacteria form spores within their cells (endospores) which can survive conditions that destroy the vegetative cells. Thirteen genera of spore-forming bacteria have been described [54]. Of these, members of the genera *Bacillus* and *Clostridium* are the most important with respect to the safety and spoilage of foods. The mature endospore is comprised of a central dehydrated protoplast, a peptidoglycan cortex, and outer proteinaceous coats [55]. The conditions under which spores form can affect their resistance to environmental stresses. For example, spores formed at higher temperatures generally have protoplasts that are more dehydrated and more resistant to heat than spores formed at lower temperatures [56]. Spores are far more resistant than the parent microorganisms to a wide range of environmental stresses [57]. As spores can survive for many years in natural environments [58] and clostridial spores in fruits and vegetables have been reported to be unaffected by freezing [59], it appears that spores will generally not be inactivated by freezing.

Although most bacteria do not form spores, all become more resistant to environmental stresses when they enter the stationary phase. The stationary phase condition involves not only the cessation of growth but also substantial physiological changes induced by nutrient starvation [60]. During entry into the stationary phase, some of the proteins that are synthesized in response to starvation are common for all growth-limiting nutrients, while the induction of others vary with the conditions of starvation [61]. Moreover, physiological changes continue after the cessation of growth, as some proteins that enhance viability are synthesized during the first few hours that bacteria are in the stationary phase [62]. Stationary phase cells of any bacterium are generally considerably more resistant to the effects of freezing than cells of the same organism that are growing exponentially [63].

However, general and specific stress proteins are also induced in growing bacteria by various inimicable environmental conditions. The induction of a core group of stress proteins in response to other environmental stresses as well as starvation gives rise to the phenomenon of cross protection [64]. That is, the development of resistance to one stress confers resistance to other stresses that need not be obviously related. Thus, cryotolerance can be induced in various bacteria not only by cultivation at temperatures near their minima for growth [65] but also by osmotic, oxidative, or other stresses [66].

In addition to injury from freezing being variable with the physiological condition of bacteria, there are wide variations in susceptibility to freezing injury between different species of bacteria and between different strains of the same organism [67]. In general, Gram-positive bacteria appear to be less susceptible to freezing injury than Gram-negative organisms [68]. Thus, staphylococci and listerias appear to be relatively resistant to the adverse affects of freezing [69,70], while slow freezing of meat results in enrichment of the spoilage flora for Gram-positive organisms such as lactobacilli and *Brochothrix thermosphacta* at the expense of the Gram-negative organisms that usually predominate on meat stored in air [71]. However, other Gram-positive organisms, such as some lactic acid bacteria and vegetative cell of some clostridia, are reported to be highly susceptible

TABLE 4.5
Decreases in Numbers of Exponential or Stationary Phase *E. coli* and *Salmonella* Frozen in Nutrient Broth with or without Additional Mg^{2+}

	Decrease (log cfu/ml)			
	E. coli		Salmonella	
Growth Phase	**Broth**	**Broth + Mg^{2+}**	**Broth**	**Broth + Mg^{++}**
Exponential	4.5	2.9	3.3	2.5
Stationary	1.2	0.9	1.4	0.8

Source: From MG Smith. *Journal of Food Science*, 60:509–512, 1995.

to the lethal effects of freezing [72,73], while relatively rapid, cryogenic freezing of ground beef has little effect on the Gram-negative coliforms in the meat microflora [74].

In some studies, freezing has been shown to destroy large fractions of Gram-negative pathogenic or spoilage bacteria in laboratory media or foods [75–78]. However, extensive reduction of numbers in some circumstances may not be replicated in others. For example, the numbers of *Escherichia coli* 0157:H7 frozen in peptone water declined by >4 log units during prolonged frozen storage [79], but the same organism in ground beef was little affected by storage for a similar time at the same temperature [80]. It is therefore evident that the susceptibility of vegetative bacterial to lethal injury from the freezing of foods is likely to vary widely with the circumstances under which freezing occurs (Table 4.5).

C. Fungi

Both unicellular and filamentous fungi, that is, yeasts and molds, produce spores during sexual reproduction and molds can also produce spores asexually [81,82]. However, only the asexual forms of the organisms, and thus only asexual or no spores, are known for many molds or yeasts, respectively [83]. In general, sexually produced spores are likely to be more resistant to freezing damage than spores that are produced asexually, with the former possibly surviving for years at freezing temperatures [84] while the latter may rapidly lose viability during the first hours after freezing [85,86]. However, the viability of frozen fungal spores can be dependent on the conditions under which they are formed and frozen, and fungal spores are commonly preserved as freeze-dried preparations [87,88].

Lethal effects are difficult to quantify for the filamentous molds, so the effects of freezing on vegetative fungi have been investigated mostly with yeasts. As with bacteria, the susceptibility of vegetative cells to freezing damage varies widely with the growth phase of the cells, the conditions under which they are cultivated, exposure to other stresses before freezing, and the presence or lack of cryoprotectants in the freezing medium. Yeast cells in the stationary phase are more resistant to freezing damage than are cells that are growing logarithmically [89]. Cells of *Saccharomyces cerevisiae* that were growing using ethanol as the carbon source were found to be less susceptible to the lethal effects of freezing than cells that were utilizing glucose or galactose [90]. Exposure of yeasts to osmotic stress or heat shock before freezing can lead to the accumulation of compatible solutes such as glycerol and trehalose [91], and substantially enhanced resistance to freezing damage (Table 4.6). As with other microorganisms, freezing damage of yeasts is greatly reduced by inclusion either or both cell penetrating or nonpenetrating cryoprotectants in the growth medium [93]. Cryoprotectants such as milk proteins and sugars can be highly effective for enhancing the survival of frozen yeasts [47].

TABLE 4.6
Survival of Exponential Phase Cells of *Geotrichum candidum* Frozen Without or After Exposure to Various Concentrations of NaCl for 2 h

NaCl Concentration (M)	Survival (%)
0	41
0.5	49
1.0	62
1.5	62
2.0	72

Source: From S Dubernet, JM Panoff, B Thammavongs, M Gueguen. *International Journal of Food Microbiology*, 76:215–221, 2002.

D. Protozoa

Protozoan parasites that can contaminate water and be found in foods include both organisms with relatively simple life cycles that are completed in a single host, and organisms with complex life cycles that involve intermediate hosts as well as the definitive host in which sexual reproduction occurs [94]. However, most have resistant forms which are variously spores, cysts, or oocysts. These are spread with faeces to contaminate the environment and spread infections to new hosts. Some protozoan parasites such as *Toxoplasma gondii* also form cysts in host tissues, which serve to transmit the parasite when the tissues are consumed by another host.

The tissue cysts of *T. gondii* were inactivated when infected muscle tissue was frozen to −12°C, but survived for increasing times up to about 3 weeks at increasingly higher freezing temperatures [95]. Oocysts of cryptosporidia have also been reported to survive longer at higher than at lower freezing temperatures, being inactivated by freezing to −70°C and surviving for only a few hours at −20°C, but surviving for over 12 weeks at −4°C [96,97]. As with the cryptosporidia, spores of a coccidian parasite of chickens, *Eimeria acervulina* were rapidly inactivated by freezing at −18°C [98]. In contrast, spores of a protozoan that causes disease in oysters were found to be highly resistant to freezing, surviving for over 7 months at −70°C [99], while cysts of *Giardia intestinalis* were inactivated within 7 days at −4°C [97].

However, the conditions under which freezing occurs may greatly affect the survival of infectious forms of protozoan parasites as it has been reported that oocysts of cryptosporidia remained viable for more than 4 weeks when frozen slowly at −22°C [100], and were more resistant to freezing when suspended in salt solutions rather than water [101]. Spores of microsporidia that infect insects have remained viable during storage in liquid nitrogen for 25 year, with survival being enhanced by suspension in cryoprotective glycerol media [102].

E. Multicellular Parasites

Various helminthic parasites of humans have infective forms that can be present in foods. The most important of those foodborne parasitic diseases are trichinella and tape worm infections acquired from red meats [103]. Other infections may be acquired from raw fish or shellfish, contaminated vegetables or accidentally ingested invertebrates [104]. The life cycles of helminthic parasites are generally complex and mostly involve intermediate as well as definitive host species. Although infections may occur from the consumption of eggs or other forms of a parasite that are usually involved in the infection of intermediate hosts, human infections are usually due to the consumption

TABLE 4.7
Times and Temperatures for Inactivation of *T. spiralis* Larvae by Freezing under Experimental or Commercial Conditions

Experimental Conditions		Commercial Conditions	
Temperature (°C)	Time	Temperature (°C)	Time
−10	4 days	−15	20 days
−15	64 min	−23	10 days
−20	8 min	−29	6 days

Sources: From KD Murell. *Food Technol.*, 39 (3):65–68, 101–111, 1985; PC Beaver, RC Jung, EW Cupp. In: *Clinical Parasitology*, Phildelphia: Lea & Febiger, 1984, pp. 231–252.

of muscle tissues of mammals, finfish, or shellfish which contain encysted or unencysted larvae of round or tape worms or metacercariae of flukes.

Larvae of helminths in muscle tissue are inactivated by freezing, with the rate of inactivation increasing with decreasing temperature (Table 4.7). Trichinosis in humans is commonly due to the consumption of pork infected with the larvae of *Trichinella spiralis*. Freezing of larvae of *T. spiralis* from pork muscle, at −20°C under well-controlled conditions, rendered the larvae uninfective after only a few minutes [105]. With storage under commercial conditions at the same temperature, several days have been found to be required for inactivation of the larvae in blocks of pork muscle tissue, but complete inactivation has been achieved without further holding when the temperature at the centre of pork muscle was reduced to −35°C [106]. However, such findings for *T. spiralis* from pork may not be true for all the organisms that cause trichinosis. There is continuing uncertainty as to whether *Trichinella* found in various hosts in different regions are strains of *T.* spiralis or are distinct species [107]. Whichever is the case, larvae of a strain or species termed *Trichinella nativa* from bear, fox, and dog in arctic and temperate regions have been found to remain viable after freezing for several months [108,109]. In contrast, larvae of *Trichinella papuae*, a strain or species from animals in tropical regions that was used to experimentally infect foxes, were found to be highly susceptible to inactivation by freezing [110].

Larvae (or cysticerci) of *Taenia saginata* and *Taenia solium*, the tape worms acquired by humans from beef and pork, respectively, are inactivated by freezing under commercial conditions similarly to the larvae of *T. spiralis* in pork [111,112] Larvae of the various helminths that may be present in fish muscle are also inactivated by commercial freezing [113,114]. However, data on the effects of freezing on eggs and metacercariae of parasites that can be infectious for humans appear to be lacking.

Despite the apparently usual lethal effects of freezing on larvae of parasitic helminths, some at least can be preserved by controlled freezing in cryoprotective media [115].

V. FREEZING OF FOODS IN RELATION TO HACCP SYSTEMS

Foods are usually frozen to prevent their spoilage by microorganisms and to preserve their desirable eating qualities [116], while the purpose of HACCP systems is to assure the safety of foods [117]. Thus, for most foods the only risk from microbiological hazards that may increase with freezing, and which is likely to be recognized in and controlled through a HACCP system, is the proliferation of pathogens during freezing or frozen storage. Generally, freezing at rates that preserve desirable eating qualities is likely to prevent any substantial growth of pathogenic organisms, while pathogens cannot grow at the temperatures now usually maintained for frozen foods. Thus, risks from microbiological hazards are likely to increase only when freezing is delayed or prevented by overloading of freezer facilities, improper stacking that insulates some product from freezing air, or failure of refrigeration equipment.

HACCP systems generally do not take account of reductions in microbiological risks that might be obtained with freezing. An exception to that generalization is the use of freezing to ensure the safety of beef and pork with respect to the larvae of helminthic parasites that may be present in flesh, which has long been a treatment regarded as effective by regulatory authorities [118,119]. More recently, some regulatory authorities have mandated the freezing of fish to destroy parasitic larvae, if it is intended that the meat be consumed while it is raw [120]. Also, in some countries, freezing has been used as a means of reducing the numbers of *Campylobacter* on poultry carcasses [121], as those Gram-negative, pathogenic bacteria are very susceptible to the lethal effects of freezing and other environmental stresses [122,123].

Freezing has also been shown to substantially reduce the numbers of various other pathogens in foods as, for examples, pathogenic vibrios in oysters [124,125] or vegetative *Clostridium perfringens* in various, comminuted meats [126]. In contrast, some pathogens have been found to be little affected by the freezing of foods, as, for examples, *L. monocytogenes* in ice cream [127] or *Salmonella typhimurium* in cooked meats [128].

Evidently, freezing may be used, and be recognized in a HACCP system, as a means of reducing pathogens in or eliminating them from some frozen foods. However, in view of the many biological, chemical, and physical factors that can affect the survival of all types of pathogens in frozen foods, prediction of the destruction of any pathogen of concern in a particular food subject to a specific freezing process would in most instances be highly uncertain. Therefore, freezing could usually be accepted in a HACCP system as a decontaminating treatment for a food only with direct evidence that the pathogen or pathogens of concern were in fact destroyed in the food in question during its freezing in the process being operated under HACCP control.

VI. CONCLUSIONS

The effects of freezing on microorganisms in foods vary greatly with the type of microorganism, the physiological state or stage in the life cycle of the microorganism, the composition of the food, and the rates of freezing and thawing. In general, viruses, bacterial spores, and sexual spores of fungi are likely to be preserved by freezing, irrespective of the composition of the food and the rates of freezing and thawing. Other microorganisms are likely to be damaged by freezing, but the extent to which freezing and subsequent frozen storage reduces the numbers of any organism may be trivial if the population is in a resistant physiological state or stage of the life cycle, the food contains cryoprotective substances, and freezing and thawing are rapid. Thus, except with large larval or adult forms of multicellular parasites, it cannot be safely assumed that freezing will destroy large numbers of any microorganism which may be present in a food. Despite that, in some circumstances freezing of a food will cause extensive inactivation of at least some microorganisms. However, substantial reductions in the numbers of viable organisms of concern would have to be validated for specific foods in specific processes if freezing is to be a recognized decontaminating treatment for a food.

REFERENCES

1. EO Caul. Foodborne viruses. In: BM Lund, TC Baird-Parker, GW Gould, Eds., *The Microbiological Safety and Quality of Foods*. Gaithersburg, MD, USE: Aspen Publishers, 2000, pp. 1437–1489.
2. CW Kim, Helminths in meat. In: MP Doyle, LR Beuchat, TJ Montville, Eds., *Food Microbiology Fundamentals and Frontiers*. Washington, DC, USA: American Society for Microbiology Press,1997, pp. 449–462.

3. EG Hayunga. Helminths acquired from finfish, shellfish and other food sources. In: MP Doyle, LR Beuchat, TJ Montville, Eds., *Food Microbiology Fundamentals and Frontiers*. Washington, DC, USA: American Society for Microbiology Press, 1997, pp. 463–477.
4. CA Speer. Protozoan parasites acquired from food and water. In: MP Doyle, LR Beuchat, TJ Montville, Eds., *Food Microbiology Fundamentals and Frontiers*. Washington, DC, USA: American Society for Microbiology Press, 1997, pp. 478–493.
5. CO Gill. Microbial control with cold temperatures. In: VK Juneja, JN Sofos, Eds., *Control of Foodborne Microorganisms*. New York: Marcel Dekker, 2002, pp. 55–74.
6. L Bogh-Sorensen, M Jul. Effects of freezing/thawing on foods. In: RK Robinson, Ed., *Microbiology of Frozen Foods*. London: Elsevier Applied Science, 1985, pp. 41–82.
7. L Vanden Berg. Physiochemical changes in foods during freezing and subsequent storage. In: J Hawthorne, EJ Rolfe, Eds., *Low Temperature Biology of Foodstuffs*. Oxford: Pergamon Press, 1968, pp. 205–219.
8. O Fennema. Frozen foods: Challenge for the future. *Food Australia*. 45:374–381, 1993.
9. O Fennema. In: LB Rockland, GF Stewart, Eds., *Water Activity: Influences on Food Quality*. New York: Academic Press, 1981, pp. 504–550.
10. L Leistner, NJ Russell. Solutes and low water activity. In: NJ Russell, GW Gould, Eds., *Food Preservatives*. Glasgow, UK: Blackie,1991, pp. 111–134.
11. L Leistner, W Rodel, K Krispien. Microbiology of meat and meat products in high and intermediate moisture ranges. In: LB Rockland, BF Stewart, Eds., *Water Activity: Influences on Food Quality*, New York: Academic Press, 1981, pp. 885–916.
12. PH Calcott. Manifestations of freeze-thaw injury. In: JG Cook, Ed., *Freezing and Thawing of Microbes*. Shildon, UK: Meadow Field Press, 1978, pp. 26–40.
13. P Mazur. Physical and chemical basis of injury in single-celled microorganisms subjected to freezing and thawing. In: HT Meryman, Ed., *Cryobiology*. London: Academic Press, 1966, pp. 213–315.
14. G Kaes. Freezer burn as a limiting factor in the storage of animal tissue. I. Experiments with liver frozen without weight loss. *Food Technology* 15 (3):122–128, 1961.
15. JC Olson Jr., PM Nottingham. Temperature, In: *International Commission on Microbiological Specifications for Foods, Microbial Ecology of Foods*, Vol. 1. New York: Academic Press, 1980, pp. 1–37.
16. AK Kraft. *Psychrotrophic Bacteria in Foods: Disease and Spoilage*. Boca Raton, FL, USA: CRC Press, 2000, pp. 147–184.
17. TA McMeekin, JN Olley, T Ross, DA Ratkowsky. In: *Predictive Microbiology: Theory and Application*. New York: Wiley, 1993, pp. 165–190.
18. T Jones, CO Gill, LM McMullen. Behaviour of log-phase *Escherichia coli* at temperatures near the minimum for growth. *International Journal of Food Microbiology* 88:55–61, 2003.
19. T Jones, CO Gill, LM McMullen. The behaviour of log-phase *Escherichia coli* at temperatures that fluctuate about the minimum for growth. *Letters in Applied Microbiology*. 39:296–300, 2004.
20. M Sasajima. Studies of psychrotolerant bacteria in fish and shellfish V. The growth or viability of trimethylamineoxide reducing psychrotrophic bacteria and their activity at subzero temperatures. *Bulletin Japanese Society for Scientific Fisheries* 40:625–630, 1974.
21. JM Larkin, JL Stokes. Growth of psychrophilic microorganisms at subzero temperatures. *Canadian Journal of Microbiology*. 14:97–101, 1968.
22. PD Lowry, CO Gill. Temperature and water activity minima for growth of spoilage moulds from meat. *Journal Applied Bacteriology* 56:193–199, 1984.
23. W Schmidt-Lorenz, J Gutschmidt. Mikrobielle und sensorische Veranderungen gefrorener Brathahnchen und Poularden bei Lagerung im Temperaturbereich von $-2.5°$ bis $-10°C$. *Fleiswirtschaft* 49:1033–1041, 1969.
24. PD Lowry, CO Gill. Development of yeast microflora on frozen lamb stored at $-5°C$. *Journal Food Protection* 47:309–311, 1984.
25. PD Lowry, CO Gill. Mould growth on meat at freezing temperatures. *International Journal Refrigeration* 7:133–136, 1984.
26. M Jul. In: *The Quality of Frozen Foods*. London: Academic Press, 1984, pp. 156–208.
27. P Mazur. Cryobiology: the freezing of biological systems. *Science* 168: 939–949, 1970.

28. BM Mackey. Lethal and sublethal effects of refrigeration, freezing and freeze-drying on microorganisms. In: MHE Andrew, AD Russel, Eds., *The Revival of Injured Microbes*, New York: Academic Press, 1984, pp. 45–75.
29. K Yamanaka, L Fang, M Inouye. The CspA family in *Escherichia coli*: multiple gene duplication for stress adaptation. *Molecular Microbiology* 27:247–255, 1998.
30. J Christophersen. Effect of freezing and thawing on the microbial population of foodstuffs. In: J Hawthorn, EJ Rolfe, Eds., *Low Temperature Biology of Foodstuff*. Oxford: Pergamon Press, 1968, pp. 251–269.
31. M Ingram, BM Mackey. Inactivation by cold. In: FA Skinner, WB Hugo, Eds., *Inhibition and Inactivation of Vegetative Microbes*. New York: Academic Press, 1976, pp. 111–151.
32. HT Meryman. The exceeding of a minimum tolerable cell volume in hypertonic suspension as a cause of freezing injury. In: GEW Wolstenholme, M O'Connor, Eds., *The Frozen Cell*. London: J&A Churchill, 1970, pp. 51–65.
33. GG Litvan. Mechanism of cryoinjury in biological systems. *Cryobiology* 9:182–191, 1972.
34. P Mazur. Kinetic of water loss from cells at subzero temperatures and the likelihood of intracellular freezing. *Journal of General Physiology* 47:347–369, 1964.
35. H Bank, P Mazur. Visualization of freezing damage. *Journal Cell Biology* 37:729–742, 1973.
36. GJ Morris, D Smith, GE Coulson. A comparative study of the changes in the morphology of hyphae during freezing and viability upon thawing for twenty species of fungi. *Journal of General Microbiology* 134:2897–2906, 1988.
37. R Fayer, T Nerad. Effect of low temperatures on viability of *Cryptosporidium parvum* oocysts. *Applied Environmental Microbiology* 62:1431–1433, 1996.
38. R Davies, A Obafemi. Response of micro-organisms to freeze-thaw stress. In: RK Robinson, Ed., *Microbiology of Frozen Foods*. London: Elsevier Applied Science, 1985, pp. 83–107.
39. PH Calcott, RA Macleod, The survival of *Escherichia coli* from freeze-thaw damage: permeability barrier damage and viability. *Canadian Journal Microbiology* 21:1724–1732, 1975.
40. SJ James, C James. In: *Meat Refrigeration*, Cambridge, UK: Woodhead Publishing, 2002, pp. 137–158.
41. ML Speck, B Ray. Effects of freezing and storage on microorganisms in frozen foods: a review. *Journal Food Protection* 40:333–336, 1977.
42. AA Kraft, JC Ayres, KF Weiss, WW Marion, SL Balloun, RH Forsythe. Effect of method of freezing on survival of microorganisms on turkey. *Poultry Science* 42:128–137, 1963.
43. EH Marth. Behaviour of food microorganisms during freezing-preservation. In: OR Fennema, WD Powrie, EH Marth, Eds., *Low Temperature Preservation of Foods and Living Matter*. New York: Marcel Dekker, 1973, pp. 382–435.
44. B Ray, ML Speck. Freeze-injury in bacteria. *Critical Reviews Clinical Laboratory Science* 4:161–213, 1973.
45. AD Brown. In: *Microbial Water Stress Physiology: Principles and Perspective*. Chichester, UK: Wiley, 1990, pp. 241–279.
46. SE El-Kest, EH Marth. Injury and death of frozen *Listeria monocytogenes* as affected by glycerol and milk components. *Journal Dairy Science* 74:1201–1208, 1991.
47. M Abadias, A Benabarre, N Teixido, J Usall, I Vinas. Effect of freeze drying and protectants on viability of the biocontrol yeast *Candida sake. International Journal Food Microbiology* 65: 173–182, 2001.
48. NH Bean, PM Griffin. Foodborne disease outbreaks in the United States 1973–1987: pathogens, vehicles and trends. *Journal Food Protection* 53:804–817, 1990.
49. DO Cliver. Foodborne viruses. In: MP Doyle, LR Beuchat, TJ Montville, Eds., *Food Microbiology Fundamentals and Frontiers*. Washington, DC: American Society for Microbiology, 1997, pp. 437–446.
50. DO Cliver. Viruses: introduction. In: RK Robinson, CA Batt, PD Patel, Eds., *Encyclopedia of Food Microbiology*, Vol. 3. London: Academic Press, 2000, pp. 2264–2267.
51. RK Lynt. Survival and recovery of enteroviruses from foods. *Applied Microbiology* 14:218–222, 1966.
52. MJ Eyles, Assessment of cooked prawns as a vehicle for transmission of viral disease. *Journal Food Protection* 42:426–428, 1983.

53. AS Khan, CL Moe, RI Glass, SS Monroe, X Jiang, J Wang, MK Estes, Y Seto, C Humphrey, E Pon, JK Iskander. Norwalk virus-associated gastroenteritis traced to ice consumption aboard a cruise ship in Hawaii: comparison and application of molecular method-based assays. *Journal Clinical Microbiology* 32:318–322, 1994.
54. RCW Berkeley, N Ali. Classification and identification of endospore-forming bacteria. *Journal Applied Bacteriology* Symposium Supplement, 76:15–85, 1994.
55. J Errington. *Bacillus subtilis* sporulation: regulation of gene expression and control of morphogenesis. *Microbiological Reviews* 57:1–33, 1993.
56. P Gerhardt, RE Marquis. Spore thermoresistance mechanisms. In: I Smith, R Slepecky, P Setlow, Eds., *Regulation of Prokaryotic Development*. Washington, DC: American Society for Microbiology, 1989, pp. 43–63.
57. GW Gould. Mechanisms of resistance and dormancy. In: A Hurst, GW Gould, Eds., *The Bacterial Spore*, Vol. 2, London: Academic Press, 1983, pp. 173–210.
58. P Setlow. I will survive: protecting and repairing spore DNA. *Journal Bacteriology* 174:2733–2741, 1992.
59. GI Wallace, SE Park. Microbiology of Frozen Foods V. The behaviour of *Clostridium botulinum* in frozen fruit and vegetables. *Journal Infectious Diseases* 52:150–156, 1933.
60. R Kolter, DA Siegele, A Tormo. The stationary phase of the bacterial life cycle. *Annual Reviews Microbiology* 47:855–874, 1993.
61. A Martin, EA Auger, PH Blum, JE Schultz. Genetic basis of starvation survival in nondifferentiating bacteria. *Annual Reviews Microbiology* 43:293–316, 1989.
62. MP Spector, JW Foster. Starvation-stress response (SSR) of *Salmonella typhimurium*: gene expression and survival during nutrient starvation. In: S Kjelleberg, Ed., *Starvation of Bacteria*. New York: Plenum, 1993, pp. 201–224.
63. MG Smith. Survival of *E. coli* and *Salmonella* after chilling and freezing in liquid media. *Journal Food Science* 60:509–512, 1995.
64. JW Foster, MP Spector. How *Salmonella* survive against the odds. *Annual Reviews Microbiology* 49:145–174, 1995.
65. JM Panoff, B Thammavongs, M Guélguen, P Boutibonnes. Cold stress responses in mesophilic bacteria. *Cryobiology* 36:75–83, 1998.
66. CER Dodd, RL Sharman, SF Bloomfield, IR Booth. GSAB Stewart, Inimical processes: bacterial self destruction and sub-lethal injury. *Trends Food Science Technology* 8:238–241, 1997.
67. SE El-Kest, EH Marth. Strains and suspending menstrua as factors affecting death and injury of *Listeria monocytogenes* during freezing and frozen storage. *Journal Dairy Science* 74:1209–1213, 1991.
68. DI Georgala, A Hurst. The survival of food poisoning bacteria in frozen foods. *Journal Application Bacteriology* 26:346–358, 1963.
69. PH Demchick, SA Palumbo, JL Smith. Influence of pH on freeze-thaw lethality in *Staphylococcus aureus*. *Journal Food Safety* 4:185–189, 1982.
70. SE El-Kest, EH Marth, Freezing of *Literia monocytogenes* and other microorganisms: a review. *Journal Food Protection* 55, 639–648, 1992.
71. PD Lowry, CO Gill. Microbiology of frozen meat and meat products. In: RK Robinson, Ed., *Microbiology of Frozen Foods*. London: Elsevier Applied Science, 1985, pp. 109–168.
72. WS Kim, NW Dunn. Identification of cold shock genes in lactic acid bacteria and the effect of cold shock on cryotolerance. *Current Microbiology* 35:59–63, 1977.
73. EM Barnes, JE Despaul, M Ingram. The behaviour of food poisoning strains of *Clostridium welchii* in beef. *Journal Applied Bacteriology* 26:415–427, 1963.
74. CO Gill, K Rahn, K Sloan, LM McMullen. Assessment of the hygienic performances of hamburger patty production processes. *International Journal Food Microbiology* 36:171–178, 1997.
75. NJ Stern. Enumeration and reduction of *Campylobacter jejuni* in poultry and red meats. *Journal Food Protection* 48:606–610, 1985.
76. DW Cook. Refrigeration of oyster shellshock: conditions which minimize the outgrowth of *Vibrio vulnifircus*. *Journal Food Protection* 60:349–352, 1997.
77. J Emborg, BG Laursen, T. Rathjen, P Daalgard. Microbial spoilage and formation of biogenic amines in fresh and thawed modified atmosphere-packed salmon (*Salmo salar*) at 2°C. *Journal Applied Microbiology* 92:790–799, 2002.

78. R Foschino. Freezing injury of *Escherichia coli* during the production of ice cream. *Annual Microbiology* 52:39–46, 2002.
79. JJ Semanchek, DA Golden. Influence of growth temperature on inactivation and injury of *Escherichia coli* 0157:H7 by heat, acid, and freezing. *Journal Food Protection* 61:395–401, 1998.
80. MP Doyle, JL Schoeni. Survival and growth characteristics of *Escherichia coli* associated with hemorrhagic colitis. *Applied Environmental Microbiology* 48:855–856, 1984.
81. PHB Talbot. *Principles of Fungal Taxonomy*. London: MacMillan, 1971.
82. NJW Kreger-van Rij. Classification of yeasts. In: AH Rose, JS Harrison, Eds., *The Yeasts*, 2nd ed. London: Academic Press, 1987, pp. 5–61.
83. JI Pitt, AD Hocking. In: *Fungi and Food Spoilage*. Gaithersburg, MD: Aspen Publishers, 1999, pp. 59–170.
84. PG Sanderson, SN Jeffers. Collection, viability and storage of ascospores of *Monilinia oxycocci*. *Phytopathology* 82:160–163, 1992.
85. PM Robinson, R Kennedy. Survival of arthrospores of *Geotrichum candidum* at sub-zero temperature. *Transactions of British Mycological Society* 82:697–705, 1984.
86. SB Alves, RM Percira, JL Stimac, SA Vieira. Delayed germination of *Beauveria bassiana* conidia after prolonged storage at low, above freezing temperatures. *Biocontrol Science Technology* 6:575–581, 1991.
87. S. Cliquet, MA Jackson, Influence of culture conditions on production and freeze-drying tolerance of *Paecilomyces fumosoroseus* blastospores. *Journal Industrial Microbiology and Biotechnology* 23:97–102, 1999.
88. JF Berny, GL Hennebert. Viability and stability of yeast cells and filamentous fungus spores during freeze-drying: effects of protectants and cooling rates. *Mycologia* 83:805–815, 1991.
89. GJ Morris, GE Coulson, KJ Clarke. Freezing injury in *Saccharomyces cerevisiae*: the effect of growth conditions. *Cryobiology* 25:471–482, 1988.
90. C-G Hounsa, EV Brandt, J Thevelein, S Hohmann, BA Prior. Roll of trehalose in survival of *Saccharomyces cerevisiae* under osmotic stress. *Microbiology* 144:671–680, 1988.
91. JG Lewis, RP Learmonth, K Watson. Induction of heat, freezing and salt tolerance by heat and salt shock in *Saccharomyces cerevisiae*. *Microbiology* 141:687–694, 1995.
92. S Dubernet, JM Panoff, B Thammavongs, M Gueguen. Nystatin and osmotica as chemical enhancers of the phenotypic adaptation of freeze-thaw stress in *Geotrichum candidum* ATCC 204307 *International Journal Food Microbiology* 76:215–221, 2002.
93. E Breierova, A Kockova-Kratochvilova, Cryoprotective effects of yeast extracellular polysaccharides and glycoproteins. *Cryobiology* 29:385–390, 1992.
94. CA Speer. Protozoan parasites acquired from food and water. In: MP Doyle, LR Beuchat, TJ Montville, Eds., *Food Microbiology Fundamentals and Frontiers*. Washington, DC: American Society for Microbiology, 1997, pp. 478–493.
95. AW Kotula, JP Dubey, AK Sharar, CD Andres, SK Shen, DS Lindsay. Effect of freezing on infectivity of *Toxoplasma gondii* tissue cysts in pork. *Journal Food Protection* 54:687–690, 1991.
96. R Fayer, T Nerad. Effects of low temperatures on viability of *Cryptosporidium parvum* oocysts. *Applied Environmental Microbiology* 62:1431–1433, 1996.
97. ME Olson, J Goh, M Phillips, N Guselle, TA McAllister. *Guiardia* cyst and *Cryptosporidium* oocyst survival in water, soil, and cattle feces. *Journal Environmental Quality*. 28:1991–1996, 1999.
98. MB Lee, EH Lee. Coccidial contamination of raspberries: mock contamination with *Eimeria acervulina* as a model for decontamination treatment studies. *Journal Food Protection* 64, 1854–1857, 2001.
99. RD Adlard, RJG Lester. Survival of spores of the oyster pathogen *Marteilia sydneyi* (Protozoa, Paramyxea) as assessed using fluorogenic dyes. *Diseases Aquatic Organisms* 36:221–226, 1999.
100. LJ Robterson, AT Campbell, HV Smith. Survival of *Cryptosporidium parvum* oocysts under various environmental pressures. *Applied Environmental Microbiology* 58:3494–3500, 1992.
101. M Walker, K Leddy, E Hager. Effects of combined water potential and temperature stress on *Cryptosporidium parvum* oocysts. *Applied Environmental Microbiology* 67:5526–5529, 2001.
102. JV Madox, LF Solter. Long-term storage of infective microsporidian spores in liquid nitrogen. *Journal Eukaryote Microbiology* 43:221–225, 1996.
103. KD Murrell, R Fayer, JP Dubey. Parasitic organisms. In: AM Pearson, TR Dutson, Eds., *Advances in Meat Research*. Vol. 2, Westport, CT: AVI Publishing Co.,1986, pp. 311–377.

104. EG Hayunga. Helminths acquired from finfish, shellfish, and other food sources. In: MP Doyle, LR Beuchat, TJ Montville, Eds., *Food Microbiology Fundamentals and Frontiers*. Washington, DC: American Society for Microbiology 1997, pp. 463–477.
105. AW Kotula, AK Sharar, E Paroczay, HR Gamble, KD Murrell, L Douglass. Infectivity of *Trichinella spiralis* from frozen pork. *Journal Food Protection* 53:571–573, 626, 1990.
106. KD Murrell. Strategies for the control of human trichinosis transmitted by pork. *Food Technology* 39 (3):65–68, 110–111, 1985.
107. PC Beaver, RC Jung, EW Cupp. In: *Clinical Parasitology*, 9th ed., Philadelphia: Lea & Febiger, 1984, pp. 231–252.
108. TA Dick. Species and intraspecific variation. In: WC Campbell, Ed., *Trichinella* and *Trichinosis*, New York: Plenum Press, 1983, pp. 31–73.
109. J Cui, ZQ Wang. Outbreaks of human trichinellosis caused by consumption of dog meat in China. *Journal de la Societé Francaise de Parasitologie* 8 (2):S74–S77, 2001.
110. P Webster, A Malakauskas, CMD Kapel. Infectivity of *Trichinella papuae* for experimentally infected red foxes (*Vulpes vulpes*) *Veterinary Parasitology* 105:215–218, 2002.
111. RW Hilwig, JD Cramer, KS Forsyth. Freezing times and temperatures required to kill cysticerci of *Taenia saginata* in beef. *Veterinary Parasitology* 4:215–221, 1978.
112. JTR Robinson, PG Chambers. Observations on cysticercosis in Rhodesia II. The survival time for *Cysticercus cellulosae* cysts at low temperatures. *Rhodesian Veterinary Journal* 7:32–40, 1976.
113. TL Deardorff, R Throm, Commercial blast freezing of third-stage *Anisakis simplex* larvae encapsulated in salmon and rockfish. *Journal Parasitology* 74:600–603, 1988.
114. AM Adams, KD Murrell, JH Cross. Parasites of fish and risks to public health. *Revue Scientifique de l'Office International des Epizooties* 16:652–660, 1997.
115. LC Wang, D Chao, CM Yen, ER Chen. Improvements in the infectivity of cryopreserved third-stage larvae of *Angiostrongylus cantonensis* using a programmable freezer. *Parasitology Research* 85:151–154, 1999.
116. B Hendley. Developing products for the market. In: CP Mallett, Ed., *Frozen Food Technology*, London: Blackie Academic and Professional, 1993, pp. 141–167.
117. DA Corlett. *HACCP User's Manual*, Gaithersburg, MD: Aspen Publishers, 1998.
118. JC Leighty. Regulatory action to control *Trichinella spiralis*. *Food Technology* 37 (3):95–97, 1983.
119. GR Snyder, KD Murrell. Bovine cysticercosis. In: G Woods, Ed., *Practices in Veterinary Public Health and Preventative Medicine*, Ames, IA: Iowa State University Press, 1983, pp. 161–170.
120. PM Schantz. The dangers of eating fish raw. *New England Journal Medicine* 320:1143–1145, 1989.
121. NJ Stern, KL Hiett, GA Alfredsson, KG Kristinsson, J Reiersen, H Hardardottir, H Briem, E Gunnarsson, F Georgsson, R Lowman, E Berndtson, AM Lammerding, GM Paoli, MT Musgrove. *Campylobacter* spp. in Icelandic poultry operations and human disease. *Epidemiology Infection* 130:23–32, 2003.
122. NJ Stern. Enumeration and reduction of *Campylobacter jejuni* in poultry and red meats. *Journal Food Protection* 48:606–610, 1985.
123. B Thurston Solow, OM Cloak, PM Fratamico. Effect of temperature on viability of *Campylobacter jejuni* and *Campylobacter coli* on raw chicken or pork skin. *Journal Food Protection* 66:2023–2031, 2003.
124. H Johnson, J Liston. Sensitivity of vibrio *parahaemolyticus* to cold in oysters, fish fillets and crab meat. *Journal Food Science* 38:437–471, 1973.
125. RW Parker, EM Maurer, AB Childers, DH Lewis. Effect of frozen storage and vacuum packaging on survival of *Vibrio vulnificus* in Gulf Coast oysters (*Crassostrea virginica*). *Journal Food Protection* 57:604–606, 1994.
126. SP Trakulchang, AA Kraft. Survival of *Clostridium perfringens* in refrigerated and frozen meat and poultry items. *Journal Food Science* 42:518–521, 1977.
127. AM Lammerding, MP Doyle. Stability of *Listeria monocytogenes* to non-thermal processing conditions. In: AJ Miller, JL Smith, GA Somkuti, Eds., *Foodborne Listeriosis*. Amsterdam: Elsevier Science, 1990, pp. 195–202.
128. RAE Barrell. The survival and recovery of *Salmonella typhimurium* phage type U285 in frozen meats and tryptone soya yeast extract broth. *International Journal Food Microbiology* 6:309–316, 1988.

5 Thermophysical Properties of Frozen Foods

Lijun Wang and Curtis L. Weller
University of Nebraska-Lincoln, Lincoln, Nebraska, USA

CONTENTS

I. INTRODUCTION

Thermophysical properties control thermal energy transport, energy storage, and phase transformations in food materials. Thermophysical properties of frozen foods are used to estimate the rate of heat transfer and to calculate the heat load in processes such as freezing and thawing. In early calculations and analyses associated with freezing and thawing, constant and uniform thermophysical properties were primarily used. Those calculations and analyses were typically oversimplified and inaccurate. Numerical analyses such as finite element and finite difference methods begun to be used widely to analyze thermal food processes. Modern numerical analysis can account for nonuniform and varying thermophysical properties, which change with time, temperature, and location during food processing. The benefits of modern numerical analyses increase the demand for more accurate thermophysical properties of frozen foods.

The large number of food products available today creates a great demand for knowledge of the thermophysical properties. The chemical composition and structure of foods vary widely with products. It is impossible to compile an exhaustive database for all foods. Although there is a fairly large amount of data for some particular foods, the data are sometimes contradictory. This is due to the different conditions at which the properties were measured and the differences among the same foods based on origin, composition, and structure. Therefore, it is most common to use prediction models of a more general nature for estimating the thermophysical properties of frozen foods. Meanwhile, the necessary background and sophisticated measurement technologies are provided so that judgments can be made on the applicability and confidence of a prediction model for specific thermophysical properties of any frozen food.

As foodstuffs are composite materials, the thermophysical properties are clearly functions of the components. The magnitude of a thermophysical property will be a function of the temperature through the temperature dependence of the property for each component in the foodstuffs. Water is usually the main component of a food material and it can be either in liquid or in solid form in frozen foods. Particular attention must be paid to the frozen foods because of the dual states of water that can exist. Major general reviews of research on thermophysical properties of foods have been provided by Mellor [1–4], Heldman and Singh [5], Murakami and Okos [6], Sweat [7], Singh [8], Urbicain and Lozano [9], Roos [10], and Rahman [11,12].

In this chapter, the concepts of freezing point, ice content, enthalpy, specific heat, latent heat, thermal conductivity, density, and thermal diffusivity of frozen foods are introduced. A study was then conducted on prediction models and measurement technologies to determining the thermophysical properties of frozen foods.

II. FREEZING POINT

A. Definition and Applications

The equilibrium or initial freezing point is one of the most important physical properties of a food material because of the discontinuity of thermophysical properties exhibited at the initial freezing point. Therefore, the initial freezing point is required for the prediction of thermophysical properties of frozen foods. Freezing point data also play a key role in the design of methods such as hydrofluidization freezing and high-pressure-assisted freezing.

Figure 5.1 gives a phase diagram of a solution showing the freezing curve at constant pressure [12]. The freezing curve represents the equilibrium between the solution and formed ice. The freezing point of the solution is any point on the freezing curve. It can be seen from Figure 5.1 that the freezing point decreases with increasing solute concentration. The temperature at which ice crystals form within the solution is always depressed below that of a pure water system where the mass fraction of solute is zero.

Although any solid food is not a solution, the aqueous component in the frozen solid food can be considered as a mixture of ice and a solution of nonaqueous components in liquid water. As water in the food freezes into pure ice during freezing, the remaining solution becomes more and more concentrated. Thus, depression of the freezing point of the food occurs and continues as concentration increases. The net effects of dynamic freezing point depression are that the initial freezing point of the food is below 0°C. The freezing process occurs over a range of temperatures, which is different from the freezing process of pure water at a unique temperature [13].

Owing to the freezing point depression of a solution, it is possible to produce a low-temperature solution, which has an initial freezing temperature much lower than that of foods. The unfreezable solution can be used as the refrigerating media in an emerging fast-freezing technology called hydrofluidization freezing. During hydrofluidization freezing, a low-temperature liquid

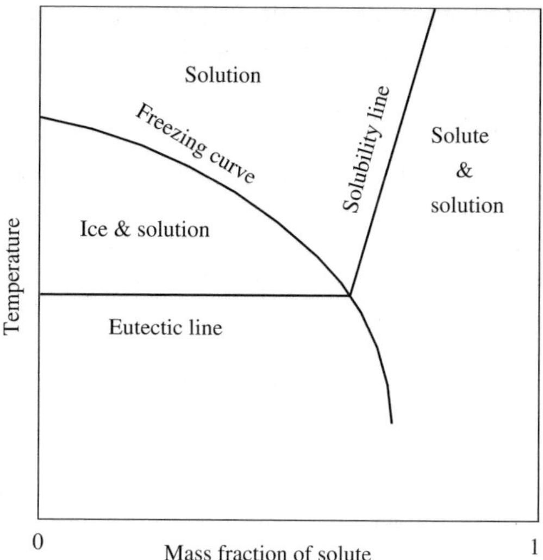

FIGURE 5.1 Phase diagram showing a typical freezing curve for a food material at constant pressure where temperature at which water freezes is a function of its solute concentration.

refrigerating media such as an aqueous solution is pumped by a recirculating system upward, forming a fluidized bed and moving products. The maximum surface heat-transfer coefficient achieved by hydrofluidization exceeded 900 W/m^2 °C, compared with 378 W/m^2 °C for immersion freezing [14,15].

Hydrostatic pressure has a significant influence on the phase transition of water by depressing the freezing or melting point and reducing the latent heat of fusion. Figure 5.2 is a pressure–temperature diagram for water showing freezing curves at different pressures [16]. The freezing point of water decreases as pressure increases in the region of ice I from atmospheric pressure of 0.1 to 200 MPa. At the pressure of 207.5 MPa, the freezing point of water decreases to a minimum of −22°C [17]. Therefore, water in a product can be kept in the liquid state at a temperature much lower than 0°C while under pressure during high-pressure-assisted freezing. When pressure is decreased, the freezing point of water moves to the left along the phase equilibrium line and the freezing point of the product increases, resulting in a high temperature gradient

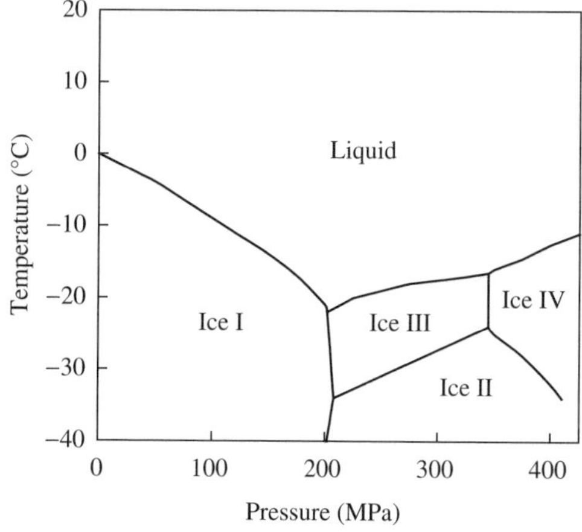

FIGURE 5.2 Phase of water under various pressure–temperature combinations.

between the supercooled temperature of the product and the new freezing point. The product thus freezes quickly and many small ice crystals are formed [18].

B. Freezing Point Measurements and Data

A freezing curve can be used to determine an initial freezing point for a food material. A typical freezing curve for most foods is shown in Figure 5.3. The abrupt rise in temperature from point A to point B in Figure 5.3 due to the liberation of the heat of fusion after initial supercooling indicates the onset of ice crystallization. Cooling below the initial freezing point of a food material without formation of ice crystal nuclei is defined as supercooling. Once the critical number of nuclei has been reached, the growth of ice crystals will begin at point A in Figure 5.3. At this point, latent heat is being released faster than heat is being removed from the system. So the temperature of the material increases instantly to the initial freezing temperature at point B. The temperature history of a sample food product as illustrated in Figure 5.3 can be recorded using a temperature sensor such as thermocouple, thermistor, or thermometer. The freezing point is derived from the relatively long temperature plateau which follows supercooling on the freezing curve [11].

Alternatively, the onset, peak, and end of freezing can be determined from a heat flow exotherm, which can be developed using differential scanning calorimetry (DSC). A typical melting curve developed for a food material using DSC analysis is given in Figure 5.4. T_m is defined as the melting point in Figure 5.4 [10]. More details about measurements using DSC analysis are given in the literature [12].

Initial freezing points of selected foods are given in Table 5.1 [6,19,20]. As the theoretical values of initial freezing point often are of dubious origin, the data should be used with caution. Typical values of initial freezing temperature for fish, meats, fruits, and vegetables are in the range from -2.0 to -0.5°C. For high-moisture foods (>55% water, wet basis), use of -1.0°C is recommended as a first approximation if a better estimation is not available [13].

C. Prediction Models of Initial Freezing Point

The initial freezing point of foods varies with water content, other nonwater components, component molecular weight, component interaction, and water-binding characteristics. It is obvious that the magnitude of freezing point depression is a function of product composition. The relationship between product composition and freezing temperature is most often explained in terms of the

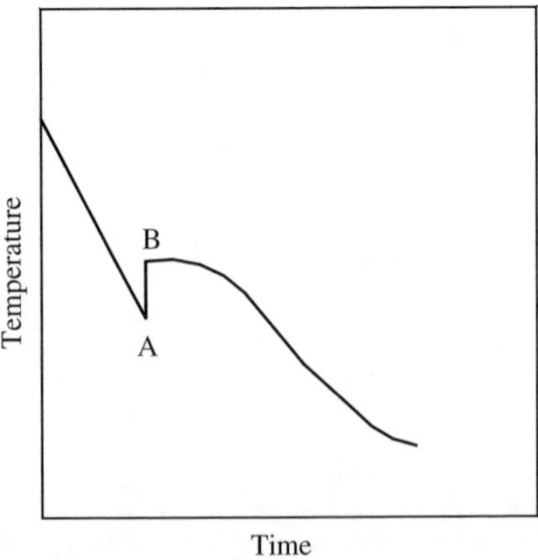

FIGURE 5.3 Typical temperature over time curve for a food material as it undergoes cooling and freezing where A is the ice crystallization temperature and B is the equilibrium freezing point.

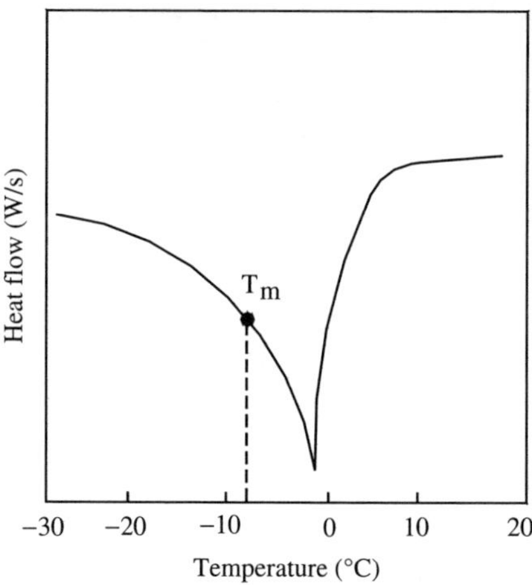

FIGURE 5.4 Typical heat flow vs. temperature curve for a food material indicating point of phase change or melting T_m.

freezing temperature depression for a solution. The initial freezing point of a solution can be calculated with use of the Clausius–Clapeyron equation, which is given by [11,21]

$$\frac{1}{T_F} = \frac{1}{T_W} - \frac{R}{M_w \lambda_w}\left[\ln\left(\frac{X_{w0}/M_w}{X_{w0}/M_w + \sum_{j=1}^{n}(X_{j0}/M_j)}\right)\right] \quad (5.1)$$

For an ideal dilute solution, the freezing point depression can be calculated by

$$\Delta T_F = \frac{RT_w^2}{M_w \lambda_w}\left[\ln\left(\frac{\sum_{j=1}^{n}(X_{j0}/M_j)}{X_{w0}/M_w + \sum_{j=1}^{n}(X_{j0}/M_j)}\right)\right] \quad (5.2)$$

For a nonideal solution, the freezing point depression can be calculated by

$$\Delta T_F = \frac{RT_w^2}{M_w \lambda_w}\left[\ln\left(\frac{\sum_{j=1}^{n}(\tau_j X_{j0}/M_j)}{(X_{w0} - X_{uw})/M_w + \sum_{j=1}^{n}(\tau_j X_{j0}/M_j)}\right)\right] \quad (5.3)$$

An empirical correlation is often used to determine the freezing temperature depression of a food material, when it is difficult to obtain the molecular weight of each food component. As total water content is usually the main part of a food and an easily measured property, most empirical correlations are thus based on the water content. Generalized correlations between initial freezing point and water content have been expressed as linear and nonlinear equations in literature [11]. A number of these correlations for several types of food materials are listed as follows.

For meat products [22–25]:

$$\Delta T_F = 1.9 - 1.4X_{w0} \quad (5.4)$$

$$\Delta T_F = \frac{X_{w0} - 1}{0.069 - 0.439X_{w0}} \quad (5.5)$$

$$\Delta T_F = \frac{X_{w0} - 1}{0.072 - 0.488X_{w0}} \quad (5.6)$$

$$\Delta T_F = \frac{X_{w0} - 1}{0.078 - 0.140X_{w0}} \quad (5.7)$$

TABLE 5.1
Initial Freezing Point of Select Meat Cuts, Fish, Vegetables, and Fruits at Various Moisture Contents [6,19,20]

Product	X_{w0}	T_c(°C)	References
	Meats		
Beef muscle	0.800	−0.73	[20]
Beef muscle	0.740	−0.99	[20]
Beef muscle	0.500	−3.63	[20]
Beef muscle	0.261	−13.46	[20]
Beef muscle	0.740	−1.75	[6]
Beef round	0.745	−1.75	[6]
Beef flank	0.745	−1.75	[6]
Beef veal	0.745	−1.75	[6]
Beef sirloin	0.740	−1.00	[6]
Lamb muscle	0.745	−1.74	[6]
Lamb muscle	0.740	−1.00	[6]
Lamb loin	0.649	−0.90	[20]
Lamb loin	0.524	−0.84	[20]
Lamb kidney	0.798	−0.96	[20]
Pork muscle	0.745	−1.75	[6]
Pork muscle	0.740	−1.00	[6]
Tukey	0.740	−2.80	[6]
	Fish		
Haddock	0.836	−0.89	[20]
Haddock	0.803	−1.00	[6]
Cod	0.820	−0.90	[20]
Cod	0.780	−2.20	[6]
Cod	0.803	−0.91	[20]
Cod	0.500	−3.57	[20]
Catfish	0.803	−1.00	[6]
Red fish	0.803	−1.00	[6]
Salmon	0.670	−2.20	[6]
Sea perch	0.791	−0.86	[6]
	Vegetables and fruits		
Grape juice	0.800	−2.90	[19]
Tomato juice	0.850	−1.60	[19]
Yeast	0.720	−1.37	[19]
Green peas	0.760	−1.74	[19]
Spinach	0.800	−0.55	[19]

For vegetable and fruit products [22]:

$$\Delta T_F = -14.46 + 49.19X_{w0} - 37.07X_{w0}^2 \tag{5.8}$$

For juice products [22, 26]:

$$\Delta T_F = 152.63 - 327.35X_{w0} + 176.49X_{w0}^2 \tag{5.9}$$

$$\Delta T_F = 10X_{s0} + 50X_{s0}^3 \tag{5.10}$$

III. ICE CONTENT

A. Ice Content, Freezable Water, and Unfreezable Water

The phase change of water, between liquid and ice, in foods dominates the changes in thermophysical properties of foods during freezing and thawing. The total water content in frozen foods consists of three fractions: freezable liquid water, unfreezable bound liquid water, and frozen water (ice), which can be expressed as

$$X_{w0} = X_{fw} + X_{uw} + X_i \tag{5.11}$$

The total liquid water in a food material during freezing is the sum of freezable and unfreezable water:

$$X_w = X_{fw} + X_{uw} \tag{5.12}$$

Methods to determine different-state water fractions in the food are required to predict thermal properties.

Unfreezable water is the unfrozen fraction left in a food material at $-40°C$ [6]. Unfreezable water fraction, X_{uw}, is usually measured and expressed in terms of the ratio of kg of unfreezable water to kg of total dry solids, B:

$$X_{uw} = BX_{s0} = B(1 - X_{wo}) \tag{5.13}$$

The values of B for different foods are given in Table 5.2 [6,20,27,28]. If better data are not available for a special food, the use of $B = 0.25$ is suggested [13].

Riedel has provided the basic data of dynamic changes of unfrozen water during freezing for a wide range of frozen foods, and a part of the data is given in Table 5.3 [8,21]. Figure 5.5 illustrates the typical change in fraction of ice with temperature for a high-moisture food during freezing [13]. The ice fraction in a food material as shown in Figure 5.5 increases gradually as temperature drops below the initial freezing temperature. The transformation of water from liquid to solid in the food occurs over a range of temperatures below the initial freezing point.

B. Prediction Models of Ice Content

The accurate prediction of the temperature-dependent ice fraction in frozen foods is critical for accurately determining their various thermophysical properties. The ice content can be predicted

TABLE 5.2
Ranges of *B* (Ratio of Mass of Unfreezable Water to Mass of Total Dry Solids in a Food Material) Values for Calculating Unfreezable Water in Select Food Materials [6]

Food Materials	*B* Range	References
Meat, fish	0.14–0.32	[20]
Sucrose	0.30	[27]
Glucose	0.15–0.20	[27]
Egg	0.11	[20]
Bread	0.11–0.14	[28]
Orange juice	0.00	[27]
Vegetables	0.18–0.25	[27]

TABLE 5.3
Enthalpy and Percentage of Unfrozen Water in Select Frozen Foods over Various Temperatures[a] [8,21]

Product	Water content (wt%)	Mean specific heat[b] 4–32°C (kJ/kg°C)		Temperature (°C)																	
				−40	−30	−20	−18	−16	−14	−12	−10	−9	−8	−7	−6	−5	−4	−3	−2	−1	0
Fruits and vegetables																					
Applesauce	82.8	3.73	Enthalpy (kJ/kg)	0	23	51	58	65	73	84	95	102	110	120	132	152	175	210	286	339	343
			% water unfrozen[c]	—	6	9	10	12	14	17	19	21	23	27	30	37	44	57	82	100	—
Asparagus, peeled	92.6	3.98	Enthalpy (kJ/kg)	0	19	40	45	50	55	61	69	73	77	83	90	99	108	123	155	243	381
			% water unfrozen	—	—	—	—	—	5	6	—	7	8	10	12	15	17	20	29	58	100
Bilberries	85.1	3.77	Enthalpy (kJ/kg)	0	21	45	50	57	64	73	82	87	94	101	110	125	140	167	218	348	352
			% water unfrozen	—	—	—	7	8	9	11	14	15	17	18	21	25	30	38	57	100	—
Carrots	87.5	3.90	Enthalpy (kJ/kg)	0	21	46	51	57	64	72	81	87	94	102	111	124	139	166	218	357	361
			% water unfrozen	—	—	—	7	8	9	11	14	15	17	18	20	24	29	37	53	100	—
Cucumbers	95.4	1.02	Enthalpy (kJ/kg)	0	18	39	43	47	51	57	64	67	70	74	79	85	93	104	125	184	390
			% water unfrozen	—	—	—	—	—	—	—	—	5	—	—	—	—	11	14	20	37	100
Onions	85.5	3.81	Enthalpy (kJ/kg)	0	23	50	55	62	71	81	91	97	105	115	125	141	163	196	263	349	353
			% water unfrozen	—	5	8	10	12	14	16	18	19	20	23	26	31	38	49	71	100	—
Peaches without stones	85.1	3.77	Enthalpy (kJ/kg)	0	23	50	57	64	72	82	93	100	108	118	129	146	170	202	274	348	352
			% water unfrozen	—	5	8	9	11	13	16	18	20	22	25	28	33	40	51	75	100	—
Pears, Bartlett	83.8	3.73	Enthalpy (kJ/kg)	0	23	51	57	64	73	83	95	101	109	120	132	150	173	207	282	343	347
			% water unfrozen	—	6	9	10	12	14	17	19	21	23	26	29	35	43	54	80	100	—
Plums without stones	80.3	3.65	Enthalpy (kJ/kg)	0	25	57	65	74	84	97	111	119	129	142	159	182	214	262	326	329	333
			% water unfrozen	—	8	14	16	18	20	23	27	29	33	37	42	50	61	78	100	—	—
Raspberries	82.7	3.73	Enthalpy (kJ/kg)	0	20	47	53	59	65	75	85	90	97	105	115	129	148	174	231	340	344
			% water unfrozen	—	—	7	8	9	10	13	16	17	18	20	23	27	33	42	61	100	—
Spinach	90.2	3.90	Enthalpy (kJ/kg)	0	19	40	44	49	54	60	66	70	74	79	86	94	103	117	145	224	371
			% water unfrozen	—	—	—	—	—	—	6	7	—	—	9	11	13	16	19	28	53	100
Strawberries	89.3	3.94	Enthalpy (kJ/kg)	0	20	44	49	54	60	67	76	81	88	95	102	114	127	150	191	318	367
			% water unfrozen	—	—	5	—	6	7	9	11	12	14	16	18	20	24	30	43	86	100
Sweet cherries without stones	77.0	3.60	Enthalpy (kJ/kg)	0	26	58	66	76	87	100	114	123	133	149	166	190	225	276	317	320	324
			% water unfrozen	—	9	15	17	19	21	26	29	32	36	40	47	55	67	86	100	—	—
Tall peas	75.8	3.56	Enthalpy (kJ/kg)	0	23	51	56	64	73	84	95	102	111	121	133	152	176	212	289	319	323
			% water unfrozen	—	6	10	12	14	16	18	21	23	26	28	33	39	48	61	90	100	—

Tomato pulp	92.9	4.02	Enthalpy (kJ/kg)	0	20	42	47	52	57	63	71	75	81	87	93	103	114	131	166	266	382
			% water unfrozen	—	—	—	—	5	—	6	7	8	10	12	14	16	18	24	33	65	100
Egg products																					
Egg white	86.5	3.81	Enthalpy (kJ/kg)	0	18	39	43	48	53	58	65	68	72	75	81	87	96	109	134	210	352
			% water unfrozen	—	—	10	—	—	—	—	13	—	—	—	18	20	23	28	40	82	100
Egg yolk	40.0	2.85	Enthalpy (kJ/kg)	0	19	40	45	50	56	62	68	72	76	80	85	92	99	109	128	182	191
			% water unfrozen	20	—	—	22	—	24	—	27	28	29	31	33	35	38	45	58	94	100
Whole egg with shell[d]	66.4	3.31	Enthalpy (kJ/kg)	0	17	36	40	45	50	55	61	64	67	71	75	81	88	98	117	175	281
			% water unfrozen																		
Fish and meats																					
Cod	80.3	3.69	Enthalpy (kJ/kg)	0	19	42	47	53	66	74	79	84	89	96	105	118	137	177	298	323	
			% water unfrozen	10	10	11	12	12	13	14	16	17	18	19	21	23	27	34	48	92	100
Haddock	83.6	3.73	Enthalpy (kJ/kg)	0	19	42	47	53	59	66	73	77	82	88	95	104	116	136	177	307	337
			% water unfrozen	8	8	9	10	11	11	12	13	14	15	16	18	20	24	31	44	90	100
Perch	79.1	3.60	Enthalpy (kJ/kg)	0	19	41	46	52	58	65	72	76	81	86	93	101	112	129	165	284	318
			% water unfrozen	10	10	11	12	12	13	14	15	16	17	18	20	22	26	32	44	87	100
Beef, lean fresh[e]	74.5	3.52	Enthalpy (kJ/kg)	0	19	42	47	52	58	65	72	76	81	88	95	105	113	138	180	285	304
			% water unfrozen	10	10	11	12	13	14	15	16	17	18	20	22	24	31	40	55	95	100
Beef, lean dried	26.1	2.47	Enthalpy (kJ/kg)	0	19	42	47	53	62	66	70	—	74	—	79	—	84	—	89	—	93
			% water unfrozen	96	96	97	98	98	100	—	—	—	—	—	—	—	—	—	—	—	—
Breads																					
White	37.3	2.60	Enthalpy (kJ/kg)	0	17	35	39	44	49	56	67	75	83	93	104	117	124	128	131	134	137
Whole wheat	42.4	2.68	Enthalpy (kJ/kg)	0	17	36	41	48	56	66	78	86	95	106	119	135	150	154	157	160	163

[a]Above −40°C.

[b]Temperature range limited to 0–20°C for meats and 20–40°C for egg yolk.

[c]Total weight of unfrozen water = [(total weight of food)(%water content/100)]/(water unfrozen/100).

[d]Calculated for a weight composition of 58% white (86.5% water) and 32% yolk (50% water).

[e]Data for chicken, veal, and venison very nearly matched the data for beef of the same water content.

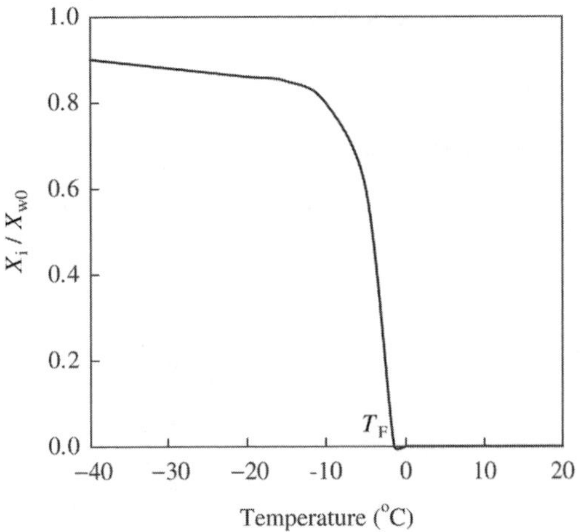

FIGURE 5.5 Typical change in fraction of ice (X_i) with temperature for a high-moisture food during freezing.

using Clausius–Clapeyron relationship and Raoult's law [29]. For a dilute solution, the ice content can be determined by a simple method based on Raoult's law [27].

During the freezing process, as the sum of ice and liquid water fractions remains constant, the ice fraction can be calculated from the mass fraction of water based on simple mass balance given by Equation (5.11). If the initial freezing point T_F is known, the mass fraction of water below the freezing point can inversely be estimated at any desired temperature based on Equation (5.1). The expression for determining the dynamic mass fraction of water in frozen foods, which is derived from Equation (5.1), is given by [29]:

$$X_w = X_{w0}\left(\frac{F_F - F_W}{F - F_W}\right) \tag{5.14}$$

where

$$F = F(T) = \exp\left(\frac{\lambda}{RT}\right) \tag{5.15}$$

The above equation does not account for the effect of unfreezable water content in a food material. For foods containing significant amount of unfreezable water, Equation (5.14) can be modified as [27,29,30]:

$$X_w = X_{uw} + (X_{w0} - X_{uw})\left[\frac{F_F - F_W}{F - F_W}\right] \tag{5.16}$$

As the latent heat of ice fusion decreases with decreasing temperature during freezing, the latent heat in Equation (5.15) can linearly be calculated by

$$\lambda = \lambda_0 + \lambda_1 T \tag{5.17}$$

Equation (5.16) is thus rewritten as

$$X_w = X_{uw} + (X_{w0} - X_{uw})\left[\frac{F'_W - F'_W}{F' - F'_W}\right] \tag{5.18}$$

where

$$F' = F'(T) = T^{-(\lambda_1/R)} \exp\left(\frac{\lambda_0}{RT}\right) \tag{5.19}$$

For very dilute solutions, Equation (5.14) to Equation (5.19) can be approximated with reasonable accuracy by [19,27]:

$$X_i = X_{w0}\left(1 - \frac{T_F}{T}\right) \tag{5.20}$$

Experimental data showed that the above equation underestimates slightly the real mass fraction of ice contents for temperatures near the initial freezing point and overestimates the values with decreasing temperature. Therefore, the above equation can be further modified as [27]:

$$X_i = (X_{w0} - X_{uw})\left(1 - \frac{T_F}{T}\right) \tag{5.21}$$

IV. ENTHALPY

A. Enthalpy and Measurement

Enthalpy is the heat content per unit mass of a food material with typical units of J/kg. Because it is difficult to define the absolute value of enthalpy, a zero value is usually arbitrarily defined at −40°C, 0°C, or any other convenient temperature. It is very convenient to use enthalpy for quantifying energy in frozen foods because it is difficult to separate latent and sensible heats in frozen foods as some unfrozen water exists in the foods even at very low temperature.

The calorimetric method has been the most popular method to determine the enthalpy of frozen foods. However, the enthalpy measurement probably depends on freezing rate for at least some foods. Enthalpy may change in a frozen food during storage even at a constant temperature if the unfrozen water percentage changes. Riedel has compiled the basic data of enthalpy for a wide range of frozen foods and a part of the data is given in Table 5.3 [8,21]. In Table 5.3, enthalpy was assumed to be zero at −40°C. Typical change in enthalpy with temperature for a high-moisture food is given in Figure 5.6 [13].

B. Prediction Models of Enthalpy

Using the information of freezing point depression and the mass fractions of ice and liquid water in a food material, several researchers have presented models for calculating enthalpy. A number of models for predicting enthalpy of frozen foods are presented here.

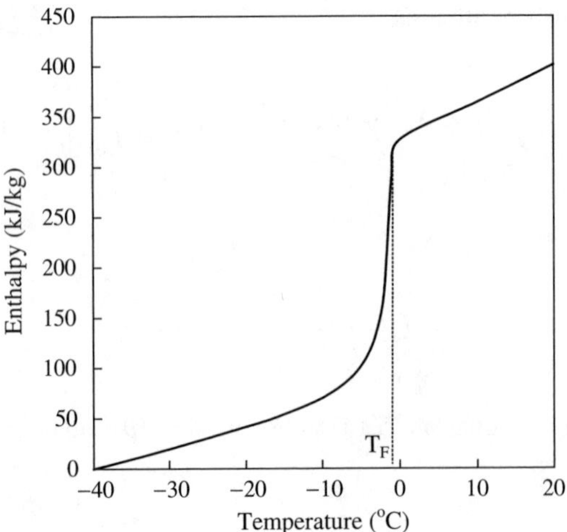

FIGURE 5.6 Typical change in enthalpy with temperature for a high-moisture food during freezing.

Schwartzberg [27]:

$$H_f = (T - T_D)\left[c_{uf} + (X_{uw} - X_{w0})(c_w - c_i) + (1 - X_{w0})\frac{M_w}{M_s}\left\{\frac{RT_P^2}{(T_P - T)(T_P - T_D)}\right\} - 0.8(c_w - c_i)\right] \tag{5.22}$$

Mannapperuma and Singh [29]:

$$\begin{aligned} H_f = {} & (1 - X_{w0})c_s(T - T_D) + X_{w0}\left[c_{i0}(T - T_D) + \frac{1}{2}c_{i1}(T^2 - T_D^2)\right] \\ & + \left[(X_{w0} - X_{uw})\left(\frac{F'_F - F'_P}{F' - F'_P}\right) + X_{uw}\right](\lambda_0 + \lambda_1 T) \\ & - \left[(X_{w0} - X_{uw})\left(\frac{F'_F - F'_P}{F'_D - F'_P}\right) + X_{uw}\right](\lambda_0 + \lambda_1 T_D) \end{aligned} \tag{5.23}$$

$$H_{uf} = H_F + (c_w X_w + c_s X_s)(T - T_F) \tag{5.24}$$

where the function F' was defined in Equation (5.19).

Chen [30]:

$$\frac{H_f}{\psi} = (T - T_D)\left[0.37 + 0.3X_s + X_s\frac{RM_w T_P^2}{M_s(T - T_P)(T_D - T_P)}\right] \tag{5.25}$$

$$\frac{H_{uf}}{\psi} = H_F + (T - T_F)(1 - 0.55X_s - 0.15X_s^3) \tag{5.26}$$

where ψ is a conversion factor of units (=4184 J/cal).

V. SPECIFIC HEAT

A. Specific Heat and Measurement

Specific heat is defined as the amount of heat necessary to raise the temperature of a unit mass of a food material by a unit degree. The SI unit for specific heat is therefore typically J/kg K. Specific heat is independent of mass density. Knowing the specific heat of each component of a mixture is usually sufficient to predict the specific heat of the mixture. However, specific heat depends on the nature of the process such as a constant-pressure process or a constant-volume process. For almost all liquid and solid foods, specific heat at constant pressure, which is normally denoted by c_p, is enough for analyzing food processes. The pressure dependence of specific heat for solids and liquids is very small until extremely high pressure is involved. However, for a gas, it is necessary to distinguish between the specific heat at constant pressure, c_p, and the specific heat at constant volume, c_v.

Specific heat is often measured by calorimetric methods. However, these methods are more useful in determining specific heat when a phase change occurs at a fixed temperature. The phase change during food freezing occurs over a range of temperature. As a result, calorimetric methods have a limited application in determining the specific heats of frozen foods.

An alternative approach involves experimentally determining the enthalpy values of frozen foods for a range of temperatures. An apparent specific heat can then be calculated from the data of enthalpy and temperature using differential thermal analysis of the enthalpy.

Figure 5.7 gives the typical change in specific heat with temperature for a high-moisture food during freezing [13]. The apparent specific heat of a frozen food increases with increasing temperature until reaching the initial freezing temperature. At the initial freezing temperature, the apparent specific heat reaches its maximum value and then decreases to a value for unfrozen food as the temperature climbs beyond the initial freezing temperature. During freezing, there is a rapid decease in the apparent specific heat as temperature just passes the initial freezing temperature because major portions of the latent heat of fusion are removed from the food material in the region near the initial freezing point.

B. Prediction Models of Specific Heat

For temperatures above its freezing point, if the composition of a food material is known, the specific heats of all individual components can be summed in gravimetric proportions to be the

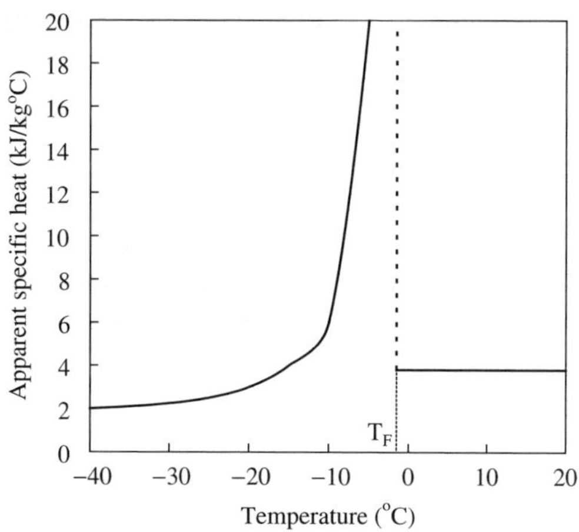

FIGURE 5.7 Typical change in specific heat with temperature for a high-moisture food during freezing.

specific heat of the food material. It can be expressed as

$$c = \sum_{j=1}^{n} c_j X_j \tag{5.27}$$

The specific heat of each component can be determined using theoretical values as functions of temperature based on the regression of experimental data, which are given in Table 5.4 [31].

However, Equation (5.27) is valid in a temperature range where there is no phase change. If there is a phase change such as freezing or thawing, the latent heat due to the phase change must be incorporated. In this case, a new term called the apparent specific heat is introduced. The apparent specific heat is obtained by differentiating the enthalpy of the frozen food with repect to temperature. The relationship between apparent specific heat and enthalpy can be expressed as

$$H = \int_{T_1}^{T_2} c_e \, dT \tag{5.28}$$

or

$$c_e = \frac{dH}{dT} \tag{5.29}$$

Therefore, the following expressions for the prediction of apparent specific heat can be derived from the expressions for the prediction of enthalpy.

Schwartzberg [27]:

$$c_f = c_{uf} + (X_{uw} - X_{w0})(c_w - c_i) + (1 - X_{w0})\frac{M_w}{M_s}\left[\frac{RT_P^2}{(T_P - T)^2} - 0.8(c_w - c_i)\right] \tag{5.30}$$

TABLE 5.4
Equations for Estimating Specific Heats of Major Food Component as Functions of Temperature [31]

Major Component	Equation
Carbohydrate[a]	$c = 1.5488 + 1.9625 \times 10^{-3}\,T - 5.9399 \times 10^{-6}\,T^2$
Fiber[a]	$c = 1.8459 + 1.8306 \times 10^{-3}\,T - 4.6509 \times 10^{-6}\,T^2$
Protein[a]	$c = 2.0082 + 1.2089 \times 10^{-3}\,T - 1.3129 \times 10^{-6}\,T^2$
Fat[a]	$c = 1.9842 + 1.4733 \times 10^{-3}\,T - 4.8008 \times 10^{-6}\,T^2$
Ash[a]	$c = 1.0926 + 1.8896 \times 10^{-3}\,T - 3.6817 \times 10^{-6}\,T^2$
Water[b]	$c = 4.0817 - 5.3062 \times 10^{-3}\,T + 9.9516 \times 10^{-6}\,T^2$
Water[c]	$c = 4.0817 - 5.3062 \times 10^{-3}\,T + 9.9516 \times 10^{-6}\,T^2$
Ice	$c = 2.0623 + 6.0769 \times 10^{-3}\,T$

[a] *T*: − 40 to 150°C.
[b] *T*: −40 to 0°C.
[c] *T*: 0 to 150°C.

Mannapperuma and Singh [29]:

$$c_{\mathrm{f}} = (1 - X_{\mathrm{w0}})c_{\mathrm{s}} + X_{\mathrm{w0}}[c_{\mathrm{i0}} + c_{\mathrm{i1}}T] + (X_{\mathrm{w0}} - X_{\mathrm{uw}})\left(\frac{F'(\lambda_0 + \lambda_1 T)^2}{RT^2(F' - F'_{\mathrm{P}})} + \lambda_1\right)\frac{F'_{\mathrm{F}} - F'_{\mathrm{P}}}{F' - F'_{\mathrm{P}}} + X_{\mathrm{uw}}\lambda_1 \quad (5.31)$$

$$c_{\mathrm{uf}} = c_{\mathrm{s}}X_{\mathrm{s}} + c_{\mathrm{w}}X_{\mathrm{w}} \quad (5.32)$$

$$c_{\mathrm{i}} = c_{\mathrm{i0}} + c_{\mathrm{i1}}T \quad (5.33)$$

where the function F' was defined in Equation (5.19).

Chen [30]:

$$\frac{c_{\mathrm{f}}}{\psi} = 0.37 + 0.3X_{\mathrm{s}} + M_{\mathrm{s}}\left(\frac{RM_{\mathrm{w}}T_{\mathrm{P}}^2}{M_{\mathrm{s}}(T - T_{\mathrm{P}})^2}\right) \quad (5.34)$$

$$\frac{c_{\mathrm{f}}}{\psi} = 1 - 0.55X_{\mathrm{s}} - 0.15X_{\mathrm{s}}^3 \quad (5.35)$$

where ψ is a conversion factor of units (=4184 J/cal).

VI. LATENT HEAT

Latent heat is defined as the amount of heat released or absorbed at a specific temperature when a unit mass of food material is transformed from one state to another. The SI unit for latent heat is typically J/g. The latent heat of fusion of ice decreases as the temperature at which freezing occurs decreases. The temperature dependence of latent heat of fusion can be expressed as [11]:

$$\lambda_{\mathrm{w}} = 334.2 + 2.12T + 0.0042T^2 \quad (5.36)$$

VII. THERMAL CONDUCTIVITY

A. Definition of Thermal Conductivity

The thermal conductivity of a food material is a measure of its ability to conduct heat. According to Fourier's law of heat conduction, the amount of heat, Q, that flows through a slab of material can be expressed as

$$Q = kA\left(\frac{T_1 - T_2}{l}\right) \quad (5.37)$$

where A is the surface area of the material normal to the direction of heat flow, T_1 and T_2 are the surface temperatures of the material, and l is the thickness of the material.

In Equation (5.37), the proportionality constant, k, is the thermal conductivity of the food material. The typical SI unit for thermal conductivity is W/m°C.

B. Thermal Conductivity Measurement and Data

Measurement techniques for thermal conductivity can be grouped into steady state, transient, and quasi-steady state [7,9,12,32–35]. Guarded hot plate and heated probe are two popular experimental instruments used for the measurement of thermal conductivity. As food materials normally have low conductivities, it takes longer time (e.g., 12 h) to reach the steady state, resulting in moisture migration and property changes due to long exposure at high temperature. Therefore, some well-established standard techniques for measuring thermal conductivity such as the guarded hot plate work well for nonbiological materials but are not well suited for foods because of long time for temperature equilibration, moisture migration in the sample, and the need for large sample size.

Transient and quasi-steady-state techniques are more popular for measuring the thermal conductivity of food materials because they require short measurement times and relatively small samples [7,36]. The line heat source thermal conductivity probe, which is based on the transient techniques, is recommended for most food applications. A typical probe is shown in Figure 5.8 [7]. The probe has an insulated heater wire inside the needle tubing, running from the handle to the tip and back. An insulated thermocouple is also inserted in the tubing, with the junction located halfway between the probe handle and the needle tip. The needle, thermocouple, and heater wire are all electrically insulated from one another by plastic tubing.

During measurement, the line heat source probe is inserted into a food sample that is initially at a uniform temperature. The probe is heated at a constant rate, and the temperature adjacent to the line heat source is monitored. After a brief transient period, the plot of the natural logarithm of time versus the monitored temperature is linear and the slope is $Q/4\pi k$. Therefore, thermal conductivity can be written as [7]:

$$k = Q\frac{\ln[(t_2 - t_0)/(t_1 - t_0)]}{4\pi(T_2 - T_1)} \tag{5.38}$$

where k is the thermal conductivity of the sample, Q the power generated by the probe heater, t_0 a time correction factor, and T_1 and T_2 are the temperatures of the probe thermocouple at times t_1 and t_2, respectively.

The length of the thermal conductivity test varies from 3 sec for liquids to 10–12 sec for most solid foods. Power levels from 5 to 30 W/m of wire have been used. Materials having higher conductivity require higher power levels to obtain sufficient temperature increases. The time correction factor, t_0, has been found to be negligible for probes with small diameters (e.g., 0.66 mm). For high accuracy, it is necessary to have a probe length-to-diameter ratio greater than 25 and to have an adequate sample diameter. The line heat source probe is not well suited for nonviscous fluids because of convection currents that arise during probe heating. A longer probe and larger sample diameter may be needed for measuring thermal conductivity of ice because of the high thermal diffusivity of ice [7].

Variation of thermal conductivity of frozen foods as a function of moisture and temperature is given in Table 5.5. The thermal conductivity of frozen foods can be as high as 1.5 W/m °C or higher. The thermal conductivity of ice, which is usually the main component of a frozen food, is about 2 W/m °C. For unfrozen food materials, thermal conductivity varies between 0.02 W/m °C for air and 0.62 W/m°C for water. Figure 5.9 illustrates a typical change in thermal conductivity with temperature for a high-moisture food material (e.g., >75%, wet basis) [13]. It can be seen from Figure 5.9 that the thermal conductivity of frozen food decreases gradually

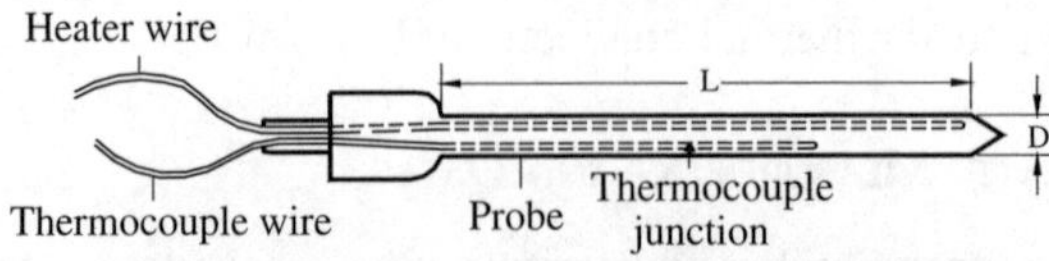

FIGURE 5.8 Cross-section of a line heat source thermal conductivity probe indicating length (L) and diameter (D).

TABLE 5.5
Observed Thermal Conductivity Values for Frozen Fish as a Function of Temperature and Composition [12]

	Thermal Conductivity (W/m °C)					
	Composition			Temperature (°C)		
Fish Variety	X_w	X_f	X_p	−20	−10	0
Saman	0.7730	0.0046	0.2020	1.428	1.334	0.409
Sol	0.7760	0.0046	0.2046	1.365	1.294	0.398
Black bhitki	0.7770	0.6000	0.1945	1.563	1.487	0.416
Black pomphret	0.7500	0.0476	0.1791	1.315	1.244	0.412
Mackerel	0.7735	0.0058	0.1867	1.237	1.088	0.416
Red bhitki	0.7953	0.0042	0.1689	1.358	1.207	0.413
Singara	0.7787	0.0057	0.2012	1.284	1.254	0.420
Hilsa	0.7470	0.0513	0.1722	1.328	1.234	0.409
Surama	0.7800	0.0077	0.1882	1.327	1.067	0.415
White pomphret	0.7483	0.0410	0.1856	1.395	1.118	0.390
Malli	0.7820	0.0045	0.7935	—	1.356	0.411
Rohu	0.7520	0.0057	0.2148	1.535	1.284	0.421

Note: X_w, water mass fraction; X_f, fat oil mass fraction; X_p, protein mass fraction.

from about 1.5 W/m °C at −40°C to 0.5 W/m °C at the initial freezing temperature. There is a sharp decrease in the thermal conductivities for food materials near their initial freezing temperatures due to the melting of ice.

C. Prediction Models of Thermal Conductivity

Thermal conductivity of a food material depends on its chemical composition, the physical arrangement or structure of each chemical component, and the temperature of the material. The structure of a food material has a significant effect on its thermal conductivity. Foods that contain fibers exhibit different thermal conductivities parallel to the fibers compared with conductivities

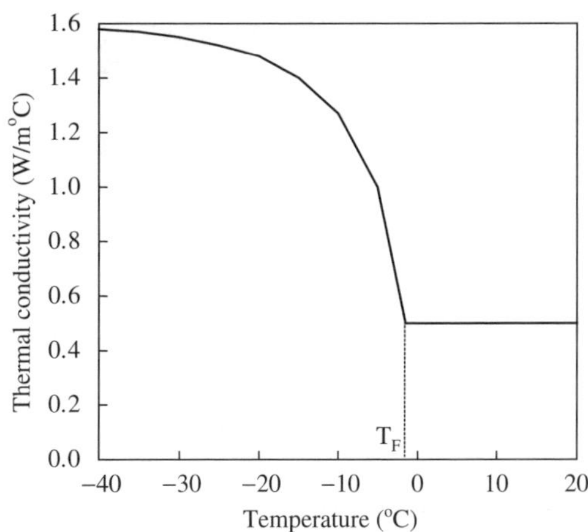

FIGURE 5.9 Typical change in thermal conductivity with temperature for a high-moisture food during freezing.

perpendicular to the fibers. Porosity has a major influence on thermal conductivities of foods. As food systems have wide variation in composition and structure, it is difficult to find a very accurate model for predicting thermal conductivity of a broad range of foods. Existing models can generally be divided into four groups: (1) models assuming structural arrangement; (2) models using a structural parameter; (3) empirical correlations for a specific food; and (4) generalized correlations.

Most models are based on volume rather than mass fractions. Volume fractions of each component are calculated by

$$\varepsilon = \frac{\rho X_j}{\rho_j} \tag{5.39}$$

The presence of ice in frozen materials has been found to greatly influence the thermal conductivities of the materials. As the thermal conductivity of ice (2.0 W/m °C) is about four times that of liquid water (0.5 W/m °C), the accuracy of a thermal conductivity model largely depends on the accuracy of ice content prediction. Murakami and Okos [6] considered various models for different nonporous foods. Above the freezing point, the simple parallel model was recommended as the best, which can be expressed as

$$k = \sum_{j=1}^{n} k_j \varepsilon_j \tag{5.40}$$

Below the freezing point, a combined parallel–series model was suggested [6]. The structural model has the non-water component arranged in parallel. Then the nonwater and water components are in perpendicular to each other. The parallel–series model is given by

$$\frac{1}{k} = \frac{1 - \varepsilon_{\mathrm{w}}}{k_{\mathrm{s}}} + \frac{\varepsilon_{\mathrm{w}}}{k_{\mathrm{w}}} \tag{5.41}$$

where

$$k_{\mathrm{s}} = \sum_{j=1}^{n} k_j \varepsilon_j \tag{5.42}$$

For nonporous meat products, the Maxwell–Eucken model was found to be most accurate [37]. In the Maxwell–Eucken model, a food material can be considered to be consist of a continuous phase and a dispersed phase. The thermal conductivity can thus be calculated by

$$k = k_{\mathrm{c}} \left[\frac{1 - 2\xi \varepsilon_{\mathrm{d}}}{1 + \xi \varepsilon_{\mathrm{d}}} \right] \tag{5.43}$$

where

$$\xi = \frac{k_{\mathrm{c}} - k_{\mathrm{d}}}{2k_{\mathrm{c}} + k_{\mathrm{d}}} \tag{5.44}$$

In the above equations, k_{d} is the thermal conductivity of the dispersed phase, k_{c} the thermal conductivity of the continuous phase, and ε_{d} the volume fraction of the dispersed phase. For unfrozen foods, the nonwater food constituents are considered as the dispersed phase and water the continuous phase. For frozen foods, a two-stage model can be used to estimate the thermal conductivity.

In the first stage, ice was considered to disperse in water, and in the second stage, a food solid matrix was considered to disperse in an ice–water mixture. The freezing phenomena may significantly alter the porosity of a food material. For porous frozen foods, Miles et al. [38] recommended the parallel model but found that similar accuracy can be achieved using the Maxwell–Eucken equation.

Equation (5.40) to Equation (5.44) require thermal conductivity values of components. Thermal conductivity prediction equations of major food components are given in Table 5.6 [31].

VIII. DENSITY

Density is defined as the mass per unit volume with typical SI unit of kg/m^3. As food products have different shapes and sizes, it may be difficult to accurately measure the food volume. A recommended procedure for solid foods is to add a known mass of sample to a calibrated volume flask, completing the volume with distilled water at 22°C, and calculating density by the following equation [9]:

$$\rho = \frac{m}{V_s} = \frac{m}{V - V_w} \tag{5.45}$$

In this above equation, m and V_s are the mass and volume of the sample, respectively. V_s is calculated from the total volume of the flask V, and the added water volume V_w.

Figure 5.10 gives typical change in density with temperature for a high-moisture food material during freezing [13]. The densities of unfrozen and totally frozen foods are relatively constant. Heldman [21] reported that the overall change in the density of strawberries during freezing between 5 and −40°C was less than 10%. However, as shown in Figure 5.10, there is a dramatic decrease in density at and just below the initial freezing temperature as the fraction of ice in the product increases.

The density of a frozen food can be determined by [39,40]:

$$\frac{1}{\rho} = \frac{X_{uw}}{\rho_{uw}} + \frac{X_s}{\rho_s} + \frac{X_i}{\rho_i} \tag{5.46}$$

where ρ is the density, ρ_{uw}, ρ_s, and ρ_i are the densities of unfrozen water, product solid, and ice, respectively, and X_{uw}, X_s, and X_i are the mass fraction of unfrozen water, product solid, and ice, respectively.

TABLE 5.6
Equations for Estimating Thermal Conductivities of Major Food Components as Function of Temperature over the Range −40°C to 150°C [31]

Major Component	Equation
Carbohydrate	$k = 0.20141 + 1.3874 \times 10^{-3}\,T - 4.3312 \times 10^{-6}\,T^2$
Fiber	$k = 0.18331 + 1.2497 \times 10^{-3}\,T - 3.1683 \times 10^{-6}\,T^2$
Protein	$k = 0.17881 + 1.1958 \times 10^{-3}\,T - 2.7178 \times 10^{-6}\,T^2$
Fat	$k = 0.18071 + 2.7604 \times 10^{-3}\,T - 1.7749 \times 10^{-6}\,T^2$
Ash	$k = 0.32962 + 1.4011 \times 10^{-3}\,T - 2.9069 \times 10^{-6}\,T^2$
Liquid water	$k = 0.57109 + 1.7625 \times 10^{-3}\,T - 6.7036 \times 10^{-6}\,T^2$
Ice	$k = 2.2196 - 6.2489 \times 10^{-3}\,T + 1.0154 \times 10^{-4}\,T^2$

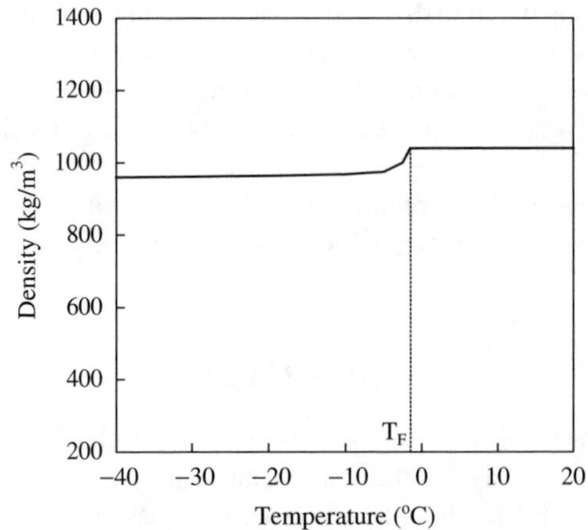

FIGURE 5.10 Typical change in density with temperature for a high-moisture food during freezing.

As the porosity of a food material can strongly influence its density, Mannapperuma and Singh [29] modified Equation (5.46) to incorporate the effect of porosity:

$$\frac{1}{\rho} = \frac{1}{1-\varepsilon}\sum_{j=1}^{n}\frac{X_j}{\rho_j} \tag{5.47}$$

where ε is the porosity and j denotes the jth component in the food.

As given in Equation (5.46) and Equation (5.47), the density is usually expressed as a function of composition. The densities of all components are further expressed in the literature as functions of temperature based on the regression of experimental data, which are given in Table 5.7 [31].

IX. THERMAL DIFFUSIVITY

Thermal diffusivity is used in the determination of heat transfer rate in a solid food material of any shape. If physical properties are grouped, the energy transport in a food material can be described

TABLE 5.7
Equations for Estimating Densities of Major Food Components as Function of Temperature over the Range of −40 to 150°C [31]

Major Component	Equation
Carbohydrate	$\rho = 1599.1 - 0.3105T$
Fiber	$\rho = 1311.5 - 0.3659T$
Protein	$\rho = 1330.0 - 0.518T$
Fat	$\rho = 925.6 - 0.417.6T$
Ash	$\rho = 2423.8 - 0.2806T$
Liquid water	$\rho = 997.2 + 3.1439 \times 10^{-3}\,T - 3.7574 \times 10^{-3}\,T^2$
Ice	$\rho = 916.9 - 0.1307T$

by the well-known Laplace's equation, which is given by

$$\frac{\partial T}{\partial t} = \alpha \nabla^2 T \tag{5.48}$$

In this equation, the proportionality constant α is the thermal diffusivity of the material. Typical SI unit for thermal diffusivity is m^2/s. Physically, it relates the ability of a material to conduct heat to its ability to store heat.

Direct measurement of thermal diffusivity is uncommon. There is also no generic model available for directly predicting thermal diffusivity below a freezing point. Apparent thermal diffusivity of a frozen food is usually calculated from known values of thermal conductivity, density, and specific heat using the following definition [41]:

$$\alpha_e = \frac{k}{\rho c_e} \tag{5.49}$$

In this equation, the values of thermal conductivity, density, and specific heat can be calculated using the expressions presented in the previous sections. Any error associated with these values can lead to erroneous predictions of thermal diffusivity. Typical average value of thermal diffusivity for unfrozen food is 1.3×10^{-7} m^2/s and for frozen foods it is 4×10^{-7} m^2/s [37].

Figure 5.11 gives a typical plot of apparent thermal diffusivity as a function of temperature [37]. There is a near discontinuity at the initial freezing temperature. The thermal diffusivity of an unfrozen food does not change substantially with temperature because the effects of any changes of thermal conductivity, density, and specific heat on the thermal diffusivity are compensated among one another. However, there is a sharp decrease in the thermal diffusivity for frozen foods near the initial freezing point. The sharp decrease is caused by the sharp increase of apparent specific heat which includes the latent heat of fusion [8].

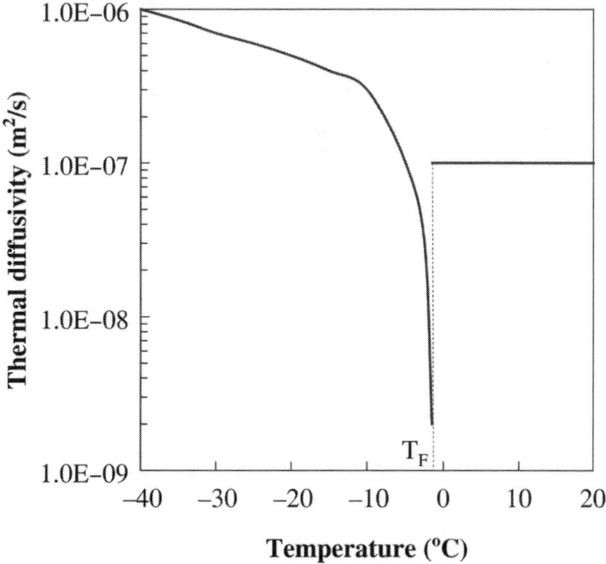

FIGURE 5.11 Typical change in thermal diffusivity with temperature for a food material during freezing.

X. CONCLUSIONS

The thermal properties of frozen foods were reviewed in this chapter. Some of the well-known measurement methods and preferred prediction models were introduced. Thermophysical properties of frozen foods in literature show wide variations due to the effects of diverse experimental methods, and variations in composition and structure of food materials. Special caution is required when using experimental data in literature. Prediction models can be used to estimate the thermophysical properties of frozen foods for a wide range of conditions and compositions with a reasonable accuracy.

With the increasing number of formulated and new frozen foods, it is desirable and warranted to develop more molds and refine the existing prediction models and experimental techniques for determining the freezing point, ice content, enthalpy or specific heat, latent heat, thermal conductivity, density, and thermal diffusivity of frozen foods. With the continued development of novel freezing and thawing technologies such as high-pressure-assisted freezing and microwave-assisted thawing, prediction models and experimental techniques are needed for determining the thermal properties of foods with phase transformation under pressure and electromagnetic field.

NOMENCLATURE

A	area (m^2)
B	unfreezable water per kg of total dry solids (kg water/kg solids)
c	specific heat (kJ/kg °C)
c_e	apparent specific heat (kJ/kg °C)
c_{i0}	parameter in Equation (5.23)
c_{i1}	parameter in Equation (5.23)
F	function defined in Equation (5.15)
F'	function defined in Equation (5.19)
H	enthalpy (kJ/kg)
k	thermal conductivity (W/m °C)
l	thickness (m)
m	mass (kg)
M	molecular weight (g/mol)
n	number of component in a food
Q	heat flux (W)
R	universal gas constant (8.314 kJ/kg mol K)
ΔT_F	freezing point depression (°C)
T	temperature (K or °C)
t	time (sec)
t_0	time correction factor in Equation (5.38) (sec)
V	volume (m^3)
X	mass fraction

Greek symbols

α	thermal diffusivity (m/sec^2)
ε	porosity or volume fraction
λ	latent heat of fusion (J/g)
λ_0	parameter in Equation (5.17)
λ_1	parameter in Equation (5.17)

ξ function defined in Equation (5.44)
ρ density (kg/m^3)
τ molecular dissociation defined in Equation (5.3)
ψ conversion factor (4184 J/cal)

Subscripts

c continuous phase
d dispersed phase
D at the datum point
e apparent or efficient value
f frozen
F at the initial freezing point of a food material
fw freezable liquid water
i ice
j the *j*th component of the food material
*j*0 the *j*th component of the food material at the initial freezing point
P pure water
s food solid parts
s0 food solid parts before freezing
uf unfrozen
uw unfreezable bound water
w water
w0 water at the initial freezing point of a food material
W at the initial freezing point of pure water

REFERENCES

1. JD Mellor. Thermophysical properties of foodstuffs. I. Introductory review. *Bulletin of International Institute of Refrigeration* 56:551–563, 1976.
2. JD Mellor. Thermophysical properties of foodstuffs. II. Theoretical aspects. *Bulletin of International Institute of Refrigeration* 58:569–584, 1978.
3. JD Mellor. Thermophysical properties of foodstuffs. III. Measurements. *Bulletin of International Institute of Refrigeration* 59:551–563, 1979.
4. JD Mellor. Thermophysical properties of foodstuffs. IV. General bibliography. *Bulletin of International Institute of Refrigeration* 60:493–515, 1980.
5. DR Heldman, RP Singh. Thermal properties of frozen foods. In: MR Okos, Ed., *Physical and Chemical Properties of Foods*. St. Joseph: ASAE, 1983, pp. 120–137.
6. EG Murakami, MR Okos. Measurement and prediction of thermal properties of foods. In: RP Singh, AG Medina, Eds., *Food Properties and Computer Aided Engineering of Food Processing Systems*. Amsterdam: Kluwer Academic Publishers, 1989, pp. 3–48.
7. VE Sweat. Thermal properties of foods. In: MA Rao, SS H Rizvi, Eds., *Engineering Properties of Foods*. New York: Marcel Dekker, 1995, pp. 99–138.
8. RP Singh. Thermal properties of frozen foods. In: MA Rao, SS H Rizvi, Eds., Engineering Properties of Foods. New York: Marcel Dekker, 1995, pp. 139–167.
9. MJ Urbicain, JE Lozano. Thermal and rheological properties of foodstuffs. In: KJ Valentas, E Rotstein, RP Singh, Eds., *Food Engineering Practice*. Florida: CRC Press, 1997, pp. 425–486.

10. YH Roos. Phase transitions and unfreezable water content of carrots, reindeer meat and white bread studied using differential scanning calorimetry. *Journal of Food Science* 51:684–686, 1986.
11. MS Rahman. Thermophysical properties of foods. In: DW Sun, Ed., *Advances in Food Refrigeration*. Leatherhead: Leatherhead Publishing, 2001, pp. 70–109.
12. MS Rahman. Food Properties Handbook. Florida: CRC Press, 1995, pp. 87–177.
13. DJ Cleland, KJ Valentas. Prediction of freezing time and design of food freezers. In: KJ Valentas, E Rotstein, RP Singh, Eds., *Food Engineering Practice*. Florida: CRC. Press, 1997, pp. 71–123.
14. AG Fikiin. New method and fluidized water system for intensive chilling and freezing of fish. *Food Control* 3 (3):153–160, 1992.
15. KA Fikiin, AG Fikiin. Individual quick freezing of foods by hydro-fluidisation and pumpable ice slurries. *AIRAH Journal* 55 (11):15–18, 2001.
16. W Wagner, A Saul, A Pruss. International equations for the pressure along the melting and along the sublimation curve of ordinary water substance. *Journal of Physical Chemistry* 23: 515–527, 1994.
17. S Denys, AM Van Loey, ME Hendrickx. Modeling heat transfer during high-pressure freezing and thawing. *Biotechnology Progress* 13:416–423, 1997.
18. JC Cheftel, J Levy, E Dumay. Pressure-assisted freezing and thawing: principles and potential applications. *Food Reviews International* 16:453–483, 2000.
19. KA Fikiin. Ice content prediction methods during food freezing: a survey of the eastern European literature. *Journal of Food Engineering* 38:331–339, 1998.
20. QT Pham. Calculation of bound water in frozen food. *Journal of Food Science* 52:210–212, 1987.
21. DR Heldman. Food freezing. In: DR Heldman, DB Lund, Eds., *Handbook of Food Engineering*. New York: Marcel Dekker, 1992, pp. 277–315.
22. HD Chang, LC Tao. Correlations of enthalpies of food systems. *Journal of Food Science* 46:1493–1497, 1981.
23. PD Sanz, M Dominguez, RH Mascheroni. Equations for the predictions of thermophysical properties of meat products. *Latin American Applied Research* 19:155–160, 1989.
24. MS Rahman. The accuracy of prediction of the freezing point of meat from general models. *Journal of Food Engineering* 21:127–136, 1994.
25. MS Rahman, RH Driscoll. Freezing points of selected seafoods (invertebrates). *International Journal of Food Science and Technology* 29:51–61, 1994.
26. L Riedel. The refrigeration required to freeze fruits and vegetables. *Refrigeration Engineering* 59:670, 1951.
27. HG Schwartzberg. Effective heat capacities for the freezing and thawing of foods. *Journal of Food Science* 41:152–156, 1976.
28. DR Heldman. Factors influencing food freezing. *Food Technology* 37:103–109, 1974.
29. JD Mannapperuma, RP Singh. Developments in food freezing. In: H Schwartzberg, A Rao, Eds., *Biotechnology and Food Process Engineering*. New York: Marcel Dekker, 1990.
30. CS Chen. Thermodynamic analysis of freezing and thawing of foods: enthalpy and apparent specific heat. *Journal of Food Science* 50:1158–1162, 1985.
31. Y Choi, MR Okos. Effects of temperature and composition on the thermal properties of foods. In: ML Maguer, P Jelen, Eds., *Food Engineering and Process Applications*, Vol. 1, *Transport Phenomena*. New York: Elsevier, 1986, pp. 93–101.
32. NN Mohsein. *Thermal Properties of Foods and Agricultural Products*. New York: Gordon and Breach, 1980.
33. NP Nesvadba. Methods for the measurement of thermal conductivity and diffusivity of foods. *Journal of Food Engineering* 1:93–113, 1982.
34. M Kent, K Christiansen, A van Haneghem, E Holtz, MJ Morley, P Nesvadba, KP Poulsen. COST 90 collaborative measurements of thermal properties of foods. *Journal of Food Engineering* 3:117–150, 1984.
35. GD Saravacos, ZB Maroulis. Transport properties of foods. New York: Marcel Dekker, 2001, pp. 269–358.
36. VE Sweat, CG Haugh. A thermal conductivity probe for small food samples. *Transaction of the ASAE* 17:56–58, 1974.

37. QT Pham. Prediction of thermal conductivity of meats and other animal products from composition data. In: WEL Speiss, H Schubert, Eds., *Engineering and Food*, Vol. 1. London: Elsevier, 1990, pp. 408–423.
38. CA Miles, G van Beek, CH Veerkamp. Calculation of thermophysical properties of foods. In: R Jowitt, Ed., *Physical Properties of Food*. London: Applied Science Publishers, 1983, pp. 269–313.
39. RC Hsieh, LE Lerew, DR Heldman. Prediction of freezing times for foods as influenced by product properties. *Journal of Food Process Engineering* 1:183, 1977.
40. DR Heldman. Food properties during freezing. *Food Technology* 36:92–96, 1982.
41. RP Singh. Thermal diffusivity in food processing. *Food Technology* 36:87–91, 1982.

6 Freezing Loads and Freezing Time Calculation

Gauri S. Mittal
University of Guelph, Guelph, Ontario, Canada

CONTENTS

I. INTRODUCTION

The purposes of food freezing are: (i) preservation of food; (ii) reducing the activity of enzymes and microorganisms; (iii) reducing the amount of liquid water for microbial growth; and (iv) reducing water activity (a_w) of foods. Many types of freezers are used for this purpose. Some of these are: (i) air blast freezers, batch, or continuous; (ii) still air freezers; (iii) belt freezers; (iv) spiral belt freezers; (v) fluidized bed freezers; (vi) plate freezers — a series of flat plates kept cool by circulating a coolant; (vii) liquid immersion freezers — chilled brine or glycol is used, can also be sprayed; and (viii) cryogenic freezers — liquid N_2 or liquid CO_2 is used (the boiling point for N_2 is -196°C and for CO_2 is -79°C).

Product quality is influenced by ice-crystal size and configuration during the freezing operation. The advantages of fast freezing operation can be lost during the storage because of the formation of large ice crystals by joining small crystals. Hence, complete product freezing in the freezer is more important [1]. The process of ice-crystal formation is based on two operations: (1) nucleation or crystal formation — it influences the type of crystal structure formed in a food product and ice-crystal nucleation is created by supercooling below initial freezing point, similar to crystallization process; (2) rate of crystal growth — is also supercooling-driven, which depends on (i) diffusion rate of water molecules from the unfrozen solution to the crystal surface, (ii) the rate at which heat is removed, and (iii) temperature of the solution.

II. FREEZING LOAD

A. Calculation

Freezing load or enthalpy change (ΔH) to reduce the product temperature (T_i) from some level above the freezing point (T_F) to some desired final temperature (T) is given by

$$\begin{aligned}\Delta H = {} & \text{sensible heat removed from the product solids } (\Delta H_\mathrm{s}) \\ & + \text{sensible heat removed from unfrozen water } (\Delta H_\mathrm{u}) \\ & + \text{enthalpy change due to latent heat } (\Delta H_\mathrm{L}) \\ & + \text{sensible heat removed from the frozen water } (\Delta H_\mathrm{I})\end{aligned}$$

$$\Delta H_\mathrm{s} = M_\mathrm{s}\, C_\mathrm{ps}(T_\mathrm{i} - T_\mathrm{F}) + \int_{T_\mathrm{F}}^{T} M_\mathrm{s} C_\mathrm{ps}\, \mathrm{d}T \quad (6.1\mathrm{a})$$

$$\Delta H_\mathrm{u} = M_\mathrm{u}\, C_\mathrm{pu}(T_\mathrm{i} - T_\mathrm{F}) + M_\mathrm{u}\, C'_\mathrm{pu}(T_\mathrm{F} - T) \quad (6.1\mathrm{b})$$

$$\Delta H_\mathrm{L} = M_\mathrm{I}\, L_\mathrm{V} \quad (6.1\mathrm{c})$$

$$\Delta H_\mathrm{I} = M_\mathrm{I}\, C_\mathrm{pI}(T_\mathrm{F} - T) \text{ or } \int_{T_\mathrm{F}}^{T} M_\mathrm{I}\, C_\mathrm{pI}\, \mathrm{d}T \quad (6.1\mathrm{d})$$

where M_s is the mass of solids, C_ps the specific heat of solids, M_u the mass of unfrozen water, C_pu the specific heat of unfrozen water, M_I the mass of ice or unfrozen water, L_v the latent heat of freezing, C_pI the specific heat of ice, and C'_pu the specific heat of unfrozen water below T_F. Enthalpy composition charts for different food materials using experimental data were provided [2,3]. One example is given in Figure 6.1.

B. Freezing Rate and Thermal Center

The absence of a consistent definition for the freezing time is one of the problems associated with the published literature on the freezing of foods. This problem arises mainly because foods do not freeze at a distinct temperature, but rather the phase change takes place over a range of temperatures. A definition of the freezing time requires a definition of the freezing point. A variable temperature distribution exists within the food product during the freezing process, giving different freezing times depending on the point within the product where the temperature is monitored. The "thermal center", defined as the location in the material which cools most slowly, is generally used as the reference location. The effective freezing time, defined by the International Institute of

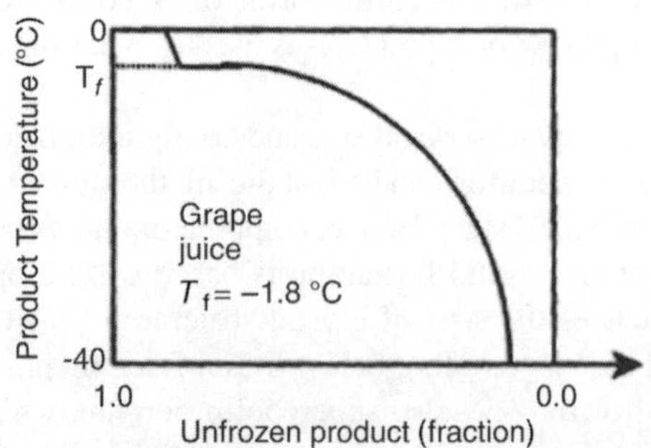

FIGURE 6.1 Riedel plot for grape juice. (From L Riedel, *Kaltetechnik* 9:38–40, 1957.)

Refrigeration [4], is the total time required to lower the temperature of a food material at its thermal center to a desired temperature below the initial freezing point. Other definitions are:

1. The time required to reduce the product temperature at the slowest cooling location from the initial freezing point to some desired and specified temperature below the initial freezing point.
2. International Institute of Refrigeration [4] definition: It is the ratio between the minimum distance from the surface to the thermal center and the time elapsed between the surface reaching 0°C and the thermal center reaching 5°C colder than the temperature of initial ice formation at the thermal center (cm/h).

III. FREEZING TIME OR RATE PREDICTION

It is important to accurately predict the freezing times of foods to assess the quality, processing requirements, and economical aspects of food freezing. A number of models have been proposed in the literature to predict freezing times. However, as the freezing process is a moving boundary problem, that is, one involving a phase change, most of the single-phase, unsteady-state solutions are unsuitable.

Foods, undergoing freezing, release latent heat over a range of temperatures. Freezing does not occur at a unique temperature. In addition, foods do not have constant thermal properties during freezing [5]. As a result, no exact mathematical model exists for predicting the freezing of foods. Researchers, who have found a solution, have either used numerical finite difference or finite element methods. So, models for predicting freezing times range from approximate analytical solutions to more complex numerical methods.

In the past, an extensive amount of work has been done to develop mathematical models for the prediction of food freezing times. The accuracy of such models is dependent on how closely the corresponding assumptions approach reality. Most of these models are usually categorized into one of two forms, analytical or numerical, with the latter generally considered as more accurate due to the inclusion of a set of assumptions and boundary conditions, which are of a more realistic nature than those pertaining to the former. Approximately, 30 different methods to predict freezing and thawing times were reviewed [6]. Details on these models are given elsewhere [7,8].

The general approach of researchers in the food-freezing field has been to seek approximate or empirical relationships, rather than to try to derive exact analytical equations. The method can be classified into two groups: (1) methods relying on analytical approximations, such as those of Refs. [9–13] or (2) methods relying on regression of computer results or experimental data, such as those of Refs. [14–18]. The methods vary considerably in complexity and accuracy, the number of arbitrary or empirical parameters used ranging from 0 to more than 50 [19].

A. Plank's Equation

Plank's equation was derived based on energy balance principle [9]. Heat condition through frozen region is written as: (Figure 6.2)

$$q = k_I A\left(\frac{T_S - T_F}{X}\right) \tag{6.2}$$

Convective heat transfer at the surface is given by:

$$q = h_c A\,(T_\infty - T_S) \tag{6.3}$$

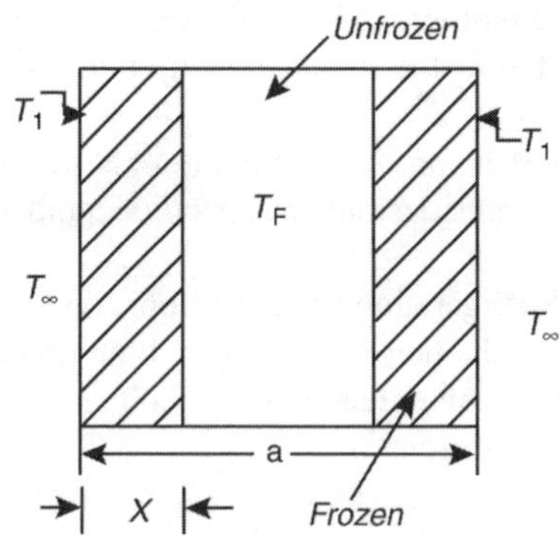

FIGURE 6.2 Diagram to derive Plank's equation.

Total resistance

$$R_t = \frac{X}{k_I A} + \frac{1}{h_c A} \tag{6.4}$$

or

$$q(\text{overall}) = \frac{\Delta T}{R_t} = \frac{T_\infty - T_F}{(X/k_I A + 1/h_c A)} \tag{6.5}$$

This heat transfer should be equal to the latent heat of freezing or

$$q = A\,\frac{dX}{dt}\rho L_V \tag{6.6}$$

$$\frac{dX}{dt} = \text{the velocity of the freezing front} \tag{6.7}$$

or

$$A\frac{dX}{dt}\rho L_V = -\frac{(T_\infty - T_F)A}{(X/k_I + 1/h_c)}(\text{negative heat transfer}) \tag{6.8}$$

or

$$\int_0^{t_F} dt = -\frac{L_V \rho}{T_\infty - T_F}\int_0^{a/2}\left(\frac{1}{h_c} + \frac{X}{k_I}\right) dX \tag{6.9}$$

or

$$t_F = -\frac{L_V\rho}{T_\infty - T_F}\left[\frac{a}{2h_c} + \frac{a^2}{8k_I}\right] = \frac{L_V\rho}{T_F - T_\infty}\left[\frac{a}{2h_c} + \frac{a^2}{8k_I}\right] \tag{6.10}$$

General form:

$$t_F = \frac{\rho L_V}{T_F - T_\infty}\left[\frac{Pa}{h_c} + \frac{Ra^2}{k_I}\right] \tag{6.11}$$

where T_F is the initial freezing point of the product, T_S the surface temperature, k_I the thermal conductivity of frozen food, X the thickness of frozen food, h_c the convective heat transfer coefficient, A the surface area, T_∞ the ambient temperature, L_v the latent heat of freezing, and ρ the food density.

P and R values for different shaped foods are:

	Infinite slab	Infinite cylinder	Sphere
P	1/2	1/4	1/6
R	1/8	1/16	1/24

For brick-shaped material, Figure 6.3 provides P and R for different β_1 and β_2 values [20].

Example 1: Lean beef block with dimensions of $1\,\text{m} \times 0.25\,\text{m} \times 0.6\,\text{m}$, $h_c = 30\,\text{W/(m}^2\,\text{K)}$, $T_0 = 5°\text{C}$, $T = -10°\text{C}$, $T_\infty = -30°\text{C}$, $\rho = 1050\,\text{kg/m}^3$, $L_V = 333.22\,\text{kJ/kg}$, m.c. $= 74.5\%$, $k_I = 1.108$ (W/m K), $T_F = -1.75°\text{C}$. Find freezing time using Plank's equation.

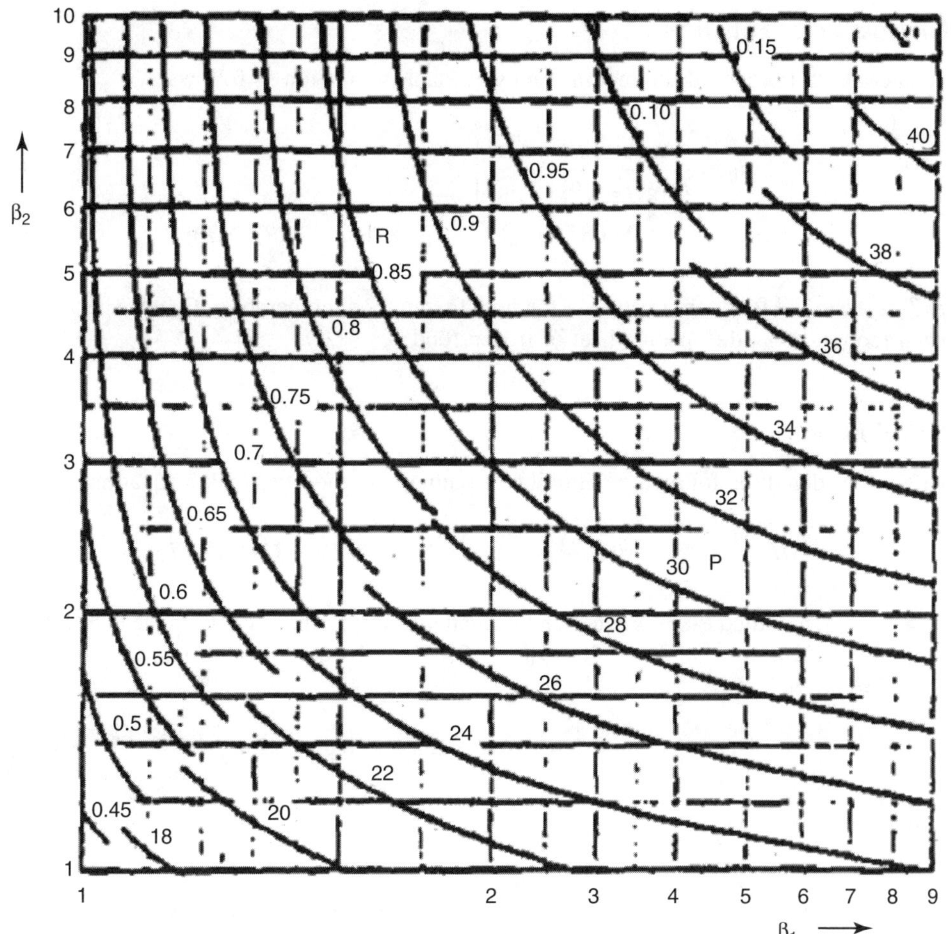

FIGURE 6.3 *P* and *R* values for different β_1 and β_2. (From AJ Ede. Modern Refrigeration 52:52-55, 1949.)

Solution:

$$\beta_1 = \frac{0.6}{0.25} = 2.4, \quad \beta_2 = \frac{1}{0.25} = 4, \quad \therefore P = 0.3, \quad R = 0.085$$

$$t = \frac{(1050)(333.22 \times 0.745)(1000\,\text{J/kJ})}{[-1.75 - (-30)]3600\,\text{s/h}} \left[\frac{0.3(0.25)}{30} + \frac{0.085(0.025)^2}{1.108} \right] = 18.7\,\text{h}$$

The limitations of Plank's equation are as follows:

1. It neglects the time required to remove sensible heat above the initial freezing point.
2. It does not consider the gradual removal of latent heat over a range of temperatures during the freezing process.
3. Constant thermal conductivity assumed for frozen material.
4. It assumes the product to be completely in liquid phase.

Many modifications were suggested on Plank's equation to improve its accuracy. Some of these are given in the subsequent sections.

B. Nagaoka et al. Equation

Nagaoka et al. [21] proposed the modifications of Plank's equation as follows:

$$t_F = \frac{\Delta H' \rho}{T_F - T_\infty} \left[\frac{Pa}{h_c} + \frac{Pa^2}{k_I} \right] \tag{6.12}$$

$$\Delta H' = (1 + 0.008T_i)[C_{pu}(T_i - T_F) + L_V + C_{PI}(T_F - T)] \tag{6.13}$$

where T_i is the initial food temperature, T the final frozen food temperature, C_{pu} the specific heat of unfrozen food, and C_{PI} the specific heat of frozen food.

C. Levy Equation

Levy [22] considered the following definition of enthalpy to modify Plank's equation:

$$\Delta H' = (1 + 0.008(T_i - T_F))[C_{pu}(T_i - T_F) + L_V + C_{PI}(T_F - T)] \tag{6.14}$$

Example 2: Use modified Plank's equation to calculate the freezing time for the lean beef block of $1\,\text{m} \times 0.6\,\text{m} \times 0.25\,\text{m}$, $h_c = 30\,\text{W/(m}^2\text{K)}$, $T_0 = 5°\text{C}$, $T = -10°\text{C}$, $T_\infty = -30°\text{C}$, $\rho = 1050\,\text{kg/m}^3$, $T_F = -1.75°\text{C}$, $t_F = ?$

$$\Delta H = 333.22\ \text{kJ/kg}\ (0.745\,\text{m.c.}) = 248.25\ \text{kJ/kg},$$

Solution:

$$\beta_2 = \frac{0.6}{0.25} = 2.4, \quad \beta_1 = \frac{1}{0.25} = 4, \quad \therefore P = 0.3, \quad R = 0.085$$

$$C_{pu} = 3.52\,\text{kJ/kg K}, \quad C_{PI} = 2.05\ \text{kJ/kg K}, \quad k_I = 1.108\,\text{W/m K}$$

$$\Delta H' = \{1 + 0.008[5 - (-1.75)]\}[3.52(5 - (-1.75)) + 248.25 + 2.05(-1.75 - (-10))]$$
$$= 297.59\,\text{kJ/kg}$$

$$t_\text{f} = \frac{(1050)(297.59)(1000)}{[-1.75 - (30)]3600}\left[\frac{0.3(0.25)}{30} + \frac{0.085(0.025)^2}{1.108}\right] = 22.41\,\text{h}$$

D. Cleland and Earle Equation

Cleland and Earle [23] modified Plank's equation using the nondimensional numbers as follows:

$$N_\text{Ste} = \text{Stefan number} = \frac{C_\text{PI}(T_\text{F} - T_\infty)}{\Delta H_\text{ref}} \tag{6.15}$$

$$t_\text{F} = \frac{\rho \Delta H_\text{ref}}{E(T_\text{F} - T_\infty)}\left[\frac{Pa}{h_\text{c}} + \frac{Ra^2}{k_\text{I}}\right]\left(1 - \frac{1.65 N_\text{Ste}}{k_\text{I}}\ln\left[\frac{T - T_\infty}{T_\text{ref} - T_\infty}\right]\right) \tag{6.16}$$

$$N_\text{PK} = \text{Plank's number} = \frac{C_\text{pu}(T_\text{i} - T_\text{F})}{\Delta H_\text{ref}} \tag{6.17}$$

$$P = 0.5[1.026 + 0.5808 N_\text{PK} + N_\text{Ste}(0.2296 N_\text{PK} + 0.105)] \tag{6.18}$$

$$R = 0.125[1.202 + N_\text{Ste}(3.410 N_\text{PK} + 0.7336)] \tag{6.19}$$

where T_ref is the reference temperature and E is 1 for an infinite slab, 2 for an infinite cylinder, and 3 for a sphere. T_ref is taken as $-10°\text{C}$ and ΔH_ref is enthalpy change from T_F to T_ref. $0.15 \le N_\text{Ste} \le 0.35, 0.2 \le N_\text{Bi} \le 20$, and $0 \le N_\text{PK} \le 0.55$.

Example 3 (Cleland and Earle [23] approach): Lamb steak (slab) 0.025 m thick, $T_\text{i} = 20°\text{C}$, $T = -10°\text{C}$, $T_\infty = -30°\text{C}$, $\rho = 1050\,\text{kg/m}^3$, $T_\text{F} = -2.75°\text{C}$, $k_\text{I} = 1.35\,\text{W/m K}$, $h_\text{c} = 20\,\text{W/(m}^2\,\text{K)}$, $E = 1$ for slab, $t_\text{F} = ?$ $C_\text{pu} = 3\,\text{kJ/kg K}$, $C_\text{PI} = 1.75\,\text{kJ/kg K}$, $\Delta H = 240\,\text{kJ/kg}$.

Solution:

$$\Delta H_\text{ref} = 240 + 1.75(-2.75 + 10) = 252.7\,\text{kJ/kg}$$

$$N_\text{Ste} = \frac{C_\text{PI}(T_\text{F} - T_\infty)}{\Delta H_\text{ref}} = \frac{1.75(-2.75 + 30)}{252.7} = 0.189$$

$$N_\text{PK} = \frac{C_\text{pu}(T_\text{i} - T_\text{F})}{\Delta H_\text{ref}} = \frac{3(20 + 2.75)}{252.7} = 0.270$$

$$P = 0.5[1.026 + 0.5808(0.270) + 0.189(0.2296(0.270) + 0.105)] = 0.607$$

$$R = 0.125[1.202 + 0.189(3.410(0.270) + 0.7336)] = 0.189$$

$$t_F = \frac{\rho \Delta H_{ref}}{T_F - T_\infty}\left[\frac{Pa}{h_c} + \frac{Ra^2}{k_I}\right]\left(1 - \frac{1.65\,N_{Ste}}{k_I}\ln\left[\frac{T - T_\infty}{T_{ref} - T_\infty}\right]\right)$$

$$= \frac{252.7(1000)(1050)}{(-2.75 + 30)3600}\left[\frac{0.607(0.025)}{20} + \frac{0.189(0.025)^2}{1.35}\right](1) = 2.289\,\text{h}$$

E. Cleland et al. Method

Cleland et al. [24,25] method is based on Calvelo [26] approach, which is given below:

$$t_f = \frac{1.3179\rho C_{PI} a^2}{k_I E}\left[\frac{0.5}{N_{Bi}N_{Ste}} + \frac{0.125}{N_{Ste}}\right]^{0.9576} N_{Ste}^{0.0550}\,10^{0.0017N_{Bi}+0.1727N_{PK}} \times \left[1 - \frac{1.65N_{Ste}}{k_I}\ln\frac{T - T_\infty}{T_{ref} - T_\infty}\right] \tag{6.20}$$

T_{ref} is also $-10°C$. N_{Bi} is given by ha/k_I.

F. Pham Method

Pham method [13] involves total of precooling, phase change, and tempering times.

$$t_f = \frac{1}{E}\sum_{i=1}^{3} \Delta H_i a \frac{(1 + N_{Bi_i}/a_i)}{2\Delta T_i h_c} \tag{6.21}$$

where

$$\Delta H_1 = C_{pu}(T_i - T_{f,ave}) \tag{6.22}$$

$$\Delta T_1 = \frac{(T_i - T_\infty) - (T_{f,ave} - T_\infty)}{\ln((T_i - T_\infty)/(T_{f,ave} - T_\infty))}, \quad a_1 = 6 \tag{6.23}$$

$$N_{Bi_i} = 0.5\left(\frac{h_c a}{k_I} + \frac{h_c a}{k_u}\right) \tag{6.24}$$

$$\Delta H_3 = C_{PI}(T_{f,ave} - T_{ave}), \quad N_{Bi_3} = N_{Bi_2} \tag{6.25}$$

$$\Delta T_3 = \frac{(T_{f,ave} - T_\infty) - (T_{ave} - T_\infty)}{\ln((T_{f,ave} - T_\infty)/(T_{ave} - T_\infty))} \tag{6.26}$$

$$\Delta H_2 = \Delta H_f, \quad \Delta T_2 = T_{f,ave} - T_\infty, \quad N_{Bi_2} = \frac{h_c a}{k_I}, \quad a_2 = 4 \tag{6.27}$$

$$T_{ave} = T - \frac{T - T_\infty}{2 + 4/N_{Bi_3}}, \quad a_3 = 6, \quad T_{f,ave} = T_F - 1.5 \tag{6.28}$$

where k_W is the thermal conductivity of unfrozen food.

G. Modified Pham Method

This modified method of Pham [19] was given after summing precooling, phase change, and tempering times. E is given by the literature [27]. This method is to calculate the freezing and thawing time for finite size objects of any shape by approximating them to be similar to an ellipsoid. The following assumptions were used in developing this method: (i) uniform initial product temperature, T_i; (ii) uniform and constant ambient conditions; (iii) a fixed value of final product temperature, T; and (iv) convective surface heat transfer is following Newton's law of cooling.

For infinite slab, the freezing time (t_{slab}) is given by

$$t_{slab} = \frac{\rho a}{2h_c}\left[\frac{\Delta H_1}{\Delta T_1} + \frac{\Delta H_2}{\Delta T_2}\right]\left[1 + \frac{N_{Bi}}{4}\right] \tag{6.29}$$

Equation (6.21) is valid for the following ranges: $0.02 < N_{Bi} < 11, 0.11 < N_{Ste} < 0.36$, and $0.03 < N_{PK} < 0.61$.

The thawing time is given by for thawing to $T_f = 0°C$:

$$t_{slab} = \frac{1.4921 C_{pu} a^2}{k_u}\left[\frac{0.5}{N_{Bi} N_{Ste}} + \frac{0.125}{N_{Ste}}\right]^{1.0248} N_{Ste}^{0.2712} N_{PK}^{0.061} \tag{6.30}$$

Equation (6.30) is valid for the following ranges: $0.3 < N_{Bi} < 41$, $0.08 < N_{Ste} < 0.77$, and $0.06 < N_{PK} < 0.27$.

where

$$\Delta H_1 = C_{pu}(T_i - T_3) \tag{6.31}$$

$$\Delta H_2 = \Delta H + C_{PI}\,(T_3 - T) \tag{6.32}$$

$$\Delta T_1 = \frac{T_i - T_3}{2} - T_\infty, \quad \Delta T_2 = T_3 - T_\infty, \quad \Delta T_3 = 1.8 + 0.263T + 0.105T_\infty \tag{6.33}$$

$$N_{Bi} = h_c a / k_I \tag{6.34}$$

$$N_{Ste} = \frac{C_{PI}\,(T_F - T_\infty)}{\Delta H_{ref}} \tag{6.35}$$

$$N_{PK} = \frac{C_{pu}\,(T_i - T_F)}{\Delta H_{ref}} \tag{6.36}$$

where C_{pu} is the specific heat of unfrozen product (J/(kg K)), C_{PI} the specific heat of the frozen product (J/(kg K)), h_c the convective heat transfer coefficient (W/(m^2 K)), T_∞ the ambient temperature (°C), T_F the initial freezing temperature (°C), ρ the product density (kg/m^3), k_I the thermal conductivity of frozen product (W/(m K)), k_u the thermal conductivity of unfrozen product (W/(m K)), ΔH the enthalpy change due to freezing = (moisture content) (333 220) (J/kg), R the characteristic dimension (m), that is radius of cylinder of sphere or half thickness of slab or other geometries, N_{Bi} the Biot number, N_{Ste} the Stefan number, N_{PK} the Plank number, and T the final product temperature (°C).

For other shapes than infinite slab, the following modification is used:

$$t_{ellipsoid} = \frac{t_{slab}}{E} \tag{6.37}$$

$$E = 1 + \frac{1 + 2/N_{Bi}}{\beta_1^2 + 2\beta_1/N_{Bi}} + \frac{1 + 2/N_{Bi}}{\beta_2^2 + 2\beta_2/N_{Bi}} \tag{6.38}$$

where V is the volume (m^3) and A the smallest cross-sectional area that incorporate R (m^2).

For an infinite slab, E is 1, for an infinite cylinder E is 2, and for a sphere E is 3.

$$\beta_1 = \frac{A}{\pi R^2} \quad \text{and} \quad \beta_2 = \frac{V}{\beta_1(4/3\pi R^3)} \tag{6.39}$$

Notes:

1. For Equations (6.11), (6.12), (6.16), (6.20), (6.21), (6.29), (6.30), and (6.34), a is slab thickness or diameter of cylinder or sphere, or the smallest dimension of brick-shaped or dissimilar products.
2. ΔH = (moisture content) (latent heat of fusion); $\Delta H'$ for modified Plank's equations such as Levy's [22] and Nagaoka et al. [21], and $\Delta H_{10} = \Delta H + C_{\mathrm{PI}}(T_{\mathrm{F}} - 10)$.

Example 4: Beef slab of $1\,\mathrm{m} \times 0.6\,\mathrm{m} \times 0.25\,\mathrm{m}$, $R = 0.25/2 = 0.125\,\mathrm{m}$, $h_c = 30\,\mathrm{W/(m^2\,K)}$, $\rho = 1050\,\mathrm{kg/m^3}$, $T_i = 5°\mathrm{C}$, $T = -15°\mathrm{C}$, $T_\infty = -30°\mathrm{C}$, $C_{\mathrm{PI}} = 2.5\,\mathrm{kJ/(kg\,K)}$, $C_{\mathrm{pu}} = 3.52\,\mathrm{kJ/(kg\,K)}$, moisture content = 74.5% wet basis, $\Delta H = 333.22(0.745) = 248.25 = \mathrm{kJ/kg}$, m.c. = 74.5%, $k_{\mathrm{I}} = 1.108(\mathrm{W/mK})$, $T_{\mathrm{F}} = -1.75°\mathrm{C}$, $t = ?$

Solution:

$$N_{\mathrm{Bi}} = \frac{h_c a}{k_{\mathrm{I}}} = \frac{30(0.125)}{1.108} = 3.3845$$

$$N_{\mathrm{Ste}} = \frac{C_{\mathrm{PI}}(T_{\mathrm{F}} - T_\infty)}{\Delta H_{\mathrm{ref}}} = \frac{2.05(-1.75 + 30)}{248.25} = 0.234$$

$$N_{\mathrm{PK}} = \frac{C_{\mathrm{pu}}(T_i - T_{\mathrm{F}})}{\Delta H_{\mathrm{ref}}} = \frac{3.52(5 + 1.75)}{248.25} = 0.0955$$

$$T_3 = 1.8 + 0.263T + 0.105T_\infty = 1.8 + 0.263(-15) + 0.105(-30) = -5.295$$

$$\Delta H_1 = C_{\mathrm{pu}}(T_i - T_3) = 3520(5 + 5.295) = 36238.4\,\mathrm{J/kg}$$

$$\Delta H_2 = \Delta H + C_{\mathrm{PI}}(T_3 - T) = 248250 + 3520(-5.295 + 15) = 282411.6\,\mathrm{J/kg}$$

$$\Delta T_1 = \frac{T_i - T_3}{2} - T_\infty = \frac{(5 + 5.295)}{2} - (-30) = 35.1475$$

$$\Delta T_2 = T_3 - T_\infty = -5.295 + 30 = 24.705$$

The freezing time (t_{slab}) is given by

$$t_{\mathrm{slab}} = \frac{\rho R}{h_c}\left[\frac{\Delta H_1}{\Delta T_1} + \frac{\Delta H_2}{\Delta T_2}\right]\left[1 + \frac{N_{\mathrm{Bi}}}{2}\right] = \frac{1050(0.125)}{30}\left[\frac{36238.4}{35.1475} + \frac{282411.6}{24.705}\right]\left(1 + \frac{3.3845}{2}\right)$$

$$= 146789.45\,\mathrm{s} = 40.775\,\mathrm{h}$$

$$\beta_1 = \frac{A}{\pi R^2} = \frac{0.25(0.6)}{\pi(0.125)^2} = 3.056$$

$$\beta_2 = \frac{V}{\beta_1\left(\frac{4}{3}\pi R^3\right)} = \frac{0.25(0.6)(1)}{\frac{4}{3}\pi(3.056)(0.125)^3} = 6.0$$

Actual freezing time is given by

$$t_{\text{ellipsoid}} = \frac{t_{\text{slab}}}{E}$$

$$E = 1 + \frac{1 + 2/N_{\text{Bi}}}{\beta_1^2 + 2\beta_1/N_{\text{Bi}}} + \frac{1 + 2/N_{\text{Bi}}}{\beta_2^2 + 2\beta_2/N_{\text{Bi}}}$$

$$= 1 + \frac{1 + 2/3.3845}{3.056^2 + 2(3.056)/3.3845} + \frac{1 + 2/3.3845}{6^2 + 2(6)/3.3845} = 1.4939$$

Therefore, $t = 40.775/1.4939 = 27.294$ h.

IV. THAWING TIME PREDICTION

Although thawing is the opposite process of freezing, the earlier equations on freezing time prediction cannot be readily applied to thawing process. The thawing time is given for thawing to $T_F = 0°C$, and can be calculated by one of the following methods.

1. Power law approach to modifying Plank's equation as proposed by Calvelo [26] and Cleland [28]: This and other methods are valid for the following ranges: $0.6 < N_{\text{Bi}} < 57.3$, $0.08 < N_{\text{Ste}} < 0.77$, and $0.06 < N_{\text{PK}} < 0.27$.

$$t_{\text{slab}} = \frac{1.4921\rho C_{\text{pu}} a^2}{k_u}\left[\frac{0.5}{N_{\text{Bi}} N_{\text{Ste}}} + \frac{0.125}{N_{\text{Ste}}}\right]^{1.0248} N_{\text{Ste}}^{0.2712} N_{\text{PK}}^{0.061} \quad (6.40)$$

where

$$N_{\text{Bi}} = \frac{h_c a}{k_I} \quad (6.41)$$

$$N_{\text{Ste}} = \frac{C_{\text{PI}}(T_\infty - T_F)}{\Delta H_{10}} \quad (6.42)$$

$$N_{\text{PK}} = \frac{C_{\text{pu}}(T_F - T_i)}{\Delta H_{10}} \quad (6.43)$$

Here ΔH_{10} is the enthalpy change for the temperature change from 0 to $-10°C$.

2. Linear correction [23]:

$$t = \frac{\rho C_{\text{pu}} a^2}{k_I E}\left[\frac{P}{N_{\text{Bi}} N_{\text{Ste}}} + \frac{R}{N_{\text{Ste}}}\right] \quad (6.44)$$

$$P = 0.5[0.7754 + 2.2828 N_{\text{Ste}} N_{\text{PK}}] \quad (6.45)$$

$$R = 0.125[0.4271 + 2.1220 N_{\text{Ste}} - 1.4847 N_{\text{Ste}}^2] \quad (6.46)$$

3. Three-stage calculation method [13]:

$$t = \frac{\rho}{E}\sum_{i=1}^{3} \Delta H_i a \frac{(1 + ha/4k_I)}{2\Delta T_i h_c} \quad (6.47)$$

where

$$\Delta H_1 = C_{pu}(T_{f,ave} - T) \tag{6.48}$$

$$\Delta T_1 = T_\infty - \frac{(T_i + T_{f,ave})}{2}, \quad k_1 = k_I \tag{6.49}$$

$$\Delta H_3 = C_{pu}(T_{ave} - T_{f,ave}) \tag{6.50}$$

$$\Delta T_3 = T_\infty - \frac{(T_{ave} + T_{f,ave})}{2}, \quad k_3 = k_u \tag{6.51}$$

$$\Delta H_2 = \Delta H_f \tag{6.52}$$

$$\Delta T_2 = T_\infty - T_{f,ave}, \quad k_2 = 0.25k_I + 0.75k_u \tag{6.53}$$

$$T_{f,ave} = T_F - 1.5 \tag{6.54}$$

$$\Delta T_{ave} = T - \frac{(T - T_\infty)}{2 + 4/N_{Bi}}, \tag{6.55}$$

4. Correction of Plank's equation [13]:

$$t_t = \frac{\rho C_{pu} a^2}{k_u E}\left[\frac{1}{2N_{Bi}N_{Ste}} + \frac{1}{8N_{Ste}}\right]\left[0.8941 - \frac{0.0244}{N_{Ste}} + \frac{0.6192N_{PK}}{N_{Bi}}\right] \times \left[1 + \frac{C_{pu}\,(T_{ave} - T)}{\Delta H_{10}}\right] \tag{6.56}$$

IV. CONCLUSIONS

Many equations and models have been suggested to calculate freezing time of foods. Whenever a freezing time prediction method is used, some inaccuracy will be inevitable. This may arise from one of the three sources: (a) inaccuracy in thermal data; (b) inaccurate knowledge of freezing conditions, particularly the surface heat transfer coefficient; and (c) inaccuracy arising from assumptions made in the derivation of the prediction equation. The best freezing time prediction method will be the one in which the error arising from the category (c) is the least. The method should require as few input data as possible, and preferably should avoid lengthy or complex operations or reference to grasp and table. Three important parameters affecting the freezing time prediction are L_v, h_c, and D. The parameter h_c is the most difficult one to measure accurately, and therefore, is a major source of error.

NOMENCLATURE

A	smallest cross-sectional area that incorporate R (m^2)
A	surface area (m^2)
C_{PI}	specific heat of the frozen product (J/(kg K))
C_{ps}	specific heat of solids (J/(kgK))
C'_{pu}	specific heat of unfrozen water below T_F (J/(kg K))
C_{pu}	specific heat of unfrozen product (J/(kg K))

E shape factor
h_c convective heat transfer coefficient (W/(m^2 K))
ΔH enthalpy change due to freezing = (moisture content) (333220 J/kg)
ΔH freezing load or enthalpy change (J/kg)
ΔH_I sensible heat removed from the frozen water (J/kg)
ΔH_L enthalpy change due to latent heat (J/kg)
ΔH_s enthalpy change of product solids (J/kg)
ΔH_u sensible heat removed from unfrozen water (J/kg)
ΔH_{10} enthalpy change for the temperature change from 0 to −10°C J/kg)
k_I thermal conductivity of frozen product (W/(m, K))
k_u thermal conductivity of unfrozen product (W/(m, K))
k_W thermal conductivity of unfrozen food (W/(m, K))
L_v latent heat of freezing (J/kg)
M_I mass of ice or unfrozen water (kg)
M_S mass of solids (kg)
M_u mass of unfrozen water (kg)
N_{Bi} Biot number
N_{PK} Plank number
N_{Ste} Stefan number
R characteristic dimension (m), that is radius of cylinder of sphere or half thickness, of slab or other geometries
t_{slab} freezing time (s)
T final frozen product temperature (°C)
T_F initial freezing temperature (°C)
T_i initial food temperature (°C)
T_{ref} reference temperature (°C)
T_S surface temperature (°C)
T_∞ ambient temperature (°C)
V volume (m^3)
X thickness of frozen food (m)
ρ product density (kg/m^3)

REFERENCES

1. DJ Cleland, KJ Valentas. Prediction of freezing time and design of food freezers. In: KJ Valentas, E Rotstein, RP Singh, Eds. *Handbook of Food Engineering Practice*. Boca Raton, FL: CRC Press, 1997, pp. 71–124.
2. L Riedel. Calorimetric investigations of the freezing of fish meat. *Kaltetechnik* 8 (12):374–377, 1956.
3. L Riedel. Calorimetric investigations of the meat freezing process. *Kaltetechnik* 9:38–40, 1957.
4. IIR, *Recommendations for the Processing and Handling of Frozen Foods*. International Institute of Refrigeration, Paris, France, 1972.
5. EJ Rolfe. The chilling and freezing of foodstuffs. In: N Blakebrough, Ed., *Biochemical and Biological Engineering Science*, Vol. 2, London, UK: Academic Press, 1968, pp. 137–208.
6. K Hayakawa. Estimation of heat transfer during freezing or defrosting of food. In: *Freezing, Frozen Storage, and Freeze Drying, Bulletin of the International Institute of Refrigeration*. 1:293–301, 1977.
7. YC Hung. Prediction of cooling and freezing times. *Food Technology* 44 (5):137–153, 1990.
8. HS Ramaswamy, MA Tung. Review on predicting freezing times of foods. *Journal of Food Process Engineering* 7:169–203, 1984.

9. R Plank. Beitrage zur Berechrung und Bewertung der Gefrigeschwindikeit von Lebensmittelm zeitschrift fur die gesamte kalte Industrie. *Beih Rcihe* 3 (10):1–16, 1941.
10. AK Fleming. Immersion freezing small meat products. In: *Proceedings of the 12th International Congress of Refrigeration*. Madrid, 2:683–694, 1967.
11. RH Mascheroni, A Calvelo. A simplified model for freezing time calculation in foods. *Journal of Food Science* 47:1201–1207, 1982.
12. QT Pham. An approximate analytical method for predicting freezing times for rectangular blocks of food stuffs. *International Journal of Refrigeration* 8:43–47, 1985.
13. QT Pham. Extension to Plank's equation for predicting freezing times of foodstuffs of simple shapes. *International Journal of Refrigeration* 7:377–383, 1984.
14. AC Cleland, RL Earle. A comparison of analytical and numerical methods of predicting the Freezing times of foods. *Journal of Food Science* 42:1390–1395, 1977.
15. AC Cleland, RL Earle. Predicting freezing times of food in rectangular packages, *Journal of Food Science* 44:964–970, 1979.
16. YC Hung, DR Thompson. Freezing time prediction for slab shape foodstuffs by an improved analytical method. *Journal of Food Science* 48:555–560, 1983.
17. J Succar, K Hayakawa. Parametric analysis for predicting freezing time of infinitely slab shaped food. *Journal of Food Science* 49:468–477, 1984.
18. C Lacroix, F Castaigne. Simple method for freezing time calculations for infinite flat slabs, infinite cylinders and spheres. *Canadian Institution of Food Science and Technology Journal* 20:251–259, 1987.
19. QT Pham. Simplified equations for predicting the freezing times of foodstuffs. *Journal of Food Technology* 21:209–219, 1986.
20. AJ Ede. The calculation of the freezing and thawing of foodstuffs. *Modern Refrigeration* 52:52–55, 1949.
21. J Nagaoka, S Takagi, S Hotani. Experiments on the freezing of fish by air blast freezer. *Journal of Tokyo University of Fisheries* 42 (1):65–73, 1956.
22. FL Levy. Calculating freezing time of fish in air blast freezers. *Journal of Refrigeration* 1:55–58, 1958.
23. AC Cleland, R L Earle. Freezing time prediction for different final product temperatures. *Journal of Food Science* 49:1230–1232, 1984.
24. DJ Cleland, AC Cleland, RL Earle. Prediction of freezing and thawing times for multidimensional shapes by simple formulae. I. Regular shapes. *International Journal of Refrigeration* 10:156–164, 1987.
25. DJ Cleland, AC Cleland, RL Earle. Prediction of freezing and thawing times for multidimensional shapes by simple formulae. II. Irregular shapes. *International Journal of Refrigeration* 10:234–240, 1987.
26. A Calvelo. Recent studies on meat freezing. In: R Lawrie, Ed., *Developments in Meat Science–2*. London: Applied Science, 1981, pp. 125–158.
27. MdM Hossain, DJ Cleland, AC Cleland. Prediction of freezing and thawing times for foods of 3-dimensional irregular shape by using a semi-analytical geometric factor. *International Journal of Refrigeration* 15:241–246, 1992.
28. AC Cleland, *Food Refrigeration Processes – Analysis, Design and Simulation*. New York: Elsevier Applied Science, 1990, pp. 79–152.

7 Mathematical Modeling of Freezing Processes

Q. Tuan Pham
School of Chemical Engineering and Industrial Chemistry, University of New South Wales, Sydney, Australia

CONTENTS

I. INTRODUCTION

Practical food freezing is a complex problem, involving several simultaneous physical phenomena: heat transfer, mass transfer, nucleation, crystal growth, volume change, mechanical strains, and stresses. Analytical methods can only deal with a few idealized cases, and for the vast majority of situations, some numerical model must be used.

The mathematical modeling of food freezing and associated phenomena poses special challenges. Around the freezing point, there are large and sudden variations in thermophysical properties such as specific heat and thermal conductivity. This leads to a highly nonlinear partial differential equation (PDE), which is difficult to solve. In complex-shaped objects, the progress of the freezing front can be highly unpredictable. Freezing is also associated with volume change, mass transfer (in the form of moisture diffusion, water vapor diffusion, and perhaps solute diffusion, in the case of immersion freezing of unwrapped foods), stress and cracking, cellular dehydration, supercooling, ice nucleation and propagation phenomena, which are only now beginning to be studied in detail.

II. GOVERNING EQUATIONS

First, we consider the simple case of a water-containing solid food wrapped in an impervious skin, undergoing refrigeration. At each point in the solid, the *Fourier heat conduction equation* applies [1]:

$$\rho c \frac{\partial T}{\partial t} = \nabla(k\nabla T) + q \tag{7.1}$$

with initial condition

$$T_{(t=0)} = T_0(\mathbf{r}) \tag{7.2}$$

On each part of the solid surface, either the temperature may be specified (*Dirichlet boundary condition*):

$$T = f(\mathbf{r}, t) \tag{7.3}$$

or the temperature gradient, and hence heat flux, may be given (*Neumann boundary condition*):

$$\frac{\partial T}{\partial n} = f(\mathbf{r}, t, T) \tag{7.4}$$

where $\mathbf{n}$ is the normal outward unit vector. In Equation (7.1), the heat generation term q is zero in most if not all, freezing applications but not necessarily in thawing operations (e.g., in microwave thawing).

III. METHODS FOR DISCRETIZATION OF THE PDE

The numerical solution of the PDE governing heat flows involves two steps: discretizing the space domain to obtain a set of ordinary differential equations (ODEs) relating a finite number of nodal temperatures and then solving this set of ODEs. It will be seen that these ODEs can be written in matrix form as:

$$\mathbf{C}\frac{\mathrm{d}\mathbf{T}}{\mathrm{d}t} + \mathbf{KT} = \mathbf{f} \tag{7.5}$$

where $\mathbf{T}$ is the vector of nodal temperatures, $\mathbf{C}$ the *capacitance matrix* (containing the specific heat c), $\mathbf{K}$ the *conductance matrix* (containing the thermal conductivity k), and $\mathbf{f}$ the *forcing matrix* (containing terms arising from heat generation or heat fluxes from boundaries). There are three commonly used methods for discretizing space: finite difference (FDM), finite element (FEM), and finite volume (FVM).

A. Finite Difference Method

FDM is the oldest of the three discretization methods (although FVM may have been used informally earlier) and is the most convenient and efficient for problems involving simple geometries such as slabs, cylinders, or brick-shaped objects. For shapes that do not deviate very much from a regular geometry, the use of a boundary fitted orthogonal grid [2] extends the use of FDM, which is much faster than the unstructured meshes of FEM or FVM. FDM involves superimposing a grid of structured (arrayed) *nodes* on the calculation domain and calculating the space derivatives of temperature around these nodes by *central differences.*

For example, for one-dimensional heat transfer across the thickness of a slab, Equation (7.1) can be written as:

$$\rho c \frac{\partial T}{\partial t} = \frac{\partial}{\partial x}\left(k \frac{\partial T}{\partial x}\right) + q \tag{7.6}$$

If an array of nodes with spacing Δx is imposed on the slab, the temperature gradient between nodes i and $i+1$ can be written as:

$$\left(\frac{\partial T}{\partial x}\right)_+ = \frac{T_{i+1} - T_i}{\Delta x} \tag{7.7}$$

Similarly, for the segment between nodes $i-1$ and i,

$$\left(\frac{\partial T}{\partial x}\right)_- = \frac{T_i - T_{i-1}}{\Delta x} \tag{7.8}$$

Equation (7.1) then becomes

$$\rho c \frac{\partial T_i}{\partial t} = \frac{k_+(T_{i+1} - T_i) - k_-(T_i - T_{i-1})}{\Delta x^2} + q \tag{7.9}$$

or

$$\rho c \frac{\partial T_i}{\partial t} = \frac{k_- T_{i-1} - (k_- + k_+)T_i + k_+ T_{i+1}}{\Delta x^2} + q \tag{7.10}$$

When all such equations are written down for nodes 1, 2, and so on, we obtain the matrix Equation (7.5), where **C** is the diagonal matrix containing terms such as ρc, **K** the (tridiagonal) matrix containing terms such as $-k_-/\Delta x^2$, $(k_- + k_+)/\Delta x^2$, and $-k_+/\Delta x^2$, and **f** the forcing matrix containing terms arising from heat generation or heat fluxes from boundaries.

With FDM, there are problems in discretizing a Neumann boundary condition. The surface flux applies at the surface node itself, whereas the flux on the inner side of the surface node applies at a point $\Delta x/2$ away, causing an asymmetric situation. For similar reasons, if unequally spaced grids are used, it is important to ensure that the changes in spacing are gradual and that a sufficiently large number of nodes is used to minimize the inaccuracy. Another practical problem is that with very high heat transfer coefficients, the temperature of the surface nodes may be subjected to instabilities due to the excessive surface heat flux.

The boundary condition is handled better by a *control-volume* formulation of finite differences, which implements the conservation law rigorously at all nodes. In this version, the solid is first divided into control volumes of thickness Δx (equal volumes are being considered for simplicity, although this is not a requirement). The nodes are placed in the middle of each control volume

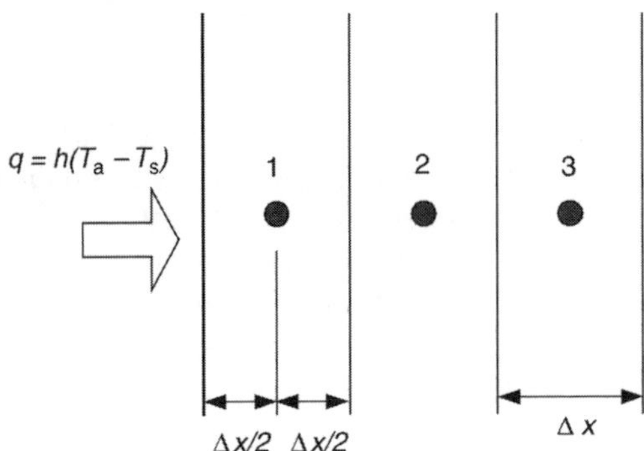

FIGURE 7.1 Control volume discretization of a slab.

(Figure 7.1), with the result that the boundary node is not at the surface boundary but some distance into the solid. The thermal resistance between the surrounding and node 1 is $1/h + (\Delta x/2/k)$. The boundary condition is taken into account by the thermal balance of the first control volume:

$$\rho c \frac{\partial T_1}{\partial t} = \frac{k_+((T_2 - T_1)/\Delta x) - (T_1 - T_a)/(1/h + \Delta x/2k_-)}{\Delta x/2} + q \tag{7.11}$$

where $\Delta x/2$ is the distance from the surface to node 1. This restores second-order accuracy to the (near-)boundary nodes. For the Dirichlet boundary condition, $1/h$ is set to 0. T_1 represents the temperature at a point below the surface, hence the surface temperature, if needed, must be calculated by interpolation between T_a and T_1.

The one-dimensional formulation mentioned earlier can be extended to infinite cylinders and spheres (including hollow ones), as in these geometries the heat equation can be written as:

$$\rho c \frac{\partial T}{\partial t} = \frac{1}{r^n} \frac{\partial}{\partial r}\left(k r^n \frac{\partial T}{\partial r}\right) + q \tag{7.12}$$

where $n = 1$ for cylinders and $n = 2$ for spheres. In the control-volume formulation, the nodes can be visualized as being leek- or onion-like layers.

B. Time Stepping

Having obtained a set of ODEs in time relating the nodal temperatures (7.5), solution will proceed in a series of time steps starting from the known initial conditions:

$$\mathbf{C} \frac{\mathbf{T}^{\text{New}} - \mathbf{T}}{\Delta t} = -\mathbf{K}\bar{\mathbf{T}} + \bar{\mathbf{f}} \tag{7.13}$$

On the right-hand side of Equation (7.13), some time-averaged value of the nodal temperatures must be used, such as:

$$\bar{\mathbf{T}} = \alpha \mathbf{T}^{\text{New}} + (1 - \alpha)\mathbf{T} \tag{7.14}$$

where the superscript "New" refers to temperatures at the end of the time step. α is a parameter varying between 0 and 1. For $\alpha = 0$ (the Euler method or explicit method), the last computed (known) temperature field T_i is used to calculate the gradients. The nodal temperatures can therefore be easily computed one by one. For example, for the internal nodes in an equally spaced grid:

$$T_i^{\text{New}} = T_i + \frac{k_+(T_{i+1} - T_i) - k_-(T_i - T_{i-1})}{\rho c \Delta x^2}\Delta t + \frac{q}{\rho c}\Delta t \tag{7.15}$$

Although easy to implement (and therefore recommended for one-off programs or small problems), the Euler method is only first-order accurate and is unstable for

$$\frac{k\Delta t}{\rho c \Delta x^2} > \frac{1}{2} \tag{7.16}$$

In two and three dimensions, the restriction is more stringent still: the upper limit becomes $\frac{1}{4}$ and $\frac{1}{6}$, respectively. Even when stable, the Euler method rapidly loses accuracy as Δt increases. The critical time interval decreases as the square of the space interval Δx, hence the solution is very time consuming when the domain is finely divided.

For $\alpha = 1$ (*backward difference*), the new (and yet unknown) temperatures are used to compute the gradients. For $\alpha = 0.5$ (central difference or *Crank–Nicholson*), the arithmetic mean of old and new is used [3]. The latter is the most popular method in terms of its combination of unconditional stability and second-order accuracy. Even then, the Crank–Nicholson scheme will be subject to large, slow decaying oscillations when $k\Delta t/(\rho c \Delta x^2)$ is large.

For all values of $\alpha \neq 0$, unknown nodal temperatures appear on both sides of Equation (7.13) and this set of equations has to be rearranged to bring all the unknown temperatures to the left-hand side. Fortunately, because each equation involves only three neighboring temperatures T_{i-1}, T_i, and T_{i+1}, the resulting matrix equation is tridiagonal and can be easily solved by the *tridiagonal matrix algorithm* [4].

Another popular time-stepping procedure in the food freezing literature is the *Lee's three-level scheme* [5]: the temperature gradients are calculated from the mean of the temperatures at the present time step, the previous time step, and the next time step. The resulting heat accumulation is used to calculate the temperature change between the previous time step and the next time step (instead of between the present time step and the next time step). In theory, this will allow rapid property changes to be better accounted for; however, the author has not found any advantage of this scheme over the Crank–Nicholson scheme [6].

In regular shapes with two or three dimensions (finite cylinders, rods, or brick shape), the FDM is implemented by applying an orthogonal grid. For any time-stepping method except Euler's, the resulting set of equations is no longer tridiagonal as the heating rate of each node is influenced by all the temperatures around it (four neighboring nodes in two dimensions and six in three dimensions). Nowadays, this is not a problem with poweful computer hardware and software that can solve very large matrix equations, but it is still time consuming to do so. Instead, the *alternating direction method* is often used [7,8]. This method involves sweeping in each of the x, y, and z directions independently in a series of pseudo-one-dimensional solutions. An example in two dimensions using the Crank–Nicholson scheme illustrates the procedure (Figure 7.2). In the first (x) sweep, we consider one row of nodes at a time and calculate a set of intermediate temperatures $\mathbf{T}^*$, using a time step of $\Delta t/2$. The (unknown) intermediate values $\mathbf{T}^*$ are used to express temperature gradients in the x direction and the (known) present values $\mathbf{T}$ to express temperature gradients in the y direction:

$$\mathbf{T}_{ij}^* = T_{ij} + \frac{\Delta t}{2}\left(\partial_x \mathbf{T}^* + \partial_y T + \frac{q}{\rho c}\right) \tag{7.17}$$

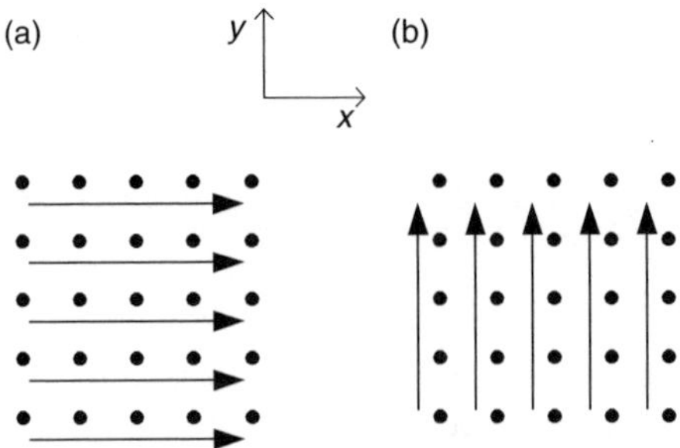

FIGURE 7.2 Alternating direction method: (a) sweeps in x direction and (b) sweeps in y direction.

where $\partial_x T$ is shorthand for $(k_+(T_{i+1,j} - T_{i,j}) - k_-(T_{i,j} - T_{i-1,j}))/\rho c \Delta x^2$ (the heating rate due to heat flow along the x direction) and similarly for $\partial_y T$. It is to be noted that the heat fluxes in the x direction are computed using the intermediate temperature field (i.e., backward differences), whereas those in the y direction are computed using the existing temperature field (i.e., explicit). This yields, for each row, a tridiagonal matrix equation that can be solved for $\mathbf{T}^*$. In the second (y) sweep, we consider one column of nodes at a time and write down the discretized Fourier equation for the next $\Delta t/2$, using the (unknown) new values to express temperature gradients in the y direction and the (known) intermediate values $\mathbf{T}^*$ to express temperature gradients in the x direction:

$$\mathbf{T}_{ij}^{\text{New}} = \mathbf{T}_{ij}^* + \frac{\Delta t}{2}\left(\partial_x \mathbf{T}^* + \partial_y \mathbf{T}^{\text{New}} + \frac{q}{\rho c}\right) \tag{7.18}$$

which again yields a tridiagonal matrix equation for each column that can be solved for $\mathbf{T}^{\text{New}}$.

C. Finite Element Method

For shapes that cannot be represented by a regular orthogonal grid, FEM and FVM are more flexible than FDM. In FEM, the object is divided into elements, which share certain nodes. Within each element, the temperature field at $\mathbf{x}$ is approximated by interpolation:

$$T(\mathbf{x}, t) = \mathbf{N}^{\text{T}}(\mathbf{x})\mathbf{T}_{\mathbf{N}}(t) \tag{7.19}$$

where $\mathbf{T}_{\mathbf{N}}(t)$ are the vector of the temperatures at the nodes and $\mathbf{N}(\mathbf{x})$ is the vector of "shape functions." The shape functions are position-dependent factors, that allow the (approximate) temperature at each location within the element to be found by interpolation between the temperatures at the nodes. For example, in a linear one-dimensional element (a segment enclosed by two nodes), the temperature T at any point P in the element is obtained by linearly interpolating between the vertices A and B:

$$T = (1 - \xi)T_A + \xi T_B \tag{7.20}$$

hence

$$\mathbf{N} = \begin{pmatrix} 1 - \xi \\ \xi \end{pmatrix} \tag{7.21}$$

where $\xi = AP/AB$ is the relative distance from point P to node A (Figure 7.3). It is easy to see that N_i must be 1 at node i itself and that $\Sigma N_i = 1$ everywhere in the element.

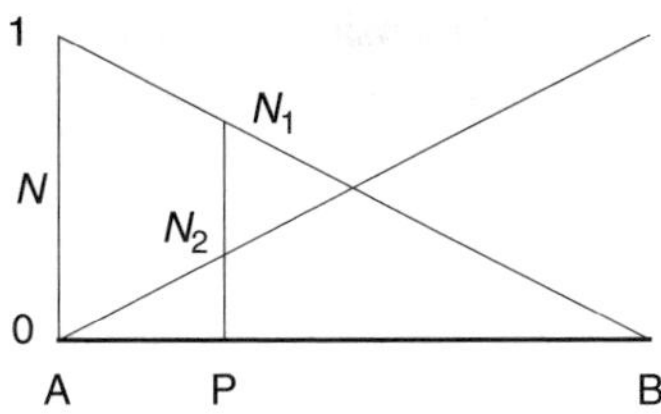

FIGURE 7.3 Shape functions in a linear one-dimensional element.

Because the temperature field is only approximate, the heat flow equation (7.1) will generally not hold exactly at every point in the element, but it is reasonable to require that energy would be conserved over the element as a whole (i.e., energy is conserved "on average"). This could be done by integrating the *residual* $\rho c(\partial T/\partial t) - \nabla(k\nabla T) - q$ over the whole element and setting it to zero, but that would give only one equation, which is not enough to solve for all the nodal temperatures. We need as many equations as there are nodes. This is obtained by requiring that the integrated *weighted residuals* should also be zero when the residual is weighted toward each node (by some function which is maximal at the node and decreases gradually with distance). In the *Galerkin FEM*, the shape functions, which have this property, are also used as weighting functions:

$$\int_{\Omega} \mathbf{N}\left[\rho c \frac{\partial T}{\partial t} - \nabla(k\nabla T) - q\right] \mathrm{d}\Omega = 0 \tag{7.22}$$

As $\Sigma N_i = 1$ at all points in the element, by summing the terms of the earlier vector equation, it will be seen if Equation (7.22) is obeyed, energy will be conserved over the element as a whole as well. Substituting Equation (7.19) into Equation (7.22) and solving it yields a relationship between the nodal temperatures which is of the following form:

$$\mathbf{C}^{(e)} \frac{\mathrm{d}\mathbf{T}^{(e)}}{\mathrm{d}t} + \mathbf{K}^{(e)}\mathbf{T}^{(e)} = \mathbf{f}^{(e)} \tag{7.23}$$

where

$$\mathbf{C}^{(e)} = \int_{\Omega} \rho c \mathbf{N}\mathbf{N}^{\mathrm{T}}\, \mathrm{d}\Omega \tag{7.24}$$

and (for a convective boundary condition)

$$\mathbf{B} = \nabla^{\mathrm{T}}\mathbf{N} = \left[\frac{\partial \mathbf{N}}{\partial x} \frac{\partial \mathbf{N}}{\partial y} \frac{\partial \mathbf{N}}{\partial z}\right] \tag{7.25}$$

$$\mathbf{K}^{(e)} = \int_{\Omega} k\mathbf{B}\mathbf{B}^{\mathrm{T}}\, \mathrm{d}\Omega + \int_{S} h\mathbf{N}\mathbf{N}^{\mathrm{T}}\, \mathrm{d}S \tag{7.26}$$

$$\mathbf{f}^{(e)} = \int_{S} hT_{\mathrm{a}}\mathbf{N}\, \mathrm{d}S \tag{7.27}$$

$\mathbf{T}^{(e)}$ is the vector of nodal temperatures, $\mathbf{C}^{(e)}$ the capacitance matrix, $\mathbf{K}^{(e)}$ the conductance matrix, and $\mathbf{f}^{(e)}$ the forcing vector (containing all terms that are independent of nodal temperatures, such as those arising from the boundary conditions and heat generation). Ω is the element domain and S its boundary. The superscript (e) indicates that this is a relationship between the nodes in one element only. For example, with the linear one-dimensional element, and assuming an internal element ($\mathbf{f} = 0$) with constant properties, substituting for $\mathbf{N}$ from Equation (7.21) into Equation (7.23) to Equation (7.27) gives:

$$\rho c l \begin{bmatrix} \frac{1}{3} & \frac{1}{6} \\ \frac{1}{6} & \frac{1}{3} \end{bmatrix} \frac{\partial}{\partial t} \begin{pmatrix} T_A \\ T_B \end{pmatrix} + \frac{k}{l} \begin{bmatrix} 1 & -1 \\ -1 & 1 \end{bmatrix} \begin{pmatrix} \bar{T}_A \\ \bar{T}_B \end{pmatrix} = 0 \tag{7.28}$$

or

$$\begin{aligned} \rho c l \left(\frac{1}{3} \frac{\partial T_A}{\partial t} + \frac{1}{6} \frac{\partial T_B}{\partial t} \right) &= k \frac{T_A - T_B}{l} \\ \rho c l \left(\frac{1}{6} \frac{\partial T_A}{\partial t} + \frac{1}{3} \frac{\partial T_B}{\partial t} \right) &= -k \frac{T_A - T_B}{l} \end{aligned} \tag{7.28a}$$

where l is the element's length. Details of how to obtain the matrices for more general situations can be found in the literature on FEM [9, 10].

Although the term on the right-hand side of Equation (7.28) is the same as that found in FDM, the capacitance terms on the left is less intuitively obvious and arise from the fact that in FEM, heat capacity is distributed over the element. Indeed, it has often been said (especially by engineers) that the FEM is an esoteric mathematical device, but a rough physical interpretation is possible and helps to understand how the method works. Each term of the $\mathbf{K}$ matrix, K_{ij}, represents the conductance between nodes i and j of the material within the element. If two elements share two nodes, there are two parallel conduction paths and the conductances from the elements can be added together. Within an element, the thermal energy $\rho c T$ at each point is in some sense attributed to the nodes according to the shape function, that is, more toward the nearest node and less toward the farthest. When the temperature at a node i is changed, this affects the temperature profile and hence the thermal energy throughout the element, and therefore, C_{ij} represents the effect of a change in T_i on the thermal energy attributed to node j.

In the *lumped capacitance* version of FEM, all the thermal energy change due to a change in T_i is attributed to node i itself, hence $\mathbf{C}$ becomes a diagonal matrix. For example, Equation (7.28) for a linear one-dimensional element becomes

$$\rho c l \begin{bmatrix} \frac{1}{2} & 0 \\ 0 & \frac{1}{2} \end{bmatrix} \frac{\partial}{\partial t} \begin{pmatrix} T_A \\ T_B \end{pmatrix} + \frac{k}{l} \begin{bmatrix} 1 & -1 \\ -1 & 1 \end{bmatrix} \begin{pmatrix} \bar{T}_A \\ \bar{T}_B \end{pmatrix} = 0 \tag{7.29}$$

or

$$\begin{aligned} \frac{\rho c l}{2} \frac{\partial T_A}{\partial t} &= k \frac{T_A - T_B}{l} \\ \frac{\rho c l}{2} \frac{\partial T_B}{\partial t} &= -k \frac{T_A - T_B}{l} \end{aligned} \tag{7.30}$$

which is more intuitively obvious. It happens to be identical with the finite-difference formulation, but the resemblance is only for the linear one-dimensional element. In other words, the mass of each element is assumed to be concentrated at the nodes instead of being distributed over the element. This formulation has some advantages over the Galerkin formulation in terms of simplicity and stability [11] and is particularly useful for dealing with the latent heat peak during freezing, as will be seen later.

With the earlier interpretation, it is easy to see how we can obtain we can obtain Equation (7.5) again, which relates all the n nodes in the domain considered. **C**, **K**, and **f** in Equation (7.5) are the global capacitance matrix, global conductance matrix, and global forcing factor, respectively, to distinguish them from the elemental variety. K_{ij} is obtained by adding up all the conductance terms relating nodes i and j from each elemental $\mathbf{K}^{(e)}$, and C_{ij} and f_i are obtained in a similar manner. Most terms, C_{ij} and K_{ij}, will be zero and only when nodes i and j are connected by one or more element will these terms be nonzero.

Equation (7.5) can then be solved by any of the time-stepping methods mentioned earlier for FDM: Euler, Crank–Nicholson, backward difference, Lee's three-level scheme, and so on. Because a sparse matrix equation has to be solved, FEM requires powerful solution procedures and is more time-consuming than FDM, which only has tridiagonal matrices to deal with. Note that there is no advantage in using the Euler method with the Galerkin FEM, as **C** is not a diagonal matrix and hence the nodal temperatures cannot be found one by one. In the lumped capacitance formulation of FEM, however, **C** becomes diagonal and thus the new temperatures can be calculated individually and explicitly.

As with all numerical approximations, care must be taken with the design of the FEM grid to maximize accuracy. The grid must be denser where thermal gradients are steep, and the shape of the elements should not be overly skewed or elongated. Nowadays, commercial FEM software can normally ensure this by automatic grid adaptation.

D. Finite Volume Method

Although the terminology is relatively new, FVM in some primitive form has been used by engineers for a long time (even before the days of computers) in view of its conceptual clarity. Indeed, every PDE used in engineering is derived by taking the infinitesimal limit of some finite-volume model. The underlying principle of FVM was also used in the control-volume version of FDM. Nevertheless, this chapter will use the term "control-volume FDM" rather than "FVM" for this submethod, because of the wide differences in applicability and speed. FDM uses orthogonal structured arrays and generates tridiagonal matrices, whereas FVM does not have to, and is thus applicable to irregular geometries. In FVM, the domain considered is divided into control volumes, each associated with a node at the center. The control volumes and nodes do not have to be in a regular array, and therefore, there is a great flexibility in dealing with complex shapes (as in FEM). The heat conservation equation is assumed to hold over each control volume as a whole:

$$\int_{\Omega}\left[\rho c\frac{\partial T}{\partial t} - \nabla(k\nabla T) - q\right]\mathrm{d}\Omega = 0 \tag{7.31}$$

Using the divergence theorem, the second term in the volume integral can be transformed into an integrated surface flux over the volume's boundaries:

$$\int_{\Omega}\left(\rho c\frac{\partial T}{\partial t} - q\right)\mathrm{d}\Omega - \int_{S}\mathbf{n}(k\nabla T)\mathrm{d}S = 0 \tag{7.32}$$

with dS representing a surface element and **n** the outward normal to it. The first term is the rate of enthalpy gain of the control volume and can be approximated by:

$$\int_{\Omega}\rho c\frac{\partial T}{\partial t}\mathrm{d}\Omega \approx \delta V\rho_{\mathrm{m}}c_{\mathrm{m}}\frac{\partial T_i}{\partial t} \tag{7.33}$$

where δV is the volume of the control volume, ρ_{m} its mean density, c_{m} its mean specific heat, and T_i is the temperature of the node associated with it.

The second term of Equation (7.32) is the sum of all the heat fluxes into the control volume through its boundaries. At each surface, the heat flux can be calculated from the (mean) temperature gradient normal to the surface. Various methods have been proposed to estimate this mean temperature gradient, all of which result in a linear expression involving the nodal temperatures in the vicinity of the surface in question. Thus, Equation (7.32) can be replaced by:

$$\delta V \rho c \frac{\partial T_i}{\partial t} = \sum_{j=1}^{N} k B_{ij} T_j + q \delta V \qquad (7.34)$$

where B_{ij} are coefficients, which depend on the nodal arrangement.

One such equation is available for each node i, and it can be easily seen that this procedure again yields Equation (7.5). As with FDM or lumped capacitance FEM (but unlike Galerkin FEM), in this case, **C** in Equation (7.5) is a diagonal matrix, which will present some advantages. Solution is by the same time-stepping techniques as for FDM and FEM.

IV. DEALING WITH CHANGES IN THERMAL PROPERTIES

A. Dealing with Latent Heat

1. Classification of Methods

In the "classical" freezing problem, also known as the *Stefan problem*, the freezing process is governed entirely by heat transfer. There is no retardation of freezing due to diffusion or nucleation phenomenon. The major issue in the numerical solution of this problem is in dealing with the large latent heat, which evolves over a very small temperature range. For a few idealized situations, analytical solutions are available such as that of Plank [12], but for all realistic situations, a numerical solution is necessary. Except where indicated, the techniques described here apply equally to FDM, FEM, and FVM.

Voller [13] gave a comprehensive review of special techniques developed to deal with phase change. These methods can be divided into *fixed grid methods* and *moving grid methods*. In the latter, the object is divided into a frozen zone and an unfrozen zone. Some nodes, element boundaries, or control-volume boundaries are put on the freezing front itself and allowed to move with it. The front is tracked by calculating its precise position at every time-interval, using the following heat-balance equation

$$[k\nabla T]_{\mathrm{u}}\mathbf{n} - [k\nabla T]_{\mathrm{f}}\mathbf{n} = W L_{\mathrm{V}} \mathbf{v}_{\mathrm{f}} \mathbf{n} \qquad (7.35)$$

where $\mathbf{v}_{\mathrm{f}}$ is the velocity of the freezing front and the subscripts u and f refer to the heat fluxes on the frozen and unfrozen sides of the front.

Moving grid methods can give precise, nonoscillating solutions for the temperature and ice-front position. However, they are less flexible than fixed grid methods because most foods do not have a sharp phase-change temperature but freeze gradually, hence it is not clear how the freezing front should be defined. With foods of complex shape, the calculation of the front's position and subsequent grid adjustment can be a complicated issue. Therefore, this chapter will concentrate on fixed grid methods. If desired, the freezing front can still be located in these methods by carrying out an interpolation to locate the position where the freezing temperature or enthalpy applies [14,15].

Fixed grid methods have the disadvantage that they produce unrealistic stepwise temperature histories and/or ice-front position, due to a whole control-volume freezing at the same time [13]. However, this is much less apparent in foods with their gradual phase change than in a pure substance, and the stepwise characteristic can in any case be minimized by choosing small space intervals or by some interpolation method.

Of the fixed methods, some researchers treat latent heat as a source term, separate from the specific heat, and include it in the term q in Equation (7.1) [16]. This approach is not suitable for most foods for which latent heat is evolved over a wide range of temperatures (icecream, butter, salted meats) and is thus hard to distinguish from sensible heat. Therefore, they will not be considered further here. The remaining methods can be classified into *apparent specific heat methods* and *enthalpy methods*. Generally, they are the most useful methods as they can deal with sharp as well as gradual phase-change. A sharp phase-change problem (such as the freezing of pure water) can be converted into gradual phase-change by the simple expedient of assuming that the phase change happens over a narrow temperature range, say between -0.001 and $+0.001$°C.

2. Apparent Specific Heat Methods

In the apparent specific heat methods, latent heat is merged with sensible heat to produce a specific heat curve with a large peak around the freezing point (Figure 7.4). Because of the large variations in specific heat, an iteration must be carried out at every step: the specific heat at each node is estimated (say from the present temperatures) and used to calculate the **C** matrix, Equation (7.13) is solved for the new nodal temperatures, the mean temperature over the most recent time step is calculated, then the specific heat is re-estimated from the specific heat–temperature relationship, and so on. It is difficult to obtain convergence with this technique, and there is always a chance that the latent heat is underestimated ("latent heat peak jumping"). This happens when the temperature at a node steps over the peak in the apparent specific heat curve: the mean specific heat between the initial and final temperatures is then always less than the peak, and the temperature change will therefore be overestimated. For this reason, the apparent specific heat method is not recommended.

Many approximate methods have been proposed to calculate the effective specific heat near the freezing point [17–21] but none is entirely satisfactory [22]. For this reason, the enthalpy formulation has become more popular.

3. Enthalpy Methods

The basic conduction equation, Equation (7.1), is written in the form:

$$\rho \frac{\partial H}{\partial t} = \nabla(k\nabla T) + q \tag{7.36}$$

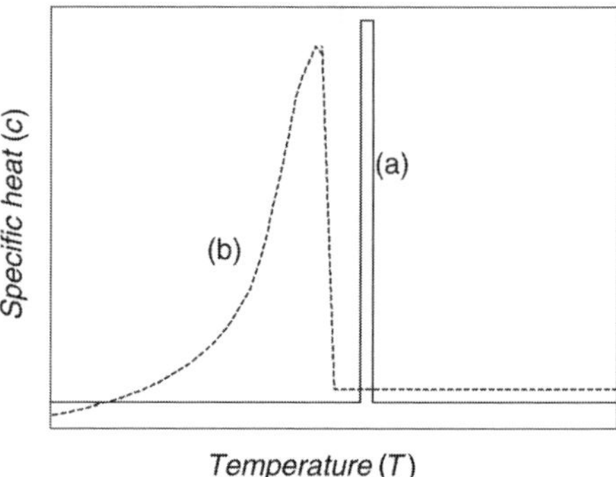

FIGURE 7.4 Apparent specific heat for (a) a material with sharp phase-change and (b) a material with gradual phase-change.

where H is the (specific) enthalpy, defined by

$$H \equiv \int_{T_{\mathrm{REF}}}^{T} c_{\mathrm{p}}\, \mathrm{d}\theta \tag{7.37}$$

and T_{REF} is an arbitrarily chosen reference temperature. After the usual FDM, FEM, or FVM manipulations, we obtain the matrix equation:

$$\mathbf{M}\frac{\mathrm{d}\mathbf{H}}{\mathrm{d}t} + \mathbf{KT} = \mathbf{f} \tag{7.38}$$

where $\mathbf{M}$ is the *mass matrix*, which remains constant with time, and $\mathbf{H}$ is the vector of nodal enthalpies.

To apply the enthalpy method, the functional relationship $T(H)$ must be available and programmed into the computer. In FDM, FVM, and lumped capacitance FEM, the solution of Equation (7.38) by Euler's method (explicit time stepping) is very simple, as $\mathbf{M}$ is a diagonal matrix, and Equation (7.38) becomes

$$H_i^{\mathrm{New}} = H_i + \frac{\delta t}{M_{ii}}\left[\sum_{j=1}^{N}(K_{ij}T_j) + f_i\right] \quad i = 1\text{–}N \tag{7.39}$$

where all the terms on the right are known present values. This method was first proposed for FDM by Eyres et al. [23]. The new nodal enthalpies are calculated one by one from the present temperature field, then the new nodal temperatures are calculated from the H–T relationship, and so on.

To obtain an exact solution to any implicit ($\alpha > 0$) solution of Equation (7.38), an iteration must be carried out at every time-step. The enthalpy change vector $\Delta\mathbf{H} \equiv \mathbf{H}^{\mathrm{New}} - \mathbf{H}$ over the present time step is iteratively adjusted until the residual vector $\mathbf{r} = \mathbf{M}(\Delta\mathbf{H}/\Delta t) + \overline{\mathbf{KT}} - \bar{\mathbf{f}}$ becomes zero to within an acceptable tolerance. A *successive substitution* scheme such as Gauss Seidel can be used: $\mathbf{H}^{\mathrm{New}}$ is calculated from $\mathbf{H} + \Delta t\mathbf{M}^{-1}(\overline{\mathbf{KT}} - \bar{\mathbf{f}})$, then $\mathbf{T}^{\mathrm{New}}$ is calculated from $\mathbf{H}^{\mathrm{New}}$, and so on. Convergence with this type of scheme tends to be very slow, and various over-relaxation schemes have been proposed [24]. A better approach is to use a Newton–Raphson iteration, where the following equation is solved iteratively for the enthalpy-change vector $\Delta\mathbf{H}$ (m being the iteration counter):

$$\mathbf{J}\Delta\mathbf{H}^{(m+1)} = \mathbf{J}\Delta\mathbf{H}^{(m)} - \mathbf{r}^{(m)} \tag{7.40}$$

$$\mathbf{J} = \frac{\mathrm{d}\mathbf{r}}{\mathrm{d}\mathbf{H}^{\mathrm{New}}} = \frac{\mathbf{M}}{\Delta t} + \mathbf{K}\frac{\partial\bar{\mathbf{T}}}{\partial\mathbf{H}^{\mathrm{New}}} - \frac{\partial\bar{\mathbf{f}}}{\partial\mathbf{H}^{\mathrm{New}}} \tag{7.41}$$

If the Crank–Nicholson scheme is used, where $\bar{\mathbf{T}} = (\mathbf{T} + \mathbf{T}^{\mathrm{New}})/2$, $\bar{\mathbf{f}} = (\mathbf{f} + \mathbf{f}^{\mathrm{New}})/2$ the Jacobian $\mathbf{J}$ becomes

$$\mathbf{J} = \frac{\mathbf{M}}{\Delta t} + \frac{1}{2}\left(\mathbf{K}\frac{\partial\mathbf{T}^{\mathrm{New}}}{\partial\mathbf{H}^{\mathrm{New}}} - \frac{\partial\mathbf{f}^{\mathrm{New}}}{\partial\mathbf{H}^{\mathrm{New}}}\right) \tag{7.42}$$

Other iterative solution methods are available [4].

The discretization of highly nonlinear problems such as the phase-change problem with Galerkin FEM poses serious difficulties from the fundamental point of view. The Galerkin approach assumes that temperature is distributed over the element according to the shape function, i.e., $T = \mathbf{N}^{\mathrm{T}}\mathbf{T}$, where $\mathbf{T}$ is the vector of nodal temperatures. As H is a nonlinear function of T, it

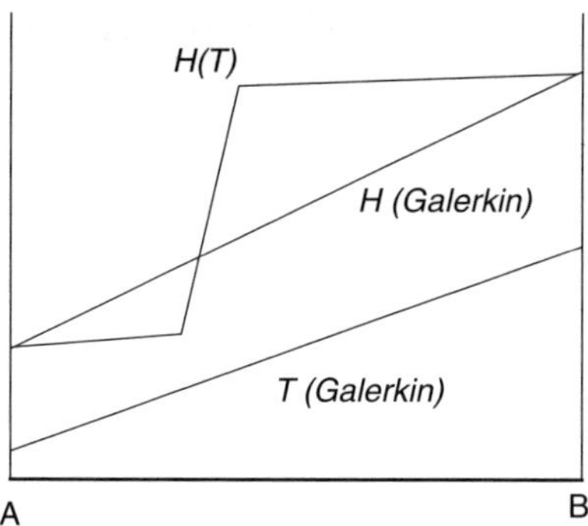

FIGURE 7.5 Temperature and enthalpy approximations in an one-dimensional element around the freezing point.

cannot be assumed that H is distributed according to $H = \mathbf{N}^T\mathbf{H}$ as well, where $\mathbf{H}$ is the vector of nodal enthalpies. In fact, this interpolation will be very inaccurate around the freezing point (see Figure 7.5 where the nodal temperatures in an one-dimensional element are just above and below the freezing point). However, in the enthalpy method, this assumption has to be made when transforming Equation (7.36) into a Galerkin FEM equation. In the effective specific heat method, the Galerkin FEM user is faced with how to calculate an effective specific heat over both time (Δt) and space (the element's domain): numerical averaging methods (which uses some sampling procedure over the element's domain) fail when $c(T)$ has a very sharp peak (Figure 7.5). For these reasons, the use of lumped capacitance FEM is highly recommended over Galerkin FEM.

4. Quasi-Enthalpy Method

Pham [25] proposed a simple correction to the specific heat formulation which, like enthalpy methods, is effective in dealing with the latent heat peak, but avoids the need for iteration. The method was first applied to FDM but was subsequently extended to lumped capacitance FEM [14] and to Galerkin FEM [17]. The method essentially consists of adding a specific heat estimation step and a temperature correction step.

(a) *Specific heat estimation step*: For each time step, the nodal enthalpy changes are first estimated from the incoming heat fluxes, using the present temperature field:

$$\frac{\mathbf{\Delta H} = \mathbf{M}^{-1}\mathbf{KT} - \mathbf{f}}{\mathbf{\Delta t}} \tag{7.43}$$

and an effective specific heat over the time interval can be estimated:

$$T_i^{\text{New}} = T(H_i + \Delta H_i) \tag{7.44}$$

$$c_i = \frac{\Delta H_i}{T_i^{\text{New}} - T_i} \tag{7.45}$$

These effective specific heats are then substituted into Equation (7.13), which is then solved once only per time step.

(b) *Temperature correction step*: To avoid latent heat peak jumping, the new nodal temperatures are then corrected according to

$$H_i^{\text{New}} = H(T_i) + c_i(T_i^{\text{New}} - T_i) \tag{7.46}$$

$$T_i^{\text{Corrected}} = T(H_i^{\text{New}}) \tag{7.47}$$

Further investigations [13,22] indicate that the temperature correction is the crucial step in this method, whereas the preliminary specific heat estimation step is of lesser importance. The physical basis of the temperature correction is illustrated in Figure 7.6. It can be seen from the last equation that this is basically an enthalpy method, because it is really the nodal enthalpy changes that are calculated at each time step. As in other enthalpy methods, the functional relationship $T(H)$ is needed.

Using well-known test problems, Pham [22] compared ten of the most advanced fixed-grid FEMs to date (after eliminating several others) in terms of accuracy, time interval for convergence, *heat balance error* (percentage difference between heat flows through boundaries and total heat gain of product — a measure of whether the latent heat load peak has been missed), and computing time as measured by the number of matrix inversions required. The test problems use both materials with a sharp phase change (heat released over 0.01 K) and a material with food-like properties. He concluded that the (noniterative) lumped capacitance FEM with Pham's quasi-enthalpy method performed best in terms of most of the earlier mentioned criteria. In his comprehensive 1996 review, Voller [13] also concluded that this method is an "excellent scheme" for fixed grids.

Because no iteration is involved, strict energy conservation is not attained in the above methods (heat fluxes are calculated using the uncorrected temperatures, whereas nodal enthalpy changes are based on the corrected temperature). This apparent defect can be turned to advantage in estimating the accuracy of the results: the overall heat balance (relative difference between the boundary flux, integrated over all boundaries and entire freezing time, and the total change in the heat content of the food) can serve as a useful indication of whether the time step is sufficiently small: a heat balance error of less than 1% generally indicates that convergence has been reached. *Iterative enthalpy methods* ensure a good heat balance at all time steps, but this does not guarantee an accurate solution because at large time intervals the nonlinearities are "smoothed over."

In summary, the methods mentioned here for dealing with latent heat can be classified and compared, as in Table 7.1.

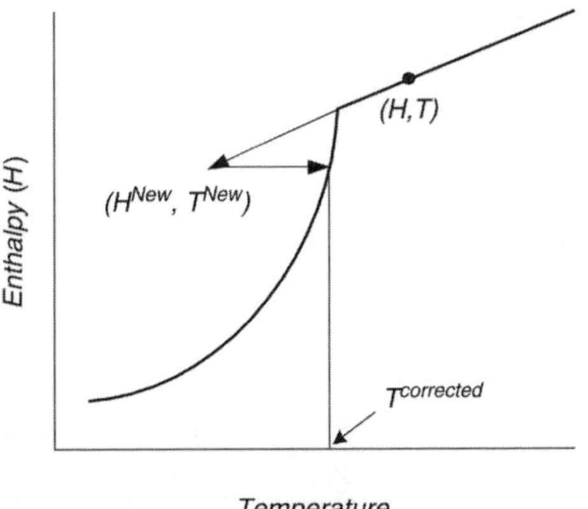

FIGURE 7.6 Illustration of Pham's temperature correction step.

TABLE 7.1
Summary of Methods for Dealing with Latent Heat

Method	Time Stepping	Material Applicability	Progamming Difficulty	Accuracy	Speed
Moving grid		Well-defined freezing range	Not considered here		
Fixed grid					
Source methods		Well-defined freezing range	Not considered here		
Apparent specific heat methods					
Euler		Gradual phase change	Very easy	Poor	Slow
Implicit non-iterative		Gradual phase change	Easy	Poor	Slow
Implicit iterative		Gradual phase change	Easy	Poor	Slow
Enthalpy methods					
Euler		All materials	Very easy	Good	Slow
Implicit iterative		All materials	Hard	Good	[a]
Pham's quasi-enthalpy method					
Implicit non-iterative		All materials	Easy	Good	Fast

[a] Slow to fast depending on iteration method.

B. Dealing with Variable Thermal Conductivity

The rapid change in thermal conductivity around the freezing point contributes to the difficulty in the numerical modeling of phase change. In computing the heat flux $k(T_{i+1} - T_i)$ between nodes $i + 1$ and i, it is unclear as to what value should be used for k: the thermal conductivity calculated at the mean temperature $(T_{i+1} + T_i)/2$, the mean value of thermal conductivities $(k_{i+1} + k_i)/2$, or some other combination such as the series model:

$$\left(\frac{1}{k_{i+1}} + \frac{1}{k_i}\right)^{-1}$$

A more rigorous formulation is obtained by using the *Kirchhoff transformation* [26,27]:

$$u = \int_{T_{\mathrm{REF}}}^{T} k\,\mathrm{d}\theta \tag{7.48}$$

or

$$\mathrm{d}u = k\,\mathrm{d}T \tag{7.49}$$

which when substituted into the Fourier equation, Equation (7.1), gives

$$\frac{\rho c}{k}\frac{\partial u}{\partial t} = \nabla^2 u + q \tag{7.50}$$

The ratio $\rho c/k$ is a property of the material which depends on temperature, and therefore, on u only. This technique groups all the nonlinearities into a single factor, after which the equation can be solved by FDM, FEM, or FVM using the apparent specific heat method as described earlier (with T replaced by e and c replaced by c/k). Alternatively, the left-hand side is written in terms of enthalpy [28]

$$\rho \frac{\partial H}{\partial t} = \nabla^2 u + q \tag{7.51}$$

which can be then solved by any enthalpy-type method or Pham's quasi-enthalpy method, with T replaced by u. Scheerlinck [29,30] found that the Kirchhoff transformation leads to a significant reduction in computation time when using an iterative method, because the **K** matrix becomes a constant and does not have to be recomputed.

With composite materials, the Kirchhoff transformation may cause some problems in the modeling of boundaries between different materials, particularly with FEM. For example, when two adjacent elements made of different materials share the same nodes, the values of u at these nodes will be different depending on whether they are viewed from one element or the other. The elemental Equation (7.23) cannot be assembled into a global matrix equation in the usual manner. Instead, each node that is shared by two materials must be treated as two separate nodes.

V. COUPLED HEAT AND MASS TRANSFER

In food freezing, heat transfer is always accompanied by mass transfer and the latter may have important implications on weight loss and product quality. We will concentrate only on the transfer of moisture, which is the most common situation, although solute transfer also happens in immersion freezing.

When mass transfer occurs, conduction is not the only mode of heat transfer. Thermal energy is also conveyed by the diffusing substance, necessitating the addition of a second transport term. This can most easily be expressed with the enthalpy form of the heat transport equation:

$$\rho \frac{\partial H}{\partial t} = \nabla(k \nabla T) + \nabla(H_{\mathrm{w}} \dot{m}_{\mathrm{w}}) \tag{7.52}$$

$\dot{m}_{\mathrm{w}}$ is the mass flux and H_{w} where the enthalpy of the diffusing substance. The mass flux is assumed to follow Fick's law:

$$\dot{m}_{\mathrm{w}} = D_{\mathrm{w}} \nabla W \tag{7.53}$$

where W is the mass concentration of the diffusing substance (kg water/kg dry solid) and D_{w} its (effective) diffusivity. The governing equation for mass transfer is therefore

$$\frac{\partial W}{\partial t} = \nabla(D_{\mathrm{w}} \nabla W) \tag{7.54}$$

Mechanical effects (gravity and pressure gradient) have been ignored, as well as the mass diffusion due to temperature gradient (Soret effect) and heat diffusion due to concentration gradient (Dufour effect). The second term in the heat transport equation can be usually neglected in dense foods due to the very slow moisture diffusion rate, but not in porous foods.

In the freezing of meat and other dense (nonporous) foods, water evaporates from the surface and is replenished by deep water diffusing toward the surface, until freezing occurs. Thereafter, the water sublimes from the ice front at or near the surface, but there is no significant water movement in the food. Mass transfer occurs in a rather thin layer near the surface only, in contrast to heat transfer which happens throughout, and the main problem is how to deal with the different scales effectively. In the freezing of porous foods such as bread and dough, moisture movement continues right through the freezing process deep inside the food. Here, the heat and mass transfer scales are similar. Yet, another type of mass transfer occurs in whole biological tissue between the intra- and extracellular spaces. Each situation presents a different set of challenges that may require a special modeling approach.

A. Mass Transfer during the Freezing of Dense Foods

Moisture in foods can simultaneously exist in several phases: vapor, "free" liquid, and various types of "bound" moisture, each of which has its own diffusion rate but, because the phases are in intimate contact, all of them must be in thermodynamic equilibrium with each other. However, owing to lack of data, it is commonly assumed that moisture movement in foods can be described by a single-phase diffusion equation, Equation (7.54), with an effective diffusivity D_w. This equation is of the same form as the heat conduction equation, Equation (7.1), and can be solved by the same methods (FDM, FEM, or FVM). In dense food, moisture diffusion is very slow and its contribution to heat transport can be neglected, hence the second term in Equation (7.52) can be neglected, except at the evaporating surface itself.

The problems to be considered are the changes in boundary conditions and the differences in scale between heat and mass transfer. During the precooling phase (prior to surface freezing), water evaporates from the surface and is replenished by moisture diffusing from the inside:

$$\rho_s D_w \left(\frac{\partial W}{\partial n} \right)_S = -k_g (P_s - P_a) \tag{7.55}$$

where ρ_s is the density of the dry solid component, $\mathbf{n}$ the unit normal vector, k_g the mass transfer coefficient, and P_s and P_a the water partial pressure at the food surface and in the surroundings, respectively. P_s is related to the surface moisture by:

$$P_s = \alpha_w(T_s, W_s) P_{sat}(T_s) \tag{7.56}$$

where $P_{sat}(T_s)$ is the saturated water vapor pressure at the surface temperature T_s and α_w the surface water activity as a function of surface temperature and moisture, respectively, and T_s and W_s. The latent heat of vaporization L_V must be taken into account in the boundary condition of the heat conduction equation:

$$-k \left(\frac{\partial T}{\partial n} \right)_S = h_{eff}(T_s - T_a) + L_V k_g (P_s - P_a) \tag{7.57}$$

where h_{eff} is an effective heat transfer coefficient, which may include radiation effects. The boundary conditions for both heat and mass transfer becomes nonlinear as they contain a term P_s, which is a nonlinear function of temperature and moisture. If the Euler time stepping is used, this poses no particular problem, as all variables are calculated explicitly from known conditions. For any other stepping method, P_s can be linearized around the present temperature and moisture value [31]:

$$P_s = a_1 + b_1 T_s \tag{7.58}$$

$$P_s = a_2 + b_2 W_s \tag{7.59}$$

and the terms $b_1 T_s$ and $b_2 W_s$ brought into the vector **KT** or **KW** of the discretized Equation (7.5). Alternatively, an iteration can be carried out at every time-step.

Inside the food, heat and mass transfer are also coupled (but to a weaker degree than at the surface), as the diffusive properties of moisture and heat depend on both temperature and moisture content. As long as a small enough time-step is used (which can be found by trial-and-error), accuracy is not greatly affected by not iterating at every time-step for the inner nodes.

A further complication is caused by the differences in scale between heat and mass transfer. Moisture diffusivity in dense foods is typically of the order $D \approx 10^{-10}$ m^2 s^{-1}, whereas thermal diffusivity is of the order $k/\rho c \approx 10^{-6}$ m^2 s^{-1}, which means that by the time the freezing process is completed, only a very thin layer near the surface has undergone change in moisture. Modeling moisture movement accurately would require an extremely fine grid, which (because of the high thermal diffusivity) would in turn require extremely small time intervals to avoid a large value of $\rho c \Delta t/\Delta x^2$, which would cause excessive oscillation and inaccuracies in the temperature field. Pham and Karuri [32] proposed resolving this difficulty by using a two-grid method, where a second, very fine, one-dimensional finite-volume surface grid is used to model the mass transfer. At each time step, the heat flow equation is solved first using the first grid, then the mass transfer equation using the second grid. The approach was successfully implemented by Trujillo [33] in modeling the chilling of a beef side, using the FVM-based computational fluid dynamics (CFD) software FLUENT. In this case, the mass transfer grid was incorporated as a user-defined function.

Once the surface has frozen (at an initial freezing point determined thermodynamically by the-surface water activity), water becomes immobilized and internal diffusion stops. Moisture then sublimes, at first from the surface, then through a layer of dessicated food that gradually thickens as the ice front recedes, at a rate determined by [34,35]:

$$\dot{m} = \frac{P_{\text{sat}}(T_{\text{s}}) - P_{\text{a}}}{1/k_{\text{g}} + \delta/D_{\delta}} \tag{7.60}$$

where δ is the dessicated thickness and D_δ the diffusivity of water through it. The problem was modeled for one-dimensional geometry using a front-tracking FDM [36]. The dehydrated zone is modeled by a flexible grid with distance increments increasing proportionately to the depth of the freezing zone. The undehydrated zone (both frozen and unfrozen) was modeled by a fixed grid, except that the last node moves with the sublimating interface (and hence the last space increment of the undehydrated zone decreases with time). An apparent heat capacity method appeared to have been used to deal with the freezing front.

Because the desiccated layer is normally very thin in the freezing of dense foods, modeling it numerically requires a very thin grid. In fact, the moment freezing starts, δ may be zero, therefore, an infinitely fine grid is required, which would cause some difficulty. Unfortunately, Ref. [36] did not mention how this was handled. In the author's view, when the rate of sublimation is very slow and the desiccated layer is very thin (thinner than, say, half the thickness of a control volume), we can assume pseudo-steady state (i.e., the water vapor profile in the dessicated layer is as though the sublimating front was stationary), and it is sufficiently accurate to use an ODE approach:

$$\frac{\mathrm{d}\delta}{\mathrm{d}t} = \frac{\dot{m}_{\text{w}}}{\rho_{\text{s}} W(\delta)} \tag{7.61}$$

where $W(\delta)$ is the moisture content at depth δ. This equation can be integrated over each time-interval and the resulting value of δ substituted into Equation (7.61) to calculate the surface mass flux, which can then be used in the boundary condition for the heat and mass transfer PDEs. For porous foods, however, the thickness of the desiccated layer may be much thicker and so it would be more appropriate to model it with a FDM, FEM, or FVM grid.

B. Mass Transfer during Freezing of Porous Foods

In porous food, mass diffusion is much faster than that in dense foods due to vapor diffusion in the pores. Water evaporates from the warm inner parts of the food and diffuses toward the outside. When the freezing point is reached, the vapor condenses into ice. This situation has been modeled by van der Sluis [37] and Hamdami et al. [38,39] for bread freezing, in view of the suspected influence of ice formation under the crust on crust detachment.

1. Evaporation–Condensation Model

One way to handle the effect of mass transfer on heat transfer is the evaporation–condensation model [40–42]. Evaporating moisture absorbs latent heat, and when it recondenses, this heat is released. This evaporation–condensation mechanism is very efficient in transferring heat — in fact, it is the principle behind heat pipes. As diffusion takes place only in the void fraction, the flux of water vapor $\dot{m}$ will be (Figure 7.7):

$$\dot{m} = \varphi \frac{D_v}{1 - y_v} \nabla \rho_v \qquad (7.62)$$

where $(D_v/1 - y_v)\nabla\rho_v$ is the unrestricted flux through stationary air [43], φ a factor to take into account the void fraction and tortuosity of the diffusion path, ρ_v the mass concentration of water vapor in the pores, y_v the mole fraction of water in the pores, and D_v the diffusivity of water vapor in air. The cross flux due to ∇T in Equation (7.54) is usually assumed to be negligible during freezing.

Owing to condensation, the flux will vary with position and the local rate of condensation will be ∇N_v. The latent heat released by condensation will be therefore

$$q_{\text{condens}} = L_V \nabla N_v \qquad (7.63)$$

where L_V is the latent heat of vaporization per kilogram.

Using the isotherm relationship for vapor concentration, we obtain:

$$y_v = \frac{a_W P_{\text{sat}}}{P_{\text{atm}}} \qquad (7.64)$$

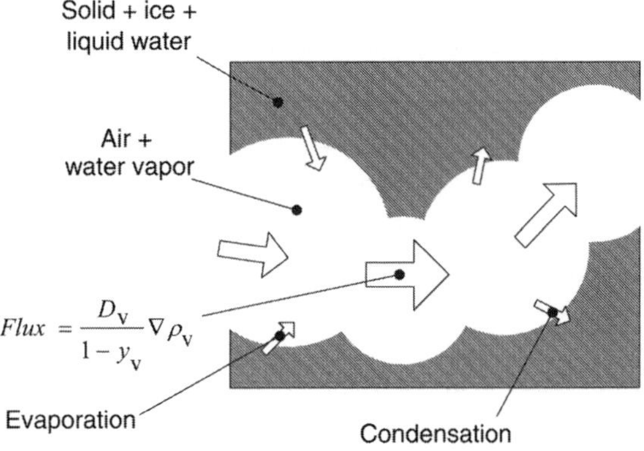

FIGURE 7.7 Diffusion and evaporation–condensation in porous material.

and from the ideal gas law

$$\rho_v = \frac{M_w a_w P_{sat}}{RT} \tag{7.65}$$

Substituting into Equation (7.63), we obtain the expression for latent heat release due to condensation:

$$L_V \nabla N_v = L_V \nabla \left(\varphi D_v \frac{M_w a_w}{RT} \frac{P_{atm}}{P_{atm} - a_w P_{sat}} \frac{\partial P_{sat}}{\partial T} \nabla T \right) \tag{7.66}$$

This term can be added to the right-hand side of heat conduction Equation (7.1) as a heat source, giving:

$$\rho c \frac{\partial T}{\partial t} = \nabla([k + k_{eva-con}]\nabla T) \tag{7.67}$$

where $k_{eva-con}$ is an effective thermal diffusivity due to evaporation–condensation:

$$k_{eva-con} = \varphi \frac{L_V M_W D_v a_w}{RT} \frac{P_{atm}}{P_{atm} - a_w P_{sat}} \frac{dP_{sat}}{dT} \tag{7.68}$$

2. A General Model for the Freezing of Porous Foods

$k_{eva-con}$ contains the latent heat of condensation, L_V, which will vary depending on whether the vapor condenses into or is formed from liquid or ice. The evaporation–condensation model is therefore not convenient to use when this is not known (e.g., when the material is partly dessicated so that its local freezing point is below the local temperature). In this case, it is better to calculate nodal enthalpies and moisture contents from the temperature and moisture fields, via Equation (7.52)–Equation (7.54). The new nodal temperatures can then be found from H and W by inverting the function $H(T, W)$. In calculating H from T and W, we must know how much of the moisture is in the vapor, liquid, and ice phases. The total moisture content in (possibly porous) food can be divided into bound moisture, (mobile) liquid moisture, water vapor, and ice:

$$W = W_b + W_l + W_v + W_{ice} \tag{7.69}$$

Bound water is assumed to remain constant. Vapor and liquid always exist and are in equilibrium, and because the vapor pressure in the pores is at temperature T and partial pressure $a_w(W_{ice}, T)P_{sat}(T)$:

$$W_v = \psi \frac{P_{sat}(T) a_w(W_l, T)}{T} \tag{7.70}$$

In contrast, ice appears only when the freezing point of the solution is equal to the local temperature — in other words, when the liquid moisture content has reached the liquid–ice equilibrium value (or maximum possible value) $W_{l,max}$ for that temperature. Any excess moisture will then condense as ice. The equilibrium value is given by [44]:

$$W_{l,max} = (W_0 - W_b) \frac{T_f - 273.15}{T - 273.15} \tag{7.71}$$

where W_0 is the initial moisture content and T_f the initial freezing point, before any moisture has been lost or gained. Thus, the procedure for calculating the vapor, liquid, and ice content from the total moisture content W is:

(a) If $W \leq W_{l,max} + W_v(W_{l,max})$, then the material is at or above its freezing point and hence

$$W_{ice} = 0 \tag{7.72}$$

$$W = W_l + W_v \tag{7.73}$$

W_v is a function of W_l (liquid–vapor equilibrium), hence W_l can be found by iteration. In practice, $W_v \ll W$ so convergence will be rapid.

(b) If $W > W_{l,max} + W_v(W_{l,max})$, then the material is below its freezing point and hence

$$W_l = W_{l,max} \tag{7.74}$$

$$W_v = W_v(W_{l,max}) \tag{7.75}$$

$$W_{ice} = W - W_l - W_v \tag{7.76}$$

Equation (7.72)–Equation (7.76) may be difficult to implement in commercial FDM, FEM, or FVM software due to the inflexibility of the latter. Often, they do not allow algebraic equations (such as an equilibrium relationship) for field variables. A practical work around is to replace an algebraic equation such as $W_l = W_{l,max}$ by a source term $(W_l - W_{l,max})/\tau$ in the differential equation for W_l. τ is an arbitrary time constant, chosen to be small enough to ensure (near-)equilibrium, but not so small as to create instability in the calculations.

C. Mass Transfer between Intra- and Extracellular Spaces

Although neglected by food engineers, this topic is of potential importance in predicting the quality of foods, especially in conjunction with the modeling of intracellular nucleation. The phenomenon has been modeled in the field of cryosurgery and a numerical model of Devireddy et al. [45], will be described later in the section on nucleation modeling.

VI. SUPERCOOLING AND NUCLEATION EFFECTS

Thus far, we have assumed that the freezing process is entirely governed by heat transfer. However, in many cases, the dynamics of nucleation and mass transfer has observable effects. It is well known that foods almost never start to freeze at their thermodynamic freezing points. When water is cooled below the freezing point, it remains liquid until the temperature is low enough for stable ice crystals to form and grow. Above this nucleation temperature, any ice crystal that might form will lose molecules faster than it gains due to surface energy effects (the curved surface of a crystal has higher free energy than a flat surface and therefore tends to lose molecules faster than it gains). There are two types of nucleation: *homogeneous* and *heterogeneous*. Homogeneous nucleation happens only in pure water, in the absence of any foreign material, at a temperature of about $-40°C$. In food freezing, heterogeneous nucleation is the prevailing mechanism. It is caused by contact with a foreign material or with impurities, on which crystals form and grow. Heterogeneous nucleation happens at a higher temperature than homogeneous nucleation because the foreign material enables water molecule to form clusters on its surface with a large radius of curvature, thus lessening the surface energy. If water or a water-rich material is cooled very quickly, nucleation does not have time to occur and, at the glass-transition temperature, the liquid in the food becomes an amorphous solid or glass, a process known as vitrification. Vitrification requires extremely fast cooling rates that are never obtained in ordinary food freezing. Even in cryogenic freezing, only a very thin layer (of the order of a few microns) is susceptible to vitrification. The glass-transition temperature of pure water is very low (about 135 K), but those of most foods are considerably higher.

Some degree of supercooling is observed in most food freezing processes, where the surface dips briefly below freezing point before suddenly coming up to the freezing temperature. Once nucleation has occurred, ice crystals will grow, and in most industrial freezing processes, the freezing process reverts to being heat transfer controlled. Pham [46] has modeled this type of behavior with a finite difference model using the quasi-enthalpy method and validated the model with data from Ref. [47]. Miyawaki et al. [48] independently used the apparent specific heat technique to solve the same problem. To simulate supercooling, the specific heat (and thermal conductivity) above freezing is assumed to continue to apply below the initial freezing point, until the coldest node reaches nucleation temperature. At that point, the normal time-stepping solution is momentarily stopped and all the nodes that have an enthalpy value H below freezing are assumed to freeze instantaneously, releasing enough latent heat for the node to warm up to the equilibrium temperature $T(H)$ (Figure 7.8). Incidentally, Figure 7.8 resembles Figure 7.6, which illustrates Pham's temperature correction step [25]. Calculations continue normally throughout the material from that point onward. Pham [46] found that for the amount of supercooling that is commonly observed (a few degrees), supercooling has negligible effect on freezing time.

However, this conclusion may not hold for all types of foods. When water is held as small droplets in an emulsion, such as in ice cream or butter, each ice crystal cannot grow beyond its droplet and each droplet has to crystallize separately, a probabilistic phenomenon. In such cases, the freezing process may be very gradual and a freezing plateau may not even be present, as was observed experimentally [49]. This may have important implication on the heat load in freezers, which will be overestimated, and cold stores, which will be underestimated. The product may also undergo internal warming due to gradual latent heat release during cold storage.

Another important reason to model nucleation and crystal growth is its effect on food quality, cellular damage, and drip loss. Maximum drip loss in meat is believed to happen when a large intracellular crystal forms in each cell, which causes maximal distortion and damage to the cell wall [50]. This happens at an intermediate freezing rate, as faster freezing causes the formation of multiple small intracellular crystals, the slower freezing leads to extracellular freezing. Devireddy et al. [45] developed a finite-volume model to predict the formation of intracellular ice in biological tissues in the context of cryosurgery. The material is divided into two phases, extra- and

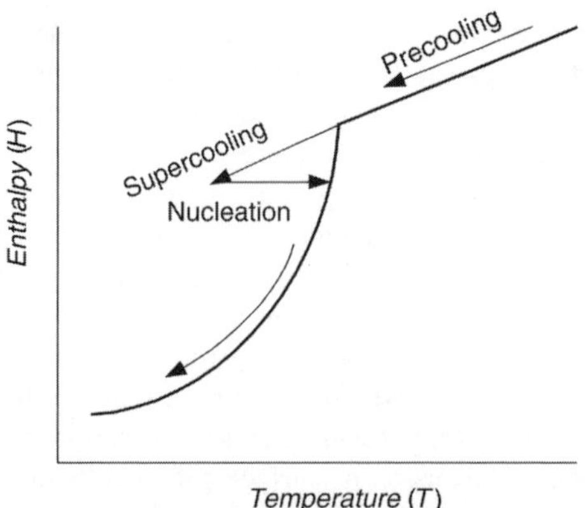

FIGURE 7.8 Modeling supercooling and nucleation on the H–T diagram.

intracellular. Extracellular liquid is assumed to freeze without supercooling. As it does so, the extracellular solute concentration increases, causing water to diffuse across the cell membrane and the cell volume to shrink at a rate [51]:

$$\frac{dV}{dt} = -\frac{L_pART}{v_w}\left\{\ln\left[\frac{V - V_b}{V - V_b + \phi v_w n_s}\right] - \frac{L}{R}\left[\frac{1}{T_{REF}} - \frac{1}{T}\right]\right\} \tag{7.77}$$

where L_p is the cell membrane's permeability, A the surface area of the cell, L the molar latent heat of freezing, R the gas constant, v_w the molar volume of water, ϕ the salt's dissociation constant, n_s the moles of salt in the cell, T_{REF} the reference temperature (273.15 K), V the cell volume and V_b the bound water volume in a cell. Owing to this diffusion, intracellular solute concentration increases, but at a slower rate than extracellular fluid. However, the temperature in the cell falls just as fast as that of the extracellular medium. There is thus some degree of supercooling inside the cell. The probability of intracellular nucleation P_{if} is obtained from the amount of supercooling, using a model by Toner [52]:

$$P_{if} = 1 - \exp\left\{-\int_{T_{seed}}^{T} A\Omega_0\left(\frac{T}{T_{f0}}\right)^{1/2}\frac{\eta_0}{\eta}\frac{A}{A_0}\exp\left[\frac{-\kappa_0(T_f/T_{f0})^{1/4}}{(T - T_f)^2T^3}\right]\left(\frac{dT}{dt}\right)^{-1}dt\right\} \tag{7.78}$$

where T_{seed} is the initial freezing temperature of the extracellular fluid, A the cell membrane area, T_f is the equilibrium phase change temperature of the intracellular fluid, η the intracellular viscosity, and κ and Ω cell type-dependent parameters. The subscript 0 refers to isotonic conditions.

Because of the mass transfer process between extra- and intracellular spaces and the supercooling of the latter, Equation (7.1) cannot be solved directly as in heat-transfer-controlled freezing. Instead, an iterative procedure has to be carried out at every time-step to satisfy the heat balance as well as the intra- and extracellular mass balances:

Guess the rate of ice formation in each control volume.
Then, carry out the following iteration procedure:
- Calculate the new nodal temperatures throughout the domain from the heat conduction equation (7.1) and the amount of ice formed over the time step.
- Calculate the extracellular ice from the temperatures, assuming thermodynamic equilibrium in the extracellular space.
- Calculate the probabilities of intracellular nucleation, that is, the increase in the number of cells with internal ice, from Toner's equation (7.78).
- Calculate the amount of intracellular ice.
- Calculate the total latent heat released by extra- and intra-cellular ice

until convergence is reached.

At every time step, material balances have to be set up to keep track of the amount of ice and unfrozen water inside and outside of the cells. The model requires many parameters that are not yet available about foodstuffs, such as cell size, cell surface area, cell membrane permeability, extracellular volume, and parameters related to the onset of nucleation. As far as the hypothesis of Bevilacqua et al. [50] about cell damage is concerned, it still does not allow the prediction of how many intracellular crystals will occur and how large.

In a liquid food, as the ice grows, solute is excluded from it and a concentration gradient will form in the liquid in front of it. This will increase the freezing point of the remaining water next to the freezing front (constitutional supercooling). If, due to irregularities in the front, part of the front protrudes into the unfrozen zone, it will be in contact with liquid at a lower concentration and hence lower freezing point, which will then tend to freeze before the liquid in other areas (Figure 7.9).

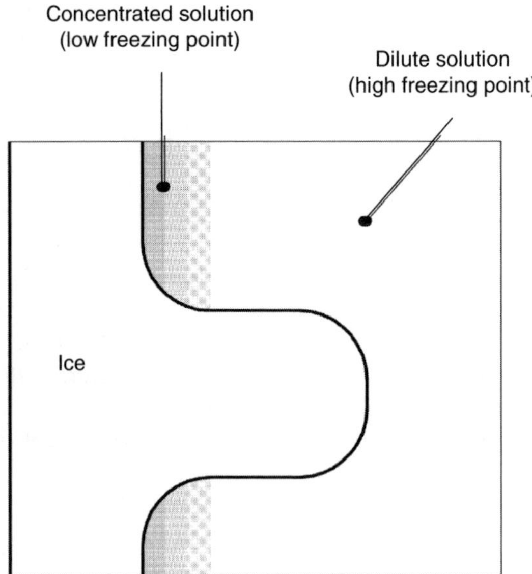

FIGURE 7.9 Dendritic growth due to constitutional supercooling.

The protrusion will therefore continue ahead of the rest of front, resulting in the formation of a dendrite. At very fast freezing rate, solute diffusion may become significant in controlling the progress of the freezing front.

In the freezing or freeze concentration of fruit juice, the rate of freezing affects the velocity of dendrite growth, which in turn influences the degree of separation of solute and ice. At low dendrite growth rate, solutes have time to diffuse away from the freezing front and there is good separation, but at high growth rate, solutes and suspensions become embedded in the ice matrix [53]. In solid foods, mass diffusion is usually very slow and therefore the solution becomes more and more concentrated locally as ice forms, lowering the freezing point and causing the characteristic gradual enthalpy–temperature curve.

To date, the phenomena of nucleation, crystal growth, and vitrification in foods and their effect on the quality of frozen foods has not received adequate attention from numerical modelers. The main reason is probably the lack of data on the parameters involved. An approximate analytical equation for predicting crystal size from dendritic growth theory was presented in Refs. [54–56], assuming a Neumann boundary condition, and validated against data from agar gel freezing. Udaykumar et al. [15] presented and validated a finite-volume technique for computing dendritic growth of crystals from pure melts, assuming diffusion control. An entirely different modeling approach is the use of cellular automata or hybrid automata [57,58], where the material is modeled as a collection of microscopic elements that change phase stochastically depending on the state of the surrounding elements.

VII. MODELING OF HIGH PRESSURE FREEZING AND THAWING

High pressure freezing, and particularly pressure shift freezing, is gaining attention as a freezing method for high quality or freeze-sensitive foods [59]. In pressure shift freezing, the food is cooled under high pressure to subzero temperatures. Because the freezing point decreases with pressure, phase change does not take place. When the product temperature has more or less equilibrated, pressure is released suddenly. The food is now supercooled by several degrees and nucleation takes place spontaneously throughout the supercooled product, causing an instantaneous temperature rise. There may be a short period (a few seconds) of equilibration where some water remains supercooled [60], but this can be probably neglected in heat transfer modeling and thermodynamic equilibrium may be assumed for modeling purposes. The uniform nucleation ensures evenly small crystal size and minimal textural damage. High-pressure thawing has also been investigated as a fast thawing method (because of the lowering of the freezing point, the difference between product and ambient temperatures is increased, hence a larger heat flux is obtained).

In high-pressure thawing and freezing, the effect of pressure on thermal properties (latent heat, freezing point, and thermal conductivity) must be taken into account. Freezing point is decreased by pressure according to Clapeyron's equation. Latent heat is also decreased. Thermal conductivity below zero will also be different, as there is no ice.

Chourot et al. [61] modeled high-pressure thawing of an infinite cylinder of pure water, using FDM with Crank–Nicholson stepping and the apparent specific heat approach. The latent heat is assumed to contribute a triangular peak spanning 1 K at the base. Thermal conductivity is assumed to be constant above and below the phase change range and vary linearly over this range. The total latent heat and mean phase-change temperature are given as polynomial functions of pressure. The entire thawing process takes place under pressure.

Denys et al. [62] modeled pressure shift freezing using FDM with explicit stepping and apparent specific heat formulation. At the moment of pressure release, the temperature rise from T_i to T_{New} is calculated by an enthalpy balance. The product is assumed to be at uniform temperature when pressure is released; however, in a subsequent paper [63], this restriction is relaxed and the energy balance is carried out node by node. Calculations continue normally from there on.

In the light of what has been discussed earlier on the handling of latent heat and the shortcomings of the apparent specific heat approach, it can be seen that a simpler, more efficient, and flexible program, which can handle any temperature and pressure regime, can be written by using the enthalpy or quasi-enthalpy formulations, that is, at every time step:

1. Calculate nodal enthalpies from Equation (7.38) (enthalpy method) or Equation (7.43)–Equation (7.47) (quasi-enthalpy method)
2. Calculate nodal temperatures from nodal enthalpies and pressure

Figure 7.10 shows a pressure shift process on the enthalpy–temperature diagram. The food is cooled at high pressure from A to B, pressure is released along BC causing partial freezing, then freezing is completed along CD. From the programming point of view this presents no extra complication over a "standard" freezing program, apart from the need to solve for T from H and P instead of from H alone. If only a finite number of pressure levels are used (usually two), the

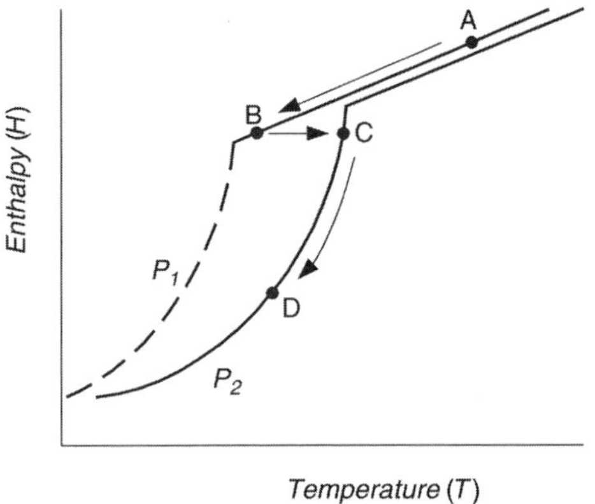

FIGURE 7.10 Pressure shift freezing on the enthalpy–temperature diagram.

property subroutine needs only contain the H–T relationship at these pressures. However, it is usually found that due to expansion during phase change, pressure will fluctuate after release, hence a continuous functional relationship would be more desirable. In addition, if it is not intended that nucleation happens during the high-pressure stage, there is no need to know the frozen section (dotted curve) of the H–T curve for high pressure.

VIII. MODELING OF THERMOMECHANICAL EFFECTS DURING FREEZING

Water expands by about 9% by volume when turning into ice, causing considerable stresses in foods during freezing. In cryogenic freezing, this expansion is followed by a significant thermal contraction, of the order of 0.5% in linear terms or 1.5% in volumetric terms [64]. Frozen food is brittle and these stresses may cause cracking in the food, especially at high cooling rates such as in cryogenic freezing. Rubinsky et al. [65] carried out an approximate analytical analysis of thermal stresses during the freezing of organs but neglected phase-related volume change. Rabin and Steif [66] calculated thermal stresses in freezing a sphere, taking both phase-change expansion and thermal contraction into account, but assumed that the unfrozen food is liquid and neglected the property changes due to freezing. Shi et al. [67,68] carried out thermal strain and stress calculations using the commercial software ABAQUS (ABAQUS Inc., Rhode Island, USA). They used both elastic and viscoelastic models, but neglected thermal contraction.

At moderate values of strains and stresses, thermophysical properties can be assumed to be unaffected, therefore, the analysis can be carried out in two stages: the thermal history is calculated first, using any of the methods listed earlier, followed by stress and strain calculations. This procedure is not necessarily valid when stress values are very large, such as when the volume is constrained, generating very high pressures and consequent thermal property changes. The stress analysis assumes that total strain is the sum of thermal strain (due to temperature change) and mechanical strain (due to mechanical stresses):

$$\varepsilon_{ij} = \varepsilon_{ij}^{(\mathrm{m})} + \varepsilon_{ij}^{(\mathrm{T})} \tag{7.79}$$

where the thermal strain can be calculated as a function temperature and should include both the phase change expansion and thermal contraction of ice:

$$\varepsilon_{ij}^{(\mathrm{T})} = \delta_{ij} \int_{T_{\mathrm{REF}}}^{T} \beta \, \mathrm{d}T \tag{7.80}$$

To solve for the mechanical strains, some constitutive relationships for the material must be assumed. In an elastic model, strains are linearly related to the (present) stresses:

$$\sigma_{ij} = \delta_{ij} K e_{kk}^{(\mathrm{m})} + G e_{ij}^{(\mathrm{m})} \tag{7.81}$$

whereas in a viscoelastic model, strains depend on stress history and vice versa:

$$\sigma_{ij} = \delta_{ij} \int_0^t K(t-\tau) \frac{\partial e_{kk}^{(\mathrm{m})}}{\partial \tau} \mathrm{d}\tau + \int_0^t G(t-\tau) \frac{\partial e_{ij}^{(\mathrm{m})}}{\partial \tau} \mathrm{d}\tau \tag{7.82}$$

where e_{ij} are the deviatoric strains:

$$e_{ij} = \varepsilon_{ij} - \delta_{ij} \frac{\sum_k \varepsilon_{kk}}{3} \tag{7.83}$$

The mechanical stress thus consists of an expansion component and a shear or deviatoric component. K is the bulk modulus and G is the shear modulus of the material, that is, the stresses caused by a unit step in the strain. For elastic materials, both K and G are constant, whereas for viscoelastic materials, they are functions of time. The integrals in Equation (7.82) are termed "hereditary integrals." They are obtained by assuming that an arbitrary strain pattern can be decomposed into a series of steps $\delta\varepsilon$, each occurring at time τ, and the stress at time t resulting from each step, $\delta\varepsilon K(t-\tau)$ and $\delta\varepsilon G(t-\tau)$, can be summed up or integrated.

The various components of stress and strain are not independent but are related by equilibrium relationships

$$\sum_j \frac{\partial \sigma_{ij}}{\partial x_j} + F_i = 0 \tag{7.84}$$

where F_i is the ith component of body force, and compatibility conditions arising from degrees of freedom considerations:

$$\frac{\partial^2 \varepsilon_{ij}}{\partial x_k \partial x_l} + \frac{\partial^2 \varepsilon_{kl}}{\partial x_i \partial x_j} - \frac{\partial^2 \varepsilon_{lj}}{\partial x_k \partial x_i} - \frac{\partial^2 \varepsilon_{ki}}{\partial x_l \partial x_j} = 0 \tag{7.85}$$

In general, it can be said that strain and stress calculations are a specialized field best left to experts and specialized software. However, in the case of spherical and cylindrical foods, which are of considerable practical interest, the problem is greatly simplified by the disappearance of most of the terms in the stress and strain equations. In spherical coordinates, for example, most of the terms in the stress tensor disappear, leaving only two: the radial stress σ_r and the tangential (or circumferential or azimuthal) stress σ_t. The compatibility equations reduce to

$$\varepsilon_r = \frac{du}{dr} \tag{7.86}$$

$$\varepsilon_t = \frac{u}{r} \tag{7.87}$$

where u is the radial displacement. By considering the forces acting on a thin shell, the equilibrium relationships reduce to

$$\frac{d\sigma_r}{dr} + \frac{2}{r}(\sigma_r - \sigma_t) = 0 \tag{7.88}$$

and the system of equations can be easily solved once the constitutive relationships and thermal fields are known.

IX. CONCLUSIONS

The numerical modeling of the classical "pure thermal" freezing problem can be considered as solved in principle. An enthalpy or quasi-enthalpy method is recommended, in conjunction with control-volume FDM, lumped capacitance FEM, or standard FVM. Explicit time stepping is recommended for small or one-off problems, Pham's quasi-enthalpy method for those who want speed as well as uncomplicated programming. Iterative enthalpy methods are useful from a mathematician's point of view to provide rigorous second-order results, which guarantee strict energy balance at all time steps. However, if a commercial piece of software such as FEMLAB, ABAQUS, Fluent, or CFX is used, it may be difficult for the user to apply the enthalpy or quasi-enthalpy methods as they are not a standard option or may not be an option at all. This needs to be looked at by the commercial software providers.

Hardly, any freezing problem is "thermal only," and more attention will be devoted to solving for the effects of parallel coupled physical processes: mass transfer, nucleation, crystal growth, mass transfer across cell membrane, vitrification, thermal expansion, mechanical strain and stress, and cracking. Even in normal freezing, internal pressure may have some effect on the freezing point that has up till now been neglected. The modern food engineer is no longer interested only in freezing times or heat loads, but also in food quality factors: drip, color, texture, flavor, distortion and cracks, and microbial growth (especially during thawing). To predict these factors, detailed modeling is needed on physical processes other than heat transfer.

Although conventional FDM, FEM, and FVM can deal with any continuous deterministic phenomenon that can be described by PDEs, some phenomena such as nucleation, crystal growth, crack initiation, and crack growth are by nature discrete and stochastic and thus the PDE approach may need to be augmented by another modeling approach altogether, such as cellular automata or hybrid automata models.

To model nonthermal phenomena successfully, data on some food properties hitherto neglected by food technologists (moisture diffusivity, absorption isotherm, nucleation parameters, cell size, cell membrane permeability, viscoelastic properties, tortuosity factor in porous foods, etc.) will have to be collected. For the modeling of high pressure freezing and thawing, there is a need for more data and prediction methods for thermal properties as a function of pressure. More and better data on heat transfer coefficients and thermal properties and better methods for their prediction will also be an ongoing area of research. The use of CFD for calculating heat transfer coefficients in food refrigeration is increasingly popular, but the lack of a satisfactory turbulence model for many practical situations (circulating flows, natural or mixed convection) means that CFD results cannot yet be completely trusted.

Nowadays, computers are so fast that practically any food freezing problem (except CFD) can be simulated within reasonable time by using an Euler stepping method. It could be believed that there is no need to search for more efficient methods. However, freezing models are more useful if they can be incorporated in larger programs such as models of whole food plants. Furthermore, the food engineer does not model for the fun of it but with the ultimate objective of being able to optimize products and processes. Computer optimization involves running the model hundreds or thousands of times (in the case of stochastic optimization methods such as genetic algorithms, even tens of thousands). Models are also used in the determination of product properties and other parameters by error minimization, where they have to be run a similar number of times. Therefore, the search for more efficient algorithms will continue, even in this day and age of fast computers.

NOMENCLATURE

c	specific heat, J kg^{-1} K^{-1}
$\mathbf{B}$	$\nabla^{\mathrm{T}}\mathbf{N}$
$\mathbf{C}$	capacitance matrix
D_{w}	diffusivity of water, m^2 s^{-1}
$D_{T\mathrm{w}}$	heat diffusivity due to concentration gradient, W m kg^{-1}
$D_{\mathrm{w}T}$	mass diffusivity due to temperature gradient, kg s^{-1} m^{-2} K^{-1}
D_{δ}	effective diffusivity of water vapor in dessicated layer, m^2 s s^{-1}
e	deviatoric strain
E	Young's modulus, Pa
G	shear modulus, Pa
$\mathbf{f}$	forcing vector
h	heat transfer coefficient, W m^{-2} K^{-1}
H	specific enthalpy, J kg^{-3}

$\mathbf{H}$	vector of nodal enthalpies
i, j	node number
$\mathbf{J}$	Jacobian matrix
k	thermal conductivity, W m^{-1} K^{-1}
$k_{eva-con}$	thermal conductivity due to evaporation–condensation, W m^{-1} K^{-1}
k_g	mass transfer coefficient, kg m^2 s^{-1} Pa^{-1}
l	length of one-dimensional element, m
K	bulk modulus, Pa
$\mathbf{K}$	conductance matrix
L_V	latent heat of vaporization, J kg^{-1}
$\dot{m}$	mass flux, kg m^{-2} s^{-1}
$\mathbf{M}$	mass matrix
M_w	molecular mass of water, kg $kmol^{-1}$
$\mathbf{n}$	unit vector normal to surface
$\mathbf{N}$	vector of shape functions
P	partial pressure of water, Pa
P_{atm}	atmospheric pressure, Pa
P_{if}	nucleating probability in a single cell over a time interval
q	heat generation, J m^{-3}
$\mathbf{r}$	position vector
r	radial coordinate, m
R	universal gas constant, 8314.4 J $kmol^{-1}$ K^{-1}
S	boundary surface
t	time, s
Δt	time step, s
T	temperature, K
T_f	initial freezing temperature, K
$\mathbf{T}$	nodal temperature vector, K
u	Kirchhoff transform, $u = \int_{T_{REF}}^{T} k\, d\theta$, W m^{-1}
$\mathbf{v}_f$	velocity vector of freezing front, m s^{-1}
V	cell volume, m^3
δV	control volume, m^3
W	total water concentration, kg m^{-3}
W_g, W_l, W_{ice}	water vapor, liquid moisture, and ice concentrations, respectively, kg m^{-3}
$\mathbf{W}$	vector of nodal moisture concentrations
x	position, m
Δx	node spacing, m
y_v	mole fraction of water vapor
α	weighting coefficient for time stepping
α_w	water activity
β	thermal expansion coefficient, K^{-1}
δ	thickness of desiccated layer, m
δ_{ij}	Kronecker delta (=1 if $i = j$, else = 0)
ε_{ij}	strain
φ	factor for diffusion in porous media
ν	Poisson ratio
ρ	density, kg m^{-3}
ρ_s	bulk density of dry solid, kg m^{-3}
ρ_v	concentration of water vapor in air, kg m^{-3}
σ	stress, Pa

ρ	density, kg m^{-3}
τ	time, s
ξ	relative position in element
Ω	element domain

Subscripts

a	value in bulk air
b	bound water
eff	effective
i, j	node number
ice	ice
l	liquid
m	mean
s	value at food surface
sat	saturation value for pure water
r	radial
t	tangential
v	vapor

Superscripts

(e)	belonging to one element
New	value at the end of time step
(T)	thermal
(m)	mechanical

REFERENCES

1. HS Carslaw, JC Jaeger. *Conduction of Heat in Solids*, 2nd ed. Oxford: Clarendon Press, 1959, p. 10.
2. AN Califano, NE Zaritzky. Simulation of freezing or thawing heat conduction in irregular two dimensional domains by a boundary fitted grid method. *Lebensmittel-Wissenschaft und-Technologie* 30:70–76, 1997.
3. J Crank, P Nicolson. A practical method for numerical integration of solutions of partial differential equations of heat conduction type. *Proceedings of Cambridge Philosophical Society* 43:50–67, 1947.
4. WH Press, BP Flannery, WT Vetterling, SA Teukolky. *Numerical Recipes in C*, 2nd ed. Cambridge: Cambridge U.P., 1986, pp. 50–51.
5. M Lees. A linear three-level difference scheme for quasilinear parabolic equations. *Mathematics of Computation* 20:516–522, 1976.
6. QT Pham. A comparison of some finite-difference methods for phase-change problems. *Proceedings 17th International Congress of Refrigeration*, Vol. B, Vienna, 1987, pp. 267–271.
7. DW Peaceman, HH Rachford. The numerical solution of parabolic and elliptic differential equations. *Journal of the Society for Industrial and Applied Mathematics* 3 (1):28–41, 1955.
8. AC Cleland, RL Earle. Prediction of freezing times for foods in rectangular packages. *Journal of Food Science* 44:964–970, 1979.
9. L Segerlind. *Applied Finite Element Analysis*, 2nd ed. New York: John Wiley & Sons, 1984.
10. OC Zienkiewicz, RL Taylor. *The Finite Element Method*. London: McGraw-Hill, 1991.
11. J Banaszek. Comparison of control volume and Galerkin finite element methods for diffusion-type problems. *Numerical Heat Transfer* 16:59–78, 1989.
12. R Plank. Beitrage zur Berechnung und Bewertung der Gefriergeschwindigkeit von Lebensmitteln. *Zeitschrift fur die gesamte Kalte Industrie Reihe 3 Heft* 10:1–16, 1913.

13. VR Voller. An overview of numerical methods for solving phase change problems. In: WJ Minkowycz, EM Sparrow, Eds., *Advances in Numerical Heat Transfer*, Vol. 1. London: Taylor & Francis, 1996, pp. 341–375.
14. QT Pham. The use of lumped capacitances in the finite-element solution of heat conduction with phase change. *International Journal of Heat and Mass Transfer* 29:285–292, 1986.
15. HS Udaykumar, L Mao, R Mittal. A finite-volume sharp interface scheme for dendritic growth simulations: comparison with microscopic solvability theory. *Numerical Heat Transfer, Part B* 42:389–409, 2002.
16. VR Voller, CR Swaminathan. General source-based method for solidification phase change. *Numerical Heat Transfer Part B* 24:161–180, 1991.
17. G Comini, S Del Giudice, O Saro. Conservative equivalent heat capacity methods for non-linear heat conduction. In: RW Lewis, K Morgan, Eds., *Numerical Methods in Thermal Problems*, Vol. 6, Part 1. Swansea: Pineridge Press, 1989, pp. 5–15.
18. G Comini, S Del Giudice. Thermal aspects of cryosurgery. *Journal of Heat Transfer* 98:543–549, 1976.
19. EC Lemmon. Phase change technique for finite element conduction code. In: RW Lewis, K Morgan, Eds., *Numerical Methods in Thermal Problems*. Swansea: Pineridge Press, 1979, pp. 149–158.
20. DJ Cleland, AC Cleland, RW Earle, SJ Byrne. Prediction of rates of freezing, thawing and cooling in solids of arbitrary shape using the finite element method. *International Journal of Refrigeration* 7: 6–13, 1984.
21. K Morgan, RW Lewis, OC Zienkiewicz. An improved algorithm for heat conduction problems with phase change. *International Journal of Numererical Methods in Engineering* 12:1191–1195, 1978.
22. QT Pham. Comparison of general purpose finite element methods for the Stefan problem. *Numerical Heat Transfer Part B — Fundamentals* 27:417–435, 1995.
23. NR Eyres, DR Hartree, J Ingham, R Jackson, RJ Sarjant, JB Wagstaff. The calculation of variable heat flow in solids. *Transactions of the Royal Society* A240:1, 1946.
24. VR Voller, CR Swaminathan, BG Thomas. Fixed grid techniques for phase change problems: a review. *International Journal of Numererical Methods in Engineering* 30:875–898, 1990.
25. QT Pham. A fast, unconditionally stable finite-difference method for heat conduction with phase change. *International Journal of Heat and Mass Transfer* 28:2079–2084, 1985.
26. G Kirchhoff. Vorlesungen uber die Theorie der Warme, 1894. In: H.S. Carslaw, J.C. Jaeger, Eds. *Conduction of Heat in Solids*, 2nd ed. Oxford: Clarendon Press, p. 11, 1986.
27. KA Fikiin. Generalised numerical modeling of unsteady heat transfer during cooling and freezing using an improved enthalpy method and quasi-one-dimensional formulation. *International Journal of Refrigeration* 19:132–140, 1996.
28. VR Voller, CR Swaminathan. Treatment of discontinuous thermal conductivity in control volume solution of phase change problems. *Numerical Heat Transfer Part B* 19:175–189, 1991.
29. N Scheerlinck. Uncertainty Propagation Analysis of Coupled and Non-Linear Heat and Mass Transfer Models. Ph.D. thesis, Katholieke Universteit Leuven, 2000, pp. 180–181.
30. N Scheerlinck, P Verboven, KA Fikiin, J de Baerdemaeker, BM Nicolai. Finite-element computation of unsteady phase change heat transfer during freezing or thawing of food using a combined enthalpy and Kirchhoff transform method. *Transactions of the American Society of Agricultural Engineers* 44:429–438, 2001.
31. LM Davey, QT Pham. Predicting the dynamic product heat load and weight loss during beef chilling using a multi-region finite difference approach. *International Journal of Refrigeration* 20:470–482, 1997.
32. QT Pham, N Karuri. A computationally efficient technique for calculating simultaneous heat and mass transfer during food chilling. *20th International Congress of Refrigeration*, Sydney, 19–24 September, 1999.
33. F Trujillo. A Computational Fluid Dynamic Model of Heat and Moisture Transfer during Beef Chilling. Ph.D. thesis, School of Chemical Engineering and Industrial Chemistry, University of New South Wales, Sydney, Australia, 2004.
34. QT Pham, R Mawson. Moisture migration and ice recrystallization in frozen foods. In: MC Erickson, YC Hung, Eds., *Quality in Frozen Foods*. New York: Chapman and Hall, 1997, pp. 67–91.

35. QT Pham, J Willix. A model for food dessication in frozen storage. *Journal of Food Science* 49:1275–1281, 1294, 1984.
36. LA Campañone, VO Salvadori, RH Mascheroni. Weight loss during freezing and storage of unpackaged foods. *Journal of Food Engineering* 47:69–79, 2001.
37. SM Van der Sluis. *Cooling and Freezing Simulation of Bakery Products*, International Institute of Refrigeration Meeting Commissions B1, B2, D1, D2/3, Palmerston North, New Zealand, 1993.
38. N Hamdami, JY Monteau, A Le Bail. Heat and mass transfer simulation during freezing in bread. *Proceedings ICEF9 9th International Congress on Engineering and Foods*, Montpellier (CDROM), 2004.
39. N Hamdami, JY Monteau, A Le Bail. Simulation of coupled heat and mass transfer during freezing of a porous humid matrix. *International Journal of Refrigeration* 27:595–603, 2004.
40. DA de Vries. The thermal conductivity of granular materials. *Annex 1952-1 Bulletin of the International Institute of Refrigeration*, 1952, pp. 115–131.
41. DA de Vries. Simultaneous transfer of heat and moisture in porous media. *Transactions of the American Geophysical Union* 39:909–916, 1958.
42. DA de Vries. The theory of heat and moisture transfer in porous media revisited. *International Journal of Heat and Mass Transfer* 30:1343–1350, 1987.
43. RB Bird, WE Stewart, EN Lightfoot. *Transport Phenomena*. New York: John Wiley & Sons, 1960, p. 523.
44. HG Schwartzberg. Effective heat capacities for the freezing and thawing of food. *Journal of Food Science* 41:152–156, 1976.
45. RV Devireddy, DJ Smith, JC Bischof. Effect of microscale mass transport and phase change on numerical prediction of freezing in biological tissues. *Journal of Heat Transfer* 124:365–374, 2002.
46. QT Pham. Effect of supercooling on freezing times due to dendritic growth of ice crystals. *International Journal of Refrigeration* 12:295–300, 1989.
47. FC Menegalli, A Calvelo. Dendritic growth of ice crystals during freezing of beef. *Meat Science* 3:179–199, 1978.
48. O Miyawaki, T Abe, T Yano. A numerical model to describe freezing of foods when supercooling occurs. *Journal of Food Engineering* 9:143–151, 1989.
49. A Nahid, JE Bronlund, DJ Cleland, DJ Oldfield, B Philpott. Prediction of thawing and freezing of bulk palletised butter. *Proceedings of the 9th International Congress on Engineering and Foods (ICEF 9)*, Montpellier (CDROM), 2004, paper 668.
50. A Bevilacqua, NE Zaritsky, A Calvelo. Histological mesurements of ice in frozen beef. *Journal of Food Technology* 1:237–251, 1979.
51. P Mazur. Kinetics of water loss from cells at subzero temperature and the likelihood of intracellular freezing. *Journal of General Physiology* 47:347–369, 1963.
52. M Toner. Nucleation of ice crystals inside biological cells. In: P Steponkus, Ed., *Advances in Low Temperature Biology*. London: JAI Press, 1993, pp. 1–52.
53. O Miyawaki, L Liu, GX Ye, T Suzuki, R Takai, K Kagitani. Progressive freeze-concentration: Principle and applications. *Proceedings of the 9th International Congress on Engineering and Foods (ICEF 9)*, Montpellier (CDROM), 2004, paper 469.
54. DJ Jarvis, SGR Brown, JA Spittle. Modeling of non-equilibrium solidification in ternary alloys: comparison of 1D, 2D and 3D cellular automaton-finite difference simulation. *Materials Science and Technology* 16:1420–1424, 2000.
55. SGR Brown. Simulation of diffusional composite growth using the cellular automation finite difference (CAFD) method. *Journal of Materials Science* 33:4769–4773.
56. D Raabe. Cellular automata in materials science with particular reference to recrystallization simulation. *Annual Review of Materials Research* 32:53–76, 2002.
57. B Woinet, J Andrieu, M Laurent. Experimental and theoretical study of model food freezing, Part 1: heat transfer modeling. *Journal of Food Engineering* 35:381–393, 1998.
58. B Woinet, J Andrieu, M Laurent, SG Min. Experimental and theoretical study of model food freezing, Part 2: characterization and modeling of ice crystal size. *Journal of Food Engineering* 35:394–407, 1998.
59. A Le Bail, D Chevalier, DM Mussa, M Ghoul. High pressure freezing and thawing of foods: a review. *International Journal of Refrigeration* 25:504–513, 2002.

60. L Otero, P Sanz. High pressure shift freezing, Part 1: amount of ice instantaneously formed in the process. *Biotechnology Progress* 16:1030–1036, 2000.
61. JM Chourot, L Boillereaux, M Havet, A Le Bail. Numerical modeling of high pressure thawing: application to water thawing. *Journal of Food Engineering* 34:63–75, 1997.
62. S Denys, AM VanLoey, ME Hendrickx. Modeling heat transfer during high-pressure freezing and thawing. *Biotechnology Progress* 13:416–423, 1997.
63. S Denys, AM Van Loey, ME Hendrickx. Modeling conductive heat transfer during high-pressure thawing processes: determination of latent heat as a function of pressure. *Biotechnology Progress* 16:447–455, 2000.
64. Y Rabin, MJ Taylor, N Wolmark. Thermal expansion measurement of frozen biological tissues at cryogenic temperatures. *ASME Journal of Biomechanical Engineering* 120:259–266, 1998.
65. B Rubinsky, EG Cravalho, B Mikic. Thermal stresses in frozen organs. *Cryobiology* 17:66–73, 1980.
66. Y Rabin, PS Steif. Thermal stresses in a freezing sphere and its application in cryobiology. *Transactions of the ASME* 65:328–333, 1998.
67. X Shi, AK Datta, Y Mukherjee. Thermal stresses from large volumetric expansion during freezing of biomaterials. *Transactions of the ASME* 120:720–726, 1998.
68. X Shi, AK Datta, Y Mukherjee. Thermal fracture in a biomaterial during rapid freezing. *Journal of Thermal Stresses* 22:275–292, 1999.

8 Innovations in Freezing Process

Da-Wen Sun and Liyun Zheng
National University of Ireland, Dublin, Ireland

CONTENTS

I. INTRODUCTION

Freezing is an excellent preservation method for foods [1,2]. The quality of frozen foods is closely related to the size and distribution of ice crystals. Existence of large ice crystals within the frozen food tissue could result in mechanical damage, drip loss, and thus reduction in product quality.

The rate of freezing strongly affects the size and distribution of ice crystals [2,3]. Rapid freezing intends to produce small and even ice crystals, whereas large ice crystals are normally formed during slow freezing. Therefore, the rate of freezing and the formation of ice crystals are critical to the quality of frozen foods.

Currently, air-blast, plate, immersion, and cryogenic freezing are the most common methods used in the food industry [4]. However, food products have low thermal conductivity with typical values of about 0.5–1.5 W/mK, which highly limits the achievable freezing rates. Therefore, chemical and physical aids are researched and developed in order to accelerate the freezing process. As a result, numerous innovations have taken place in achieving rapid freezing processes. These innovations include high-pressure shift freezing (PSF) [5], ultrasonic-assisted freezing [6], dehydrofreezing [7], and applications of antifreeze protein (AFP) [8] and ice nucleation protein (INP) [9]. Through these innovations, significant improvements in product quality have been achieved.

II. THE FREEZING PROCESS

Freezing is the process of removing sensible and latent heat in order to lower product temperature generally to $-18^{\circ}C$ or below [1,2]. Figure 8.1 shows a typical freezing curve. It can be seen that during freezing, the temperature in a particular location of a food product consists of three distinct phases: precooling or chilling phase, phase change period, and subcooling or tempering phase [1,2].

In the precooling or chilling phase, only the sensible heat of the product is removed, the product temperature is lowered from the initial temperature to the temperature at which water crystallization is about to begin. Then, if the temperature is further reduced, the free water in the food will start to crystallize to form ice crystals. This is the phase change period and a freezing plateau can be seen in Figure 8.1. In the process of phase transition, latent heat of fusion is released and removed as ice is formed. In the temperature range over which water crystallization occurs, there is also removal of sensible heat of other food components [5]. Although the temperature is almost constant as water is the main component of a food item, it falls slightly because of solute concentration [10]. This phase change period is very important to product quality due to the formation of the ice crystals. In order to achieve high quality of frozen foods, effective control of ice structure in this period is highly critical. When most of the freezable water is converted to ice, subcooling or tempering follows [11]. Besides water, food also contains protein, fat, carbohydrate, and other components. As food is a multicomponent system, the freezing curve of food is different from those of pure substances.

A. Formation of Ice Crystals

The formation of ice crystals during freezing consists of the initial nucleation and subsequent crystal growth. In nucleation, molecules are combined into an ordered particle of a size sufficient to survive, which also serves as a site of crystal growth. The very fine particle formed is called nucleus or seed. In crystal growth, the nucleus is simply enlarged by orderly addition of more molecules [3].

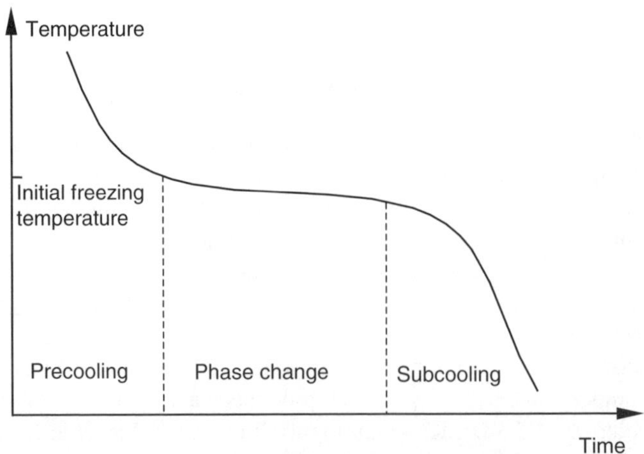

FIGURE 8.1 Typical freezing curve showing a freezing process consisting of three periods: precooling or chilling, phase change, and subcooling or tempering.

1. Nucleation of Ice

Nucleation of ice occurs when the temperature of a food is lowered to the initial freezing point. The nucleus formed is a minute crystal, which is in equilibrium with the surrounding water. Ice nucleation can take place in two ways depending on the purity of water: homogeneous nucleation and heterogeneous nucleation. The former occurs only in extremely pure water, where an ice nucleus is formed by the random orientation and combination of water molecules, while the latter is more likely to take place in foods, and it occurs when water molecules aggregate in a crystalline arrangement on nucleating agents or nucleation activators such as suspended foreign particles, surface films, or walls of containers [12,13].

2. Growth of Ice Crystal

The growth of ice crystal largely depends on the removal of latent heat released. As ice nucleation rate is faster when heat removal is more efficient. Therefore, it is well known that rapidly frozen foods contain small and numerous ice crystals. However, it should be noted that different frozen foods have different sizes of ice crystals even when the foods undergo the same freezing rate and have the same dimensions. This is due to the difference in the amount of free water available in different foods [11,12].

B. Food Microstructure during Freezing

Ice crystallization strongly affects the structure of tissue foods, which in turn damages the palatable attributes and consumer acceptance of the frozen products. The extent of these damages is a function of the size and location of the crystals formed and therefore depends on freezing rate, nature of the cells [3], and permeability of the cells [14].

In traditional freezing processes such as air-blast, plate contact, and cryogenic freezing, when a food is in contact with the freezing medium, ice nucleation occurs in the region next to the frozen border and is controlled by the magnitude of supercooling reached in this zone. Supercooling is the difference between the actual temperature of the sample and the expected solid–liquid equilibrium temperature at that pressure, and is the driving force of ice nucleation, which is also an important parameter in controling the size and number of ice crystals [15]. For example, when pure liquid water is cooled at atmospheric pressure, it does not freeze spontaneously at 0°C because of the existence of supercooling. For pure water without any foreign particles, the supercooling temperature or the freezing temperature can be as low as −40°C [16]. For each degree kelvin of supercooling reached, there is about tenfold increase in the ice-nucleation rate [17]. Between an interior point of the food and surface, there exist thermal gradients, which decrease toward the center of the product and are particularly important in large volume products [18]. In the inner regions of the product, because of the thermal gradients, supercooling to produce ice nucleation is not achieved, resulting in the growth of large ice crystals.

In slow freezing, extracellular ice crystals are formed predominantly. During the process of converting water in the extracellular fluid into ice, the intracellular fluid still remains in some supercooled condition. Therefore, the vapor pressure of the intracellular fluid is higher than that of the extracellular fluid and ice crystals. This pressure difference causes the moving of the intracellular water from cells to deposit on the extracellular ice crystals and thus the growing of large ice crystals [19]. This extensive migration of water would dehydrate the cells, and cause a shrunken appearance of cells in the frozen state. Figure 8.2 illustrates this effect. When thawing, it will increase the drip loss and reduce the overall quality of the food. Extracellular ice crystals can grow to large sizes with the internal transfer of water. If the ice crystal is large enough, it can deform the cells, or even rupture the cells permanently, and subsequently reduce the product quality [20].

In rapid freezing, both extracellular crystallization and intracellular crystallization occur simultaneously. As heat transfer is very rapid in this case, supercooling exists in the intracellular fluid,

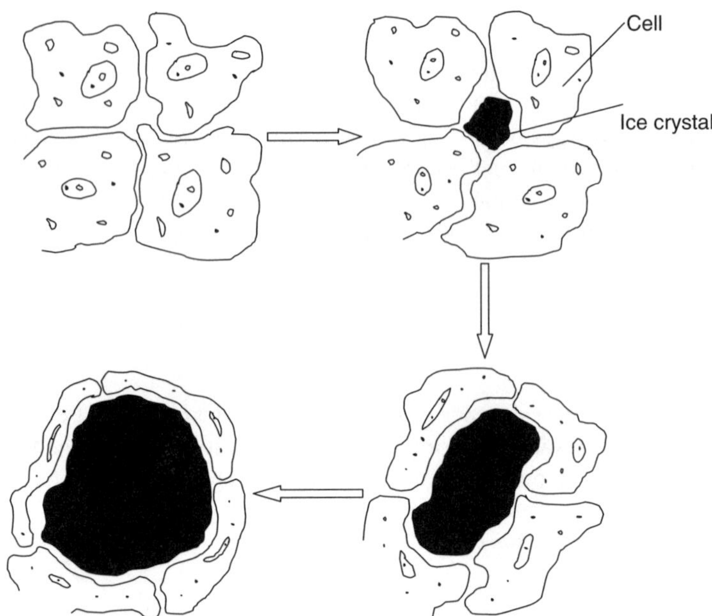

FIGURE 8.2 Development of ice crystals in tissue during slow freezing. (Adapted from HT Meryman. *Federation Proceedings*, 22 (1 P1):81, 1963.)

which initiates nucleation. Therefore, numerous small ice crystals are formed both inside and outside the cells. Unlike slow freezing, water migration is low, therefore, a frozen appearance similar to the original unfrozen appearance is obtained and the high quality of the frozen foods is maintained. Figure 8.3 shows the case of fine and uniform ice crystals distribution in internal and external cells in rapid freezing, providing a product with good quality [20]. The phenomenon in Figure 8.3 only occurs in tissues with very low permeability. If tissue permeability is high, moisture is rapidly transferred from the inside of the cell and the concentration of the intracellular fluid is increased, leading to the depression in the freezing point in the concentrated fluid, thus the intracellular fluid is undercooled and does not freeze. Therefore, even at high freezing rate, dehydration in the cells may still independently occur [14].

The resistance to freezing damage of animal cells is different from that of plant cells. Animal cells have a flexible structure of cell membrane, which can resist rupture during freezing. As the

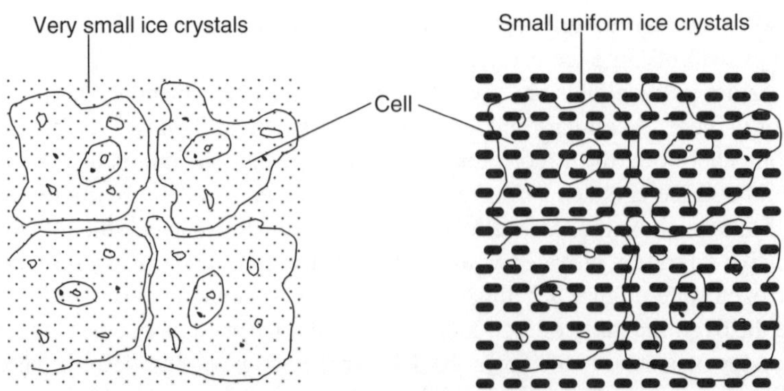

FIGURE 8.3 Development of ice crystals in tissue during rapid freezing (Adapted from HT Meryman. *Federation Proceedings*, 22 (1 P1):81, 1963.)

membrane is less effective against the propagation of ice, intercellular crystallization occurs more often [21]. On the other hand, plant cells possess a semi-rigid cell structure, which is composed of cell membrane and cell wall. As the cell wall is less flexible under stress, the tissues take more risk of freezing damage, depending on the size and locations of ice crystals.

C. Factors Affecting Freezing Rate

In order to maintain the high quality of frozen foods, high freezing rate is desirable. Many factors affect the freezing rate. These factors include temperature difference between freezing medium and food, effective heat transfer coefficient, product shape and size, and physical properties of the food system [10,22].

During freezing, heat transfer occurs from the surface to the freezing medium by convection, and within the food by conduction.

For heat convection, the heat transfer rate per unit area can be calculated by

$$\frac{q}{A} = h(T_s - T_f) \tag{8.1}$$

where T_f (K) is the temperature of the freezing medium, h (kW/m^{2}K) the convective heat transfer coefficient, T_s (K) the surface temperature, A (m^2) the surface area, and q (kW) the rate of heat flow.

For heat conduction, the heat transfer rate per unit area is given by the following equation

$$\frac{q}{A} = k\frac{T - T_s}{X} \tag{8.2}$$

where T (K) is the temperature of a given location within the food, X (m) the distance between that location and the surface, and k (kW/mK) the thermal conductivity of the food.

1. Temperature Difference between Freezing Medium and Food Product

The temperature difference between the freezing medium and food is the driving force for removal of heat as indicated in Equation (8.1). If the freezing temperature is lowered, the freezing rate will always increase. Therefore, lowering the freezing temperature is one method to accelerate the freezing process. For instance, cryogenic refrigeration based on the evaporation of the refrigerant used can provide much lower T_f (about $-78.5°C$ for CO_2 and $-196°C$ for N_2) than that of mechanical refrigeration where the feasible temperature is about $-40°C$ [22]. However, the minimum temperature that a particular freezing medium can reach is limited by the properties of the refrigerant itself.

2. Surface Heat Transfer Coefficient

As indicated in Equation (8.1), if the heat transfer coefficient h is high, more heat can be transferred. The h value is affected by many factors such as freezing medium, flow velocity, type of contact, size, and spatial distribution of the product, existence of packaging and air space between product and packaging [23]. For obtaining high heat transfer coefficient, high freezing medium velocity or turbulence flow is required. The existence of packaging provides a resistance to the transfer of heat from a product to the freezing medium. The influence of packaging on h is dependent on the packaging materials, their thickness, and the presence of air space. Therefore, in order to achieve high freezing rate, food products should be packaged after freezing.

3. Shape and Size of Food Product

The size of food significantly affects the rate of freezing as shown in Equation (8.2). In plate freezing, Woinet et al. [24] indicated that the mean ice crystal size in model food of gelatin gel grew proportionally with the distance from the cold plate. This means that the freezing rate is not uniform within the food, but depending on the distance that the heat must travel. For regular shaped foods, the thermal center is their geometry center where the temperature changes most slowly. Therefore, ice crystal grows larger in the thermal center. For obtaining rapid freezing, small size products are better. In many cases, it is not feasible to change the dimensions of the products, however, it is possible to arrange the products in a single layer rather than in multiple layers of greater thickness in order to improve heat transfer.

For the shape of products, if the thickness of an infinite slab, the diameter of an infinite cylinder, and the diameter of a sphere have the same value, and these three items are exposed to the same freezing conditions, their freezing times are calculated approximately in the ratio of 6 : 3 : 2. It means that a sphere will freeze in two thirds the time of a cylinder and in one third the time of a slab [1,25].

4. Thermophysical Properties of Food

The thermal conductivity k of food is the most influential property in the freezing rate. The thermal conductivity changes with the phase change. The k value for ice is nearly fourfold higher than that for unfrozen water. Furthermore, thermal conductivity is affected by the direction of fiber of foods. Heat transfers faster along the direction of muscle fibers than across the fibers, therefore, the difference in thermal conductivity parallel and perpendicular to the food fibers should be taken into account in evaluating the freezing rate [25,26].

Besides thermal conductivity, product density changes significantly during freezing, however, the influence of density changes on freezing rate is not significant [27]. In the meantime, the specific heat of a food item at a temperature above its initial freezing point is much lower than that below the freezing point, as during freezing, both sensible and latent heat must be removed and latent heat is released over a range of temperatures.

III. HIGH-PRESSURE SHIFT FREEZING

The effects of pressure on the phase diagram of water is shown in Figure 8.4 [28]. At atmospheric pressure, when water is frozen, its volume increases. This increase in volume is because of the formation of ice, which uniquely has a lower density than liquid water, resulting in a volume increase of about 9% on freezing at 0°C and about 13% at −20°C [5]. This volume increase is the main reason of tissue damage during freezing. However, under high pressure, several types of ice (ice II–IX) are formed with different chemical structures and physical properties. The densities of high-pressure ice are greater than that of water. Therefore, during phase transition, high-pressure ice (ice II–IX) do not expand in volume, which may reduce tissue damage.

The frozen preservation of food can take advantage of the above phase change diagram of water to achieve the changing of the physical state of food using external manipulation of pressure [12]. As shown in Figure 8.4, there exists a nonfrozen region of water below 0°C under high pressure. The freezing point of water can be reduced to a minimum of −22°C at 207.5 MPa. When pressure is released, a very high supercooling can be obtained, as a result the ice-nucleation rate is greatly increased. At a very high pressure of 900 MPa, ice VI having a density of 1.31×10^3 kg/m^3 may be formed at room temperature, which means that foods can be frozen without any form of cooling under very high pressure. However, very few experiments have been carried out in this area, owing to the very high pressure required [5]. It is noted from Figure 8.4 that during reduction of pressure, water state changes, for example, as pressure is reduced from 600 MPa to atmospheric pressure, at −20°C, water state will pass ice VI, ice V, ice III, then liquid and finally ice I.

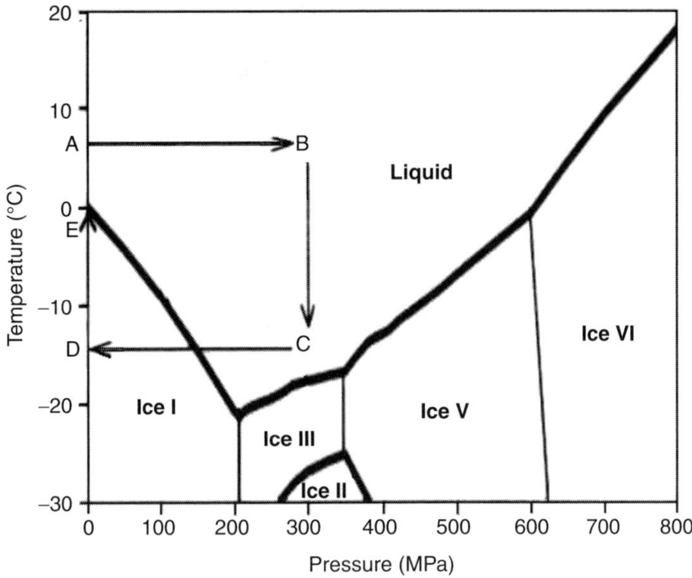

FIGURE 8.4 Principle of high PSF. (Adapted from A Le Bail, D Chevalier, DM Mussa, M Ghoul. *International Journal of Refrigeration* 25:504–513, 2002.)

With a good understanding of the water phase change, high pressure can be used to greatly aid the freezing process and to improve product quality. The process of reducing the temperature well below 0°C under pressure and then quickly releasing the pressure to initiate rapid ice nucleation is called pressure shift freezing (PSF), which is marked as A → B → C → D → E in Figure 8.4. In PSF, the food products are initially kept in the liquid state at subzero temperature with high-pressure (Figure 8.4: A → B → C). Phase transition then occurs as a result of a pressure release (Figure 8.4: C → D) that promotes high supercooling (Figure 8.4: D → E) [17]. Because of this large degree of supercooling, the initial formation of ice is instantaneous and homogeneous throughout the whole volume of the product and not only on the surface [13,18,29]. In practice, PSF is carried out in a high-pressure vessel with its temperature being regulated at subzero temperatures. Food samples are cooled to a typical low temperature of −20°C at 200 MPa in the vessel where the high-pressure maintains the water within the food in a liquid state. The pressure is then rapidly released to the atmosphere and the samples undergo a sudden temperature rise up to the phase change temperature at the existing pressure. Partial freezing is initiated during this pressure release because of high supercooling of the samples. The temperature in the samples increases according to the temperature–pressure equilibrium relationship of liquid water and ice I [15]. Only partial freezing can be obtained during PSF as experiments show that the ice to water ratio can only reach 0.36 for a sample of pure water at the end of the pressure release step. Freezing must thereafter be completed at atmospheric pressure. Therefore, PSF technology can be especially useful to freeze large items of foods in which a uniform ice crystal distribution is required. Such large items are difficult to freeze even using efficient classical freezing methods including cryogenic freezing [13,29], as under very low freezing medium temperature, thermal gradients within the foods are pronounced and damage of freeze-cracking would be possible. The use of high pressure facilitates supercooling, promotes uniform and rapid ice nucleation and growth, thus producing smaller crystals [5]. From the microstructural point of view, damage to cells is minimized because of the small size of ice crystals, resulting in a significant improvement in product quality [13,15,30–32]. Another high-pressure freezing process which is distinguished and essentially different from PSF is called high-pressure-assisted freezing (PAF). In PAF, phase transition takes place under constant high pressure when the temperature is lowered to corresponding freezing

TABLE 8.1
Experimental Studies Available in Literature on High-Pressure Freezing of Foods

Food System	Process	Conditions	Analysis	Comparison	Ref.
Tofu	PAF/PSF	100, 200, 340, 400–600, 700 MPa; −19°C	Microstructure	AB: −20°C, −30°C, −80°C	[30]
Tofu	PSF	200 MPa; −18°C	Microstructure	AB: −10°C, −18°C	[33]
Carrot	PAF/PSF	100–700 MPa; −18 ∼ −20°C	Texture Microstructure Pectin content		[34]
Carrot	PAF/PSF	Various conditions at 50–400 MPa; −15 ∼ −28°C	Microstructure		[35]
Chinese cabbage	PAF/PSF	100–700 MPa; −18 ∼ −20°C	Microstructure Pectin content	AB: −30°C	[31]
Potato	PSF	400 MPa; −15°C	Texture Color Microstructure	AB: −30°C	[36]
Eggplant	PSF		Texture Microstructure Drip losses	AB	[37]
Pork	PSF	200 MPa; −20°C	Microstructure		[13]
Pork, beef	PSF	200 MPa; −20°C	Protein stability Microstructure		[38]
Norway lobsters	PSF	200 MPa; −18°C	Protein stability Microstructure	AB: −30°C	[15]
Protein gels	PAF/PSF	200 MPa; −33°C 270 MPa; −20°C	Texture Microstructure Drip losses	SA: −43°C LN: −80°C	[39]
Cheeses	PSF	200 MPa; −20°C	Texture Rheology		[40]
Peach, Mango	PSF	200 MPa; −20°C	Microstructure		[29]

AB: air-blast freezing; SA: slow air freezing; LN: liquid nitrogen freezing; PAF: pressure-assisted freezing; PSF: pressure shift freezing.

point. Therefore, ice I or other forms of ice can be obtained. Cooling of the sample occurs from the surface to center as in a normal freezing process at atmospheric pressure [17].

PSF technology has recently attracted a greater attention. Table 8.1 lists the studies available in the literature on high-pressure freezing of various food systems. Most of the studies investigate the effect of PSF on microstructure of frozen foods. Martino et al. [13] conducted experiments on PSF of large pork pieces. Comparing with traditional air-blast and liquid freezing, it was found that high-pressure frozen samples of uniform, small-sized ice crystals both at the surface and at the central zones, whereas air-blast and cryogenic fluid freezing, having thermal gradients, showed nonuniform ice crystal distribution [13]. Chevalier et al. [15] and Kanda et al. [33] also confirmed that high-pressure freezing resulted in a reduction in the size of ice crystals and in a much preserved microstructure in comparison with air-blast frozen samples, however, the operation of high-pressure freezing should adopt lower pressure levels so that the effect of pressure on proteins can be minimized.

To study the effect of PSF on textural changes of foods, experiments on high pressure freezing of carrots were carried out and the results showed that high-pressure freezing at 200, 340, and 400 MPa would be effective in improving both the texture and histological structure of frozen carrots [34,35]. This result was generally in agreement with the findings on high-pressure-frozen Chinese cabbage and tofu [30–33]. Comparison between PSF-treated potatoes and untreated samples also revealed that PSF preserved the texture well, whereas air-blast freezing resulted in a reduction of the rupture strength [36]. Pork samples also had less damage with PSF in terms of structural preservation compared to classical air-blast or cryogenic freezing methods [13]. However, it was also reported that the textural and structure of konnyaku frozen at 0.1–700 MPa then thawed at atmospheric pressure changed greatly, but high-pressure freezing was ineffective in improving the texture of the frozen konnyaku [41].

Peach and mango were also cooled under high pressure at 200 MPa to $-20°C$ without ice formation, then the pressure was released to 0.1 MPa. By scanning electron microscopy (SEM), it was observed that the cells were arranged adjacently without clear breakage, indicating that PSF was the freezing method that best preserved their vegetal microstructures [29].

Most of the earlier studies showed that PSF-treated foods have a much preserved texture. Microscopic observations also indicate that the microstructures of foods can be preserved as long as no ice is formed before depressurization. However, texture measurements have revealed that denaturation might occur under pressure, resulting in a modification of the texture of foods containing high protein. Drip loss reduction is variable from one product to another. Further studies on potential microbial destruction of microorganisms in PSF-treated foods are needed [28].

At the commercial level, Japan is at the forefront of PSF application for food processing. The United States and Europe are also exploring the commercialization of this technology [42]. In the commercialization of PSF technology, the biggest obstacle is the high capital costs [43]. As the operation of high-pressure equipment is at subzero temperature, the use of special steel is needed for vessel design and suitable pressure transmission fluid is required. Furthermore, precise monitoring is also necessary for improving product quality and stability of the operation.

IV. ULTRASONIC FREEZING

Although the use of power ultrasound to assist food-freezing is a relatively new subject, recent research advances indicate that its potential is promising [6,44]. The beneficial use of the ultrasound is realized through the mechanical and physical effects that it generates upon the medium through which it transmits. The basic components of a freezing system contain ice crystals and aqueous phase. When it is subjected to the action of the acoustic energy, the compression and rarefaction of the sound waves can cause the occurrence of cavitation in the aqueous phase [45]. Cavitation leads to the production of gas bubbles that will continue to grow and act as nucleating agents to promote nucleation [46]. Experiments with concentrated sucrose solution showed that the number of nuclei increased with the application of power ultrasound [47]. Microstreaming is another significant acoustic phenomena associated with cavitation, which occurs because of the vigorous circulatory motion of the cavitation bubbles in the sonic field [48]. The violent agitation that ultrasound provides can benefit in increasing heat and mass transfer rate [6,49], which can therefore accelerate the freezing process [6]. Similar to other dense and incompressible materials, ice crystals will fracture when they are subjected to the alternating acoustic stress [50]. This will consequently lead to products with smaller crystal size distribution. An experimental study shows that for a sucrose solution treated with power ultrasound, 32% of the water exists as crystals with a diameter of 50 μm or larger, in comparison with 77% for the one without acoustic treatment [50].

Resulting from the above acoustic effects, power ultrasound has been widely used to initiate ice nucleation [47]. It can also be applied to modify and control crystal size distribution during the solidification of liquid food [50]. The cleaning action of cavitation can also effectively prevent encrustation on freezing surface [51]. As the area of concern of this chapter is freeze preservation of fresh foodstuffs, only the most relevant study conducted by Li and Sun [6,44] will be discussed in detail. In their experimental investigation, power ultrasound was applied intermittently during immersion freezing of potato slices. Their results showed that ultrasound with a power level of 15.85 and 25.89 W can both lead to noticeable increase of freezing rate, as indicated by a more rapid reduction of product temperature (Figure 8.5).

Furthermore, they found that the ability of power ultrasound in accelerating the freezing process is dependent on the level of acoustic power applied as well as the duration of the acoustic treatment [6]. As shown in Figure 8.5, ultrasound of 7.34 W did not cause any obvious change in freezing rate. However, when the acoustic power was increased to either 15.85 or 25.89 W, freezing rate under both conditions was observed to increase yet at different patterns. A more rapid temperature reduction is observed for the latter than the former during the phase changing period, although towards the end of the phase changing period, temperature of product treated at 25.89 W of acoustic power declines more slowly than the one at 15.85 W (Figure 8.5). The reason for this is not yet fully understood. Li and Sun [6] suggested that this might be associated with the accumulated thermal effect that power ultrasound might have upon the product, as reported by other researchers that depending on the nature of the medium, sound waves can be absorbed by the medium of transmission and converted into heat [52]. In this case, as acoustic energy transmitted through the refrigerant before it reached the product, it can therefore be lost either to the refrigerant or the product; neither of them benefits the freezing process. The lower freezing rate for acoustic power of 25.89 W during the later stage of freezing seemed to imply that thermal effect is proportional to the amount of acoustic power applied.

The results from Li and Sun [6] also indicated that the duration of the power ultrasound has a similar effect on the freezing rate as the level of acoustic power. As shown in Figure 8.6, total acoustic treatment of 1 min only resulted in very slight change in the freezing rate. During the phase-changing period, as duration of power ultrasound was increased from 1.5 to 2.5 min, freezing rate was observed to increase. Towards the end of the phase-changing period, possibly because of a similar thermal effect as discussed earlier, temperature reduction for potatoes treated for 2.5 min was slower than that for 2 min.

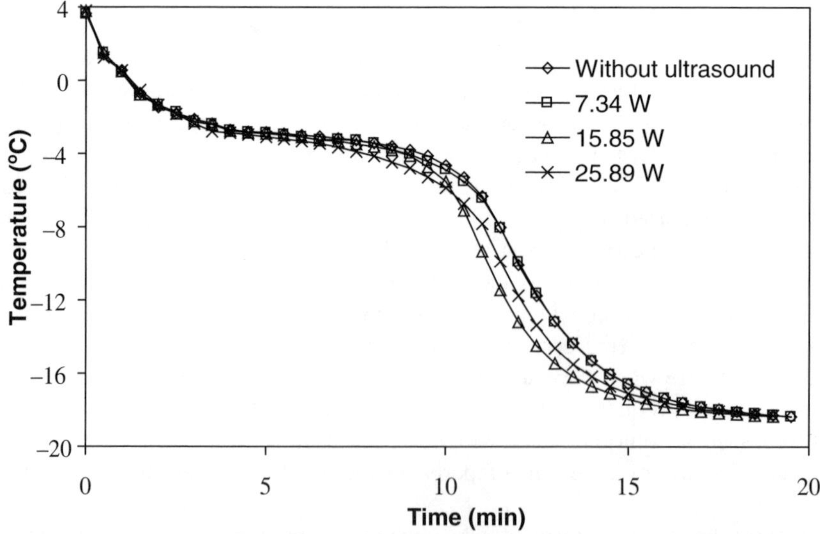

FIGURE 8.5 Effect of acoustic power on the ability of power ultrasound in accelerating the food-freezing process. (Adapted from B Li, D-W Sun. *Journal of Food Engineering*, 55 (3):277–282, 2002.)

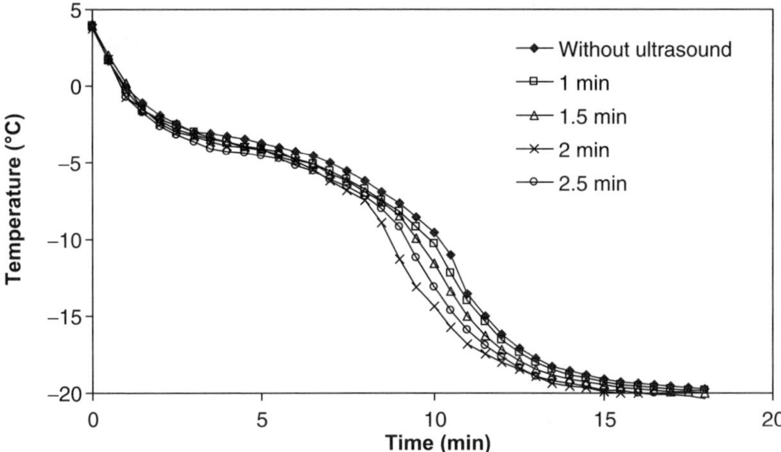

FIGURE 8.6 Effect of acoustic duration on the ability of power ultrasound (power level = 15.85 W) in accelerating the food-freezing process. (Adapted from B Li, D-W Sun. *Journal of Food Engineering* 55 (3):277–282, 2002.)

In addition to its ability to accelerate the freezing process, power ultrasound can also improve product quality [44]. The SEM of photos showed that plant tissues of ultrasound-assisted frozen potatoes exhibit a better cellular structure than those without acoustic treatment as less extracellular void and cell breakage/disruption were observed [44]. This is attributable to several factors. First, it might be because of the higher freezing rate resulting from the acoustic treatment, as fast-freezing has been widely proven to be one of the most effective methods for the production of high-quality frozen food [2]. Crystal fragmentation is another possibility, which can reduce the size of crystals inside the product. As small crystals execute less mechanical stress upon the cell membrane, consequently there will be less cell breakage and deformation [53]. Finally, cavitation bubbles might also initiate intracellular nucleation, which usually might not occur without the acoustic treatment owing to insufficient degree of supercooling [3]. Cavitation can also contribute to a higher nucleation rate in the extracellular region. Both of them will help the frozen product to have a similar appearance to its original unfrozen shape and thus achieve higher product quality.

The use of power ultrasound in assisting food-freezing is promising, which not only enhances the freezing rate, but also leads to a product of better quality. The future development of this technology is still strongly related to the availability of cost-effective and easily operated equipment. Instead of being a new freezing technique, the ultrasound is an aid to existing freezing process. Therefore, it is preferred that ultrasonic device can be designed in such a way that it can be easily connected to existing freezing equipment, which still requires further research effort. In general, although commercial application of this new technology is yet to be realized, research hurdles do not appear to be unsurpassable and its benefits to food manufacturers are obvious.

V. DEHYDROFREEZING

Dehydrofreezing is a well-established commercial method to reduce cost of shipping, handling, and storage of fruits and vegetables. During dehydrofreezing, food is dehydrated first to desirable moisture and then frozen [7,54]. Most fresh fruits and vegetables contain more water than meats, and their cellular structure of cell wall is less elastic than cell membrane and could be susceptible to large ice crystals formed during freezing. Therefore, in commercial freezing, the presence of large amount of water in fruits and vegetables could inevitably cause tissue damage. Dehydrofreezing provides a promising way to preserve fruits and vegetables as part of water is removed from the foods prior to freezing [7,55,56]. A reduction in moisture content also reduces the amount of water to be frozen, therefore, refrigeration load needed during freezing can be lowered. Furthermore, dehydrofrozen products can reduce cost of packaging, distribution and storage, and maintain product quality comparable to conventional products [55]. Figure 8.7 shows the comparison between the conventional freezing and dehydrofreezing processes of fruit and vegetable products [57].

Figure 8.7 shows that besides typical stages such as product preparation, treatment, packaging and storage, the additional stage in dehydrofreezing is the partial dehydration, which influences the subsequent freezing process and quality of final products. Air-drying and osmotic dehydration are the common methods used to remove part of the water. The efficiency of dehydration process is evaluated in terms of rate and extent of water removal [58]. Therefore, osmotic dehydration is a more popular method as it has advantages over convective hot-air-drying, such as adaptability to a wider variety of products and lower energy requirement, better texture, taste, and final appearance of the food. However, when using osmotic dehydration, care should be taken in choosing the aqueous solution of high osmotic pressure as solute uptake often leads to substantial modification of the product composition with a negative impact on sensory characteristics [59]. Sucrose is often used as an osmotic agent for osmotic dehydration of fruits. However, it is not suitable for vegetables because of excessive sweetness from sucrose uptake. For vegetables, sodium chloride is commonly used. Significant changes in vitamin C content can be prevented by adding ascorbic acid to the osmotic drying solution. Other osmotic agents include glucose, fructose, lactose, maltodextrin, corn syrup, and so on [55,60].

Dehydrofreezing has been successfully applied to fruits and vegetables. Robbers et al. [7] dehydrofroze samples of fresh kiwi by first immersing them in 68% (w/w) aqueous sucrose solution to dehydrate for 3 h, then freezing in an air-blast freezer with an air velocity of 3 m/s at about $-3°C$.

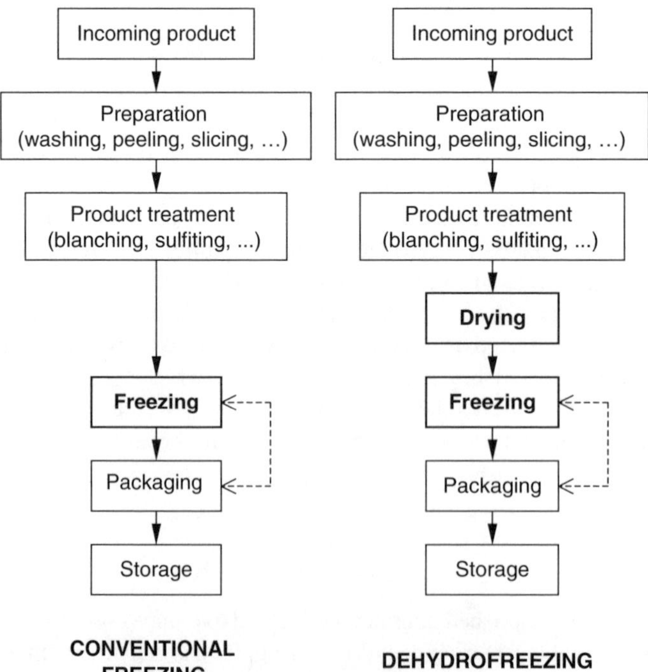

FIGURE 8.7 Principle of dehydrofreezing process for fruit and vegetables. (Adapted from KS Suslick. In: KS Suslick, Ed., *Ultrasound*. New York: VCH, 1988, pp. 123–163.)

The experiment showed that freezing began at a lower temperature in the dehydrated product and the temperature of the dehydrated samples was reduced to −18°C in 19–20 min, which was about 20–30% faster compared to untreated kiwi, which required the freezing time of 23–24 min. Generally speaking, lower water content of dehydrated food always induces a lower freezing point and a shorter freezing time as there is less water to freeze and consequently less heat to remove [54]. Garrote and Bertone [56] dehydrofroze strawberry halves by osmotically treating them in the presence of solution of glycerol, glucose, and sucrose of varying concentrations and then freezing with liquid refrigerant, and found that the dehydrofrozen strawberry halves sustained a significantly smaller exudates loss, although untreated fresh strawberry halves produced a larger amount of exudates. Similarly, melon samples which were dehydrated longer produced less exudates, which was in agreement with the lower water content and presumably lower extent of structural damage caused by freezing [54]. Furthermore, research on dehydrofreezing of muskmelon also confirmed the reduction of exudate loss on thawing. In the meantime, analysis of exudates loss, texture, color, aroma composition, and sensory characteristics of dehydrofrozen muskmelon indicated that the cultivar also had a greater influence on the quality of the end-products [61].

For frozen fruits and vegetables, except freezing rate and exudates, sensory characteristics and textures are important quality factors affecting the acceptability by consumers. Hardness, taste, and overall acceptability were evaluated for dehydrofrozen green beans, which were dehydrated by soaking in NaCl–water solution and then frozen in an air blast freezer [55]. Sensory analysis suggested that osmotically dehydrated frozen green beans were as good and equally acceptable as conventionally frozen green beans. Apples, peaches, and apricots were either dehydrated osmotically or by combining osmosis with air-drying and then frozen, color, texture, and sensory evaluation indicated that osmotically dehydrated frozen fruits were organoleptically acceptable [62].

With the recognition of the advantage of using dehydrofreezing technology, perhaps the most important point for further research is that some efficient systems for water removal need to be developed. For example, hot-air-drying could be equipped with some heat recovery devices in order to reduce the overall energy cost. As dehydrofreezing of foods would be expected to require rehydration prior to consumption, establishing correlation between processing conditions and rehydration characteristics is also important in order to maintain good flavor and texture of the dehydrofrozen foods [57].

VI. ANTIFREEZE PROTEINS (AFP) AND ICE NUCLEATION PROTEINS (INP)

The size of ice crystals in frozen food significantly affects the quality of the products. Therefore, controlling the growth of ice crystals is one of the primary concerns of food technologists [63]. Antifreeze proteins (AFP) and ice nuclear proteins (INP) can both influence ice crystal development and can be directly added to food to interact with ice. Although both AFP and INP influence ice crystal size and crystal structure within the food, they are very different substances in structure and they function distinctly and oppositely during freezing [9]. AFPs are used to lower the freezing temperature and retard recrystallization on frozen storage; however ice nucleation proteins raise the temperature of ice nucleation and reduce the degree of supercooling [8,64]. Although both proteins show opposite effects on controling ice crystallization, they are potentially added in food and therefore attract much attention from food technologists.

A. Antifreeze Proteins

The discovery of AFP was based on the observation of fishes that inhabit in polar and northern coastal waters whose freezing point is close to −1.9°C or about 1°C below the plasma freezing point of the fishes [65]. AFP were first identified by DeVries and Wohlschlag in 1969 in the blood of these fishes [66]. The proteins obviously served to lower the freezing point of the blood of the fishes to below the freezing point of seawater, without significantly increasing the osmotic pressure of the plasma [67]. Since then, AFPs have been identified in a wide range of fish in areas susceptible to ice formation; they have also been reported to be present in many invertebrates including most insects and in higher plants as well as in fungi and bacteria [68,69].

The most studied proteins with antifreeze activity are from fish. Based on the presence or absence of carbohydrates, AFP are classified into two main types: glycoproteins and nonglycoproteins [9]. Antifreeze glycoproteins (AFGP) mainly consist of repeating units of two amino acids, in which one of them is glycosylated [8]. For convenience, nonglycoproteins are still called AFP', which can be further subdivided into four distinct antifreeze subtypes: the alanine-rich AFP of right eye flounders and sculpins (type I), the cystine-rich AFP of sea raven smelt and herrins (type II), an AFP (type III) found in ocean pout and eelpout wolfish, and the glutamine and glutamate-rich AFP of long horn sculpin (type IV) [67]. Table 8.2 lists the characteristics of these AFPs present in fish [67]. As the function of the AFP is to lower the point at which ice crystals grow and to modify the ice habit and growth rate [9], so that smaller crystals and crystals of different shapes are formed, it is generally accepted that AFP functions by binding to ice and interfering with water molecule propagation to crystal surface. Different AFP's obviously show preference for different crystal planes [70]. Wen and Laursen [65] proposed a two-step process for the binding of AFP to ice surface: at low concentration, AFP molecules bind individually to the surface; at sufficiently high concentration, AFP molecules pack together in a cooperative manner to exert maximal activity [65]. They also suggested a model for the inhibition of ice crystal growth. In this model, patches or aggregates of AFP molecules are assumed to bind tightly to the ice surface, so that the ice lattice is only allowed to grow in the spaces between AFP molecules, hence decreasing the stability of the surface at the ice water interface. Therefore, the addition of water to ice surface is unfavorable, and the growth of the crystal is inhibited. Moreover, when the AFP is adsorbed to ice surfaces, it tends to bind to ice prism faces. The dipole nature of the AFPs might account for their preferential binding [71]. It is postulated that the dipole field of the α-helix would align dipole moments of individual water molecules in the ice crystals, therefore, a dipole–dipole interaction is induced between the protein molecule and the ice crystals. These interactions would lead to specific adsorption on the prismatic facets of ice. As a result, the ice habit is modified. Harrison et al. [72] also presented the selective growth facet action of AFGP.

The potential for the application of AFP in foods to suppress freezing point and inhibit recrystallization during freezing is very promising. AFP could be genetically introduced into foodstuffs or could be synthesized genetically or chemically and added to the food produced [8]. However, practical applications are still seldom reported. One possible application of AFP is to inhibit recrystallization of ice in dairy products such as ice cream and deicing agents, as the ice crystal existing in ice cream is a very important factor to preserve the smooth and creamy texture of ice cream. However, recrystallization occurs inevitably if temperature fluctuates during storage or in transit, resulting in coarse texture of ice cream and damage in quality. In a patent reported by Warren et al. [73], AFP was added to food product, which is a composite of a root beet shell with a heart of vanilla ice cream. By adding a small quantity of AFP, the sample was then frozen at about $-80°C$, and stored at -6 to $-8°C$. Very little ice crystal growth was observed after 1 h of storage in the sample with AFP, whereas the control sample showed a definite increase in the size of ice crystals.

The function of AFP in inhibiting recrystallization may also be very useful in maintaining the high quality of chilled and frozen meats, as in slow freezing large ice crystals may form intracellularly, resulting in drip loss of nutrition during thawing. Bovine and ovine muscles meats have been experimented by soaking in solutions of up to 1 mg/ml type I AFP or AFGP prior to freezing at $-20°C$ and the results showed evidence of reduced ice crystal size [74]. Even amount of AFGP as small as 0.5 $\mu g/ml$ can give extensive inhibition of ice crystal growth [8]. AFP could be incorporated into live animals and still have effects on meat quality during freezing and thawing. In the study of Payne and Young [75], AFGP isolated from Antarctic cod was injected intravenously into lambs at various times prior to slaughter. Samples of meat were then vacuum-packed and stored at

TABLE 8.2
Characteristics of the Five Types of AFP Present in Fish

Characteristics	AFGP	Type I AFP	Type II AFP	Type III AFP	Type IV AFP
Molecular mass (Da)	2600–33,000	3300–4500	11,000–24,000	6500–14,000	12,300
Primary structure	(Alanine–alanine–threonine)$_n$ disaccharide	Alanine-rich multiple of 11 aminoacid repeats	Cystine-rich, disulfide linked	No dominant aminoacid(s) or repeat units (e.g., as in AFGP)	Glutamine- and glutamate-rich (26%)
Glycoprotein	Yes	No	No (exception: smelts have $<3\%$ carbohydrate)	No	–
Secondary structure	Expanded	α-Helical amphiphilic	β-Sheet	β-Sandwich	α-Helix
Tertiary structure	Not determined	100% Helical	Globular c-type lectin fold	Globular	Four-helix bundle
Protein components	8	7	2–6	12	1
Gene copies	Not determined	80–100	15	30–150	Not determined
Natural source	Antarctic notothenioids, northern cods (Atlantic cod, Greenland cod)	Right-eyed flounders (winter flounder), shorthorn sculpin	Sea raven, smelt, herring	Ocean pout, eelpout wolffish	Longhorn sculpin

Source: Adapted from RWR Crevel, JK Fedyk. *Food and Chemical Toxicology,* 40:899–903, 2002.

−20°C for 2–16 weeks. The injection of AFGP at either 1 or 24 h before slaughter was reported to reduce drip loss and ice crystal size. Particularly, ice crystals were the smallest in the lambs injected at 24 h before slaughter with a final concentration of 0.01 μg/kg AFGP [75].

The commercial application of AFP in foods most probably will depend on the cost of the proteins [63]. Although commercial products of AFP or AFGP are currently available, they are mainly for research or special uses because of their high price. Chemical synthesis and genetic engineering may be a solution to produce cost-effective AFP, hence to promote their applications in frozen food products.

B. Ice Nucleation Proteins

As discussed previously, the formation of ice crystals during freezing is initiated by ice nucleation, which is promoted by foreign particles. These foreign particles are generally termed as ice nucleation activators (INAs). There exist various types of ice nucleation activators of biogenic origin in plant bacteria, insects, intertidal invertebrates, plants, and lichen [9], the INP from some ice nucleating bacteria are the highest level of INA [76]. The functions of bacterial INA are to reduce the degree of supercooling and catalyze ice formation. Figure 8.8 shows two freezing curves, one with the addition of INP and the other without INP. It can be seen that with the addition of INP, supercooling is significantly reduced, and ice nucleation occurs much earlier, thus leading to shortening of freezing time. The most common species of ice nucleation active bacteria that have been found to produce INA belongs to genera *Pseudomonas, Erwinia*, and *Xanthomonas.* Some strains of *Fusarium* and related genera of fungi are also active during ice nucleation [16]. These bacteria can catalyze ice formation at temperatures as high as −2 to −3°C, resulting in frost damage of many important crops [76]. However, because not all natural strains exhibit ice nucleation activity, those which produce INA substance are called Ina^+, and those, which do not are Ina^-. The study by Phelps and coworkers [77] on isolation of cell-free ice nuclei from *Erwinia herbicola M1* followed

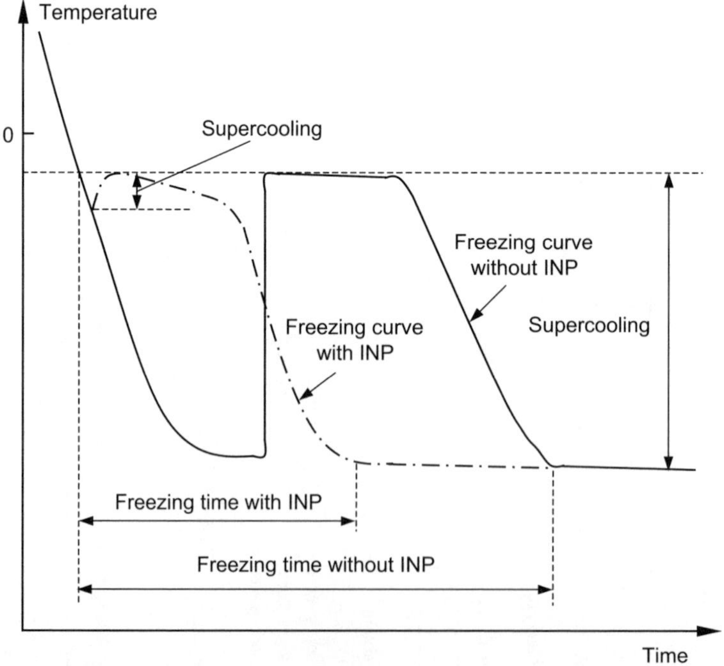

FIGURE 8.8 Typical freezing curves with and without addition of INP.

by a series of treatments suggested that cell-free INA substance is associated with outer membrane vesicles. The bacterial ice nucleation phenotype is very sensitive to proteases and sulfhydryl-modifying chemicals, indicating that a protein is required for icenucleating activity [78,79]. Phospholipid is also a requirement for expression of ice nucleating activity in *Pseudomonas syringae* [80]. Li and Lee [81] summarized that each protein consists of three distinguishable domain structures: a N-terminal domain, which is relatively hydrophobic; a very hydrophilic C-terminal domain; and a central repeating domain which is hydrophilic and particularly rich in alanine, glycine, serine, and threonine.

Ina^+ bacterial cells and their products such as INP have great potential for their applications in the freezing of foods. They elevate the temperature of ice nucleation, shorten freezing time, increase freezing rate, and change the texture of frozen foods, thus decreasing refrigeration cost and improving the quality. Therefore, INP have been added to frozen products in various experiments. Arai and Watanabe [82] froze samples of egg whites at −10°C and observed supercooling lower than −6°C; however, when INA bacterial cells (*Erwinia ananas*) were added, the samples only underwent a slight degree of supercooling [82]. Similar results were also found in the case of using INA bacterial cells entrapped in calcium alginate gel [83]. Ina^+ *P. syringae* cells were also added to sucrose (10% w/w), egg white (9% w/w), safflower oil (20% w/w), and salmon muscle which were subjected to freezing at −6 or −5°C (for salmon muscle sample); they significantly elevated ice nucleation temperature from nonfreezing, −5.1, −6.0 and −4.9 to −1.8, −0.6, −0.8 and −1.5°C, respectively, as a result, 12–33% reduction of total freezing time was achieved [81]. Li and Lee [64] also reported that bacterial extracellular ice nucleation (ECIN) from *E. ananas* were used for efficient freezing and texture modification. Different food samples, such as liquid (milk, oil, cream, and juice), semi-solid (ice cream and yogurt) and solid (ground beef, rice flour paste, and tilapia fish fillet), were investigated. When 700 units of ECINs (70 g protein) were added to 10 ml liquid samples freezing at −6°C, the degree of supercooling was reduced significantly. Table 8.3 lists some samples whose ice nucleation temperatures were significantly increased by adding ECINs [64].

Research has also been carried out to investigate the effect of INP on ice formation patterns which greatly affect the quality such as texture of frozen foods. Arai and Watanabe [82] proposed

TABLE 8.3
Effect of ECINs on Nucleation Temperature (°C) of Various Foods Freezing at −6°C

Sample	Control	+ECIN
Sucrose (10%)	Did not freeze	−1.3 ± 0.3
Egg white (9%)	Did not freeze	−1.7 ± 0.1
Safflower oil (20%)	−4.3 ± 0.5	−0.8 ± 0.2
Whole milk	Did not freeze	−1.9 ± 0.3
1% milk	Did not freeze	−1.6 ± 0.1
2% milk	Did not freeze	−1.2 ± 0.2
Heavy cream	Did not freeze	−1.5 ± 0.1
Non-dairy cream	Did not freeze	−0.6 ± 0.1
Vanilla ice cream	Did not freeze	−4.7 ± 0.3
Chocolate ice cream	Did not freeze	−5.2 ± 0.4
Starch gel (5%)	Did not freeze	−0.6 ± 0.2
Starch gel (10%)	Did not freeze	−1.8 ± 0.3
Yogurt (banana)	Did not freeze	−1.5 ± 0.2
Yogurt (strawberry)	Did not freeze	−3.3 ± 0.2

Source: Adapted from J Li, TC Lee. *Journal of Food Science*, 63 (3):375–381, 1998.

a process to use *E. ananas* bacterial cells as heterogeneous ice nuclei to form anisotropically texture products. Egg white samples without INA bacterial cells froze at temperature as low as −15°C and formed many small ice crystals. On the other hand, the samples with the bacterial cells that stored at subzero temperatures to allow slow freezing at −3°C or slightly lower formed large and long ice crystals in a mutually parallel directions, which suggested that egg white could be textured into a film or flake state. Such directional textures were also obtained by adding INA bacterial cells to isotropic aqueous dispersions or hydrogels of proteins and polysaccharides, such as bovine blood, 5–15% soybean protein isolate, soybean curd, milk curd, 0.5–2% agar, 5–20% corn starch paste, and 0.5–2% glucomannan [82]. As the mechanical and sensory properties of foods are closely related to their textures, Li and Lee [64] further reported that addition of ECINs obviously affected ice formation patterns. Ice crystals formed at −10°C in the absence of ECINs seemed to be smooth, with no directionality and consisted of very fine particles. In contrast, in the presence of ECINs, supercooling was greatly reduced by 7°C and ice crystals formed at −3°C appeared to be ordered, with a defined directionality and uneven surfaces. Rice flour pasta with ECINs freezing and then thawing showed a higher degree of hardness and was easily fractured [64]. With ECINs, unique ice formation patterns could be obtained. Desirable fiber-like texture for some foods, such as tofu and alkali-extracted red meat or poultry proteins, can be produced using INA bacterial cells and their products [64]. Some other proteinaceous foods such as egg white, bovine blood, soy protein isolate, and milk can be also textured by freezing with whole cells of *E. ananas* at −5°C [82].

In the food industry, there is a trend for developing rapid freezing techniques to preserve high quality of the frozen foods. The addition of INA bacterial cells and their products in food products can elevate ice nucleation temperature, thus reducing freezing time and improving the cost effectiveness of the rapid freezing process. However, one major concern to their applications in the food industry is that bacterial ice nucleators must be robust, environmentally safe, nontoxic, nonpathogenic, and palatable [81]. If whole bacterial cells are used, it is very important to make sure that inedible microorganisms are killed completely before the food is consumed.

VII. CONCLUSIONS

Freezing processes are complex, involving heat transfer and possibilities of a series of physical and chemical changes, which may greatly affect product quality. From energy saving or quality improving point of view, new methods are necessary. The novel methods of high-pressure freezing, ultrasound-assisted freezing, and dehydrofreezing accelerate freezing process, thus forming small and uniform ice crystals. The use of AFP and INP improve freezing process directly by interacting with ice crystals formed. Although the potential of the application of these innovative methods in the food industry for both improving product quality and increasing process efficiency is promising, further research is needed to develop them into cost-effective and highly efficient methods.

REFERENCES

1. AE Delgado, D-W Sun. Heat and mass transfer models for predicting freezing processes – a review. *Journal of Food Engineering* 47 (3):157–174, 2001.
2. B Li, D-W Sun. Novel methods for rapid freezing and thawing of foods — a review. *Journal of Food Engineering* 54 (3):175–182, 2002.
3. OR Fennema, WD Powrie, EH Marth. *Low-Temperature Preservation of Foods and Living Matter*. New York: Marcel Dekker, 1973, pp. 354–380, 598–599.
4. D-W Sun, Ed., *Advances in Food Refrigeration*. Surrey: Leatherhead Publishing. LFRA Ltd., 2001.
5. MT Kalichevsky, D Knorr, PJ Lillford. Potential food applications of high-pressure effects on ice-water transitions. *Trends in Food Science and Technology* 6:253–258, 1995.

6. B Li, D-W Sun. Effect of power ultrasound on freezing rate during immersion freezing. *Journal of Food Engineering* 55 (3):277–282, 2002.
7. M Robbers, RP Singh, LM Cunha. Osmotic-convective dehydrofreezing process for drying kiwifruit. *Journal of Food Science* 62 (5):1039–1042, 1047, 1997.
8. RE Feeney, Y Yeh. Antifreeze proteins: properties, mechanism of action, and possible applications. *Food Technologies* 47:82–88, 90, 1993.
9. CL Hew, DSC Yang. Protein interaction with ice. *European Journal of Biochemistry* 203:33–42, 1992.
10. P Fellows. *Food Processing Technology — Principles and Practice*, 2nd ed., Chichester: Ellis Horwood, 2000, pp. 418–440.
11. PO Persson, G Londahl. Freezing technology. In: CP Mallett, Ed., *Frozen Food Technology*. Glasgow: Blackie Academic & Professional, 1993, pp. 20–58.
12. D Knorr, O Schlueter, V Heinz. Impact of high hydrostatic pressure on phase transitions of foods. *Food Technologies* 52 (9):42–45, 1998.
13. MN Martino, L Otero, PD Sanz, NE Zaritzky. Size and location of ice crystals in pork frozen by high-pressure-assisted freezing as compared to classical methods. *Meat Science* 50 (3):303–313, 1998.
14. DS Reid. Basic physical phenomena in the freezing and thawing of plant and animal tissues. In: Mallett, Ed., 2nd ed. *Frozen Food Technology*. Glasgow: Blackie Academic & Professional, 1994, pp. 1–19.
15. D Chevalier, M Sentissi, M Havet, A Le Bail. Comparison of air-blast and pressure shift freezing on Norway lobster quality. *Journal of Food Science* 65 (2):329–333, 2000.
16. AC Rubiolo. Evaluation of freezing and thawing processes using experimental and mathematical determinations. In: Lozano, Anon, Parada-Arias, and Barbosa-Canovas, Eds., *Trends in Food Engineering*. Lancaster: Technomic Publishing Company, 2000, pp. 179–190.
17. L Otero, PD Sanz. Modelling heat transfer in high pressure food processing: a review. *Innovative Food Science and Emerging Technologies* 4:121–134, 2003.
18. PD Sanz, L Otero, CD Elvira, JA Carrasco. Freezing processes in high-pressure domains. *International Journal of Refrigeration* 20 (5):301–307, 1997.
19. ME Sahagian, HD Goff. Fundamental aspects of the freezing process. In: Jeremiah, Ed., *Freezing Effects on Food Quality*, New York: Marcel Dekker, 1996, pp. 1–50.
20. HT Meryman. Preservation of living cells. *Federation Proceedings* 22 (1P1):81, 1963.
21. M Jul. *The Quality of Frozen Foods*. London: Academic Press, 1984.
22. RH Mascheroni. Engineering trends in food freezing. In: Lozano, Anon, Parada-Arias and Barbosa-Canovas, Eds., *Trends in Food Engineering*, Lancaster: Technomic Publishing Company, 2000, pp. 165–177.
23. AM Tocci, RH Mascheroni. Heat and mass transfer coefficients during the refrigeration, freezing and storage of meats, meat products and analogues. *Journal of Food Engineering* 26:147–160, 1995.
24. B Woinet, J Andrieu, B Laurent, SG Min. Experimental and theoretical study of model food freezing. Part II. Characterisation and modelling the ice crystal size. *Journal of Food Engineering* 35:395–407, 1998.
25. BR Becker, BA Ficke. Freezing times of regularly shaped food items. *International Communication Heat and Mass Transfer* 26:617–626, 1999a.
26. (a) X Zhu, D-W Sun. The effects of thermal conductivity calculation on the accuracy of freezing time predicted by numerical methods. *AIRAH Journal* 55 (10):32–34, 2001. 93. (b) Al De Vries, DE Wohlschlag. Freezing resistance in some Antarctic fishes. *Science*, 163:1074–1075, 1969.
27. YC Hung. Prediction of cooling and freezing times, *Food Technology* 44(May):137–144, 227, 1990.
28. A LeBail, D Chevalier, DM Mussa, M Ghoul. High pressure freezing and thawing of foods: a review. *International Journal of Refrigeration* 25:504–513, 2002.
29. L Otero, M Martino, N Zaritzky, M Solas, PD Sanz. Preservation of microstructure in peach and mango during high-pressure-shift freezing. *Journal of Food Science* 65 (3):466–470, 2000.
30. A Fuchigami, A Teramoto. Structural and textural changes in kinu-tofu due to high-pressure-freezing. *Journal Food Science* 62 (4):828–837, 1997.
31. M Fuchigami, N Kato, A Teramoto. High-pressure-freezing effects on textural quality of Chinese cabbage. *Journal of Food Science* 63 (1):122–125, 1998.

32. M Fuchigami, A Teramoto, N Ogawa. Structural and textural quality of kinu-tofu frozen-then-thawed at high-pressure. *Journal of Food Science* 63 (6):1054–1057, 1998.
33. Y Kanda, M Aoki, T Kosugi. Freezing of tofu (soybean curd) by pressure-shift: freezing and its structure. *Journal of Japanese Society of Food Science and Technology* 39 (7):608–614, 1992.
34. M Fuchigami, N Kato, A Teramoto. High-pressure-freezing effects on textural quality of carrots. *Journal of Food Science* 62 (4):804–808, 1997.
35. M Fuchigami, K Miyazaki, N Kato, A Teramoto. Histological changes in high-pressure-frozen carrots. *Journal of Food Science* 62 (4):809–812, 1997.
36. H Koch, I Seyderhelm, P Wille, MT Kalishevsky, D Knorr. Pressure-shift freezing and its influence on texture, colour, microstructure and rehydration behaviour of potato cubes. *Nahrung* 40:125–131, 1996.
37. L Otero, MT Solas, PD Sanz, C de Elvira, JA Carasco. Contrasting effects of high-pressure assisted freezing and conventional air-freezing on eggplant microstructure. *Zeitschrift fuer Lebensmittel Untersuchung und Forschung* 206 (5):338–342, 1998.
38. FM Fernandez, L Otero, MT Solas, PD Sanz. Protein denaturation and structural damage during high-pressure-shift freezing of porcine and bovine muscle. *Journal of Food Science* 65 (6):1002–1008, 2000.
39. H Barry, EM Dumay, JC Cheftel. Influence of pressure assisted freezing on the structure, hydration and mechanical properties of a protein gel. In: NS Isaacs, Ed., *High Pressure Food Science, Bioscience and Chemistry*. London: Royal Society of Chemistry, 1998, pp. 343–353.
40. DE Johnson. The effects of freezing at high pressure on the rheology of Cheddar and Mozzarella cheeses. *Milchwissenschaft* 55 (10):559–562, 2000.
41. A Teramoto, M Fuchigami. Changes in temperature, texture and structure of konnyaku (konjac glucomannan gel) during high-pressure-freezing. *Journal of Food Science* 65 (3):491–497, 2000.
42. RJ Swientek. High hydrostatic pressure for food preservation. *Food Processing* 53:90–91, 1992.
43. B Mertens, G Deplace. Engineering aspects of high-pressure technology in the food industry. *Food Technology* 47 (6):164–169, 1993.
44. D-W Sun, B Li. Microstructural change of potato tissues frozen by ultrasound-assisted immersion freezing. *Journal of Food Engineering* 57:337–345, 2003.
45. M Ashokkumar, F Grieser. Ultrasound assisted chemical process. *Review of Chemical Engineering* 15 (1):41–83, 1999.
46. TJ Mason, L Paniwnyk, JP Lorimer. The use of ultrasound in food technology. *Ultrasonic Sonochemistry* 3:S253–S256, 1996.
47. KS Suslick. Chemical, biological and physical effects. In: KS Suslick, Ed., *Ultrasound*. New York: VCH, 1988, pp. 123–163.
48. G Scheba, RB Weige, JR O'Brien. Quantitative assessment of germicidal efficiency of ultrasonic energy. *Applied Environmental Microbiology* 57:2079–2084, 1991.
49. SK Sastry, GQ Shen, JL Blaisdell. Effect of ultrasonic vibration on fluid-to-particle convective heat transfer coefficients. *Journal of Food Science* 54:229–230, 1989.
50. E Acton, GJ Morris. Method and apparatus for the control of solidification in liquids, W.O. 99/20420, USA Patent Application, 1992.
51. TJ Mason. Power ultrasound in food processing — the way forward. In: MJW Povey, TJ Mason, Eds., *Ultrasound in Food Processing*. Glasgow: Blackie Academic & Professional, 1998, pp. 104–124.
52. JD Floros, HH Liang. Acoustically assisted diffusion through membranes and biomaterials, *Food Technology* 48(December):79–84, 1994.
53. BWW Grout, GJ Morris, MR McLellan. The freezing of fruits and vegetables. In: WB Bald, Ed., *Food Freezing: Today and Tomorrow* Berlin: Springer, 1991, pp. 113–123.
54. EA Spiazzi, I Raggio, KA Bignone, RH Mascheroni. Experiments on dehydrofreezing of fruits and vegetables: mass transfer and quality factors. *Advances in the Refrigeration Systems. Food Technologies and Cold Chain, IIF/IIR* 6:401–408, 1998.
55. RN Biswal, K Bozorgmehr, FD Tompkins, X Liu. Osmotic concentration of green beans prior to freezing. *Journal of Food Science* 56 (4):1008–1011, 1991.
56. RL Garrote, RA Bertone. Osmotic concentration at low temperature of frozen strawberry halves. Effect of glycerol, glucose and sucrose solution on exudate loss during thawing. *Food Science and Technology* 22:264–267, 1989.
57. CC Huxsoll. Reducing the refrigeration load by partial concentration of foods prior to freezing. *Food Technology* 36 (5):98–102, 1982.

58. HN Lazarides, NE Mavroudis. Freeze/thaw effects on mass transfer rates during osmotic dehydration. *Journal of Food Science* 60 (4):826–828, 857, 1995.
59. GM Dixon, JJ Jen. Changes of sugars and acids of osmovacuum-dried apple slices. *Journal of Food Science* 42:1126–1127, 1977.
60. J Hawkes, JM Flink. Osmotic concentration of fruit slices prior to freeze dehydration. *Journal of Food Processing and Preservation* 2:265–284, 1978.
61. A Maestrelli, RL Scalzo, D Lupi, G Bertolo, D Torreggiani. Partial removal of water before freezing: cultivar and pre-treatments as quality factors of frozen muskmelon (*cucumis melo*, cv reticulatus Naud.). *Journal of Food Engineering* 49:255–260, 2001.
62. D Torreggiani, E Maltini, G Bertolo, F Mingardo. Frozen intermediate moisture fruits: studies on techniques and product properties. In: *Proceedings of the International Symposium on Progress in Food Preservation Processes*. Brussels: CERIA, 1988, Vol. 1, pp. 71–72.
63. RE Feeney, Y Yeh. Antifreeze proteins: current status and possible food uses. *Trends in Food Science and Technology* 9:102–106, 1998.
64. J Li, TC Lee. Bacterial extracellular ice nucleator effects on freezing of foods. *Journal of Food Science* 63 (3):375–381, 1998.
65. D Wen, RA Laursen. Structure-function relationships in an antifreeze polypeptide. *Journal of Biological Chemistry* 268 (22):16401–16405, 1993.
66. Al De Vries, DE Wohlschlag. Freezing resistance in some Antarctic fishes. *Science* 163:1074–1075, 1969.
67. RWR Crevel, JK Fedyk, MJ Sprugeon. Antifreeze proteins: characteristics, occurrence and human exposure. *Food and Chemical Toxicology* 40:899–903, 2002.
68. M Griffith, KV Ewart. Antifreeze proteins and their potential use in frozen foods. *Biotechnology Advance* 13 (3):373–402, 1995.
69. GL Fletcher, SV Goddard, Y Wu. Antifreeze proteins and their genes: from basic research to business opportunity. *Chemtech* 29 (6):17–28, 1999.
70. J Barrett. Thermal hysteresis proteins. *International Journal of Biochemistry and Cell Biology* 53:105–107, 2001.
71. DSC Yang, M Sax, A Chakrabartty, CL Hew. Crystal structure of an antifreeze polypeptide and its mechanistic implications. *Nature* 333:232–237, 1988.
72. K Harrison, J Hallett, TS Burcham, RE Feeney, WL Kerr, Y Yeh. Ice growth in supercooled solutions of antifreeze glycoproteins. *Nature* 328:241–243, 1987.
73. CJ Warren, CM Mueller, RL Mckown. Ice crystal growth suppression polypeptides and methods of preparation, US Patent 5:118,792, 1992.
74. SR Payne, D Sandford, A Harris, OA Young. The effects of antifreeze proteins on chilled and frozen meat. *Meat Science* 37:429–438, 1994.
75. SR Payne, OA Young, Effect of pre-slaughter administration of antifreeze proteins on frozen meat quality. *Meat Science* 41:147–155, 1995.
76. H Kawahara. The structures and functions of ice crystal — controlling proteins from bacteria. *Journal of Bioscience and Bioengineering* 94 (6):492–496, 2002.
77. P Phelps, TH Giddings, M Prochoda, R Fall. Release of cell-free ice nuclei by *Erwinia herbicola*. *Journal Bacteriology* 167 (2):496–502, 1986.
78. SE Lindow. The role of bacterial ice nucleation in frost injury to plants. *Annual Review Phytopathology* 21:363–384, 1983.
79. LM Kozloff, MA Schofield, M Lute. Ice-nucleating activity of *Pseudomonas syringae* and *Erwinia herbicola*. *Journal of Bacteriology* 153:222–234, 1983.
80. AG Govindarajan, SE Lindow. Phospholipid requirement for expression of ice nuclei in *Pseudomonas syringae* and in vitro. *Journal of Biology Chemistry* 263 (19):9333–9338, 1988.
81. J Li, TC Lee. Bacterial ice nucleation and its potential application in the food industry. *Trends in Food Science Technology* 6:259–265, 1995.
82. S Arai, M Watanabe. Freeze texturing of food materials by ice-nucleation with the bacterium *Erwinia ananas*. *Journal of Biology Chemistry* 50 (1):169–175, 1986.
83. M Watanabe, J Watanabe, K Kumeno, N Nakahama, S Arai. Freeze concentration of some foodstuffs using ice nucleation-active bacterial cells entrapped in calcium alginate gel. *Journal of Biology Chemistry* 53 (10):2731–2735, 1989.

Part II

Facilities for the Cold Chain

9 Freezing Methods and Equipment

Mike F. North and Simon J. Lovatt
AgResearch Ltd, Hamilton, New Zealand

CONTENTS

I. INTRODUCTION

It is common for foods to have their temperature reduced by a freezer. However, the best rate for this to occur, the importance of mass (e.g., water vapor) transfer processes in freezing, the shape of the product, its thermal properties, and other parameters all affect the choice of freezing method, and consequently the equipment used. This means that different freezing methods are suitable for different types of food products. This chapter focuses on freezing methods and freezers for foods where the water in the unfrozen phase is effectively immobilized and the food may be considered solid. Freezers for liquid foods are not considered.

II. FREEZING METHODS

A. Natural Convection Freezing

Natural convection refers to the natural fluid flow that arises when a heat source is placed in a fluid. When the fluid surrounding the heat source is warmed, the fluid around the heat source becomes less dense (and more buoyant) than the bulk of the fluid. The warm fluid rises, causing the cooler fluid in the bulk to sink, thereby creating natural convection currents that continually supply cool fluid to the heat source. In *natural convection freezing*, the food product is the heat source and the fluid is usually cold air. In the earliest freezers, the air was cooled by blocks of ice placed near the ceiling of the chiller; however, coils of tube filled with evaporating refrigerant are usual in modern systems. Unfortunately, natural convection freezing is slow. It is also relatively uncontrollable because the rate of cooling is determined only by the thickness of the product and the temperature of the air.

B. Forced Convection Freezing

Forced convection freezing relies on cooling a food product in air and removing the heat from the air by passing it over a refrigerant evaporator, as with natural convection freezing. However, in forced convection freezing, fans are used to increase the velocity of the air. This makes the process more controllable than natural convection freezing because it can achieve a more uniform air temperature throughout the freezer and the air velocity can be altered to vary the heat transfer coefficient at the surface of the product. For thin products with large surface-to-mass ratios, increasing the heat transfer coefficient may lead to substantial reductions in the freezing time. However, for products with lower surface-to-mass ratios, the freezing time is limited more by the rate of heat conduction through the thickness of the product than by the rate of heat transfer at the surface.

When moist foods are not wrapped during air cooling, the moisture on the surface and within the outer few millimeters of the food can evaporate into the air because the absolute humidity of the bulk air is usually lower than that of the air at the surface of the food. Although moisture evaporation increases the effective surface heat transfer coefficient compared with pure convection, any significant weight loss and degradation of the product surface appearance is frequently undesirable and should be avoided.

C. Liquid Immersion Freezing

In *liquid immersion freezing*, the food product is immersed in a cold liquid, generally brine, glycol, or a sugar solution, which is cooled by a refrigerant evaporator. This method results in considerably higher heat transfer coefficients, and therefore shorter freezing times, than most air-freezing methods. The product may be left bare, but it is usually frozen in liquid-tight packaging in situations where uptake of the cooling liquid by the product or cross-contamination between products is undesirable.

D. Contact Freezing

The need to transfer heat from the product to air or liquid, and then from the fluid to the refrigerant evaporator is a drawback of any air- or liquid-based cooling process. This double heat transfer step can be replaced by a single step if the food can be placed in direct contact with the refrigerant evaporator. *Contact freezing* processes achieve this by placing food products between two metal plates filled with evaporating refrigerant. In addition to providing the high rate of heat transfer that exists when good solid-to-solid contact is achieved, eliminating one heat transfer step means that for a given refrigerant evaporation temperature, the cooling medium temperature is colder than in an air or liquid immersion freezer. Further efficiencies are gained because contact freezers do not require fans and they are significantly more compact than convection or immersion freezing systems. The product can be bare or packaged, however the advantage of a higher heat transfer coefficient and a lower cooling medium temperature can be lost if a significant amount of packaging is placed between the product surface and freezer plates. Ideally, the product would have flat parallel surfaces with which the plates can make contact.

E. Cryogenic Freezing

Cryogenic freezing involves either spraying liquid nitrogen or carbon dioxide "snow" onto the food surface or immersing the food product directly into the liquid cryogen. This method also results in much shorter freezing times than air or liquid immersion freezing, owing to the large initial temperature difference between the cryogen and the food product, and as a result of the high rate of surface heat transfer that occurs when the cryogen changes phase. As with other freezing methods, the benefit gained by a higher surface heat transfer coefficient becomes smaller as the thickness of the product increases. Cryogenic freezing does not require any refrigeration machinery, therefore the capital cost of a cryogenic freezing system is comparatively low; for example, only a tank to contain the cryogen and a suitable spraying arrangement may be required. However, the cost of the cryogen is relatively high in most locations, which results in high operating costs that often make the method too costly for freezing large quantities of food. For short trial production runs, for very high value products that must be rapidly frozen, or for emergencies and overloads where no other form of refrigeration is available, cryogenic freezing may be useful [1]. As a further advantage, the low temperature of cryogens can bring about rapid freezing of the food product surface, thereby preventing excessive moisture loss. The rapid surface freezing possible with cryogens is sometimes utilized to "crust freeze" the surface of high value, wet, sticky, or delicate products, which are then fully frozen using a different type of freezer [1].

III. FREEZER TYPES

A. Batch Air Blast Freezers

Batch air-blast freezers are the simplest common form of *forced convection freezer*. They typically consist of an insulated room containing fans that force air over refrigerant evaporator coils and then circulate it over the food products. The product items are usually hung or stacked in racks to ensure that the cold air can pass over the surface of each individual item. It is important to direct the air, and position the product items so that the air flow is as even as possible across the product stow and so that short-circuiting (cold air that returns to the evaporator coil without contacting the product) is minimized. This helps avoid large variations in cooling rate between different parts of the freezer and leads to a more consistent product quality. In batch blast freezers, the air flow can be directed by the use of turning vanes or slotted ceilings, which spread the air flow evenly across the product stow, as shown in Figure 9.1 [2].

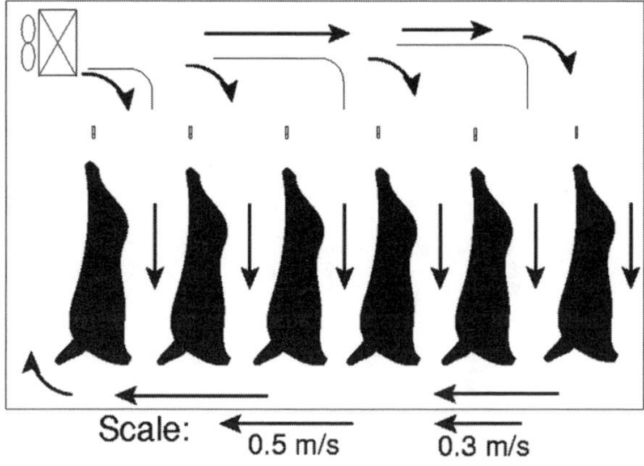

FIGURE 9.1 Batch air blast freezer for carcasses, turning vanes fitted. (From SJ Lovatt, J Willix, QT Pham. A physical model of air flow in beef chillers, in *Proceedings of International Institute of Refrigeration*, Commissions B1, B2, D1, D2/3, Palmerston North, New Zealand, 1993, pp. 199–206. With permission.)

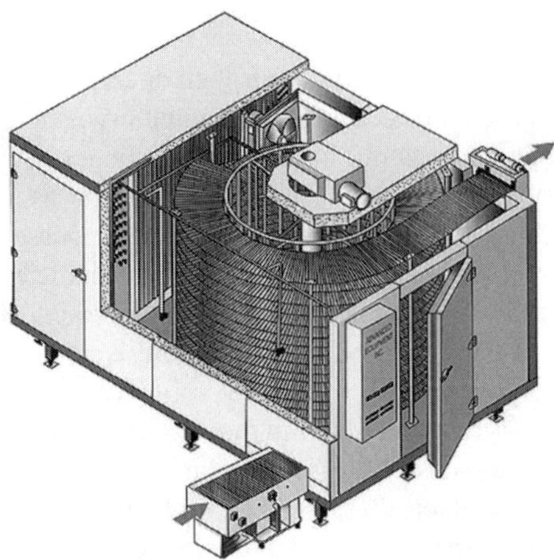

FIGURE 9.2 Spiral air blast freezer. (Courtesy of Advanced Equipment Inc., Canada.)

B. Continuous Air Blast Freezers

Continuous air-blast freezers frequently include a moving belt system that transports the food products through an environment containing air moving at high velocity. The continuous movement of the product through the freezer means that the freezing process can be integrated into the food production line and it also helps ensure that each individual item is subject to a consistent rate of cooling. Designs vary from simple one-pass belt systems to more elaborate multipass, spiral belt (Figure 9.2) or variable retention time systems (Figure 9.3).

C. Fluidized Bed Freezers

A *fluidized bed freezer* involves a continuous process that forces cold air up under the product at a high enough velocity to "fluidize" the product, as shown in Figure 9.4. The air acts as both the cooling medium and the transport medium, thus products suited to fluidized bed freezing are small and uniform in size (to aid in fluidization) and are not prone to damage caused by the high velocity mixing that occurs in a fluidized bed. Common examples are vegetables such as peas, corn kernels, and diced carrots. The high air velocity and small product size that is associated with fluidized bed freezing typically results in freezing times of less than 10 min [1].

FIGURE 9.3 Air blast freezer with variable retention time. (Courtesy of IBEX Technologies Ltd., New Zealand.)

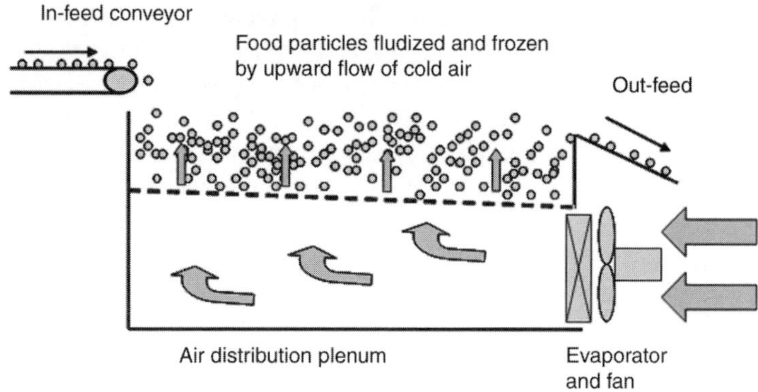

FIGURE 9.4 Fluidized bed freezer.

D. Impingement Freezers

Impingement freezers use numerous jet nozzles to direct air onto the surface of food products at a very high velocity. The process is usually continuous with air jets positioned above and below a mesh belt conveyor system. The airflow direction is usually perpendicular to the product surface, which disrupts the boundary layer surrounding the product and thereby increases the surface heat transfer coefficient. Significant improvements in freezing time resulting from the high heat transfer coefficients in impingement freezers are only achieved for thin products, so impingement freezing is generally not cost effective for thick products.

E. Liquid Immersion Freezers

Liquid immersion freezers can be designed to operate in a batch or a continuous mode. In a typical batch operation, products are placed into baskets and these baskets are then immersed in a bath of the cold liquid. The baskets are removed from the liquid after the product is frozen. Although batch systems are simple and easy to instal, they are more labor-intensive than continuous systems. Continuous immersion freezers use some type of conveyor system to move the product through the cooling liquid. Existing conveyor designs include auger systems (Figure 9.5) and solid or

FIGURE 9.5 Liquid immersion auger freezer. (Courtesy of MSC Engineering Ltd., New Zealand.)

mesh belts that may incorporate baffles to ensure that all products are continuously moved through the freezer. The choice of conveyor system depends on the buoyancy of the product in the cooling medium because, for example, products that float cannot be conveyed on a belt system placed at the bottom of a liquid immersion tank. It is often undesirable for the product to remain wet following its removal from a liquid immersion freezer, so most commercial systems either allow the cooling liquid to drain off or use air knives to blow the liquid off the product.

F. Batch Plate Freezers

Batch plate freezers consist of several layers of plates and are available in horizontal or vertical arrangements. In horizontal systems, the bottom layer is loaded first, either manually or automatically (Figure 9.6). Once the bottom layer is full, the next plate up is lowered to clamp the product into place and the next layer of product is then loaded on top of this plate. In a vertical plate freezer (Figure 9.7), the plates lie vertically and the product is loaded from the top. Once frozen, the plates are unclamped and the product falls out the bottom of the *plate freezer*. Vertical plate freezers are mainly used for bare product, whereas horizontal systems are used with packaged or bare products. In addition to the rapid rates of cooling obtained in plate freezers, the pressure of the plates against the product surfaces during freezing minimizes the bulging that often occurs in air-blast freezing systems, making the product squarer and easier to stack [1].

G. Continuous Plate Freezers

Continuous plate freezers can be classified into two types — drum and belt systems. Drum systems consist of continually rolling plates, in the form of two conveyor belts, with the product held between them. Belt systems come in two main configurations. The first consists of two long, stationary refrigerated plates between which the product is conveyed by continuously moving plastic films. The second type of belt system, although not strictly a plate freezer because it does not place the product in direct contact with a refrigerant evaporator, uses two continuously moving, solid, stainless steel belts which are typically air-cooled [1]. Continuous plate freezers are commonly used for thin, flat-sided products with short freezing times, such as hamburger patties or fish fillets. The systems may be used to completely freeze the product or simply to

FIGURE 9.6 Horizontal plate freezer with automatic loading system. (Courtesy of RealCold Milmech Pty Ltd.)

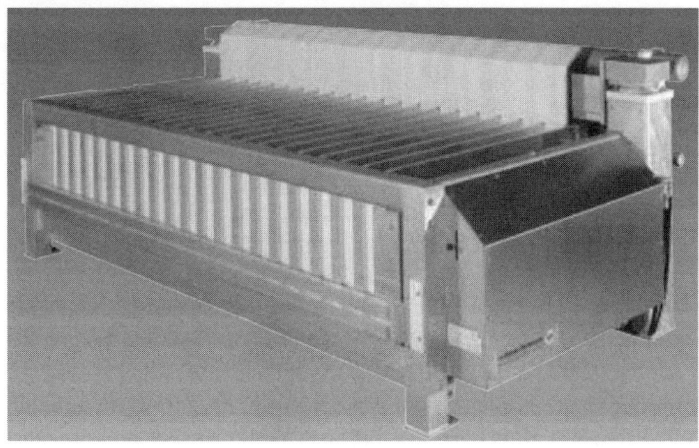

FIGURE 9.7 Vertical plate freezer. (Courtesy of A/S Dybvab Stål Industri (DSI), Denmark.)

"crust freeze" the first few millimeters of the product surface to aid subsequent processing of the product.

H. Liquid Nitrogen Freezers

The most common arrangement for a *liquid nitrogen freezer* is a continuous belt that conveys the product through a tunnel. Liquid nitrogen at approximately −196°C is usually fed onto the product toward the end of the tunnel. As the liquid nitrogen removes heat from the product, it boils and the cold nitrogen vapors are directed toward the start of the tunnel, thereby cooling the incoming product [1]. Other liquid nitrogen freezer arrangements include batch cabinets, multipass and spiral belt systems, and full immersion baths. Depending on the composition and initial temperature of the food product, between 0.3 and 1.5 kg of liquid nitrogen may be required to fully freeze 1 kg of a product.

I. Carbon Dioxide Freezers

Carbon dioxide freezers are configured in a similar way to liquid nitrogen freezers. However, carbon dioxide acts very differently from liquid nitrogen in the freezer. At atmospheric pressure, carbon dioxide sublimes instead of passing through the liquid phase. Therefore, carbon dioxide must be transported and piped into the freezer as a high-pressure liquid. Once the carbon dioxide exits the applicator nozzle in the freezer, it immediately expands into a mixture of gas and small "snow" particles at a temperature of −79°C. The mixture of cold vapor and solid removes heat from the product surface upon contact, which causes the solid carbon dioxide particles to sublime directly into a gas. The vapor generated in a *carbon dioxide freezer* does not provide as much refrigeration effect as the vapor generated in a liquid nitrogen freezer because it is at a much higher temperature. Therefore, in a carbon dioxide freezer, it is the sublimation of the solid particles that provides the majority of the refrigerating effect. To maximize the contact time between the solid carbon dioxide particles and the product, carbon dioxide is usually sprayed throughout the entire length of the freezer tunnel. Even though subliming carbon dioxide removes (per unit mass of cryogen) about 50% more heat from the product than boiling liquid nitrogen, the higher temperature of the carbon dioxide vapors usually mean that a carbon dioxide freezer has a slightly higher cryogen consumption rate than a liquid nitrogen freezer.

IV. EQUIPMENT DESIGN AND OPERATION

A. Product Suitability

The first consideration for freezer design is the selection of a freezing process that is suitable for the product. For example, high value, wet, sticky, or delicate products such as individual quick-frozen (IQF) strawberries or shrimp may require extremely rapid surface freezing to form a solid ice "crust" that allows the product to be handled without severe damage or clumping. Such rapid surface freezing is probably only achievable in a cryogenic freezer. However, if desired, the internal part of the product may be frozen at a slower rate using a different type of freezer.

Although rapid freezing maximizes the quality of many food products, this is not the case for meat. Slaughtered animals undergo a natural biochemical process that converts their muscle into meat. If carcasses are cooled too quickly after slaughter, the conversion process may not reach completion and the resulting meat will be of poor quality. Similarly, a certain amount of moisture loss from the surface of a carcass is often desirable because a drier surface makes it easier to manipulate during cutting operations. Excessive moisture loss is not desirable, however, because this will affect the product quality and yield. These product requirements make air-blast freezing well suited to the cooling of carcasses because freezing is not too rapid to allow muscle to be transformed into meat, and some surface evaporation occurs, but usually not so much that a badly "dried-out" appearance results [3].

Table 9.1 shows commonly used freezing methods for various types of food products.

TABLE 9.1
Commonly Used Freezing Methods for Various Foods

Product Type	Commonly Used Freezing Methods
Meat, carcasses	Air blast
Meat, cartons	Air blast, plate
Meat, large individually wrapped cuts	Liquid immersion, air blast
Meat, small or diced pieces	Cryogenic, liquid immersion
Meat, cured or processed products	Air blast, liquid immersion, cryogenic
Meat, hamburger patties	Plate, impingement, air blast, cryogenic
Poultry, whole bird or pieces	Air blast, often preceded by liquid immersion or cryogenic, plate (in packages)
Poultry, processed or breaded products	Cryogenic, impingement, air blast, plate (in packages)
Fish, whole or eviscerated	Air blast, cryogenic, plate, liquid immersion (particularly aboard fishing vessels)
Fish, fillets or small diced pieces	Plate, air blast, cryogenic, impingement
Fish, minced blocks	Plate
Fish, processed or breaded products	Cryogenic, impingement, air blast, plate (in packages)
Shellfish	Air blast, cryogenic, plate (in packages)
Prawns and shrimp	Cryogenic, impingement, air blast
Fruits, small size (whole)	Air blast, cryogenic, plate (in packages), fluidized bed
Fruits, large size (sliced)	Air blast, plate (in packages)
Fruits, purée or pulp	Air blast, plate
Vegetables, small size (e.g., peas)	Fluidized bed, air blast, cryogenic, plate (in packages)
Vegetables, medium size (e.g., corn cobs)	Air blast, plate (in packages)
Vegetables, leafy (e.g., spinach)	Air blast, plate (in packages)
Cheese and butter	Plate, air blast
Dough, bread, and baked products	Air blast, plate
Pre-cooked ready meals	Air blast, plate (in packages)

B. Predicting Food Freezing Times and Heat Loads

In the past, many freezing processes were successfully designed by engineers who applied their experience or used their judgment to scale up or down from a similar process. Increasingly, designers are using mathematical models as part of their design tool kit. A range of mathematical models exists for calculating the temperature profile of a food product over time during a freezing process [4]. From this information, the prediction of microbial growth and product quality at the end of the process is often possible. Models are also available for predicting the heat load that must be removed by the refrigeration system [5]. Although the heat removed from the product is important, there are often many other heat loads that must be considered. These include heat loads from equipment (e.g., fans, lights, warm structures, and forklifts) or people in the freezer, and heat infiltrating into the freezer through the walls, ceiling, floor, and openings (e.g., doors).

C. Economics

There are often situations where more than one freezer type could be used to achieve a certain product specification. Consequently, selection of the best freezer for the application frequently depends on the economics of the individual options. A careful comparison of possible freezing methods must consider both capital and operating costs.

The *capital cost* of freezers is generally high when compared with other processing equipment, and the refrigeration system often comprises much of that capital cost. However, for plate freezers or sophisticated continuous systems, the cost of materials, installation, and complex conveying and control systems can be considerable. An exception is cryogenic freezers, which tend to have low capital costs because they do not require a mechanical refrigeration system. When calculating the capital cost, it is also important to consider the space taken up by the freezer. Although a batch air-blast freezer may not be as expensive as a plate freezer of the same throughput, it will take up considerably more space, which may represent a significant cost.

Freezers often represent a small fraction of the *operating cost* in a food processing operation, when compared for example, with labor or raw material purchase costs. Again, an exception is cryogenic freezing, which has a high operating cost because the refrigerant is continuously used up during operation. The operating costs of freezers vary with the product throughput and include electricity, labor, maintenance, and cleaning costs. It is also important to consider mass and quality loss from the product when comparing the operating cost of freezers. Mass loss can occur in many different ways depending on the product and freezer used, but common mass losses include evaporation or drip from the product surface, product sticking to conveyor belts, and product breakages. Quality losses that may cause the product value to be downgraded include dehydration, color changes, and the buildup of unwanted ice crystals in packages. Large variations in product exit temperatures can also cause inconsistent product quality and can lead to the downgrading of product.

D. Control and Operation

At the most basic level, freezer temperatures are typically regulated to a given set-point temperature through the use of *feedback control* systems. These systems measure a temperature in the freezing environment and adjust either the rate at which refrigerant enters the evaporator or the pressure of the evaporating refrigerant. The measured temperature can be the refrigerant evaporating temperature or (more useful in most cases) the temperature of the fluid (e.g., air) used to transfer heat from the product to the evaporator. Traditionally, the control mechanisms used were of an analog type, using, for example, the temperature-dependent expansion of a vapor, vapor–liquid mixture, or a coiled strip of metal to drive the movement of a valve. In modern installations, feedback controllers are usually constructed from digital electronics, and the temperatures are measured by thermistor, thermocouple, or resistance temperature detector (RTD).

Simple feedback temperature control alone may be satisfactory for simple freezing regimes, where the freezer is to operate at a single temperature throughout the freezing cycle. For sophisticated freezing regimes, it may be necessary to change the freezer temperature according to a time schedule, according to measurements beyond those of the refrigerant or heat transfer fluid (e.g., those made by temperature probes inserted in product items), or for some other reason. In those cases, the more sophisticated decisions required to select the freezer temperature may be programmed into a programmable logic controller (PLC) or a supervisory control and data acquisition (SCADA) system.

In practice, it is often not a good strategy to control a freezer directly from measurements of product item temperatures. In addition to the damage that can be incurred by inserting temperature probes in product items, the temperature that would be measured at the slowest cooling point in a product item of significant thickness may not be very helpful in controlling the freezer. For example, if a freezer is unloaded when the temperature probe inserted into the slowest cooling point of a product item reaches $-12°C$, the measured point in the product item will usually continue to cool for some time after the product is unloaded, thereby cooling the product below the target temperature of $-12°C$. In addition, the measured product item may not be representative of the other items in the freezer. Many of those items may be warmer or cooler than the measured item and may therefore be insufficiently or excessively frozen.

An alternative control method used in some cases has been to use a mathematical model of the cooling product, calibrated to the conditions in the freezer. Unlike a temperature probe, a mathematical model can indicate that temperature a representative product item should reach at some time in the future. Thus, the model can be used to predict when the product should be unloaded under given conditions or what freezing conditions would be required to ensure that the product load was appropriately frozen by a specified time and that product quality targets are met [6]. This method has been called *model-based control*. For air-blast freezers, the most convenient freezing condition to control with a model-based controller can sometimes be the air velocity, by modulating the speed of the freezer fans. This has the additional advantage of reducing the energy consumption of the freezer, if the fan speed is reduced significantly [7].

For freezers that operate in batch mode, a part of the freezing cycle is allocated to *loading* the product items at the start and *unloading* them at the end of the cycle. For small freezers where the batch size is a few tens of kilograms, loading and unloading may be done manually, but larger freezers are loaded with trolleys carrying product items, or racks containing the product are carried into the freezer using a forklift truck or specially designed automatic loading mechanisms. Except where loading is automatic, therefore, one or more staff would typically be required for this task.

While loading or unloading an air-blast batch freezer, it is usual to halt or reduce the speed of freezer fans. In addition to making the working conditions more comfortable and safe for the loading staff, this can reduce the flow of warm air into the freezer through the open freezer doors and hence the heat load on the freezer. Air flowing into a freezer will also carry moisture that will freeze on cold surfaces (such as the evaporators), causing frost, or ice, which can be unsightly can reduce heat transfer at the evaporator surface or even be hazardous to staff working in the freezer. In cases where the initial rate of freezing is limited by product quality requirements (e.g., for meat, if time must be allowed for aging to take place), it can be necessary to raise the temperature of the freezer above 0°C during loading and for a period afterward. This may require installation of a heating system in the freezer or the use of hot-gas bypass to heat the freezer through the evaporator coils. This type of freezing regime also requires careful design during construction, because frequent cycling between temperatures above and below 0°C can risk freeze or thaw damage to structural components of the freezer — especially concrete floors and pillars.

When loading takes some time (e.g., as may occur if the loading rate is limited by the rate at which the product is packaged), it is often desirable to load the freezer in stages, with the doors closed and the fans running between each stage to ensure that the product already loaded starts to cool before the freezer is full. This is particularly important when product cooling is required to minimize microbial growth and hence ensure that the resulting product is safe to consume.

E. Reliability and Maintenance

By their nature, freezers operate in harsh conditions for extended periods of time. It is therefore critical that freezers be designed and constructed to achieve high reliability. There is often no back-up system for the freezer in a food processing operation, so regular checking and *maintenance* are essential to minimize the chance of breakdowns, which may cause the entire process line to stop and may result in downgrading or loss of a significant amount of product. As with many items of refrigeration equipment, there is often a trade-off between the effort required for maintenance and the energy efficiency of the equipment. Thus, equipment that is designed to minimize maintenance can have higher energy costs and *vice versa.* It is important to make this trade-off consciously during the freezer design process to ensure that the best compromise is achieved for a given installation.

F. Cleanability and Defrosting

Freezers may be cleaned manually or with automated clean in-place (CIP) systems. Either procedure can be made difficult because the use of water-based *cleaning* materials usually requires the freezer to be warmed above 0°C for the period of cleaning. Fortunately, the low temperatures at which freezers operate restrict or (usually) completely prevent the growth of the bacteria and molds that cleaning is intended to eliminate. It is therefore possible to clean freezers infrequently, in comparison with chillers, for example. In addition, for cases where food products are enclosed in packages, there can be little or no opportunity for biological material to pass from the freezer to the food, and cleanliness may be more an esthetic than a food safety requirement.

One area where cleanliness is particularly important in a freezer relates to *defrosting*. Frost will build up on cold surfaces whenever moisture-containing air can enter the freezer or whenever moisture evaporates into the freezer environment from exposed food product items. Although small amounts of frost can be tolerated, defrosting must be carried out periodically to prevent frost building up to an unacceptable level. If most of the frost collects on the evaporators (as would normally be the case), it can be removed by heating the evaporator surface above 0°C for long enough for the frost to melt. In an air-blast freezer, this should be done with the fans switched off to avoid meltwater being sprayed throughout the freezer. Normally, the meltwater should drip into a well-drained tray below the evaporator. It is often desirable to insulate and heat the drain to ensure that it is not blocked by ice when the freezer is operating. Common methods of heating the evaporator surface include electrical resistance heating, bypassing hot refrigerant gas through the evaporator and (less desirably in most cases because of the risk of water getting into the rest of the freezer room) water sprays. Once parts of the freezer are damp and above 0°C, there is an opportunity for microbial growth to occur, so it is important that the areas of the freezer affected by defrosting receive particular attention during cleaning and that they are designed to dry quickly after defrosting or cleaning is complete.

V. CONCLUSIONS

On the basis of a few fundamental methods of freezing, the requirements of different food products, economics, design, and operational trade-offs have resulted in a wide range of freezing equipment being available for use in food processing operations. Judicious selection of this equipment and its operating procedures are necessary to ensure that the resulting frozen food has the required quality at an acceptable cost. It is important that this is done when the equipment is specified and designed, because there are few opportunities to change most of these attributes during operation.

REFERENCES

1. ASHRAE. ASHRAE Handbook: Refrigeration, SI Edition. Atlanta, GA: American Society of Heating, Refrigerating and Airconditioning Engineers, 2002, pp. 15.1–15.6.
2. SJ Lovatt, J Willix, QT Pham. A physical model of air flow in beef chillers. In: *Proceedings of International Institute of Refrigeration*, Commissions B1, B2, D1, D2/3, Palmerston North, New Zealand, 1993, pp. 199–206.
3. CE Devine, RG Bell, SJ Lovatt, BB Chrystall, LE Jeremiah. Chapter 2: Red Meats. In: LE Jeremiah, Ed., *Freezing Effects on Food Quality*. New York: Marcel Dekker, 1995, pp. 51–84.
4. AC Cleland and S Ozilgen. Thermal design calculations for food freezing equipment — past, present and future, *International Journal of Refrigeration* 21:359–371, 1998.
5. SJ Lovatt, I Merts, Recommended methods for food refrigeration process design, *Proceedings of the 20th International Congress of Refrigeration*, Sydney, Australia, 1999, Paper 350.
6. MPF Loeffen, A Carrie. Computerized freezer control to protect meat tenderness, *Transactions of the Institution of Professional Engineers*, 13:113–118, 1986.
7. J Walford and DT Lindsay. A model-based controller for batch-loaded, air-blast, lamb carcass freezers. In: PR Johnstone, Ed., *Proceedings of the 28th Meat Industry Research Conference*. Auckland, New Zealand, 1994, pp. 317–322.

10 Cold Store Design and Maintenance

Laurence Ketteringham and Stephen James
Food Refrigeration and Process Engineering Research Centre (FRPERC),
University of Bristol, Langford, UK

CONTENTS

I. INTRODUCTION

The purpose of a cold store for frozen food is to maintain the temperature of the previously frozen food below a set value. This temperature depends on the type of food being stored and the desired storage life. As long as the temperature remains below -12°C, there will be no growth of pathogenic microorganisms, so the food will remain safe [1]. Frozen *storage life* will be limited by enzymic reactions, which affect the taste of the frozen product. The rates of these reactions are a function of temperature, so the storage life will be generally longer at lower temperatures. Publications such as that in the International Institute of Refrigeration [2] provide data on the storage life of many foods at different temperatures. Storage lives can be as short as 3 to 4 months for individually quick frozen, polybag-packed shrimps at -18°C [3]. In contrast, lamb stored at -25°C can be kept for over 2.5 years [4].

Temperature fluctuations during frozen storage have little effect on storage life for many foods unless the temperature rises above -12°C. The cold store set-point temperature and control differential are therefore governed by the overall economics of the operation. Storage temperatures for high value tuna may be as low as -80°C, whereas -18°C will be adequate for short term (3 to 6 months) storage of red meat.

II. TYPES OF STORAGE ROOMS

A. Bulk Storage Rooms

Most food is stored frozen in large, forced air circulation rooms. In a small and diminishing number of cases, the food is frozen unwrapped or packaged in nonmoisture-proof materials. To minimize *weight loss* and appearance changes associated with desiccation, air movement around an unwrapped product should be the minimum required to maintain a constant temperature. Low air velocities are also desirable with wrapped products to minimize energy consumption. However, many storage rooms are designed and constructed with little regard to air distribution and localized velocities over products. Horizontal throw refrigeration coils are often mounted in the free space above the racks or rails of product, and no attempt is made to distribute the air evenly around the products.

Using a false ceiling or other forms of ducting to distribute the air throughout the storage room can substantially reduce variations in velocity and temperature. It is claimed that an even air distribution can be maintained using air socks, with localized velocities not exceeding 0.2 m/s.

B. Jacketed Cold Stores

Cooling the walls, floor, and ceiling of a store produces very good temperature control in the enclosed space with the minimum of air movement. A refrigerated jacket can be provided either by embedding pipe coils in the structure or by utilizing a double skin construction, through which refrigerated air is circulated.

Although a refrigerated jacket is efficient in absorbing any heat from the environment around the store before it reaches the food, the lack of air circulation within the enclosed space indicates that heat removal from the product is very limited. Therefore, care must be taken to:

- Attain the desired storage temperature throughout the product before storing
- Minimize any heat loads produced during loading and unloading
- Provide the supplementary refrigeration required for any products which respire

C. Tempering Rooms

There is no exact definition for the word "tempering" in the food industry. In practice, it is a process by which the temperature of the product is either raised or lowered to a value that is optimal for the next processing stage.

Burgers (patties), sausage, canned meats, pet foods, frozen prepared foods, and portion of controlled steaks and specialities rely heavily on frozen ingredients. Much of this frozen raw material is tempered rather than thawed before processing. Tempering, as an alternative to thawing, eliminates the accompanying problems of drip loss, bacterial growth, and other adverse changes.

In a tempering room, the temperature of the frozen product is raised from the long-term storage temperature of colder than $-18°C$ to the -5 to $-2°C$ required for further processing.

An increasing proportion of bacon and cooked meat is presliced and packed before it is delivered to wholesalers and retailers. Slicers have to be operated at very high speeds to achieve the required throughput. Maximizing the yield of high-quality slices from these high-speed slicers requires the meat to be in a semifrozen, tempered state before slicing. Obtaining the correct temperature, which can range from -3 to $-11°C$, throughout the product is crucial for a high yield of undamaged slices [5]. In this case, the tempering system has to lower the product temperature to its optimum value before slicing.

Traditionally, tempering rooms operated at the desired final product temperature and the product took a long time, up to 2 weeks, to reach the desired value. The design of the room was a compromise between the need to rapidly change the temperature of new product placed in the room and the need to maintain the temperature of the already tempered product. Two-stage processes are increasingly being adopted, where the desired amount of heat addition or subtraction is carried out before the product is placed in the tempering room to stabilize at the correct temperature. The tempering room can be then designed as a frozen storage system with little or no air movement over the product.

III. DESIGN OF FROZEN STORAGE ROOMS

The function of the equipment must be absolutely clear when specifying refrigeration equipment; either the food passing through the process should be maintained at its entry temperature or the temperature should be changed. These two functions require very different equipment. Theoretically, a frozen storage room should always be loaded with food at the desired storage temperature. In practice, though unfrozen or partially frozen food is often loaded into a frozen storage room, a temperature change is required. If a room is to serve several functions, then each function must be clearly identified. The optimum conditions needed for the function must be evaluated and a clear compromise made between the conflicting uses. The result will likely be a room that does not perform any of the functions effectively.

There are three stages in obtaining the correct refrigeration system for a specific process. The first is determining the process specifications, the second is converting these into engineering specifications (that a refrigeration engineer can handle), and the third is procurement of the plant.

A. Process Specification

Poor design in existing frozen stores is often due to a mismatch between what the room was originally designed to do and how it is actually used. The first task in designing such plant is therefore the preparation of a clear specification by the user of how the room will be used. In preparing this specification, the user would do well to consult with all parties concerned: these may be officials enforcing legislation, customers, other departments within the company, and engineering consultants or contractors. The user's decisions alone should form the basis of the specification.

1. Throughput

The throughput must be specified in terms of the food to be stored and whether it is wrapped or unwrapped in boxes, on pallets, and so on. If more than one product is to be stored, then separate specifications must be made for each product. If fish or meat in carcass form is to be stored, then average and maximum weights and dimensions are required.

A throughput profile is needed. Few food stores handle the same type and quantity of product each day and therefore the average throughput is not adequate in the specification. The maximum *product load* must be catered for and the store should also be designed to perform adequately and economically at all other throughputs.

2. Temperature Requirements

The range of temperature requirements for each product must also be clearly stated. Several other requirements must be considered in deciding on the range of temperature requirements. First of all, what legislative requirements (e.g., the EEC quick frozen food regulations) are there? What customer requirements are there? These requirements may come from your existing customers or from future customers who you are hoping to attract. What are your personal requirements? Some companies sell a quality product under their own brand name, which should include a cooling specification. Finally, one must decide to what extent the earlier mentioned requirements may be compromised. The reason for this will become apparent later. Many companies will compromise on their requirements to some extent. This can lead to poor quality and those that get caught, and are called to task for this, can lose orders or have their production disrupted. Other firms know to what extent they can push the inspectors or their customers and ensure that they stay within accepted limits.

3. Weight Loss

If unwrapped or permeably wrapped food is to be stored in the room, then reducing weight loss can be critical to the economics of its operation. The rate of sublimation of ice from a frozen surface is considerably slower than the rate of evaporation from a moist surface, and the ability of air to hold water rapidly diminishes as its temperature falls below 0°C. However, as clearly demonstrated in Figure 10.1, a warmer storage temperature and moving air can substantially increase weight loss. The relatively small increase in capital and operating costs resulting from designing and operating the cold store to maintain a lower temperature would be rapidly recovered from the reduction in weight loss.

4. Change of Use

All the information collected so far, and the decisions taken, will be based on existing production. Another question that needs to be asked is "Will there be any changes in the use of the frozen store in the future?" In practice, the answer to this question is almost always yes. Very few food processors will carry out exactly the same processing within the life of a cold store, which can be between 10 and 50 years (judging by present stores). Changes should be envisaged and quantified in as much detail as possible.

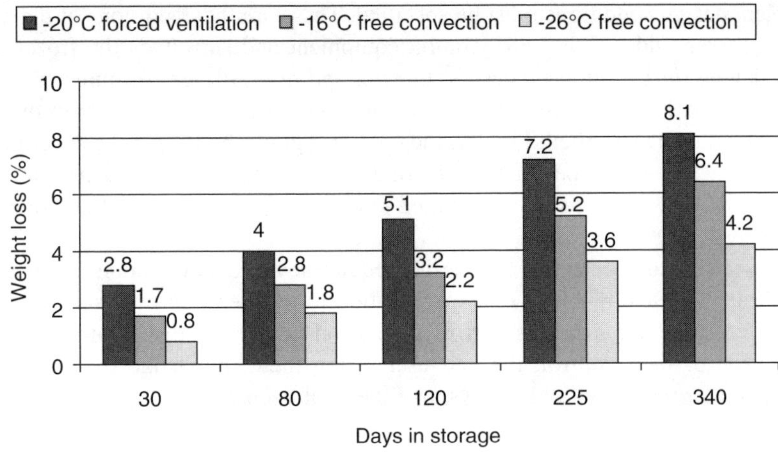

FIGURE 10.1 Weight loss from unwrapped hams in frozen storage. (From R Malton, SJ James. Using refrigeration to reduce weight loss from meat. In: *Proceedings of the Symposium on Profitability of Food Processing* — 1984 Onwards — The Chemical Engineers Contribution, Bath, 1984, pp. 207–217.) With permission.

It is still not possible at this stage in the design to finalize the layout and operation of the store. However some idea of the position in which it will be operated, how it will be laid out, its size, and so on must be made now. This must be kept flexible until the engineering specification has been formulated (explained later). It is common practice that the decisions are made in advance of producing a store's specification. Lack of flexibility in changing them is often responsible for poor performance once the installation is completed.

5. Plant Layout

Frozen storage is one in a sequence of operations. It influences the whole production system and interacts with it. An idea must be obtained of how the room is loaded, unloaded, and cleaned, and these operations must always be intimately involved with those of the rest of the operation. It is important to know where the frozen food will be produced and where it will be sorted for orders. There is often a conflict of interest within a frozen store. In practice, the frozen store is often used as a marshaling area for sorting orders. If it is intended that this operation is to take place in the store, the design must be made much more flexible to cover the conditions needed in a marshalling area.

Products must be loaded into and out of the frozen store and the process may be continuous, batch, or semicontinuous. In the case of batch and semicontinuous processes, holding areas may be required to even-out flows of material from adjacent processes.

The earlier mentioned specifications should enable the conditions within the storage room in terms of air temperatures, velocities, and possibly relative humidity to be specified. If there is a requirement to freeze or complete freezing within the store, then this extra load can also be calculated, although it may be complex to calculate due to the interactions between this stage and previous processes. Where design data exist, they should be utilized to specify the product load.

Other refrigeration loads also need to be specified. Many of these, such as ambient air infiltration through door openings and gaps in the structure, the use of lights, machinery and people working in the refrigerated space, and refrigeration system defrosts, are all under the control of the user and must be specified so that the final design can cope with the heat load created by them. Ideally, all of the loads should be summed together where they interact on a time basis, to produce a load profile. It is important to know the load profile over time to achieve an economic solution, if the refrigeration process is to be incorporated with all other processes within a plant.

The ambient design conditions must be specified. These are the conditions of the air that cools the hot, high-pressure side of the refrigeration equipment and surrounds the frozen store on all sides. In standalone refrigeration processes, the wet and dry bulb temperatures of the outside of air will be significant. If the process is to be integrated with heat reclamation, then the temperature of the heat sinks must be specified. Finally, the defrost regime should also be specified. There are times in a process where it may be undesirable for defrosts to take place (e.g., sometimes during the loading of a room), and so they should be timed to clear the coil of frost before commencing this part of the process.

The end user should specify all the earlier requirements. It is common practice throughout European industry to leave much of this specification to refrigeration contractors or engineering specialists. Often, they are in a position to give good advice on this. However, as all of the above are outside of their control, the end user, using their knowledge of how well they can control their overall process, should always take the final decision.

B. Engineering Specification

The ideal engineering specification will represent the process requirements in a form that any refrigeration engineer can use to design the system without knowledge of the way the room will be used. If the first part of the process specification has been completed, then the engineering specification will be largely in place. It consists of the environmental conditions within the refrigerated enclosure (air temperature, velocity, and humidity), the way in which the air will move within the refrigerated enclosure, the capacity of the refrigeration equipment, the refrigeration load profile, the ambient design conditions, and the defrost requirements. The final phase of the engineering specification should be drawing up a schedule for testing the engineering specification prior to handing over the equipment. This test will be in engineering, not product, terms.

During this process, the user must play an active role because a number of the decisions taken in this stage will affect other aspects of his operation. The specification produced should be the document that forms the basis for quotations and, finally, the contract between the user and contractor. It must be stated in terms that are objectively measurable once the chiller is completed. Arguments can result from an unclear, ambiguous, or unenforceable specification and so clarity and accuracy are essential. These disagreements and any legal cases that result will be expensive to all parties involved.

1. Environmental Conditions

Usually, a frozen store will be designed to keep product at a constant temperature. As previously discussed, the products to be stored and the storage time will determine this temperature. If the store has to cater for a range of products, then the product that requires the lowest storage temperature should determine the correct storage temperature. If the room will be only used for storage of prefrozen food, then the air velocity over the products should be the minimum required to maintain the air temperature.

A very low air movement is especially important for any storage room that may contain unwrapped products. To further reduce weight loss, a high relative humidity is also required.

If it is known that the room will be loaded with food that is warmer than the storage temperature, then provision must be made to maintain higher (>1 m/s) air movement over the warm product. It is not recommended to design a room to both extract heat (lower food temperatures) and maintain food temperatures. If it has to be achieved, then variable or dual speed evaporator fans are required. Alternatively, auxiliary fans may be installed within the room to generate the required air movement over the warm product. In either case, the heat generated by the fans must be included in the heat load calculations.

2. Room Construction and Size

The size of the room can be determined using throughput information from the user specification. The desired racking, stacking, and loading system for the frozen store will determine the total internal space required. Fuller [6] and Trott [7] describe in detail the effect of different stacking patterns on room size. Frozen stores vary from small rooms with manually loaded racking either side of a central aisle to large, automatically loaded systems with movable racking. High rooms that will allow up to ten pallets to be stacked vertically cost less per unit volume than systems with lower ceiling heights. However, the height may be unacceptable to planning authorities and the pallet loading and retrieval will take longer.

3. Floors

Although cold store floors appear simple, they are multilayer construction with each layer performing a specific function [6]. The main components are:

- Sub base — to provide flexible but firm support
- Base slab — to provide main support for whole structure
- Heaters and screed — to prevent freezing below the insulation
- Vapour barrier — essential to prevent moisture ingress into the insulation
- Insulation — to prevent heat infiltration into the room (especially from the heaters)
- Slip layer — protects insulation and allows relative movement
- Wear floor — strong, hygienic, and wear resistant working surface

4. Walls and Ceilings

The walls and ceilings of modern frozen stores consist mainly of insulating panels and their supporting structures. The panels are traditionally plastic-coated steel sheets bonded to an insulating core of expanded polystyrene, polyurethane, or mineral fiber. Increasingly, insurance requirements are making it mandatory for the structure to be fire retardant.

One of the most important features of the walls and ceiling is a continuous and effective vapor barrier on the outer surfaces of the insulation to prevent moisture ingress into the insulation.

In any cold store, especially if mechanical handling is used, internal protective rails are required to prevent damage to the structure.

5. Doors

Air infiltration through doorways into cold storage rooms during loading and unloading is by far the largest source of heat ingress. The ingress of warm, moist air causes many problems to a store's operators, including increased running costs [8] and defrost requirements together with ice buildup, which can lead to accidents involving personnel. Information from the UK Health and Safety Executive showed that 3% of all accidents reported to them from the distribution industry were in cold stores [9].

The size of doors, their position, the length of time they are open, and the infiltration protection fitted to them (if any) will all affect the calculations used to produce the engineering specification. Positioning doors in opposite walls, which creates through air movements when both are open or poorly fitted, should always be avoided. All doors should be fitted with the most appropriate method of reducing infiltration.

The traditional and most common method of reducing infiltration is by fitting a transparent PVC strip curtain. The ability of devices such as strip curtains to reduce infiltration is specified in terms of their infiltration reduction effectiveness. An effectiveness of 100% means that infiltration is completely removed and 0% means that the infiltration is equal to that which would occur with an unprotected door. The static (no traffic moving through the entrance) effectiveness of strip curtains has been measured to be between 90 and 96% [10]. However, they are "generally considered as unsafe, not particularly efficient, unhygienic and requiring much maintenance and it is possible that they may be banned in the future" [11]. Vestibules (air locks) and flexible, fast opening doors, often in combination, are other methods employed to reduce infiltration. Vestibules are very effective, but restrict access too much for some operators and are difficult to fit to existing sites because of the amount of space that they occupy. Flexible, fast opening doors suffer from the disadvantages of heavy maintenance requirements and lack of vision for forklift truck operators.

Air curtains reduce infiltration without taking up much space or impeding traffic. The origin of the air curtain dates back to a patent by Theophilus van Kennel in 1904 and they have been popular for around 50 years. Air curtains consist of a fan unit that produces a sheet of air, which forms a moving barrier to heat, moisture, dust, odors, insects, and so on. In the case of cold room air curtains, the fan unit is situated either above the door, blowing a jet vertically down, or at the side of the door, blowing a jet horizontally across the door.

Air curtains vary from simple single jets to dual or triple air jets [11]. Some air curtains recirculate their air via a return duct, whereas others do not. The static effectiveness of air curtains as-found (before improvements were made) have been measured between −44 (actually detrimental) and 78% [10]. This was improved to 42 to 80% after adjustments were made to the air curtains. Studies at FRPERC also showed that testing and adjustment of air curtains could greatly improve performance above that obtained under a standard installation.

6. Refrigeration Load

Refrigeration load calculations can now be performed, leading to a load profile for the room. If the store is only loaded with fully frozen product, the infiltration load is likely to be the most important. When loading or unloading a frozen store, the doors may be left open for long periods, which can allow a fully established air flow to take place between the room and ambient air, either from buoyant flow by a single door or from through flow of air if more than one door is open. Designers often decide that the door will only be open for short periods and that fully established airflow will never occur. A clear process specification will show whether this assumption is valid.

The heat load through the structure is usually much smaller than that through the doors. The same applies to the heat load imposed by people, machinery, and lighting in the store. Unfortunately, these loads are normally concurrent with the infiltration load and must therefore be added to these to calculate the total peak load.

The evaporator fans can also produce high heat loads. At this point, in the design, an approximate figure for evaporator fan power must be used but when the final design is completed, and more accurate data are available, this must be substituted and the calculations reworked. A contingency or safety factor is often added to the earlier mentioned calculations, to allow for errors.

The heat load calculations are more complicated if the room is expected to cope with product that is loaded at a warmer temperature than the storage temperature. Product at the center of a shrink-wrapped pallet will change temperature very slowly. This is an advantage if the correct temperature has been achieved before loading the product into the frozen store, but a problem otherwise. The process specification should clearly define the maximum acceptable time to reduce the food temperature to that required for further storage. The conditions required to reduce the temperature at this time, if practically possible, must be determined and incorporated in the engineering specification.

In some cases, this may simply require the provision of more powerful fans to achieve the required airflow and more powerful refrigeration to meet both the peak product heat load and that from the added fans. In other cases, the required time may not be achievable without changing the configuration of the product itself and hence the process specification.

7. Refrigeration Plant Capacity

The capacity of the refrigeration plant must now be decided. If there is a large difference between the peak and average load, the specification should be made to meet the peak heat load. If the refrigeration system can meet the peak heat load, the planned cooling times and the agreed specification can be achieved. However, the peak heat load may be over a very short time and so the plant will be larger than required for most of the running time. Larger refrigeration plant working at low load is much less efficient than smaller plant working at full or nearly full load and is therefore far more costly to run; it is also more expensive to purchase.

There are some possible solutions to the designer's dilemma. If refrigeration capacity is demanded elsewhere on site, but at different times, the provision of a central plant serving both facilities can make use of this diversity. It is therefore important at this stage to look at the refrigeration load profile for the entire plant or site — there may be blast freezers which are only operated well after the time that the peak cooling load has passed, and by careful refrigeration design, plant may be installed and shared between both facilities. However, this only applies to a part of the refrigeration plant (the compressors and condensers) and not to the evaporators. Another option is to spread the loading time of the room over a longer period, and so reduce the peak product loads.

Whatever decision is taken, the peak product load that the refrigeration plant is expected to accommodate should be clearly stated in the agreed engineering specification. A load profile should also be given to ensure that the refrigeration designer provides a plant that will run as efficiently as possible over the entire product load range.

8. Ambient Design Conditions

The conditions of the air outside the frozen store must be defined in the engineering specification. Both the air infiltration and fabric loads are dependent on the outside temperature, which therefore has an important effect on the capacity of the refrigeration plant. Ambient temperature also affects the capacity of the refrigeration plant because heat must be given up to this air to condense the refrigerant in a cooling tower or condenser. If it is intended that the room should function under all possible ambient conditions, very high ambient wet and dry bulb temperatures must be specified. However, these normally occur only during exceptional circumstances and only briefly at or soon after midday. For design purposes, temperatures that are not exceeded for more than 2.5% of the total time in the year are normally acceptable and often a figure of 5% is used. Both wet and dry bulb temperatures should be specified, as this may allow the designer the option of using an evaporative condenser or cooling tower for heat rejection to the atmosphere, which leads to a more efficient and smaller cooling plant.

9. Defrosts

The defrost events should be specified to avoid the peak heat load periods while still ensuring that during these peak periods, the evaporator is clear of ice. The defrost system and timings should be well designed so that the ice is fully removed by defrosts, as gradual buildup can lead to blocked coils, which seriously impede the refrigeration performance.

10. Engineering Design Summary

The engineering specification should, therefore, include each of the items shown below:

1. Store air temperature, air speed, and relative humidity for each product specification (covering complete range) and the time that each of these periods will be operating.
2. The ambient air temperature: both wet and dry bulbs.
3. The peak and average heat loads.
4. Infiltration load, that is, the number of door openings and the time they will remain open, under what circumstances and conditions.
5. Evaporator and condenser temperatures.
6. All the conditions laid down in the engineering specification can be measured and therefore do not depend upon variation in usage or even abuse of the store and should therefore form the basis for a contract.

IV. MAINTENANCE OF FROZEN STORAGE ROOMS

The maintenance of frozen storage rooms can be broken down into the structure of the store and the refrigeration system. Monitoring is the key to maintaining the optimum operation of a frozen store.

A. Structure

The structure of many existing frozen storage rooms has survived for 30 to 50 years without any routine maintenance. Over that period, the insulating properties of the wall materials will gradually deteriorate. Any breaks in damp-proof membranes will lead to moisture ingress into the insulation, which will seriously impair its insulating properties, especially if it turns to ice. Consequently, any damage to the structure needs to be repaired and resealed immediately. However, in many cases, breaks are not obvious until too late when the expanding ice produces noticeable signs and damage.

Routine monitoring of heater circuits in the doors and floors should reveal any breakdowns before serious ice buildup occurs. Replacing heaters that are embedded in concrete screed may be a very difficult and costly operation.

Poor sealing of doors and failure of door protection systems cause the most common problems. The ingress of warm damp air will initially increase the operating costs of the store. Ice buildup on the floors and walls can result in accidents, whereas buildup on the coils can cause a reduction in refrigeration capacity or refrigeration failure. Routine checking and repair or replacement of door seals, PVC strip curtains, and any other door infiltration protection devices are essential for efficient operation. Air curtains are more problematic as it is not always obvious that they are not working in an optimal manner. Poorly installed air curtains or ones that become poorly adjusted can actually increase the infiltration over that of an open, unprotected door. Systems should be routinely checked for correct operation.

B. Refrigeration System

Increasingly, refrigeration plant is becoming fully automatic in operation. However, there is still a need for routine cleaning of filters and strainers, attention to oil and lubricant levels, and so on.

The correct and efficient operation of any plant requires full flow through heat exchangers. Finned coils on condensers and evaporators need to be routinely cleaned. Studies have shown that many refrigerated stores are not routinely cleaned, and that in many cases, large numbers of bacteria can be found on and around evaporator coils [12]. Evaporator coils are usually designed to give good heat transfer rather than be easily cleaned, thus they are often situated in difficult places to reach.

Poor design and operation of defrost cycles can often result in coils becoming totally blocked by ice. Careful monitoring of ice buildup is essential for the maintenance of optimal performance of the storage room.

C. Monitoring

Continuous monitoring of key temperatures is one method of maintaining system performance and identifying potential problems before they affect food temperatures. Recording and plotting temperatures after successful commissioning of a frozen store will establish the pattern of temperature cycling that occurs when the system is performing as designed. Any deviation from that pattern will either indicate that something has changed in the refrigeration system or it is being used in a different manner to that specified. Sensors near the door openings will show any changes in the door opening schedules. Sensing the air on and off evaporator coils will immediately indicate any problems with ice buildup or the refrigeration system itself. However, monitoring is only of use if the data produced are analyzed and actions taken when any deviations are found.

V. COLDROOM MODEL

Recently, FRPERC has produced the user-friendly model 'ColdRoom' to help in the design and operation of chilled and frozen storage rooms.

The main project objectives were to improve the safety, quality, and economics of chilled and frozen storage by closer control of food temperature. This was achieved by developing a user-friendly model to predict food temperatures in chilled and frozen storage rooms under real operating conditions. The model allows:

1. Cold room operators, contractors, and manufacturers to specify and design cold rooms to keep food at optimum temperatures under actual working conditions.
2. Users to rapidly predict the effect of operating conditions and loading patterns on performance and identify how they can avoid unacceptable food temperatures.

A. Mathematical Model

The core of the mathematical model consists of an iterative solution method that repeatedly assembles simultaneous equations that represent heat flows in the room, solves these equations to give the temperatures, and then increments the model through time. The heart of the model is the air within the cold room. This air is subdivided into blocks over the width, height, and depth of the room. These air blocks exchange heat by convection with each other and surface heat transfer with any solid objects that they are adjacent to (walls or food). Wall blocks are layered around these room air blocks and food can be placed within the room air blocks. Features that add or subtract energy from the room, such as evaporators, air and moisture ingress through door openings and people, machinery, and lighting in the room, are incorporated.

The model subdivides the air of a cuboid room into $3 \times 3 \times 3$ blocks. Energy and mass balances are carried out on each block, allowing each to exchange heat and mass with the air blocks adjacent to it and heat with the wall blocks adjacent to it or food blocks within it. Each block is a cuboid with six faces, so six energy and mass balances are necessary for each block. If food is within the block, a further term is necessary for this heat transfer. Boundary conditions are set for air temperatures outside each of the main faces of the walls, roof and floor, in front of the door, and onto the evaporator. A constant heat transfer coefficient between the outer surfaces of the room and the surrounding air is assumed.

All of these heat transfer terms are converted to a set of linear equations, which can be reduced to coefficients of the unknown future temperatures and of known temperatures. In this way, they can be represented in matrix form and solved to find the future temperatures at each time step in the model using a matrix solution method. The fastest solution method that was found was an optimized matrix solution method, which takes advantage of the sparsity of the matrix to eliminate many of the calculations. To increase model speed further, variable time stepping was incorporated into the model solution.

B. Room Air Movement

The flow into and out of each of the cells, of which there are few compared with, say, a computational fluid dynamics (CFD) mesh, is calculated from a simple function or algorithm. Using CFD modeling to develop equations and algorithms, work was carried out to create a set of rules and equations that define the flow of air around the inside of room. To solve the flow regime analytically, the model was divided into an aisle portion (two-dimensional flow as a vortex, forced by the fan) flanked by food storage portions (porosity based). CFD showed that with no temperature gradient or buoyancy, the throw (distance of the far side of the vortex from the front of the fan) is a function of the ratio of the height of the fan to the height of the room and the entrainment coefficient ($\sim$0.1 in most cases) and is largely independent of the speed and quantity of air leaving the evaporator fan. An analytical solution was developed for this problem and incorporated into the aisle portion of the model. The flow through the food storage portions is modeled using a porosity model, the inputs being the pressures created by the velocity profile in the aisle vortex. The porosity of the food storage portions is dependent on the amount of product placed therein.

C. Infiltration and Door Openings

It is assumed that all of the air infiltration occurs through the doorway. The user enters the specification of the door and the conditions outside. A variable amount of air is exchanged through the door depending on whether the door is closed, open, or open but protected by such devices as air curtains or strip curtains. As examples of the data input windows, and the data input methods required by the user, screenshots of the door opening scheduler, and the food loading and unloading order windows are shown in Figure 10.2 and Figure 10.3.

The model predicts the mass of air that would enter the blocks adjacent to the door. The store will lose or gain heat depending on the temperature of the air outside the room, compared to that inside the room, adjacent to the top of the door.

The air infiltration calculations were based on equations recommended by studies undertaken by FRPERC, which carried out experimental and CFD work on the infiltration of air through door entrances [13,14].

D. Defrosts

The air that passes through the door will carry moisture with it. If this air carries more moisture than the air inside the room, the model considers that the moisture accumulates in the room air and then condenses onto the evaporator coil, where it freezes if the coil is below 0°C.

If the coil is above 0°C when the water condenses, it is not cold enough to freeze the water, so the model considers it to drop from the coil and pass out of the room. The same will happen during a

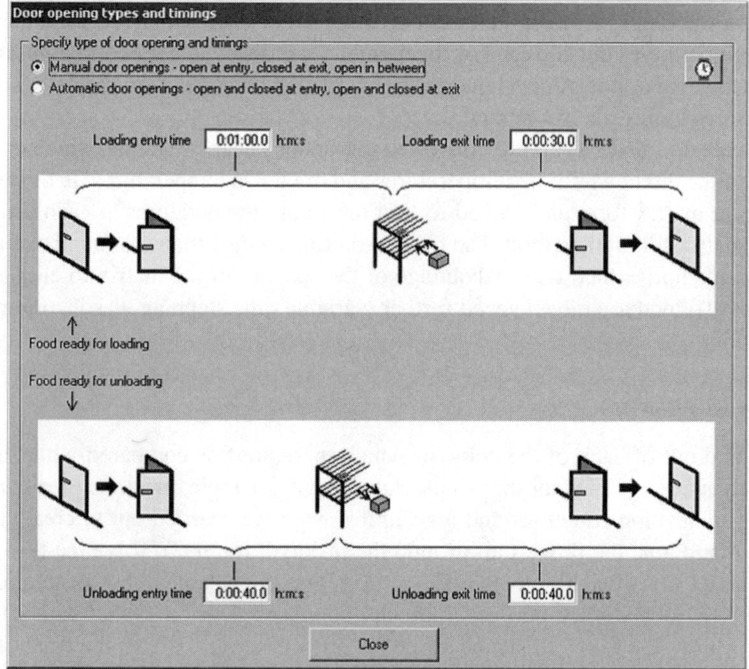

FIGURE 10.2 Screenshot of the door-opening window for ColdRoom program.

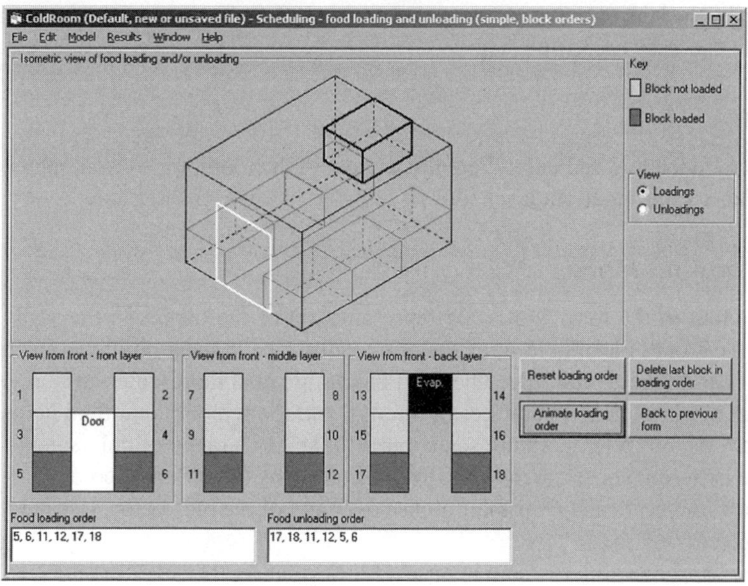

FIGURE 10.3 Screenshot of the food loading (and unloading) window for ColdRoom program.

scheduled defrost or if the refrigeration system remains off for long enough for the room air to heat the coil and mass of ice above 0°C as in an off-cycle defrost.

E. Refrigeration System

The model of the refrigeration system was developed to remove energy from the room air in a manner similar to that of a refrigeration system. This module uses the thermophysical properties (enthalpy, entropy, specific volume, and saturation conditions) of the chosen refrigerant, which varies with pressure and temperature as they pass around the refrigeration system. It relies on lookup tables and interpolation therein to model the state of the refrigerant, adding or losing heat as it flows through the different parts of the system.

The refrigeration module makes some initial estimates regarding the state of the refrigerant in the system and then uses these guesses, the current room and ambient air temperatures, and flow-rates and some other known values regarding the refrigeration system to work around an iterative loop, recalculating the guesses when possible, and so converging on a set of conditions that balance.

Once all of the estimated properties have converged, the refrigeration model is balanced at a steady-state condition and the module can pass the necessary information, such as evaporator and condenser duties, to the calling procedure, which can in turn incorporate the duties into the model solution matrix.

F. People and Machinery

Heat inputs are entered into the model for people and machinery that will enter the room when the product loading occurs. These heat loads are entered into the center, bottom of the room for the duration of the food loading period.

The heat load set for the machinery assumes that the machinery that is used for loading the food is the same at each food-loading event.

The heat given out by people is related to the temperature of their surroundings at the times they are in the room. The model calculates the heat output by multiplying the number of people who enter the room by the heat output per person. As the room temperature can change at each time step, the heat output can vary each time step.

G. Scheduling

Scheduling of food loading and unloading into and out of the room, food thermophysical properties and loading density are input via a scheduling window in the user interface.

H. Output from the Model

A graphical output in the form of a scale representation of the blocks in the walls and room as viewed from the front is available after the temperatures have been predicted in the model core and written to tab-delimited text file. The food blocks are also represented in this viewer.

The viewer displays the temperatures of the wall and room blocks and food items and layers in two ways. First, there is a temperature scale that is linked to a range of colors, so that differences in temperature between blocks can be visualized instantly by the differing colors. For a more accurate temperature, the user can hover their mouse over any of the blocks to show the temperature to two decimal places.

The viewer has the facility to zoom in on any blocks that are too small to see when all the blocks are fitted to scale on the screen (such as thin wall claddings) and center or resize the representation to fit the screen using simple controls. The room block representations can be dragged across the screen to pan the view across the room.

The room layers can be stepped through from front to back, starting from the outer layer of the front wall, going through the front wall into the room, where the room air and the side walls, roof and floor can be seen, into the back wall and through to the outer layer of the back wall.

I. Verification of the Model

The model was verified against data provided for a chilled cold room operating at approximately 3.5°C. The room was modeled in ColdRoom using data provided about the room construction and refrigeration system. In all the food blocks, the measured center temperature of the food was compared to the temperatures predicted by ColdRoom. The temperatures predicted by ColdRoom were within 0.5°C of that measured at any time during the whole test period in all 18 of the food items in the room. There was no overall trend for the model to over or under predict the real temperatures. Over the whole test period, the model predicted 41% of the center food temperatures to within 0.2°C and 65% of the center food temperatures to within 0.3°C of the real data. Ninety-four percent of the center food temperatures were within 0.4°C of the real data.

J. Use of ColdRoom Model

The model can be used in the design and specification of new storage rooms and to indicate potential problems in operation. An example comparison is detailed subsequently, illustrating the operational problems that arise from loading food into the room at the wrong temperature. The first scenario has food loaded into the room at the correct storage temperature (−18°C), whereas the second has the food loaded much warmer (−8°C).

A frozen store was modeled with a large, unprotected door opening, operating with a thermostat set point of −18°C and a ±2°C control differential. The room was surrounded by high ambient air temperatures (25°C) with high humidity air at the door entrance (50%). The defrosts were set to start 1 h into the run and every 6 h thereafter. Each defrost would run for a maximum of 30 min, but would be terminated earlier if the evaporator coil went above 5°C (the defrost termination temperature, set in the model). A delay of 2 min was scheduled after each defrost termination to simulate a "drain down time," where the refrigeration system remains off to allow water on the coil to drain

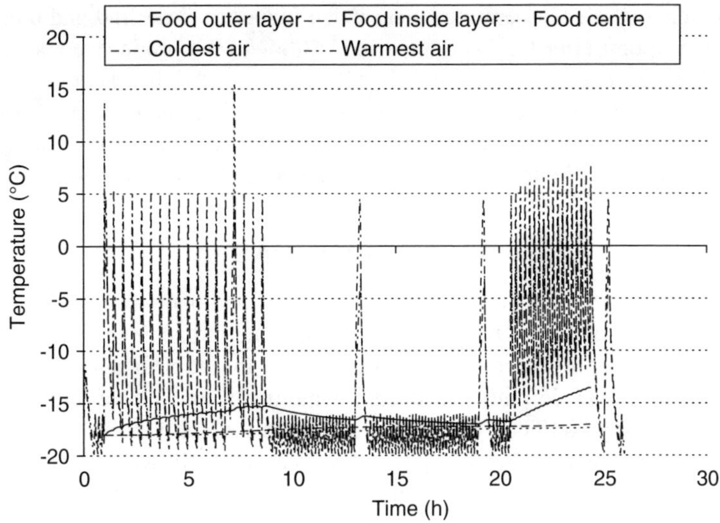

FIGURE 10.4 Loading with food at $-18°C$.

off, so it is not refrozen on restart. The temperature peaks resulting from the defrost events can be seen in the figures explained subsequently.

The food was loaded into the store at $-18°C$ over an 8 h period with regular door openings, each of 2 min, 40 sec, held for 12 h, then unloaded over 4 h. This created large changes in both air and product temperatures as can be seen in Figure 10.4. However, this severe usage pattern still results in changes in food temperatures that would have little, if any effect, on overall storage life or quality of most frozen foods.

The problems associated with loading food that is not at its desired storage temperature become clear in the second simulation, the results of which can be seen in Figure 10.5. The food, loading patterns, and all other factors were identical to that previously modeled with the exception of loading the food at $-8°C$.

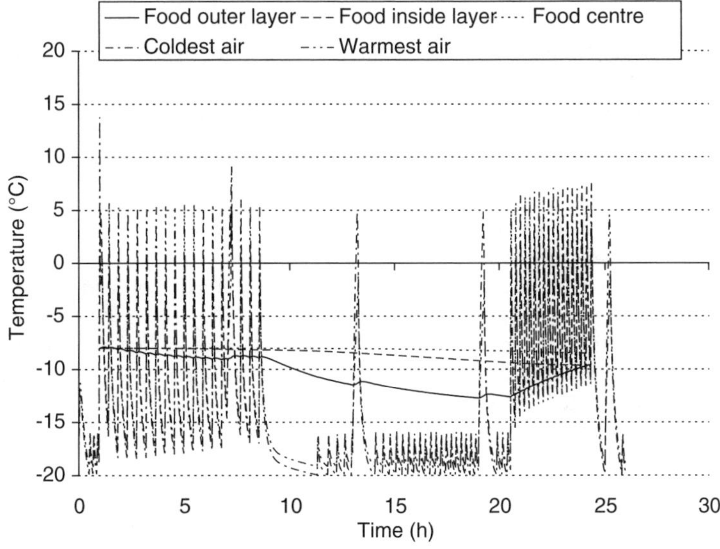

FIGURE 10.5 Loading with food at $-8°C$.

During food loading, air temperatures were much higher than desired and took almost 3 h to return to the control point after the loading period. Although the product temperature fell during the holding period, the core temperature was still above $-9°C$ by the time the product was unloaded and the surface had warmed back up to $-10°C$. The predictions clearly demonstrate the problems of loading food into frozen storage rooms above their desired storage temperature and show how simple it can be to demonstrate this using the ColdRoom model.

VI. CONCLUSIONS

Many factors have to be taken into account when designing and operating a frozen food store. The key task is to define the conditions that have to be maintained to achieve the desired high-quality storage in the products being stored. It is most important that a clear and unambiguous process throughout the specification is initially produced. Changing this into a quantifiable engineering specification will identify any conflicts in the required process, which will have to be resolved.

ColdRoom, a predictive, user-friendly model developed by FRPERC, can be a substantial aid to optimizing the design and operation of frozen cold stores.

REFERENCES

1. SJ James, C James. Microbiology of refrigerated meat. In: *Meat Refrigeration*. Cambridge, England: Woodhead Publishing Limited, 2001, pp. 3–19.
2. Anonymous, *Recommendations for the Processing and Handling of Frozen Foods*, 3rd ed., International Institute of Refrigeration, Paris, France, 1986.
3. G Löndahl, CE Danielsson. Time temperature tolerances for some meat and fish products. In: *Proceedings of the International Institute of Refrigeration Commission C2*, Warsaw, 1972.
4. SJ James, JA Evans. Frozen storage of meat and meat products, FAIR Concerted Action PL95-1180, 1997.
5. SJ James, C Bailey. Bacon tempering for high speed slicing. In: *Proceedings of the XVIIth International Congress of Refrigeration C*, Vienna, C2-1, 1987.
6. R Fuller. Storing frozen food: cold store equipment and maintenance. In: CJ Kennedy, Ed., *Managing Frozen Foods*. Cambridge, England: Woodhead Publishing Ltd, 2000, pp. 213–232.
7. AR Trott. Cold store construction. In: *Refrigeration and Air-Conditioning*. London: Butterworths, 1984, pp. 143–157.
8. P Chen, DJ Cleland, SJ Lovatt, MR Bassett. Air infiltration into refrigerated stores through rapid-roll doors. In: *Proceedings of the 20th International Congress of Refrigeration*, Vol. 4, 19–24 September, 1999, pp. 925–932.
9. MFG Boast. Frost free operation of large and high rise cold storage. In: *Proceedings of the Institute of Refrigeration*, Vol. 6, 2002, pp. 1–11.
10. CC Downing, WA Meffert. Effectiveness of cold-storage door infiltration protective devices, *ASHRAE Transactions* 99 (2):356–366, 1993.
11. PJJH Ligtenburg, DJ Wijjfels. Innovative air curtains for frozen food stores. In: *Proceedings of the 19th International Congress on Refrigeration*, 1995, pp. 420–437.
12. JE Evans, SL Russell, C James, JEL Corry. Microbial contamination of food refrigeration equipment. *Journal of Food Engineering*, 62 (3):225–232, 2004.
13. AM Foster, R Barrett, SJ James, MJ Swain. Measurement and prediction of air movement through doorways in refrigerated rooms. *International Journal of Refrigeration* 25 (8):1102–1109, 2002.
14. AM Foster, MJ Swain, R Barrett, SJ James. Experimental verification of analytical and CFD predictions of infiltration through cold store entrances, *Intertational Journal of Refrigeration* 26 (8):918–925, 2003.

11 Transportation of Frozen Foods

Silvia Estrada-Flores
Food Science Australia, North Ryde, NSW, Australia

CONTENTS

I. INTRODUCTION

About 650 million tons of food are shipped every year worldwide [1]. Most of these shipments are via maritime transport, but high-value frozen foods are being increasingly shipped by air. In this scenario, supply chain management has emerged as an integrated approach, which evaluates the effect of variables such as logistics, distribution, technology, quality, safety, costs, and times in the overall efficiency of a particular commercial operation. The transportation of goods is now seen as a part of a "system," rather than as an isolated event within the commercial operations required to position a product in the market. The analysis of the supply chain components for chilled and frozen foods needs to be particularly meticulous, to minimize negative economic, legal, and moral consequences associated with the loss of quality and integrity of products.

The transport of frozen foods offers a number of formidable challenges to the supply chain manager. McKinnon and Campbell [2] offered some examples:

- Frozen distribution systems are capital- and energy-intensive.
- The industry is often diverse, comprising small and large firms; in the latter case, the size of the shipments makes it difficult to achieve a quick delivery response to the changing demands of the market.
- The flow of transport of frozen foods is highly affected by seasonality, marketing strategies, demand, and competition.
- The value of frozen products is relatively low and the demand is price-sensitive.

In contrast, the transport of frozen goods has been traditionally viewed as a less technically demanding task than the transport of respiring, chilled products. No serious quality or safety issues have been associated with subtle losses of temperature control, as long as product temperatures remain below $-18^{\circ}C$. Only product temperatures above $-12^{\circ}C$ have been associated with an increase in bacterial counts and loss of quality [3].

The selection of transport modes in the distribution of frozen products significantly affects the profitability of the freezing industry. Figure 11.1 shows the costs of operation for shipping a 40 ft (12.2 m) container traveling from the American East Coast to Rotterdam [4,5]. Transport times and loading and unloading operations represent a large share of transportation costs. The latter two are fixed costs. Hence, one of the key areas of focus to decrease the overall cost is to expedite these operations [5] and increase the cost-effectiveness of the transport operation such as increasing the energy efficiency of the operation. Emerging regulations on the use of energy in transport and the impact of emissions on the environment have also fueled the investigation of low-energy, low-polluting refrigeration systems.

It has been estimated that the distribution of frozen foods requires 70% more energy than that of products at ambient temperature [4], but transportation of frozen cargo requires less energy than chilled cargo. Andersen [6] analyzed the energy consumption required to transport the aquacultural fish production from Norway to overseas markets during 1994, in the forms of frozen and fresh cargo. In the case of frozen fish, the modes of transport were trucks from Bergen to Oslo and refrigerated vessels from Olso to the importing country. Fresh fish was transported using trucks

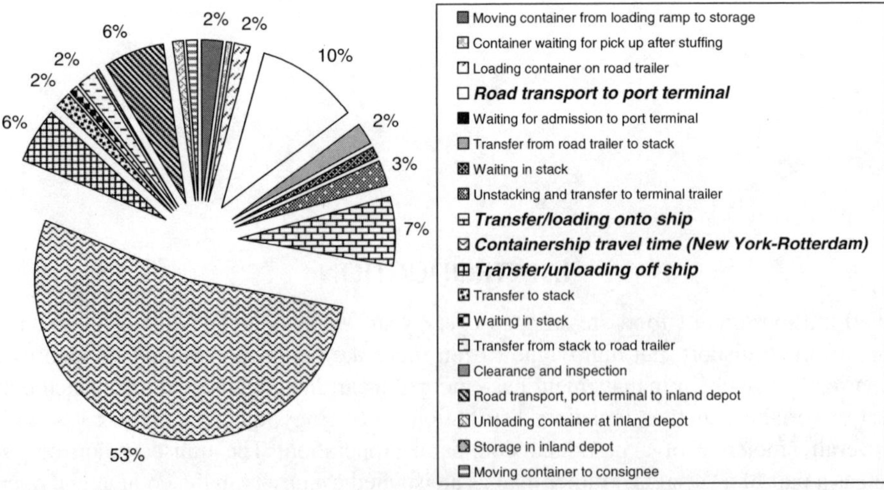

FIGURE 11.1 Cost structure of moving a 40 ft (12.2 m) container between the American East Coast and Western Europe. (Data from W Coyle, W Hall, N Ballenger. Transportation technology and the rising share of US perishable food trade. In: A Regmi, Ed., *Changing Structure of Global Food Consumption and Trade*. Report WRS 01-1. USDA Agriculture and Trade, 2001, pp. 31–40.)

from Bergen to Frankfurt and airplanes from Frankfurt to the importing country. Andersen [6] found that the energy used during the transport of frozen cargo was 2.1 kW h/kg, whereas the transport of fresh fish required 21.6 kW h/kg. Although road and air transport may present lower energy efficiencies than rail and sea transport, fresh products require fast distribution channels to avoid significant quality losses and to take care of safety issues. In contrast, the supply chain for frozen products has longer time frames and can benefit from more cost and energy-effective transport options.

II. QUALITY AND SAFETY RISKS DURING THE TRANSPORT OF FROZEN FOODS

Unpleasant smells, stained packaging, or large ice crystals around the product after transport are all tell-tale signs of a disrupted cold chain during the transport of frozen products [7].

Several guidelines and codes of practice for refrigerated foods have been published [8–12]. Most of these acknowledge that the following quality changes in frozen products may appear during transport:

- Partial melting of ice crystals and moisture migration within the product due to temperature fluctuations during transport. A growth of larger ice crystals at the expense of smaller ones, a change in shape,[1] and changes of crystal orientation may also be present.
- Formation of ice or frost on the surface of packages and pallets due to: (i) entry of warm, moist air to the cargo space (either during the loading of the transport, during door openings, or through the door seals of the insulated body) and (ii) migration of vapor from the product to the internal surface of the package.

Some physical issues are:

- Mechanical damage due to acceleration effects, motion of the transport system, vibration and vertical impacts during accidental falls, collapsing of the package or pallet, and rough handling during loading and unloading.
- Contamination of the frozen cargo due to the use of the same transport to carry substances incompatible with the present cargo, the presence of residues and odors from previous cargoes, the use of cleaning agents or pesticides, unsuitable or contaminated dunnage, and securing materials or materials used in the construction of the transport.

Transient situations leading to loss of temperature control (e.g., defrost, temporary loss of refrigeration power, loading and unloading operations between transport modes and between cold storage, and sudden change of external environmental conditions, among others) are often unavoidable. Nevertheless, the impact of these factors can be minimized by: (a) ensuring that the refrigerated transport is operating optimally; (b) selecting a route that does not compromise the integrity of the product; (c) providing training to staff involved in the logistic operations; (d) selecting adequate packaging; (e) communicating in a continuous manner with the carriers; and (f) designing contingency plans to implement when faced with undesirable scenarios that may compromise the quality of the frozen cargo.

Even though frozen products are more resistant to mild temperature abuse than chilled products, the former also requires the continuity of transport operations expected for chilled cargo. Various anecdotes and reported cases exist for frozen cargo that was unloaded and left at ambient temperature for extended periods of time (sometimes overnight), waiting to be loaded into the next transport [13]. Mason and Wallace [14] investigated the performance of vehicles carrying frozen food from manufacturer's warehouses to bulk or retail warehouses in Queensland. They reported that insufficient precooling of trucks and lack of air distribution systems

[1] Sharp surfaces will have a tendency to become smoother over time.

(e.g., ceiling ducts and floor channels) resulted in excessive warming of the products, in some cases exceeding the maximum recommended temperature of $-18^{\circ}C$ for frozen goods. The most expensive blunders in the cold chain of frozen products seem to arise from human errors and congestion of appropriate facilities in the logistics operations. Thus, a sound knowledge of (a) the issues likely to appear in the logistics of frozen foods, (b) the physical and chemical characteristics of the product to be transported, and (c) the characteristics (e.g., cost, duration, and demand) of the distribution channels available will contribute significantly toward decreasing quality risks during the transport of frozen products.

III. DESIGN AND OPERATIONAL FACTORS AFFECTING TEMPERATURE UNIFORMITY DURING TRANSPORT

There are some common features that need to be considered during the design and operation of refrigerated transport equipment for frozen products. These are discussed below.

A. Initial Temperature of Product

The extra heat load generated by product entering at temperatures above the set-point of a refrigerated vehicle can significantly impair its thermal performance. Temperature differences in boxes of product before loading can persist until the end of the transport period. Moreover, during the transport of mixed loads of products, which may encompass items highly responsive to temperature changes (e.g., prepacked frozen meals with significant air spaces enclosed in the packaging materials) and products with a slow response to temperature changes (e.g., ice cream transported in bulk containers), the thermal response of the cargo space will be largely determined by the predominant thermal mass (e.g., bulk containers) [15]. A refrigerated loading dock should be used to minimize the rise of product temperature during loading and unloading and to avoid the entrance of warm, moist ambient air.

B. Airflow Management and Loading Conditions

The most common air circulation pattern in refrigerated trucks is top-air delivery, lengthwise, front-to-rear [16]. Bottom-air delivery is more common in refrigerated containers [17]. The performance of a refrigerated vehicle can be greatly improved by generating adequate air movement within the cargo space and by avoiding the entry of warm air during product deliveries (in the case of multi-delivery logistics). Recommended measures to achieve effective air circulation include: installing solid return air bulkheads, providing uniform, solid block stowage, securing the load away from doors and sidewalls, providing space underneath the load to create effective return paths for the air, and ensuring that cargo has been loaded below the red line marking the space required for unrestricted airflow back to the unit (in the case of bottom-air delivery equipments) or from the unit (for top-air delivery units). Plastic or air curtains for doors can greatly decrease the entrance of warm air during transport that requires multiple deliveries [18]. In nondedicated containers that may be used to transport either chilled or frozen products, the fresh air ventilators must always be closed and the humidity indicator should be in the OFF position when frozen products are transported [19].

C. Type of Packaging

Packaging designs for frozen products do not require the presence of perforations to aid convective heat transfer between the cold air and the product. Aspects such as structural integrity of the packages and pallets and heat transmission between the package and the surrounding air are more important for frozen foods.

D. Refrigeration System and Temperature Control

Mechanical refrigeration systems coupled with thermostats sensing the air return temperatures are the most common means of temperature control in the transport of frozen products. In vehicles designed to transport both chilled and frozen cargo, the control system usually includes a second sensor in the air delivery. The sizing of the refrigeration unit is critical and should account for extra heat loads such as the heat conducted through the insulated walls, heat contributed by defrost systems and fan motors, and door openings. Control systems can vary from simple ON/OFF strategies, commonly used in long-distance transport of frozen foods, to proportional integral derivative (PID) controls [17]. Temperature differentials lengthwise and along the height of the cargo space may be minimized with adequate control systems, correct placement of temperature sensors, and efficient insulation.

E. Door Seals, Insulation, and Aging of Unit

The possibilities of water-damaged insulation and faulty door seals increase in old vehicles for which maintenance has been insufficient. High air leakage rates and high heat infiltration will lead to loss of temperature control, unless the refrigeration unit has been sized to take these into account. The deterioration of the insulation materials due to aging is estimated to be about 5% of the insulating quality per year [16]. Regarding the refrigeration plant and control components, Jiang and Wang [20] suggested an average lifetime of 7 years for marine reefers before the refrigeration plant is in need of major repairs.

Figure 11.2 summarizes some important characteristics of refrigerated road, rail, sea, and air transport modes. Detailed technical descriptions of refrigerated trucks, containers, rail boxcars, and air shipping containers commonly used to transport chilled and frozen products can be found elsewhere [16,17,21]. Some particular characteristics and operating conditions for frozen products transported in these vehicles will be examined in the following sections.

IV. CONTAINERIZED SEA TRANSPORT

The development and expansion of containerization has been recognized as a significant factor that has contributed to the steady reduction of transportation costs worldwide since 1950 [22]. Containers can be used in maritime, road, and railway transports. The main types of marine containers are porthole containers and integral (reefer) containers.

Porthole (or Con-Air) containers were developed in the late 1960s to facilitate the transport of chilled and frozen products on fixed routes between terminals with similar refrigeration systems [23]. These same design features are now hindering their flexibility in modern shipping ports, and porthole units are likely to be totally replaced by integral containers in the near future. Porthole containers do not have an in-built source of refrigeration, but receive ducted air through two sealable portholes on the front bulkhead [16,17]. The cold air produced is distributed by a central refrigeration system in the ship, which serves several porthole units at the same time. When the containers are being used off the ship, the units are cooled by electrically driven or autonomous clip-on refrigeration units [17,24].

Integral containers accounted for 82% of the world production of 20 ft (6.1 m) equivalent units (TEUs) during 1998 [25]; during 2003, the production of integral containers increased to 99%. The rise in the global production of integral reefers can be attributed to the phasing out of obsolete 20 ft (6.1 m) porthole units, the aging of existing Con-Air vessels (most of these over 30 years old) and the recent increase in containerized reefer traffic [26]. Over 90% of all current reefer production encompasses 40 ft (12.2 m) high-cube units.

Integral containers have their own in-built refrigeration unit that usually runs from a three-phase electrical power supply, generated onboard the ship or by an independent diesel generator

TRANSPORT MODE	TYPICAL VOYAGE TIMES	INSULATED BODIES COMMONLY USED	REPRESENTATIVE TYPES OF TRANSPORT	TYPICAL DIMENSIONS: Size (external dimensions)	L	W	H	Typical loading capacity	TYPICAL COOLING SYSTEM[a]: M/D	M/E	E/D	EP	C
VAN OR COMPARTMENT IN MULTI-TEMPERATURE TRUCK (URBAN/SHORT-DISTANCE DISTRIBUTION)	1 TO 12 HRS	(1) Normal van with sidewalls and doors foamed with polyurethane or polystyrene (2) Insulated box specially designed to fit inside a van	Panel van		3.265 m	1.736 m	1.633 m	13.4 m³	✓		✓	✓	✓
		(3) Non-insulated van carrying (a) product packaged in insulating materials and aided by phase change materials or (b) insulated boxes with refrigeration coupled with van's engine	Van truck		5.805 m	2.155 m	2.180 m	20.6 m³	✓	✓	✓	✓	✓
TRUCK/TRAILER/SEMI-TRAILER (SHORT/LONG-DISTANCE DISTRIBUTION)	12 HRS TO 3 DAYS	(1) Integral container: the refrigeration unit is fitted in the insulated body; the unit may run by means of a separate engine or it may be coupled to the truck's engine	Highway and intermodal trailers	*20 ft* *40 ft* *45 ft High Cube*	6.096 m 12.192 m 13.710 m	2.438 m 2.438 m 2.438 m	2.438 m 2.438 m 2.890 m	33.0 m³ 54.9 m³ 75.4 m³		✓			✓
		(2) Insulated box: polyurethane or polystyrene layers reinforced with fiberglass or steel	Truck	*45 ft High Cube* *48 ft*	13.710 m 14.500 m	2.438 m 2.580 m	2.890 m 2.920 m	75.4 m³ 88.0 m³	✓	✓	✓	✓	✓
RAIL (LONG-DISTANCE DISTRIBUTION)	3 TO 16 DAYS	(1) Non-mechanically refrigerated insulated cars: cooling by (a) dry ice or other cryogenic substances or (b) air cooled by ice/water mixtures	Double-axle insulated wagon	*WAI28B*	16.500 m	2.700 m	2.550 m	84.8 m³	Typical system comprises two ice tanks and electric fans				
		(2) Mechanically refrigerated cars: run on electricity supplied by a central power unit on the train or by an autonomous motor	Double-axle wagon	*WAF26*	15.500 m	4.230 m	2.400 m	74.7 m³	✓	✓			
			Four-axle wagon A	*WAF36*	21.480 m	4.060 m	2.400 m	107.1 m³					
			Four-axle wagon B	*WAF34.2*	21.040 m	4.060 m	2.400 m	107.1 m³					
		(3) Multi-modal container	integral containers, semi-trailers and swap bodies (see intermodal trailers)										
AIRPLANE (LONG-DISTANCE DISTRIBUTION)	1 TO 100 HRS	(1) Passive insulated system: Sizes vary depending on specific needs of product.	B 727-100	Main hold Forward hold Aft hold	2.240 m	3.180 m	2.500 m	82 - 88 m³ 11.9 m³ 13.3 m³					✓
		(2) Active system: Envirotainer ™ is the mayor manufacturer of these units. All models have a similar cooling range of -20 to +20 °C.	B 757-200	Main hold Forward hold Aft hold	2.240 m	3.180 m	2.500 m	155 - 178 m³ 19.8 m³ 30.3 m³					
		Capacities are: (a) model RAP =8.2 m³; (b) model RKN =2.9 m³ (c) model CLD =0.164 m³	B 767-300	Main hold Forward hold Aft hold	2.240 m	3.180 m	2.500 m	474 - 479 m³ 41 - 47 m³ 31 - 37 m³					
			B 747-100	Main hold Forward hold Aft & Bulk hold	2.240 m	3.180 m	2.500 m	301 - 344 m³ 59 m³ 47.5 & 28.3 m³					
SHIP (LONG-DISTANCE DISTRIBUTION)	3 DAYS TO 2 MONTHS	(1) Porthole container: Containers with portholes, through which cold air from the ship's refrigeration plant can be delivered to the cargo space	1st gen converted cargo vessel	Years 1956-1970	135 -200 m	Draft <9 m	For container sizes see intermodal trailers	TEU[b] 500-700	✓	✓			
			1st gen converted tanker	Years 1956-1970	135 -200 m	Draft <9 m		TEU 800					
		(2) Integral container: the refrigeration unit is fitted in the insulated body; the unit may use energy supplied by the on-board electricity system, an independent diesel engine or cryogenic systems	2nd gen cellular containership	Years 1970-1980	215 m	Draft 10 m		TEU 1000-2500					
			3rd gen Panamax class	Years 1980-1988	250-290 m	Draft 11-12 m		TEU 3000-4000					
		(3) Refrigerated vessel: All or part of the holds are insulated and have ducting to channel cold air from evaporators using expanded refrigerant or brine	4th gen post Panamax	Years 1988-2000	275-305 m	Draft 11-13 m		TEU 4000-5000					
			5th gen post Panamax plus	Years 2000-?	335 m	Draft 13-14 m		TEU 5000-?					

[a] M/D: Mechanical/Diesel; M/E: Mechanical/Electricity; E/D: Vehicle's engine or alternator; EP: Eutectic plates; C: Cryogenic;
[b] TEU= twenty-foot equivalent unit; a TEU is a measure of containerized cargo equal to one standard 20 ft container.

FIGURE 11.2 Important characteristics of the most common modes of transport of frozen products. (*Sources*: Anonymous. Guide to Refrigerated Transport. IIR TFF-TFA. Paris: International Institute of Refrigeration, 1995, pp. 70–130; J. Frith. *The Transport of Perishable Foodstuffs.* Cambridge: Ship Owners Refrigerated Cargo Research Association, 1991, pp. 7–29; Anonymous. *ASHRAE Handbook, Refrigeration, Systems and Applications.* Atlanta, GA: American Society of Heating, Refrigerating, and Air-Conditioning Engineers, 1994, pp. 1–30.4; Y. Wild. Refrigerated containers and CA technology. In: *Container Handbook. Cargo Loss Prevention Information for German Marine Insurers, 2003.* Berlin: Transport and Loss prevention Department.)

[16,17,25]. These containers are generally carried on deck or in a particular area below deck that can be cooled sufficiently to dissipate the heat rejected by the container's condensers.

In most containers, air flows from the bottom of the unit and is delivered to the cargo area through a T-section floor. The depth of the floor channels must allow the flow of air under the cargo and toward the doors. Finally, the air returns to the refrigeration unit via the space between the top of the cargo and the ceiling. The frozen cargo is stacked solidly in the center of the container, with no gap between the boxes and ensuring that a space is left between the pallet and the walls of the container. This space provides some protection to the frozen cargo against heat infiltrated through the insulated walls into the container [17].

Temperature control in refrigerated containers carrying frozen products is based on the measurement of the temperature of air returning to the evaporator. The cargo temperature is often different to the air temperature, due to the thermal inertia of the frozen product. Hence, a $\pm 3°C$ tolerance band around the set-point temperature (usually $-18°C$) is generally used to account for those deviations, unless specific regulations for a particular product apply.

Some particular recommendations to ensure a successful transit of containerized frozen products are given as follows:

- Whenever possible, the use of containers older than 7 years should be avoided, to decrease the impact of aging defects (e.g., quality of insulation and reliability of mechanical parts).
- Door and portholes seals are particularly sensitive to "wearing and tearing" effects and should be inspected before loading.
- Containers should be clean and free of odors.
- A pretrip inspection should be carried out by the carrier. The container should be precooled at the required set-point temperature (equal or below $-18°C$) before loading.
- It is important to keep temperature records of the container at all times (i.e., during transport to the port, while on the port of departure, during the voyage, and while on the port of arrival). Standard procedures are required to recover and store temperature information logged in the temperature recorder of the container. Temperature records (e.g., circular charts, often referred to as Partlow charts) are maintained by the staff onboard the ship. However, these records are often kept confidential by the shipping company and they are not readily available in the case of a legal dispute that involves temperature management practices.
- Precise instructions about the location of the refrigerated container (e.g., "under the deck" or "on deck" stowage) need to be discussed with the carrier, to avoid delays in switching of refrigeration (in the case of porthole containers) and unnecessary exposition of the containers to direct sunlight.

V. VESSEL SEA TRANSPORT

Conventional refrigerated ships are designed to hold frozen product at $-20°C$. Ships intended for transporting break-bulk or palletized cargo have holds divided by decks 2–3 m high, providing spaces with perforated deck grating [27]. Refrigeration can be achieved by direct expansion or brine systems. In both cases, the coils transporting the coolant may form part of a normal evaporator with fans to mobilize the air or they may be distributed within the walls and ceiling of the hold [28]. Cold air is blown into the ducts at the end of the hold or through side ducts. The air ascends through the cargo and through the duct at the opposite end of the hold, and then returns to the refrigeration unit via spaces above the cargo. Two between-deck spaces can be combined to form a common air space, although such combination risks incomplete refrigeration at the higher levels of the hold. The holds are usually insulated with polyurethane, although polystyrene and expanded polyvinyl chloride are also used [16].

The loading of frozen cartons has proved particularly challenging in the shipping industry. At loading, pallets are removed from the truck trailer and placed on the dock, where they are subsequently lifted into the hold using the ship's gear such as slings, lifting platforms, or flying forks. In the hold, lift trucks engage the pallets and transport them to locations near where the cartons will be stowed. Stevedores then manually remove the cartons from the pallets and stack them for shipping. The pallets are then returned to the square of the hatch and are stacked to be hoisted out of the hold and back onto the dock by the ship's gear. Delays in bringing a sufficient quantity of product to the dock and in unloading the cartons from the truck can increase the time needed to load the vessel [29]. Space restrictions in the vessel also limit the number of workers stowing the cartons. Excessive delays in loading result in cartons being left on the dock or in the truck, allowing the product to warm up. It may also result in increased condensation of moisture on the cartons, which can complicate the handling process. As the frozen food industry is seeking to use less wax on the cartons and to utilize paper-coated boxes, the damaging effect of condensation and internal thawing on the boxes increases. Overall, high costs, significant expenditure of manual labor and bottlenecks that slow the loading process may result in product degradation or spoilage are still frequent in loading and unloading operations of ships. Technical solutions have been offered [29] but these have not been widely implemented. Thus, transhipment should be avoided whenever possible. Even though freight cost may decrease by using this option, the risk of losing cold chain integrity may offset this advantage.

VI. ROAD TRANSPORT

Numero and Jones [30] are credited with the invention of the first practical refrigeration unit used in trucks and railcars. Since its introduction, refrigerated trucks have become a necessary link in the supply chain for frozen products.

The growth of online (the Internet) shopping has changed the panorama of urban distribution of foods. It is expected that by 2007, the value of online grocery shopping will reach US$ 85 billion in the United States alone, with 25% of U.S. households making use of the Internet to obtain their food supplies [31]. This trend has led to an increase in the use of small, multitemperature, multicompartment vans, to deliver a mixture of frozen and unfrozen goods. Vans typically have a working schedule of 8–12 h, including multiple loading and unloading of small parcels with several door openings and evaporator defrosts occurring during that time.

Panel vans have been successfully adapted to work as either single-temperature or multitemperature delivery units. As an example, normal Sprinter vans converted to refrigerated vans have achieved a heat leakage value (also known as *K*-value) below 0.4 $W\ m^{-2}\ K^{-1}$ thus complying with the European statutory regulations for the transport of chilled produce [32]. The refrigeration plant in panel vans is usually driven by the vehicle's engine, but an electric motor that draws energy from the vehicle's alternator and an auxiliary battery can also be used. In larger multitemperature vans carrying chilled and frozen cargo, refrigeration systems comprising two or more evaporators that share both compressor and condenser are common. When the chiller compartment requires cooling, the supply of refrigerant to the frozen compartment is suspended and the demand of the chiller is attended [33]. The size of the compartments can be varied by means of removable partitions. Diesel, electricity, or a combination of both are commonly used to run multitemperature refrigeration units, but eutectic plates for cooling are also commonly used.

Eutectic systems can only hold product for a period of less than 10 h [16], depending on the heat load entering the cargo space. The holding temperatures for frozen products range from −20°C to −30°C. The performance of eutectic compartments greatly depends on the sound design of the system (e.g., amount of solution, size and position of plates within the insulated compartment) and, once installed, in the complete freezing of the solution during recharging periods, usually occurring overnight in a normal urban distribution chain.

Long-distance road transport relies on the use of highway and intermodal trailers and trucks [21]. Long-distance vehicles withstand continuous operation periods of about 1000 h, while downtime due to maintenance needs to be kept to a minimum; thus, the following aspects are more significant for vehicles that are used a few hours per day:

- Efficiency of the insulated body. This parameter is often referred to as a *K*-value ($W\ m^{-2}\ K^{-1}$), or the transfer of heat through the structure of the insulated body, measured under certain test conditions, divided by the product of the mean surface area of the body and the difference between the external and internal air temperatures.
- Airtightness of the unit, measured as air leakage under a pressurized condition ($m^3\ h^{-1}$).
- Time to pull temperatures down to the required set-point temperature.
- Steadiness of the cargo space temperature during normal operating conditions.
- Steadiness of the cargo space temperature when an extra heat load is applied in excess of the normal heat load (either simulating the effect of door openings or additional heat infiltration through the walls).
- Temperature variations within the cargo space with respect to voyage time, also related to airflow distribution.
- Effect of defrost systems.

Some examples of international regulations covering the aspects mentioned earlier are given below.

1. The Agreement on the International Carriage of Perishable Foodstuffs and on the Special Equipment To Be Used for Such Carriage (ATP Agreement) [8]. This agreement provides a number of classification temperatures that need to be assessed under an ambient temperature of +30°C, among other experimental conditions. The relevant classifications for the transport of frozen products are:

- Class C: mechanically refrigerated equipment fitted with a refrigerating appliance operating at a temperature between +12 and −20°C, inclusive.
- Class F: mechanically refrigerated equipment fitted with a refrigerating appliance operating at a temperature below −20°C.

The vehicles are tested in terms of their insulation capacity and the efficiency of the refrigeration unit under steady conditions. The *K*-value is required to be less than or equal to $0.4\ W\ m^{-2}\ K^{-1}$. To account for door openings and other sporadic cooling demands, the refrigeration system is required to have an extra capacity of 35% over the expected cooling requirements at normal conditions, when operating at its minimum classification temperature.

2. The Australian Standard 4982-2003 [34]. This standard follows closely the requirements of the ATP agreement. However, the external ambient temperature used to test the vehicles is more stringent (+38°C). The classifications pertaining to frozen products are:

- Class C: the mean inside temperature needs to be assessed at −18°C.
- Class D: the mean inside temperature needs to be assessed at −28°C.

The standard does not specify a maximum *K*-value, but the unit needs to operate at the stated classification temperature for at least 8 h. Furthermore, the vehicle must keep the classification temperature for 4 h, with an extra heat load equivalent to 35% of the heat infiltrating the envelope.

Thermal imaging and full mapping of temperature variation in the cargo space may be useful to test vehicles used for the transport of frozen products [35].

Shortages of storage space for holding frozen products may encourage more efficient inventory management, but often have a negative impact on the handling of frozen products. Transport fleets are being increasingly used as a flexible resource to compensate for the lack of storage space. In a recent survey in the U.K., McKinnon and Campbell [2] found that trucks spent nearly 10 h out of 48 h fully loaded and stationary. Holding frozen products in trucks is less energy-efficient than using a cold store; however, standard distribution practices frequently demand preloading of trucks and synchronization of the loading of inbound and outbound cargoes to achieve a more efficient distribution of workload and staff.

VII. RAIL TRANSPORT

Rail transport costs for frozen products have been compared with road transport [36]. Even though rail maintenance and construction costs are high and these impact on the rates offered to the users, costs can be offset if transhipments (e.g., loading and unloading) are minimized, transit occurs between distant geographical points, and the size of the cargo is significant.

Railcars used for the transport of frozen goods may be grouped into three categories [16,21]:

1. Refrigerated wagons are insulated bodies using a nonmechanical source of cold (e.g., ice with or without the addition of salt; eutectic plates; dry ice, with or without sublimation control; liquefied gases, with or without evaporation control). According to the ATP agreement, a wagon class C carrying frozen products should be able to maintain a temperature of −20°C with a mean outside temperature of +30°C.
2. Mechanically refrigerated wagons are insulated wagons either fitted with their own refrigeration plant, or served jointly with other wagons by a central mechanical refrigeration unit. According to the ATP Agreement, a wagon class C carrying frozen products should be able to maintain a temperature between +12°C and −20°C, inclusive, with a mean outside temperature of +30°C.
3. Multi-modal (road–rail) units are particularly suited to carrying large volumes of product over long distances. These units can be integral containers (as discussed in Section IV), semi-trailers (either carried in flat cars or bogies), refrigerated swap bodies, or large-capacity containers.

The use of the frozen product thermal mass as a means to maintain temperatures below the recommended −18°C temperature guideline has been attempted [37], but isolated warm spots (usually in the top corners of the stow) will appear in these scenarios [38]; conductive heat transfer within the cargo is too slow to even out temperature differentials.

VIII. AIR TRANSPORT

Airfreight is often perceived as a safe, reliable, and fast transport mode for frozen products. However, products may have waiting times of several hours before being loaded into the aircraft, with no additional cooling but any autonomous, in-built source (e.g., phase change materials) provided in the packaging or container. Temperature-controlled cargo spaces are frequently limited and few airlines have adequately trained personnel in good transport practices for temperature-sensitive products [39]. Delays in air transport are also more common due to the recent implementation of tight security measures.

Having said this, some factors have led to an increase in the use of airfreight for the delivery of temperature-sensitive goods, such as the separation of freight from passenger services [40] and the delivery of specialized services for chilled and frozen products [41].

Frozen products may be air freighted by means of:

- Passive insulating shipping systems, also known as thermal packages. These mainly consist of cardboard or plastic boxes insulated with polystyrene, polyurethane, or vacuum panels.

Dry ice or eutectic plates filled with phase change materials (such as salt solutions) are used to provide low, noncontrolled temperatures during the voyage.
- Active shipping systems, such as insulated containers with in-built refrigeration systems and means for temperature control. Most of these systems are battery-powered and use dry ice as a refrigerant.

Some recommendations apply to the transport of frozen products by air:

- Air cargo containers usually benefit from "thermal blankets," which are multilayered covers that protect the pallet from radiant and convective heat.
- Airfreight should be booked in advance, with specific departure and arrival times negotiated beforehand.
- Air forwarders specialist in perishable products and with worldwide representations and connections are recommended over more generic firms with more limited resources.
- The perishable nature of the products carried should be clearly stated on the "Bill of Landing," air waybill, or consignment note, along with instructions for keeping the product in a temperature-controlled facility at the correct temperature.
- Documentation should be expedient and the product should be transported immediately to appropriate coldstorage facilities.

IX. MONITORING AND CONTROL DURING THE DISTRIBUTION OF FROZEN PRODUCTS

A. Regulatory Approaches

Although most current regulations on food product safety during refrigerated transport are self-regulatory measures developed by the food industry, the implementation of public temperature performance standards to improve food safety has been suggested before [42]. Two examples of current private and public regulations related to safe temperatures for the transport of frozen products are described below.

- In Australia, the Australian Cold Chain Guidelines [10], a self-regulatory guideline, provides recommendations for the safe transport of perishable products. These specify that frozen products should be handled never warmer than $-18^{\circ}C$.
- In Europe, the Quick Frozen Foodstuffs (Amendment) Regulations [9], the British interpretation of EEC directive 92/1/EEC, establishes the use of recording instruments to monitor the air temperatures to which quick frozen foods are subjected. Quick frozen foods are required to be stored and transported at $-18^{\circ}C$.

It is worthwhile noting that a Transportation Technical Analysis Group (TAG) was established in 1995 in the U.S.A. The group identified a number of issues during the road transport of perishable products and drafted a proposed temperature regulation for the transport of perishables, including frozen foods [43]. The proposed regulation has not yet been implemented.

Thermal performance and safe temperature regulations have the common mission of ensuring optimum temperature management during the transport of perishable products. Even though there also seems to be mutual exclusivity of parties developing public thermal performance standards and those undertaking food safety regulations [15], temperature monitoring is the most important step to enforce both types of regulations.

B. Technology for Temperature Monitoring During Transport

A variety of temperature and time–temperature indicators/integrators (TTIs) as well as temperature data loggers, with and without display, are available from various suppliers. Some characteristics required from these systems for the transport industry are: sensitivity, satisfactory accuracy for the intended purpose, robustness to withstand harsh conditions (including vibration), traceability, ease of use, small space requirements and low unit, and total cost per monitored shipment [44].

Temperature-measuring devices may be based on physicochemical properties (e.g., melting point, thermal expansion, emissivity, diffusion, solidification temperature, and viscoelastic properties) or chemical reactions (e.g., electrochemical corrosion, enzymatic reactions, and polymerization). Two popular temperature-measuring systems are described below.

- Graphic recorders are used for monitoring air temperatures in containers, airplane holds, railcars, and trucks. The device has a bimetal coil as a sensing element; the coil expands or contracts depending on the surrounding temperature. A stylus attached to the coil creates a "temperature line" in the paper, which correlates with a range of temperatures, typically from −28.9 to +37.8°C.
- Electronic data loggers are now replacing graphic recorders. One reason is that an electronic format of temperature data is more suitable for quality assurance analyses and records. Flexibility of electronic data loggers is a second reason, as these can accompany the product itself and are not necessarily part of the transport system. Electronic loggers have a sensing element, which changes its electric resistance in response to the temperatures sensed. The resistance is translated into temperatures by internally built software or by an external computer. The results are stored in an internal memory and data can be transferred afterward to a computer for analysis.

Some temperature-monitoring devices do not present a display of temperatures, but undergo physical changes as a consequence of temperature changes. The British Standard 4908 [45] classifies these temperature indicators in the following groups:

- Temperature indicators with ascending function, thaw, or threshold indicators. These typically measure temperatures in the range −20 to 30°C.
- Temperature indicators with descending function. These typically measure temperatures in the range 0 to −6°C.
- Partial TTIs. These need to hit a temperature threshold to change their properties and signal temperature abuse.
- Full TTIs. These indicate temperature changes over the full temperature range.

Some examples of TTIs are shown in Table 11.1.

Currently, the value of full-history TTIs range from US$ 1.00 to 4.00. More sophisticated approaches, such as radio frequency identification devices (RFIDs), have a value of US$ 2.50 to US$ 5.00 [44]. An RFID-based temperature-measuring system generally encompasses a sensor, a tag, and a reader that communicate with one another by means of radio transmission. Active RFID systems are battery-powered, which allow them to be independent of a common energy source and can, thus, be used for transport applications. The information collected by the tag, such as temperature, identification code, or others, can be obtained at real-time, and sudden situations that endanger the integrity of the frozen goods can be addressed promptly. New developments in the field of TTIs include the combination of RFID technology and enzyme-based technologies, opening the possibility of tracking shelf-life of frozen products remotely [46].

TABLE 11.1
Summary of Specifications of some Commerical TTIs

Name and Type	Change in TTI Noticed by User	Principle	Temperature or Time Limits: Upper Limit	Temperature or Time Limits: Lower Limit	Dimensions
3M Monitor Mark® 9860A	Diffusing blue front along the length of a porous wick	Diffusion of colored substance if temperature measured is higher than melting point of octyl octanoate	−15°C	−20°C	95 mm × 19 mm; thickness = 2 mm
VITSAB CheckPoint™ labels	Color change of label, caused by a decrease in acidity of active substance	Enzymatic hydrolysis of a lipid substrate (occurs at −18°C)	4 days	Variable[a]	22 mm × 36 mm; thickness = 0.8 mm
WarmMark™ 51034	Color change of label, caused by a decrease in acidity of active substance	Enzymatic hydrolysis of a lipid substrate (occurs at −18°C)	12 hr	Variable[a]	19 mm × 46 mm; thickness = 1.5 mm

[a]Depends on the severity of warming during temperature abuse.

Sources: PS Taoukis, TP Labuza. In: CM Bourgeois, TA Roberts, Eds., *Predictive Microbiology Applied to Chilled Food Preservation. Proceedings of the International Symposium,* Quimper, France, June 16–18, 1997. Refrigeration Science and Technology Proceedings Series. Paris, France: International Institute of Refrigeration, 1997; 3M MonitorMark™ Time Temperature Indicators brochure, 2004; VITSAB™ Time Temperature Indicators brochure, 2004; Delta Track Thermolabels brochure, 2004.

X. FUTURE TECHNOLOGIES

McKinnon and Campbell [2] found that some transport users were dissatisfied with the outdated materials-handling systems during distribution of frozen foods and the scarcity of technical innovations in the logistics of frozen foods. Competitive pressures in the refrigerated industry are also encouraging manufacturers to include criteria such as environmental issues, food safety regulations, and costs in the development of new transport technologies. Some of these new technologies, as detected by various authors [47–50], include:

1. Sanitation aspects.
2. Multicompartment, multitemperature vehicles with features that increase the flexibility to change the vehicle's capacity and its use (e.g., switching from a frozen application to a chilled application) according to the product demand.
3. New technologies for traceability, control, and prediction of shelf-life during transport.
4. Labeling and other means of information to the consumer.
5. Management of interfaces during the cold chain (e.g., from cold store to transport, from transport to retail cold store, and others).

6. Improvement of the airflow distribution in refrigerated transport by means of air ducting.
7. Alternative refrigeration cycles and systems.

Recent patents [49,51] refer to cryogenic technologies in the transport of frozen goods. Some cryogenic systems have been tested with mixed success in railcars and containers, indicating the need for finetuning these systems to achieve full commercialization.

In terms of energy efficiency, Van Gerwen et al. [52] estimated that, in refrigerated road vehicles, between 10 and 40% of the total energy consumption is related to the refrigeration system. There is a considerable room for improvement in this area. Current regulations aimed to ensure a certain level of thermal performance for refrigerated vehicles will need to encompass energy efficiency standards in the future.

Systematic risk assessments of temperature abuse during transport and the relative significance of this operation on the final quality of frozen products have not been undertaken. The frozen food industry relies on enforcing temperatures deemed to be safe for storage conditions, rather than specific transport guidelines for frozen products.

It is important to keep in mind that new developments in the transport of frozen products are motivated by (a) compliance with current and future regulations in terms of energy efficiency and environmental impact, (b) product quality and safety, and (c) cost reduction. Novel designs for refrigerated vehicles need to address all these areas to be successfully marketed.

XI. CONCLUSIONS

The transport of frozen foods has been traditionally viewed as a less technically demanding task than the transport of chilled products. However, the industry still faces challenges related to the high use of energy during transport, the capital investment required to establish distribution channels for frozen foods, the diversity of the industry, the introduction of e-commerce worldwide, and the price-sensitive demand of frozen products, among others. A sound knowledge of the issues likely to appear in the logistics of frozen foods, the physicochemical characteristics of the products transported, and the characteristics of the distribution channels (e.g., cost, duration, and demand) is required to overcome the aforementioned challenges.

International thermal performance and safe temperature regulations have been traditionally developed and implemented using separate regulatory structures, even though both types of policies have the common goal of ensuring optimum temperature management during the transport of frozen products. Given that traceability, safety, and shelf-life have been detected as key areas for the development of new technologies in the perishables transport industry, there is a need to homogenize the regulatory efforts for the transport of frozen products and perishables in general.

REFERENCES

1. B Halweil. Food trade slumps. In: L Starke, Ed., *Vital Signs 2001*; Worldwatch Institute. New York: WW Norton & Company, 2001, 62 pp.
2. B McKinnon, J Campbell. Quick-response in the frozen food supply chain: the manufacturer's perspective. Christian Salvesen Research Paper 2, 1998, pp. 3–36.
3. Anonymous. *Recommendations for the processing and handling of frozen foods*, 3rd ed. Paris: International Institute of Refrigeration, 1986, pp. 52–54.
4. Anonymous. Congestion Points Study (Phase III). *Best Practices Manual and Technical Report*, Vol. 2. *Sea Transport*. Asia-Pacific Economic Cooperation, 1997, 105 pp.
5. W Coyle, W Hall, N Ballenger. Transportation technology and the rising share of U.S. perishable food trade. In: A Regmi, Ed., *Changing Structure of Global Food Consumption and Trade*. Report WRS 01-1. USDA Agriculture and Trade, 2001, pp. 31–40.

6. O Andersen. Transport of fish from Norway: energy analysis using industrial ecology as the framework. *Journal of Cleaner Production* 10 (1):581–588, 2002.
7. Anonymous. Thawing problems. The Navigator. April, 2001 Auckland, New Zealand: International Marine Insurance Agency Ltd., 2001, pp. 1–2.
8. Anonymous. Agreement on the International Carriage of Perishable Foodstuffs and on the Special Equipment to be used for such carriage. Working Party on the Transport of Perishable Foodstuffs. Geneva: Economic Commission For Europe. Inland Transport Committee, 1970 (including amendments thereto up to Sept 2000), pp. 11–12.
9. Anonymous. The Quick-Frozen Foodstuffs (Amendment) Regulations. On the monitoring of temperatures in the means of transport, warehousing and storage of quick-frozen foodstuffs intended for human consumption, 1994. Commission Directive 92/1/EEC 298, 28 pp.
10. Anonymous. The Australian Cold Chain Guidelines for the handling, storage and transport of frozen foods, ice cream and chilled foods for retail sale and in food service outlets. Australian Food & Grocery Council, Australian Supermarket Institute and Refrigerated Warehouse & Transport Association, 1999, pp. 1–6.
11. G Eksteen. Minimum requirements for vessels carrying containers and equipment for the export of perishable products. Perishable Products Export Control Board Protocol HP20, 2002, 1 pp.
12. G Bruwer. Loading deep frozen cargo into sea containers. Perishable Products Export Control Board Protocol HP13, 2003, pp. 1–2.
13. M Jul. *The Quality of Frozen Foods*. London: Academic Press, 1984, pp. 174–180.
14. RL Mason, RB Wallace. The cold hard facts: a survey of the Queensland frozen food industry. *Food Australia* 45 (11):532–537, 1993.
15. S Estrada-Flores. Safe temperature regulations during the road transport of fresh-cuts. In: *Proceedings of the International Congress of Refrigeration*, August 17–22, 2003, Washington, DC, 2003 (CD ROM).
16. Anonymous. *Guide to Refrigerated Transport*. IIR TFF-TFA. Paris: International Institute of Refrigeration, 1995, pp. 70–130.
17. J Frith. *The Transport of Perishable Foodstuffs*. Cambridge: Ship Owners Refrigerated Cargo Research Association, 1991, pp. 7–29.
18. CP Tso, SCM Yu, HJ Poh, PG Jolly. Experimental study on the heat and mass transfer characteristics in a refrigerated truck. *International Journal of Refrigeration* 25 (3):340–350, 2002.
19. Anonymous. Frozen products. In: *Cool Facts*. Copenhagen K, Denmark: A.P. Møller, 2004, pp. 1–41.
20. YQ Jiang, SL Wang. Statistical analysis of reliability of container refrigeration units. *International Journal of Refrigeration* 19 (6):407–413, 1996.
21. Anonymous. *ASHRAE Handbook, Refrigeration, Systems and Applications*. Atlanta, GA: American Society of Heating, Refrigerating, and Air-Conditioning Engineers, 1994, pp. 1–30.4.
22. N Ballenger, W Coyle, W Hall, RG Hawkins. Transportation technology eases the journey for perishables going abroad. Agricultural Outlook Jan.–Feb. Economic Research Service USDA, 1999, pp. 18–22.
23. R Silliars. Reefer containers below deck. *Marine Bulletin, Lloyd's Register Marine Business* 1 (26): 1–2, 2001.
24. Y Wild. Refrigerated containers and CA technology. In: *Container Handbook. Cargo Loss Prevention Information for German Marine Insurers*. Berlin: Transport and Loss Prevention Department, 2003.
25. Y Wild. Comparison between porthole and integrated reefer system containers from technical, operational and economic aspects. In: *Proceedings of the IIF-IRF Commission D2/3*, Gdansk, September 29–October 1 1994, pp. 1–7.
26. Anonymous. Reefer market stays hot. *World Cargo News*, Leatherhead, U.K.: WCN Publishing, 2004.
27. R Nordstrom, H Nurminen. Cargo carrier refrigeration system. U.S. Patent 6,230,640, May 15, 2001.
28. Anonymous. Refrigerating installation (reefer). In: *Rules for the Classification of Steel Ships*, 2000, Bureau Veritas, Paris. NR469.3 DTM ROOE, pp. 30–35.
29. SW Coblenz. Method and apparatus for handling, transporting, pallet removal and loading cartons of frozen animal products onto vessels. US Patent 6,375,407, April 23, 2002.
30. JA Numero, FM Jones. Air conditioner for vehicles. US Patent 2,303,857, December 1, 1942.
31. H Tat Keh, E Shieh. Online grocery retailing: success factors and potential pitfalls. *Business Horizons*, June/July 2001, pp. 73–83.

32. Anonymous. *Kerstner Cooling Van brochure*. Groß-Rohrheim, Germany: Werner-von-Siemens, 2002.
33. B Chopko, A Stumpf, B Valentin. Multi-temperature transport unit refrigeration design. In: *Proceedings of the International Congress of Refrigeration*, August 17–22, 2003, Washington, DC, 2003 (CD ROM).
34. Anonymous. Thermal performance of refrigerated transport equipment — specification and testing. Committee ME-006 AS 4982-2003, Standards Australia, 2002, pp. 1–20.
35. S Estrada-Flores. Current Australian regulations for refrigerated road transport of temperature-sensitive products. In: *Proceedings of the Cold Chain Distribution Conference*, Sydney, Australia, June, 23–25 2004 IQPC Australia, (CD ROM).
36. PO Roberts. Logistics and freight transport: review of concepts affecting bulk transportation. *Presentation to the World Bank*. February 10, 1999, pp. 30–31.
37. L Tyree. Dry ice rail car cooling system. US Patent 5,979,173, November 9, 1999.
38. J Middlehurst, NS Parker, MF Coffey. Holding of frozen cartoned meat in insulated shipping containers. Meat research Newsletter. North Ryde, NSW: CSIRO Div. of Food Research, 1978, pp. 1–2.
39. Anonymous. Instructions for perishable products "Blue Book." In: *Perishable Products Export Control Board Directory 2004*, Plattekloof, Capetown, SA.
40. Anonymous. Freight logistics and transport systems in Europe. Final report. Paris, France: European Council of Applied Sciences and Engineering, 2004, pp. 49–50.
41. H Chong. Cold chain distribution. In: *Proceedings of the Cold Chain Distribution Conference*, June 23–25, 2004 Sydney, Australia: IQPC Australia(CD ROM).
42. PS Taoukis, TP Labuza. Chemical time–temperature integrators as quality monitors in the chill chain. In: CM Bourgeois, TA Roberts, Eds., *Predictive Microbiology Applied to Chilled Food Preservation. Proceedings of the International Symposium*, Quimper, France, June 16–18, 1997. Refrigeration Science and Technology Proceedings Series. Paris, France: International Institute of Refrigeration. 1997.
43. USA proposed rulemaking in addition to 21 CFR Part 110. *Federal Register* 61 (227):59372–59382, 1996.
44. K Romann. Cold chain temperature monitoring: overview of latest technology for optimal temperature monitoring. In: *Best Practice for Cold Chain Distribution*. U.K.: IQPC, 2003, (CD ROM).
45. British Standard Institution. Packaging temperature and time–temperature indicator — performance specification and reference testing. BSI 7908, 1999, pp. 1–3.
46. Anonymous. *Bioett Brochure*. Lund: Bioett AB, 2004.
47. R Heap. Refrigerated transport: progress achieved and challenges to be met. 16th Informatory Note. *Bulletin of the International Institute of Refrigeration* 84 (1):27–33, 2004.
48. J Guilpart. Refrigerated road transport in Europe. In: *Proceedings of the International Congress of Refrigeration*, August 17–22, 2003, Washington, DC, 2003 (CD ROM).
49. P Franklin Jr. Insulated freight container with recessed CO_2 system. US Patent 6,109,058, August 29, 2000.
50. MJ Murdock. Transport refrigeration, food safety and quality. In: *Proceedings of the International Congress of Refrigeration*, August 17–22, 2003, Washington, DC, 2003 (CD ROM).
51. R Garlov, V Saveliev, K Gavrylov, L Golovin, H Pedolsky. Refrigeration of a Food Transport Vehicle Utilising Liquid Nitrogen. US Patent 6,345,509, February 2002.
52. RJM Van Gerwen, SM van der Sluis, H Schiphouwer. Energy efficiency in refrigerated transport. IIF-IIR Comm D2/3 and D1. 2, Cambridge, 1998, pp. 39–49.

12 Retail Display Equipment and Management

Giovanni Cortella and Paola D'Agaro
University of Udine, Italy

CONTENTS

I. INTRODUCTION

In retail premises, display cabinets play an important role in persuading the customer to buy food products, while satisfactorily preserving them. This is a complex task, because these two purposes are in some way conflicting.

The display function is assured by making the product clearly visible and easy to reach to persuade potential customers. However, such features lead to food temperature fluctuations because of its exposure to heat sources, for example, warm ambient air and radiative heat from lighting. Unfortunately, temperature fluctuations are recognized to be one of the primary causes of quality and safety loss in frozen foods. Therefore, the requisite of an optimal product display is in conflict with the need to maintain the required storage conditions [1–5].

Compliance with the preserving function is regulated by various standards, both for the manufacturer and for the shop manager. In general terms, as regards the manufacturer, the display cabinet must be designed to keep the products at the prescribed temperature when the cabinet is operated at a certain climatic class defined by the ambient conditions. The manufacturer must therefore find a right balance between the preserving and display functions through an optimization process at the design stage of the cabinet. The shop manager must in turn install, load, and maintain the cabinet following the prescriptions of the manufacturer: exposure to air conditioning or warm air streams and to lighting must be carefully evaluated; heat exchangers must be kept clean; the refrigerated volume shall not be overloaded.

In the past few years, the reduction of energy consumption has become a new priority, especially in such applications where significant results can be achieved. Supermarkets are one of these applications, because they are subject to a great energy demand, a great portion of which is due to the refrigeration load. As an example, the use of a centralized control system with electronic expansion valves can lead to major benefits in terms of energy saving when compared with the use of traditional thermostatic expansion valves.

In this chapter, a brief description about the design, installation, operation, monitoring, and maintenance of display cabinets to be operated in supermarkets for retail sale is given.

II. DESIGN

First of all, it has to be pointed out that retail display cabinets are only intended for displaying products at the correct storage temperature, that is to say they are neither designed to freeze food nor to reduce its temperature. The load volume is kept refrigerated usually through forced circulation of cold air. One or more heat exchangers are provided for air cooling, and the refrigerating power is supplied by a self-contained or a remote unit. Self-contained cabinets are used exclusively for particular applications in small shops (e.g., ice creams), whereas remote unit cabinets are most widely used in supermarkets.

A. Classification

Display cabinets for frozen food are only for self-service operation; therefore, their opening always faces the customer. They are usually classified depending on their geometry. Typically, two great classes are known, the horizontal open-top cabinet and the vertical multideck cabinet. Both can be equipped with glass doors for the sake of energy saving and better temperature control [6–8].

1. Horizontal Open-Top Cabinet

Horizontal cabinets are preferred among the open cabinets for frozen food because of their particular ability to preserve the load volume from warm ambient air infiltration. This behavior is due to cold-air stratification, and the most important effects are low-energy consumption and a better food temperature control. Two examples of horizontal open-top cabinets are represented in Figure 12.1 and Figure 12.2. In Figure 12.1, a wall-site unit is shown, which is designed to be positioned against a shop wall or a similar unit (back-to-back.) In Figure 12.2, a island-unit is represented, which is designed to be accessed from all sides.

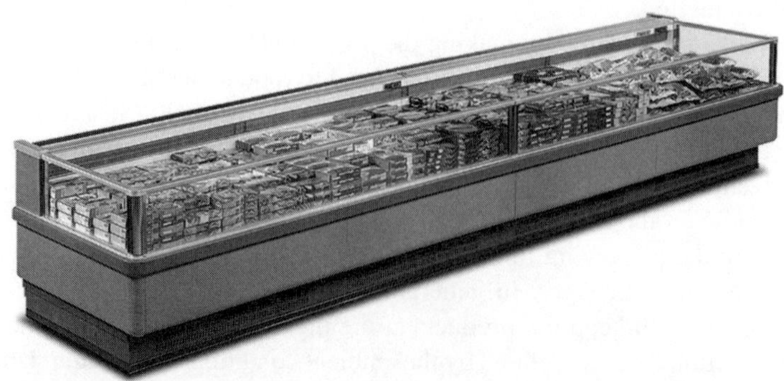

FIGURE 12.1 Horizontal open-top display cabinet, wall-site unit. (Courtesy of Arneg S.p.A., Italy.)

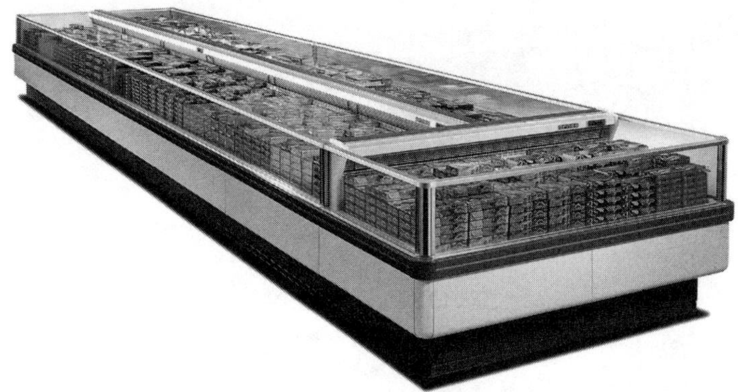

FIGURE 12.2 Horizontal open-top display cabinet, island unit. (Courtesy of Arneg S.p.A., Italy.)

Load is kept refrigerated through cold-air circulation, which is usually forced by fans in the case of frozen food cabinets, to remove the high heat flux. Air circulates in a cavity surrounding the whole load volume apart from the open-top, where a horizontal air-curtain is established to reduce the effect of radiative heat on the load surface. For further improvement in performance, covers are provided for night operation, when the shop is closed, whereas sliding glass doors are sometimes provided for shop opening hours operation to reduce radiative heat exchange and ambient-air infiltration.

Therefore, horizontal cabinets are considered to be among the most effective cabinets in terms of the preserving function, especially when sliding doors are employed. However, they are not so effective in terms of the display function because the absence of shelves makes only the top layer of products visible. They can be successfully employed in the case of large turnover products when the storage of a great quantity of merchandise can be helpful to the retailer.

2. Vertical Multideck Cabinet

Vertical multideck cabinets are preferred for their ability to save floor space because of the presence of three to six shelves that improve their display function. Two examples of vertical multideck cabinets are represented in Figure 12.3 and Figure 12.4, with and without glass doors, respectively. In open cabinets, the load volume is kept refrigerated through cold-air-forced circulation. At the opening, one or more parallel air curtains are necessary to reduce warm air entrainment from the ambient, which could result in difficult food temperature control and high energy consumption. Air curtains improve temperature control but fail to reduce energy consumption. Doors are much more effective for this purpose, particularly when the glass is treated with a reflective layer preventing radiative heat. Closed cabinets should be used whenever possible, even if the display function is strongly affected for two reasons: the shopper must open the door to reach the product and the door, once it has been opened, is subject to water vapor condensation on the internal side of the glass, thus reducing product visibility. To overcome this problem, doors are usually triple glazed and have electric heaters with a demisting function.

Finally, the earlier mentioned types of cabinets can be combined for a better exploitation of the floor space, as shown, for example, in Figure 12.5.

B. Heat Balance

A deep knowledge of the heat balance is necessary for a correct design of a refrigerated display cabinet [9,10]. Heat fluxes originate both from the external ambient and from the internal sources.

FIGURE 12.3 Vertical multi-deck closed display cabinet. (Courtesy of Arneg S.p.A., Italy.)

1. Heat Fluxes from the External Ambient

In a display cabinet, heat gains from the external ambient to the load can be classified depending on whether they take place for conduction, radiation, or warm air entrainment.

Conductive heat transfer takes place from the ambient through the unit walls and doors, if any. As such, walls are made of well-insulated sandwich panels and the transparent doors of triple glazed glass, and the contribution to the global heat gain is usually of minor importance. Furthermore, conductive heat transfer influences the refrigerating power and hardly influences the load temperature, because the load volume is almost completely surrounded by the refrigerated air ducts.

FIGURE 12.4 Vertical multi-deck open display cabinet. (Courtesy of Arneg S.p.A., Italy.)

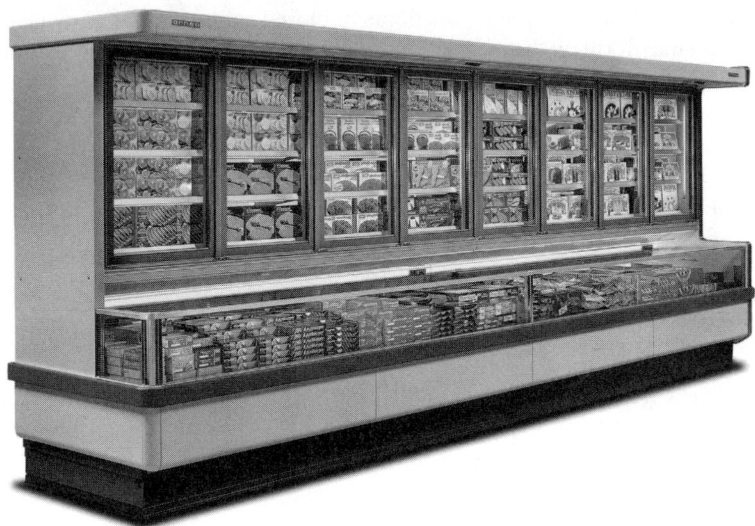

FIGURE 12.5 Combination of a horizontal open-top and a vertical multideck display cabinet. (Courtesy of Arneg S.p.A., Italy.)

Radiative heat transfer takes place through the openings of the display cabinet and through the glass doors. It is due to the difference of temperature between the room ceiling or the lighting appliances and the load. Products placed in the upper layer of open-top horizontal cabinets or in the more external position in multideck vertical cabinets can reach temperatures of 5 to 7°C higher than products not exposed to radiation. This is due to the high emissivity (i.e., about 0.9) of the emitting surfaces of both the ambient and the food packaging. Radiative heat transfer can therefore be effectively reduced by controlling the temperature and emissivity of the emitting surfaces and the emissivity of product packaging [11]. For packaging, it has been measured that the use of low emissivity materials such as aluminum instead of paper allows a decrease in food temperature of up to 5°C in the uppermost layer.

Warm air entrainment takes place continuously in open cabinets and during door openings in closed cabinets. In open cabinets, a single or multiple air curtain is created to separate the refrigerated load from the ambient air. Such a device is effective in open-top horizontal cabinets, where stratification takes place and helps to reduce warm-air infiltration, but in vertical cabinets, the creation of a barrier to the incoming warm and humid air remains a crucial problem.

2. Heat Fluxes from Internal Sources

Heat gains from internal sources are primarily due to lighting, defrosting, and demisting devices.

The refrigerated compartment has to be adequately lit, but high-intensity lighting raises product temperature by several degrees. Especially in vertical multideck cabinets, where lights are usually placed under the extremity of each shelf, a concentrated heating effect on the edge packages takes place, which are at the same time also affected by radiative heat transfer. It is necessary to use fluorescent lights or other high-efficiency lights with low surface temperature. The use of LED lights has been tested, showing a more uniform lighting with a similar energy consumption [12].

Defrosting devices are used to keep the air ducts and cooling coils clear of ice. Regardless of the defrosting method employed (electrical or hot gas), during the defrosting period, the refrigerating power is missing and a huge amount of heat is applied to the cooling coil and sometimes to the air ducts. Fans are usually stopped to avoid warm-air circulation over the load; however, the product is subject to a temperature rise, which might be significant in some cases.

Demisting heaters are used to keep glass doors of closed cabinets clear of condensation. Water vapor condensation may form on the internal surface of glass doors when these are opened or on the external surface of glass doors when the external climate is humid and hot. Combined use of triple glazed doors and electric heaters prevents condensation on the external side and ensures demisting on the internal side in an acceptable time.

Finally, further sources of heat can be found in additional electric heaters fitted into the rim around the top of open-top chests, which might come into contact with the customers' hands and feel unpleasant if particularly cold.

C. Air Distribution

Display cabinets for frozen food are cooled by forced air circulation, because a large amount of refrigerating power is required and this would be difficult to transfer by air circulating merely by natural convection.

In both horizontal and vertical cabinets, air is refrigerated in a finned coil, situated underneath the bottom of the load compartment, and then is forced to circulate through air ducts, which surround the load volume. The circulating cold air refrigerates the compartment and, in the case of open cabinets, helps in limiting the infiltration of air from the room through the establishment of the so-called air curtains.

In the case of an open-top horizontal cabinet at a temperature varying from -30 to -25°C, only a single air curtain is employed. Air velocity is very important in this case, because if air velocity is too low, the necessary refrigerating effect cannot be guaranteed. On the contrary, if it is too high, the air stream becomes more turbulent, thus increasing heat and mass transfer with the environment. Figure 12.6 shows a schematic cross section of a horizontal display cabinet with the air circulation arrangement, whereas in Figure 12.7, the temperature distribution in the air curtain over the load is shown, which was calculated with the computational fluid dynamics (CFD) technique [13].

A single air curtain is also employed in the case of a closed vertical cabinet. The air curtain operates with open and closed doors, with the main purpose of guaranteeing that only a minimal amount of ambient air is entrained when the door is opened. After each door opening, the air

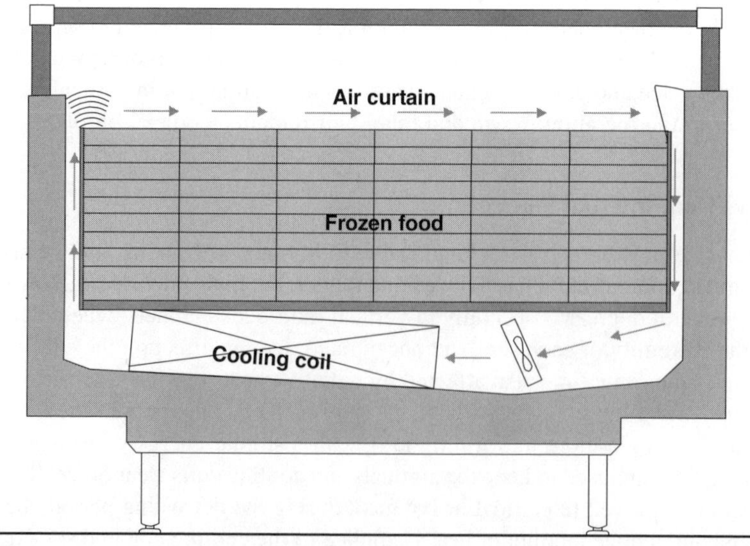

FIGURE 12.6 Schematic cross section of a horizontal open-top display cabinet. (Reprinted from G Cortella. *Computers and Electronics in Agriculture* 34:43–66, 2002. With permission.)

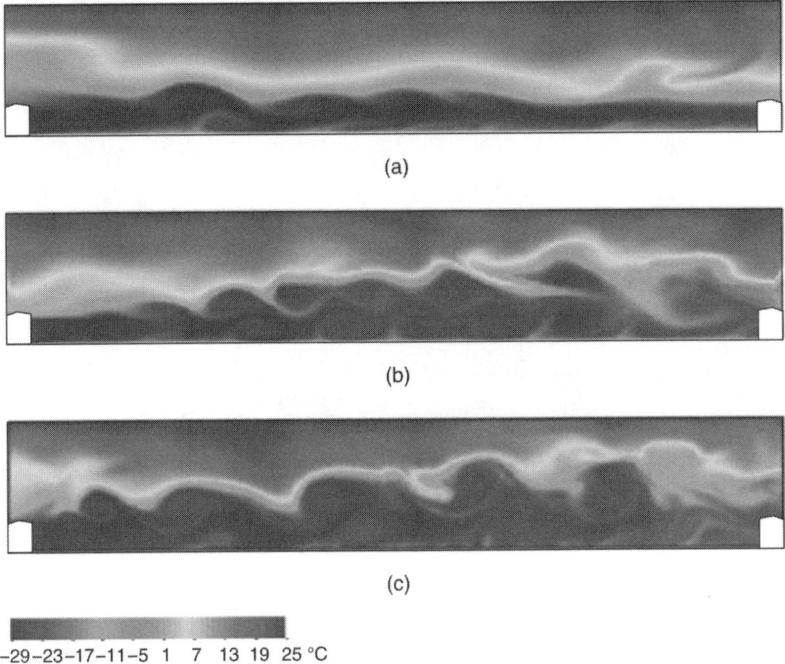

FIGURE 12.7 Calculated temperature maps of the air curtain in a horizontal display cabinet at various value of the air velocity: (a) 0.2 m/s, (b) 0.3 m/s, and (c) 0.4 m/s. (Reprinted from G. Cortella. *Computer and Electronics in Agriculture* 34:43–66, 2002. With permission.)

curtain helps glass demisting by supplying dry air on its surface. Demisting is the result of a combined action of the electrical heaters and the air curtain. The air curtain alone could not be effective because its low temperature would result in icing condensed water.

In the case of open vertical cabinets, the air circulation is different. At least two parallel air curtains are employed. The internal curtain is maintained at low temperature (i.e., about −30°C) to keep the load refrigerated while removing all the heat gains (radiation, lights, and defrosting). The internal air curtain temperature rises to about −18°C at the return opening. The external curtain is at higher temperature (i.e., about 0°C) and acts as a vertical barrier to warm-air infiltration from the room into the refrigerated compartment. The two curtains move downwards at the front of the cabinet until a strong fluctuation periodically entrains warm room air, which has to be cooled and dehumidified once it reaches the cooling coil. For this reason, often a third air curtain is employed, which is made of ambient air taken at the top of the cabinet and not collected at the return grill. The third air curtain helps to prevent strong fluctuations due to turbulence and protects the refrigerated curtains from the effects of ambient air movements. Figure 12.8 shows a schematic cross section of a vertical open display cabinet with the air circulation arrangement. The three air curtains form a very complex system, an example of which is shown in Figure 12.9, and only a deep understanding of its fluid dynamics leads to the awareness of the correct thermal energy balance.

A large amount of important improvements have been made in the last two decades on the design of commercial retail cabinets, and most of them are direct consequence of the introduction of CFD [13, 14–20]. As an example, it is now possible to perform a three-dimensional analysis of the air curtains flow pattern in front of a vertical open display, as can be seen in Figure 12.10. This kind of analysis demonstrates the presence of extremity effects due to the side walls: in the case of Figure 12.10, two large-scale vortices form close to each of the walls, and by the depression produced by these vortices, the main fraction of the overspilled cold air exits the cabinet.

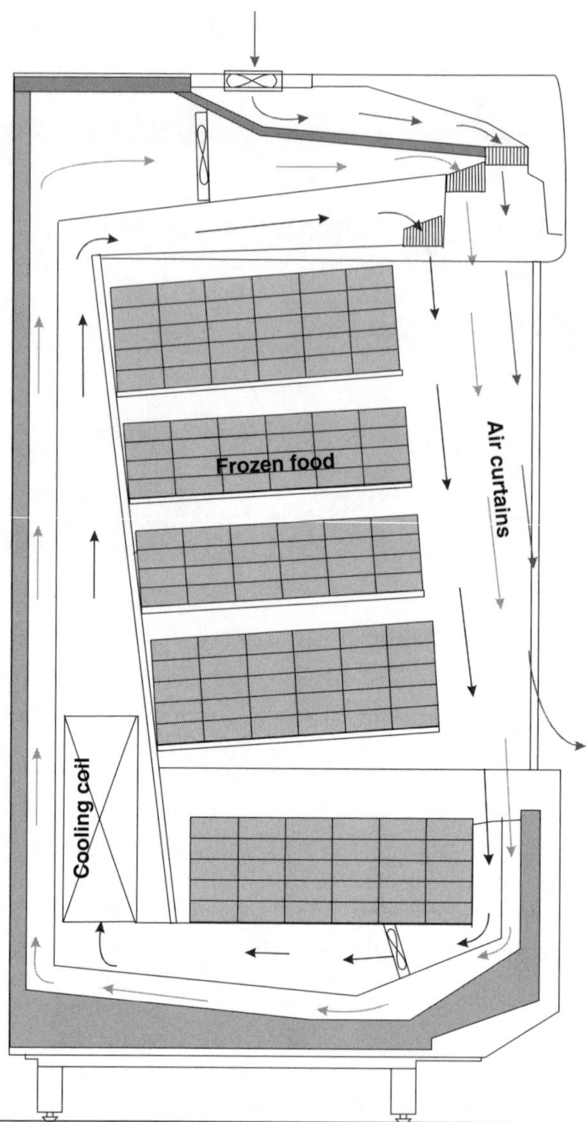

FIGURE 12.8 Schematic cross section of a vertical open display cabinet, with the air circulation arrangement. (Reprinted from G Cortella. *Computer and Electronics in Agriculture* 34:43–66, 2002. With permission.)

In any case, it must be pointed out that experimental tests must always support and validate the theoretical calculations due to the many factors related to cabinet operation and to ambient conditions that act simultaneously and affect each other.

D. Refrigerating Equipment

Refrigerated display cabinets have to perform a heavy refrigeration duty to counterbalance the several earlier mentioned heat gains. A vapor compression refrigeration cycle is always employed, and they can be classified as standalone units and remote condensing units, depending on where the compressor and condenser are located.

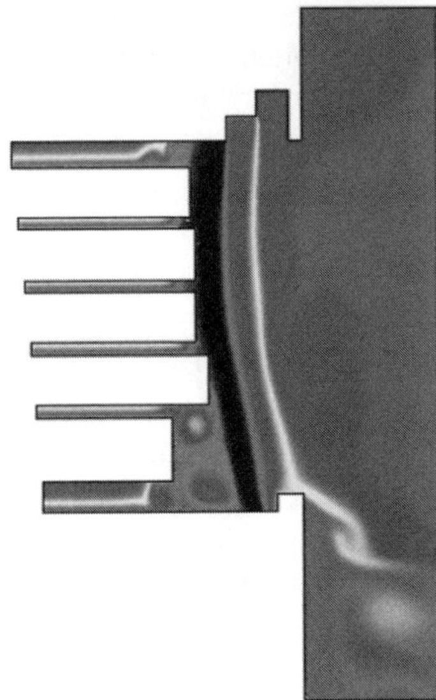

FIGURE 12.9 Calculated two-dimensional air flow pattern from the temperature distribution calculated in a vertical open display cabinet.

In a standalone unit, the whole refrigerating plant is enclosed at the bottom of the cabinet, thus requiring only a power supply plug. This kind of unit is available only for small cabinets, usually dedicated to spot merchandising.

In remote condensing units, both the compressor and the condenser are situated far away from the display cabinet, often in a dedicated room or outside of the shop. The two sections of the refrigerating plant are connected by pipes carrying refrigerant, thus requiring sometimes a huge quantity

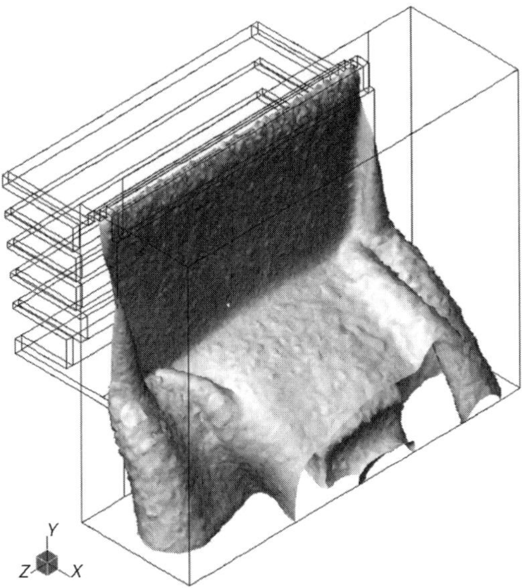

FIGURE 12.10 Calculated three-dimensional air flow pattern from the isothermal surface at $t = 25°C$ calculated in front of a vertical open display cabinet.

of fluid. Centralized multicompressor plants are preferred when a large number of cabinets is utilized with a computerized control, which often allows heat recovery for space or water heating.

The use of secondary fluids may be necessary in the case of toxic or flammable refrigerants such as ammonia or hydrocarbons or advisable with the aim of reducing the total amount of refrigerant [21–23].

E. Defrosting

During the operation of a display cabinet, frosting occurs on the evaporator, on the return duct, and on many cold surfaces, due to the entrainment of air from the ambient. This is a major problem for the correct operation of the unit because the deposit of frost leads to the obstruction of the cooling coil and the air ducts and to loss in performance of the fans. Consequently, both the refrigerating power and the air flow rate reduce, thus causing insufficient food cooling and air curtains instability. Defrosting is therefore necessary to remove ice from all the surfaces. It can be performed by reversing the refrigerating cycle in the standalone units ("hot-gas" defrosting) or by means of electrical heaters. New methods are being evaluated, such as the use of the liquid refrigerant instead of the hot superheated gas [24,25]. Rise in food temperature is however unavoidable, depending on the duration of the defrosting operation and especially on the food packages that are more exposed to radiative heating and air circulation. Figure 12.11 shows food temperature measured in the upper layer of an open-top horizontal cabinet, in the case of two defrostings per day. It is clear that it takes a rather long time for the food temperature to return to the correct value. The choice of the defrosting frequency is very important and can be managed by a smart control device [26,27]. As shown in Figure 12.11, a long interval between defrosting cycles causes ice building up and, therefore, a longer and more difficult defrosting operation. On the contrary, frequent defrosting cycles are more efficient but may expose food to undesired recurrent temperature fluctuations.

Defrosting can be considerably reduced for open cabinets by using night covers or plastic curtains during shop closing time, thus preventing warm and humid air entrainment.

F. Future Trends

There is room for improvement in the design of display cabinets, through the employment of new numerical techniques such as the CFD and experimental techniques such as the laser doppler

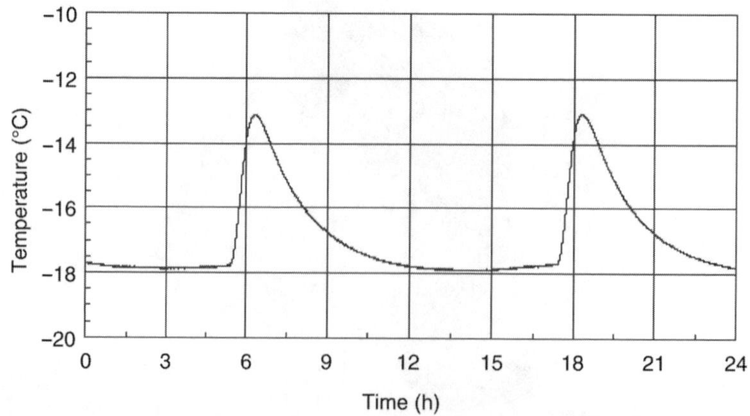

FIGURE 12.11 Influence of the defrosting operations on food temperature (experimental values, upper layer of an open-top horizontal cabinet). (Adapted from G Cortella. *Computers and Electronics in Agriculture* 34:43–66, 2002. With permission.)

anemometry (LDA) and particle image velocimetry (PIV), all aimed at the best exploitation of the air curtains [28]. As regards the refrigerating equipment, room for improvement comes from the change to new refrigerants (due to the environmental impact of the old ones) and from the availability of complex control systems [29–35]. The whole supermarket is being considered as a unique system including refrigerating units, ambient air conditioning/heating, and hot-water production, with the aim of significantly reducing energy consumption [36–38].

III. SUITABILITY FOR DIFFERENT PRODUCTS

The choice of the retail display cabinet for frozen food is not particularly influenced by the different products displayed, but in some way by their packaging. In fact, frozen food is always packaged and soft-packaged food is best stacked in horizontal cabinets. Another criterion of choice could be the turnover of the product. Vertical cabinets are preferred for low turnover products, as they display small quantities of many different products well. On the contrary, the great load volume of horizontal cabinets may be helpful for high turnover products, thus reducing the number of weekly reloading. Finally, food whose shelf life is more affected by temperature fluctuations should be displayed in closed cabinets, where the presence of doors limits heating due to warm-air entrainment or radiative exchange and reduces the need for defrosting operations [6,7].

IV. INSTALLATION

Display cabinets may be tested according to different standards, but in all cases, after having stated the ambient conditions, they are usually classified by "climatic classes." A display cabinet whose performance is certified for a certain climatic class should be installed in ambient conditions as close as possible to test conditions. It is worth noting, however, that the standards are used for the comparison of various cabinet performance in specified ambient conditions through the measurement of test packages temperature. Such packages are made up of a particular mixture of water and cellulose in compliance with the EU standards or by boxes containing a sponge immersed in water and propylene glycol in compliance with the American Society of Heating, Refrigerating, and Air-conditioning Engineers (ASHRAE) standards. Therefore, the compliance of a display cabinet with a certain standard does not guarantee correct food storage temperature due to the different thermophysical properties of food and test packages.

For a correct installation, it is important for the temperature, relative humidity, and air velocity of the environment to be similar to those prescribed for the corresponding climatic class and test conditions [39–41]. Use of air conditioning systems is suggested where possible, so as to maintain controlled conditions. Ambient temperature and relative humidity control is important for the types of display cabinets considered here. High ambient temperature causes food temperature fluctuations and increased energy consumption. High ambient humidity provokes frost formation on the cooling coil and the air ducts, leading to poor refrigeration performance, high energy consumption, and high defrosting frequency [42]. Ambient air velocity control is particularly important for open cabinets because the air curtains flow pattern can be significantly influenced by ambient air movement. Great temperature fluctuations or great variation in food temperature on different sections of the same cabinet can be encountered due to external air velocity. European standards require 0.1 to 0.2 m/s transversal air velocity; good refrigeration practice demands for lower values whenever possible. For this reason, it is becoming common practice in large stores to separate the air conditioning plant dedicated to the frozen and chilled food sale zone from the rest of the shop. In this area, air should be distributed at low velocity, with an efficient temperature control and with the lowest possible relative humidity. To reduce customers' discomfort, air conditioning return grids could be positioned on the floor close to the cabinets, so as to collect the warm air exiting the standalone units or the cold air spilling over the open cabinets.

Some further requirements are also important from the radiative heat gain view point. Cabinets should be installed as far as possible from direct sunlight, windows, or any high-temperature surface. Heat from lighting can be reduced by replacing incandescent floodlights with fluorescent tubes and by restricting illumination to a maximum value of 600 to 700 lx. Reflective surfaces placed over the open-top cabinets are suggested by some authors [6] to reduce the radiative heat exchange with the room walls and ceiling.

Finally, customers should be encouraged to the good practice of picking frozen food as the last item by placing cabinets close to the shop tills.

V. OPERATION

In discussing the operation, the behavior of both the retailer and the customer becomes important [1,2].

From the retailer point of view, the cabinet has to be loaded with products at the correct storage temperature because the refrigerating unit is not designed to cool the goods, but just to maintain them at the desired temperature. Furthermore, the cabinet should be filled by minimizing the time spent for the operation and respecting the "load limit line," which indicates the maximum capacity of the load volume. Packages placed beyond the load limit line will not be stored at the required temperature, and they will probably impede the correct arrangement of the air curtain, with harmful effects on the whole load. The need for defrosting can be appreciably reduced by applying night covers or plastic curtains on open cabinets during shop closing hours. Finally, it has to be remarked that food in retail cabinets is inevitably subject to temperature fluctuations, which severely affect its shelf life. Therefore, it is important that a first-in–first-out logic is applied when loading the cabinet, placing new goods behind the existing.

From the customer point of view, food storage conditions can be significantly improved by limiting the number of door openings (if any) or avoiding unnecessary handling of the goods.

VI. MONITORING

Monitoring the performance of a display cabinet should strictly mean to check the goods temperature and verify them with the required value. However this is a destructive measurement, which cannot be taken into consideration for obvious reasons. Furthermore, load temperature is particularly uneven, especially in cabinets with an ample display area where temperature differences of up to 15°C can be encountered between the warmest and coldest packages. This makes it impossible to use the so-called "plastic chickens," which are data loggers enclosed in containers with dimensions and thermal properties similar to those of food. Being so difficult to say whether all the goods are stored at or below the required temperature, the only way for monitoring the display cabinet is a continuous check of its refrigerating performance. The most important point of measurement is the temperature of the returning air. Every fault in the air curtains, excessive radiative heat, defective door closing, or frost buildup on the evaporating coil, and air ducts immediately cause a rise in the return air temperature, which can be easily detected. A prompt solution can then be applied before load temperature increases the hazard for food safety. A more complete control could check the temperature difference between the inlet and the return temperature, so as to understand whether the return air temperature increases due to any refrigerating circuit faults. This control might also prove useful for an automatic defrosting device, as the reduction in the air mass flow rate due to frost buildup results in an increase in such temperature difference.

Smart monitoring systems are being developed, including control functions on the refrigerating circuit, to achieve the best performance both in terms of food storage and energy consumption.

VII. MAINTENANCE

Any fault in a display cabinet operation results in significant loss to the retailer due to missed display area and food detriment. A proper maintenance is important to maintain the unit in ideal working conditions. Daily inspections should check the temperature recordings, if any, or temperature measurements, in looking for any defect in the air curtains or in the refrigerating circuit. The inlet and return air ducts must be kept clean and free of any obstacle to air circulation. Load should be frequently rearranged after customer handling, to comply with the load limit line. The whole refrigerated compartment, the air ducts, and the cooling coils should be periodically cleaned from dust and ice, and conveniently sanitized.

The maintenance of the refrigerating unit should also be performed by trained personnel. In standalone units, the air-cooled condensers must be kept clear from dust.

In the regrettable event of a breakdown, all the precautions to avoid food heating must be carried out. The thermal capacity of the load ensures a few hours of storage at sufficiently low temperature in the case of short-term failure, provided that night covers are applied or doors are locked and that lighting is switched off.

VIII. CONCLUSIONS

Refrigerated display cabinets are a critical link in the frozen food cold chain. The requisite of an optimal product display is in fact in conflict with the need to maintain the required storage conditions. With the aim of improving the display function, the control of food temperature becomes a crucial problem, as temperature fluctuations happen frequently in spite of the higher energy consumption.

Such problems are shown to be recurrent with open cabinets, which rely on the effectiveness of one or more air curtains for the control of ambient air entrainment. For this reason, open vertical multideck display cabinets are rarely used for frozen foods, although they are widely used for chilled foods. On the contrary, open-top horizontal cabinets are extensively employed, especially because they are efficient and effective in terms of their preserving function, but less so in terms of their display function.

Careful design and proper installation are the keys for the best results. There is room for improvement in the design of display cabinets through the application of advanced methodologies. Numerical techniques such as the CFD and experimental techniques such as the LDA and PIV are being employed for the best exploitation of the air curtains. The application of the restrictions to the use of chlorofluorocarbons and hydrochlorofluorocarbons as refrigerants gives room for improvement of the refrigerating units. If natural refrigerants such as carbon dioxide or ammonia are used, the refrigerating plant must be completely redesigned. In addition, the availability of complex control systems gives interesting options for a significant reduction in both energy consumption and temperature fluctuations.

There is also room for improvement in the food temperature monitoring. New instrumentation is available at low cost and should be used to enhance temperature control and prevent undesired temperature rise. Nowadays, only the return air temperature has to be monitored in the European countries and it is well known that this information is not adequate. Owing to radiative heating, excessive ambient air movement, or unnecessary door openings, food temperature can rise well above the prescribed values. It has to be mentioned that after the recent introduction of the Hazard Analysis and Critical Control Points (HACCP), the display cabinet could be identified as a "critical control point," in which the food temperature should be monitored.

REFERENCES

1. Anonymous. *Control of the Cold Chain for Quick-Frozen Foods Handbook.* IIR — International Institute of Refrigeration, Paris, 1999.
2. Anonymous. *The Refrigerated Food Industry Confederation, Guide to the Storage and Handling of Frozen Foods (The Gold Book)*, British Frozen Food Federation, Grantham, UK, 1994.
3. S Bobbo, G Cortella, M Manzan. The temperature of frozen foods in open display freezer cabinets. In: *Proceedings of the 19th International Congress of Refrigeration.* International Institute of Refrigeration, Paris, Vol II, 1995, pp. 697–704.
4. WEL Spiess, T Boehme, W Wolf. Quality changes during distribution of deep-frozen and chilled foods: distribution chain situation and modeling considerations. In: IA Taub, RP Singh, Eds., *Food Storage Stability.* Boca Raton, FL, USA: CRC Press, 1997, pp. 399–417.
5. WEL Spiess, T Grünewald, M Hefft. Residence time behaviour of deep frozen food in the frozen food chain. In: M Le Maguer, P Jelen, Eds., *Food Engineering and Process Application.* Vol. 2, London: Elsevier Applied Science, 1986, pp. 67–77.
6. A Gac, W Gautherin. Le Froid dans les Magasins de Vente de Denrées Périssables. Paris: Pyc Livres, 1987.
7. G Rigot. Meubles et Vitrines Frigorifiques. Paris: Pyc Livres, 1990.
8. C Morillon, F Penot. La modélisation: une aide à la conception thermoaéraulique des meubles frigorifiques de vente, Revue Générale du Froid, 968, 48–53, 1996.
9. D Clodic, X Pan. Energy balance, temperature dispersion in an innovative medium temperature open type display case. In: *Proceedings of the IIF-IIR Commission D1, B1 Meeting: New Technologies in Commercial Refrigeration.* International Institute of Refrigeration, Paris, 2002, pp. 191–199.
10. F Billiard, W Gautherin. Heat balance of an open type freezer food display cabinet. In: *Proceedings of the IIF-IIR Commission B1, B2, D1, D2/3 Meeting: Cold Chain Refrigeration Equipment by Design.* International Institute of Refrigeration, Paris, 1993, pp. 322–332.
11. P Nesvadba. Radiation heat transfer to products in refrigerated display cabinets. In: *Proceedings of the IIF-IIR Commission C2, D3 Meeting.* International Institute of Refrigeration, Paris, 1985, pp. 323–329.
12. N Narendran, R Raghavan. Solid-state lighting for refrigerated display cases. In: *Proceedings of the IIF-IIR Commission D1, B1 Meeting: New Technologies in Commercial Refrigeration.* International Institute of Refrigeration, Paris, 2002, pp. 66–71.
13. G Cortella. CFD aided retail cabinets design. *Computers and Electronics in Agriculture*, 34, 43–66, 2002.
14. AM Foster, GL Quarini. Using advanced modelling techniques to reduce the cold spillage from retail display cabinets into supermarket stores. In: *Proceedings of the IRC/IIR Conference 'Refrigerated Transport, Storage and Retail Display'.* International Institute of Refrigeration, Paris, 1998, pp. 217–225.
15. H van Oort, RJM van Gerwen. Air flow optimisation in refrigerated cabinets. In: *Proceedings of the 19th International Congress of Refrigeration.* International Institute of Refrigeration, Paris, 1995, Vol. II, pp. 446–453.
16. JN Baleo, L Guyonnad, C Solliec. Numerical simulation of air flow distribution in a refrigerated display case air curtain. In: *Proceedings of the 19th International Congress of Refrigeration.* International Institute of Refrigeration, Paris, 1995, Vol. II, pp. 681–688.
17. F Penot, C Morillon, S Mousset. Analyse et numérisation d'images d'écoulements pour l'étude des meubles frigorifiques de vente. In: *Proceedings of the 19th International Congress of Refrigeration*, International Institute of Refrigeration, Paris, 1995, Vol. II, pp. 106–113.
18. D Stribling, SA Tassou, D Marriott. A two-dimensional computational fluid dynamic model of a refrigerated display case. *ASHRAE Transactions*, 103 (1):88–94, 1997.
19. G Cortella, M Manzan, G. Comini. Computation of air velocity and temperature distributions in open display cabinets. In: *Proceedings of the IIF-IIR Commission C2, D1 Meeting: Advances in the Refrigeration Systems, Food Technologies and Cold Chain.* International Institute of Refrigeration, Paris, 1998, pp. 617–625.
20. P Schiesaro, G Cortella. Optimisation of air circulation in a vertical frozen food display cabinet. In: *Proceedings of the 20th International Congress of Refrigeration.* International Institute of Refrigeration, Paris, 1999, Paper 174.

21. A Aittomäki, J Kianta. Designing of indirect cooling systems. In: *Proceedings of the IIF-IIR Commission D1, B1 Meeting: New Technologies in commercial refrigeration*. International Institute of Refrigeration, Paris, 2002, pp. 132–140.
22. J Nyvad, S Lund. Indirect cooling with ammonia in supermarkets. In: *Proceedings of the IIF-IIR Commission B1, B2, E1, E2 Meeting: Applications for Natural Refrigerants*. International Institute of Refrigeration, Paris, 1996, pp. 207–217.
23. M Verwoerd, N Liem, R van Gerwen. Performance of a frozen food display cabinet with potassium formate as secondary refrigerant. In: *Proceedings of the 20th International Congress of Refrigeration*. International Institute of Refrigeration, Paris, 1999, Paper 552.
24. VD Baxter, VC Mei. Warm liquid defrosting technology for supermarket display cases. In: *Proceedings of the IIF-IIR Commission D1, B1 Meeting: New Technologies in Commercial Refrigeration*. International Institute of Refrigeration, Paris, 2002, pp. 147–151.
25. C Gage, G Kazachki. Warm-liquid defrost for commercial food display-cases: experimental investigation at 32.2°C condensing. In: *Proceedings of the IIF-IIR Commission D1, B1 Meeting: New Technologies in Commercial Refrigeration*. International Institute of Refrigeration, Paris, 2002, pp. 169–178.
26. D Datta, S Tassou. Implementation of a defrost on demand control strategy on a retail display cabinet. In: *Proceedings of the IIF-IIR Commission D1, B1 Meeting: New Technologies in Commercial Refrigeration*. International Institute of Refrigeration, Paris, 2002, pp. 218–226.
27. D Datta, SA Tassou, D Marriott. Experimental investigations into frost formation on display cabinet evaporators in order to implement defrost on demand. In: *Proceedings of the International Refrigeration Conference*. Purdue, West Lafayette, IN, USA, 1998, pp. 259–264.
28. B Field, R Kalluri, E Loth. PIV investigation of air-curtain entrainment in open display cases. In: *Proceedings of the IIF-IIR Commission D1, B1 Meeting: New Technologies in Commercial Refrigeration*. International Institute of Refrigeration, Paris, 2002, pp. 72–82.
29. P Schiesaro, H Kruse. Development of a two stage CO_2 supermarket system. In: *Proceedings of the IIF-IIR Commission D1, B1 Meeting: New Technologies in Commercial Refrigeration*. International Institute of Refrigeration, Paris, 2002, pp. 11–21.
30. Y Mao, W Terrell, P Hrnjak. Performance of a display case at low temperatures refrigerated with R404A and secondary coolants. In: *Proceedings of the IIF-IIR Commission D1, D2/3 Meeting: Refrigerated Transport, Storage and Retail Display*. International Institute of Refrigeration, Paris, 1998, pp. 181–189.
31. G Eggen, K Aflekt. Commercial refrigeration with ammonia and CO_2 as refrigerant. In: *Proceedings of the IIF-IIR Commission B1, E1, E2 Meeting: Natural Working Fluids*. International Institute of Refrigeration, Paris, 1998, pp. 281–292.
32. CA Infante Ferreira, S Soesanto. CO_2 in comparison with R404A. In: *Proceedings of the IIF-IIR Commission B1, E1, E2 Meeting: Heat Transfer Issues in Natural Refrigerants*. International Institute of Refrigeration, Paris, 1998, pp. 141–149.
33. KG Christensen. Use of CO_2 as primary and secondary refrigerant in supermarket applications. In: *Proceedings of the 20th International Congress of Refrigeration*. International Institute of Refrigeration, Paris, 1999, Paper 375.
34. SL Russel, P Fitt. The application of air cycle refrigeration technology to supermarket retail display cabinets. In: *Proceedings of the IIF-IIR Commission D1, D2/3 Meeting: Refrigerated Transport, Storage and Retail Display*. International Institute of Refrigeration. Paris, 1998, pp. 199–207.
35. S Russell, A Gigiel, SJ James. Progress in the use of air cycle technology in food refrigeration and retail display. In: *Proceedings of the IIF-IIR Commission C2, D1 Meeting: Advances in the Refrigeration Systems, Food Technologies and Cold Chain*. International Institute of Refrigeration, Paris, 1998, pp. 137–145.
36. JA Evans, SM van der Sluis, AJ Gigiel. Energy labelling of supermerket refrigerated cabinets. In: *Proceedings of the IIF-IIR Commission D1, D2/3 Meeting: Refrigerated Transport, Storage and Retail Display*. International Institute of Refrigeration, Paris, 1998, pp. 252–263.
37. GG Maidment, X Zhao, SB Riffat, G Prosser. Application of combined heat-and-power and absorption cooling in a supermarket. *Applied Energy* 63:169–190, 1999.
38. GG Maidment, RM Tozer. Combined cooling heat and power in supermarkets. *Applied Thermal Engineering* 22:653–665, 2002.

39. M Axell, P Fahlén. Climatic influence on display cabinet performance. In: *Proceedings of the IIF-IIR Commission D1, B1 Meeting: New Technologies in commercial refrigeration*. International Institute of Refrigeration, Paris, 2002, pp. 181–190.
40. W Gautherin and S Srour. Effect of climatic conditions on the operation of refrigerating equipment in a hypermarket. In: *Proceedings of the 19th International Congress of Refrigeration*. International Institute of Refrigeration, Paris, Vol. II, 1995, pp. 705–712.
41. A Tassone. Hypermarchés: le casse-tête de la climatisation. Revue Pratique du Froid, 845:35–38, 1997.
42. RH Howell, L Rosario, D Riiska, M Bondoc. Potential savings in display case energy with reduced supermarket relative humidity. In: *Proceedings of the 20th International Congress of Refrigeration*. International Institute of Refrigeration, Paris, 1999, Paper 113.

13 Household Refrigerators and Freezers

Rodolfo H. Mascheroni and Viviana O. Salvadori
CIDCA (CONICET – UNLP), La Plata, Buenos Aires, Argentina

CONTENTS

I. INTRODUCTION

Household refrigerators and freezers are intended for keeping small volumes of many different foods and drinks simultaneously at low temperatures, as can be found in any household. These items require diverse storage temperatures (and relative humidities in many cases) and have different storage lives. With the exception of air conditioning, domestic refrigeration systems are primarily used for food storage [1].

The use of these appliances by untrained people implies that their operation and maintenance must be very simple and, in fact, it is probable that their owners perform no maintenance at all over long-time periods. In addition, domestic refrigerators must withstand frequent use and provisions must be made for economical and effective servicing in case of malfunction or damage [2].

These features condition the design and operation of household refrigerators and freezers, which have little in common to those of commercial storage chambers and industrial freezers. The design of domestic refrigerators must also deal with the overall appearance of the appliance (size, shape, color, and surface finishing) and provide special-purpose storage compartments such as vegetable crispers, meat keepers, high-humidity compartments, and butter keepers. In many cases, modern refrigerators and freezers also include additional facilities such as automatic icemakers, crushers, and dispensers or chilled water and juice dispensers [2].

In addition, these apparatus must make use of nontoxic and nonflammable refrigerants; their operation must be as silent as possible and must comply with energy consumption regulations.

Household refrigerators appeared with many of their distinctive characteristics at the beginning of the third decade in the 20th century and since then they have undergone continuous improvements parallel to those of other home appliances [3].

As outlined in this brief introduction, household refrigerators and freezers have many distinctive characteristics, and this chapter intends to provide information on their design, suitability for different products, operation, and performance.

II. EVOLUTION OF DESIGN AND CHARACTERISTICS OF MATERIALS AND COMPONENTS

A. Historic Evolution

The first household refrigerator was developed in 1803 by Thomas Moore, a farmer from Maryland (USA), and consisted of a box cooled by a mixture of salt and ice. In fact, domestic cooling capacities in the shape of iceboxes began to find frequent use more than 20 years later. At that time, only cool larders were used for domestic food storage. In 1826, the Societé d'Encouragement pour la Industrie proposed a prize of 2000 French francs for the development of a proceeding to maintain the ice stored in household iceboxes as long as possible. At the same time in the USA, Frederic Tudor, from New York, proposed the use of natural ice in domestic iceboxes, which became a common practice during the next decades and lasted even well within the past century. Production of natural ice developed to an organized activity with standard ice block sizes, harvesting methods and storage and distribution facilities, which concluded in the ice man that made the household distribution of ice pieces [3,4].

It was in the 20th century that the first successful attempts to develop domestic refrigerators using vapour-compression systems were made. Ammonia, methyl chloride, sulfur dioxide, propane, and isobutane were used as refrigerants in the first prototypes [4]. All of these refrigerants were dangerous, flammable and toxic, for their use in household refrigerators. Of them, propane and isobutene, although flammable, proved to be very safe, probably due to the very low volumes charged in those refrigerators. A proof of their safety is their revalorization in recent years and the actual frequent use of isobutane in household refrigerators.

The true evolution of domestic refrigeration took place in the third and fourth decades of the 20th century, helped by the design and development of fractional horsepower motors and with the introduction of fully sealed systems, which eliminated the belts [5]. The development of modern refrigerants, basically dicholorodifluoromethane (R12), provided household appliances with a safe and efficient, in every sense, refrigerant, the ideal for any design engineer. This led to the general adoption of rotary compressors instead of reciprocating ones and the replacement of expansion valves by capillary tubes [5].

Domestic two-temperature refrigerators (the first refrigerator-freezers) were developed later, by 1940, consisting of two separate compartments. Only 50 years ago, they began to be an increasingly popular household appliance. These refrigerators used mostly chlorodifluoromethane (R22) as the refrigerant, which is nontoxic and nonflammable as well as thermodynamically efficient.

Different distinctive features of the evolution of domestic refrigerators, mainly those of refrigerants, materials, and design, deserve a brief individual look.

1. Refrigerants

A special mention must be made of the ongoing evolution of refrigerants. During 50 years, different CFCs and HCFCs were tested, selected, and widely used for each application in industrial and domestic refrigeration. The recent need for the replacement of CFCs (and HCFCs) involved the development of new refrigerants and the testing of many new and old of them, pure or in mixtures for their use in different applications of industrial and domestic refrigeration.

For the specific use in household refrigeration, tests determined that, in the first stage, the best single substance for replacement of R12 was 1,1,1,2-tetrafluoroethane (R134a). Its main disadvantages are the relatively poor performance at low evaporating temperatures (with respect to R12) and the need to use synthetic lubricants [4]. The great majority of household appliances run nowadays with R134a.

Among the choices of researchers as alternatives to CFCs and HCFCs, many hydrocarbons, pure or in mixtures, have been tested. Examples of this are propane [6], isobutane [7–10], cyclopropane [11], mixtures of propane/isobutane [12], propane/butane/isobutane [6], propane/butane/R134a [13], and LPG [14], among many others. Of them, isobutane (R600a) is, by far, the best characterized and has found important applications for compressors of household refrigerators. It requires a large volumetric flow, but this drawback is balanced by a high latent heat of vaporization, low cost of refrigerant and lubricant, negligible sensitivity to moisture content, and silent operation, leading to highly efficient systems. Although it is flammable, its safety record has been excellent. As a result, hydrocarbon refrigerants have become dominant, in some countries, for domestic refrigerators that have no automatic defrost, mainly in Northern Europe and also in India, China, and South America [4,8].

2. Materials

Many of the first designs had wooden cabinets and iron tubing, belt drivers, expansion valves, and other features that evolved within a short time toward the actual components of household appliances.

As mentioned, in the 1930s, the adoption of R12 and hermetic rotary low-power compressors led to the replacement of iron for nonferrous tubing, expansion valves for capillary tubes, and the elimination of belt drivers. The refrigerators were then built of steel, self-contained, and with better insulation. Better design led to the placement of compressor at the bottom of the unit.

A special mention needs to be made of the evolution of insulants. Urethane foam was developed in the 1950s and evolved to rigid foam in the 1960s, using fluorinated-expanding agents, mainly R11. Foam in-place insulation had an important influence due to its superior insulant capacity that permitted to use lower foam thickness; besides the foam's rigidity contributes to reinforce the structure of the appliances. Therefore, this product became the standard insulating material for refrigerators and freezers until the appearance of the environmental concerns on CFCs [5]. At present, the main option is the use of cyclopentane as blowing agent. An alternative, already in use, is vacuum insulation, which has several technological concerns.

The external surface of cabinets is normally made of steel in the shape of a structure that supports all the components of the appliance (door, refrigeration system, and inner food compartments).

The main evolution of refrigerator cabinets during the last 50 years has been on its shape and size, the introduction of different specialized compartments, and the best use of the inner space available. The continuous developments in plastics enabled the inclusion of different compartments in the inner door liners, to support thousands of door openings without breakage, even when heavily loaded, for example, with bottles of different drinks.

In brief, domestic refrigerators changed considerably from their early designs during the first part of the second half of the 20th century. Multiple compartments made them more complex, the design being refined to meet each time more stringent customer demands.

B. Refrigerating Systems

1. Vapour-Compression Cycle

Nowadays the vapour-compression cycle is almost universally used for household refrigerators and freezers. Where compared with other practical, electrically powered refrigerating systems, it is the most efficient: 1.39 W/W against 0.09 W/W of thermoelectric system and 0.44 W/W of the absorption system [2].

The principal components of the refrigeration circuit are a compressor, a condenser, a capillary tube, and an evaporator (Chapter 3). In domestic refrigerators and freezers, all systems are hermetically sealed, and in theory, they do not require replenishment of refrigerant during their useful life.

The heat transfer through the walls, door, and gasket and the individual characteristics of each component, all affect the energy consumption and the efficiency of household refrigerators. The principal characteristics of each component are as follows.

The compressors used in domestic refrigerators are displacement compressors (with reciprocating piston or rotary mechanisms), with capacities from 90 to 600 W. The compressor and the motor are mounted together and hermetically sealed. The compressor is cooled rejecting heat to the surroundings.

The condenser is the main heat-rejecting component. In the majority of the appliances, it consists of a flat steel-tubing serpentine, placed on the back wall of the cabinet and cooled by natural air convection. Sometimes, the condenser tubing is attached to the inside surface of the cabinet; the problem in this type of component is external sweating. Some special applications have the condenser cooled by forced air (with a small fan incorporated) and may also have a section for compressor cooling.

The capillary tube connects the outlet of the condenser to the evaporator inlet. It regulates the flow and liquid condition of the refrigerant toward the evaporator, and then regulates the performance of the condenser. It is very simple, without moving parts, and may be soldered to the suction line. A recent innovation in this area is the replacement of the capillary tube by a turbo expander, which generates additional mechanical work [15]. This work can be used to drive a fan to force air convection on heat exchangers (evaporator or condenser), increasing the coefficient of performance (COP) and allowing for energy savings.

There are three types of evaporators: manual defrost, cycle defrost, and no-frost. The manual defrost system is seldom used in new appliances, as the cooling effect is generated by natural convection of air over a refrigerated surface. This evaporator is usually a box with three or four sides refrigerated, and the refrigerant is carried in tubes brazed to the walls of the box or the box is made up of two sheets of a metal (usually aluminium) bonded together with internal channels for the circulation of the refrigerant. The evaporator is sized using empirical correlations that relate the evaporator area and the refrigerator internal volume. The performance of the evaporator depends on channel pattern, channel internal diameter, evaporator position, refrigerant pressure drop, and overall heat transfer coefficient.

The refrigerators with cycle defrost use two evaporators, one for fresh food compartment and the other for the freezer. The first evaporator defrosts during each compressor cycle and is designed for natural defrost operation; this first evaporator is usually a vertical plate, which consists of bonded sheets of metal (normally aluminium) with internal channels for refrigerant flow or a serpentine with or without fins. The freezer evaporator does not require defrosting.

As an example of the characteristics of these types of evaporators, the heat transfer performance of vertical plate-type evaporators was studied [16]; the results show a weak dependence of the heat transfer coefficient on the evaporator position. Three positions were studied: original position (inclined forward 3.2°), inclined forward 8°, and inclined backward 2.4°. The evaporator heat transfer rate was obtained from the readings of flux meters placed on 20 test points on both sides of the evaporator. The global heat transfer coefficient U varied from 7.69 to 8.16 $\mathrm{Wm^{-2}\,K^{-1}}$.

The no-frost evaporator is a forced-air fin and tube arrangement, which minimizes the effect of frost accumulation. In this, the air is forced by a fan and the evaporator is defrosted by an electric heater or by hot refrigerant gas. Some equipment includes adaptive defrost, that is, the period of defrosting is set according to the evaporator condition.

To reduce the thermodynamic irreversibilities resulting from an inefficient operation of the Carnot cycle and to introduce environmentally safe refrigerants, many researchers have worked on developing alternative vapor compression refrigeration cycles [5], which are detailed subsequently.

The Lorenz–Meutzner Cycle. It exploits the thermodynamic advantages of the temperature glide of zeotropic mixtures; a 9% of reduction in energy consumption is obtained with a mixture of R22/R123 when compared with conventional refrigerant.

Dual-Loop System. It employs two separate refrigeration cycles, which cool the freezer and fresh compartments independently.

Two-Stage System. It consists of one condenser, two compressors, two evaporators, and at least one suction-line heat exchanger. There are two patented versions [5]. This system promises an improvement of 48.6% over a single stage system.

Control Valve System. It includes two evaporators, one compressor, and one condenser. Two capillary tubes and a control valve are installed between the food and freezer evaporator inlets and the condenser outlet.

Ejector Refrigerator. The energy wasted in the capillary tube is partially used by an ejector to raise the suction pressure entering the compressor.

Tandem System. This is a conventional system that controls the temperature of each compartment by using a thermodamper.

2. Alternative Refrigeration Systems

a. Stirling Cycle

It is an alternative refrigeration method that uses an inert gas, such as air, helium, and so on, as the refrigerant, so this cycle has no associated environmental problems.

This cycle is potentially the most energy-efficient refrigeration process.

The Stirling cycle differs from conventional refrigeration cycles in that the working fluid remains in the gas phase throughout. It has two moving parts: a compressor piston and a displacer, and the cycle has two heat exchangers of small area associated to it. The design of these heat exchangers is crucial to the performance of the refrigerator.

In this respect, EA Technology and Oxford University have jointly developed a concept domestic freezer employing the Stirling cycle [17]. This freezer used a conventional 80 L cabinet (from a frost-free vapor compression freezer) set at $-20°C$ (internal temperature), with a cooling duty of 60 W. The cabinet was modified to accommodate the Stirling cycle cooler unit and the external heat exchangers.

The hot end thermosyphon is a boiling cell with a compact natural convection condenser, which uses water at subatmospheric pressure. The cold thermosyphon is a standard evaporator and uses isobutane at the working temperature of $-30°C$. The fan selected is typical for cooling electronic equipment, with less power input than fans used in standard refrigerators. This feature is demonstrated to have a significant impact on energy consumption.

The freezer was tested at standard conditions (internal temperature of $-20°C$ and ambient temperature of 25°C). The overall energy consumption was 1.48 kWh/24 h when compared with 1.8 kWh/24 h measured for the equivalent vapor compression cycle freezer. Therefore, the Stirling cycle permits an energy saving of 17%. Other tests indicate high-energy savings at low cooling duty ($<$20 W). COP is over 0.8 at cooling duties less than 50 W; this value is 80% higher than that of a conventional compressor rated at the same duty.

Other example is given by Berchowitz et al. [18], who tested two domestic refrigerators equipped with a free piston Stirling cooler, with well-insulated cabinet, and with plastic heat exchangers to cool the walls. The data of one of the tested systems show a 30% improvement in energy consumption, when compared with a conventional Rankine cycle.

b. Absorption Cycle

This is similar to the vapor compression; the difference is that the compressor is replaced by an absorption system. More details are given in Chapter 3.

The advantages of this system are the lack of moving parts, the lack of noise and vibration, and the operation without electric power input. Recently, the role of this cycle has been reconsidered because it is free of CFC. Nevertheless, being 30% less efficient than vapor-compression systems, this cycle has been used for long times to take advantage of solar heating and as an alternative in the absence of electrical supply.

c. Thermoelectric Cycle

This is a small heat pump without moving parts that uses the Peltier effect to transfer heat from a cold refrigerated space to the ambient air (electrons absorb energy and pass it from one semiconductor to another). To increase the heat transfer rate between the thermoelectric cell and the refrigerated space, a coolant (ethylene glycol) flowing through a heat exchanger is used and a fan forces the air over the heat exchanger. This is repeated on heat rejection side of the cell.

The COP is defined as [20]:

$$\mathrm{COP} = \frac{T_{\mathrm{c,co}} - T_{\mathrm{c}}}{(T_{\mathrm{h,ci}} - T_{\mathrm{h}})C_{\mathrm{r}} - (T_{\mathrm{c,co}} - T_{\mathrm{c}})} \tag{13.1}$$

where T_c is the mean temperature of the cold fluid, $T_{c,co}$ the temperature of cold fluid at outlet of Peltier cell, T_h the mean temperature of the hot fluid, $T_{h,ci}$ is the temperature of hot fluid at inlet of Peltier cell, and C_r the heat capacity ratio.

Because of its very low efficiency and its high price, this system is used only in specific applications (military, aerospace, and medical instruments).

III. CHARACTERISTICS OF APPLIANCES AND RECOMMENDATIONS FOR FOOD STORAGE

A. Refrigerators

The main purpose of a domestic refrigerator is to provide the adequate conditions for the simultaneous short time chilled storage of different types of foods that need diverse temperature and humidity conditions for keeping their quality. This primary function is usually complemented with several other simultaneous capabilities such as drink and water chilling, ice making, frozen food storage, and so on.

In many cases, modern refrigerators also include additional facilities such as automatic ice makers, crushers and dispensers, chilled water, and juice dispensers, and quick-freezing and quick thawing capabilities [2,19].

This wide field of possibilities parallels the variety of models of refrigerators to be found in the market. Irrespective of specialized types like under-counter and compact refrigerators, most appliances can be classified into three broad types: single door refrigerators, top-mount combination, and side-by-side combination.

1. Single-Door Refrigerators

They are intended for refrigerated storage of all type of foods and drinks. The frozen food space is not sufficiently cold to provide freezing capacity and is only good for short time keeping of already frozen foods (temperature in this compartment is between -6 and -12°C).

The evaporator is placed in the (upper) frozen food storage volume. An insulated baffle separates this volume from the (lower) refrigerated zone, but allowing the sufficient air to pass around it, by natural convection, so as to maintain sufficiently low temperatures in the refrigerated compartment.

Most of the refrigerators are of the manual defrost type (nonautomatic), although most modern appliances include automatic defrosting at predetermined periods.

2. Top-Mount Combination or Fridge–Freezer

This is a combination of refrigerator and freezer, each with independent compartments (with separate exterior doors) and sometimes also with independent compressors; the freezer usually is mounted on top of the refrigerator. Different models have different temperatures in the freezer zone, leading to a classification based on freezer performance (discussed later).

Most combination refrigerator–freezers use the no-frost system. They are equipped with a fan that blows over a concealed evaporator. Hence, the evaporator collects almost all the frost and is automatically defrosted by an electrical heater or by hot gas, according to different design policies. Very little frost is formed on the frozen foods or on the walls of the freezer compartment. It has the drawback that the circulating air increases water evaporation in the refrigerated zone. Owing to this fact, it is usually suggested that foods be stored packaged or in closed vessels.

3. Side-by-Side Combination

In fact, they are two upright independent appliances (one refrigerator and one freezer) with a common wall. Each side is a complete system by itself. Normally, these are the most expensive refrigerators of a particular brand and include facilities such as ice and refrigerated water or drinks dispensers and even fast freezing and rapid thawing capacities.

All of them are of the no-frost type. Some appliances include microprocessor-based control that allows for adaptive defrost. Certain parameters are monitored and the control software determines the moment of defrosting. Adaptive defrosting assures energy efficiency and better quality of stored foods.

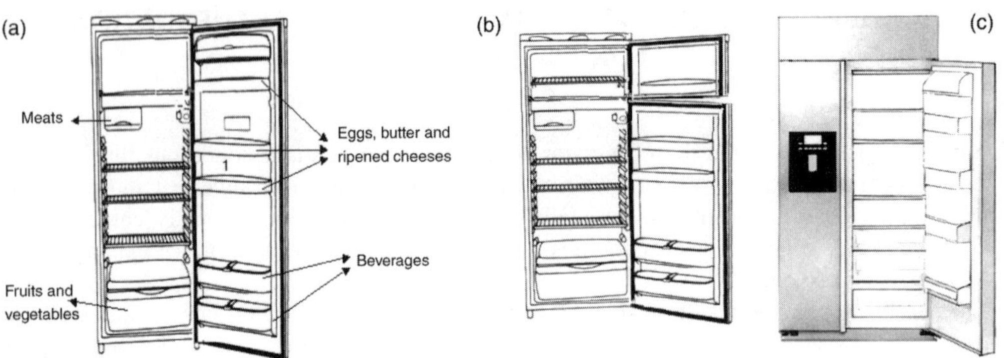

FIGURE 13.1 Scheme of different types of refrigerators: (a) single door, (b) top mount, (c) side-by-side.

Irrespective of refrigerator type, all of them are provided with shelves devised for specific types of foods and drinks. The most usual are below the freezing compartment for meats because meats need the lower temperatures of the refrigerated enclosure, on the lower part of the cabinet for fruits and vegetables, because this zone has higher temperatures, and in the door for eggs, butter, cheeses, and bottles. Figure 13.1 shows a sketch of the three types of refrigerators, illustrating the distribution of the different compartments in the one-door model.

There is a classification of freezer capabilities of refrigerators based on the number of "stars" that are assigned to the freezer. A "one-star" refrigerator reaches −6°C in the freezer compartment, a "two-star" reaches −12°C, and a "three-star" reaches −18°C. A "four-star" refrigerator has "true" freezing capabilities with freezer temperature ranging between −24 and −30°C.

According to ambient temperatures, refrigerators may be designed as "tropical" (T) for temperatures up to 43°C, "semitropical" (ST) for temperatures of up to 35°C, and "normal" (N) for ambient temperatures up to 32°C.

When considering the most adequate refrigerated storage temperature range for different types of foods, designers work on the basis of four ranges or "zones" in the refrigerated volume:

1. The colder zone (next to the evaporator) in the range of 0 to 4°C, which is mainly devoted to meats and meat products, fresh and cooked sausages, fishes, pasterized milk, creams, fresh desserts and cheeses, fresh fruit juices, salads, home-made prepared foods with sauces or creams, foods in thawing process, and so on
2. Intermediate temperature zone (the central volume of the refrigerator) in the range of 4 to 6°C, which is adequate for homemade preparations (as cooked meats), cut fruits and vegetables, yogurts, and so on
3. In the vegetable holder (temperature about 6°C) for fresh fruits and vegetables and ripening cheeses (packaged)
4. In the shelves of the door for eggs, butter, ripened cheeses, esterilized milk and fruit juices, seasonings (mayonaisse, ketchup, and mustard), and also beverages (according to personal temperature preferences).

B. FREEZERS

The characteristics of domestic freezers have much less variations than those of refrigerators. Basically, there are two broad types of design: upright (vertical) and chest (horizontal) freezers. All of them assure at least −18°C storage temperature and a minimum daily freezing capacity of about 10% of the holding capacity (as specified in user's manual). An important issue is not to exceed the design freezing capacity when introducing unfrozen foods. In a contrary case,

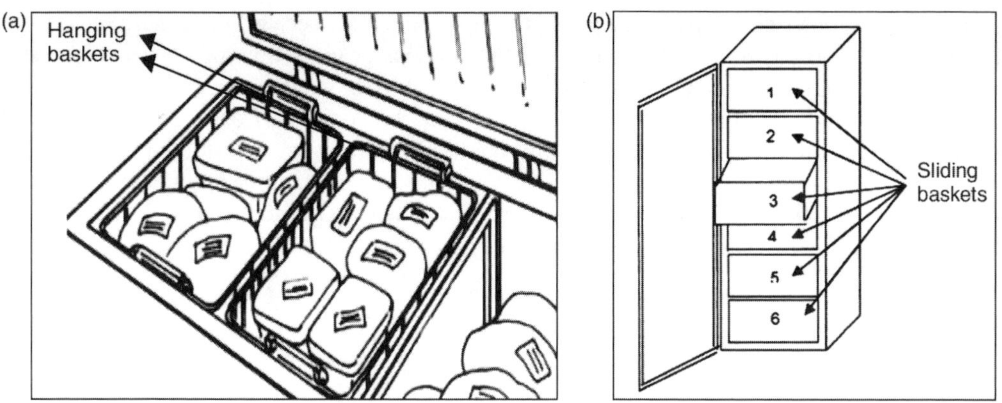

FIGURE 13.2 Scheme of different types of freezers: (a) chest and (b) upright.

temperature of stored goods would increase and the freezing of the new charge will last much more than convenient for keeping adequate quality.

As its use is not as continuous as the refrigerators, usually all models include a power supply indicator light or some type of thermometer with external dial or display for warning of high storage temperature.

Upright freezers are normally intended for more frequent use (loading and discharge of food), because the access to goods is easy. They enable to maintain foods classified by type or date of storage, using different compartments. Normally, compartments are of the sliding type, to facilitate access, and have a plastic front to help reduce temperature increase during openings. Some appliances have a "quick freeze"compartment for freezing newly stored fresh foods.

Meanwhile, *chest freezers* are more ample and best suited for big or irregular-shaped pieces of food and for long-term storage. As a drawback, the access to foods stored in the lower levels of the freezer can be quite complicated. In many cases, freezers include hanging baskets and vertical divisions to facilitate the storing and searching for frozen goods. Some models have a "fast freeze" setting, by which the compressor runs continuously, to be used for the freeezing of recently charged unfrozen food.

In many cases, there is a possibility of setting the appliance as "bottle refrigerator," with temperatures of about 0°C. Figure 13.2 presents a sketch of horizontal and vertical freezers.

IV. OPERATION AND PERFORMANCE OF REFRIGERATORS AND FREEZERS

In the previous section, some indications about recommended temperatures for refrigerated storage of different foods were given. In addition, the consumers are provided with practical guides about the good practices to ensure that the foods stored in their refrigerators or freezers keep their quality characteristics until the moment of consumption. However, in spite of the manufacturers' indications, there is no control on consumer attitudes, and probably food is stored at temperatures too high in domestic refrigerators, with undesirable consequences. Different studies on consumer attitudes on handling of chilled foods and performance of domestic refrigerators at home permit us to describe some general trends.

Various extensive surveys were performed in different countries: 252 refrigerators in the United Kingdom [21], 150 in Northern Ireland [22], 125 in the Netherlands [23], 50 in New Zealand [24], 136 in Greece [25], and finally, 143 in France [26]. These studies analyze different aspects of the performance of the refrigerators, which will be discussed in the following paragraphs. Other authors studied the temperature performance of different appliances working at laboratory under controlled conditions [27,28].

Literature surveys on the performance of freezers are scarce, because the condition of frozen storage by itself usually assures food security. Fluctuations or the setting of freezer temperature (always below 0°C) affect mainly food quality (texture, drip on thawing, overall appearance, and maybe taste) but normally not its sanitary condition. In addition, frozen foods are usually cooked after thawing, so most of quality differences originating on diverse storage conditions are not perceived by the end consumer. Probably, thermal abuse during in-shop storage and at transport to the home may be much more quality detrimental than what could happen during domestic storage [19,29].

A. Results of the Surveys

1. Transport from Supermarket to Household Refrigerator

As the conditions and length of transport from shop to home strongly influence the initial conditions of food in domestic storage, James and Evans [21] measured the thermal response of 19 refrigerated products during 1 h transport at an ambient temperature of 23 to 27°C. The temperatures of foodstuffs at parking were between 4 and 20°C (these high values were determined in thin sliced products because of heating in the shopping trolley). Upon arrival to home, the temperature ranged between 5 and 18°C in the products placed in a cool box or container and from 18 to 38°C in the products placed in the boot of the car without protection. Thin-sliced products (smoked salmon trout) showed the highest temperature variation and thicker products (cooked chicken) were less influenced. This demonstrates that the temperature of chilled foods can reach unacceptably high values during transport. After being placed in the refrigerator, these foods required approximately 5 h to reduce the temperature to below 7°C.

2. Refrigerator Characteristics

The different studies carried out within the home provide statistical data about type, age, and other characteristics (placing, seals, and setting temperature) of the refrigerators.

a. Type

In the United Kingdom [21], the most popular design was the two-door refrigerator or "fridge–freezer" (49.4% of the appliances) followed by the one-door refrigerator or "box-plate" (31.9%) and the larder refrigerator, with plate evaporator in the back panel and without a freezing compartment (18.7%). In Ireland [22], 77.3% of box-plate refrigerators and 22.7% of larder refrigerators were encountered with or without a freezing compartment. A more recent survey [26] indicates that 58% of the appliances were the two-door type or fridge–freezers and the other 42% was one-door refrigerators. Of the total of 143 refrigerators surveyed, 77 (54%) has static refrigeration system, 14 (10%) has ventilated ones (frost free), and the remaining 52 participants of the survey did not know the refrigeration systems of their appliance.

The type of the refrigerator influences the distribution of temperature inside it. Box-plate refrigerators had the lowest average temperature with an even temperature distribution, whereas the other two types showed higher temperature values at the upper limit of up to 5°C. The position of the plate evaporator in the larder refrigerator does not affect the vertical temperature gradient, but it may affect the temperature across the shelves. Laguerre et al. [28] compared the performance of two types of refrigerators in laboratory tests: in one of them air is forced by a fan (hidden to the consumer) placed under the evaporator and is focused to the cabinet through a conduct. The second one is a conventional static system equipped with a fan attached to the back panel of the cabinet. In spite of the differences in the air pattern, the average temperatures were similar. The coldest zone is near the air exit and the warmest near the door in the first refrigerator, and in the second one, the coldest place was near the evaporator and the warmest near the lateral panels.

b. *Age*

The majority of the appliances were less than 10 years old.

c. *Place*

Among 395 refrigerators, 110 (27.8%) of them were located near a heat source (oven, dish washing machine, etc.).

d. *Seals*

In 292 of 402 appliances, the condition of the seal was good or excellent.

3. Refrigerator Temperature

The safety of the chilled foods depends principally on temperature throughout all stages from production to consumption; in the cold chain, the domestic refrigerator is an important link that probably had the worst control. ISO Standard 7371 recommends 5°C or less to control the growth of food-poisoning microorganisms. In spite of this, the majority of consumers were unable to state the temperature at which they attempted to operate their refrigerator.

To analyze the distribution of the temperature inside the refrigerator cabinet, three air (top, middle, and bottom) and one or two product temperatures were measured during several days. Analysis of variance of the results obtained by different authors [21,26] revealed that generally the average temperature and the values corresponding to the different positions are not statistically different.

Only 15% of the consumers in the UK survey [21] kept a thermometer in the refrigerator and used this information to set the refrigerator temperature, and only one consumer kept an integral thermometer that enabled temperatures to be read while the door was closed. In the Northern Ireland survey [22], none of the participants of the study had a thermometer in their refrigerator.

a. *Influence of Ambient Temperature*

The ambient temperatures in the majority of the houses visited in the different surveys were between 17 and 23°C, which is the range of design of refrigerators and so this temperature is expected not to influence the refrigerator performance. In the experimental work of Laguerre et al. [28], three ambient temperatures were studied (16, 25, and 32°C); the influence of high ambient temperature is significant on energy consumption (with an increase of 60%) but not on average air temperature because of thermostat control.

b. *Average Temperatures*

In the study performed in the UK [21], the overall mean temperature of each refrigerator was 6.0°C, with a minimum value of -0.9°C and a maximum value of 11.4°C. On an average, 29.9% of the tested appliances operated below 5°C and 66.7% operated below 7°C. Only 7.3% operated on an average above 9°C.

In an empty box-plate refrigerator, average temperatures were between 0.5 and 1.5°C on the shelves and just above 3°C in the door, with a cycle of less than 0.5°C [26]. However, in a fridge–freezer, the average temperatures were less uniform, with values between 14.7°C at the top and 2.1°C at the bottom.

A study in China [30] showed that only 2.3% of domestic refrigerators operate at temperatures lower than 6°C, 34.1% operate within 8 to 12°C, 34.1% operate within 12 to 14°C, and 29% operate at temperature higher than 14°C.

In Northern Ireland [22], 71.3% of all the refrigerators had an average internal temperature above the recommended value of 5°C, with the mean value of 6.54°C, the minimum of 0.8°C, and the maximum of 12.6°C.

In other study carried out in France [26], 119 sets of temperatures over a total of 143 refrigerators surveyed indicate that the global temperature was above 5°C in 80% of the refrigerators and over 8°C in 26% of the total. The mean temperature was 6.6°C, with a minimum of 0.9°C and a maximum of 11.4°C.

c. Distribution and Range of Temperatures

The difference between the highest and the lowest temperatures over the whole refrigerator was between 4.5 and 30.5°C [21], the highest temperature occurs at top in 70% of 252 appliances and the lowest temperature is measured near the middle of the cabinet in 45%. In empty refrigerators, the difference between maximum and minimum temperatures were between 4 and 8°C in spite of forced air system [28].

Recently, thermal distribution inside the cabinet has been studied by means of computational fluid dynamics (CFD). The numerical analysis [31] shows that in larder refrigerators with conventional air-supply systems, there are high-temperature regions above the top shelf and the top door pocket and there is a low-temperature region in the center of each shelf. Thermal uniformity can be improved by modifying the air-supply system, adding a blower, jet slots, and grills for air return.

d. Effect of Loading

The response of two experimental refrigerators to loading with precooled and warm foods indicates that loading reduces the average temperature by 0.5 to 2°C, depending on the refrigerator type. A minimum cooling time of approximately 2 h was required in either appliance to cool the warm products, even when the thermostat was set at its lowest value [27]. The loading of the refrigerator with new products results the increase of the temperature of the stored foods by 1 or 2°C when the tax of charge increases from 15 to 30%. The warmest products were placed near the lateral panels, far from the cool air-supply system [28].

e. Effect of Door Openings

The recovery to within 1°C of the original temperature occurred in 2 to 10 min, depending on the type of refrigerator and the time the door was open (10 s to 1 min). With multiple door openings, the time increases and the degree of temperature recovery is reduced [27]. The global temperature is higher in those refrigerators with a high frequency of door openings, more than 20 times per day [26].

f. Thermostat Setting

Despite the advances in refrigerator design (more compact appliances, chilled drink, and ice dispensers), only the more expensive refrigerators are equipped with more sophisticated temperature controls.

The different studies [22,28] show that there was no correlation between temperature setting and refrigerator temperature. Some refrigerators, with their thermostat set at the coolest value, had an internal temperature above 5°C, indicating an inefficient operation of the refrigerator. Besides, few families use a thermometer to control temperature and a great number of consumers set their refrigerator to the less cold value to reduce the energy consumption, without consciousness of food safety. In empty refrigerators, the temperature may be 2°C above the set value [28].

g. Microbiological Aspects

The relationship between temperature distribution and frequency of microbial contamination was also investigated [21,22,25]. Predictions of bacterial growth (*Pseudomonas* and *Listeria*) during an 1 h transport were made using a mathematical model [21]. The increase of microorganisms was between 0.4 and 2 generations depending on the thermal protection of the foods during their transport (cool box and without protection, respectively).

In a study conducted by Flynn et al. [22], the temperatures were recorded and an assessment of the effect of cabinet temperatures on the microbiological quality of chilled foods was made.

In the survey performed in Greece [25], from a total of 136 domestic refrigerators, samples from the walls, shelves, and cheese compartments were taken and the presence of *Listeria* spp. and *L. monocytogenes* was investigated. *L. monocytogenes* was detected in only two cases, in spite of the high mean temperature encountered (55% of the refrigerator were over 9°C). Other researchers obtained similar results [25].

B. Experimental Results of Tests on a One-Door Refrigerator

Tests were run in our laboratory on a brand-new one-door refrigerator Electrolux Model R310 (unpublished data). Its main features are as follows: loading capacity, 294 dm^3; overall height, 1499 mm; width, 601 mm; depth, 618 mm; defrosting, semiautomatic; special characteristics, upper freezing compartment; shelves, below the freezing compartment for meats, on the lower part of the cabinet for fruits and vegetables, and door for eggs and bottles. Figure 13.1 shows a scheme of the refrigerator. Tests were done on the empty refrigerator at two levels of setting of the thermostat and on minced meat slabs placed in different shelves of the partially filled refrigerator.

1. Empty Refrigerator

The settings used were level 7 (maximum cold recommended for ambient temperatures higher than 28°C) and level 3 (normal, for ambient temperatures between 18 and 28°C). Results are given in Table 13.1 for an average ambient temperature of 16°C.

TABLE 13.1
Results of Tests on an Empty One-Door Domestic Refrigerator

	Maximum Cold	Normal Cold
	Average temperatures	
Freezing compartment	−17.0°C	−11.0°C
Upper shelf	−1.8°C	0.2°C
Lower shelf	−2.6°C	−0.8°C
Fruit and vegetable compartment	0.9°C	3.5°C
Door shelf [(1) in Figure 13.1]	−2.2°C	0.4°C
	Recovery time to within 1°C after a 1 min door opening	
Freezing compartment	8 min	11 min
Upper shelf	13 min	15 min
Lower shelf	8 min	9.5 min
Fruit and vegetable compartment	Temperature rise lower than 1°C	Temperature rise lower than 1°C
Door shelf [(1) in Figure 13.1]	18 min	25 min

These results show a temperature profile within the cabinet of about only 4°C in both the regimes, with average temperatures at each location about 2°C lower for maximum cold regime. As expected, recovery times were shorter for the maximum cooling regime. These results are in accordance with the previously cited data from literature.

2. Loaded Refrigerator

Another test was done with a half-loaded refrigerator (with food on all shelves and compartments), working at normal regime. Three plastic trays each loaded with 750 g of minced meat were placed in the freezer compartment, meat shelf, and fruit and vegetable compartment, respectively. Average registered temperatures were −12.2°C in the center of the dish and −13.2°C on the surface of the sample in the freezer compartment. The amplitude of temperature fluctuations was minimal, ranging between 0.30 and 0.35°C. In meat compartment, temperatures were of −4.4°C in the center and −4.6°C on the surface, with negligible fluctuations. In the vegetables compartment, the temperature in the center of the meat slab was 2.3°C, again with minimal fluctuations. These figures are lower than those of the empty refrigerator working at the same regime (Table 13.1). Measured data is in accordance with literature [28], which reports lower average temperatures in filled refrigerators.

C. Experimental Results of Tests on Domestic Freezers

1. Upright Freezer

We tested at our laboratory an Electrolux R22 freezer (unpublished data). Its main characteristics are as follows: six sliding compartments, with one fast freeze compartment (the second if numbering from the upper shelf). Thermostat is in the lower rear part (out of the compartment). It has no indication of temperature setting; only refrigeration level that is, from (−) to (+). The equipment was tested during normal use at our laboratory, with all the compartments filled with vessels containing frozen samples.

Figure 13.3 shows the temperatures registered in the air within the different compartments, including the thermal response to a 1-min door opening. The setting of thermostat was in (−) (lowest temperature). As can be seen, temperatures are in the order of −25 to −28°C. Door opening implies an increase of air temperature of 10 to 15°C for the three upper levels, but of only 1.5 to 3°C for the lower ones. Time to recover the initial temperature after the 1-min door opening ranged between 25 and 45 min for different crispers.

In the same conditions, two plastic trays wrapped in polystyrene, each with 800 g of minced meat were frozen one on top of the other in the fast freeze compartment (lower tray is numbered "1" and upper is "2"). Figure 13.4 presents the thermal history measured at different locations in each tray (lower face, middle of the height, and upper face). As it can be noticed, freezing of the samples begins after 2 h of being charged. The total freezing time (up to −18°C) is in the order of 15.5 h.

The same procedure was repeated with the freezer at its lowest cooling capacity (highest temperature). In this case, freezing times increased to reach about 18 h, as shown in Figure 13.5.

In both situations (maximum and minimum settings), freezing times exceed 15 h, which can be clearly classified as "slow freezing."

2. Chest Freezer

The appliance tested at our laboratory (unpublished data) was a freezer Whirlpool model AFG 145, with a useful volume of 0.32 m^3 and a daily freezing capacity of 16 kg. The equipment was tested during normal use at our laboratory, with all the compartments almost completely filled with vessels or packages containing frozen samples. Tests were done by placing thermocouples at different locations within the volume of the freezer as shown in the scheme of Figure 13.6.

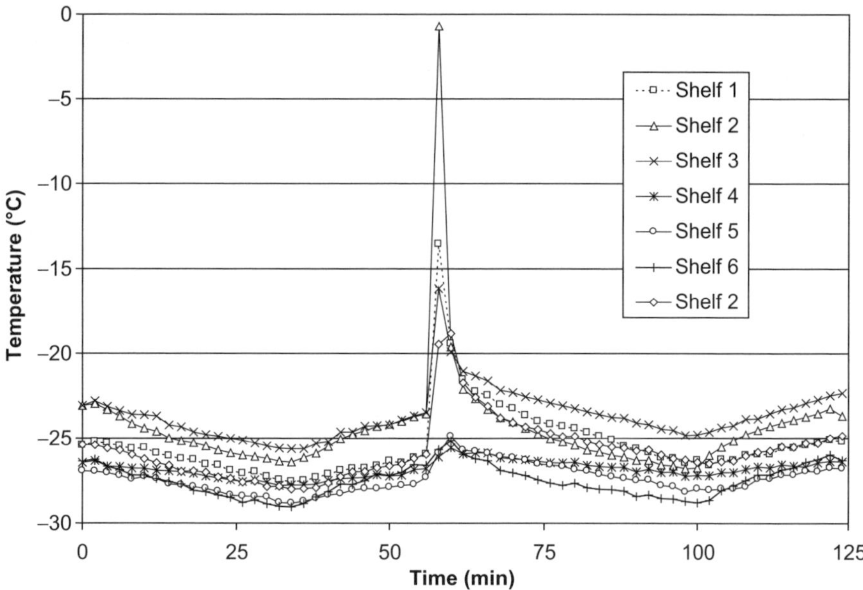

FIGURE 13.3 Air temperatures in an upright freezer, including the thermal response to a 1 min door opening.

During normal use (without door openings) at the lower temperature setting, there are fluctuations in the order of 5°C in air temperature at locations 4 and 5 (near the evaporator), the fluctuations decrease to 1.5 to 2°C in other locations. Average temperatures at different locations are: 1, −21.9°C; 2, −18.8°C; 3, −17.6°C; 4 and 5, −22.8°C; and 6, −19.5°C. Obviously, thermocouples near the evaporator show the lower temperatures and those in the upper baskets show the higher ones.

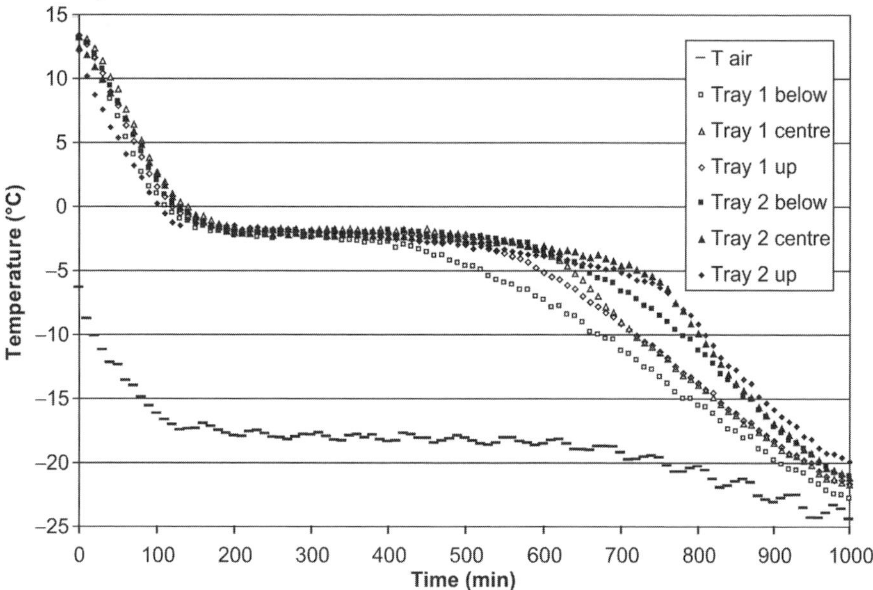

FIGURE 13.4 Thermal histories of two trays of minced meat, placed in the fast freeze compartment of an upright freezer. Tray "1" is below tray "2". The freezer was set at its highest cooling capacity.

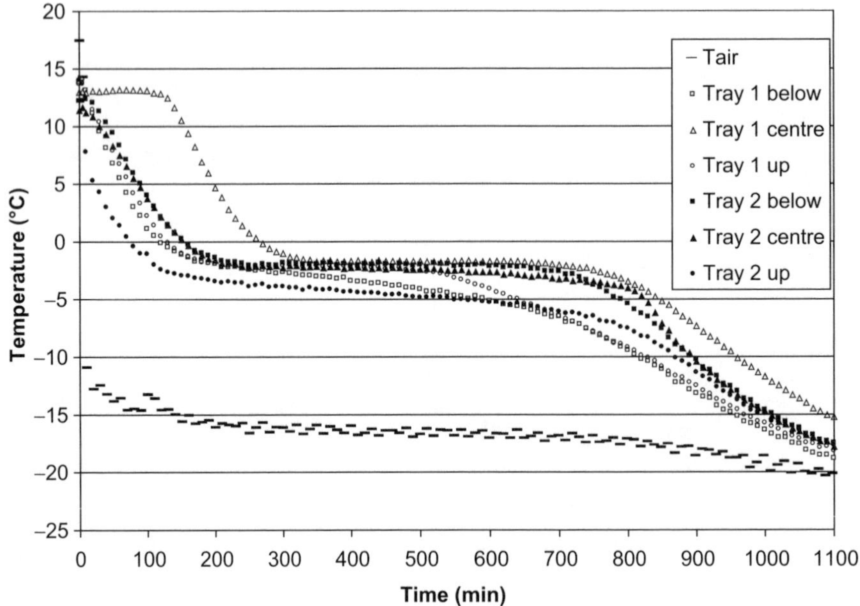

FIGURE 13.5 Thermal histories of two trays of minced meat, placed in the fast freeze compartment of an upright freezer. Tray "1" is below tray "2". The freezer was set at its lowest cooling capacity.

The 1-min door opening ensures that thermocouples on baskets reach 0°C, recovering their original temperature in about 30 min. Temperature in locations 4 and 5 reach about −11°C and goes back to its original value in about 18 min.

To determine freezing capacity, four trays each with 750 g of minced beef were set on locations 1, 2, near 5 (but not against the wall), and 6. Measured freezing times (up to −18°C in the center of the dish) were 640, 1050, 1300, and 810 min, respectively. These results show a great dispersion in freezing rate, depending on position. All the cases can be classified as 'slow freezing.'

D. Energy Consumption

Energy efficiency of refrigerators and freezers is receiving attention lately because there is a potential of substantial energy savings. Household refrigerators and freezers are probably the largest single end users of electricity in the residential sector due to their continuous operation and widespread use. Therefore, there is a potential of substantial energy savings, improving energy efficiency. Nowadays, there are many test standards to measure the energy consumption of refrigerators. Sometimes the results provided by the different standards differ significantly and it is difficult to compare the performance of different appliances.

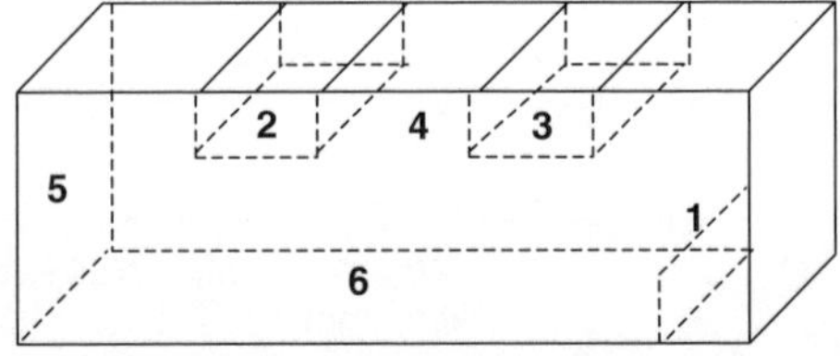

FIGURE 13.6 Position of the thermocouples in the chest freezer: 1, over the compressor compartment; 2 and 3, in the baskets; 4 and 5, on the rear and left walls (near the evaporator); 6, on the floor.

TABLE 13.2
Some Important International Test Standards

Standards	Region
ISO 5155 for freezers, ISO 7371 for all refrigerators; ISO 8187 for refrigerator or freezers, and ISO 8561 for forced air frost free units. Probably the four standards will be combined into only one	ISO
ANSI/AHAM HRF-1-2001	North America
JIS C 9607-1986	Japan
CNS 9577 (general) and C3164–1989	China and Korea
GOST 16317-87, 1991	Russia
AS/NZS 4474.1, 1997	Australia and New Zealand
IRAM 2120 (Parts 1, 2, 3, 4, and 5) and IRAM 2404 (Parts 1, 2, and 3)	Argentina
NBR 7070 to 7093 and NBR 12863 to 12886	Brazil

TABLE 13.3
Values of Parameters Measured in the ISO Standards (the Values in Brackets Correspond to Testing Conditions of the Cited Property or Parameter in Other Currently Operative Standards Listed in Table 13.2)

Parameter	Condition
Ambient temperature	25/32 ± 0.5°C [15°C; 25°C; 30°C; 32°C]
Relative humidity	45 to 75%
Number of measure points	3 [1; 2]
Location of measure points	350 mm from the walls [between 250 and 350]
Fresh food compartment temperature	5°C [3°C; 3.3°C; 5°C; 7.2°C]
Number of measure points	3
Frozen food compartment	
Freezer one-star temperature	−6°C [−9°C; −9.4°C]
Freezer two-star temperature	−12°C [−15°C]
Freezer three-star temperature	−18°C [−18°C]
Number of measure points	4 to 6 [3 to 12]
Loading of test packages	Yes [no; 75% filled]
Volume of all compartments	Storage [Gross]
Volume adjustment of freezer compartment	2.15 [1.6; 1.63]
Energy adjustment of separate freezer	None [0.7; 0.85]
Anti-sweat heaters	When needed on [always on; average on and off]
Door openings	Only in the JIS standard
Period of energy measurement	≥24 h [less than 1 kWh; 16 h; 3 h $< t <$ 24 h]
Installation of the refrigerator	On a wooden platform and next to a rear wall at the minimum allowable distance. [Each norm has its own condition.]

Bansal [32] reviews the principal test standards, identifying the major differences among the conditions of the parameters to be measured. The most important standards are listed in Table 13.2.

By summarizing the requirements for the different tests, one can find that the following parameters need to be measured to determine energy consumption: ambient temperature, relative humidity, cabinet (freezer or refrigerator) temperature, freezer compartment volume, the existence of antisweat heaters, the effect of door openings, the period of energy measurement, the loading (of test packages), the place of installation of the refrigerator, the operation of automatic defrost, and so on.

Table 13.3 summarizes the parameter values of the ISO standards and gives the values of the other cited tests (between brackets).

As there are so many standards currently operative, it is not clear how a refrigerator tested in a country would operate in any other country, which is tested with a different standard. The problem is great when the refrigerators are exported to countries with different climate conditions.

V. CONCLUSIONS

During the last 40 to 50 years, the evolution of refrigerators and freezers was based not on new types of equipment or refrigerants but mainly on best control of temperature and humidity through design and operation, new insulants, lining and finishing materials, the combination of refrigerating and freezing capabilities, the inclusion of a variety of special-purpose storage compartments and additional facilities such as ice makers and dispensers or chilled water and juice dispensers.

The actual trends put their focus on new refrigerants and blowing agents, energy efficiency, and on each time more sophisticated additional features.

NOMENCLATURE

COP	coefficient of performance
C_r	heat capacity ratio
N	normal (refrigerator type)
ST	semitropical (refrigerator type)
T	tropical (refrigerator type)
T_c	mean temperature of the cold fluid
$T_{c,co}$	temperature of cold fluid at outlet of Peltier cell
T_h	mean temperature of the hot fluid,
$T_{h,ci}$	temperature of hot fluid at inlet of Peltier cell
U	Global heat transfer coefficient

REFERENCES

1. SJ James. Developments in domestic refrigeration and consumer attitudes, *IIR Bulletin*, 5:5–17, 2003.
2. Anonymous. Household refrigerators and freezers. In: *ASHRAE Refrigeration Handbook*. Chapter 49, Atlanta: ASHRAE, 2002, pp. 49.1–49.12.
3. F Beltrán Cortés. Apuntes para una historia del frío en España, Consejo Superior de Investigaciones Científicas, Madrid 1983.
4. SF Pearson. Refrigerants past, present and future. *IIR Bulletin* 3:5–25, 2004.
5. R Radermacher, K Kim. Domestic refrigerators: recent developments. *International Journal of Refrigeration* 19:61–69, 1996.
6. MA Hammad, MA Alsaad. The use of hydrocarbon mixtures as refrigerants in domestic refrigerators. *Applied Thermal Engineering* 19:1181–1189, 1999.
7. H Iz, T Yilmaz, Y Tanes. Experimental results of the safety tests on domestic refrigerators for refrigerant R600a. In: *Proceedings of IIR Meeting of Comissions B1, B2, E1 and E2*. Aarhus, 1996, pp. 321–328.

8. RS Agarwat. Isobutane as refrigerant for domestic refrigeration in developing countries, In: *Proceedings of IIR Meeting of Comissions B1, B2, E1 and E2*. Aarhus, 1996, pp. 75–86.
9. RH Pereira, MA Lunardi, JL Driessen, MR Thiessen. Hydrocarbon refrigerants as substitutes for CFC-12 in domestic refrigeration system. In: *Proceedings of Refrigeration Science of Technology*. Germany, 1994, pp. 561–569.
10. DR Riffe. Isobutane as a refrigerator freezer refrigerant. In: *Proceedings of International Refrigeration Conference*. Purdue, 1994, pp. 245–254.
11. LJM Kuijpers, JA de Wit, AAJ Benschop, MJP Janssen. Experimental investigation into the ternary blend HCFC22/124/152a as a substitute in domestic refrigeration. In: *Proceedings of International Refrigeration Conference*. Purdue, 1990, pp. 314–324.
12. D Jung, ChB Kim, K Song, B Park. Testing of propane/isobutane mixture in domestic refrigerators. *International Journal of Refrigeration* 23:517–527, 2000.
13. B Tashtoush, M Tahat, MA Shudeifat. Experimental study of new refrigerant mixtures to replace R12 in domestic refrigerators. *Applied Thermal Engineering* 22:495–506, 2002.
14. BA Akash, SA Said. Assessment of LPG as a possible alternative to R-12 in domestic refrigerators. *Energy Conversion and Management* 44:381–388, 2003.
15. A Zoughaib, D Clodic. A turbo expander development for domestic refrigeration appliances. In: *Proceedings of the 21st International Congress of Refrigeration*, IIR/IIF. Washington, 2003, 1–8 Paper ICR0144.
16. L Silva, C Melo, RH Pereira. Heat transfer characteristics of plate-type evaporators. In: *Proceedings of the 20th International Congress of Refrigeration*. IIR/IIF, Sydney, 1999, Vol. III, pp. 495–502.
17. RH Green, PB Bailey, L Roberts, G Davey. The design and testing of a Stirling cycle, In: *Proceedings of IIR Meeting of Commissions B1, B2, E1 and E2*. Aarhus, 1996, pp. 153–161.
18. DM Berchowitz, DE Kiikka, BD Mennink. Tests results for Stirling cycle cooled domestic refrigerators. In: *Proceedings of IIR Meeting of Commissions B1, B2, E1 and E2*. Aarhus, 1996, pp. 133–141.
19. BA Anderson, S Sun, F Erdogdu, RP Singh. Thawing and freezing of selected meat products in household refrigerators. *International Journal of Refrigeration* 27:63–72, 2004.
20. PK Bansal, A Martin. Comparative study of vapour compression, thermoelectric and absorption refrigerators. *International Journal of Energy Research* 24:93–107, 2000.
21. S James, J Evans. Consumer handling of chilled foods: temperature performance, *International Journal of Refrigeration* 15:290–306, 1992.
22. O Flynn, I Blair, D McDowell. The efficiency and consumer operation of domestic refrigerators. *International Journal of Refrigeration* 15:307–312, 1992.
23. PAL Coulander. Koelkast temperature thuis, Report of the Regional Inspectorate for Health Protection, Leewarden, 1994.
24. GD O'Brien. Domestic refrigerator air temperatures and the public's awareness of refrigerator use. *International Journal of Environmental Health Research* 7:141–148, 1997.
25. D Sergelidis, A Abrahim, A Sarimvei, C Panoulis, P Karaioannoglou, C Genigeorgis. Temperature distribution and prevalence of *Listeria* spp. in domestic, retail and industrial refrigerators in Greece. *International Journal of Food Microbiology* 34:171–177, 1997.
26. O Laguerre, E Derens, B Palagos. Study of domestic refrigerator temperature and analysis of factors affecting temperature: a French survey. *International Journal of Refrigeration* 25:653–659, 2002.
27. S James, J Evans. The temperature performance of domestic refrigerators. *International Journal of Refrigeration* 15:313–319, 1992.
28. O Laguerre, J Gahartian, S Srour. Etude de la performance des réfrigérateurs domestiques. Revue General du Froid, 1037: 22–28, 2003.
29. RC Martins, CLM Silva. Frozen green beans (*Phaseolus vulgaris*, L.) quality profile evaluation during home storage. *Journal of Food Engineering* 64:481–488, 2004.
30. B Shixiong, X Jing. Testing of home refrigerators and measures to improve their performance. In: *Proceedings of IIR Meeting of Commissions B2, C3, D1, D2/3*. Dresden, 1990, pp. 411–415.
31. K Fukuyo, T Tanaami, H Ashida. Thermal uniformity and rapid cooling inside refrigerators. *International Journal of Refrigeration* 26:249–255, 2003.
32. PK Bansal. Developing new test procedures for domestic refrigerators: harmonisation issues and future R&D needs — a review. *International Journal of Refrigeration* 26:735–748, 2003.

14 Monitoring and Control of the Cold Chain

Maria C. Giannakourou and Petros S. Taoukis
National Technical University of Athens, Athens, Greece
George-John E. Nychas
Agricultural University of Athens, Athens, Greece

CONTENTS

I. INTRODUCTION

Modern lifestyle and the evolution of consumer requirements over the past decade have led to significant increase in demand for frozen foods, which offers the advantages of easy and quick preparation of a "fresh-like" meal. In Europe and the United States, the market of frozen foods has been expanded so as to include a variety of products, aimed at all of the segments' consumers. Frozen food purchase represents a significant percentage of the total food expenses. For example, in France frozen food consumption reaches 45% of the total food purchase [1].

The mass consumption of frozen foods and the new consumer patterns, that is, reduced cooking times for minimal quality loss, microwave cooking, have accentuated the need for constant and systematic control of the temperature handling of frozen foods throughout their distribution in the cold chain, from the point of production to their final consumption. Several studies have been recently carried out to assess the importance of low-temperature handling of frozen food, focusing on the effect of temperature fluctuations or temperature abuses during handling on product quality [2–4]. It is recognized that, even at adequately low freezing temperatures, physicochemical and biochemical mechanisms take place, degrading the product quality [3].

All these modes of deterioration and potential hazards are exaggerated usually by the fluctuating time–temperature environment during storage, especially when freeze–thaw cycles occur [5]. Additionally, glass transition temperature (T_g) in relation to the stability of frozen foods has been widely discussed recently [6–8], focusing on the accelerated quality loss at temperatures above the T_g. In contrast, investigations in the logistics of the food chain reveal the existence of weak links, such as loading or unloading operations, temporary storage, retail outlet, and consumer transport that "break" the low-temperature continuity and accelerate frozen food deterioration dramatically. One should also consider that frozen products follow complex circuits, subjected not only to varying environmental parameters, but also to intrinsic constraints, such as transport media liability, suitability of cooling equipment and controlling methods, and so on. In any case, transfer points, that is, points where frozen products are moved from a cold area to another, are known to be frequent black spots for temperature abuse and detrimental fluctuations, as well as mishandling [9].

When temperature fluctuations occur during frozen storage, the amount of ice in a system will generally remain constant, but the number of ice crystals will be reduced and the average size will increase [10]. This is mainly due to a natural tendency for reduced surface area, both at constant temperatures and at fluctuating time–temperature profiles. Especially when temperature is not constant, recrystallization takes place, increasing the size of ice crystals. The growth in size of ice crystals can significantly influence the quality loss during frozen storage and handling of perishable foods. Additionally, temperature variation within a product can cause moisture migration, relocating the water within the product so as to move toward surfaces and to leave the denser regions of the product. So, when there is void space around a product in a package, moisture will transfer into this space and tend to accumulate on the surface of the product and the internal surface of the package. One common example of this is the significant in-package desiccation, frequently observed in sales cabinets for products that are loosely packaged in plastic bags [11]. Unfortunately, most studies are performed under constant temperature conditions, and the effect of temperature fluctuations is only theoretically addressed [12].

In this context, the required temperature conditions need to be maintained all the way from the producer to the consumer, assuring a maximum low temperature of $-18°C$, a limit set by the majority of international and national regulations. Any increase in the temperature of the environment in which the product is held above that marginally accepted temperature is proven to have a significant adverse effect on the quality, and sometimes even on the safety of the product. Especially when the food is inadvertently thawed, microbiological issues become serious and may lead to food rejection [13,14]. To put that in practical perspective, when a frozen product is held, even for a few minutes, in warmer than $-18°C$ air, it will start to thaw, despite its "frozen-like" appearance. Restoring the temperature at the appropriate levels will lead the product to slower freezing because the equipment in the cold chain is designed to maintain the product at $-18°C$ and not freeze product down to that temperature.

A considerable amount of work on the effect of temperature on the quality degradation of different food products has been published in the earlier and recent literature [15–19]. Many studies are focused on the nutritional degradation, described by vitamin loss [20–23], or sensory deterioration [24,25] of frozen fruits and vegetables, pointing out, in most cases, the stability obtained when low temperatures are preserved. However, there is a change of attitude recently toward the recommended temperature for frozen storage and handling. In the earlier days, a very low temperature (of about $-30°C$) was recommended for the earlier stages of the distribution chain (producer and wholesale stage) and $-18°C$ for retail outlet and home freezers. Taking into consideration the energy issues involved and the optimization of the frozen storage and handling process desired by the food producer, the two contradictory requirements of low temperature and energy saving should be compromised in an effective management scheme.

Considering the multiple parameters that affect the efficiency of the current cold chain and the importance of a steady and adequately cold logistic path for the product acceptance, in terms of both safety and quality, it becomes evident that monitoring and control of the cold chain is a prerequisite for reliable quality management and optimization [26–28]. Good temperature control is essential in all sections of the frozen food chain and can be obtained through improved equipment design, through quality assurance systems application, and by an increased operator awareness. The current philosophy, however, for food quality optimization is to introduce temperature monitoring in an integrated, structured quality assurance system, based mostly on prevention, through the entire lifecycle of the product [29,30].

II. MONITORING THE CONTROL CHAIN

The process of constantly monitoring and keeping records of temperature handling has been recently a key point for the frozen food sector. This is mainly due to stricter legislation, an increased concern for more effective management, and improvements of available monitoring equipment [31].

According to a definition provided by the Concerted Action FAIR-CT 96-1180 [2] sponsored by the European Commission "the cold chain is the part of the Food Industry which deals with the transport, storage, distribution, and selling of frozen foods. It includes equipment and the operation of that equipment to maintain frozen food in a fully frozen condition at the correct temperature." Alternatively, according to the International Institute of Refrigeration, the "cold chain" refers to the continuity of frozen distribution, that is, the means successively employed to ensure the frozen preservation of perishable foodstuffs from the production to the consumption stage.

A. Requirements and Control of the Stages of the Cold Chain

1. Cold Store

During its journey from the producer to the home freezer, and final consumption, the frozen product is stored at different points of the chain inside chambers of different characteristics and performance. The size of the cabinet, initial temperature of the incoming food, temperature required, temperatures of the surroundings, mechanical characteristics (location of refrigeration machinery, compressors, ventilation, and insulation) and energy and cost are issues of first priority when considering cold store requirements. In any case, an effective stock rotation and a safe stacking within any storage area are of cardinal importance for an optimized frozen food management. Until now, it is always stated that the "first-in–first-out" management approach must be strictly adhered to in all stages of the freezer chain [11] through fully automatic handling procedures in the freezer storage rooms. As it will be discussed in Section IV, a more sophisticated product circulation system (least shelf-life first out, LSFO), based on the real quality status of the product, would reduce the possibility of out-of-date stock, leading to a cost-effective product management.

Regarding temperature requirements during frozen storage, according to EU Directive 89/108 (Quick Frozen Food Directive, QFF) [32], after quick freezing, the product temperature should be maintained at $-18^\circ C$ or colder after thermal stabilization. Some frozen foods, for example, beef, broilers, butter, have a fairly long storage life even at $-12^\circ C$, whereas foods such as lean fish require storage temperature around $-28^\circ C$ to reduce the quality loss and prolong their storage life [9].

In the United States, a temperature of $-18^\circ C$ or colder is recommended, adding that some products, for example, ice cream and frozen snacks require $-23^\circ C$ or colder. The EU Directive

92/1 [33] requires that a temperature-recording device must be installed in each storage facility to register and store for at least a year the temperature data of air surrounding the perishable food.

2. Transport

The different points of transport, from the cold store to the retail outlet and then to the consumer's refrigerator, are critical points for the product's overall quality and safety. A significant factor is the temperature inside the transport vehicles, and the fluctuations occurring during transit. The vehicle must be provided with a good refrigerating system, operating constantly during transportation to maintain the product frozen. Another important issue is to avoid undesirable heat infiltration, which may occur due to hot weather, sunny conditions, inadequate insulation, or air leakage. When taking precautions to avoid these, it should be possible to achieve good-quality, healthy, and safe frozen food products.

Legislation on control of transport equipment and temperatures during transport has been increasingly stricter, especially for intra-European transport of frozen foods. The Quick Frozen Food (QFF) Directive requires that "the temperature of quick frozen food must be maintained at −18°C or colder at all points in the product, with possible brief upward fluctuations of no more than 3°C during transport (Article 5) [32]." Directive 92/1 [33] requires that transport equipment must have installed an appropriate temperature-recording device, which should be approved by the authorities in the EU member state, where the vehicle is registered. The temperature data should be dated and stored for at least a year by a responsible person.

The Agreement on Transport of Perishables, the so-called ATP agreement [34], has been ratified by about 30 countries, mainly in Europe, but also by Russia, United States, and other countries. In cold transport between countries participating in ATP agreement, special equipment must be used, which should be inspected and tested for compliance with the standards in Annex 1, Appendices 1, 2, 3, and 4 [34]. In ATP, Annex 2, it is stated that "for the carriage of frozen and quick frozen foodstuffs, the transport equipment has to be selected and used in such a way that during carriage the highest temperature of the foodstuff in any point of the load does not exceed −20°C (for ice cream), −18°C (for quick frozen food, frozen fish, etc.), −12°C (for all frozen foodstuffs, except butter), and −10°C (for butter) [34]."

The ATP agreement includes precise and strict requirements on the technical properties of transport equipment (quality of insulation, construction, etc.). In most EU countries, these rules are not enforced, allowing for the transport and distribution of frozen foods to occasionally take place in unsuitable equipment, that is, inadequate insulation, insufficient cooling capacity of the refrigeration machinery, etc. In France, however, ATP-certified equipment must be used for the transport of frozen foods, prescribing the exact ATP category for different groups of foods.

The U.S. Code of Recommended Practices [35] suggests that temperature should be measured in an appropriate place and recorded in vehicles used for frozen food transport.

According to a definition assigned by the U.K. authority, local distribution is the part of the distribution chain in which the product is delivered to the point of retail sale, including sale to a catering establishment [9]. In France, local distribution is limited to 8 h, and the U.S. Code of Practice recommends that a frozen food measured with a temperature above −12°C should be rejected, or, at least, examined for acceptable quality prior to being offered for sale.

Finally, one should not overlook the fact that one of the weakest links in the distribution chain is the transport period from the product purchase to the consumer's domestic freezer. When this time period is not part of the thawing process, meaning that the product will not be immediately consumed but it will be stored in home freezer, the effect of this time might be significant for product quality and wholesomeness. According to the results of a consumer survey conducted in Greece, 26% of people need more than 20 min to carry food from the point of purchase to home freezer, with a 2% exceeding 45 min. Considering the usual temperatures during summer months (>32°C), this temperature abuse might lead frozen food to significant thawing, and consequently to major deterioration.

3. Retail Display

In current practice, several types of cabinets are used, such as: (a) vertical multideck with or without glass doors, using refrigerated air circulated by fans throughout the cabinet and (b) open top cabinets, which lower food temperature by forced air circulation, and natural convection. A common display cabinet consists of a thermally insulated body that will bear the food load and the cooling equipment. The refrigeration unit may be totally within the cabinet (integral cabinet) or partially situated in a remote location, with only the heat-exchanging coils and the fan inside the cabinet.

Temperature conditions within the retail cabinets play a significant role in the product's final quality status, and several surveys published [36,37] show a wide variation in product temperatures, with a significant percentage (>20%) of recorded closed vertical freezers in Spain and Portugal exhibiting an average temperature higher than −12°C. Considering a common distribution scenario for any frozen food, it could be stated that an important percentage of the total quality degradation is due to the conditions experienced during its storage in retail freezers. Another important issue during this part of the freezer chain is the temperature fluctuations occurring due to automatic defrosting of equipment, variations within the cabinet depending on the location of products, consumer handling of products, and regular replenishment with new products.

As far as regulation is concerned, according to QFF EU Directive [Article 5.2(b)] "tolerances in the product temperature in accordance with good practice are permitted. These tolerances may reach 3°C (to a product temperature of −15°C), if and to the extent that the Member States so decide. The Member State shall select the temperature in the light of stock or product rotation in the retail trade. The Commission shall be informed of the measures taken" [32]. According to Directive 92/1, temperature recording is not mandatory and the temperature is measured by at least one clearly visible thermometer, which in open (gondola-type) cabinets must indicate the temperature of the return air at the load-line level [33].

Cabinets must be installed in positions having suitable climate conditions. Cooling equipment is certified by the manufacturer to comply with European Standard EN441 for a specific "climatic class" [2]. Direct exposure to sunlight and draughts must be avoided. The required cabinet performance will only be achieved if the ambient conditions are cooler and less humid than limits specified for the climatic class shown in the nameplate. Air conditioning is advisable, if proper conditions cannot be guaranteed. In the United States, display cabinets should have the capacity to maintain constantly a product temperature of −18°C or colder with the exception of the defrosting cycle and short periods of loading. For newly introduced equipment, there is a suggestion that it bears an audible or visual alarm, which will activate when refrigeration failure occurs, providing a rapid response to adverse temperature conditions.

In Australia, the temperature requirements for frozen foods in retail cabinets are set since 1983 by the Australian standard AS 1731, according to which a food business displaying frozen (and potentially hazardous) foods must ensure that the food remains frozen when displayed [2].

It is important that the cabinet is only loaded with products at −18°C or below, following the foodmaker's instructions. As discussed by Jul [11], a systematic stock rotation and maintenance of the first-in–first-out principle could contribute more to product quality than expensive cabinet modifications. As will be discussed later, an innovative stock management system, based on product actual time–temperature history (LSFO), could further optimize the distribution chain, minimizing the unacceptable products. Similarly, loading procedures and handling of products before stocking in freezer cabinets are points of potential improvement.

4. Home Storage

The last part of the freezer chain is the least studied stage of frozen distribution, probably due to difficulties in data collection, concerning temperature conditions in domestic freezer, consumer habits, and approximate storage periods before consumption. However, when addressing the quality issue of frozen foods from production to final consumption in an integrated and structured way, such a period should be included in the evaluation of quality losses in the freezer chain. In a survey conducted in 100 Greek households, with miniature dataloggers (COX TRACER, Belmont, NC), recording temperature every 10 min for a period of 10 days, it was found that of the households >10% operate at temperatures higher than −10°C [38]. Another important observation of this survey is that, in many cases, there was an important temperature fluctuation recorded throughout the day, possibly due to door opening, product replenishment, or inefficient refrigeration system, that allows wide temperature variations (even of ±3°C) (Figure 14.1). In Figure 14.1, the fluctuating performance of three representative domestic freezers is illustrated, ranging from unduly low temperatures to inadequate, elevated freezing temperatures.

5. Transfer Points

Transfer points, that is, points where frozen products are moved from a cold area to another, are known to be frequent black spots for temperature abuse and mishandling. At these points, the control of temperature conditions is frequently lost and there is a change of the responsible personnel, leading sometimes to severe violation of handling requirements. As Jul describes [11], a frequent occurrence is that a truck has to be emptied completely to gain access to a particular shipment due to ineffective loading. It is then almost certain that there will be an undue delay in placing the rest of the shipment back, and restoring the appropriate temperature conditions. There are also many cases where frozen products transported by sea are left on the pier due to delays, subjected possibly to abusive temperature conditions.

A first necessary step for minimizing or even eliminating the detrimental effects of these points is the identification, the recording, and the evaluation of the potential hazard that transfer points may represent. The personnel involved in this relay path should be trained to ensures the continuity of the freezer chain and a fast handover from one point to the other. Finally, a reporting system should be introduced so that any temperature abuse is reported to both ends of the commercial food chain, that is, the producer and the retailer to ensure that remedial action can be taken where necessary.

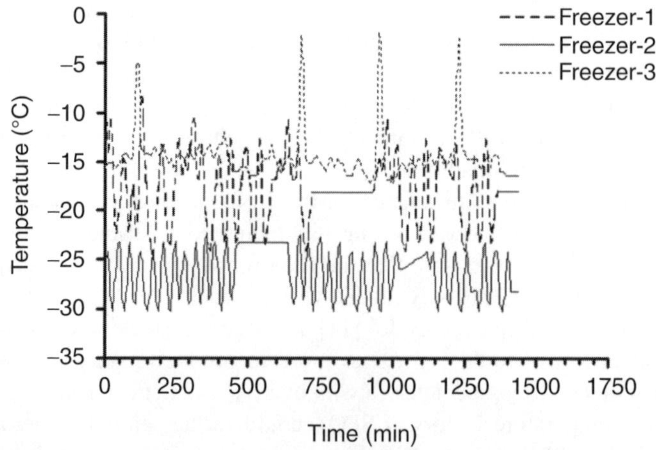

FIGURE 14.1 Temperature conditions recorded in three domestic freezers in a 10-day period.

B. Temperature Control

The sampling procedure that should be followed in the official control of the temperature of frozen foods is described in EU Directive 92/2, Annex 1 [39]. According to this Directive, the packages selected for temperature measurement shall be such that their temperature is representative of the warmest point of the load. For instance, in storage rooms, samples should be chosen from critical points, for example, near the doors, in the center of the room, and in the air returning to the refrigeration unit. During transport and local distribution, four samples must be selected from the following critical points: top and bottom of the load near the doors, top rear points of the load (farthest away from the refrigeration unit), center of the load, and center of the front surface of the load (closest to the return air intake of the refrigeration unit).

In ATP agreement, Annex 2, Appendix 2 [34], the procedure for the sampling and measurement of temperatures for carriage of frozen and quick frozen perishable foodstuffs is described in a similar way to Directive 92/2, omitting however the word "critical" from the text. According to ATP agreement, when a load has been selected for temperature control, a nondestructive measurement (between-case or between-pack) should be used at first. Only when the results do not conform to the prescribed temperatures of the temperature-monitoring devices (taking into account allowable tolerances), destructive product measurements should be carried out.

Directive 92/2, Annex 2 [39] describes also the reference destructive method to be used in the official control, providing specifications about the temperature-measuring instruments.

C. Equipment for Temperature Monitoring

To obtain an efficient record of the product history, temperature of both the food and its surroundings should be monitored. Additionally, multiple measurements at different locations should be taken, in case of a large batch or varying conditions in the chamber. Measurements can be realized either by a mechanical or an electronic equipment, with or without the potential of recording and maintaining an electronic file of data.

1. Sensors

The three principal types of sensors commercially available are thermocouples, platinum resistance, and semiconductors (thermistor). The choice depends on requirements for accuracy, speed of response, range of temperatures to be monitored, robustness, and cost [31]. The predominant types of thermocouples are of type K (with nickel–chromium and nickel–aluminum alloy wires) and type T (with copper–nickel alloy). The main advantages of the thermocouples are their low cost, facility to be hand-prepared, and a very wide range of temperatures measured (from -184 to 1600°C). Errors in the use of thermocouples are due to induced voltage from motors or transmitters, moisture and thermal gradients in other junctions, and can be increased when the ambient temperature varies widely. Each sensor and instrument used for monitoring throughout the cold chain has to be frequently checked to ensure that it meets the specification and obtains an accuracy within preset tolerances. This is reliably achieved by calibration equipment that measures the sensor temperature for a range of applied temperatures and based on the results, allows for correcting the actual reading.

2. Read Out and Recording Systems

In this category, the most common device is the electronic digital readout instrument, which is powered by batteries and allows for storing and printing out, or even an alarm notification when the temperature goes outside a preset limit. The recent miniaturization of circuit systems has produced some compact and powerful data logging systems, which can potentially follow the food, within the food case or pallet throughout all stages of the cold chain (e.g., COX Tracer, Cox

Recorders, Belmont, NC; "Diligence," Comark, Hertfordshire, England; KoolWatch, Cold Chain Technologies, MA, U.S.A.; DL200-T, Telatemp, California, U.S.A.; i-Button, Dallas semiconductor, Maxim, Texas, U.S.A.; Dickson TK-500, Dickson Addison, IL). Woolfe [31] reports another type of data logger recently developed (called "electronic chicken") that monitors the display cabinet. This type of logger is placed on a shelf and records temperatures from a food simulant included within the logger, which has the same thermal properties as the food-target displayed on the retail shelf. The device is equipped with an alarm light that instantly notifies any case of abuse. Development in this area is oriented in further decrease of the data logger size, to have the opportunity to monitor the actual temperature of foods by placing the logger between food packs.

3. Temperature Monitoring with Time–Temperature Integrators

An alternative, cost-effective way to individually monitor the temperature conditions of food products throughout the chain is the use of time–temperature integrators (TTIs). The essence in TTI application lies in the use of a physicochemical mechanism and a measurable change to display (a) the current temperature, (b) the cross-over of a preset temperature, or (c) the integrated time–temperature history of the frozen food. TTI complying with the third requirement are devices with an easily measurable response that reflect the accumulated time–temperature history of the product on which they are attached. Their operation is based on irreversible reactions that are initiated at the time of their activation and proceed with an increasing rate as temperature is elevated in a manner that resembles the temperature dependence of most quality loss reactions of foods [29,40]. These devices are attached on the food itself or outside the packaging and actually follow the food during its circuit from manufacturer to final consumption. The ultimate purpose of their application is the "translation" of their reading to the quality status of the food through the appropriate algorithm as discussed previously. Such TTI devices are MonitorMark, FreezeWatch (3M, StPaul, Minnesota), Fresh-Check® (Temptime, previously Lifelines, Morristown, NJ), ColdMark (Cold Chain Technologies, MA, U.S.A.), VITSAB (Malmo, Sweden), and Freshpoint (Tel Aviv, Israel).

Frozen foods had been a primary target of TTI application at the early stages of development, and several studies correlating frozen food quality with TTI were published [41–48]. Due to difficulties related to kinetic modeling of frozen foods and response of TTI in the subfreezing range, there was a switch of focus to application of TTI to chilled foods [49,50], and only a few studies tested TTI at the temperature range of interest for frozen storage [23,51].

III. TEMPERATURE EFFECT ON SHELF-LIFE

The shelf-life of a frozen food is a complex concept that depends on the characteristics of the food product and the environmental conditions to which the food is exposed after being subjected to the freezing process. For frozen foods which are usually distributed packaged, the packaging also plays an important role in maintaining the quality and shelf-life of foods by serving as an integral part of the preservation system applied [52]. When addressing shelf-life of frozen food, the International Institute of Refrigeration (IIR) has recommended two different definitions: (1) high-quality life (HQL) is the time from freezing of the product to the development of a just noticeable sensory difference (70–80% correct answers in a triangular sensory test) and (2) practical storage life (PSL) is the period of proper frozen storage after freezing of an initially high-quality product during which the organoleptic quality remains suitable for consumption or for the process intended [53]. Just like any other foods, frozen foods deteriorate during storage by different mechanisms, mostly affecting quality of frozen foods, because in the low temperatures of proper frozen storage, microbes cannot grow [5].

A. Quality Modeling of Frozen Foods

Quality is a dynamic, complex attribute of food, which influences the degree of its acceptability by the consumer, and is postprocessing constantly moving toward lower levels [53]. Kinetics of chemical reactions, microbiological, or physical phenomena that lead to quality changes play an important role for food shelf-life [54]. Through a careful study of the food components and the process, the change of food quality can be represented by the loss of one or more quantifiable quality indices by the formation of an undesirable product. The order of a chemical reaction is defined by the following equation, which describes the change in concentration C of a chemical compound (or in a more generalized sense, a representative quality factor) during a chemical reaction at time t:

$$-\frac{\mathrm{d}C}{\mathrm{d}t} = kC^n \tag{14.1}$$

where k is the reaction rate constant and n the apparent order of the reaction. The use of the term "apparent" indicates that Equation (14.1) does not necessarily describe the mechanism of the measured phenomenon. The reaction rate constant k is a measure of the reactivity and shows the quality loss of the frozen food [55]. A general equation describing the loss of the quality factor C in a food system may be expressed as

$$f_{\mathrm{q}}(C) = k(I_i, E_j) \cdot t \tag{14.2}$$

where f_{q} is defined as the quality function of the food and k the apparent reaction rate constant is a function of intrinsic factors I_i, such as concentration of reactive compounds, inorganic catalysts, enzymes, reaction inhibitors, pH, water activity, as well as microbial populations, and of environmental factors E_j, such as temperature, relative humidity, total pressure and partial pressure of different gases, light and mechanical stresses. The methodology for the determination of the apparent reaction order and reaction rate constant is described by Taoukis et al. [53].

The value of the quality index C_{θ_s} that signals or corresponds to the limit of acceptability of the food can be translated to a value of the quality function, $f_{\mathrm{q}}(C_{\theta_s})$. The time to reach this value at specified conditions, that is, the shelf-life θ_s, is inversely proportional to the rate constant at these conditions:

$$\theta_{\mathrm{s}} = \frac{f_{\mathrm{q}}(C_{\theta_s})}{k} \tag{14.3}$$

Kinetic equations for shelf-life estimation are specific to the food studied and the environmental conditions used. Among the environmental factors considered, the one being invariably introduced in the shelf-life model is temperature. Scientific knowledge has established its predominant effect on post-processing reaction rates, emphasizing the impact of abusive or fluctuating conditions. Additionally, temperature of frozen food cannot be controlled by means of initial processing and food packaging during subsequent handling, distribution, and storage and therefore high-quality preservation cannot be ensured throughout the cold chain.

The temperature dependence of reactions and quality loss rates is often described by the following Arrhenius-type relation, derived from thermodynamic laws and statistical mechanics principles [56]. The Arrhenius relationship is the most frequently used equation to model the temperature dependence of various quality changes in foods:

$$k = k_{\mathrm{o}} \exp\left(\frac{-E_{\mathrm{A}}}{RT}\right) \tag{14.4}$$

where k_0 is the frequency factor and E_A in J/mol is defined as the activation energy, that is, the excess energy barrier that quality parameter C needs to overcome to proceed to degradation products. R is the universal gas constant (8.3144 J/mol K). To estimate the effect of temperature on the reaction rate of a specific quality deterioration mode, values of k are estimated at different temperatures in the range of interest and ln(k) is plotted against $1/T$ in a semilog graph. A straight line is obtained with a slope of $-E_A/R$ from which the activation energy is calculated.

Instead of using the parameter k_0 in the Arrhenius equation, which is of no practical interest, the use of a reference temperature T_{ref} is alternatively recommended, corresponding to a representative value in the temperature range of the process or storage of study. Equation (14.4) is then mathematically transformed as follows:

$$k = k_{ref} \exp\left[\frac{-E_A}{R}\left(\frac{1}{T} - \frac{1}{T_{ref}}\right)\right] \quad (14.5)$$

where k_{ref} is the rate constant at the reference temperature T_{ref} ($T_{ref} = -18$ or -20°C for frozen foods).

In the recent literature, there are numerous shelf-life models based on the Arrhenius equation that describes the temperature dependence of several chemical or microbiological reactions in foods [57,58]. However, due to difficulties encountered at the low temperatures of frozen foods, there are few cases reported that undertake a full kinetic study of quality degradation in a wide subfreezing zone, including temperatures of possible abusive handling in the cold chain [18,59]. Additionally, there are few studies that validate the developed shelf-life models under fluctuating temperature conditions within the temperature range of interest (−5 to −30°C) to be able to predict quality loss in the real distribution chain [38,51].

An empirical approach in studies of temperature-dependent kinetics of quality loss is through Q_{10} concept, a tool of practical importance to the food industry. Q_{10}, which has been used in the early food science and biochemistry literature, is the ratio of the reaction rate constants at temperatures differing by 10°C or, equivalently, it shows the reduction of shelf-life θ_s when the food is stored at a temperature higher by 10°C:

$$Q_{10} = \frac{k_{(T+10)}}{k_{(T)}} = \frac{\theta_s(T)}{\theta_s(T+10)} \quad (14.6)$$

The Q_{10} approach in essence introduces a temperature dependence equation in the form of the following form:

$$k(T) = k_o\, e^{bT} \Longrightarrow \ln k = \ln k_o + bT \quad (14.7)$$

which implies that if ln(k) is plotted against temperature (instead of $1/T$ of the Arrhenius equation), a straight line is obtained. Alternatively, shelf-life (θ_s) can be plotted against temperature in the following equation:

$$\theta_s(T) = \theta_{so}\, e^{-bT} \Longrightarrow \ln(\theta_s) = \ln(\theta_{so}) - bT \quad (14.8)$$

where the outcoming plots are often called shelf-life plots, where b is the slope of the shelf-life plot and θ_{so} the intercept. Shelf-life plots are practical and easier to understand as one can read directly the shelf-life of the food at any storage temperature. These plots are true straight lines only for narrow temperature ranges of 10–20°C. Within this interval, Q_{10} and b are functions of temperature, correlated to the activation energy of the food quality deterioration reaction, given as:

$$\ln Q_{10} = 10b = \frac{E_A}{R} \frac{10}{T(T+10)} \tag{14.9}$$

Despite the extended application of the Arrhenius law, there are factors related to frozen food structure and characteristics that can cause significant deviations from an Arrhenius-like behavior [60]. In frozen foods, the freeze-concentration effect is predominant on the reaction rate in the immediate subfreezing temperature range and the observed rate increase is especially notable for reactants of low initial concentration. Consequently, the Arrhenius plot would show an abrupt change in this range and a single Arrhenius line should not be used in case the freezing point is crossed within the temperature range studied [53].

The effects of storage temperature on food stability of frozen foods is of prime importance due to the impact on deteriorating reaction rates that result in loss of nutritional and other quality characteristics [61,62]. In the recent research concerning the long-term stability of frozen foods, focus has also been directed to glass transition phenomena, found to affect the physical properties, the translational mobility, and consequently the stability of the frozen matrix [63]. Glass transition is related to dramatic changes of food mechanical properties and molecular mobility and may occur in carbohydrate-containing foods when storage conditions are suddenly modified, such as during rapid cooling or solvent removal. The focal point of this "glass-dynamics" approach is that at temperatures below the glass transition temperature (T_g or T_g'), the extremely high viscosity of the glassy state hinders the mobility and the diffusion of water molecules, leading to significant improvement of food stability [61,64–66]. So, in such systems, due to drastic acceleration of the diffusion-controlled reactions above T_g, the dependence of the rate of a food reaction on temperature cannot be described by a single Arrhenius equation.

It has also been proposed that frozen food stability, particularly the effect of temperature on reaction rates of the unfrozen phase can be described by an alternative equation, the following Williams–Landel–Ferry (WLF) expression that empirically models the temperature dependence of mechanical and dielectric relaxations in the range $T_g < T < T_g + 100$:

$$\log\left(\frac{k_{ref}}{k}\right) = \frac{C_1(T - T_{ref})}{C_2 + (T - T_{ref})} \tag{14.10}$$

where k_{ref} is the rate constant at the reference temperature T_{ref} ($T_{ref} > T_g$) and C_1 and C_2 are system-dependent coefficients. Williams et al. [67], assuming $T_{ref} = T_g$ and applying WLF equation for data available for various polymers, estimated mean values of the coefficients $C_1 = -17.44$ and $C_2 = 51.6$. However, the uniform application of these constants is often problematic [68–70] and the calculation of system-specific values, whenever possible should be preferred.

Besides questions of theoretical validity of the Arrhenius equation in wide temperature ranges that include phase transition phenomena, most notably the frozen range, cautious application even on empirical basis and within well-defined temperature limits of practical significance serves as a useful tool for shelf-life calculations and predictions. The Arrhenius parameters such as the E_A value give a well-comprehended measure of temperature dependence comparable to the respective ample information existing from kinetic modeling at most food systems in the frozen and nonfrozen temperature ranges [54,71].

B. Microbiology of Frozen Foods

Spoilage of food can be considered as an ecological phenomenon that encompasses the changes of the substrata available (e.g., low-molecular-weight compounds), during the proliferation of the bacterial population that consists of the microbial association of the stored food products.

The prevailing of a particular microbial association of products depends on factors that persist during processing, transportation, and storage in the market. It is well established that in food systems five categories of ecological determinants (e.g., intrinsic, extrinsic, processing, implicit, and emergent factors) influence the establishment of the particular microbial association and determine the rate of attainment of a climax population, the so-called "specific spoilage microorganisms" (SSO) although their domination is rather ephemeral and characterizes only a particular processing condition [72]. These microorganisms survive or even grow because of the variety of ecological strategies that they are able to adopt [73].

Among the factors mentioned earlier, the extrinsic factors, that is, temperature, relative humidity, and the composition of the gaseous atmosphere obtained during distribution and storage, seems to contribute significantly to the selection of the particular (SSO) microbial flora [74]. Specifically the temperature is of great importance. For example, freezer temperatures can be used as the only hurdle. The key problem in the frozen products is the enumeration of microbial population of such ecosystems. It is widely recognized that microorganisms are injured by exposure to reduced temperatures, leading to sublethal damage in microbial populations. The effects of sublethal damage in microbial populations include (a) increased lag times and (b) inability to develop quantitatively on selective media that do not exert any inhibitory effect on undamaged populations of the same taxon. This phenomenon, and especially the prolonged lag phase, is less noticeable when the food ecosystem is refrozen and again analyzed after a short period of storage. Studies on the effect of different environmental stresses (among these freezing) including on the enumeration and the recovery of microorganisms is centered, however, on pathogenic microorganisms [75]. In this case, the important feature is to ascertain the presence or absence of the pathogenic bacterium. The importance of the results obtained obviously has a cardinal role in the evaluation of microbiological hazards.

It is important to note that, except for the resuscitation of microorganisms using artificial laboratory ecosystems prior to the enumeration in the final ecosystem, resuscitation of the injured flora may take place in the natural ecosystem. This occurs after the removal of the determinant which imposed the damage to the microbial population. It is evident from the above that the population of a food ecosystem and, in particular, its enumeration is a matter of creating appropriate laboratory ecosystems in which the selected determinants applied will contribute to the recovery of the natural one present therein. The aforementioned feature was suggested by the observation that when the ecosystem was exposed to conditions for microbial growth, after a long lag phase (2–3 days) during which resuscitation took place, the pattern of its subsequent decomposition was analogous to that of ecosystems with an uninjured flora.

A large number of microorganisms have been reported to grow at or below 0°C, especially yeasts and molds, rather than bacteria. However, food-related bacteria have been reported to grow at −20°C and around −12°C [76]. The lowest recorded temperature of growth for a food-related microorganism is −34°C for a pink yeast. Growth of *Vibrio* spp. has been reported at −5°C, of *Yersinia enterocolitica* at −2°C, of *Brochothrix thermospacta* at −0.8°C and of *Aeromonas hydrophila* at −0.5°C.

Microbial growth at and below freezing temperatures depends on nutrient content, pH, and the availability of liquid water [77]. Foods that are likely to suffer from microbial degradation at subfreezing temperatures include fruit juice concentrates, bacon, ice cream, and certain fruits.

Studies of microbial growth at subfreezing temperatures clearly indicate that microbial growth does not occur in food ecosystems with a temperature of ≤−8°C. Freezing temperatures have been shown to influence the killing of certain microorganisms, with cocci being generally more resistant than gram-negative rods. Of the food pathogens, *Salmonellae* are less resistant than *Staphylococcus aureus* or vegetative cells of *Clostridia*, whereas endospores and food-poisoning toxins are apparently unaffected by low temperatures. In any case, freezing and subsequent frozen storage should not be considered as a microbial destructive preservation method. Frozen foods are assigned a specific shelf-life that is mostly based on texture, flavor, tenderness, color, and nutritional quality requirements rather than microbiological issues [77].

Thus, the main determinant for the storage period of a properly frozen food ecosystem is the physical, chemical, or biochemical changes that are unrelated to microbiological proliferation [72].

C. Physicochemical Indices of Quality of Frozen Food

The principal physical change that occurs in frozen foods is moisture migration, that can be manifested as moisture loss by sublimation, moisture absorption and redistribution in foods, ice recrystallization, or drip loss during thawing [78]. An important problem frequently encountered during frozen storage is desiccation, causing a significant weight loss and affecting consequently the manufacturer profitability [16]. Fluctuations in storage temperatures magnify moisture migration, having in some cases a serious effect on product appearance known as "freezer burn." The inability to sustain a continuous, adequately cold distribution chain also leads to detrimental ice recrystallization, which signals loss of quality for the frozen product.

Another important problem associated with frozen storage is protein denaturation that has been attributed to the formation of ice crystals, dehydration, and concentration of solutes in the tissue or protein solution. Freeze-induced protein denaturation, subsequent enzyme inactivation, and related functionality losses are frequently observed in frozen fish, meat, poultry, egg products, and doughs [79].

In contrast, frozen storage at low temperatures, though effective in adequately controlling microbial spoilage, can slow down but does not inactivate tissue enzymes. Many undesirable changes in frozen foods are due to enzymatic activities that, despite being retarded due to low temperatures, are also strongly influenced by the freeze-concentration. For example, in meat products, hydrolytic rancidity, textural softening, color loss, or acceleration in lipid oxidation are consequences of hydrolytic enzyme activity [80]. Kinetics of nonenzymatic reactions are also influenced by the freeze-concentration phenomenon, which tends to override the stabilizing effect of low temperatures [62].

During frozen storage, numerous chemical changes can take place, leading to product deterioration. A large number of studies has shown the effect of constant or fluctuating low-temperature storage on lipid oxidation [16,81,82], enzymatic browning [51,80,83] and degradation of chlorophyll, other pigments, and vitamins [19,23]. The rate enhancement of these reactions due to freeze-concentration cause irreversible nutritional and sensory damage to the product. Ascorbic acid loss in fruits and vegetables, formation of rancid off-flavors in frozen meats and fish, and color loss are some of the direct consequences of frozen storage, which are of major concern to both the consumer and the manufacturer. Singh and Wang [16] point out that the important role of ascorbic acid due to its protective role for the other nutrients is frequently considered as a representative index of quality loss for frozen fruits and vegetables.

Temperature fluctuations, especially when they involve unduly high temperatures, can produce cumulative adverse effects on frozen food quality. Singh and Wang [16] summarize the conclusions of studies on the effect of nonisothermal storage, pointing out the significance of fluctuations of the order of 10°C that usually occur at loading or transfer points of the cold chain. The negative effect is mostly pronounced in physical properties of the products (moisture loss and desiccation), whereas chemical properties may not be significantly damaged.

D. Shelf-Life Prediction under Fluctuating Temperature Conditions

A thorough kinetic study of the quality deterioration of a frozen food requires as the first step the selection of a representative quality index and the monitoring of its change during isothermal frozen storage in a wide range of subfreezing temperatures. Then, the temperature dependence of the quality loss function is estimated through the Arrhenius equation and the obtained model is validated under nonisothermal conditions. The use of this model would finally allow for a reliable estimation of the quality loss of the product in question at time–temperature conditions that differ from the experimental ones. This methodology has been effectively used in some studies [23,38].

Considering the current cold chain as discussed earlier, frozen food products are bound to be exposed to a variable temperature environment that does not infrequently include stages of abusive storage or handling conditions. In a general form, the value of the quality function [Equation (14.2)] at time t is calculated by the following integral, with $T(t)$ describing the change of temperature as a function of time:

$$f_q(C) = \int k[T(t)]\,dt \tag{14.11}$$

To represent the integrated effect of the temperature variability on product quality degradation, the term of the effective temperature T_{eff} can be introduced. T_{eff} is defined as the constant temperature that results in the same quality value as the variable temperature distribution over the same time period. This approach that equals the overall effect of a nonisothermal handling with a single, constant value simplifies Equation (14.11) to the following expression:

$$f_q(C) = \int_0^{t_{tot}} k(T(t))\,dt = k_{eff}t_{tot} \tag{14.12}$$

where k_{eff} is the value of the rate of the quality loss reaction at the effective temperature. If the $T(t)$ function can be described by a step sequence, or equivalently can be discretized in small time increments t_i of constant temperature T_i (with $\Sigma t_i = t_{tot}$), then Equation (14.12) is modified assuming the applicability of Arrhenius equation:

$$k_{ref}\sum_i\left(\exp\left[-\frac{E_A}{R}\left(\frac{1}{T_i}-\frac{1}{T_{ref}}\right)\right]t_i\right) = k_{eff}t_{tot} \tag{14.13}$$

From Equation (14.13), the value of k_{eff} can be estimated, and subsequently from the Arrhenius model of Equation (14.5), the effective temperature T_{eff} can be calculated.

Based on the aforementioned approach, in the real distribution chain of frozen products that include several stages of storage, transport, and handling, one can estimate the extent of quality loss of a product when its quality function and its time–temperature history are known.

To calculate the fraction of shelf-life consumed at the end of each stage f_{con}, the time–temperature–tolerance (TTT) approach [5,15] can equivalently be used. According to this methodology, the f_{con} is estimated as the sum of the times at each constant temperature segment t_i divided by the shelf-life at that particular temperature θ_I, that is,

$$f_{con} = \sum_i \frac{t_i}{\theta_i} \tag{14.14}$$

where index i represents the different time–temperature steps within the particular stage of study. The remaining shelf-life of products can be calculated at a reference temperature, representative of their storage conditions after each stage as $(1 - \Sigma f_{con})\theta$, where θ is the shelf-life at that reference temperature.

This approach is seeking to use the developed validated kinetic models of quality loss of frozen food during the real, nonisothermal handling of products, mimicking the actual distribution of frozen foods [23]. As an indicative case study, the quality loss of frozen peas was assessed, described adequately by green color change, and L-ascorbic oxidation of frozen vegetables [51] by the following equations. The kinetic results of the use of the Arrhenius equation in the range between −3 and −20°C are shown in Table 14.1 and the shelf-life, based either on nutritional (L-ascorbic acid loss) or sensory (color degradation) criteria is estimated in the range of interest

TABLE 14.1
Arrhenius Parameters, Shelf-Life Endpoint, Statistics, and Q_{10} Value for the Quality Loss of Frozen Green Peas

	Kinetic Parameters	
	L-Ascorbic Acid loss	**Color Loss**
Apparent reaction order	First-order	Zero-order
E_A (kJ/mol)	136.8 ± 20.5[a]	79.2 ± 19.2
k_{ref} (d^{-1}) ($T_{ref} = -18°C$)	0.00102	0.0269
R^2	0.993	0.983
Endpoint of shelf-life[b]	$\left(\frac{[L-asc]}{[L-asc]_0}\right) = 60\%$	$\Delta C = 10$
Q_{10} (in the range −15 to −5°C)	10.8	4.0

[a] 95% confidence intervals based on the statistical variation of the kinetic parameters of the Arrhenius model (regression analysis).

[b] The endpoint is based on nutritional criteria (L-ascorbic acid), or on sensory tests for color acceptability (ΔC reduction)

in the shelf-life plot shown in Figure 14.2.

$$C_L = C_{L_0}\, e^{-k_L t} \Longrightarrow \ln\left(\frac{C_L}{C_{L_0}}\right) = -k_L t \quad (14.15a)$$

$$\Delta C = \sqrt{(a - a_0)^2 + (b - b_0)^2} = k_C t \quad (14.15b)$$

where C_L and C_{L_0} are the concentrations of L-ascorbic acid at time t and zero, ΔC describes the *chroma* change, estimated by the CIELab values (subscript 0 refers to time zero), and k_C and k_L are the apparent reaction rates of color loss and L-ascorbic acid oxidation, respectively.

Similar to the methodology used for frozen green peas shelf-life study, the gradual change of color of frozen slices of white mushroom during frozen storage was studied and the results are summarized in Table 14.2. As stated by Giannakourou and Taoukis [51], the main deterioration mode

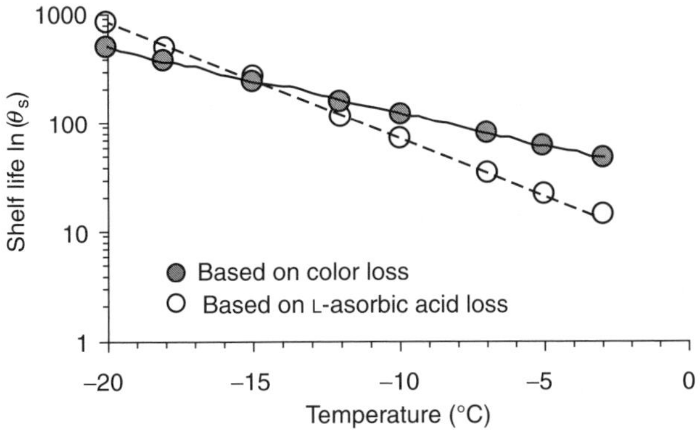

FIGURE 14.2 Shelf-life plot for frozen green peas, based on sensory ($\Delta C_{final} = 10$) or nutritional (40% L-ascorbic acid loss) criteria.

TABLE 14.2
Arrhenius Parameters, Statistics and Q_{10} Value, and Shelf-Life at Four Temperatures in the Frozen Storage Range for the Lightness Loss of frozen Mushrooms

	Kinetic Parameters
	Lightness loss
E_A (kJ/mol)	155.2 ± 60.3[a]
k_{ref} (d^{-1}) ($T_{ref} = -18°C$)	9.9×10^{-4}
R^2	0.957
Q_{10} (in the range −15 to −5°C)	14.8
Temperature (°C)	Shelf-life (days)[b]
−5	10
−10	39
−15	153
−20	637

[a]95% confidence intervals based on the statistical variation of the kinetic parameters of the Arrhenius model (regression analysis).

[b]Shelf-life is based on 70% loss of the initial white color (L_0), set by a sensory panel.

observed is related to the loss of the initial bright white color expressed by the change of the *L*-parameter (CIE Lab scale), which is found to follow a first-order reaction:

$$L = L_0\, e^{-k_m t} \Longrightarrow \ln\left(\frac{L}{L_0}\right) = -k_m t \tag{14.16}$$

where L and L_0 are values of color parameter L at time t and zero and k_m the apparent reaction rates of color lightness.

Assuming a dynamic temperature profile for frozen peas, it is possible to calculate the fraction of shelf-life consumed at any point of their lifecycle, that is, f_{con} in Equation (14.14). Real time–temperature scenarios, including all stages of the cold chain of a commercial frozen vegetable were assumed. These profiles were actually selected from a survey of frozen vegetables handling throughout the distribution, realized in the context of a large field test where 100 packages of frozen green peas and frozen mushrooms with attached dataloggers followed the whole marketing route [38]. For each of these scenarios representing an alternative path of frozen peas distribution of total duration of 135 days, f_{con} was estimated based on the real time–temperature history of the product and the shelf-life kinetics and compared to the f_{con} "expected" when the $T(t)$ profile of the product is presumed to be at a constant temperature of −18°C (food label declaration). In Figure 14.3, the application of this TTT approach is shown both for L-ascorbic and color loss of frozen green peas, for two different time–temperature scenarios illustrated in the interior of Figure 14.3. Comparing the value of f_{con} for L-ascorbic loss for the real dynamic temperature conditions (solid line in Figure 14.3) with the "presumed" f_{con} under "ideal" constant conditions of −18°C (dotted line), the importance of considering the actual temperature exposure of the food (usually quite different from the recommended) becomes evident.

If, however, ideal conditions are simplistically assumed for both modes of quality deterioration of frozen peas (L-ascorbic loss and color change), the error in f_{con} and subsequently in the remaining shelf-life prediction of each product at any point of the cold chain may significantly increase

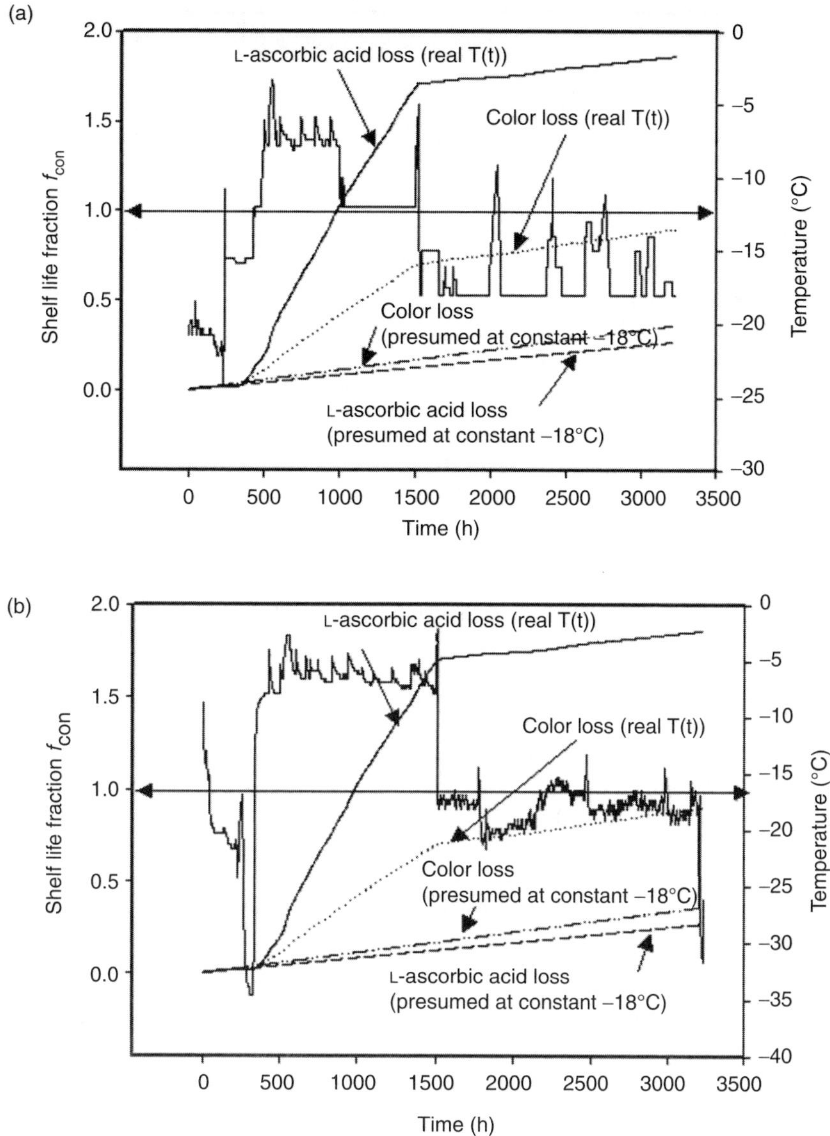

FIGURE 14.3 Fraction of remaining shelf-life (f_{con}) of frozen green peas, based on color loss ($\Delta C = 10$) and L-ascorbic acid loss (40% loss), estimated by the actual time–temperature profile throughout the illustrated distribution scenarios (a) and (b). The respective change of f_{con} is also shown, if an isothermal proper storage at $-18°C$ is assumed.

(Figure 14.3). Chill chain management decisions based on this assumption and ignoring real temperature history would be seriously ineffective.

If the same scenario of Figure 14.3a is applied for frozen mushrooms, the value of f_{con} estimated for the real dynamic temperature conditions is significantly higher than the corresponding "presumed" f_{con}, revealing a more severe deterioration than expected (Figure 14.4). In the case of color change of mushrooms, which is even more temperature-dependent (very high E_A value), the error introduced by the simplistic assumption of a proper handling under constant temperatures ("presumed" f_{con}) is even greater than that in the case of frozen green peas.

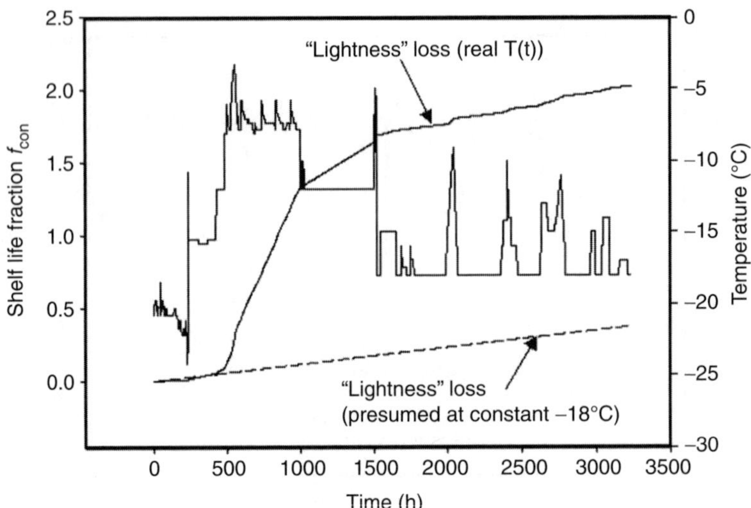

FIGURE 14.4 Fraction of remaining shelf-life (f_{con}) of frozen mushrooms, based on color loss (30% loss), estimated by the actual time–temperature profile, throughout the illustrated distribution scenario. The respective change of f_{con} is also shown, if an isothermal proper storage at −18°C is assumed.

E. Application of TTI as Monitoring and Prediction Tools

The basic principles of TTI modeling and application for quality monitoring are detailed in the literature [40,84,85]. Combining Equation (14.2) and Equation (14.5), loss of shelf-life of a food (based on the deterioration of the selected index C) can be expressed as

$$f_q(C) = k_{C_{ref}} \exp\left(-\frac{E_A}{R}\left(\frac{1}{T} - \frac{1}{T_{ref}}\right)\right)t \tag{14.17}$$

where E_A is the activation energy of the reaction that controls quality loss. Similar to Equation (14.17), a response function $F(X)$ can be defined for TTI such that $F(X) = k_I t$, with k_I an Arrhenius function of T.

The value of the functions $f_q(C)_t$ at time t after exposure at a known variable temperature exposure $T(t)$ can be found by Equation (14.12), when the term of the effective temperature T_{eff} is introduced. For a TTI exposed to the same temperature fluctuations $T(t)$ as the food product, and corresponding to an effective temperature T_{eff}, the response function can be similarly expressed as

$$\begin{aligned} F(X) &= k_{I_{ref}} \int_0^t \exp\left(-\frac{E_{A_I}}{R}\left(\frac{1}{T} - \frac{1}{T_{ref}}\right)\right) dt \\ &= k_{I_{ref}} \exp\left(-\frac{E_{A_I}}{R}\left(\frac{1}{T_{eff}} - \frac{1}{T_{ref}}\right)\right)t \end{aligned} \tag{14.18}$$

where $k_{I_{ref}}$ and E_{A_I} are the Arrhenius parameters of the TTI.

Thus, the basic elements for a TTI-based food quality monitoring scheme are (a) a well-established kinetic model to describe quality loss of the food, (b) the response function of the TTI, and (c) the temperature dependence of both food quality loss and TTI response rate expressed by the respective values of the activation energies. The essence in TTI implementation algorithm lies in the calculation of the T_{eff} of the exposure [Equation (14.18)], based on the TTI response reading (Figure 14.5), which is assumed to describe the integrated effect of temperature history on food quality loss. This assumption requires that food quality degradation and TTI response rate are similarly affected by temperature, that is, the activation energies of the two phenomena do not differ by more than 25 kJ/mol. Under these conditions, the application scheme would reliably provide the extent of the quality deterioration of the food and a prediction of the remaining shelf-life at any assumed average storage temperature.

To assess this application scheme for the distribution scenarios illustrated in Figure 14.3, the fraction of shelf-life consumed f_{con} or equivalently the remaining shelf-life is calculated by means of a TTI attached on the food, following the same marketing path in the cold chain. The principle of TTI use is based on the "translation" of its response, through the appropriate kinetic models [38,51] to the corresponding T_{eff}, and then to the value of the f_{con}. So, the prediction obtained by the indicator is compared with the "actual" f_{con} (illustrated as a solid line in Figure 14.3) based on the real time–temperature integral. For the case of frozen green peas, TTI of enzymatic type M2-21 (VITSAB AB, Malmo, Sweden) were used to "mimic" frozen peas loss of quality [51]. Their response studied under isothermal experiments and using the Arrhenius equation was found to have an activation energy of 92.2 $\pm$ 18.7 kJ/mol ($R^2 = 0.955$). The developed model was validated under nonisothermal conditions, so that it can be reliably applied in the real distribution chain. In the time–temperature scenario illustrated in Figure 14.3b, the remaining shelf-life of frozen green peas can be estimated at designated points of the cold chain, based on the actual time–temperature of products and the developed kinetic models for their quality loss, and compared with the prediction based on TTI reading. Table 14.3 shows the aforementioned comparison at times that refer to warehouse storage of frozen products (20 days after production), to stocking at a central distribution center (42 days after production), to retail display in commercial freezers (62 days after production), and to the endpoint of its lifecycle, before final consumption (135 days after production).

From Table 14.3, it is obvious that TTI prediction is adequately accurate, when color loss is considered as the main criterion of frozen green peas rejection. In contrast, if L-ascorbic loss signaled the end of shelf-life, TTI predictions would differ significantly from the "real" remaining shelf-life, demonstrating the importance of the requirement shown in Figure 14.5 [E_A(food) $\cong$ E_A(TTI)]. The latter, however, is a theoretical hypothesis because in the range of temperatures considered, the endpoint of color unacceptability ($\Delta C = 10$) is actually reached before 40% L-ascorbic acid loss. Therefore, chroma change is the criterion that sets the shelf-life limit for frozen peas.

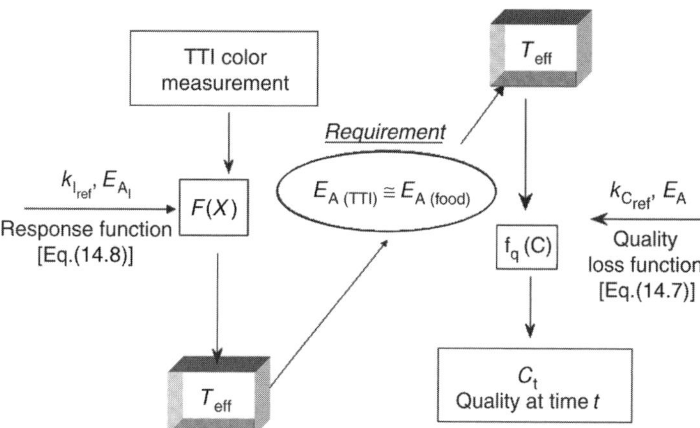

FIGURE 14.5 Application scheme of TTI as quality monitors and tools for predicting food remaining shelf-life. All kinetic data necessary as input for food quality loss and TTI response are also shown.

TABLE 14.3
Comparison of the Remaining Shelf-Life Calculated, Either Directly from the Real Time–Temperature History of Frozen Green Peas or Through the Attached TTI Response, Using the Developed Shelf-Life Models

Criterion of Quality Loss of Frozen Green Peas	Remaining Shelf-Life (*d* at −18°C)					
	Green Color Loss			L-Ascorbic Acid Loss		
Time of Measurement	Based on Real *T*(*t*)	Based on TTI[a] Reading	Presumed[b]	Based on Real *T*(*t*)	Based on TTI[a] Reading	Presumed[b]
20 days (industry warehouse stocking)	355	355	352	480	471	474
42 days (distribution center)	305	303	330	161	212	459
62 days (retail outlet)	246	242	309	67	121	438
135 days (domestic storage—end of lifecycle)	81	72	237	Rejected	38	366

Note: The presumed remaining shelf-life, based solely on the expiration date and the time spent in distribution was also calculated at the different stages of distribution.

[a]TTI of type M2-21 was used.

[b]The "presumed" remaining shelf-life is based on the assumption of uniform, proper handling throughout the cold chain (i.e., isothermal conditions of ≅ −18°C) and is calculated to be 372 days for color loss and 500 days for L-ascorbic acid loss.

Nevertheless, in any case, TTI predictions are of high practical value, offering a substantial improvement to the erroneous value of remaining shelf-life based on the expiration date label. Table 14.3 shows that the presumption of uniform, ideal handling leads to values that are very inaccurate in the real cases of fluctuating time–temperature conditions.

In a similar way, Table 14.4 illustrates the effective prediction of the remaining shelf-life of frozen mushroom by TTI at different points of distribution for the rotation scenario shown in Figure 14.4. The potential of TTI as monitors of frozen mushrooms handling and means for predicting their quality status is highlighted when compared with the grossly overestimated remaining shelf-life expected from the final consumer whose judgment is based merely on the expiration label. In the case of frozen mushroom, a different enzymatic TTI of type *L* was used, which was found to be more sensitive to temperature ($E_A = 160$ kJ/mol), following closely the behavior of white color of the frozen products.

IV. OPTIMIZED MANAGEMENT WITH TTI

The results in Table 14.3 and Table 14.4 showed the effectiveness of TTI as monitoring and controlling tools for the real distribution of frozen vegetables. TTI applicability was further validated through a controlled, representative field test for frozen green peas and mushrooms following the same methodology as for Figure 14.3 and Figure 14.4 and Table 14.3 and Table 14.4 [38]. Thus, TTI considered as temperature history recorders could be reliably used to indicate quality and remaining shelf-life, and potentially to introduce an optimized cold chain management system. This improved system, coded as LSFO, which has been previously assessed by Taoukis et al. [49] for chilled products has recently been evaluated for frozen vegetables [38].

This novel approach proposed as an alternative to FIFO policy is based on the assortment of products according to their quality status, as it is predicted by the attached TTI at designated points of

TABLE 14.4
Comparison of the Remaining Shelf-Life Calculated, Either Directly from the Real Time–Temperature History of Frozen Mushrooms or Through the Attached TTI Response, Using the Developed Shelf-Life Models

	Remaining Shelf-Life (d at −18°C)		
Time of Measurement	Based on Real $T(t)$	Based on TTI[a] Reading	"Presumed"[b]
20 days	327	327	340
25 days	200	198	335
33 days	38	37	327
35 days	1	Rejected (−1d)	325

[a]TTI, of type *L* was used.

[b]The "presumed" remaining shelf-life is based on the assumption of uniform, proper handling throughout the cold chain (i.e., isothermal conditions of ≅ −18°C) and is calculated to be 360 days for "lightness" loss.

Note: The presumed remaining shelf-life, based solely on the expiration date and the time spent in distribution was also calculated at the different stages of distribution.

their marketing route. The principles that lie behind this management system are illustrated in Figure 14.6, showing a case study where a batch of frozen green pea products arrives at the retail warehouse, where it is supposed to be split into three groups for successive stocking of the retail freezer cabinets, after 15, 30, and 45 days, respectively. TTI application allows for a classification based on real quality criteria (loss of green color) instead of a random split, according to the FIFO approach. With LSFO, products are advanced successively to the retail shelves (with a replenishment period of 15 days) according to their quality classification, sold and transported to the domestic freezers, where they would be stored for 45 days before final consumption. The distribution

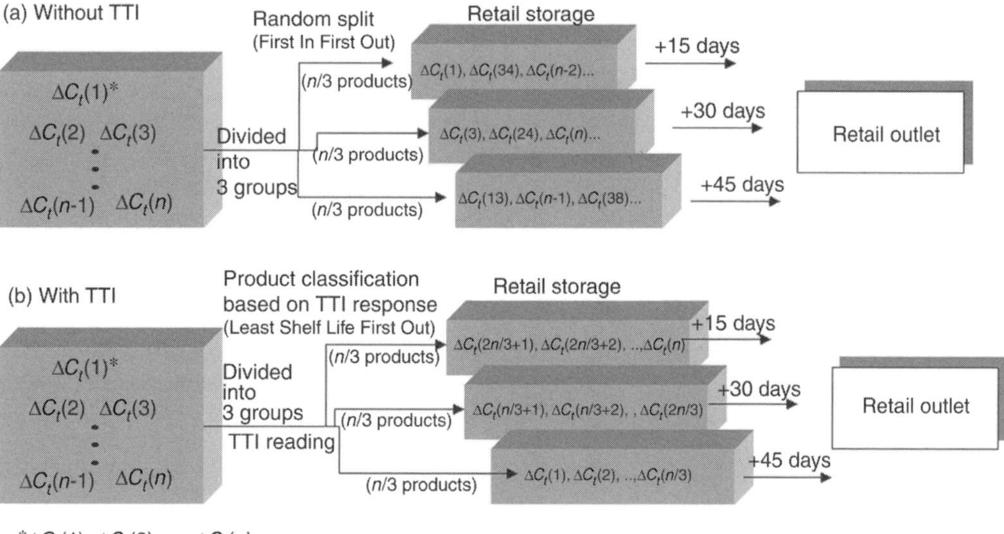

FIGURE 14.6 Schematic illustration of the decision-making principles of the TTI-based management system, at the point of retail display of frozen green pea products compared with the traditional FIFO practice.

considered in this case, shown in Figure 14.7, is based on the real data of the cold chain obtained from the extended field test in [38] and agrees with the structure of the food chain applied in Germany [86], representing therefore a realistic scenario for LSFO application. At the stage of retail display and domestic storage, temperature data from surveys illustrated in Figure 14.8 was used. The novelty in LSFO algorithm is that, at decision points, it classifies products and forwards the ones closer to their actual expiration. The key point in LSFO implementation is the estimation of each product time–temperature history through the reading of the attached TTI. This is accomplished by translating TTI response through the algorithm of Figure 14.5 to the respective temperature integral, or the corresponding T_{eff}. Overall, this system would optimize the current inventory management system, leading to products of more consistent quality and nutritional value at the time of consumption.

To validate the effectiveness of the proposed frozen food management system, the Monte Carlo numerical simulation technique was used [87], allowing for the study of numerous, alternative distribution scenarios. It is based on the repetitive simulation of the marketing route of frozen peas, with temperature taking each time a different value, out of a given distribution (Figure 14.8). Instead of unrealistically assuming a single-point, fixed estimate for temperatures at different stages throughout the cold chain, temperature variation (or uncertainty) is taken into account [29,41,87] to reflect the actual conditions in the frozen distribution. Eventually, this numerical procedure leads to a new distribution of the output of interest (remaining shelf-life based on color loss of frozen green peas), instead of a single-point estimate, as shown in Figure 14.3 and Figure 14.4, and Table 14.3 and Table 14.4, where a specific scenario was applied. Any point in the final frequency curve reflects the possibility of a product to have a certain value of remaining shelf-life ($\pm$30 days), or alternatively the percentage of products out of the same batch that are of the same quality. Negative values of remaining shelf-life correspond to products that have exceeded the limit of acceptability before reaching the time to final consumption.

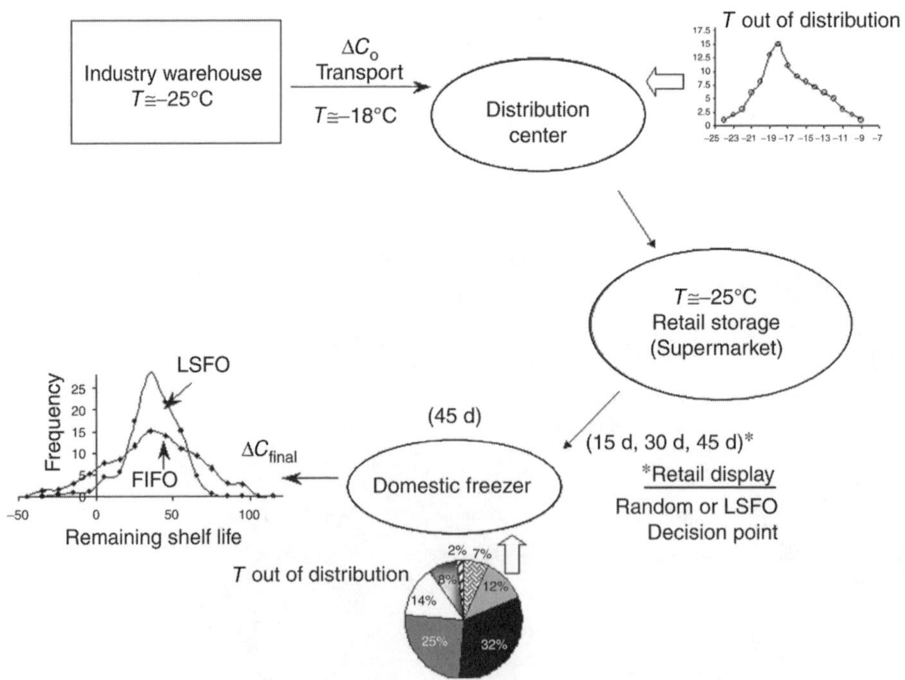

FIGURE 14.7 Schematic illustration of the marketing route of frozen green peas, assumed for the evaluation of LSFO vs. FIFO policies at the final consumption time. Temperature distributions and the LSFO decision point are also shown.

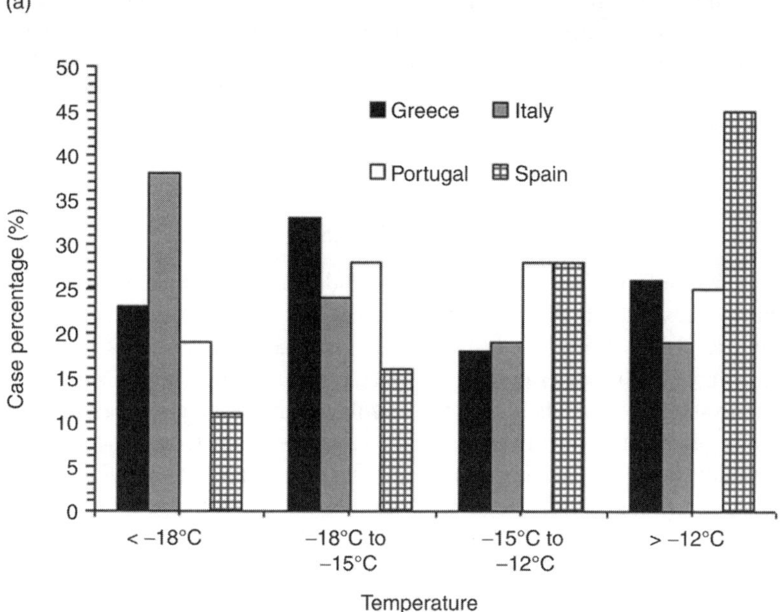

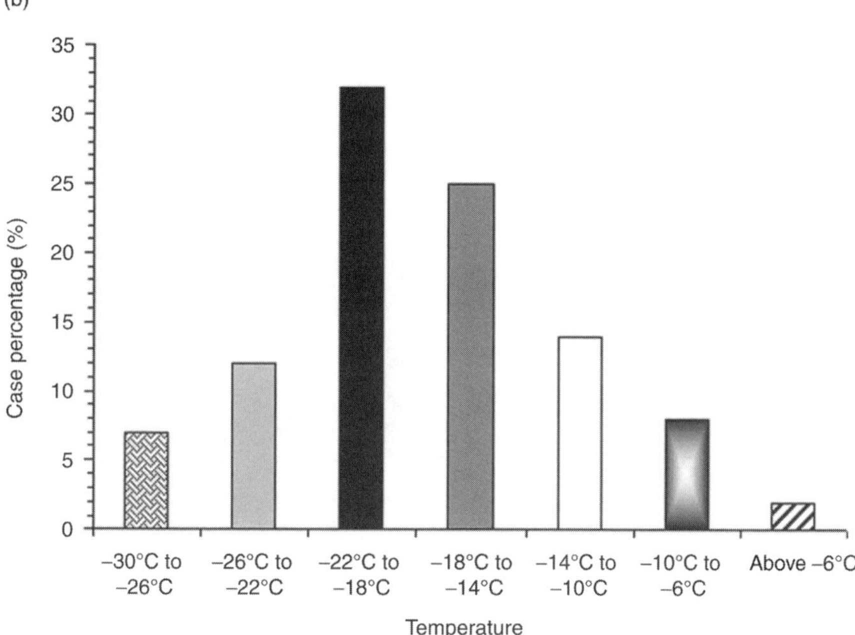

FIGURE 14.8 Temperature distribution in (a) retail display in open, horizontal commercial freezers of four Mediterranean countries. (From Anon. Final Report of the European Commission sponsored Research project in 4 EU member states (Greece, Italy, Spain and Portugal). Contract number: EC 1080/94/000069, 1995, 24 pp. With permission.) on frozen product temperature and (b) domestic freezers, from a survey in 100 households. (From MC Giannakourou, PS Taoukis. *Journal of Food Science* 68 (1):201–209, 2003. With permission.)

In our case study, the distribution management of frozen green peas based on LSFO instead of FIFO policy leads to a more consistent quality, reducing significantly the "tails" of the obtained distribution as clearly depicted in Figure 14.9. With FIFO system, 15.1% of products were beyond acceptable quality at the time of consumption, whereas LSFO implementation reduced this percentage to 5.8%.

V. CONCLUSIONS

In the distribution of frozen food, the conditions of the cold chain, that is, the deviations from the recommended temperatures, the observed fluctuations, and the possible "breakages" of the low-temperature continuity, play a decisive role for preserving food quality and assuring its safety. Recognizing the inevitable quality deterioration of frozen products during frozen storage, transport and handling, accurate shelf-life models for the whole subfreezing range are necessary to describe this quality loss. Additionally, the effect of temperature should be evaluated and quantified in terms of mathematical models, not only under isothermal frozen storage at low temperatures, but also under the dynamic, frequently fluctuating conditions of the real cold chain. Using such well-established, validated models, shelf-life prediction of frozen products is then possible at any point of its actual marketing route from the producer to the final consumer.

Recognizing that consumer acceptance is of prime concern for any food marketing policy, a reliable prediction of the end of shelf-life of frozen foods has become an economic decision-making process. So, to reliably predict the shelf-life, one needs to assess the integrated impact of temperature on the quality loss of the frozen product. For this purpose, suitable TTIs can be used as reliable monitoring and controlling tools for the real distribution of frozen foods indicating,

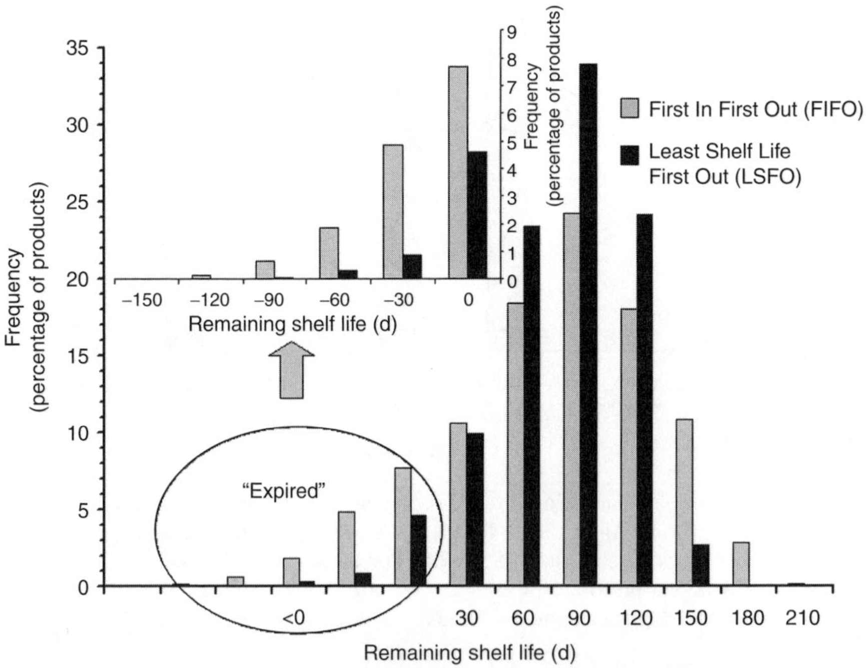

FIGURE 14.9 Distribution of quality of frozen green peas with FIFO and LSFO cold chain management, based on loss of the green color. Remaining shelf-life is the time the product would remain acceptable after the consumption time if isothermally stored at −18°C.

at any point of the cold chain, their real quality status. Based on TTI prediction, an optimized management system for frozen product distribution management and stock rotation, LSFO was developed as an alternative to the old-fashioned FIFO policy. The application of this "intelligent" management process of frozen stock was found to significantly improve quality at the point of consumption, by reducing products that have exceeded their shelf-life before reaching the consumer. TTI can be consequently considered as useful decision-support tools for the optimization of the current inventory management system to improve consumer's acceptability. With continuous improvement, this stock-rotation policy could encompass information about the initial quality variability of raw material, process parameters, and so on, in addition, to the temperature variability. The accuracy and the size of available data on cold chain diversity, cost-effective ways to monitor and register chain conditions, and further refinements of the proposed management system could potentially improve substantially the quality distribution of the final frozen products.

NOMENCLATURE

a, a_0	value of a-parameter of CIE Lab scale at time t and zero
b	slope of the semilog plot of shelf-life θ_s against temperature T
b, b_0	value of b-parameter of CIE Lab scale at time t and zero
C	concentration of a quality index (i.e., the concentration of a chemical compound)
C_L, C_{L_0}	concentration of L-ascorbic acid at time t and zero (mg L-ascorbic acid/100 g of raw product)
C_1, C_2	constants for the WLF equation
ΔC	chroma change
E_A	activation energy of a reaction (J/mol)
f_{con}	fraction of shelf-life consumed
f_q	quality function of frozen foods
$F(X)$	mathematical function of the response X of TTIs
k	reaction rate constant (h^{-1} or d^{-1})
k_C	rate constant of chroma change (h^{-1} or d^{-1})
k_L	rate constant of L-ascorbic loss (h^{-1} or d^{-1})
k_m	rate constant of color "lightness" loss (h^{-1} or d^{-1})
k_0	frequency factor (h^{-1} or d^{-1})
k_I	rate constant of TTI response (h^{-1} or d^{-1})
L, L_0	value of L-parameter of CIE Lab scale at time t and zero
n	apparent reaction order
R	universal gas constant (8.3144 J mol^{-1} K^{-1})
t_{tot}	total time of exposure in the cold chain, equal to the sum of small time increments t_i of isothermal exposure ($t_{tot} = \Sigma t_i$) (h or d)
T	temperature (K)
T_{eff}	effective temperature (K)
T_g	glass transition temperature (K)
T_{ref}	reference temperature (K)
θ_s	shelf-life (h or d)

Subscript

eff	value at the effective temperature
g	value at the glass transition temperature
ref	value at the reference temperature

REFERENCES

1. J Moureh, E Derens. Numerical modelling of the temperature increase in frozen food packaged in pallets in the distribution chain. *International Journal of Refrigeration* 23:540–552, 2000.
2. L Bøgh-Sørensen. Collaborative study on the temperature abuse of frozen foods. In: *Proceedings of the 6th Plenary Meeting of the EU Concerted Action Project "The Preservation of Frozen Food Quality and Safety throughout the Distribution Chain"* (CT96-1180). Instituto Sperimentale per la Valorizzazione Technologica dei prodotti agricoli, Milan, pp. 9–10, 1999.
3. R Gormley, T Walshe, K Hussey, F Buttler. The effect of fluctuating vs. constant frozen storage temperature regimes on some quality parameters of selected food products. *Lebensmittel-Wissenschaft und Technologie* 35:190–200, 2002.
4. E Hansen, L Lauridsen, LH Skibsted, RK Moawad, ML Andersen. Oxidative stability of frozen pork patties: effect of fluctuating temperature on lipid oxidation. *Meat Science* 68 (2):185–191, 2004.
5. B Fu, TP Labuza. Shelf life testing: procedures and prediction methods. In: MC Erickson, YC Hung, Eds., *Quality in Frozen Food*. New York: Chapman & Hall, 1997, pp. 377–415.
6. HD Goff. Low-temperature stability and the glassy state in frozen foods. *Food Research International* 25:317–325, 1992.
7. WL Kerr, MH Lim, DS Reid, H Chen. Chemical reaction kinetics in relation to glass transition temperatures in frozen food polymer solutions. *Journal of the Science of Food and Agriculture* 61:51–56, 1993.
8. YH Roos, M Karel, JL Kokini. Scientific status summary: glass transitions in low moisture and frozen foods: effects on shelf-life and quality. *Food Technology* 50 (11):95–108, 1996.
9. L Bøgh-Sørensen. Frozen food legislation. *Bulletin of the International Institute of Refrigeration* (4): 4–18, 2002.
10. DS Reid. Overview of physical/chemical aspects of freezing. In: MC Erickson, YC Hung, Eds., *Quality in Frozen Food*. New York, U.S.A: Chapman & Hall, 1997, pp. 10–28.
11. M Jul. 1984. *The Quality of Frozen Foods*. Orlando: Academic Press, 1984, pp. 112–251.
12. G Skrede. Fruits. In: LE Jeremiah, Ed., *Freezing Effects on Food Quality*. New York: Marcel Dekker, 1996, pp. 183–245.
13. CE Devine, RG Bell, S Lovatt. Red meats. In: LE Jeremiah, Ed., *Freezing Effects on Food Quality*. New York: Marcel Dekker, 1996, pp. 73–76.
14. RW Mandigo, WN Osburn. Cured and processed meats. In: LE Jeremiah, Ed., *Freezing Effects on Food Quality*. New York: Marcel Dekker, 1996, pp. 171–177.
15. WB Van Arsdel. In: WB Van Arsdel, MJ Copley, RL Olson, Eds., *Quality and Stability of Frozen Foods*. New York: Wiley/Interscience, 1969.
16. RP Singh, CY Wang. Quality of frozen foods — a review. *Journal of Food Process Engineering* 1:97–127, 1977.
17. L Jeremiah. *Freezing Effects on Food Quality*. New York: Marcel Dekker, 1996, pp. 51–400.
18. RC Martins, CLM Silva. Frozen green beans (*Phaseolus vulgaris* L.) quality profile evaluation during home storage. *Journal of Food Engineering* 64:481–488, 2004.
19. TP Labuza. Shelf-life of frozen fruits and vegetables. In: TP Labuza, Ed., *Shelf-life of Foods*. Westport, CT, U.S.A: Food & Nutrition Press, Inc., 1982, pp. 289–340.
20. DJ Favell. A comparison of the vitamin C content of fresh and frozen vegetables. *Food Chemistry* 62 (1):59–64, 1998.
21. MJ Oruña-Concha, MJ Gonzalez-Castro, J Lopez-Hernandez, J Simal-Lozano. Monitoring of the vitamin C content of frozen green beans and Padrón peppers by HPLC. *Journal of the Science of Food and Agriculture* 76:477–480, 1998.
22. MA Sahari, MF Boostani, ZE Hamidi. Effect of low temperature on the ascorbic acid content and quality characteristics of frozen strawberry. *Food Chemistry* 86:357–363, 2004.
23. MC Giannakourou, PS Taoukis. Kinetic modelling of vitamin C loss in frozen green vegetables under variable storage conditions. *Food Chemistry* 83 (1):33–41, 2003.
24. W Kmiecik, Z Lisiewska. Effect of pretreatment and conditions and period of storage on some quality indices of frozen chive (*Allium schoenoprasum* L.). *Food Chemistry* 67:61–66, 1999.
25. Z Lisiewska, W Kmiecik. Effect of storage and temperature on the chemical composition and organoleptic quality of frozen tomato cubes. *Food Chemistry* 70:167–173, 2000.

26. M Browne, J Allen. Logistics of food transport. In: R Heap, M Kierstan, G Ford, Eds., *Food Transportation*. London: Blackie Academic & Professional, 1998, pp. 22–50.
27. C Dubelaar, G Chow, P Larson. Relationships between inventory, sales and service in a retail chain store operation. *International Journal of Physical Distribution and Logistics Management* 31 (2):96–108, 2001.
28. LMM Tijkens, AC Koster, JME Jonker. Concepts of chain management and chain optimisation. In: LMM Tijkskens, MLATM Hertog, BM Nicolaï, Eds., *Food Process Modelling*. New York: CRC Press, 2001, pp. 448–469.
29. PS Taoukis. Modelling the use of time–temperature indicators. In: LMM Tijskens, MLATM Hertog, BM Nicolaï, Eds., *Food Process Modelling*. New York: CRC Press, 2001, pp. 402–431.
30. G Panozzo, G Minotto, A Barizza. Transport and distribution of foods: today's situation and future trends. *International Journal of Refrigeration* 22:625–639, 1999.
31. ML Woolfe. Temperature monitoring and measurement. In: C Dennis, M. Stringer, Eds., *Chilled Foods: A Comprehensive Guide*. Great Britain: Ellis Horwood Ltd., 1992, pp. 77–109.
32. Anonymous EU Directive 89/108 on the Approximation of the Laws of the Member States relating to Quick Frozen Foodstuffs for Human Consumption, Brussels, Official Journal L 40, 1989, pp. 34–37.
33. Anonymous EU Directive 92/1 on the Monitoring of Temperatures in the Means of Transport, Warehousing and Storage of Quick Frozen Foods Intended for Human Consumption, Brussels, Official Journal L 34, 1992, pp. 28–29.
34. Anonymous Agreement on the International Carriage of Perishable Foodstuffs and on the Special Equipment to be Used for Such Carriage (ATP) (Accord relatif aux transports internationaux de denrées périssables). United Nations, New York, E/ECE 810 Rev.1, E/ECE/TRANS/563 Rev.1, 1970.
35. Anonymous Frozen Food Roundtable. Frozen Food Handling and Merchandizing. American Frozen Food Institute, and 14 other associations/organizations in U.S.A., 1999.
36. Anonymous Deep freezing temperatures in shops — temperature control in freezers and foodstuffs. Final Report of the European Commission sponsored Research project in 4 EU member states (Greece, Italy, Spain and Portugal). Contract number: EC 1080/94/000069, 1995, 24 pp.
37. L Bøgh-Sørensen, F Bramsnaes. The effect of storage in retail cabinets on frozen foods. *Bulletin of the International Institute of Refrigeration* (Annexe 1977–1):375–381, 1977.
38. MC Giannakourou, PS Taoukis. Application of a TTI-based distribution management system for quality optimisation of frozen vegetables at the consumer end. *Journal of Food Science* 68 (1):201–209, 2003.
39. Anonymous EU Directive 92/2 on the Sampling Procedure and the Community Method of Analysis for the Official Control of the Temperature of Quick Frozen Foods Intended for Human Consumption, Brussels, Official Journal L 34, 1992, pp. 30–33.
40. PS Taoukis, TP Labuza. Time–temperature indicators (TTIs). In: R Ahvenainen, Ed., *Novel Food Packaging Techniques*. U.K.: Woodhead Publishing Limited, 2003, pp. 103–126.
41. MC Giannakourou, K Koutsoumanis, E Dermesonlouoglou, PS Taoukis. Applicability of the intelligent shelf-life decision system for control of nutritional quality of frozen vegetables. *Acta Horticulturae* 566:275–280, 2001.
42. H Schubert. Criteria for the application of T-TI indicators to quality control of deep frozen food products. I.I.F.-I.I.R.- Commissions C1/C2 Ettlingen, Germany, 1977, pp. 407–423.
43. JW Farquhar. Time/temperature monitoring of frozen and refrigerated seafood products. In: *Proceedings of XVIth International Congress of Refrigeration*, Vol. III. International Institute of Refrigeration, Commission C2, Paris, 1983, pp. 811–819.
44. KD Dolan, RP Singh, JH Wells. Evaluation of time–temperature related quality changes in ice cream during storage. *Journal of Food Process and Preservation* 9:253–271, 1985.
45. RP Singh, JH Wells. Use of time–temperature indicators to monitor quality of frozen hamburger. *Food Technology* 39 (12):42–50, 1985.
46. JH Wells, RP Singh, AC Noble. A graphical interpretation of time–temperature related quality changes in frozen food. *Journal of Food Science* 52(2):436–444, 1987.
47. DI LeBlanc. Time–temperature indicating devices for frozen foods. *Journal of the Canadian Institute of Food Science and Technology* 21 (3):236–241, 1988.
48. SH Yoon, CH Lee, DY Kim, JW Kim, KH Park. Time–temperature indicator using phospholipid–phospholipase system and application to storage of frozen pork. *Journal of Food Science* 59 (3):490–493, 1994.

49. PS Taoukis, M Bili, M Giannakourou. Application of shelf-life modelling of chilled salad products to a TTI-based distribution and stock rotation system. *Acta Horticulturae* 476:131–140, 1998.
50. PS Taoukis, K Koutsoumanis, GJE Nychas. Use of time–temperature integrators and predictive modelling for shelf-life control of chilled fish under dynamic storage conditions. *International Journal of Food Microbiology* 53:21–31, 1999.
51. MC Giannakourou, PS Taoukis. Systematic application of time–temperature integrators as tools for control of frozen vegetable quality. *Journal of Food Science* 67 (6):2221–2228, 2002.
52. MA Tung, IJ Britt, S Yada. Packaging considerations. In: NAM Eskin, DS Robinson, Eds., *Food Shelf Life Stability*. Boca Raton, FL, U.S.A: CRC Press, 2001, pp. 129–145.
53. PS Taoukis, TP Labuza, IS Saguy. Kinetics of food deterioration and shelf-life prediction. In: KJ Valentas, E Rotstein, RP Singh, Eds., *Handbook of Food Engineering Practice*. New York: CRC Press, 1997, pp. 361–403.
54. PS Taoukis, MC Giannakourou. Temperature and food stability: analysis and control. In: R Steele, Ed., *Understanding and Measuring the Shelf Life of Food*. Cambridge: CRC Press, 2004, pp. 42–68.
55. YH Roos. Water activity and plasticization. In: NAM Eskin, DS Robinson, Eds., *Food Shelf Life Stability*. Boca Raton, FL, U.S.A: CRC Press, 2001, pp. 17–22.
56. S Arrhenius. About the reaction rate of the inversion of non-refined sugar at souring. *Zeitschrift für Physikalische Chemie* 4:226–248, 1889.
57. MLATM Hertog, LMM Tijskens, PS Hak. The effects of temperature and senescence on the accumulation of reducing sugars during storage of potato (*Solanum tuberosum* L.) tubers: a mathematical model. *Postharvest Biology and Technology* 10:67–79, 1997.
58. AP Buedo, MP Elustondo, MJ Urbicain. Non-enzymatic browning of peach juice concentrate during storage. *Innovative Food Science and Emerging Technologies* 1:255–260, 2001.
59. RC Martins, CLM Silva. Modelling colour and chlorophyll losses of frozen green beans (*Phaseolus vulgaris* L.). *International Journal of Refrigeration* 25:966–974, 2002.
60. TP Labuza, D Riboh. Theory and applications of Arrhenius kinetics to the prediction of nutrient losses in food. *Food Technology* 36:66–74, 1982.
61. D Simatos, G Blond. DSC studies and stability of frozen foods. In: H Levine, L Slade, Eds., *Water Relationships in Foods*. New York: Plenum Press, 1991, pp. 139–155.
62. ME Sahagian, HD Goff. Fundamental aspects of the freezing process. In: LE Jeremiah, Ed., *Freezing Effects on Food Quality*. New York: Marcel Dekker, 1996, pp. 1–50.
63. L Slade, H Levine. Beyond water activity: recent advances based on an alternative approach to the assessment of food quality and safety. *Food Science and Nutrition* 30 (2–3):115–357, 1991.
64. D Champion, G Blond, M LeMeste, D Simatos. Reaction rate modelling in cryoconcentrated solutions: alkaline phosphatase-catalyzed DNPP hydrolysis. *Journal of Agricultural and Food Chemistry* 48:4942–4947, 2000.
65. AK Carrington, HD Goff, DW Stanley. Structure and stability of the glassy state in rapidly and slowly cooled carbohydrate solutions. *Food Research International* 29 (2):207–213, 1996.
66. CG Biliaderis, A Lazaridou, I Arvanitoyannis. Glass transition and physical properties of polyol-plasticized pullulan-starch blends at low moisture. *Carbohydrate Polymers* 40:29–47, 1999.
67. ML Williams, RF Landel, JD Ferry. The temperature dependence of relaxation mechanisms in amorphous polymers and other glass forming liquids. *Journal of Chemical Engineering* 77:3701–3707, 1955.
68. M Peleg. On the use of WLF model in polymers and foods. *Critical Reviews on Food Science* 32:59–66, 1992.
69. P Buera, M Karel. Application of the WLF equation to describe the combined effects of moisture, temperature and physical changes on non-enzymatic browning rates in food systems. *Journal of Food Processing and Preservation* 17:31–47, 1993.
70. NS Terefe, ME Hendrickx. Kinetics of the pectin methylesterase catalyzed de-esterification of pectin in frozen food model systems. *Biotechnology Progress* 18:221–228, 2002.
71. MC Giannakourou, PS Taoukis. Stability of dehydrofrozen green peas pretreated with non-conventional osmotic agents. *Journal of Food Science* 68 (6):2002–2010, 2003b.
72. GJE Nychas, E Drosinos. Meat and poultry spoilage. In: R Robinson, C Batt, P Patel, Eds., *Encyclopedia of Food Microbiology*. London: Academic Press, 2000, pp. 1253–1259.
73. L Body, JT Wimpenny. Ecological concepts in food microbiology. *Journal of Applied Bacteriology* 73:23S–38S, 1992.

74. JHJ Huis in't Veld. Microbial and biochemical spoilage of foods: an overview. *International Journal of Food Microbiology* 33:1–18, 1996.
75. A Tanghe, P Van Dijk, JM Thevelein. Determinants of freeze tolerance in microorganism, physiological importance and biotechnological applications. In: AI Laskin, JW Bennett, GM Gadd, Eds., *Advances in Applied Microbiology*, Vol. 53. Academic Press, 2003, pp. 129–176.
76. H Michener, R Elliott. Minimum growth temperatures for food poisoning, fecal-indicator, and psychrophilic microorganisms. *Advances in Food Research* 13:349–396, 1964.
77. JM Jay. *Modern Food Microbiology*. U.S.A: Aspen Publishers, 2000, pp. 323–339.
78. QT Pham, RF Mawson. Moisture migration and ice recrystallization in frozen foods. In: MC Erickson, YC Hung, Eds., *Quality in Frozen Food*. New York, U.S.A: Chapman & Hall, 1997, pp. 10–28.
79. YL Xiong. Protein denaturation and functionality losses. In: MC Erickson, YC Hung, Eds., *Quality in Frozen Food*. New York, U.S.A: Chapman & Hall, 1997, pp. 111–140.
80. RV Sista, MC Erickson, RL Shewfelt. Quality deterioration in frozen foods associated with hydrolytic enzyme activities. In: MC Erickson, YC Hung, Eds., *Quality in Frozen Food*. New York, U.S.A: Chapman & Hall, 1997, pp. 101–110.
81. MC Erickson. Lipid oxidation: flavor and nutritional quality deterioration in frozen foods. In: MC Erickson, YC Hung, Eds., *Quality in Frozen Food*. New York: Chapman & Hall, 1997, pp. 141–174.
82. JG Sebranek. Poultry and poultry products. In: LE Jeremiah, Ed., *Freezing Effects on Food Quality*. New York: Marcel Dekker, 1996, pp. 85–108.
83. OR Fennema. Freeze-preservation of foods — technological aspects. In: OR Fennema, WD Powrie, EA Marth, Eds., *Low-temperature Preservation of Foods and Living Matter*. New York: Marcel Dekker, 1973, pp. 509–550.
84. PS Taoukis, TP Labuza. Applicability of time–temperature indicators as shelf-life monitors of food products. *Journal of Food Science* 54 (4):783–789, 1989.
85. PS Taoukis, TP Labuza. Reliability of time–temperature indicators as food quality monitors under nonisothermal conditions. *Journal of Food Science* 54 (4):789–792, 1989.
86. G Zhang, W Habenicht, WEL Spieß. Improving the structure of d eep frozen and chilled food chain with tabu search procedure. *Journal of Food Engineering* 60:67–79, 2003.
87. AM Lammerdig, A Fazil. Hazard identification and exposure assessment for microbial food safety risk assessment. *International Journal of Food Microbiology* 58:147–157, 2000.

Part III

Quality and Safety of Frozen Foods

15 Quality and Safety of Frozen Meat and Meat Products

Sandra Moorhead
University of Guelph, Guelph, Ontario, Canada

CONTENTS

I. INTRODUCTION

Historically, meat has been a major component of the diet in many cultures around the world, and remains so today. However, meat is a highly perishable food commodity which unless appropriately modified or stored, will rapidly spoil from the growth and by-products of microorganisms, as well as develop unpalatable characteristics because of endogenous biochemical degradation of the meat components. Before the advent of technology enabling refrigeration and freezing as preservation methods, traditional means of meat preservation were largely either drying or salting. However, these methods change the taste and texture of the meat and meat products. Meat preservation by freezing has been used for centuries as local production of meat exceeded the immediate requirements for consumption. An approximate storage life for frozen meat is between 10 and 24 months if stored at -18°C, and between 15 and 24 months if stored at -24°C [1].

Research in the late 19th century indicated that frozen storage did not compromise the quality of red meats, and one of the first practical demonstrations of this was on 5th February, 1882, when a cargo of frozen meat was dispatched from New Zealand to England and arrived in excellent condition [2]. The retail revolution in consumer packs of frozen meats began in the 1930s in the USA, with fruit and vegetables, meats and fishes being successfully marketed. Since then, the number and variety of frozen foods has grown immensely, adapting to changing consumer needs. The range of meat cuts and meat products available to processors, retailers, and consumers is now huge, and meat may be frozen as carcasses, as packaged primal or consumer cuts, or as a range of processed meat products.

Many factors can affect the quality of frozen meat, including the state of the animal upon slaughter, the slaughter and conditioning process, processing techniques, and freezing conditions. Freezing technology includes three steps: a freezing process, storage at freezer temperatures, and a thawing process; physical, chemical, and nutritional changes can occur during each process. This chapter will discuss the quality and safety of frozen meat and meat products, resulting from preslaughter, slaughter, processing and finally freezing, frozen storage, and thawing conditions.

II. MEAT QUALITY

The quality characteristics that meat producers have to be concerned with include organoleptic and physicochemical qualities such as color, texture, flavor, exudate (drip) loss, nutrient content, and fat oxidation.

A. Prefreezing Considerations

1. Preslaughter Factors Affecting Meat Quality

One of the major preslaughter factors influencing meat quality is the concentration of muscle glycogen at slaughter [3]. Postslaughter, muscle glycogen is converted into lactic acid by way of glycolysis; gradually lowering the muscle pH to a final value is termed the ultimate pH. The ultimate pH of table quality muscles from a well-fed, rested animal is approximately 5.6. Stressful conditions in the days and hours leading to slaughter can result in low glycogen concentrations at slaughter. The resultant ultimate pH is higher than normal. When approximately more than pH 6.0 the meat takes on a dark, firm, and dry (DFD) character, with concomitant changes in flavor and keeping quality in the chilled state.

Other processing factors that may affect meat quality include the intrinsic postmortem glycolytic rate of the species concerned. Decrease in muscle pH to below 6.0 in a typical chilling environment occurs quickly in pigs (5–10 h), but more slowly in sheep (16–24 h) and cattle (up to 36 h) [4]. During this time, if the carcass is cooled too quickly, exposed muscles can "cold-shorten," toughening the meat when frozen. Electrical stimulation accelerates glycolysis, decreasing the muscle pH and hastening the onset of rigor to the point that muscles can no longer cold-shorten [5].

2. Processing Factors

Another advantage of electrical stimulation is that postslaughter animal movement is reduced, resulting in a fresher brighter meat color at 48 h [6]. The mechanism for this is believed to be destruction of mitochondrial enzyme activity such that postmortem oxidation is reduced. Concentrations of the bright red oxymyoglobin increase at the meat surface. The color of raw meat is determined by the concentration and chemical nature of the haemoproteins present [7,8]. The overall redness of the meat is because of oxymyoglobin, the oxygenated form of myoglobin, whereas the unattractive brown color after prolonged storage is because of metmyoglobin, the oxidized form of myoglobin [9–11]. With the recent trend towards centralized packaging and distribution of consumer packs to retail outlets [12], it is necessary for meat distributors to ensure that display is desirable, but only at the point of display. Color at prior times is not important (it could be argued that color is never important for eating quality, but to buy meat unseen would require consumer re-education).

Aging, the enzymatic breakdown of myofibrillar proteins resulting in tenderness, is most rapid at high temperatures. Aging can also occur during frozen storage, as free water exists in frozen water. Reactions continue, although not at the pace above freezing point [13].

Hot boning, where the meat is removed from the skeletal structure before the onset of rigor mortis, often increases the risk of muscle toughening, but this is almost certainly because of cold-shortening. Provided cooling is controlled (e.g., by immersion in temperature-controlled baths), hot-boned meat can be as tender as cold-boned meat and maybe more so.

Oxidative rancidity develops in frozen whole tissue as well as ground meats if there is excessive exposure to air. This can be achieved with modified atmospheres or vacuum packaging.

B. The Freezing Process

Freezing meat offers a product with a nutritional quality close to fresh meat, however, it may significantly affect the organoleptic properties of meat such as color [14]. Freezing of aged, well treated meat does not necessarily have any significant effect on the cooked color, flavor, odor, or juiciness of that meat, and in fact in some circumstances produces a slight tenderizing effect as in the case of frozen pork loin roasts [15], although it was also found that off-flavors developed on prolonged storage, presumably through oxidative rancidity.

1. Freezing

Freezing is the transition of water in the muscle tissue crystallizing into ice. The freezing rate refers to the rate at which any given part of the food is cooled. The undesirable changes in meat during freezing is because of mechanical damage to muscle cells from large ice crystals, and to chemical damage arising from increase in concentration of tissue solutes.

Before freezing, the meat is chilled to between typically 0 and 10°C. The changes that occur in this period have an important effect on the subsequent quality of the frozen product. They include the unwanted growth of microorganisms, leading to early onset of spoilage, as well as undesirable chemical changes such as rancidity, which if initiated during the chilled phase prefreezing, will continue at an accelerated rate after freezing. Therefore, the time meat is held chilled affects the overall quality of the meat or meat product. Moisture loss during chilling has both positive and negative effects on meat quality. The processor has to achieve enough surface drying to minimize microbial growth, while at the same time minimizing weight loss and preserving meat surface appealing to the consumer. Factors affecting moisture loss include muscle type, preslaughter factors that affect the ultimate pH, area of cut surface exposed, freezing rate, storage conditions, and thawing rate.

Petrović et al. [16] investigated a range of freezing rates from 0.22 to 5.66 cm/h on beef *M. longissimus dorsi*, studying physicochemical properties including weight loss, pH, and water-binding capacity. Mechanical tenderness, myofibrillar solubility, and sensory evaluation on the cooked product were also examined. The greatest weight losses were registered at slow freezing rates (0.22 and 0.39 cm/h), and the meat was found to be tougher and less soft. At these slow freezing rates, myofibrillar proteins were least soluble. However, very quick freezing rates (4.92 and 5.66 cm/h) caused myofibrillar damage, lower water-binding capacity, and tougher, drier meat. These authors found that a freezing rate between 3 and 4 cm/h had the least effect on muscle structure and the state of myofibrillar proteins, and cooked muscles from samples frozen at these rates were evaluated as the most tender [16].

In another study, slowly frozen meat resulted in more drip than fast frozen meat but otherwise freezing rate had no effect on the functional attributes of protein solubility, sulfydryl content, surface hydrophobicity, emulsion activity index, or meat color [17]. These authors [16,17] concluded that the current practice of blast-freezing and storage at −18 to −20°C was sufficient to maintain the quality of manufacturing beef.

2. Frozen Storage

Frozen meat storage temperatures are ideally at or below −18°C [1]. Fluctuations above this temperature, caused by a range of actions from electrical or equipment failure to movement of the product around cool-stores or during transportation, will result in a product of poorer quality. This is caused by denaturation of proteins occurring as solute concentration increases, damaging ice crystals growing in size, lipids oxidation, and dehydration of surfaces (freezer burn). These events can be partially controlled by suitable packaging and by a reduced oxygen concentration to reduce lipid oxidation.

The effect of storage time and temperature on the functional attributes of meat were assessed by Farouk et al. [17]. Storage temperatures of −18, −35, and −75°C were examined and found to affect the solubility of both myofibrillar and sarcoplasmic proteins minimally, but not other attributes such as emulsion activity and stability, water-holding capacity, color, drip, and tenderness [17]. Storage time alone had the greatest effect on thawed beef in this study. Storage times of 0, 3, 6, 9, and 12 months were studied, and increases in pH, emulsion activity, and stability were recorded, while total protein solubility decreased, indicating denaturation of proteins, responsible for loss of quality over time. Conversely, the meat became more tender over storage time, which is attributed to the breakdown of muscle structure caused by enzyme activity and ice crystal formation. These authors concluded that expensive ultra-low temperature storage of meat was unnecessary for manufacturing beef, but that extended storage times were more detrimental to meat quality.

Most nutrients were retained during freezing and subsequent storage (Table 15.1) [18]. Experiments to measure these nutrients were performed between the 1950s and 1970s, with no further work published subsequently. Soluble proteins and vitamins have been shown to be lost with drip during thawing, but the fluid lost with drip approximates the fluid lost when fresh meat is cooked, therefore the net loss of nutrients is minimal [6].

3. Thawing

Regulatory authorities advocate commercially thawing meats at low temperatures (<10°C) or that processors show equivalency for microbiological growth rates during thawing [26]. Early research focused on the effect of household thawing on the organoleptic quality of meats [27], finding that variations in thawing methods have a limited but not insignificant effect on total end product quality as shown in Table 15.2. Stoll et al. [27] found that a slow thawing process at not too low a temperature is preferable for beef.

Thawing meat results in an approximately 2–6% decrease in yield because of drip loss [28], with subsequent vitamin losses (however, the drip loss may be included in the food preparation meaning no losses). Drip loss depends on the species, the muscle, and the rate of thawing which in Table 15.3 [28] was standardized at either for 4 h at room temperature, or 20 h at 4°C. Pearson et al. [29] found a much higher nutrient loss, but thawing was at 14–15 h at 26°C, and the percent drip loss in beef was not recorded. In contrast, Ngapo et al. [30] observed only subtle differences in drip loss and in the ultra-structure of samples of pork after frozen storage and thawing.

C. Processed Meats

The product quality of processed meats is directly attributable to the quality of the raw materials. Often meat for further processing has already been frozen, amplifying the effects of further freezing, storage, and thawing. Additional ingredients are usually added which affect the quality, shelf life, and overall acceptability of these products, and the physicochemical reactions occurring during the freezing process.

TABLE 15.1
Percent Retention of Nutrients in Frozen Stored Pork and Beef (*Longissimus dorsi*)

Species	Frozen storage (days)	Temperature (°C)	Thiamin	Riboflavin	Niacin	Pantothenic Acid	Pyridoxin	Retinol	Reference
Pork	180	−18	68	66	90	nt	nt	nt	[19]
Pork	180	−26	83	97	98	nt	nt	nt	[19]
Pork	365	−18	89	144	114	106	nt	nt	[20]
Pork	180	−18	101	81	106	84	64	nt	[21]
Pork	168	−18	85	94	95	nt	nt	nt	[22]
Pork	90	−20	92	nt	nt	nt	nt	92	[23]
Pork	120	−16	78	nt	nt	nt	nt	nt	[24]
Beef	180	−18	92	91	101	91	76	nt	[25]
Beef	300	−18	98	57	96	nt	nt	nt	[25]

Note: nt, not tested.

TABLE 15.2
Effects of Various Methods of Thawing Beef Frozen in 600 g Pieces (Score: 9, highest; 5, Unsatisfactory)

Methods and Thawing Times	Organoleptic Evaluations			
	Color	Form	Flavor	Texture
Room temperature 16 h, then cooked for 2 h	7.8	7.8	7.5	7.5
Refrigerator 24 h, then cooked 2 h	7.3	7.8	8.0	7.3
Cold tap water 3 h, cooled 2 h	7.5	7.8	8.0	8.3
Pressure cooked 45 min, 8 min standing	8.0	7.8	7.5	6.8
High pressure steam cooked 1 h	5.0	6.5	7.0	7.0
Microwave 2100 W, 27 min in glass dish with gravy	6.3	6.8	6.0	5.0

Source: Reprinted from *The Quality of Frozen Foods*, M Juls (Ed.), p. 261, 1984, with permission from Elsevier.

Tempering frozen blocks of meat (partial thawing) is often employed to reduce drip loss, bacterial growth, and to facilitate comminution where this is applied. Thus tempering allows better control of quality of these processed meat products [31]. However, vitamins are likely to be lost to some extent during further processing [32].

Ashby and James [33,34] examined the effects of freezing and packaging methods on shrinkage and freezer burn in cooked hams. They found that hams frozen in a blast freezer ($-27°C$) sustained less shrinkage compared to ham frozen by forced air ($-22°C$) and still air cold rooms ($-20°C$). Forced air freezing increased amounts of freezer burn between 4 and 6 months frozen storage, compared with blast or still air-freezing methods, although differences in degree of freezer burn were lost after 6 months of frozen storage.

In another ham experiment, uncured hams were frozen on racks at $-29°C$, then stored at $-18°C$ for 3 months before various thawing and curing treatments [35]. Weight loss, raw color, aroma, and general appearance were evaluated. Table 15.4 shows the results for weight loss. From an economic consideration, Kemp et al. [35] suggested that there was a considerable advantage in curing frozen hams.

TABLE 15.3
Percent Drip Loss and B-vitamins Loss of Original Content

	Cut	% Drip	Thiamin	Riboflavin	Niacin
Pork	Tenderloin	1.4	0.9	0.70	0.72
	Loin	1.9	2.1	2.9	1.5
	Chop	2.0	2.4	3.0	1.5
Beef	Blade roast	2.2	5.1	0.4	3.1
	Rib steak[a]	6.3	16.8	6.3	1.5
	Rib steak	5.7	14.3	8.6	9.0
	Rib roast	3.2	5.3	3.2	5.3
Lamb	Leg roast	2.2	2.3	1.5	3.6

[a]Thawed at room temperature, all others thawed in refrigerator.
Source: Derived from LH Kotschevar. *Journal of the American Dietetics Association* 31:589–596, 1951.

TABLE 15.4
Percent Weight Loss because of Thawing Method of Uncured Hams After Thawing and Dry-Curing

		Treatment Preprocessing		
Processing	**Frozen**	**Cooler-Thawed, 3°C**	**Room-Thawed, 15°C**	**Water-Thawed, 38°C**
Cured	3.1	6.1	5.4	5.0
Smoked	12.9	15.7	16.1	16.8
Aged 1 month	17.5	21.0	20.6	21.4
Aged 2 months	22.0	25.5	25.3	26.0
Aged 3 months	24.9	29.3	27.8	28.7

Source: From JD Kemp, BE Langlois, JD Fox. *Journal of Food Science* 43:860–863, 1978. Reprinted with permission from The Institute of Food Technologists.

The effects of freezing and frozen storage were evaluated on all-beef and soy-extended patties by Berry [36], who reported that patties stored at −7°C had greater surface discoloration and freezer burn than patties stored at −18 or −23°C. Patties were stored for 0, 6, 9, 12, 18, or 24 months before evaluation, and results indicated that storage of patties at −7°C should be avoided owing to reduction in quality after 6 months (and longer) of storage.

III. MEAT SAFETY

A. General Considerations

One of the major advantages of freezing meats is the slowing or prevention of growth of spoilage and pathogenic microorganisms present on either the meat surface or within the meat product. However, some microbial enzymes may remain active at freezing temperatures, and cause spoilage, for example, the storage life of frozen pork is limited by rancidity caused in part by microbial or tissue lipolysis [37].

Meat, although initially a sterile surface, is microbiologically contaminated at all stages of the slaughter, dressing, chilling or freezing, storage, processing, and packaging chain [38]. Sources of contamination include the animal itself (external surfaces such as fleece, hide or skin, and the gastrointestinal and respiratory tract). The process workers and the processing environment also contribute to the microflora found on the end-product.

Contaminating bacteria are to be kept to a minimum through strict adherence to hygienic handling and processing practices. However, the same bacteria species can be isolated from beef, pork, and sheep, and numbers on fresh carcasses are commonly between 10^1 and 10^3 colony forming units (cfu)/cm^2 for beef [39–43], between 10^2 and 10^4 cfu/cm^2 for sheep [42,44–46], and between 10^3 and 10^4 cfu/cm^2 for pork [39,47]. Freezing may reduce these numbers slightly, but thawing provides the condition where proliferation of bacteria can begin. Regulatory authorities generally advocate that commercial thawing be undertaken at temperatures below 10°C to ensure that the hygienic status of the product is not compromised [26].

There are two broad groups of microorganisms of concern in meat safety, those that in large numbers cause spoilage and those that cause food-borne illness. For the latter, some species need only be present in very small numbers. Meat presents a nutritious substrate for microorganisms, and depending on the packaging used, either aerobic or anaerobic bacteria can dominate the microflora.

Freezing affects bacterial cells as much as the muscle cells of meat. Intracellular ice formation and water diffusion, resulting in a sublethal concentration of solutes is also damaging to microorganisms [48]. Ironically, freezing regimes to optimize quality in muscle cells also optimizes survival of contaminating bacterial cells [49]. For example, almost 100% of super-cooled *Saccharomyces cerevisiae* and *Escherichia coli* were able to survive frozen storage at -16 and -10°C, respectively [50]. Repeated freeze–thaw cycles can disrupt and destroy bacteria [48]; however, the effects of cyclic freezing are not well documented on most microbial pathogens.

An important consideration for microbial growth is the water activity (A_w) of the substrate, or in general terms, the available moisture. Most food spoilage bacteria require an A_w of above 0.9, with the exception of *Staphylococcus aureus* (0.86) [51]. Approximate minimum A_w values for food-related fungi are lower, but still average around 0.8. At -18°C, frozen meat containing ice in equilibrium will have an A_w of around 0.84, a condition under which few bacteria can grow [51].

Other factors that affect the survival of microorganisms during freezing and thawing include the type and strain of the microorganism, phase of growth, nutritional status, rate of cooling, substrate, the holding temperature, time frozen, and rate of thawing. Some bacteria are able to form spores as a survival mechanism, and these spores are extremely resistant to the effects of freezing and repeated freeze–thaw cycles [49].

Freezing has little effect of on viruses [52,53], but in contrast, nematode parasites are very susceptible to freezing. In fact, freezing is a regulated procedure for inactivating trichinae in pork [54].

B. Spoilage Bacteria

Spoilage bacteria include *Pseudomonas* spp., *Acinetobacter* or *Moraxella* spp., *Aeromonas* spp., *Alteromonas putrifaciens*, *Lactobacillus* spp., and *Brochothrix thermosphacta*. These psychrotrophic bacteria originate from soil, vegetation, and water on the animal hide or fleece. The processing environment can also cause contamination. The initial microflora on pork will differ from that of other red meats because of the dehairing processes. However, normal processing events recontaminate the pork with a microflora similar to that of other red meats [55].

Until spoilage is evident to the senses, the only detectable effect of bacterial growth is reduction in glucose levels, which does not alter the organoleptic qualities of meat. Once glucose supplies are depleted, microorganisms then degrade proteins, resulting in sulfides, amines, acetic acid, lactic acid, isovaleric acid, isobutyric acid, esters, and nitriles being produced. These by-products result in spoilage odors and flavors.

Growth characteristics of the spoilage bacteria on meat depend on the initial flora, level of contamination, storage time, temperature, and packaging regime. The microflora of anaerobically stored meat is usually dominated by species of *Lactobacilli*, which of necessity grow slowly and fermentatively, therefore resulting in little effect on the organoleptic quality of meat. However, spoilage flavors will develop over time because of the accumulation of volatile organic acids. Under aerobic condition, as the temperature is decreased, psychrotrophic bacteria begin to dominate the microflora, with *Pseudomonas* spp. the most dominant strain [56].

While growth of the above bacteria is stopped at temperatures below -2°C, xerotolerant molds and yeasts can continue growing. The more common meat-spoilage molds have a minimum growth temperature near -8°C [57], and cause conditions known as "black spot," "white spot," "blue-green mold," and "whiskers." At this temperature visible colonies will take several months to develop [58]. Some species of molds, for example, *Penicillium expansum*, when frozen on growth medium at a freezing rate of 2°C/min, resulted in shrinkage of hyphae, and at a faster freezing rate of 20°C/min, intracellular ice was formed [59].

C. Pathogenic Bacteria

The presence of a pathogen on frozen meat is directly affected by the prevalence of the pathogen on slaughter animals. Any organism associated with the gastrointestinal tract of slaughter animals has

TABLE 15.5
Microbiological Baseline Data of Beef Sides, Market Hogs, and Ground Beef in the United States (% positive)

Microorganism	Steers or Heifers (n = 2079)	Cows or Bulls (n = 2112)	Pork (n = 2112)	Ground Beef (n = 563)
Indicator Organisms				
Aerobic plate count ($\leq$100,000 cfu/cm^2 or cfu/g)	93.1	96.3	91.6	100
Total coliforms, ($\leq$100 cfu/cm^2 or cfu/g)	96.4	92.2	84.2	92.0
E. coli, ($\leq$10 cfu/cm^2 or cfu/g)	95.9	91.8	80.0	78.6
Pathogenic Bacteria				
C. jejuni or *coli*	2.6	1.1	31.5	0.002
E. coli O157:H7	0.2	nd	nd	nd
Salmonella spp.	1.0	2.7	8.7	7.5
C. perfringens	2.6	8.3	10.4	53.3
S. aureus	4.2	8.4	16.0	30.0
L. monocytogenes	4.1	11.3	7.4	11.7

Note: nd, not detected.

(*Sources*: From Anonymous. Nationwide beef microbiological baseline data collection program: steers and heifers. In: United Stated Department of Agriculture, Food Safety and Inspection Service, Washington, DC, 1994, pp. 1–9; Anonymous. Nationwide beef microbiological baseline data collection program: cows and bulls. In: United Stated Department of Agriculture, Food Safety and Inspection Service, Washington, DC, 1996, pp. 1–13; Anonymous. Nationwide pork microbiological baseline data collection program: market hogs, in United Stated Department of Agriculture, Food Safety and Inspection Service, Washington, DC, 1996, pp. 1–13; Anonymous. Nationwide federal plant raw ground beef microbiological survey, in United Stated Department of Agriculture, Food Safety and Inspection Service, Washington, DC, 1996, pp. 1–8.

the potential to contaminate meat from direct or indirect fecal contact. The processing environment is also a source of contamination, from equipment where biofilm-containing pathogens have become established to the hands of process workers. Table 15.5 indicates the level of contamination found in a survey of cattle and pork in the United States between 1993 and 1996 [60–63].

Most bacteria are able to survive in a nongrowing state at chilled and frozen storage temperatures, but temperatures for vegetative growth are usually higher than 0°C. For example, the minimum growth temperature for some of the more common pathogens are: *Yersinia enterocolitica*, −2°C; *Listeria monocytogenes*, 1°C, enterotoxigenic *E. coli*, 3°C; *Aeromonas hydrophila*, 0–5°C; non-proteolytic *Clostridium botulinum*, 3.3°C; *Salmonella*, 7–10°C; *Bacillus cereus*, 6–10°C; *S. aureus*, 7–10°C; proteolytic *C. botulinum*, 10°C; and *C. perfringens*, 12°C [64].

For ethical reasons, there have been limited studies on the dose–response relationship of pathogens in foods, but from results of outbreaks and a few voluntary trials, the infectious dose has been estimated for *Salmonella* spp. ($>10^5$ organisms) and *C. jejuni* (500 organisms) [65]. However, only a limited number of pathogenic bacteria genera have caused outbreaks associated with frozen foods, indicating that human pathogenic microorganisms are susceptible to freezing. This may be because of inherently low numbers present (below the infectious dose) or the inability to survive and grow on frozen meats by these organisms.

Investigations on the growth and survival characteristics of pathogenic microorganisms were performed to determine if pathogenic bacteria are susceptible to freezing and or frozen storage. The survival of *E. coli* O157:H7 strains on beef trimmings frozen to −18 or −35°C was determined in one trial. Counts on nonselective media remained constant throughout the 12-week storage trial, while counts on selective media decreased significantly, indicating a degree of sublethal injury [66]. The growth and survival of *E. coli* O157:H7 has also been examined on comminuted product.

Doyle and Schoeni [67] inoculated ground beef with *E. coli* O157:H7, formed hamburger patties, which were subsequently frozen and stored at −20°C for up to 9 months. There was inconsequential loss of viability following these procedures. Sage and Ingham [68] found a reduction in *E. coli* O157:H7 numbers in ground beef patties after frozen storage and thawing, but again, enumeration was performed on selective media. In another experiment where *E. coli* O157:H7 was inoculated onto ground beef, then formed into patties, strains survived 12 months frozen storage at −20°C, but gave a 1–2 $\log_{10}$ reduction in numbers [69].

The survival of three *Salmonella* serotypes inoculated onto beef trimmings and frozen to −18 or −35°C was investigated by Dykes and Moorhead [70] who found that bacterial numbers did not decrease over the 9-month storage trial. *Salmonella typhimurium* was shown to be sublethally damaged but not destroyed by freezing in sausage and ground beef stored at temperatures between −18 and −20°C, because more cells were recovered on nonselective media than selective [71].

With some localized exceptions relating to pork-borne *C. coli* infections, campylobacteriosis is not commonly acquired as a consequence of beef, lamb, or pork consumption, this may in part reflect the low meat pH of these species compounded by the adverse conditions for *Campylobacter* survival prevailing at the carcass surface during air-chilling. Moorhead and Dykes [72] studied the susceptibility of *C. jejuni* in beef trimmings during freezing and frozen storage at −18°C. Within 7 days, reductions in numbers from 0.6 to 2.2 $\log_{10}$ were observed, but there was no further loss of viability during subsequent storage. Earlier studies in ground beef showed that the addition of a cryoprotectant, such as 10% glycerol increased survival of *C. jejuni* during freezing [73].

Little work has been done on the effect of freezing on *Listeria* spp., but in one report, *L. monocytogenes* was shown to survive freezing well in ground beef, with the food found to contribute a protective effect [74]. Storage of raw, ground beef inoculated with *S. aureus* at −22°C for 4 months, or frozen beef for 168 days, resulted in a decrease in numbers of approximately 1 $\log_{10}$ [75,76]. Demchick et al. [77] studied the survival of *S. aureus* in lean ground beef at two pH levels, over repeated freeze–thaw cycles, and also found a 1 $\log_{10}$ reduction in numbers. Freezing did not inactivate staphylococcal enterotoxin.

Both spores and vegetative cells of *C. perfringens* are frequently found in small numbers on raw meats, and this organism often serves as an indicator for the more serious contamination of *C. botulinum*. Fresh meat products were inoculated with *C. perfringens* vegetative cells (and spores) and frozen to −27°C. The number of viable cells was reduced by 90% in 42 days [78]. These same authors [78] found that while vegetative cell numbers decreased, the numbers of spores present remained unchanged.

While these examples are not exhaustive, they do show that freezing is a poor bactericidal process.

IV. CONCLUSIONS

The quality and safety of meat and meat products is governed mainly by the presentation of slaughter stock; their state of stress and cleanliness are dominant factors. Freezing technology should be designed to ensure that the product does not significantly deteriorate in eating quality and safety from its original state. Research has shown that correct freezing regimes can produce meat of acceptable quality with regard to consumer preferences and nutritional value. However, these freezing regimes — rapid freezing followed by storage at nonfluctuating, low temperatures — may also optimize survival of contaminating bacterial cells. However, the incidence of food-borne illness from frozen meats is extremely low, indicating that there are both low numbers of pathogens that survive the antimicrobial practices in the processing system, and freezing techniques in practice are sufficient to prevent growth of any pathogens present.

REFERENCES

1. Anonymous. *Recommendations for the Processing and Handling of Frozen Foods*. 3rd ed. Paris: International Institute of Refrigeration, 1986, pp. 42–131.
2. MFG Boast. The technology of freezing. In: RK Robinson, Ed., *Microbiology of Frozen Foods*. London: Elsevier Applied Science Publishers Ltd., 1985, pp. 1–39.
3. JR Bendall. Post mortem changes in muscle. In: GH Bourne, Ed., *The Structure and Function of Muscle*. Vol. 11. New York: Academic Press, 1973, pp. 244–311.
4. W Davenport. Chilling innovations. *Meat Processing* 28:44–47, 1989.
5. BB Chrystall, CE Devine. Electrical stimulation developments in New Zealand. In: AM Pearson, TR Dutson, Eds, *Advances in Meat Research*. Vol. 1. Electrical Stimulation. Westport, CT: AVI Publishing, 1986, pp. 73–119.
6. CE Devine, RG Bell, S Lovatt, BB Chrystall, LE Jeremiah. Red meats. In: LE Jeremiah, Ed., *Freezing Effects on Food Quality*. New York: Marcel Dekker, Inc., 1996, pp. 51–84.
7. DA Ledward, Colour of raw and cooked meat. In: DA Ledward, DE Johnston, MK Knight, Eds., *The Chemistry of Muscle-based Foods*. Cambridge: Royal Society of Chemistry, 1992, pp. 128–144.
8. C Faustman, RG Cassens, The biochemical basis for discoloration in fresh meat: a review. *Journal of Muscle Foods* 1:217–243, 1990.
9. M Renerre. Review: Factors involved in the discoloration of beef. *International Journal of Food Science and Technology* 25:613–630, 1990.
10. D Cornforth. Color: its basis and importance. In: AM Pearson, TR Dutson, Eds., *Advances in Meat Research*, Vol. 9. Quality Attributes and their Measurement in Meat, Poultry and Fish Products. London: Blackie Academic and Professional, 1994, pp. 34–78.
11. DE Hunt, DH Kropf. Color and appearance. In: AM Pearson, TR Dutson, Eds., *Advances in Meat Research*, Vol. 3. Restructured Meat and Poultry Products. New York: Van Nostrand Publishers, 1987, pp. 125–159.
12. Y Tomioka. Fresh meat. In: T Kadoya, Ed., *Food Packaging*. Sandiego, CA: Academic Press, 1990, pp. 309–321.
13. CL Davey, KV Gilbert. Thaw contracture and the disappearance of adenosine triphosphate in frozen lamb. *Journal of the Science of Food and Agriculture* 27:1085–1092, 1976.
14. MC Lanari, RG Cassens, DM Schaefer, KK Scheller. Dietary vitamin E enhances color and display life of frozen beef from holstein steers. *Journal of Food Science* 58:701–704, 1993.
15. LE Jeremiah, AC Murray, LL Gibson. The effects of differences in inherent muscle quality and frozen storage on the flavor and texture profiles of pork loin roasts. *Meat Science* 27:305–327, 1990.
16. L Petrović, R Grujić, M Petrović. Definition of the optimal freezing rate-2. Investigation of the physico-chemical properties of beef *M longissimus dorsi* at different freezing rates. *Meat Science* 33:319–331, 1993.
17. MM Farouk, KJ Wieliczko, I Merts. Ultra-fast freezing and low storage temperatures are not necessary to maintain the functional properties of manufacturing beef. *Meat Science* 66:171–179, 2003.
18. PP Engler, JA Bowers. B-vitamin retention in meat during storage and preparation. A review. *Journal of the American Dietetics Association* 69:253–257, 1976.
19. WP Lehrer, AC Wiese, WR Harvey, PR Moore. Effect of frozen storage and subsequent cooking on the thiamine, riboflavin, and nicotinic acid of pork chops. *Food Research* 16:485–491, 1951.
20. BD Westerman. B-complex vitamins in meat. IV. Influence of storage time and temperature on the vitamins in pork roasts. *Journal of the American Dietetics Association* 28:331–335, 1952.
21. FA Lee, RF Brooks, AM Pearson, JI Miller, JJ Wanderstock. Effect of rate of freezing on pork quality, appearance, palatability, and vitamin content. *Journal of the American Dietetics Association* 30:351–354, 1954.
22. BD Westerman, B Oliver, DI MacKintosh. Influence of chilling rate and frozen storage on B-complex vitamin content of pork. *Journal of Agricultural and Food Chemistry* 3:603–605, 1955.
23. N Nestorov, Kozhuharova. Studies on the changes in thiamin and retinol content in pig liver and muscles after sharp freezing and long cold storage. *Proceedings of the European Meeting of Meat Research Workers*. Vol. 20, 1975, pp. 269–271.
24. JD Kemp, RE Montgomery, JD Fox. Chemical, palatability and cooking characteristics of normal and low quality pork loins as affected by freezer storage. *Journal of Food Science* 41:1–3, 1976.

25. FA Lee, RF Brooks, AM Pearson, JI Miller, F Volz. Effects of freezing rate on meat. Appearance, palatablity, and vitamin content of beef. *Food Research* 15:8–15, 1950.
26. Anonymous. Réglementation des conditions hygeniques de congélation de conservation et de décongélation des denrées animales et d'origine animales, in Ministère de l'Agriculture, Article 20, 1974.
27. K Stoll, D Dätwyler, M Fausch, T Neidhardt. Thawing of frozen foods by different methods. *Bulletin of the International Institute of Refrigeration* (Annexe 1977-1):393–397, 1977.
28. LH Kotschevar. B-vitamin retention in frozen meat. *Journal of the American Dietetics Association* 31:589–596, 1951.
29. AM Pearson, JE Burnside, HM Edwards, RS Glasscock, TJ Gunha, AP Novak. Vitamin losses in drip obtained upon defrosting frozen meat. *Food Research* 16:85–87, 1951.
30. TM Ngapo, IH Babare, J Reynolds, RF Mawson. A preliminary investigation of the effects of frozen storage on samples of pork. *Meat Science* 53:169–177, 1999.
31. A Bezanson. Thawing and tempering of frozen meat. *Proceedings of the Meat Industry Research Conference*. Chicago, Arlington, VA: American Meat Institute Foundation, 1975, pp. 51.
32. J Davídek, J Velíšek, J Pokorny. Vitamins. In: J Davídek, J Velíšek, J Pokorny, Eds., *Developments in Food Science*, Vol. 21. Chemical Changes During Food Processing, Amsterdam: Elsevier, 1990, pp. 230–293.
33. BH Ashby, GM James. Effects of freezing and packaging methods on freezer burn of hams in frozen storage. *Journal of Food Science* 38:258–260, 1978.
34. BH Ashby, GM James. Effects of freezing and packaging methods on shrinkage of hams in frozen storage. *Journal of Food Science* 38:254–257, 1978.
35. JD Kemp, BE Langlois, JD Fox. Composition, quality and microbiology of dry-cured hams produced from previously frozen green hams. *Journal of Food Science* 43:860–863, 1978.
36. BW Berry. Changes in quality of all-beef and soy-extended patties as influenced by freezing rate, frozen storage temperature, and storage time. *Journal of Food Science* 55:893–897, 1990.
37. M Ingram. Freezing, an integrated procedure. In: *Meat Freezing. Why and How*? Langford, UK: Meat Research Institute, 1974, pp. 1.1–1.4.
38. RG Bell, Chilled and frozen raw meat, poultry and their products. In: J Milner, Ed., LFRA Microbiology Handbook, Leatherhead UK, Leatherhead Food RA, 1996, pp. 1–53.
39. IB Hansson. Microbiological meat quality in high- and low-capacity slaughterhouses in Sweden. *Journal of Food Protection* 64:820–825, 2001.
40. PB Vanderlinde, B Shay, J Murray. Microbiological quality of Australian beef carcass meat and frozen bulk packed beef. *Journal of Food Protection* 61:437–443, 1998.
41. KA Murray, A Gilmour, RH Madden. Microbiological quality of chilled beef carcasses in Northern Ireland: a baseline survey. *Journal of Food Protection* 64:498–502, 2001.
42. J Sumner, E Petrenas, P Dean, P Dowsett, G West, R Wiering, G Raven. Microbial contamination on beef and sheep carcasses in South Australia. *International Journal of Food Microbiology* 81:255–260, 2003.
43. D Phillips, J Sumner, JF Alexander, KM Dutton. Microbiological quality of Australian beef. *Journal of Food Protection* 64:692–696, 2001.
44. EA Duffy, KE Belk, JN Sofos, SB LeValley, ML Kain, JD Tatum, GC Smith, CV Kimberling. Microbial contamination occurring on lamb carcasses processed in the United States. *Journal of Food Protection* 64:503–508, 2001.
45. D Phillips, J Sumner, JF Alexander, KM Dutton. Microbiological quality of Australian sheep meat. *Journal of Food Protection* 64:697–700, 2001.
46. CO Gill, J Bryant, DA Brereton. Microbiological conditions of sheep carcasses from conventional or inverted dressing processes. *Journal of Food Protection* 63:1291–1294, 2000.
47. CO Gill, J Bryant. The contamination of pork with spoilage bacteria during commercial dressing, chilling and cutting of pig carcasses. *International Journal of Food Microbiology* 16:51–62, 1992.
48. B Ray, ML Speck. Freeze injury in bacteria. *CRC Critical Reviews in Clinical Laboratory Sciences* 4:161–213, 1973.
49. BM Lund. Freezing. In: BM Lund, TC Baird-Parker, GW Gould, Eds., *The Microbiological Safety and Quality of Food*. Vol. 1. Gaithersburg MA: Aspen Publishers, 2000, pp. 122–145.

50. P Mazur. Physical and chemical basis of injury in single-celled microorganisms subjected to freezing and thawing. In: HT Meryman, Ed., *Cryobiology*. London and New York: Academic Press, 1960, pp. 214–315.
51. JHB Christian. Drying and reduction of water activity. In: BM Lund, TC Baird-Parker, GW Gould, Eds., *The Microbiological Safety and Quality of Food*. Vol. 1. Gaithersburg, MA: Aspen Publishers, 2000, pp. 146–174.
52. R DiGirolamo, J Liston, JR Matches. Survival of virus in chilled, frozen, and processed oysters. *Applied Microbiology* 20:58–63, 1970.
53. RK Lynt. Survival and recovery of enteroviruses from foods. *Applied Microbiology* 14:218–222, 1966.
54. Anonymous. Prescribed treatment of pork and products containing pork to destroy trichinae. Code of Federal Regulations. Washington, DC: US Government Printing Office, 1987, pp. 207–215.
55. GA Gardner. Microbiology of processing: bacon and ham. In: MH Brown, Ed., *Meat Microbiology*. Applied Science Publishers, London, 1982, pp. 129–178.
56. J Labadie. Consequences of packaging on bacterial growth. Meat is an ecological niche. *Meat Science* 52:299–305, 1993.
57. WJ Scott. Water relations of food spoilage microorganisms. *Advances in Food Research* 7:83–127, 1957.
58. PD Lowry, CO Gill. Temperature and water activity minima for growth of spoilage moulds from meat. *Journal of Applied Bacteriology* 56:193–1999, 1984.
59. GJ Morris, D Smith, GE Coulson. A comparative study of the changes in the morphology of hyphae during freezing and viability upon thawing for twenty species of fungi. *Journal of General Microbiology* 134:2897–2906, 1988.
60. Anonymous. Nationwide beef microbiological baseline data collection program: steers and heifers. In: United Stated Department of Agriculture, Food Safety and Inspection Service, Washington, DC, 1994, pp. 1–9.
61. Anonymous. Nationwide beef microbiological baseline data collection program: cows and bulls. In: United Stated Department of Agriculture, Food Safety and Inspection Service, Washington, DC, 1996, pp. 1–13.
62. Anonymous. Nationwide pork microbiological baseline data collection program: market hogs, in United Stated Department of Agriculture, Food Safety and Inspection Service, Washington, DC, 1996, pp. 1–13.
63. Anonymous. Nationwide federal plant raw ground beef microbiological survey, in United Stated Department of Agriculture, Food Safety and Inspection Service, Washington, DC, 1996, pp. 1–8.
64. VN Scott, L Moberg. Biological hazards and controls, In: KE Stevenson, DT Bernard, Eds., *HACCP — Establishing Hazard Analysis Critical Control Point Programs, A Workshop Manual*. Washington, DC: The Food Processors Institute, 1995, pp. 4.1–4.25.
65. MH Kothary, US Babu. Infectious dose of foodborne pathogens in volunteers: a review. *Journal of Food Safety* 21:49–73, 2001.
66. GA Dykes. The effect of freezing on the survival of *Escherichia coli* O157:H7 on beef trimmings, *Food Research International* 33:387–392, 2000.
67. MP Doyle, JL Schoeni. Survival and growth characteristics of *Escherichia coli* associated with hemorrhagic colitis. *Applied and Environmental Microbiology* 48:855–856, 1984.
68. JR Sage, SC Ingham. Survival of *Escherichia coli* O157:H7 after freezing and thawing ground beef patties. *Journal of Food Protection* 61:1181–1183, 1998.
69. SE Ansay, KA Darling, CW Kaspar. Survival of *Escherichia coli* O157:H7 in ground-beef patties during storage at 2, −2, 15 and then −2°C, and −20°C. *Journal of Food Protection* 62:1243–1247, 1999.
70. GA Dykes, SM Moorhead. Survival of three *Salmonella* serotypes on beef trimmings during simulated commercial freezing and frozen storage. *Journal of Food Safety* 21:87–96, 2001.
71. RAE Barrell. The survival and recovery of *Salmonella typhimurium* phage type U285 in frozen meats and tryptone soya yeast extract broth. *International Journal of Food Microbiology* 6:309–316, 1988.
72. SM Moorhead, GA Dykes. Survival of *Campylobacter jejuni* on beef trimmings during freezing and frozen storage. *Letters in Applied Microbiology* 34:72–76, 2002.

73. NJ Stern, AW Kotula. Survival of *Campylobacter jejuni* inoculated into ground beef. *Applied and Environmental Microbiology* 44:1150–1153, 1982.
74. SA Palumbo, AC Williams. Resistance of *Listeria monocytogenes* to freezing in foods. *Food Microbiology* 8:63–68, 1991.
75. A White, LP Hall. The effect of temperature abuse on *Staphylococcus aureus* and *Salmonellae* and in raw beef and chicken substrates during frozen storage. *Food Microbiology* 1:29–38, 1984.
76. TE Minor, EH Marth. Loss of viability by *Staphylococcus aureus* in acidified media. *Journal of Milk and Food Technology* 35:548–555, 1972.
77. PH Demchick, SA Palumbo, JL Smith. Influence of pH on freeze-thaw lethality in *Staphylococcus aureus*. *Journal of Food Safety* 4:185–189, 1982.
78. SP Trakulchang, AA Kraft. Survival of *Clostridium perfringens* in refrigerated and frozen meat and poultry items. *Journal of Food Science* 42:518–521, 1977.

16 Quality and Safety of Frozen Poultry and Poultry Products

Nahed Kotrola
Food and Beverage, Ecolab, Auburn, USA

CONTENTS

I. INTRODUCTION

In the United States, production and consumption of poultry have increased more rapidly than that of other meat sources since the 1960s. Consequently, poultry meat has become one of the meats most commonly chosen by consumers and is generally preferred by health-conscious consumers. In 1993, U.S. production of broilers reached 6.65 billion birds worth almost US$10 billion and production of turkeys totaled almost 300 million birds with a value of US$2.4 billion. Both of these statistics represent new records in production volume. The production, processing, and marketing of such large volumes of highly perishable food products make temperature control critical for this industry. Freezing and frozen storage constitutes the most effective long-term means of maintaining high quality and safety of the products for consumers.

The poultry industry has seen tremendous innovation in the past years and with this change has come alternative methods of processing, storage, and treatment of poultry meat. From the onset of refrigeration, freezing of poultry on a regular basis has been commonplace. However, with the current state of the market, in which consumers look for increasing convenience and extended preservation and safety of their food, frozen poultry has come to play a tremendous role. The modern home does not necessarily have a designated cook, as may have been the case in previous years. In dual income households, dinner needs to be easily prepared with less time involved. At the same time, consumers are demanding the same quality that the industry has come to be known for in the past. Additionally, now more than ever, food safety is at the forefront of concern for the consumer who remembers food safety failures of the past.

As frozen poultry does indeed appear to answer the needs of the common consumer, it is important to understand that this type of product carries with it several pitfalls that researchers are attempting to explore. Product tenderness is one of the primary concerns of the producer looking to freeze poultry. Compromise of the ideal consistency of the poultry product can be avoided in the freezing process. This may involve the temperature at which the meat is frozen, how quickly it is frozen, or perhaps ingredients that are added to preserve the eating quality of the meat.

Product dehydration is the other major concern of the poultry processor when attempting to freeze a poultry product. With dehydration, essentially there are problems to avoid. The first is product appearance and sensory characteristics. As we know, consumers "eat with their eyes." If the surface of the chicken product is dehydrated, it may appear discolored or washed out, making it unappealing to the consumer. Additionally, the dehydration may cause the meat to become tough or off-flavored. The second major problem with dehydration is the economic concern. As poultry is sold on a weight basis, any loss in product weight is interpreted as a financial loss in the end. This is why producers are so concerned with preserving the weight of the poultry product by preventing dehydration. It is here that producers turn to researchers to answer the question of product preservation in the realm of frozen poultry.

In addition to the meat quality concerns that producers have with frozen poultry, there is the concern of food safety. Although the simple process of freezing inhibits the growth of a large number of microorganisms, there are still a substantial number of organisms that are still present in many of these products, not only food spoilage organisms but also pathogenic microorganisms. Psychrotrophic organisms will be in the greatest abundance in this type of product, which can contribute to spoilage of food as well as human illness. It becomes apparent that with the freezing of poultry, we not only need to look at lipid oxidation as a means of spoilage but also the presence of microorganisms. Once again, it is the burden of the researcher to determine the effect of freezing on poultry safety as well as shelf life.

II. NUTRITIONAL ASPECTS OF FROZEN POULTRY AND POULTRY PRODUCTS

Although there has been much research into the nutrition of the chicken itself, there has been relatively little research on the nutritional value of poultry products for the consumer. Therefore, this section will draw inferences to the nutritional value of poultry products using information on the composition of poultry and poultry products and the nutritional requirements of adult humans. Certain nutrients known to have beneficial or detrimental effects on human nutrition will be discussed relative to their occurrence in poultry products as well.

A. Nutrient Composition of Poultry Meat

Although the requirements of humans are published by the United States Department of Agriculture (USDA) and several other sources internationally [1–6], it should be mentioned that these values are both extrapolations from a relatively small sample size and are often derived from short-term studies, which may neglect the long-term effect of a deficiency in a given nutrient. In addition, many studies use deficient populations to determine requirements, resulting in the use of subjects with altered metabolisms to determine the requirements of a nutrient for a population with normal metabolic rates. The nutrient requirements for adult humans are presented in Table 16.1. These values were obtained from the National Academy of Sciences and represent the recommended daily allowances (RDA) or acceptable macronutrient distribution range (AMDR) for macronutrients and select vitamins and minerals [1–5]. Table 16.2 shows the nutrient composition of poultry meat, both light and dark meat [6]. It should be noted that the sample size for the nutrient

TABLE 16.1
RDA of Macronutrients and Select Micronutrients for Adult Male and Female Humans

Nutrient	Adult Males (31–50 yr of age)	Adult Females (31–50 yr of age)
Protein (g/day)[a]	10–35	10–35
Digestible carbohydrates (g/day)	130	130
Fiber (g/day)	38	25
Fat (g/day)	20–35	20–35
Saturated fat (g/day)	ND	ND
Monounsaturated fat (g/day)	ND	ND
Polyunsaturated fat (g/day)	5–10	5–10
Vitamin B_{12} (μg/day)	2.4	2.4
Calcium (g/day)	1	1
Iron (mg/day)	8	18
Phosphorous (mg/day)	700	700

Note: ND refers to not determined and is specified for nutrients that are not required in the human diet.

[a]Nutrients with a range provided represent the AMDR, representing the lower limit to prevent deficiency and the upper limit to reduce the probability of developing metabolic disorders (e.g., heart disease) associated with excessive consumption of a nutrient.

Sources: Z Yang, Y Li, M Slavik. Use of antimicrobial spray applied with an inside–outside bird washer to reduce bacterial contamination on prechilled chicken carcasses. *Journal of Food Protection* 61:829–832, 1998; H Xiong, MF Slavik, JT Walker. Spraying chicken skin with selected chemicals to reduce attached *Salmonella typhimurium*. *Journal of Food Protection* 61:272–275, 1998; EC Okolocha, L Ellerbroek. The influence of acid and alkaline treatments on pathogens and the shelf life of poultry meat. *Food Control* 16 (3):217–225, 2005; G Purnell, K Mattick, T Humphrey. The use of "hot wash" treatments to reduce the number of pathogenic and spoilage bacteria on raw retail poultry. *Journal of Food Engineering* 62:29–36, 2004; RK Gast. Recovery of *Salmonella enteritidis* from inoculated pools of egg contents. *Journal of Food Protection* 56:21–24, 1993. With permission.

TABLE 16.2
Nutrient Composition of Dark and Light Cooked Chicken Meat with and Without the Skin[a]

Nutrient[b] (85.1 g portion)	Light Meat		Dark Meat	
	Skin-On	Skin-Off	Skin-On	Skin-Off
Protein (g)	26	25	22	23
Digestible carbohydrates (g)	0	0	0	0
Fiber (g)	0	0	0	0
Fat (g)	7	3	13	9
Saturated fat (g)	2	1	4	2
Monounsaturated fat (g)	3	1	5	3
Polyunsaturated fat (g)	1	1	3	2
Vitamin B_{12} (μg)	0.3	0.3	0.2	0.3
Calcium (mg)	12	13	13	13
Iron (mg)	0.9	0.9	1.2	1.2
Phosphorous (mg)	182	194	143	152

[a]Values were summarized from Appendix A in Wardlaw, 1999 [37].

[b]The standard portion size for meat samples before cooking was 3 oz or 85.1 g.

levels present is 85.1 g (3 oz), which is far less than the value that the average person would consume in a meal.

A comparison of Table 16.1 with Table 16.2 shows that poultry meat, dark or light, has benefits and shortcomings. A 85.1 g serving is more than sufficient to supply a day's worth of protein for an adult. The implication of this is that when a consumer eats a more realistic portion, they will consume protein well in excess of their requirement. As amino acids are simply broken down for energy and excreted, an excess of this nutrient is not detrimental. Fat, a nutrient that is generally considered to have detrimental effects on health, is at a low concentration in poultry meat. An 85.1 g serving of white meat without the skin has a total of 3 g of total fat, 1 g of saturated fat, and 1 g of polyunsaturated fatty acids. In contrast, dark meat cooked with the skin has 13 g of fat, 4 g of saturated fat, and 3 g of polyunsaturated fat. There is no requirement for total fat, as the body is capable of synthesizing most fatty acids; however, it is recommended that no more than 35 g of fat is consumed per day, with a maximum of ten of those grams coming from saturated fat. Poultry products, especially with the skin removed, can be easily incorporated into a diet that meets these recommendations. The amounts of calcium, phosphorous, and iron in poultry meat will contribute to the RDA but will not be sufficient to meet these requirements alone. Interestingly, the vitamin B_{12} level in poultry is such that a single serving of chicken meat is insufficient to meet the requirement. Vitamin B_{12} is only present in animal products, yet the consumption of other animal products or at least seven servings of chicken would be required to meet the daily requirement. Fortunately, vitamin B_{12} is stored in the body and the requirement does not have to be met on a daily basis.

A comparison between dark and light meat shows certain small differences. The protein level is slightly higher in white meat than in dark meat, with a correspondingly higher fat content in dark meat. This is due to the large white glycolytic fibers in the breast muscle and the combination of glycolytic and oxidative fibers in the dark meat. This fiber difference also explains the higher level of iron in dark tissue, as myoglobin is not present in high levels in glycolytic fibers. The amount of calcium and vitamin B_{12} are essentially the same in both types of meat. Comparisons of skin-on with skin-off chicken products show that fat levels are higher in skin-on products, with the fatty acid profile staying relatively constant. In addition to low saturated fat levels, cholesterol in poultry products is very low, ranging from 53 to 79 mg per 85.1 g portion [6]. There is no recommended intake of cholesterol, as it is synthesized by the body. However, in certain populations, dietary intake of cholesterol can adversely influence health [6]. To compare with cholesterol levels in other meat products, lean beef and pork cuts contain 75 mg, beef liver contains 310 mg, and kidney contains 540 mg per 85.1 g sample [6]. With moderate intake of any meat product, cholesterol levels in the plasma should not be influenced.

B. Ingredients Added to Poultry Products

The ingredients added to poultry products during further processing can affect the nutritional value of the product. Typically, these ingredients are not included at levels that would be likely to significantly reduce the nutritional value of the product. An example of a potentially harmful ingredient is salt [6]. Although sodium and chlorine are essential nutrients, the overconsumption of table salt has been linked to an increased incidence of high blood pressure. Although salt added to further processed products will add to dietary salt intake, it is unlikely that enough poultry products will be consumed to adversely affect the health of the consumer. Other ingredients, such as sodium triphosphate, may also contribute to the overall intake of sodium. Interestingly, the addition of certain nutrients to the chicken feed can increase the level of those nutrients in the meat and influence both nutritional value and shelf life. Examples of these nutrients are vitamin E, specific fatty acids, and ascorbic acid [7].

Many of the ingredients used to alter the texture or water-holding ability of the product are comprised of starches or proteins. Although it was mentioned earlier that the protein requirement for the standard adult was exceeded by an 85.1 g serving of chicken, the additional protein and starch may be of nutritional benefit and certainly will not cause harm. Organic acids that are added to the food are a source of energy, as well as a preservative, and will be of nutritional value to the consumer. Many of the ingredients that perform enzymatic functions are likely of no nutritional value.

C. Proteins

Freezing may induce some protein denaturation, as evidenced by research on freezing rates, structural changes, and drip losses. However, changes in digestibility and nutritive value of proteins, even in denatured form, are very slight and appear to be of no practical significance. The drip losses that occur during thawing and cooking can include water-soluble protein, vitamins, and minerals, but the amount lost relative to that remaining is small. Vitamin B is among the most labile nutritive components, and evidence exists that thawing and cooking may lead to significant losses of vitamin B_6, but the losses are produced largely by the subsequent thawing and cooking treatments rather than by freezing.

III. QUALITY OF POULTRY PRODUCTS

The concept of quality may be defined differently by those producing the meat than those consuming it. For the purposes of this chapter, quality will be defined by the color, texture, and tenderness of the product, the factors that most closely determine the perception of product quality. The safety of the product will be discussed subsequently. All steps of the production and processing of the meat, beginning with the transport of poultry and ending with the shipment of a consumable product, affect the quality of frozen poultry and poultry products.

A. Freezing, Packaging, and Thawing

Ideally, poultry products are frozen immediately after fabrication and held at −17.8 to −28.9°C and at approximately 85% relative humidity and retain product quality for 6–10 months [8].

Quality is maintained at temperatures below $-10^{\circ}C$ because most microbial growth, biochemical activities, and enzymatic activities are reduced to almost zero due to the fixture of cellular water molecules in a crystalline structure [9]. However, appearance and palatability are influenced by the rate of freezing, temperature, and duration of storage, packaging, and handling during and after thawing. Rapid freezing rates produce small ice crystals and result in a higher quality product. Reduced freezing rates permit larger ice crystal formation, which results in decreased product quality due to the destruction of protein structure in the muscle cell. In addition, poor or improper packaging will not protect the product surface from excessive drying and will result in freezer burn [8,9]. Lipid oxidation is a concern in certain products due to the resulting rancid off-flavors and odors but can be controlled with antioxidants and modified atmosphere packaging [10]. Proper product thawing procedures will prevent excess purge loss and reduce the risk of microbial growth. Retaining the product in the original package after thawing will prevent dehydration and drip loss. Refreezing the previously frozen products is not recommended because it drastically reduces quality and safety, causing loss of proteins, flavor, juiciness, excess drip, and increasing the likelihood of microbial growth and product deterioration [8].

B. Frozen Poultry Product Quality during Storage

The quality of frozen poultry meat may be affected by storage in a number of ways. The carcasses themselves may deteriorate externally due to dehydration, which causes the loss of the natural soft glossy appearance of the poultry meat. Additionally, with the absence of this moisture may come the hydrolysis or oxidation of the fat and protein not only on the surface of the meat but within it as well. As these things take place, the palatability of the cooked product suffers and the eating quality is subsequently lowered.

A study published as early as 1941 by the USDA [11] recognizes these effects in the quality of frozen poultry. The trend seen at this time was in contrast to the previous years. Prior to 1925, poultry carcasses were stored at temperatures typically below 10°C. As more was known about the process of freezing poultry, these temperatures tended to get lower. Processors began to store these carcasses below 0°C and, more commonly, near $-20^{\circ}C$. In the 3-year study conducted by Harshaw for the USDA [11], it was found that the poultry stored at the lower temperatures ($-20^{\circ}C$) tended to lose less weight from dehydration. Additionally, the colder carcasses also tended to have a lower incidence of off-flavors.

It is also interesting to note that in the experiment, half of the birds stored were quartered. It was found that at equally low temperatures, the quartered birds lost more weight, especially with time. This result certainly makes logical sense, as the quartered birds would have a higher surface area and thus a higher likelihood of losing moisture. This perhaps shed insight into the ideal method of storing carcasses for longer periods of time in the future. In addition, worthy of note in this experiment was the absence of quick chilling by freezing the carcasses. The time that it takes to reach the desired frozen temperature had not yet been addressed.

As it became more apparent to researchers that dehydration or drip loss during thawing was at the top of the list of concerns for frozen poultry, the next step was to attempt to prevent this phenomenon. As it is understood with the physical action of freezing a meat product, the water within the tissues begins to form crystals. With a slower freezing comes a larger crystal. The faster the freezing takes place, the smaller the ice crystals. The crystal is of concern because as it forms, it punctures the cells in the tissue. As the product thaws, the water that was within the cell is lost, which we refer to as drip loss. Therefore, the first step in attempting to retain moisture within the product is to hasten the freezing process.

The addition of nonmeat ingredients was the next logical step in attempting to retain moisture in the meat product. A study conducted by Yoon [12] examined the effect of the addition of phosphate to the product to retain water. Phosphate has been used for many years to improve the quality of finished poultry products and to provide greater processing flexibility. The addition of trisodium phosphate not only improves water binding, texture, color, and flavor [13], but it also has an antimicrobial effect [14]. The Yoon study concluded that no significant texture toughening was observed in frozen chicken breasts after 10 months of storage at frozen temperature regardless of treatment. This suggests that toughening is not a determinant factor in the quality loss of frozen chicken breasts. Instead, improving water-binding ability of chicken meat without the ice crystal formation during frozen storage is most important for preserving the eating quality of frozen chicken breasts. This result can be accomplished by treating chicken breasts with 10% trisodium phosphate or STPP before frozen storage.

In further research, it has been shown that there has been actually less drip loss from carcasses that have been stimulated before harvest and then treated with a phosphate solution before freezing. This is due to the interaction with the pH decline that takes place naturally as a part of rigor mortis. A study conducted by Young and Buhr [15] went one step further in attempting to observe this phenomenon. In this study, the carcasses were slaughtered, and the breasts were harvested 1 h *postmortem*. They were then treated with a phosphate solution and frozen. The difference in this study was that the carcasses were exposed to electrical stimulation in the slaughter process so as to rapidly decrease the pH.

In the study, electrical stimulation had no direct effect on pH, cooking loss, or shear values, whereas polyphosphate increased pH and decreased cooking loss. Polyphosphate treatment caused fillets from unstimulated carcasses to absorb more marinade and yielded more drip than those from stimulated carcasses. Fillets from stimulated carcasses marinated in NaCl solution without polyphosphate yielded less drip than those from unstimulated carcasses [15]. It becomes evident that to alter the capacity of the meat to retain water, it is important to remember the natural physiology and biochemistry of the meat as it passes through the rigor process. Freezing may trigger a moisture loss; however, there are mechanisms within the meat that can indeed hinder this from occurring.

Phosphate solution has certainly been the most popular and explored additive to improve the quality of frozen or fresh poultry products. There have been, however, other means of altering poultry quality that have been explored as well. For example, the addition of dietary tea catechins as an effect on broiler quality in the fresh or frozen state was explored [6]. The study showed that the dietary catechins inhibited lipid oxidation in long term frozen stored chicken. In another study [7], the effect of selenium, vitamin E, and ethoxyquin on lipid peroxidation was explored. It was found that dietary selenium and other factors affecting selenium status may be useful in retarding the development of oxidative rancidity in frozen poultry products.

Regardless of the approach that the researcher takes, whether it is to treat the product after harvest or to alter the dietary intake of the bird before harvest, it is apparent that many avenues have been explored to find a method to keep the quality integrity of the frozen poultry product. When considering frozen poultry, a widely overlooked aspect of the product quality discussion is the means in which it is thawed or cooked. Keeping in mind that the consumer ultimately desires a product that is convenient and quick to prepare, research does indeed need to reflect this goal.

Younathan et al. [16] had this concept in mind when they set out to determine the effect of microwave energy as a rapid thawing method for frozen poultry. There has been tremendous growth in recent years in the use of the microwave oven for practically every aspect of meal preparation. Samples in this study were frozen in a similar fashion and then thawed using a microwave oven. The samples were then compared to samples thawed overnight in a refrigerator. Microwave defrosting of frozen breasts resulted in a slightly more tender product. However, the breasts that were thawed in the microwave experienced more drip loss. Hence, it can be seen that this widely used means of preparing poultry has a mixed effect on the quality of the meat. While preserving tenderness, it actually loses moisture.

1. Carcass Changes during Storage

Fast freezing results in lighter colored carcasses than slow-frozen poultry. During slow freezing, the skin dries, shrinks, and becomes more transparent and, as a result, has a darker appearance than if the same carcass were frozen rapidly. Light-colored frozen carcasses are generally considered to be more desirable than dark-colored ones. During slow freezing, large ice crystals are formed with a resulting transparent surface layer and a darker appearance on the surface of the frozen carcass. A major portion of the darkening takes place in the skin and the remainder in the surface layer of the flesh. The rate of freezing of the flesh below the surface has no effect on the surface color. Although well-finished carcasses are desirable, fat acts as an insulator and retards the freezing rate regardless of freezing method.

Research reported that when carcasses were held for 6–24 h before freezing, the breast muscles were tendered than those of fryers held only 40 min before freezing. In addition, turkey carcasses held for 1 h and then frozen at $-2.8°C$ (27°F) for 3 days provided adequate tenderization and no adverse flavor changes were observed after 14 days at $-2.8°C$ (27°F). When the birds were frozen before the onset of rigor, wide differences in the rate of thawing did not adversely affect tenderness.

2. Bone Darkening during Storage

It is known that chilling before freezing will not prevent bone darkening. However, storing at temperatures below $-9.4°C$ (15°F) lowers or prevents bone darkening entirely. This is caused by the leaching of hemoglobin from bone marrow to adjacent muscle as result of the freeze or thaw treatment. Leaching only occurs in the carcasses from relatively young birds because the bones are not completely calcified and are more porous than in mature birds. Although product quality does not change, the appearance constitutes a negative factor in consumer acceptance. Many attempts have been made to eliminate bone darkening in young frozen birds. Although some reduction can be achieved, only removing the bone marrow or cooking before freezing will eliminate this defect.

3. Palatability Changes during Storage

Palatability changes are the most critical for producers because of the need to ultimately provide consumers with a pleasurable eating experience. Palatability differences in poultry resulting from freezing treatments are often reported as small or nonexistent. When several freezing rates were compared, it was found that surface color differed, whereas cooking yields or palatability differences were not found. Similarly, no difference in tenderness was found for chicken breasts frozen at temperatures from -18 to $-68°C$.

4. Protein Changes during Storage

Some protein denaturation and solubility changes are known to occur as a result of freezing, but the practical significance of these changes is not clear. Comparisons of thermal gelation properties of poultry dark meat frozen at different rates showed little effects of freezing or freezing rates on rheology or gel strength. Freezing, regardless of rate, resulted in somewhat greater water-holding capacity, which might be caused by increasing charged sites on the meat proteins. However, freezing does not appear to be a major detriment to processing functionality of poultry meat.

C. Intrinsic Chemical Reactions

The chemical reactions of greatest concern with frozen poultry, those involving lipid oxidation, constitute a major determinant of frozen product shelf life, because poultry lipids are relatively unsaturated and susceptibility to oxidation is high. Freezing results in concentration of solutes, which catalyzes the initiation of oxidative reactions. The greater the concentration of these catalysts, the greater the acceleration of changes. Freezing also disrupts and dehydrates cell membranes, exposing membrane phospholipids to oxidation.

1. Protein Denaturation

Chemical reactions during freezing and frozen storage may also contribute to some protein denaturation and decreased solubility. Decreases in sulfhydryl groups and ATPase activity occur during frozen storage and are indicative of protein changes. Peptides and amino acids are also increased in the drip fluid, as are nucleic acids, indicating protein changes and structural cellular damage, respectively.

2. Enzyme Activity

Biochemical reactions involving muscle enzymes may occur at very slow rates in frozen poultry, depending on storage temperature. For example, glycolysis may continue if not completed before freezing and may reach completion if enough time is available. Phosphatase activity is not completely inhibited by frozen storage. Muscle ATPase activity may continue during frozen storage, and some proteolytic enzyme activity has been suggested.

D. The Concern for Quality and Consumer Acceptance

Poultry genetics, particularly for broiler chickens, does not contribute to much variability in quality. Diets of live birds generally do not influence meat quality, with the exception of the well-recognized effect of highly unsaturated oils such as fish oils. A possible positive contribution of dietary factors may come from vitamin E (tocopherol), which has been shown to be effective in frozen and frozen or thawed beef for improving color stability and reducing oxidative changes. Similar effects of tocopherol have been observed in poultry for suppressing lipid oxidation.

Although the implementation of innovative intervention strategies are necessary in the processing plant to improve storage stability and safety, they are worthless if the desired palatability and organoleptic attributes including flavor, texture, color, and odor are compromised.

1. Preservatives

There are many antimicrobial products that may be incorporated into poultry products to increase shelf life and prevent spoilage. Salt can be added to reduce the water activity of the product and reduce the water available for the microbes to use in metabolic processes. Salt alone, however, must be added at 15% to completely prevent microbial growth [17]. This level of salt is generally not acceptable for most products, which precludes the use of salt alone as a preservative. Sodium nitrite may, as well, be used as a preservative [17–19]. The USDA has strict regulations regarding the use of nitrite in food products [17]. This is due to the potential toxic effects of nitrite itself and of nitrosamine, the product of a reaction between nitrite and amino acids when subject to high temperature. Owing to the potential formation of nitrosamines, an antioxidant must also be incorporated into the product to ensure that NO_2 is reduced to NO. This prevents any reaction with amino acids and the formation of nitrosamines.

2. Acidulants

Sodium acetate, sodium diacetate, and lactate have also been used in poultry products to reduce microbial growth. These products act as acidulants and may be incorporated at levels of 0.25% [18]. These organic acids are also commonly used in carcass rinses and chiller water to control microbial growth without incurring the flavor-altering effects of chlorine. Owing to their acidic nature, the organic acids also alter the flavor of the product, but most often the acidic flavor is desirable in the products in which they are used.

3. Antioxidants

Antioxidants can be used to preserve the product in terms of preventing lipid oxidation. Many compounds, most commonly butylated hydroxyanisole (BHA) and butylated hydroxytoluene (BHT), can be incorporated into the product [18,20]. Antioxidants are particularly important in frozen poultry products, as the rate of lipid oxidation usually determines shelf life [20]. Liquid smoke products also provide antioxidant and antimicrobial action due to the phenols that they contain. Additionally, alpha-tocopherol and ascorbic acid (vitamins E and C, respectively) can be added to the product as antioxidants. Ascorbic acid is thought to chelate iron, which is a catalyst for oxidative pathways [18]. These compounds are fat-soluble and function to quench free radicals that are the main cause of lipid oxidation. Off-flavors and rancid odors that result from lipid oxidation are often indications of impending microbial spoilage [17]. Delaying the onset of lipid oxidation can improve the shelf life of the product.

The list of ingredients discussed here is by no means exhaustive. Topics such as spices and flavors have been omitted entirely due to space considerations and the length of their potential discussions. There is a long list of ingredients that can be used to improve the water-binding ability of poultry products. Ingredients such as salt and phosphates are commonly injected into whole meat products as part of a marinade and function to increase water binding and retention of the marinade. Other products, such as the plant protein and starch ingredients described, are more useful in further processed products. The texture and water-binding properties of further processed products can be manipulated through an infinite number of combinations of the various ingredients. Careful consideration must be taken when choosing an ingredient or set of ingredients for a given product. Several antimicrobial products are also available for incorporation into poultry products. The use level of these products may be controlled by the USDA, and careful control of their incorporation and the choice of preservatives must be implemented.

4. Stability and Acceptability

The important factors in determining stability and acceptability are those that influence oxidative and organoleptic change. As discussed previously, the amount of unsaturated lipids and the oxidative or antioxidant environment surrounding membrane phospholipids are major factors. In addition, important processes are treatments such as cooking, which initiate oxidative change and decrease practical frozen storage life by about 50%. Addition of salt as well as chopping or grinding also greatly shortens frozen storage life. The effects of packaging are critical. High barrier vacuum nearly doubles storage life in most cases when compared with a low-barrier polyethylene film. The final major determinant is storage temperature. In general, intact poultry carcasses and cuts will remain acceptable for at least 12 months if they are well packaged and held at $-18°C$. Lowering storage temperature increases stability. Thus, 24 months of acceptability may be expected with temperatures of $-25°C$ or less. Cooked poultry products such as fried chicken may be expected to have a practical storage life of 6–9 months or more at $-30°C$. The stability of the frozen products can be summarized as dependent on the product characteristics, the processes used for the product, and the packaging system chosen.

E. Protecting the Quality of Frozen Poultry

Product quality can be defined using many factors including appearance, yields, eating characteristics, and microbial characteristics. The objective of the freezing process is to preserve product characteristics at a desirable level as long as possible. The use of high-quality initial materials is essential to high-quality frozen products. The USDA inspects each bird for wholesomeness and fitness for consumption and also grades ready-to-cook carcasses as A, B, or C quality. Determinants of these grade levels are conformation, fleshing fat cover, pinfeathers, and defects such as exposed flesh, discoloration, or broken bones. Freezing effects are also considered; a "freezing defects" category is used, with darkening, pockmarks, or freezer burn resulting in lower grades.

1. Product Processing

Slaughter, processing, and handling can affect carcass quality by inducing bruises, tears, or even broken bones. Most of these defects will be obvious and result in downgrading. Following slaughter and evisceration, rapid and well-controlled chilling of poultry carcasses is essential to maintaining microbiological quality and tenderness. It is undesirable to freeze poultry meat immediately following slaughter, as meat tenderness is likely to be decreased. Electrical stimulation of carcasses can be used to speed up *postmortem* tenderization if deemed necessary. Poultry must be chilled to a temperature of 4°C or less within a time limit depending on carcass weight. A large majority of poultry carcasses are chilled in ice water slush because of relatively fast chilling and because some moisture is absorbed by the carcasses, thereby increasing yields. Maximum tolerances based on carcass weight for absorbed moisture have been established by the USDA and range from 4.3% for heavy turkeys to 8% for lightweight chickens. The absorbed water is significant to the freezing process because it can influence freezing rate and the extent of ice crystal formation. This excess moisture must be drained away before packaging and freezing to prevent excessive drip or separated water from occurring during thawing. The second and perhaps more critical consideration for freezing is that adequate time before freezing must be given for the meat to achieve optimum tenderness (6–8 h for chickens and 12–24 h for turkeys).

Product preparation before freezing may include cutting, deboning, slicing, and other operations to provide greater convenience. Preparation of products by cooking before freezing is becoming increasingly more popular as a greater variety of poultry products, including breaded and fried portions, cured and smoked products, and items in marinades or broths, are offered to consumers. Poultry products, according to the USDA Food Safety Inspection Service, must be either raw or fully cooked before freezing. Fully cooked means that an internal temperature of 71°C for uncured products or 68°C for cured products must be reached. Breaded products such as patties may be fried to set the breading without further cooking of the product.

2. Product Packaging

Following product preparation, packaging requirements become a consideration. Packaging is one of the most important factors in maintaining quality during frozen storage and is especially important for cooked products. Most important are protection from exposure to oxygen and from loss of moisture. As a result, packaging must provide a good barrier to oxygen, to prevent off-flavor development, to moisture, to avoid dehydration or freezer burn. It has been suggested that the packaging system is of paramount importance to the quality of frozen products and of even greater importance than the freezing treatment itself because the stability of the product during storage is dependent on the protection provided by the package. Use of PPP (product, process, and packaging) concepts has been suggested to be the most important overall consideration for frozen product quality.

3. Product Storage

Poultry and poultry products are frequently frozen to achieve long-term storage with minimal loss of quality. Poultry accommodates freezing treatments relatively well, and the quality level at the time of freezing can be maintained as long as the freezing process and storage conditions are adequate. Research on frozen poultry generally indicates that a storage temperature of about −18°C or lower will minimize deterioration if products are well packaged and temperature fluctuation is minimal. Over a range of −10 to −30°C, a reduction of 10°C in storage will apparently double or triple the number of days that broilers retain stability and acceptability. Storage needs to be at

-18°C or less, using an appropriate high-barrier package with minimal temperature change during the storage period.

IV. SAFETY OF POULTRY PRODUCTS

In recent years, the safety of food has received an increasing amount of attention. Internationally, a Hazard Analysis and Critical Control Point (HACCP) process for the production of food has been adopted. Poultry products are no exception. The presence of physical, biological, or chemical hazards in poultry products pose health risks to the consumer and economic risks to the company. The production of wholesome, unadulterated poultry products is essential to maintain public health and to maintain product confidence. Although biological hazards are the most prevalent and dangerous of the hazards, chemical and physical hazards must also be considered in designing a HACCP plan designed to minimize these hazards. Biological hazards have received the most attention in the literature relating to poultry products. *Salmonella* and *Campylobacter* contamination remain significant concerns in the poultry industry. Physical hazards are also a significant concern in the poultry industry, although the consequences of adulteration are generally lower than for biological hazards. Chemical hazards have a low prevalence in the industry and will remain low due to sanitation standard operating procedures (SSOPs) and HACCP plans.

A. CONCERN FOR FOOD SAFETY AND QUALITY

According to the Centers for Disease Control and Prevention, 76 million illnesses, 325,000 hospitalizations, and 5200 deaths related to foodborne illness occur in the United States each year. Of these statistics, known pathogens account for an estimated 14 million illnesses, 60,000 hospitalizations, and 1800 deaths annually [21]. Consequently, the estimated annual costs of human foodborne illness in the United States range from $8.5 billion to $20 billion [22]. Mead et al. [23] reported that nearly 2.4 million cases are caused by *Campylobacter* spp., 1.4 million cases are caused by nontyphoidal *Salmonella serovars*, and 270,000 cases are caused by pathogenic *Escherichia coli* including *E. coli O157:H7.* Unfortunately, for the poultry industry, poultry and poultry products have been implicated as a major source of *Campylobacter* and *Salmonella* infections in humans [24].

A recent study conducted by Zhao et al. [25] demonstrated the prevalence of *Campylobacter* spp., *S. serovars*, and *E. coli* in raw retail meats, including chicken and turkey products, obtained from four supermarket chains in the greater Washington, D.C. area over a 14-month period. The majority (70.7%) of chicken samples ($n = 184$) were positive for *Campylobacter*, 38.7% of the chicken samples ($n = 212$) were positive for *E. coli*, and only 4.2% of the chicken samples were positive for *Salmonella.* As a result of their findings, Zhao et al. [25] suggest focusing on effective prevention strategies such as implementing on-farm practices that reduce pathogen contamination, increasing hygiene in both slaughter and meat processing, continuing implementation of HACCP systems, and increasing consumer education efforts to reduce the presence of foodborne bacterial pathogens in meat products at the retail level. Understanding that multiple entry points exist for contamination of poultry during various phases of the growth period and processing procedures is required to develop multifaceted intervention guidelines to successfully control contamination.

Although product quality is certainly an important factor when considering frozen poultry, there is also the consideration of microbiological spoilage or even pathogenic effects. As mentioned earlier, consumers "eat with their eyes," and thus any type of off-condition will affect eating experience. Dehydration can alter the external appearance of the poultry so can microbiological growth. Product safety is also at the top of the list of priorities that the modern consumer has. With the stress that producers have on biosecurity coupled to a history of food recalls, the microbiology of food products is very important.

The misconception of the common consumer is that the food is safe if it has been frozen. This is perhaps the most dangerous mistake that an individual can make. As we have come to find through research, a number of organisms grow in refrigerated and even frozen temperatures. Thus, it is crucial that processors who are looking to produce frozen poultry do not neglect the microbial safety of their product. Perhaps, the only thing more important to understand that microbes can grow in frozen conditions is knowing what type of organisms can grow under this condition.

A study conducted by Norberg [26] in 1981 set out to determine just what are the organisms growing in frozen poultry. It is the conception of the public and even some food scientists that *Salmonella* sp. is the only organism of concern in poultry. This impression is reinforced by the idea that very little growth of any organism occurs in frozen poultry. The study [26] concluded that this was certainly not the case. The investigation showed that frozen chicken does contain other pathogenic bacteria than *Salmonellae.* Namely, *Campylobacter* and *Yersinia enterocolitica* were isolated in much higher frequency. *C. jejuni* was found in 22% of the samples, *Y. enterocolitica* was found in 24.5% of the samples, and *S. typhimurium* was found in one sample. Aerobic plate counts (APCs) and numbers of coliform bacteria at 37°C were not found to be noticeably higher in samples containing pathogens than in pathogen-free samples.

B. Preventive Measures in the Processing Plant

In addition to the existing critical control points, which include proper equipment maintenance, product and washer temperature controls, water replacements, and counter-flow technology in the scalder and chiller, chemical intervention strategies have been implemented in the processing plant to control foodborne pathogens [27]. These methods include phosphate dips or sprays (e.g., trisodium phosphate and polyphosphates), ionizing radiation, ozonation, and organic acid sprays to reduce the numbers of bacterial pathogens present on raw animal products.

1. Phosphate Treatments

Trisodium phosphate (TSP) is a GRAS product approved in 1992 by the U.S. Department of Agriculture and can be used as a dip or a spray [28]. Several concurring studies proved that TSP is active against gram-negative bacteria, including salmonellae, coliforms, *E. coli O157:H7*, *campylobacters*, and *pseudomonads* on the skin of chicken carcasses [28–30]. It is anticipated that the antimicrobial mechanisms of TSP include (a) exposing microorganisms to high pH, which would affect cell membrane components; (b) enhancing detachment of bacteria from food surfaces by sequestration of metal ions; and (c) removing fat from the skin surface, thereby allowing bacteria to be more effectively washed from the food surface. Recently, Sampathkumar et al. [30] demonstrated the efficacy of high pH during TSP treatments against *S. enterica serovar enteritidis.* The process involved immersing postchill whole birds for 15 sec in a 10% solution of Av-Gard TSP and then allowing the excess TSP solution to drip from the bird. Subsequent microbiological testing demonstrated the effectiveness of TSP treatments and the pH concentrations adjusted to pH $11.0 + 0.2$. Both tests resulted in the loss of cell viability and membrane integrity, demonstrating that alkaline pH treatments permeabilize and disrupt the cytoplasmic and outer membranes of *serovar enteritidis* cells, instigating the release of intracellular contents and eventually causing cell death [30].

2. Chlorine Compounds

Chlorine is commonly used in poultry processing facilities at levels less than 50 ppm in the immersion chillers to facilitate the reduction of bacterial contaminants. The efficacy of chlorine decreases as pH, organic load, temperature, bacterial concentrations, and trace minerals change in the same environment. Acidified sodium chlorite (SanovaR) is a mixture of sodium chlorite and citric acid and results in an effectively low pH antimicrobial spray product. SanovaR is effective against *E. coli*, *Salmonella*, *Listeria*, *Campylobacter*, and other potential bacterial contaminants [31].

3. Ozonation

Ozone is a soluble and unstable blue gas commercially produced by passing electric charges or ionizing radiation through air or oxygen. The powerful compound is used to inactivate bacteria through its oxidizing properties. Ozonation is effective against bacteria including *E. coli O157:H7*, *S. typhimurium*, and *Pseudomonas fluorescens* on muscle foods [31].

4. Organic Acid Treatments

Many antimicrobial sprays have been formulated and applied with inside or outside bird washers in attempts to reduce bacterial contamination of prechilled chicken carcasses. Lactic acid has been a popular treatment applied at 1 and 2% levels. Yang et al. [32] found that a 2% lactic acid treatment was effective to reduce total aerobes by 1.03 log 10 CFU per carcass. However, this treatment caused slight chicken skin discoloration. Xiong et al. [33] also evaluated the effectiveness of a lactic acid prechill spray to reduce *S. typhimurium* attachment on chicken skin and found that a 1–2% lactic acid spray resulted in a 2.2 log 10 CFU per carcass reduction [33]. Subsequent research by Okolocha and Ellerbroek [34] reported that a 1% lactic acid dip (PuracR) was more effective than a spray, which resulted in a 0.6 log 10 CFU/ml APCs reduction, a 1.1 log 10 CFU/ml *Enterobacteriaceae* reduction, a 0.4 log 10 CFU/ml *Lactobacillus* reduction, and a 0.4 log 10 CFU/ml *Pseudomonas* reduction. Glutamal bioactiveR, a 1% formulation of active constituents (sugars, foodstuff phosphates, ascorbic/isoascorbic acid, or their inorganic salts) with lactic acid as an activator, was also more effective as a dip than a spray. Glutamal bioactiveR resulted in a 1.0 log 10 CFU/ml APC reduction, a 0.9 log 10 CFU/ml Enterobacteriaceae reduction, a 0.2 log 10 CFU/ml *Lactobacillus* reduction, and a 0.2 log 10 CFU/ml *Pseudomonas* reduction. Overall, both PuracR and Glutamal bioactiveR products effectively reduced original pathogen loads and received high acceptability scores by consumers [34].

5. Hot Wash Treatments

Recently, Purnell et al. [35] investigated the use of hot water immersion treatments to sufficiently reduce bacterial numbers during processing. The hot water immersion treatments were located after the inside or outside wash cabinets and were conducted following suitable time–temperature parameters previously determined. Purnell et al. [35] concluded that a heat treatment at 75°C for 30 sec significantly reduced levels of *Campylobacter*, Enterobacteriacae, and APC but resulted in unacceptable grades due to epidermis damage or skin tears during packaging positioning. However, at 70°C for 40 sec heat treatment resulted in both reduced bacterial levels and acceptable grades relating to surface appearance [35].

6. Temperature Treatments

Varying temperature treatment studies have reported that freezing food inactivates substantial populations of *C. jejuni* by >2 log 10 CFU/g [36,37]. The detrimental mechanisms of freezing and thawing to living cells is due to cell injury attributed to ice nucleation and dehydration. Oxidative damage to cells has also been implicated as a mechanism contributing to freeze–thaw injury, resulting from an oxidative burst occurring within cells during thawing [38].

Zhao et al. [39] conducted several trials to determine the inactivation rates of *C. jejuni* on poultry exposed to short-term and long-term freezing at different temperatures and after superchilling to an internal temperature of −3.3°C. All samples were contained in Whirl-Pak bags. Long-term holding for 8–12 weeks at −20°C resulted in a 3 log 10 CFU/g reduction when compared with short-term freezing for 72 h or less at −20°C, which resulted in a 1.7 log 10 CFU/g reduction. Superchilling only resulted in significant bacterial reductions at −196°C for 20 sec. The −196°C for 20 sec temperature treatment may have been influenced by the technique because to reach the desired temperature, submersion of the samples into liquid nitrogen was required. This technique is not likely to occur during commercial processing. The remaining superchilling temperatures at −80°C for 330 sec,

$-120°C$ for 220 sec, and $-160°C$ for 150 sec involved sample exposure to vapor-state liquid nitrogen. Therefore, the study concluded that the superchilling conditions utilized by the poultry industry are not adequate to substantially reduce *Campylobacter* populations on fresh products [39].

V. CONCLUSIONS

As has been illustrated, there has been a tremendous effort for the innovation of the frozen poultry market. There will be certainly an equal or greater push to understand the many facets of producing this product in a premium condition. With the profile of the modern consumer, the frozen poultry market will certainly continue to thrive. The need for convenience and ease of preparation will continue to be a top priority for consumers in the modern market. However, the problems that are typically associated with frozen poultry must be dealt with to produce frozen poultry in an economical and safe way.

Research thus far has certainly attempted to address these problems. In the future, there will continue to be a need for research dealing with frozen poultry. It is a relevant topic, and one that will continue to furnish abundant research endeavors. Intervention strategies protecting product quality and safety are vital during processing and are critical to subsequent steps including freezing and thawing. In addition to implementing intervention strategies, proper employee training, and consumer education are also key in the "farm-to-fork" approach. Most importantly, education is critical to reduce the incidence of foodborne illness. Education complements regulatory and research activities and instills responsibility for food safety in everyone involved.

REFERENCES

1. Anonymous. Dietary reference intakes for energy, carbohydrate, fiber, fat, fatty acids, cholesterol, protein and amino acids. The National Academies, 500 Fifth St NW, Washington DC, U.S.A., 2002.
2. Anonymous. Dietary reference intakes for calcium, phosphorous, magnesium, vitamin D and floride. The National Academies, 500 Fifth St NW, Washington DC, U.S.A., 1998.
3. Anonymous. Dietary reference intakes for vitamin C, vitamin E, selenium and carotenoids. The National Academies, 500 Fifth St NW, Washington DC, U.S.A., 1998.
4. Anonymous. Dietary reference intakes for vitamin A, vitamin K, arsenic, boron, chromium, copper, iodine, iron, manganese, molybdenum, nickel, silicon, vanadium and zinc. The National Academies, 500 Fifth St NW, Washington DC, U.S.A., 2001.
5. Anonymous. Dietary reference intakes for thiamin, riboflavin, niacin, vitamin B6, folate, vitamin B12, pantothenic acid, biotin and choline. The National Academies, 500 Fifth St NW, Washington DC, U.S.A., 1998.
6. GM Wardlaw. *Perspectives in Nutrition.* 4th ed. New York: WCB McGraw-Hill. 1999.
7. R Bou, F Guardiola, S Grimpa, A Manich, A Barroeta, R Codony. Influence of dietary fat source, α-tocopherol, and ascorbic acid supplementation on sensory quality of dark chicken meat. *Poultry Science* 80:800–807, 2001.
8. JT Keeton. Formed and emulsified products. In: AR Sams, Ed., *Poultry Meat Processing*, New York, NY: CRC Press, 2001, pp. 196–226.
9. D DeFremery, AA Klose, RN Sayre. Freezing poultry. In: NW Desrosier, Ed., Westport, CT: AVI Publishing Company, Inc., 1977, pp. 240–272.
10. CM Owens. Coated poultry products, In: AR Sams, Ed., *Poultry Meat Processing*, New York, NY: CRC Press, 2001, pp. 227–242.
11. HM Harshaw, WS Hale, TL Swenson, LM Alexander, RR Slocum. *Quality of Frozen Poultry as Affected by Storage and Other Conditions.* Technical Bulletin 768, United States Department of Agriculture, April, 1941.
12. KS Yoon. Texture and microstructure properties of frozen chicken breasts pretreated with salt and phosphate solutions. *Poultry Science* 81:1910–1915, 2002.
13. JE Steinhauer. Food phosphates for use in the meat, poultry and seafood industry. *Dairy Food Sanitation* 3:244–247,

14. RB Tompkin. Indirect antimicrobial effects in foods: phosphates. *Journal of Food Safety* 6:13–27, 1984.
15. LL Young, RJ Buhr. Effect of electrical stimulation and polyphosphate marination on drip from early-harvested, individually quick-frozen chicken breast fillets. *Poultry Science* 79:925–927, 2000.
16. MT Younathan, AJ Farr, DL Laird. Microwave energy as a rapid-thaw method for frozen poultry. *Poultry Science* 63:265–268, 1984.
17. E Dransfield, AA Sosnicki. Relationship between muscle growth and poultry meat quality. *Poultry Science* 78:743–746, 1999.
18. JT Keeton. Formed and emulsion products. In: A Sams, Ed., *Poultry Meat Quality*, Florida: CRC Press, 2001, pp. 195–226.
19. DP Smith, JC Acton. Marination, cooking and curing of poultry products. In: A Sams, Ed., *Poultry Meat Quality*, Florida: CRC Press, 2001, pp. 257–280.
20. CM Owens. Coated poultry products. In: A Sams, Ed., *Poultry Meat Quality*, Florida: CRC Press, 2001, pp. 227–242.
21. Anonymous. What is FoodNet? Centers for Disease Control and Prevention, 1600 Clifton Road, Atlanta, GA 30333, U.S.A., 2003.
22. LM Spencer. Food on the hoof: stamping out food safety concerns. *Journal of American Veterinarian Medical Association* 207:280–283, 1991.
23. PS Mead, L Slutsker, V Dietz, LF McCaig, JS Bresee, C Shapiro, PM Griffin, RV Tauxe. Food-related illness and death in the United States. *Emerging Infectious Diseases* 5:607–625, 1999.
24. JS Bailey. Detection of *Salmonella* cells within 24 to 26 hours in poultry samples with the polymerase chain reaction BAX system. *Journal of Food Protection* 61:792–795, 1998.
25. C Zhao, B Ge, JD Villena, R Sudler, E Yeh, S Zhao, DG White, D Wagner, J Meng. Prevalence of *Campylobacter* spp., *Escherichia coli* and *Salmonella* serovars in retail chicken, turkey, pork, and beef from the Greater Washington, D.C., Area. *Applied Environmental Microbiology* 67:5431–5436, 2001.
26. F Van-Immerseel, K Cauwerts, LA Devriese, F Haesebrouck, R Ducatelle. Feed additives to control *Salmonella* in poultry. *World's Poultry Science Journal* 58:501–512, 2002.
27. PL White, AR Baker, WO James. Strategies to control *Salmonella* and *Campylobacter* in raw poultry products. *Revolutionary Science Technology* 16:525–541, 1997.
28. P Coppen, S Fenner, G Salvat. Antimicrobial efficacy of AvGard carcass wash under industrial processing conditions. *British Poultry Science* 39:229–234, 1998.
29. R Capita, C Alonso-Calleja, M Sierra, B Moreno, M del Camino Garcia-Fernandez. Effect of trisodium phosphate solutions washing on the sensory evaluation of poultry meat. *Meat Science* 55: 471–474, 2000.
30. B Sampathkumar, GG Khachatourians, DR Korber. High pH during trisodium phosphate treatment causes membrane damage and destruction of *Salmonella enterica* serovar enteritidis. *Applied Environmental Microbiology* 69:122–129, 2003.
31. BM Hargis, DJ Caldwell, JA Byrd. Microbiological pathogens: live poultry considerations. In: AR Sams, Ed., *Poultry Meat Processing*, New York, NY: CRC Press, 2001, pp. 121–135.
32. Z Yang, Y Li, M Slavik. Use of antimicrobial spray applied with an inside–outside bird washer to reduce bacterial contamination on prechilled chicken carcasses. *Journal of Food Protection* 61: 829–832, 1998
33. H Xiong, MF Slavik, JT Walker. Spraying chicken skin with selected chemicals to reduce attached *Salmonella typhimurium. Journal of Food Protection* 61:272–275, 1998.
34. EC Okolocha, L Ellerbroek. The influence of acid and alkaline treatments on pathogens and the shelf life of poultry meat. *Food Control* 16 (3):217–225, 2005.
35. G Purnell, K Mattick, T Humphrey. The use of "hot wash" treatments to reduce the number of pathogenic and spoilage bacteria on raw retail poultry. *Journal of Food Engineering* 62:29–36, 2004.
36. RK Gast. Recovery of *Salmonella enteritidis* from inoculated pools of egg contents. *Journal of Food Protection* 56:21–24, 1993.
37. RK Gast. Evaluation of direct plating for detecting *Salmonella enteritidis* in pool of egg contents. *Poultry Science* 72:1611–1614, 1993.
38. RK Gast. Detection of *Salmonella enteritidis* in experimentally infected laying hens by culturing pools of egg contents. *Poultry Science* 72:267–274, 1993.
39. T Zhao, GO Ezeike, MP Doyle, YC Hung, RS Howell. Reduction of *Campylobacter jejuni* on poultry by low-temperature treatment. *Journal of Food Protection* 66:652–655, 2003.

17 Safety and Quality of Frozen Fish, Shellfish, and Related Products

Jacek Jaczynski
West Virginia University, Morgantown, USA

Angela Hunt and Jae W. Park
Oregon State University, Astoria, USA

CONTENTS

I. INTRODUCTION

It is imperative to recognize that the safety of foodstuffs must not be compromised for quality. Humans consume over 1000 species of fish and shellfish that grow in diverse habitats and geographic locations of the world. These fish and shellfish unavoidably carry a variety of microorganisms from aquatic and terrestrial sources. The high levels of free unbound water and nutrient availability, combined with the psychrophilic character of fish microflora, render fish and fishery products more susceptible to wide aspects of safety and quality issues than foods derived from terrestrial animals, often spoiling in a short period of time under refrigeration. In addition to spoilage microorganisms, fish may contain various potential human pathogens. It is often difficult to maintain the quality of fish and fishery products due to the distance between consumers and the harvesting areas, which provide opportunities for microbial growth.

The key differences between fish and terrestrial animals that directly translate into the quality issues associated with fish and fishery food products are the composition of lipids, heat-labile muscle proteins, and the abundance of proteolytic enzymes in fish tissues. These adverse biochemical reactions are temperature-dependent, and therefore, low temperature reduces quality deterioration associated with these reactions in fishery products.

Owing to the psychrophilic microflora and the adverse temperature-dependent biochemical reactions, low-temperature freezing followed by frozen storage is the single most important measure to counteract safety and quality deterioration in fishery products. To process fish and shellfish into stable products, low temperature should be applied, maintained, and strictly monitored throughout processing and distribution.

Fish is a vital component of the human diet, and therefore an enormous industry has developed to meet human nutritional needs. As a result, consumers enjoy a variety of food products that range from whole fish to pieces of fish such as cuts and fillets, canned fish in a multitude of forms, dried and cured products, fish oils and extracts, frozen portions, and complete meals as well as reformed and gelled products. The variety within one product type and the range of species used as food results in a huge matrix of possible products and potential problems.

During the last decade, the total world fish catch has stabilized at about 90 million metric tons (Table 17.1), and it is not likely to increase in the future [1]. In addition, the aquaculture sector contributes about 30 million metric tons annually, two-thirds of this being fish, the rest consisting of molluscs and crustaceans. The increased demand for fish and fishery products in the future will have to be met by further development of aquaculture and better use of available natural resources.

II. EFFECTS OF LOW TEMPERATURE ON MICROBIAL GROWTH AND RETENTION OF QUALITY ATTRIBUTES IN FISHERY PRODUCTS

The process of freezing as a method of preserving fish quality has been established since the early 1900s when Birdseye learned that fish and meats frozen in the arctic winter tasted better than those frozen in the milder spring and fall. This resulted in the proliferation of the freezing industry

TABLE 17.1
World Catches of Fishes and Aquatic Invertebrates in Million Metric Tons

	Year					
	1995	1996	1997	1998	1999	Average
Large fishes	57.3	58.6	58.2	56.6	55.9	57.3
Small pelagic fishes	22.0	22.3	21.6	16.7	22.7	21.1
Aquatic invertebrates	12.6	12.6	13.9	13.6	14.2	13.4
Total annual catch	91.9	93.5	93.8	86.9	92.9	91.8

Source: Adapted from Anonymous. *Food and Agriculture Organization Yearbook Fishery Statistics — Capture Production*, Vol. 88/1. FAO: Rome, 1999. With permission.

with the subsequent development of the food industry due to increased distribution range and shelf-life [2,3].

The rate of microbial growth and detrimental biochemical reactions in fish are a function of temperature and water availability (i.e., water activity). Therefore, the shelf-life of fishery products can be effectively extended by lowering storage temperatures and reducing free water availability due to freezing [4]. These two factors act as a double-tier hurdle to effectively hinder microbial growth. However, depending on the rate of heat removal (i.e., freezing rate), the free water may form ice crystals. Ice crystals disrupt muscle structures [5] and the increased concentration of cell electrolytes (i.e., freeze concentration) can result in the chemical denaturation of fish muscle proteins. In terms of fish quality, it is therefore advantageous to minimize the formation of ice crystals by reducing the time at the temperature zone when maximum ice crystal formation occurs.

The benefits of reducing free water availability also include lowering the water activity. To sustain growth, microorganisms require certain levels of unbound free water, which is often expressed as water activity (a_w). The overall benefits of freezing the fishery products are often outweighed by the detrimental effects of ice crystal formation, the partial dehydration of the tissue surrounding the ice crystal (i.e., sublimation — "freeze burn"), and freeze concentration.

A. Microbial Growth and Water Activity (a_w) versus Moisture Content

Microbiological growth, survival, inactivation, sporulation, germination, and toxin production are controlled by water activity (a_w) rather than total moisture. Biochemical reactions (i.e., lipid oxidation, nonenzymatic browning, and enzymatic activity) are also significantly affected by water activity (Figure 17.1). A concept of a_w as a microbial response to water availability in foods was introduced by Scott [6].

Considering a_w in relation to microbial shelf-life and food safety, the a_w values that support microbial growth of different microorganisms are of concern. Table 17.2 shows minimal a_w for foodborne pathogens at their optimum pH and temperature for growth. Corry [7] and Beuchat [8,9] provided exhaustive tables with minimum a_w values for the growth and toxin production of several pathogenic and spoilage microorganisms. The growth of the foodborne pathogens can be suppressed by controlling the a_w at or below 0.92. The exception is *Staphylococcus aureus* that can grow under aerobic conditions at a_w as low as 0.86. The two major findings of Corry [7] and Beuchat [8,9] as related to food safety and shelf-life extension can be summarized as follows: (1) at $a_w < 0.90$ most foodborne pathogens are suppressed, except *S. aureus*, which can grow at a_w 0.86 under aerobic conditions; (2) minimum a_w for growth is always equal to or lower than the minimum a_w for toxin production.

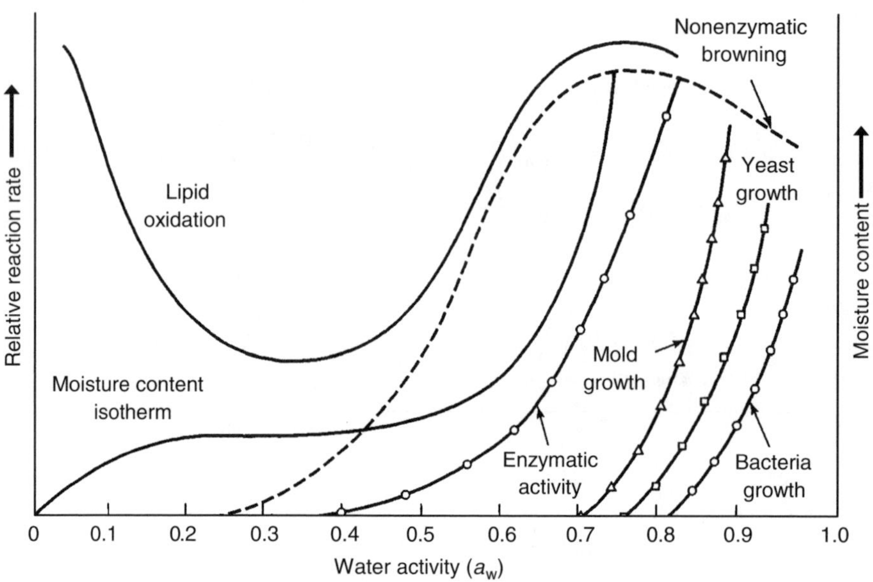

FIGURE 17.1 Effect of water activity and moisture content on relative reaction rates in food. (Adapted from J Faller. In: FJ Francis, Ed., *Wiley Encyclopedia of Food Science and Technology*, 2nd Ed., Norwich: John Wiley & Sons, 1999, pp. 959–969. With permission.)

TABLE 17.2
Minimum a_w Required for Growth of Human Food-Borne Pathogens

Bacteria	a_w
Campylobacter jejuni	0.990
Aeromonas hydrophila	0.970
Clostridium botulinum E	0.965
Clostridium botulinum G	0.965
Shigella ssp.	0.960
Yersina enterocolitica	0.960
Clostridium perfringens	0.945
Clostridium botulinum A and B	0.940
Salmonella ssp.	0.940
E. coli	0.935
Vibrio parahaemolyticus	0.932
Bacillus cereus	0.930
Listeria monocytogenes	0.920
S. aureus (anaerobic)	0.910
S. aureus (aerobic)	0.860

Source: Adapted from WJ Scott. *Advances in Food Research* 7:83–127, 1957; JEL Corry. *Progress Industrial Microbiology* 12:73–80, 1973; LR Beuchat. *Journal of Food Protection* 46:135–140, 1983. With permission.

B. Fish Microbial Flora

Low temperature, including freezing, does not cause significant microbial death. Although freezing cannot be considered as a means of microbial inactivation, freezing is the single most efficient manner of slowing microbial contamination by extending the lag phase of the microbial growth curve. Freezing is capable of some microbial lethality effect, but the extent of microbial death is practically negligible and highly variable [10]. Therefore, focusing on maintaining low initial levels of microbial flora in raw fish, processing equipment, and facilities by implementing hygienic and sanitary conditions is critical. This indicates that a preventive approach must be considered rather than corrective action after contamination reaches dangerous levels, or worse, an outbreak occurs. In general, temperatures below $-10^{\circ}C$ significantly inhibit bacterial growth, whereas fungi, including yeast and molds, cannot multiply below -12 and $-18^{\circ}C$, respectively.

Intrinsic (fish species, fishing location, catch season, etc.) and extrinsic factors (catch method, fish postharvest handling, storage time–temperature, etc.) play an important role on the composition of fish microbial flora. The concentrations of microorganisms commonly found in fish vary with catch season and fish diet [11,12]. The major species found in cold marine water fish include psychrophilic gram-negative bacteria such as *Moraxella* and *Acinetobacter* (formerly known as *Achromobacter*) [13–15], *Pseudomonas*, *Flavobacterium*, and *Vibrio* [16]. The muscle tissue of alive healthy fish is regarded as sterile, whereas the fish surface, viscera, and gills typically contain 10^2–10^7 cm^{-1}, 10^3–10^8 mL^{-1}, and 10^3–10^6 g^{-1}, respectively [11,17]. Low concentrations of aerobic species such as *Clostridium* and *Bacillus* can normally be isolated from fish intestines, but a typical bacterial flora includes *Vibrio*, *Moraxella*, *Acinetobacter*, *Pseudomonas*, and *Aeromonas.* In contrast, the major species isolated from warm-water fish include gram-positive, mesophilic bacteria such as *Corynebacterium*, *Bacillus*, *Micrococcus*, *Enterobacteriaceae*, and *Salmonella* [16].

The lethal effects of freeze-induced cellular death of microorganisms include toxic concentration of intracellular and extracellular solutes, cell dehydration, cold shock, and formation of internal ice crystals. Phospholipids in microbial membranes exhibit high polyunsaturation, which is required to provide a mobile state for cellular transport. The dehydration of the membrane phospholipids has been reported to be the major cause of freeze-induced microbial injury, leading to the disruption of cellular integrity [18].

The contamination levels of trawled fish are typically 10- to 100-fold (1–2 log) higher than those of lined fish due to passing the fish by the sea floor [16]. Careful onboard handling and avoiding physical damage as much as possible are critical. Skin and flesh damage quickly introduce microbial contaminants, resulting in cross-contamination of previously sterile muscle tissue. Eviscerated fish contain fewer bacteria than whole fish, and therefore, evisceration should immediately follow fish arrival on board [17]. During evisceration care should be taken to prevent fish from the drying effects of wind and sunlight exposure. Eviscerated fish should also be immediately chilled to approximately 0°C [19]. However, onboard evisceration is not always beneficial in all fish-harvesting operations if in-plant post-harvest processing immediately follows [20]. Not only should the fish be handled in a careful manner, but also the handling should immediately follow fish arrival on the vessel, although this is difficult in large fishing operations.

During the fish onboard transport from the time they are captured until unloading, microbial growth in fish depends on fish handling and storage time–temperature. This indicates that dual or integrated emphasis on time–temperature is needed to reduce microbial contamination. During extended fish onboard transport, slime build-up develops. Slime accumulation is a common and persistent sanitary problem for fishing vessels [21]. The persistence of slime to disinfectants and sanitizers is due to the characteristics of the bacterial biofilm. The bacteria buried in the interior of the biofilm are protected from the action of the disinfectants and sanitizers by the external layers of the biofilm. Therefore, methods that prevent biofilm formation are much more efficient. Once the biofilm forms, it is difficult to remove it from the food-contact surfaces (i.e., surfaces of the hold on the fishing vessel and processing equipment). The biofilms often harbor foodborne pathogens and serve as a reservoir of these dangerous microorganisms. On pathogen growth, they are often released from the biofilms, resulting in cross-contamination of the stored fish. The slime layer is bacterial in nature and primarily consists of pseudomonads and heavily mucoid corynebacteria, *Moraxella*, *Acinetobacter*, and *Flavobacter*. The bacterial accumulation in slime may reach concentrations as high as 10^9–10^{10} g^{-1} [21].

C. Effect of Low Temperature on Fish Microbial Flora

Although live healthy fish possess regulatory mechanisms that guard them from microbial contamination of their flesh, shortly after fish are captured cellular death occurs, leading to autolysis during rigor mortis. The autolysis yields a nutrient-rich medium for the microorganisms to thrive on, thereby, microbial contamination proceeds to the previously sterile muscle tissue.

From a microbiological point of view, captured fish should be regarded as an excellent growth medium for psychrophils and psychrotrophs. If the growth conditions for these microorganisms remain unaltered, they rapidly contaminate captured fish. As bacteria reproduce by binary fission, a simple calculation shows that a single bacterium yields a population of 10^6 CFU/g after 7 h of growth if the optimal growth conditions are provided.

Proliferating microorganisms secrete proteolytic and hydrolytic enzymes that rapidly degrade the muscle quality. As the majority of these microorganisms are psychrophilic and psychrotrophic, they readily conduct their metabolisms at low temperature [22]. There are several factors affecting microbial growth. Typical factors include onboard handling, the sanitary condition of the vessel, fish species and their physiological condition, fish size, catch method, and processing and storage conditions [20,23]. It has been estimated that approximately 10% of the annual total world catch is lost due to microbial spoilage that can be easily prevented by implementing proper onboard fish handling and storage [24].

Initial microbial contamination of chilled/fresh fillets averages 10^5 CFU/g of spoilage bacteria, mainly *Moraxella–Acinetobacter*. *Pseudomonas putrefaciens* and fluorescent pseudomonads are the primary microorganisms responsible for spoilage at refrigerated temperatures [25]. These microorganisms account for only 1% of the total initial population. Their growth to approximately 30% of the total microbial population is considered a beginning of fish spoilage despite the total microbial concentration. Even refrigerated storage near the freezing point of water (1°C) for 14 days did not suppress microbial growth, resulting in an increase of microbial population to 2.1×10^8 CFU/g [26]. When fish are stored at elevated temperatures (20°C) for 1 day, the spoilage flora will shift toward mesophilic microorganisms, such as *Alteromonas* and *Vibrionaceae*, and these species will be predominant in the microbial population [26].

P. putrefaciens, fluorescent pseudomonads, and other spoilage-causing microorganisms degrade proteins and lipids in fish muscle tissue by the proteolytic and hydrolytic enzymes secreted in their growth environment (i.e., contaminated fish tissue) [27]. Many of the products of degradation reactions contribute to the unpleasant sensory characteristics. Proteins are degraded to peptides and amino acids, and final degradation products include indole, amines, acids, sulfide compounds, and ammonia. Lipids are degraded to free fatty acids, glycerol, and other compounds. Nucleotides are decomposed into nitrogenous compounds that contribute to off-odors. Many enzymatic kits can indicate microbial spoilage in fish, including hydrogen sulfide, gelatin hydrolysis, DNase, RNase, amylase, lipase, and trimethylamine oxide.

Bacterial spores are often a problem with many other techniques that aim to reduce microbial contamination. Similar to low-temperature processing, spores are the most resistant to cold injury. Gram-negative pseudomonads are susceptible to cold injury, whereas the gram-positive bacteria including micrococci, lactobacilli, and streptococci, are more resistant. Oysters contaminated with Polio virus showed a decline in plaque-forming units at −36°C [28]. Some foodborne pathogens, such as Listeria species, have been reported to survive freezing temperatures [29]. Refrigeration considerably suppresses the growth of mesophilic microorganisms, however, at refrigeration temperatures the mesophilic flora is replaced by pshychrophils.

Cooling rate plays an important role in the survival rate of microorganisms. The maximum and minimum survival of *Escherichia coli* occurs at a cooling rate of 6 and 100°C/min, respectively [18]. Similar survival rates have been reported for *Streptococcus faecalis*, *Salmonella typhimurium*, *Klebsiella aerogenes*, *P. aeruginosa*, and *Azotobacter chroococcum* [30].

The recommended storage temperature for frozen fishery products is at or below −20°C; however the preferred temperature should be −30°C. Although cold injury usually results in cellular death, survival of microorganisms due to their repair and recovery mechanisms occurs and is greater in a supercooled environment compared with a frozen environment. It has been estimated that the number of inactivated bacterial cells subjected to frozen storage is between 50 and 90% of the initial population [18]. *Vibrio parahaemolyticus*, when inoculated in various seafoods, has been shown to survive at −15 and −30°C. However, a significant decrease of viability at frozen storage was also observed [31].

Cryoprotectants are often used in fishery products (e.g., surimi production). Cryoprotective substances counteract adverse effects of extended storage at low temperature that induce degradation of muscle proteins and other functional ingredients. However, the cryoprotectants (e.g., glycerol, egg white, carbohydrates, peptides, serum albumin, meat extract, milk, glutamic acid, dextran, glucose, polyethylene glycol, sorbitol, and erythritol) can also protect microbial cells during freezing and subsequent thawing. It has been suggested that the protective mechanism is probably due to reduced damage of the microbial cell wall and membrane [32].

D. Effect of Electron Beam on Safety and Physical Properties of Frozen Fishery Products

1. Concept of Electron Beam

Electron beam (e-beam) is a source of ionizing radiation. E-beam, in contrast to thermal and high-pressure processing, uses high-energy electrons for pasteurization or sterilization effect. Electrons are accelerated to the speed of light by a linear accelerator. Thus, electrons are passed through the product, inactivating the bacteria. The electron source is electricity and, unlike gamma radiation, e-beam does not use radioisotopes (cobalt-60 or cesium-137) [33]. The Joint Expert Committee on Food Irradiation representing Food and Agriculture Organization/International Atomic Energy Agency/World Health Organization (FAO/IAEA/WHO) concluded that irradiation of any food up to 10 kGy caused no toxicological hazards and introduced no nutritional or microbiological problems [34]. The dose is the quantity of radiation energy absorbed by the fish product as it passes through the radiation field during processing.

The dose is generally measured in grays (G) or kilograys (kGy) where 1 gray = 0.001 kGy = 1 joule (J) of energy absorbed per kilogram (kg) of irradiated fish product. Unlike gamma radiation (cobalt-60 and cesium-137), e-beam enables application of high dose rates (e-beam, 10^3–10^5 Gy/s; gamma, 0.01–1 Gy/s). Therefore, to obtain comparable levels of dose absorbed, processing time with e-beam is typically much shorter than with gamma radiation [35]. Radiation processing does not increase the temperature of processed food, and therefore, e-beam is likely to minimize the degradation of food quality commonly associated with excessive thermal processing [36]. To increase electron penetration, e-beam is usually applied to two sides of the processed product: top and bottom. This e-beam configuration is referred to as two-sided (or double-sided) e-beam.

2. Effect of e-Beam on Microbial Inactivation of Frozen and Chilled Surimi Seafood Gels

Commercial surimi seafood (crabmeat stick style) was ground, inoculated with *S. aureus*, and incubated to a concentration of 10^9 CFU/g. E-beam at 1, 2, and 4 kGy resulted in 2.9 log reduction, 6.1 log reduction, and no detectable colonies of *S. aureus* in surimi seafood at ambient temperature (23°C), respectively (Figure 17.2). Sample temperature during e-beam processing had a significant effect ($P < 0.05$) on microbial inactivation kinetics [37]. When frozen samples were subjected to

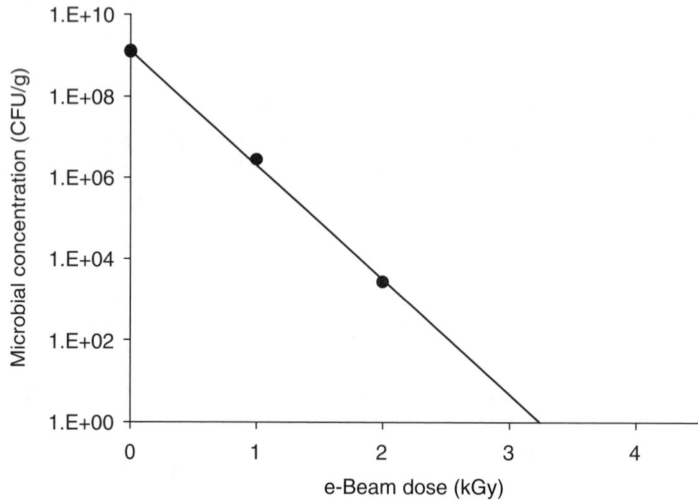

FIGURE 17.2 Survival of *S. aureus* treated with e-beam (Adapted from J Jaczynski, JW Park. *Journal of Food Science* 68(5):1788–1792, 2003. With permission.)

1 kGy radiation, microbial reduction was about 1 log lower than in unfrozen surimi seafood. E-beam at 2 kGy resulted in the best inactivation for samples at room temperature, followed by chilled, and frozen samples, respectively (Figure 17.3).

3. Effect of Electron Beam on Physical Properties of Frozen Surimi Seafood (Crabmeat)

Electrophoresis of Alaska pollock surimi and surimi gels prepared from Alaska pollock surimi was conducted under denaturing conditions of sodium dodecyl sulfate and β-mercaptoethanol, and

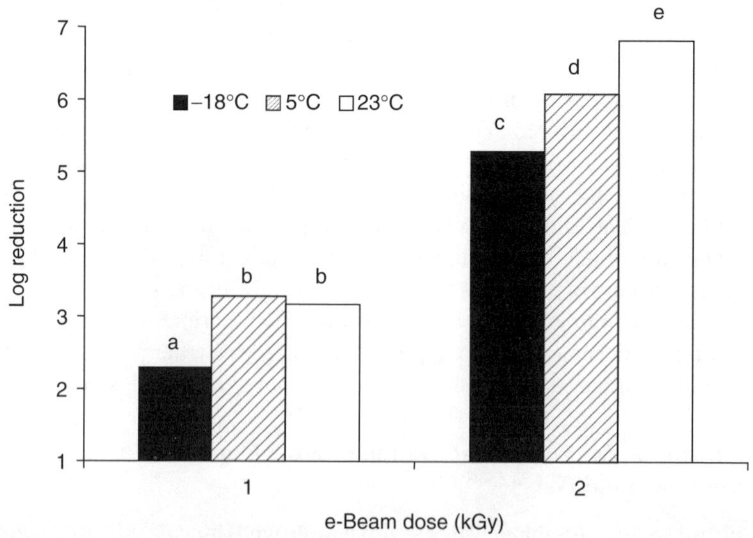

FIGURE 17.3 Effects of sample temperature on inactivation of *S. aureus* by e-beam (different letters on the bars indicate significant difference at $P < 0.05$) (Adapted from J Jaczynski, JW Park. *Journal of Food Science* 68(5):1788–1792, 2003. With permission.)

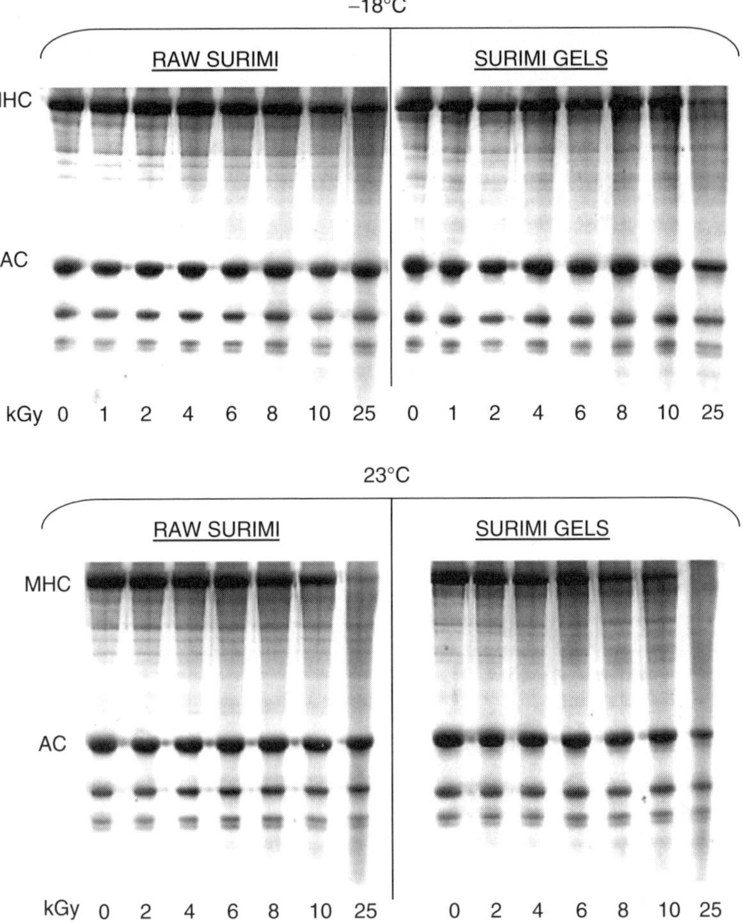

FIGURE 17.4 Effect of sample temperature during e-beam processing on MHC and AC. (Adapted from J Jaczynski, JW Park. *Journal of Food Science* 69(1):53–57, 2003. With permission.)

showed gradual degradation of myosin heavy chain (MHC) with an increase of e-beam dose (Figure 17.4) [38]. Samples subjected to 25 kGy radiation at −18°C showed a thin MHC band, suggesting slower degradation at lower temperature. Gradual disappearance of MHC resulted in an increase of smaller molecular weight proteins (200–50 kDa) in each lane below MHC (Figure 17.4). The complete disappearance of the MHC band was observed at 25 kGy for Alaska pollock surimi (23°C) and surimi gels (23°C). Actin (AC) was only slightly affected in surimi gels subjected to e-beam at 25 kGy.

Texture analysis (shear force) of surimi seafood showed that e-beam processing improved firmness when frozen samples were processed ($P < 0.05$) (Figure 17.5) [39]. This might have been caused by the formation of additional bonds and protein aggregation induced by electrons passing through frozen samples. However, the exact mechanism is not clear.

E. Selection of Raw Materials

Aquatic foods that go to processing facilities on-shore or onboard vary greatly in terms of the state, size, and form. The most widely used and accepted procedure to ensure that the raw materials pass the buyer's or processor's requirements is sensory inspection. A sensory evaluation relies on human

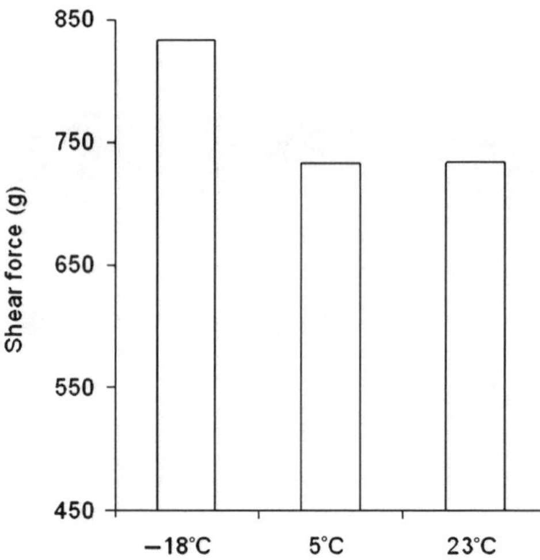

FIGURE 17.5 Effect of sample temperature during e-beam processing on shear force of commercial surimi seafood (imitation crabmeat). (Adapted from J Jaczynski, JW Park. *Journal of Food Science* 69 (1):53–57, 2003. With permission.)

senses (i.e., touch, sound, olfactory, and visual perception) to determine the acceptability of aquatic foods [40].

The inspection should also focus on the detection of microbial contamination. It is important to emphasize that foodborne human pathogens can affect human health even at very low concentrations. Unlike the presence of common fish spoilage bacteria that can be detected with sensory methods, pathogens do not manifest their presence by producing characteristic off-odors and the like. Therefore, the pathogens that pose health risks to the public may be undetected by initial inspection of raw aquatic foods. This is why it is necessary to take every feasible sanitary measure to ensure pathogen-free raw materials.

The inspection of raw aquatic foods should also focus on enzymatic degradation and other physicochemical factors that decrease the market value of the raw materials. A typical set of fresh fish characteristics includes a faint fresh, nonfishy odor; firm and elastic flesh; bright and full translucent eyes; bright pink gills; and bright and moist skin surface with no heavy deposits of slime.

Aquatic animals offer great health benefits associated with ω-3 polyunsaturated fatty acids (PUFA). Aquatic foods, unlike their terrestrial counterparts, contain higher quantities of these beneficial fatty acids, in particular eicosapentaenoic (EPA, 20:5 ω-3) and docosahexaenoic (DHA, 22:6 ω-3). Therefore, a diet rich in aquatic foods is considered beneficial to human cardiovascular status. The PUFA, owing to their high polyunsaturation level, are particularly susceptible to oxidation reactions. If fish have not been chilled immediately after capture and protected from the environment (i.e., sunlight UV and oxygen abundance in air), then they are a subject to oxidative rancidity. This is both a sensory (i.e., off-odor) and nutritional problem. Lipid oxidation from the air (and sometimes auto-oxidation within the seafood) not only does affect the odor and taste acceptability of the product, but also destroys the PUFA in the oil that are so important for human nutrition.

F. Stress at Harvest and Postharvest Handling of Raw Materials

Handling stress during harvest and the late antemortem period affect the rate and degree of muscle metabolism in the early postmortem period and consequently affects fish muscle quality [41]. In addition, the poor physiological status of fish prior to death affects fish post-harvest quality. When tuna fish are caught under stressful conditions, lactate accumulation combined with elevated

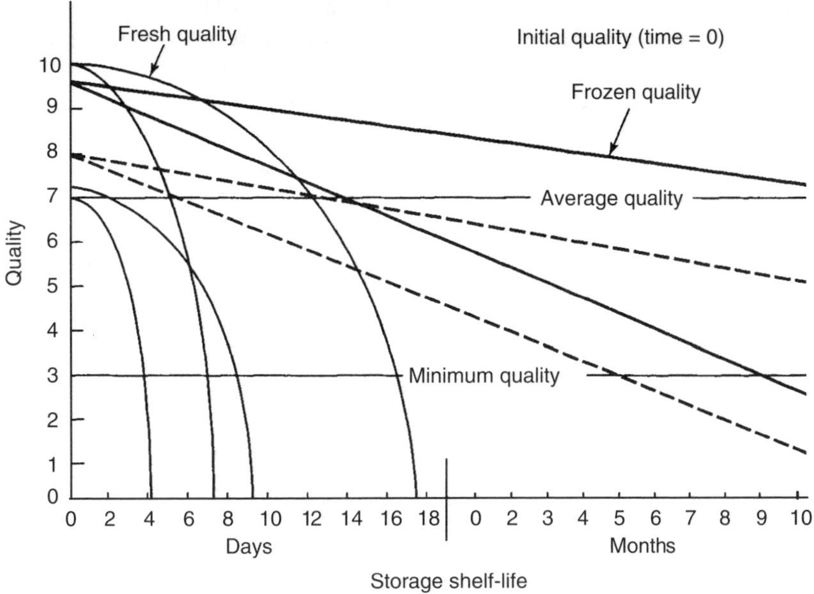

FIGURE 17.6 Relative shelf-life of fishery products as related to post-harvest handling. (Adapted from GM Pigott. In: FJ Francis, Ed., *Wiley Encyclopedia of Food Science and Technology*, 2nd Ed., Norwich: John Wiley & Sons, 1999, pp. 776–799. With permission.)

muscle temperature degrades the muscle quality. However, the tuna quality is still suitable for canning. A capture method called gill netting causes fish exhaustion, resulting in a shorter period of rigor mortis and quality deterioration of muscle tissue during subsequent storage.

The rate of onset and resolution of rigor mortis affect appearance and structure of fish muscle, thus impacting muscle quality [42]. When fish die if relaxed, creatine phosphate is degraded before the use of adenosine triphosphate (ATP). The ATP content begins to decrease when creatine phosphate reaches about the same concentration as ATP. Rigor occurs when crossbridge cycling between myosin and actin stops, and permanent myosin and actin linkages are formed [43]. Unlike beef, *postmortem* "tenderization" or autolysis of fish muscle is highly undesirable to processors and consumers [44]. Commercial markets require high flesh quality of raw products, and therefore, it is critical to control postharvest biochemical changes in fish muscle by implementing proper handling. In general, handling stress results in lower muscle creatine phosphate, ATP, and pH; a faster onset of rigor mortis; higher plasma cortisol and muscle lactate; and softer muscle texture [45–47].

Figure 17.6 summarizes the trends and tremendous variation in storage shelf-life of fresh and frozen fishery products as affected by various postharvest handling methods [48]. It has been reported that quickly chilled and carefully handled, processed, and packaged fish following capture with minimum stress results in a product that maintains quality for almost 3 weeks. At the same time, fish that have been inadequately handled onboard followed by poor processing and distribution is unacceptable after 4 days.

III. CHEMICAL REACTIONS IN FISH MUSCLE DURING FROZEN STORAGE

Frozen storage prolongs product shelf-life, however it also results in product deterioration, which depends on both extrinsic and intrinsic factors. Extrinsic factors include: enzymes, type of fatty acids in the lipid fraction, and the presence of other metabolites, which are precursors of undesirable compounds. Biochemical indicators of deterioration during frozen storage include: (1) protein denaturation (extractability, hydrophobicity, viscosity, and electrophoretic patters), (2) decrease or increase of enzyme activity, and (3) change in metabolite concentrations (amines, aldehydes, and nucleotide degradation) [49]. The physical, structural, and chemical changes in fish that result are observed by measuring protein denaturation, solubility loss, and reduced protein hydration [50–54]. In addition, adverse changes in color and flavor [55–57] as well as texture deterioration also occur [58–62]. Contributors to adverse protein changes during freezing and frozen storage include the simultaneous production of free fatty acids (FFAs) and formaldehyde (FA) [52] as well as storage time and temperature [63,64].

To a degree, deterioration of muscle protein during frozen storage is inevitable [65–68]. Sodium dodecyl sulfate polyacrylamide gel electrophoresis (SDS-PAGE) has been successfully applied to detect changes in myofibrillar proteins during frozen storage under various temperature conditions. During frozen storage, there was a decrease in the relative amount of MHC. This was possibly caused by the dissociation of MHC into the lower-molecular weight protein aggregates formed during frozen storage [69]. As concluded in the study of Davies et al. [70], myosin, not actin, is most likely to undergo modification during prolonged frozen storage.

Extended frozen storage can produce profound effects on the structural and chemical properties of muscle proteins, which subsequently influence the quality attributes of muscle foods [71]. Freeze-denaturation and aggregation of the myofibrillar proteins are responsible for the loss of protein functionality and gel-forming ability in frozen fish [67,72,73]. Loss of Ca^{2+}-ATPase activity in muscle [74] and isolated protein [75,76] can be used to detect protein changes. In addition, protein changes in frozen fish muscle are detectable through alterations in surface hydrophobicity [77–80]. Wagner and Añon [81] theorized that myofibrillar protein denaturation during frozen storage occurs in consecutive reactions. They also proposed that freezing, particularly at a slow rate, initially results in a noticeable decrease in myosin–actin affinity due to denaturation of the myosin heads. This is followed by a continued decrease in ATPase activity and myosin tail denaturation. Myosin continues to denature, although at a slower rate, and aggregation of the denatured proteins eventually occurs. Thus, solubility and viscosity also decrease.

Raman spectroscopy has provided some evidence on the structural changes during freezing and frozen storage occurring in hake muscle proteins *in situ* [82]. These include increase of the relative β-sheet content at the expense of α-helices, increase of exposure of aromatic hydrophobic residues, and increase of hydrophobic interactions of aliphatic residues. Frozen storage also affects the ultrastructure of fish muscle proteins. In transverse sections of fibers from cod muscle stored at −20°C for prolonged periods, transmission electron microscopy (TEM) showed that the spacing between thick filaments was reduced in the 10–10 crystallographic plane, and, in some areas, rows of fused thick filaments could be observed [83].

A. Protein Denaturation

The term denaturation generally describes a complex phenomenon of proteins, which involves changes in the secondary and tertiary structures. These changes result from the breakage of bonds that contribute to the stability of the native protein conformation without disrupting the covalent linkages between carbon atoms in the polypeptide chains. However, the secondary structures of the reactive groups of the extended hydrated random polypeptide coils, which lead to the formation of crosslinks, must also be considered [52].

Freeze-induced protein denaturation and related functionality losses are commonly observed in frozen muscle foods, especially fish [52,84–88]. Denaturation of muscle proteins plays a dominant role in the quality changes of frozen stored meats [67] and should correspond with a decrease in enzyme activity. Differential scanning calorimetry (DSC) is also a useful method for studying proteins [89]. According to frozen storage studies [90–92], the amount of extracted actomyosin decreased with increased storage time.

1. Mechanisms of Denaturation

The mechanism of protein denaturation caused by drying can be considered the same as the mechanism of freeze denaturation because in both processes, water molecules in the cell are removed [68]. Many hypotheses have been proposed to explain the denaturation of muscle proteins [50,52,64,93–95]. These include: (1) the effects of inorganic salts concentrated into the liquid phase of the frozen system; (2) water activity relations; (3) reactions with lipids; (4) reaction with formaldehyde derived from trimethylamine; (5) auto-oxidation; (6) surface effects at the solid–gas interface; (7) effects of heavy metals; and (8) effects of other water-soluble proteins (such as proteases). As freezing progresses, proteins are exposed to increased ionic strength in the nonfrozen aqueous phase, which then leads to extensive modification of the native protein structure [96–98]. The effect of salts on the secondary (i.e., noncovalent) forces, which stabilize the tertiary and quaternary conformation of the protein molecule, could be responsible for protein denaturation [99].

2. Factors Affecting Denaturation

Factors influencing denaturation impart either a partial or total deconformation of the native molecule. Denaturation is often attributable to a cooperative process because damage to a small fraction of the bonds of the protein molecule can result in major changes of large fragments of the molecule [52]. Factors influencing protein denaturation during freezing and frozen storage (Figure 17.7) include salt concentration, pH, ionic strength, lipid oxidation, enzymatic reaction, surface tension, and the physical effects of ice and dehydration [53,99].

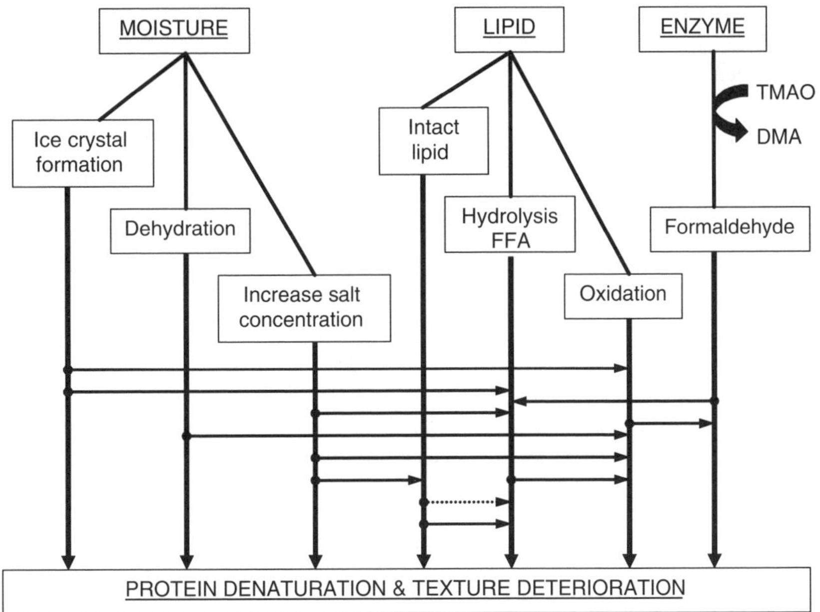

FIGURE 17.7 Factors affecting protein denaturation. Vertical pathway directly affect, whereas horizontal lines indirectly affect (solid lines positive effect, dotted line negative effect). (Adapted from SY Shenouda. *Advances in Food Research* 26:275–311, 1980. With permission.)

The disappearance of liquid water is a major factor affecting protein changes during freezing, particularly the breakdown or change of the structure of the so-called "ordered water" in the vicinity of the hydrophobic groups of the protein molecules. This reportedly leads to a change of the native protein structure. Such conformational changes may be supported by the concomitant influence of various factors such as electrolyte concentrations, semipolaric compounds like phenylalanine and phenylpyruvate, and pH changes. During frozen storage, the alterations of proteins generally increase with higher temperatures and extended storage times. For many proteins, the temperature range between -2 and $-10°C$ is particularly critical [95]. The formation of undesirable compounds related to protein changes, such as the formation of formaldehyde in Alaska pollock around $-10°C$, were maximally observed [100].

Trimethylamine oxide (TMAO) is broken down by trimethylamine oxidase (TMAO-ase) to generate formaldehyde FA and dimethylamine (DMA) during frozen storage [58,101–103]. FA is a highly reactive compound that has been shown to react easily with proteins, particularly the myofibrillar proteins [104,105] and contribute to protein denaturation in FA-forming fish [106,107]. During frozen storage of lean fish, such as gadoid species, much attention has been given to FA formation and its implication in quality loss [53,106]. FA accumulation is higher in fish dark muscle and intestines than in white muscles. Minced fish also has a higher rate of FA accumulation than intact fillets. Although the rate varies by species, it does not depend on TMAO concentration in the tissue or the size of the fish [52].

Tokunaga [108] noted that extensive accumulation of FA in frozen stored fish is accompanied by a loss of myofibrillar protein extractability. This led to the conclusion that FA is at least partially responsible for accelerating protein denturation in Alaska pollock. It was later found that regardless of storage conditions, the increase of DMA concentration was always followed by a significantly correlated decrease in protein extractability ($r = -0.77$ to -0.84). For minced Baltic cod flesh stored at $-20°C$, the correlation coefficient was found to be $r = -0.79$ [104]. In addition, Tokunaga [108] found that when DMA levels were about 6.4–7.0 mg/100 g, the yield of soluble myofibrillar protein decreased to about 50% of that from fresh fish. By assuming a linear cause-and-effect relationship between DMA levels and protein solubility, FA production was found to contribute to approximately 62% of the total variation in protein extractability ($-0.79^2 \times 100$) [109].

A factor implicated in protein denaturation in both gadoid and nongadoid fish during frozen storage is the production of FFAs by hydrolysis of muscle lipid [62]. Recent studies have demonstrated that lipid auto-oxidation is also involved in denaturation and deterioration of muscle protein functionality by causing crosslinking between proteins and lipid oxidation products [110–112]. Lipid hydrolysis and oxidation have been shown to occur during frozen storage of lean fish and influence protein denaturation, texture changes, functionality loss, and fluorescence development loss [102,107,113]. Formation of volatile lipid oxidation products during frozen storage is also evident in salmon [114,115] and trout [116]. However, deterioration in trout during frozen storage is believed to be more attributable to hydrolysis and the formation of low-volatility-free fatty acids than lipid oxidation [117]. Free fatty acids have been shown to be strongly interrelated to lipid oxidation [118,119]. The main source of free fatty acids in many fish species is phospholipids [120,121]. In tissue, free fatty acids are reported to both enhance lipid oxidation [122] and inhibit it [123]. Triglyceride hydrolysis is suggested to lead to increased oxidation, but phospholipid hydrolysis produces the opposite effect [124].

Enzymatic hydrolysis of phospholipid in frozen fish has been recognized as a major cause of quality deterioration since the late 1950s [124]. A linkage between phospholipid hydrolysis and lipid peroxidation during frozen storage is also reported for lean fish [122]. The microsomal lipid peroxidation enzyme system is active at temperatures below the freezing point of fish tissue [125]. Olley and Lovern [120] suggested that phospholipase may be activated by freezing and it would be possible that enzymatic lipid peroxidation activates phospholipase A_2 to initiate phospholipid hydrolysis in frozen fish muscle. Han and Liston [122] found that phospholipid hydrolysis was dependent on peroxidation in both microsomes and stored frozen fish muscle.

Polymerization and protein aggregation resulting from oxidized lipids in lipid–protein systems contributes to decreased protein solubility and the formation of colored complexes [126–128]. Ohta and Nishimoto [129] discovered that protein extractability could be better maintained if the lipids of frozen fish were protected during frozen storage. Although lipid oxidation still occurred, instead of forming carbonyls and other compounds contributing to rancidity, the lipids were bound in lipid–protein complexes. As a result, the texture of poorly stored frozen fish became brittle [52]. To avoid any direct or indirect effects of lipid oxidation, proper glazing and packaging of the product are necessary along with the addition of antioxidants where applicable [130,131]. Temperature also showed a preservative effect on lipid deterioration of cod (*Gadus morhua*) and haddock (*Melanogrammus aeglefinus*) during frozen storage. Lipid hydrolysis, lipid oxidation, and interaction compound formation were more pronounced at $-10°C$ than at $-30°C$ [132].

B. Protein Aggregation

Several studies have confirmed that hydrogen bonds, hydrophobic interactions, and disulfide bridges are primarily responsible for protein aggregation [61,62,64,133–143]. The involvement of each type of bond in protein aggregation during frozen storage varies for different storage conditions. Contrasting data as to the proportional involvement of each bond is likely due to differences in species, storage conditions, and methodology [61,62,133–137,144]. It was concluded that aggregates formed during frozen storage of minced cod are mostly linked by secondary interactions and disulfide bridges, and myosin and actin are the proteins mainly involved [138].

1. Mechanisms of Aggregation

On the basis of several studies conducted to determine the mechanism of heat-induced or freeze-induced aggregation, different hypotheses have been proposed. From these studies, it was concluded that myosin aggregation could be induced by:

1. Initial head-to-head interactions. Using DSC, Samejima et al. [145] reported that myosin head (S-1 fraction) has a lower denaturation temperature (43°C) than the rod fraction (55°C). The S-1 fraction showed higher heat-induced aggregation, suggesting that myosin aggregation was induced by initial head-to-head interactions. This hypothesis was supported by electron microscopy, which revealed initial head-to-head interactions followed by rod interactions [146,147].
2. Interactions between heavy meromyosin (HMM) S-2 fractions [148].
3. Side-to-side interactions, primarily involving rod fractions [148].
4. Ionic strength. At low ionic strength, myosin aggregation could be induced by side-to-side interactions, and at high ionic strength aggregation could be induced by head-to-head interactions [149].

The SDS-ME-unextractable residue can be considered the final step in the aggregation process occurring in fish muscle. Understanding the structural features of this aggregate would help elucidate the exact aggregation mechanism of frozen fish muscle [150]. It has been suggested that nondisulfide covalent bonds are involved in the formation of these aggregates [151], which could be caused by agents such as formaldehyde or oxidized lipids [53,150].

Participation of disulfide bonds in myosin aggregation has been debated. In early investigations, no significant changes in the number of —SH groups were detected. Recent studies showed the importance of this aggregation in the freeze-induced aggregation of fish myosin and actomyosin [152]. In a study conducted by Careche et al. [153], it was suggested that freeze-induced aggregation of soluble fish myosin primarily involves head-to-head interactions with a higher formation of disulfide bonds. A similar behavior has been reported for heat-induced myosin aggregation [154]. Total —SH groups decreased continuously during 15 days of storage to reach 33% of the initial value, which confirmed the importance of disulfide bonds in frozen-induced aggregation of fish myosin in solution [155].

2. Factors Influencing Aggregation

Different species exhibit different degrees of susceptibility to aggregation during frozen storage, depending on muscle intactness, storage time–temperature, and muscle composition [138,150,152,153]. Other factors influencing protein aggregation include dehydration, ice crystal formation, and salt concentration. These, combined with lipid oxidation, free fatty acid formation, and in some fish, formaldehyde formation, produce denaturation and aggregation of proteins. This in turn results in a loss of solubility and texture deterioration [53,64,151]. It has also been proposed that covalent modification of proteins during frozen storage could not be excluded as a contributor to protein aggregation in frozen fish [135].

In fish meat and other tissues, the freezing of water and the correlated concentration of polar and semipolar compounds may change the conformation in certain areas of fibrillar protein molecules, whereby particular aggregation reactions may be initiated. Solutes present in fish tissues or resulting from biochemical activity invoke conformational changes in the protein and decreased protein solubility, which are attributable to aggregation of the protein molecules [52].

C. Protein Extractability and Solubility

Extractability of proteins, usually in salt-containing solutions, is a prerequisite for several functional properties of muscle foods to occur and has long been used as an indicator of myofibrillar protein quality during frozen storage [150]. Alterations in myofibrillar proteins and their functionality have been observed in frozen muscle and isolated protein systems in terms of protein solubility [95,97,156,157] and decreased gel-forming ability [53,110,111,158] and a decrease in myofibrillar protein extractability has often been used to measure protein denaturation [159–161].

Salt solubility relates to the dissolution of myofibrillar proteins when subjected to high concentrations of salt (0.6 *M* NaCl). As frozen deterioration progresses, it has been observed that myofibrillar proteins become more resistant to solubilization [162]. Myofibrillar proteins contribute significantly to muscle functionality, including hydration and surface properties. However, these proteins are the most susceptible to changes during frozen storage, which lead to the formation of protein–protein bonds and a subsequent decrease in extractability in NaCl solutions [163].

As frozen storage progresses, the aggregates grow in size until they are no longer extractable in 0.6 *M* NaCl [164]. Myosin and actin become less extractable in 0.6 *M* NaCl after frozen storage as different kinds of protein–protein bonds form in amounts and percentages that change with storage time and result in aggregates. In a study conducted by Lim and Haard [135], disulfide bond formation was suggested as a contributing factor to loss of protein extractability of green halibut mince during frozen storage, although the kinetics of disulfide bond formation do not correspond to the time progression of protein insolubility. It is possible, however, that disulfide bond formation is a secondary event relating to protein solubility.

Hydrogen or hydrophobic bond formation has also been suggested as causing protein insolubility [52]. Decreases in protein extractability and functionality have been attributed to a fairly reliable correlation between protein extractability and meat brittleness, which seems to exist in the temperature range above −29°C [51,90]. In addition, texture deterioration and associated loss in extractable protein during frozen storage has been reported for several fish species [135]. A significant correlation between protein extractability and storage time has also been reported and explained in more detail [159,165].

The products of lipid oxidation have been well known to produce aggregation of proteins, which decreases solubility [95]. According to recent ultrastructural and chemical studies, linoleic acid hydroperoxides were about 10 times more effective in decreasing protein solubility in KCl solutions of incubated cod myofibrils than linoleic acid. This explains the rapid deterioration of the fibrillar proteins in the surface layers of frozen cod fillet blocks compared to the inner layers during extended frozen storage [95]. Dyer and Dingle [50] and Dyer [90] observed that FFA formation preceded the loss of the extractability of myofibrillar proteins and that lean fish species (i.e., cod) experienced a rapid loss of protein extractability compared with fatty fish [91,92,166]. Although release of FFA in fish from various species seems to coincide with protein deterioration during frozen storage [120], in a thorough study [121] evaluating cod, lemon sole, halibut, and dogfish, a simple connection between hydrolysis and protein denaturation could not be established.

D. ATPase Activity

Ca^{2+}-ATPase is an indicator of the structural state of myosin. Loss of Ca^{2+}-ATPase activity has been associated with denaturation of myosin [159,167]. An association between freeze-induced aggregation of myosin, measured as solubility loss, and loss of ATPase activity has been determined [168]. Freeze-induced denaturation of myofibrillar proteins can therefore be characterized by measuring ATPase activity [67,75,169–171]. However, loss of ATPase activity is not necessarily synonymous with aggregation because it is possible to have no aggregation and loose 100% activity if the active site denatures. Ca^{2+}-ATPase is affected by temperature and time of frozen storage.

The Ca^{2+} -ATPase activity of extracted Pacific whiting actomyosin showed more rapid changes at −8°C than at lower temperatures (−20, −34, and −50°C). As storage time increased, the ATPase activity of myosin isolated from frozen fish was found to proportionately decrease along with the rate of actomyosin precipitation [67,75,169–171].

E. pH

The ultimate pH of fish tissue directly impacts the texture of fresh fish and during frozen storage, and may influence the rate of textural and flavor changes. The WHC of the myofibrils is also affected by pH. The charge on the filament of the myofibrillar lattice decreases as the pH of the fillet decreases, resulting in a slightly less net negative charge. The charge repulsion between the filaments subsequently decreases, which results in contraction of the myofibrillar lattice. This contraction expels liquid from the myofibrillar proteins and increases their protein density. Consequently, if fish have high-energy reserves when caught, the final pH will tend to be lower and this may directly impact fish texture and storage stability [162].

F. Adverse Sensory Effects

After prolonged storage, even at temperatures below −20°C, significant, undesirable sensory changes occur in seafood products [52]. The changes occurring in frozen cod muscle, as shown by taste panel (texture, odor, and taste), include the development of tastelessness and loss of tenderness and the subsequent development of brittleness, off-flavors, and odors [90]. The timescale of these changes as measured by DSC also seemed to coincide with possible aggregation.

Significant texture changes during frozen storage include: changes in firmness, juiciness, and fibrousness [115]. A study by Nilsson and Ekstrand [172] reported similar changes in the texture of frozen salmon. Enzymatic hydrolysis of neutral lipids was found to be primarily responsible for the sensory deterioration of salmon during frozen storage [173]. Oil content, fatty acid composition, and concentration of astaxanthin and tocopherols were important for sensory quality as well as frozen stability of salmon during storage [174].

1. Flavor

Development of rancid off-flavors, especially unsaturated fat oxidation, is a problem for many fish species for sensory perception and shelf-life stability [51,52,175,176]. Much effort has been exerted to prevent the development of rancid off-flavors in fish species, primarily pelagics, which have a high oil content. The lipid content of these fish is greater than 5%, depending on season and the majority is stored in the tissues as triglyceride [53]. In contrast, low fat species, such as cod (*G. morhea*) and haddock (*G. aeglefinus*) store the majority of their lipid in the liver, which is easily removed in the heading and gutting process [162].

Oxidation of highly unsaturated lipids directly relates to the production of off-flavors and odors in foods [177,178]. Refsgaard et al. [115], however, demonstrated that during frozen storage of salmon, pronounced sensory deterioration was due not only to the formation of volatile oxidation products, but also to increased intensity of oil taste, bitterness, and metal taste. A substantial increase in free fatty acid content during frozen storage was instead found to contribute to the decreased sensory attributes of frozen stored salmon. By adding polyunsaturated fatty acids to fresh salmon mince that was equal to those formed during frozen storage, a sensory perception similar to that experienced after frozen storage was obtained [173]. A correlation was therefore found between sensory perception and the level of free fatty acids present [115].

2. Texture

Texture hardening, along with poor dispersibility, remains a major problem for the commercial application of frozen fish mince in formulated seafood products [179]. The causes of texture hardening in gadoid fish during prolonged frozen storage have been extensively studied. Textural changes during frozen storage are ultimately caused by changes in the myofibrils. Factors reported to cause textural deterioration are ice crystal formation [180], formaldehyde-induced intermolecular crosslinking between adjacent protein molecules [181], protein–lipid interaction [154,182], and crosslinking of sulfhydryl groups [135,183].

It is well established that fish myosins denature and aggregate during frozen storage, yielding a drier, more brittle flesh [90]. Fish from the family Gaddidae quickly become more brittle at lower storage temperatures than most other types of fish. As shown in Table 17.3, generation of

TABLE 17.3
Correlation Matrix of Chemical Data and Objective Indices of Brittleness in Raw and Cooked Hake Fillet

	DMA	HCHO	EPN	Peak Height		Peak Slope	
				Raw	Cooked	Raw	Cooked
DMA	1	0.936^b	-0.768^b	0.797^b	0.866^b	0.860^b	0.874^b
HCHO		1	-0.728^b	0.670^a	0.696^a	0.670^a	0.688^a
EPN			1	-0.744^b	-0.551	-0.680^a	-0.585
Peak height (raw)				1	0.764^b	0.910^b	0.747^b
Peak height (cooked)					1	0.899^b	0.969^b
Peak slope (raw)						1	0.896^b
Peak slope (cooked)							1

Note: DMA, dimethylamine; HCHO, formaldehyde; EPN, salt-extractable protein nitrogen.

$^aP < 0.05$.

$^bP < 0.01$.

formaldehyde and DMA from the breakdown of trimethylamine oxide (TMAO) correlates with texture changes during frozen storage [61,62,100,105,106,184,185]. DMA is therefore used as a chemical index for measuring textural deterioration [62].

The effect of formaldehyde on textural changes was demonstrated [105,177]. As formaldehyde was added to fresh cod muscle at concentrations normally found in poor-quality frozen stored fillet, the salt-extractable protein nitrogen (EPN) content diminished. In addition, evidence of brittleness was observed at the surface of tissues placed in dilute formaldehyde solutions [62]. So, the ability of formaldehyde to induce intermolecular linkages between adjacent protein molecules has been believed to cause brittleness in gadoid species [53,61,100,102,105,106,185].

Formaldehyde formation, however, is not a complete explanation for textural deterioration during frozen storage. This is evidenced by (1) mincing of haddock fillets stimulates TMAO degradation, whereas the intact fillet experiences texture deterioration without TMAO degrading to formaldehyde and (2) washing of mince to remove TMAO-ase and other water-soluble components does not stabilize the mince [162]. In addition, species not possessing the TMAO-ase enzyme still incur textural changes during frozen storage. Therefore, it seems likely that another mechanism for textural deterioration, besides formaldehyde induced crosslinking, exists [62].

The relative influence of FA and lipid degradation products in texture changes has been evaluated [77,186]. FFAs are known to cause texture deterioration by interacting with proteins and it has been proven that accumulation of FFA in frozen fish contributes to unacceptability [53,102,106]. Brittleness of hake muscle during frozen storage coincided with two separate events, loss of salt-soluble protein and WHC, both of which relate to protein denaturation. The influence of lipid hydrolysis on textural deterioration, however, is a topic of debate [62].

Sodium dodecyl sulfate (SDS) electrophoresis of hake and haddock myofibrillar proteins was carried out to gain some knowledge of the molecular events resulting in brittleness during frozen storage. This technique has found wide application in the detection of covalent interactions between food proteins. The electrophoretic patterns of freshly extracted hake myofibrils were compared with those of hake myofibrils that were soaked overnight in 1200 ppm formaldehyde, 35 days storage at -17 and $-5°C$. The myofibrillar patterns from muscle that was soaked in aqueous formaldehyde and those stored at $-5°C$ for 35 days were remarkably similar. The muscle that was stored at $-17°C$ yielded myofibrillar banding patterns somewhat intermediate between the fresh controls and the muscle stored at $-5°C$. These observations are further evidence that the changes associated with brittleness in frozen stored red hake muscle are primarily due to the covalent crosslinking of the structural proteins by formaldehyde [62].

According to Connell [156], the development of brittleness and the loss of WHC are caused by the formation of additional numbers and higher strength of existing linkages between the myofibrillar proteins. The formation of disulfide bonds, however, played an important role in textural deterioration of frozen red hake [187]. Dyer [90] stated that there is a strong correlation between protein extractability in frozen-stored fish and brittleness of the flesh as measured organoleptically after cooking. Connell [51,156,188–190] also reported on the validity of this correlation and found that in quality assessment of frozen cod, the deterioration of the texture can be measured by the decline of protein extractability. In contrast, Cowie and Little [191] obtained experimental data illustrating that in cod fillets stored for up to 82 months at $-29°C$ no correlation could be established between protein extractability, which decreased from 72 to 45%. However, a strong correlation existed between the latter and the pH of the raw muscle. At higher storage temperatures (-7 and $-14°C$), both protein changes, reflected by a decrease of extractability and the muscle pH, influenced the brittleness of the cooked flesh. Connell and Howgate [190] subsequently used pH value as a covariant in the assessment of overall quality of cod and haddock [52]. The pH of the buffer used in these extractions could decrease or promote denaturation at the homogenization stage.

Previous attempts to prevent textural deterioration in fish mince during frozen storage have included addition of amino acids [192], oil [193], hydrocolloids [194], nonfish proteins, and starch [195]. In a subsequent study, the addition of 0.4% alginate, 4% sorbitol, and 0.3% sodium tripolyphosphate (STPP) effectively controlled protein denaturation of red hake mince. Alginate appeared to be responsible for preventing muscle fiber interaction through electrostatic repulsion and chelating Ca^{2+}.

G. Biological Factors Affecting Chemical Reactions

No simple reason exists to explain the varying rates of deterioration in different fish species. This difference is often attributed to fish spawning, rough handling, and contamination of fillets with blood or viscera [162]. For spawning, the reproductive cycle of mature hake influences the metabolic state of the fish and subsequently alters its actomyosin composition [196]. It was found that fillets from prespawning fish deteriorate faster than fillets from postspawning hake during storage at −20°C [197]. Extensive formation of aggregates in prespawning hake was observed, even after only 15 days of storage. Thus, protein solubility of prespawning hake continuously decreased compared with postspawning hake where after 240 days of frozen storage only soluble aggregates were observed. It was therefore concluded that postspawning hake was desirous for the preparation of frozen products [198].

The frozen stability of fish muscle proteins also varies significantly with species [199]. Many species differ significantly in susceptibility to reactions induced by freezing and frozen storage. In nongadoid fish, such as halibut, wolfish, and ocean perch, the solubility changes of proteins during frozen storage were considerably slower and the values of the extractable proteins were not as low as in gadoid fish [105]. Differences in the stability of fish proteins also depend on the pH and ionic strength of the system [70]. Comparing the shelf-life of fatty species to lean species, significant differences are found as well. Fatty species often encounter oxidative changes of lipids and pigments, evidenced by rancid odor and flesh discoloration. In contrast, lean fish primarily suffer from severe alterations of proteins, typically denaturation or denaturation–aggregation, which significantly deteriorate texture [52]. In species that form FA during frozen storage, as the FA reacts with the myofibrillar proteins, denaturation and aggregation accelerates compared with species that do not form significant amounts of formaldehyde [78,145,200]. In addition, the proportion of protein linked by disulfide and nondisulfide covalent bonds varies by species, muscle integrity [63,146], and frozen storage temperatures [64] and increases with storage time [138,200].

The thermostability of myosin varies for the environmental temperature of the respective fish species [162,199,201,202]. A correlation has therefore been suggested between the extent of freeze-denaturation of fish muscle and the habitat temperature of the fish [203]. Evidence of this habitat dependence with myosin from cold-water species being less stable than myosin from warm-water species has been published [203–206]. The denaturation temperature of myosin from warm-water fish may be almost 20°C higher than cold water fish [202]. Howell et al. [205] studied isolated myofibrils and found that myosin from tropical species were less affected by frozen storage than cold water species. In addition, a greater degree of crosslinking occurs in the muscle of cold water fish, and the myosins of these fish are more drastically modified than myosins of tropical species. It is interesting that a significant portion of frozen warm-water fish, such as Threadfin bream, is used in surimi processing in Thailand, whereas frozen fish from cold water (i.e., Alaska pollock, Northern blue whiting, etc.) is not successfully processed into surimi.

IV. CRYOPROTECTION OF FISHERY PRODUCTS

Dyer [90] suggested that the properties of proteins in the tissues can be modified or protected by lipids, fatty acids, nucleotide compounds, or even carbohydrates [52]. The first published use of cryoprotectants for successful application in muscle proteins was a combination of sucrose (10%) and polyphosphate (0.2–0.5%), which was used to prevent denaturation of Alaska pollock muscle proteins [207–209]. Washing of the minced fish muscle prior to incorporating with cryoprotectants was also necessary to prevent denaturation during frozen storage [96].

Later it was discovered that sorbitol could replace half of the sucrose in the cryoprotectant mixture [67]. For muscle proteins, sodium tripolyphosphate and sorbitol have been found to be effective cryoprotectants [66,71]. A recent study, however, indicated that protein functionality of freeze-dried fish protein can be well preserved even without sucrose and sorbitol as long as membrane lipid was removed (Hultin and Kelleher 2000, personal communication).

Cryoprotectants prevent the drastic changes of proteins associated with freezing and thawing and subsequently, aggregation progresses at a slower rate. Cryoprotective agents include compounds such as sugars, polyalcohols [67], carbohydrates, polyols, some amino acids, and other related compounds [66,99,210]. However, many of them are not feasible for use because of high cost, food regulations, or adverse sensory properties. In a study investigating the effectiveness of several cryoprotectants in preventing freeze-induced perturbations of farmed rainbow trout (*Oncorhynchus mykiss*), Herrera and Mackie [211] found that addition of cryoprotectants maintained solubility levels significantly higher than those of the control during the entire storage period. Maintenance of the extractability of salt-soluble proteins during frozen storage has been used as an indicator of the cryoprotective action of an ingredient against protein denaturation [68].

Freezing rate and temperature control can be used to maintain protein integrity during freezing and subsequent frozen storage. For a protective effect, the rate of freezing should be high enough to prevent the formation of large ice crystals in the extracellular spaces [52]. For storage temperature, Hsu et al. [86] found that 1 month storage of Pacific whiting fillets at −50°C showed no significant difference with regard to shear strain values compared with the 0-day sample. In addition, reducing quality loss during processing and handling of fish before freezing can improve protein quality during frozen storage.

The use of vacuum and antioxidant treatment of fishery products can also be effective in maintaining protein quality during frozen storage. Vacuum packaging and antioxidant (AO) treatment dramatically affected the shelf-life stability of both hake and mackerel. The combined use of vacuum and AO treatment for long-term frozen storage could offer better protection of mackerel, whereas the absence of vacuum and AO resulted in longer shelf-life for hake. For short-term frozen storage, the use of glazing with AO for mackerel or without AO for hake seemed to offer more economical advantages compared with using nonbarrier bags with or without vacuum [212].

A. Cryoprotectants

Cryoprotectants do not prevent but rather function to minimize the negative effects of frozen storage on the physicochemical traits of myofibrillar proteins [213]. Cryoprotectants are compounds that improve the quality and extend the shelf-life of frozen foods. The term cryoprotectant includes all compounds that help to prevent deleterious changes in foods caused by freezing, thawing, and frozen storage. These substances may be added during processing and product formulation or produced naturally in the living organism from which the food is derived. Most preservative compounds that chemically stabilize food molecules at ambient temperatures can be equally effective at minimizing these same changes during freezing, frozen storage, or thawing [214].

The systematic studies on many cryoprotective substances, such as amino acids and carboxylic acids, revealed that these compounds have some common structural principles with respect to the capacity to prevent freeze denaturation of actomyosin [215–218]. The following requirements for exhibiting cryoprotective effects for fish muscle proteins have been proposed:

1. A molecule has to possess one "essential group," either —COOH, —OH, or OP_3H_2, and more than one "supplementary group," of the type —COOH, —OH, NH_2, SH, —SO_3H, and OP_3H_2.
2. The functional groups must be suitably spaced and properly oriented about each other.
3. The molecule must be comparatively small [210,219].

The selection of cryoprotectants depends on the application (i.e., comminuted, fillet, etc.). Cryoprotectants for fillet and mince include: polyphosphates, sugar, carboxylic acids, and milk proteins [72,87–90]. Sucrose/sorbitol soak provided more cryoprotection than soaking fillets in a solution of sodium lactate with regard to protecting protein structure [213]. The greatest stabilizing effect on cod-surimi proteins was obtained from carbohydrate/polyol treatments, sorbitol, glucose syrup, sucrose, and sucrose/sorbitol at a level of 8% (w/w) [220]. Commercial freezing of surimi made from Alaska pollock and Pacific whiting typically involves incorporation of sucrose (4%), sorbitol (4–5%), and polyphosphates (0.3%), which protect fish myofibrillar proteins during extended frozen storage [221]. In cod minces, however, glucose showed a significantly higher protective ability than sucrose [222].

Owing to the sweetness imparted by using sugar as a cryoprotectant, there has been much attention devoted to researching other cryoprotectants [68,71,97,157,223]. Some additives, however, like starch, successfully minimize protein denaturation during frozen storage; however they significantly increased gel strength of fish protein gels. Noguchi and Matsumoto [215] found that sodium glutamate at 0.3 *M* concentration totally prevented the loss of solubility in frozen carp myofibrillar proteins for up to 10 weeks of storage at $-20^{\circ}C$ and reduced the rate of ATPase activity loss. Compared with 1 *M* glucose, 0.3 *M* concentration of sodium glutamate was found to be equally effective [52]. In a different study, polydextrose, when compared with lactitol, glucose syrup, and the mixture of sucrose and sorbitol appeared to be the most effective cryoprotectant [222]. Antifreeze proteins are another option for cryoprotection, which involves either soaking or injecting into the muscle before freezing. Antifreeze proteins decrease drip loss and maintain the texture attributes of the proteins by controlling ice crystal size [224,225].

More recent research by Hunt et al. [226] has been conducted on glucose polymers, which are produced from corn using a patented process. These polymers were found to be 40% as sweet as sucrose and were used in the cryoprotection formulation of fish proteins for surimi manufacture as a sorbitol replacement. Compared with the commercial control, which contained 4% sucrose, 5% sorbitol, and 0.3% phosphate, the glucose polymer samples (4% sucrose, 5% glucose polymer, and 0.3% phosphate) performed similar or better with regard to salt-extractable protein (SEP), Ca^{2+}-ATPase activity, and dimethylamine (DMA) formation during 8 months of frozen storage. Evaluation of gel properties (shear stress, shear strain, and color) showed that glucose polymer samples better maintained fish proteins after 8 months of frozen storage (Figure 17.8).

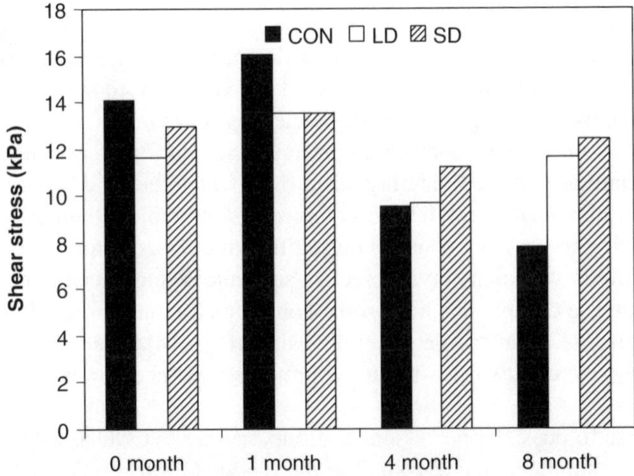

FIGURE 17.8 Shear stress values of two glucose polymers (SD and LD) compared with the commercial control sample (CON). (Adapted from A Hunt, JW Park, H Zoerb. *Proceedings of the Annual Institute of Food Technologists Meeting*, 2002. With permission.)

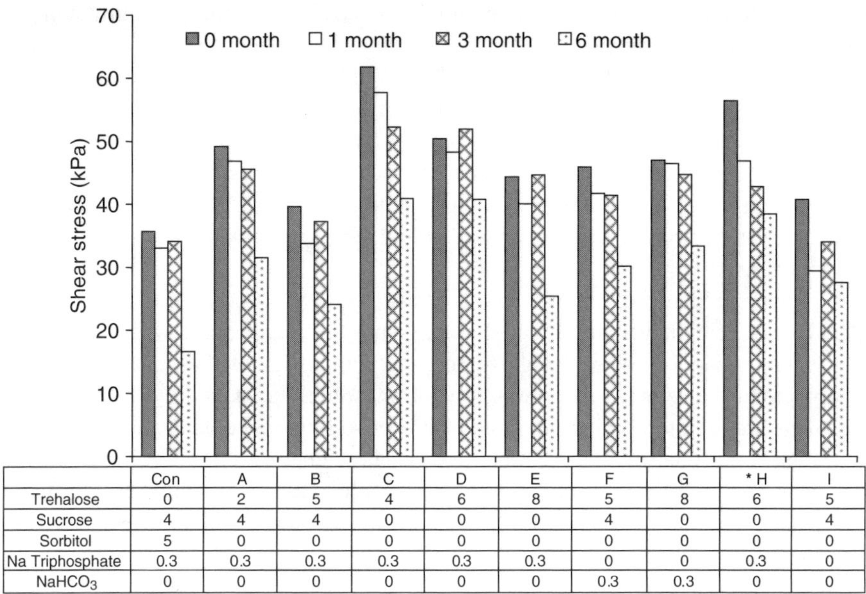

	Con	A	B	C	D	E	F	G	*H	I
Trehalose	0	2	5	4	6	8	5	8	6	5
Sucrose	4	4	4	0	0	0	4	0	0	4
Sorbitol	5	0	0	0	0	0	0	0	0	0
Na Triphosphate	0.3	0.3	0.3	0.3	0.3	0.3	0	0	0.3	0
$NaHCO_3$	0	0	0	0	0	0	0.3	0.3	0	0

FIGURE 17.9 Effects of trehalose on gel strength (shear stress) as surimi cryoprotectant during 6-month frozen storage. *Denotes regular trehalose while all others indicate crystalline trehalose. A, B, C, D, E, F, G, H, and I denote mixtures of cryoprotectants with their composition indicated in the table below Figure 17.9.

Another study by Hunt et al. [227] evaluated the performance of trehalose as a cryoprotectant of fish proteins (Figure 17.9). Trehalose is a white crystalline powder commonly found in animals and plants that is 45% the sweetness of sucrose. Under the Japan Food Sanitation Law, trehalose is regarded as a food in Japan. The freeze point depression of trehalose compared with that of sucrose showed that trehalose performed similar to sucrose. During the cryoprotection study, trehalose was used in varying concentrations alone and in combination with sucrose. With regard to gel properties, shear stress and shear strain, and dimethylamine formation, the F and G samples performed better than the commercial control (Figure 17.9) and proved effective for use as a cryoprotectant of fish proteins.

B. Phosphate

Food-grade phosphates contribute to a wide range of functions, including retention of moisture and flavor, chelating heavy metal ions to prevent lipid oxidation, shelf-life extension, and cryoprotection. For seafood, phosphates are valuable in maintaining and cryoprotecting myofibrillar proteins during freezing and frozen storage. Typically sodium tripolyphosphates or hexametaphosphates as well as blends of phosphates are used in seafood. The phosphates are applied by either dipping in or spraying fish with a phosphate solution, which is often followed by tumbling. For communited meat systems, like surimi, phosphates are typically added in a dry, powder form. The application of phosphates in seafood must be uniform and consistent. Additional benefits of phosphate addition include: retention of natural juices, prevention of fluid loss during shipment and prior to sale, and retention of positive sensory attributes, like flavor and moisture [228].

Phosphates have been reported to enhance protein functionality in fish during frozen storage [54,68,71]. Phosphates often minimize the negative effect of frozen storage on myofibrillar protein solubility, hydrophobicity, and myosin susceptibility to thermal denaturation [213]. In a different study [68], phosphate addition, alone or in combination with various sugar and polyols, appeared to have minimal protective effect on protein extractability, although addition of STP alone did seem to be slightly protective.

Tanikawa et al. [229] and Mahon and Schneider [54] used polyphosphates to minimize freezing damage and thawing drip in fish fillets. Dip treatment in tripolyphosphate effectively controlled thawing drip in several fish species and scallops. Brief treatment (30 sec to 2 min) of fish fillets in a solution of 12% sodium tripolyphosphate retained a more fresh-caught composition after frozen storage. Stabilization of mullet myofibrills during frozen storage was also better maintained when phosphate was combined with sugar and polyol rather than when phosphate or sugar and polyol was added alone [71].

Recently, chemical blends of phosphates have been evaluated for use as a cryoprotectant of fish proteins [230]. Chemical blends are obtained using a spray-drying process with a special temperature profile, which results in one molecule adjacent to another rather than one crystal adjacent to another as obtained in physical blends. Solutions of phosphoric acid, sodium hydroxide, and potassium hydroxide are used to obtain integrated blends of sodium and potassium di- and tripolyphosphates (R. Schnee, personal communication, 2004). These chemical blends are more soluble and have higher rates of dissolution compared with the phosphate blends obtained by physical mixing. The study evaluated various chemical and physical phosphate blends and the effect they have on the chemical and textural properties of fish proteins.

In general, it was concluded that the tetrapotassium pyrophosphate blend functioned the best followed by chemically blended phosphates compared with the other phosphate samples tested, including the current phosphate (a mechanical blend of sodium tripolyphosphate and tetrasodium pyrophosphate) used in industry, with regard to maintaining fish proteins during long-term frozen storage.

C. Mechanisms of Cryoprotection

The primary function of cryoprotection is to thermodynamically prevent unfolding of the protein molecules [223]. There is a striking parallel between the types of compounds that have been found to stabilize proteins against solution-induced perturbations, such as thermally induced unfolding and pH-induced dissociation, and those compounds that have been chosen as cryoprotectants for isolated proteins. There are various mechanisms to explain the cryoprotective effect of compounds, which include solute exclusion, cryostabilization, and freezing point depression. Depending on the compound, more than one mechanism can be used to explain its cryoprotective effect.

1. Solute Exclusion

A variety of compounds (i.e. glycerol, sugars, amino acids, polyalcohols, and salts) have been studied for their ability to stabilize proteins during frozen storage. The mechanism by which these compounds cryoprotect fish proteins is primarily by solute exclusion [214]. The interior of protein molecules is comprised of hydrophobic amino groups, which tend to be buried within the protein structure. In addition, large fractions of the protein surface are likewise hydrophobic given that the surface is occupied by atoms that do not have the ability to form hydrogen bonds [231]. Therefore, hydrophilic compounds, such as sugars, are excluded from the surface of the protein, thereby stabilizing the native protein structure.

During freezing, as the temperature decreases, the strength of the intramolecular hydrophobic interactions, which stabilize the native protein structure, also decreases [232]. Water preferentially hydrates the surface of the protein through hydrogen bonds as well as dipole–dipole and ion–dipole interactions [233]. The presence of water on the protein surface can induce the hydrophobic regions in the protein interior to remain buried within the structure (Figure 17.10). In addition, as pure water freezes into ice, the solute concentration and solute surface tension increase and counter

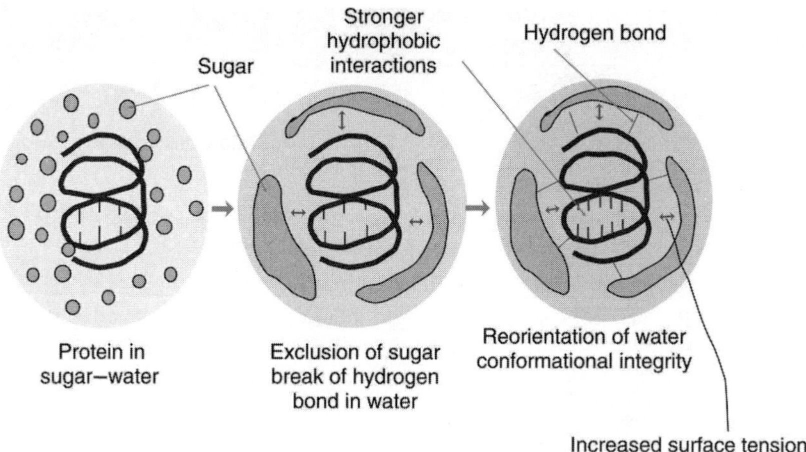

FIGURE 17.10 Solute exclusion from protein surfaces and preferential hydration of proteins by hydrophobic interaction.

the effect of weakened intramolecular hydrophobic interactions. In this way, the native conformation of the protein is maintained with cryoprotectant during freezing and frozen storage [214].

2. Cryostabilization

In contrast to low-molecular-weight polymers, which cryoprotect proteins by altering the thermodynamics of a system to favor the native protein state, high-molecular-weight polymers cryostabilize by trapping the protein into a glass where deterioration processes are significantly slowed [234]. Addition of the high-molecular-weight compounds raises the glass transition temperature, which promotes the conversion of the system to the glass state [235,236]. The higher glass transition temperature imparted by the addition of solutes effectively minimizes freeze-induced denaturation because water is immobilized in the glass structure rather than forming ice crystals, which damage the protein structure [99].

3. Freezing Point Depression

Even at low temperatures there exists a fraction of water that will not be frozen into ice. As the concentration of the solute increases, the freezing point is depressed, which leaves a fraction of unfrozen water in the system. Eventually a point is reached where the freezing point of this nonfrozen fraction equals the freezing temperature of the fish protein, resulting in an equilibrium between the ice phase and the nonfrozen fraction [162].

Cryoprotectants can depress the freezing point of the fish protein system. Evaluating the freezing point depression of variety of sugar molecules, the smaller-molecular-weight compounds further decreased the freezing point depression compared with higher-molecular-weight compounds (Figure 17.11). In addition, Matsumiya and Otake [238] observed that when the sorbitol concentration was increased from 1 to 10%, the freezing point of prepared raw surimi blended with sorbitol was depressed from 0.17 to 1.79°C.

Arctic and Antarctic fish naturally contain antifreeze glycoproteins, which are present in their blood. These glycoproteins have been found to lower the freezing temperature without affecting the osmotic pressure or melting temperature [239]. Further depression of the freezing temperature was observed when polyalcohols were added along with the antifreeze proteins. Caple et al. [240] explained that the antifreeze glycoproteins alone or with polyalcohols are thought to be specifically absorbed at the ice–water interface and this inhibits the ability of water molecules to join the ice lattice.

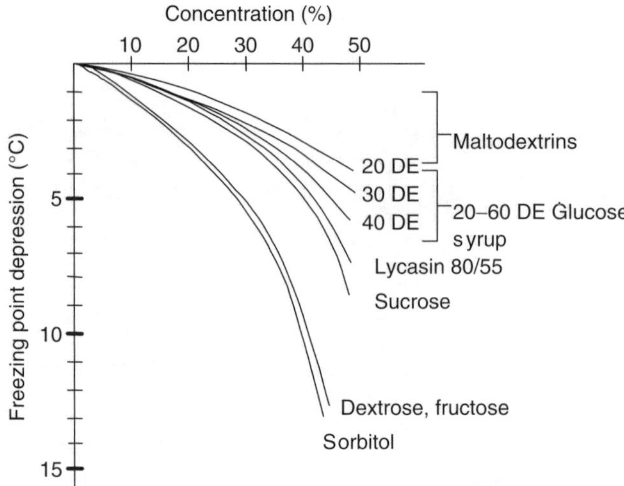

FIGURE 17.11 Freezing point depression curve of various carbohydrates (DE, dextrose equivalents). (Adapted from M Serpelloni. The food applications of sorbitol, mannitol, and hydrogenated glucose syrups. In: *Proceedings of the International Symposium of Polyols and Polydextrose*, Paris, 1985. With permission.)

V. CONCLUSIONS

Low-temperature processing, including freezing, is probably the most widely used method to slow safety and quality deterioration of aquatic foods. With the development of the refrigeration and freezing technology, consumers today can enjoy nutritional benefits of fresh-like aquatic foods. It is likely that future advances in the processing technologies as well as the development of aquaculture will further improve the safety, quality, and availability of these nutritious foods.

Freezing suppresses microbial growth, primarily by reducing water activity (a_w), whereas refrigeration slows growth by controlling the suboptimal growth temperature. Neither freezing nor refrigeration, however, can be considered as an inactivation method for microorganisms. It is necessary, therefore, that the handling of raw materials be performed in a manner that yields initial microbial loads to be as low as possible.

Fish and seafood products offer unique challenges to maintaining protein quality during freezing and frozen storage due to the inherent thermal instability of the proteins, although warm-water fish have demonstrated higher thermal stability compared with cold water fish. In addition to the chemical alterations of the protein as evidenced by protein solubility, Ca-ATPase activity, and DMA formation, protein deterioration is also evident by changes in the textural and sensory quality of fish and seafood products. Protein denaturation and aggregation occur as the proteins are damaged during freezing and frozen storage, and contribute to the resulting loss of product quality.

To counter the damaging effects of freezing and frozen storage, cryoprotectants are added to minimize protein deterioration. Typically, sugars and other carbohydrates are used. The mechanism most commonly attributed to the cryoprotective effect of these compounds is solute exclusion. Other compounds, such as phosphates, have also been found to successfully cryoprotect fish proteins when combined with sugar and sorbitol. The main challenge of cryoprotection is to find effective cryoprotectants that are less sweet than sugar and therefore impart minimal effects on the sensory quality of fish and seafood product. Recent studies involving glucose polymers and trehalose have demonstrated successful application of compounds significantly less sweet than sugar for use as fish protein cryoprotectants.

REFERENCES

1. Anonymous. *Food and Agriculture Organization Yearbook Fishery Statistics — Capture Production*, Vol. 88/1. FAO: Rome, 1999.
2. O Fennema. The U.S. frozen food industry: 1776–1976. *Food Technology* 30:56–68, 1976.
3. TP Labuza, AE Sloan. Forces of change: from osiris to open dating. *Food Technology* 35:34–43, 1981.
4. J Faller. Food processing. In: FJ Francis, Ed., *Wiley Encyclopedia of Food Science and Technology*, 2nd Ed., Norwich: John Wiley & Sons, 1999, pp. 959–969.
5. DS Reid. Physical phenomena in the freezing of tissues. In: CP Mallett, Ed., *Frozen Food Technology*. London: Blackie, 1993, pp. 1–19.
6. WJ Scott. Water relations of food spoilage microorganisms. *Advances in Food Research* 7:83–127, 1957.
7. JEL Corry. The water relations and heat resistance of microorganisms. *Progress Industrial Microbiology* 12:73–80, 1973.
8. LR Beuchat. Influence of water activity on growth, metabolic activities and survival of yeasts and molds. *Journal of Food Protection* 46:135–140, 1983.
9. LR Beuchat. Influence of water activity on sporulation, germination, outgrowth, and toxin production. In: LB Rockland, LR Beuchat, Eds., *Water Activity: Theory and Applications to Food*. New York: Marcel Dekker, 1987, pp. 137–151.
10. H Souzu, M Sato, T Kojima. Changes in chemical structure and function in *Escherichia coli* cell membranes caused by freeze thawing. II. Membrane lipid state and response of cells to dehydration. *Biochimica et Biophysica Acta* 978:112–118, 1989.
11. DC Georgala. The bacterial flora of the skin of north sea cod. *Journal of General Microbiology* 18:84–91, 1958.
12. J Listen. A Quantitative and Qualitative Study of the Bacterial Flora of Skate and Lemon Sole Trawled in the North Sea. Ph.D. thesis, Aberdeen University, Aberdeen, U.K., 1955.
13. P Baumann, M Doudoroff, RY Stanier. A study of the *Moraxella* group. I. Genus *Moraxella* and the *Neisseria catarrhalis* group. *Journal of Bacteriology* 95:58–73, 1968.
14. P Baumann, M Doudoroff, RY Stanier. A study of the *Moraxella* group. II. Oxidative-negative species (genus *Acinetobacter*). *Journal of Bacteriology* 95:1520–1541, 1968.
15. GL Gilardi. Morphological and biochemical differentiation of *Achromobacter* and *Moraxella* (DeBord's tribe Mineae). *Applied Microbiology* 16:33–38, 1968.
16. JM Shewan. Some bacteriological aspects of handling, processing and distribution of fish. *Journal of Royal Sanitary Institute* 59:394–421, 1949.
17. J Listen. Qualitative variations in the bacterial flora of flatfish. *Journal of General Microbiology* 15:305–314, 1956.
18. PH Calcott, RA Macleod. Survival of *Escherichia coli* from freeze–thaw damage: a theoretical and practical study. *Canadian Journal of Microbiology* 20:671–681, 1974.
19. DR Ward, NJ Baj. Factors affecting microbiological quality of seafoods. *Food Technology* 42:85–89, 1988.
20. WK Rodman. On board fish handling systems for offshore wetfish trawlers, work smarter not harder. In: WT Otwell, Ed., *Proceedings of the First Joint Conference of the Atlantic Fisheries Technology Society*. Florida Sea Grant: University of Florida, 1987.
21. T Chai. Usefulness of electrophoretic pattern of cell envelope protein as a taxonomic tool for fish hold slime *Moraxella* species. *Applied Environmental Microbiology* 42:351–356, 1981.
22. V Venugopal. Extracellular proteases of contaminant bacteria in fish spoilage: a review. *Journal of Food Protection* 53:341–350, 1990.
23. HC Chen, T Chai. MicrofLora of drainage from ice in fishing vessel fish holds. *Applied Environmental Microbiology* 43:1360–1365, 1982.
24. DG James. The prospects for fish for the undernourished food and nutrition. *Food and Agriculture Organization* 12:20–27, 1986.
25. T Chai. Detection and incidence of specific species of spoilage bacteria in fish. II. Relative incidence of *Pseudomonas putrefaciens* and fluorescent Pseudomonads on Haddock fillets. *Applied Microbiology* 16:1738–1741, 1968.

26. EM Ravesi, JJ Licciardello, LD Racicot. Ozone treatments of fresh Atlantic cod, *Gadus morhua. Marine Fisheries Review* 49:37–42, 1987.
27. JM Shewan, G Hobbs, W Hodgkiss. A determination scheme for the identification of certain genera of Gram-negative bacteria, with special reference to the *Pseudomonadeceae. Journal of Applied Bacteriology* 23:379–390, 1960.
28. R Digirolamo. Survival of virus in chilled, frozen, and processed oysters. *Applied Microbiology* 20:58–63, 1975.
29. SW Weagant. The incidence of Listeria species in frozen seafood products. *Journal of Food Protection* 51:655–657, 1988.
30. PH Calcott. The effect of cooling and warm rates on the survival of a variety of bacteria. *Canadian Journal of Microbiology* 22:106–109, 1976.
31. HC Johnson, J Liston. Sensitivity of *Vibrio parahaemolyticus* to cold in oysters, fillets and crabmeat. *Journal of Food Science* 38:437–441, 1973.
32. GJ Banwart. *Basic Food Microbiology*. Westport: AVI Publishing Co., 1979, pp. 781.
33. SE Luchsinger, DH Kropf, CM Garcia Zepeda, MC Hunt, JL Marsden, EJ Rubio Canas, CL Kastner, WG Kuecker, T Mata. Color and oxidative rancidity of gamma and electron beam-irradiated boneless pork chops. *Journal of Food Science* 61 (5):1000–1005, 1996.
34. World Health Organization. Wholesomeness of irradiated food. Geneva: WHO. Technical Report Series 651, 1981.
35. SJ Lewis, A Velasquez, SL Cuppett, SR McKee. Effect of e-beam irradiation on poultry meat safety and quality. *Poultry Science* 81:896–903, 2002.
36. GG Giddings. Radiation processing of fishery products. *Food Technology* 38 (4):61–65, 1984.
37. J Jaczynski, JW Park. Microbial inactivation and electron penetration in surimi seafood during electron beam processing. *Journal of Food Science* 68 (5):1788–1792, 2003.
38. J Jaczynski, JW Park. Physicochemical changes in Alaska pollock as affected by electron beam. *Journal of Food Science* 69 (1):53–57, 2003.
39. J Jaczynski, JW Park. Physicochemical properties of surimi seafood as affected by electron beam and heat. *Journal of Food Science* 68 (5):1626–1630, 2003.
40. GM Pigott. Total utilization of raw materials from the sea. *Proceedings of the Conference on Formed Foods*, Brigham Young University, Brighton, Utah, April 1–2, 1985.
41. AR Jerrett, J Stevens, AJ Holland. Tensile properties of white muscle in rested and exhausted chinook salmon (*Oncorhynchus tshawytscha*). *Journal of Food Science* 61 (3):527–532, 1996.
42. T Nakayama, T Toyoda, A Ooi. Physical property of carp muscle during rigor tension generation. *Fish Science* 60 (6):717–722, 1994.
43. EE Pate, CJ Brokaw. Cross-bridge behavior in rigor-mortis. *Biophysics of Structure and Mechanism* 7:51–54, 1980.
44. H Tsuchiya, S Kita, N Seki. Postmortem changes in actinin and connectin in carp and rainbow trout muscles. *Nippon Suisan Gakkaishi* 58:793–798, 1992.
45. T Sigholt, U Erikson, T Rustad, S Johansen, TS Nordtvedt, A Seland. Handling stress and storage temperature affect meat quality of farmed-raised Atlantic salmon (*Salmo salar*). *Journal of Food Science* 62 (4):898–905, 1997.
46. T Berg, U Erikson, TS Nordtvedt. Rigor mortis assessment of Atlantic salmon (*Salmo salar*). *Journal of Food Science* 62 (3):439–446, 1997.
47. TE Lowe, JM Ryder, JF Carragher, RMG Wells. Flesh quality in snapper, *Pagrus auratus*, affected by capture stress. *Journal of Food Science* 58 (4):770–773, 1993.
48. GM Pigott. Fin and shellfishery products. In: FJ Francis, Ed., *Wiley Encyclopedia of Food Science and Technology*, 2nd Ed., Norwich: John Wiley & Sons, 1999, pp. 776–799.
49. H Rehbein. Measuring the shelf-life of frozen fish. In: HA Bremner, Ed., *Safety and Quality Issues in Fish Processing*. Boca Raton, FL: CRC Press, 2000, pp. 407–422.
50. WJ Dyer, JR Dingle. Fish proteins with reference to freezing. In: G Borgstrom, Ed., *Fish as Food. Production, Biochemistry and Microbiology*, Vol. 1. New York: Academic Press, 1961, pp. 275.
51. JJ Connell. Fish muscle proteins and some effects on them of processing. In: HW Schultz, AF Anglemier, Eds., *Symposium on Foods: Proteins and Their Reactions*. Westport, CT: The AVI Publishing Company, Inc., 1964, pp. 255–293.

52. Z Sikorski, J Olley, S Kostuch. Protein changes in frozen fish. *Critical Reviews in Food Science and Nutrition* 8:97–129, 1976.
53. SY Shenouda. Theories of protein denaturation during frozen storage of fish flesh. *Advances in Food Research* 26:275–311, 1980.
54. JH Mahon, CG Schneider. Minimizing freezing damage and thawing drip in fish fillets. *Food Technology* 18 (12):117–118, 1964.
55. AS McGill, R Hardy, JR Burt, FD Gunstone. Hept-cis-4-enal and its contribution to the off-flavor in cold stored cod. *Journal of the Science of Food and Agriculture* 25:1477–1489, 1974.
56. AS McGill, R Hardy, FD Gunstone. Further analysis of the volatile components of frozen cold stored cod and the influence of these on flavor. *Journal of the Science of Food and Agriculture* 28:200–205, 1977.
57. MCM Banda, HO Hultin. Role of cofactors in breakdown of TMAO in frozen red hake muscle. *Journal of Food Processing and Preservation* 7:221, 1983.
58. CH Castell, W Neal, B Smith. Formation of dimethylamine in stored frozen sea fish. *Journal of the Fisheries Research Board of Canada* 10:1685–1690, 1970.
59. GH Chu, C Sterling. Parameters of texture changed in processed fish: myosin denaturation. *Journal of Texture Studies* 2:214, 1970.
60. W Mao, C Sterling. Parameters of texture change in processed fish: cross-linkage of proteins. *Journal of Texture Studies* 1:484, 1970.
61. JR Dingle, RA Keith, B Lall. Protein instability in frozen storage induced in minced muscle of flatfishes by mixture with muscle of red hake. *Canadian Institute of Food Science and Technology Journal* 10:143–146, 1977.
62. TA Gill, RA Keith, BS Lall. Textural deterioration of red hake and haddock muscle in frozen storage as related to chemical parameters and changes in the myofibrillar proteins. *Journal of Food Science* 44 (3):661–667, 1979.
63. M Tejada, P Torrejón, ML Del Mazo, M Careche. Changes in extracted natural actomyosin from frozen stored cod (*Gadus morhua*) fillets and minces. *Presented at the European Research Conference on Protein Folding and Stability*, San Feliú de Guixols, Spain, 1995.
64. M Careche, ML Del Mazo, P Torrejón, M Tejada. Importance of frozen storage temperature in the type of aggregation of myofibrillar proteins in cod (*Gadus morhua*) fillets. *Journal of Agricultural and Food Chemistry* 46:1539–1546, 1998.
65. WD Powrie. Characteristics of food myosystems and their behavior during freeze-preservation. In: OR Fennema, WE Powrie, EH Marth, Eds., *Low Temperature Preservation of Foods and Living Matter*. New York, NY: Marcel Dekker, 1973, pp. 598–634.
66. JJ Matsumoto. Denaturation of fish muscle proteins during frozen storage. In: OR Fennema, Ed., *Proteins at Low Temperatures. ACS Symposium ser 180*. Washington, DC: American Chemical Society, 1979, pp. 205–226.
67. JJ Matsumoto. Chemical deterioration of muscle proteins during frozen storage. In: JR Whitaker, M Fujimaki, Eds., *Chemical Deterioration of Proteins*. ACS Symposium ser 123. Washington, DC: American Chemical Society, 1980, pp. 95–124.
68. JW Park, TC Lanier, DP Green. Cryoprotective effects of sugars, polyols, and phosphates on Alaska pollock surimi. *Journal of Food Science* 53 (1):1–3, 1988.
69. EL LeBlanc, RJ LeBlanc. Separation of cod (Gadus morhua) fillet proteins by electrophoresis and HPLC after various frozen storage treatments. *Journal of Food Science* 54 (4):827–834, 1989.
70. JR Davies, RG Bardsley, DA Ledward, RG Poulter. Myosin thermal stability in fish muscle. *Journal of the Science of Food and Agriculture* 45:61–68, 1988.
71. JW Park, TC Lanier. Combined effects of phosphates and sugar or polyol on protein stabilization of fish myofibrils. *Journal of Food Science* 52 (6):1509–1513, 1987.
72. J Grabowska, ZE Sikorski. The gel-forming capacity of fish myofibrillar proteins. *Lebensmittel-Wissenschaft Technologie* 9:33–35, 1976.
73. T Suzuki. *Fish and Krill Protein: Processing Technology*. London: Applied Science Publishers, 1981, p. 260.
74. A Hudobro, M Tejada. Gel forming capacity of fish mince during frozen storage. In: A Lahman Bennani, D Messaho, Eds., *Contribution du Froid à la Préservation de la Qualité des Fruits, Légumes et Produits Haleutiques*. Acted Éditions. Rabat, Morocco: Institut Agronomique et Véterinaire Hassan II, 1994, pp. 203–214.

75. M Careche, M Tejada. Interaction between triolein and natural hake actomyosin during frozen storage at −18°C. *Journal of Food Biochemistry* 15:449–462, 1991.
76. M Careche, M Tejada. Hake natural actomyosin interaction with free fatty acids during frozen storage. *Journal of the Science of Food and Agriculture* 64:501–507, 1994.
77. E Li-Chan, S Nakai, DF Wood. Relationship between functional (fat binding, emulsifying) and physicochemical properties of muscle proteins. Effects of heating, freezing, pH and species. *Journal of Food Science* 50:1034–1038, 1985.
78. ML Del Mazo, A Huidoboro, P Torrejón, M Tejada, M Careche. Role of formaldehyde in formation of natural actomyosin aggregates in hake during frozen storage. *Zeitschrift fuer Lebensmittel-Untersuchung und Forschung* 198:459–464, 1994.
79. S Cofrades. Relationship between Structural and Functional Characteristics of Proteins from Different Myosystems. Influencing Factors. Ph.D. thesis, Universidad Complutense, Madrid, Spain, 1994.
80. P Torrejón. Protein Aggregation in Frozen Fish Muscle. Action of Formaldehyde. Ph.D. thesis, Universidad Complutense, Madrid, Spain, 1996.
81. JR Wagner, MC Añon. Effect of frozen storage on protein denaturation in bovine muscle. II: Influence on solubility, viscosity and electrophoretic behaviour of myofibrillar proteins. *Journal of Food Technology* 21:547–558, 1986.
82. M Careche, AM Herrero, A Rodriguez-Casado, ML Del Mazo, P Carmona. Structural changes of hake (*Merluccius merluccius* L.) fillets: effects of freezing and frozen storage. *Journal of Agricultural and Food Chemistry* 47 (3):952–959, 1999.
83. ML García, J Martín-Benito, MT Solas, B Fernández. Ultrastructure of the myofibrillar component in cod (*Gadus morhua* L.) and hake (*Merluccius merluccius* L.) stored at −20°C as a function of time. *Journal of Agricultural and Food Chemistry* 47 (9):3809–3815, 1999.
84. AJ Miller, SA Ackerman, SA Palumbo. Effects of frozen storage on functionality of meat for processing. *Journal of Food Science* 45 (6):1466–1471, 1980.
85. AJ Borderias, A Moral, M Tejada. Stability of whole, filleted, and minced trout (*Salmo irideus* Gibb) during frozen storage. *Journal of Food Biochemistry* 6 (3):187–195, 1982.
86. CK Hsu, E Kolbe, MT Morrissey, YC Chung. Protein denaturation of frozen Pacific whiting (*Merluccius productus*) fillets. *Journal of Food Science* 58 (5):1055–1056, 1075, 1993.
87. A Zotos, M Hole, G Smith. The effect of frozen storage of mackerel (*Scomber scombrus*) on its quality when hot-smoked. *Journal of the Science of Food and Agriculture* 67 (1):43–48, 1995.
88. R Hurling, H McArthur. Thawing, refreezing, and frozen storage effects on muscle functionality and sensory attributes of frozen cod (*Gadus morhua*). *Journal of Food Science* 61 (6):1289–1296, 1996.
89. RJ Hastingss, GR Rodger, R Park, AD Mathews, EM Anderson. Differential scanning calorimetry of fish muscle: the effect of processing and species variation. *Journal of Food Science* 50:503–506, 510, 1985.
90. WJ Dyer. Protein denaturation in frozen and stored fish. *Food Research* 16:522–527, 1951.
91. WJ Dyer, ML Morton. Storage of frozen plaice fillets. *Journal of the Fisheries Research Board of Canada* 13:129, 1956.
92. WJ Dyer, ML Morton, DI Fraser, EG Bligh. Storage of frozen rosefish fillets. *Journal of the Fisheries Research Board of Canada* 13:569, 1956.
93. OR Fennema, WD Powrie, EH Marth, Eds., *Low Temperature Preservation of Foods and Living Matter*. New York, NY: Marcel Dekker, 1973, pp. 577–597.
94. RM Love. The freezing of animal tissue. In: HT Meryman, Ed., *Cryobiology*. London: Academic Press, 1966, pp. 317–405.
95. W Partmann. Some aspects of protein changes in frozen foods. *Zeitschrift fuer Ernaehrungswissenschaft* 16:167–175, 1977.
96. T Fukumi, K Tamoto, T Hidesato. Mon Rep Hokkaido Municipal Fish Exp Stn 22:30. Cited from Park, Lanier, Keeton, Hamann: use of cryoprotectants to stabilize functional properties of prerigor salted beef during frozen storage. *Journal of Food Science* 52 (3):537–542, 1987.
97. JW Park, TC Lanier, HE Swaisgood, DD Hamann, JT Keeton. Effects of cryoprotectants in minimizing physicochemical changes of bovine natural actomyosin during frozen storage. *Journal of Food Biochemistry* 11 (2):143–161, 1987.
98. TM Lin, JW Park. Solubility of salmon myosin as affected by conformational changes at various ionic strengths and pH. *Journal of Food Science* 63 (2):215–218, 1998.

99. JW Park. Cryoprotection of muscle proteins by carbohydrates and polyalcohols — a review. *Journal of Aquatic Food Products Technology* 3 (3):23–41, 1994.
100. T Tokunaga. The effect of decomposition products of trimethylamine oxide on the quality of frozen Alaska pollock fillets. *Bulletin of the Japanese Society of Scientific Fisheries* 40:167–171, 1974.
101. MD Rey-Manilla, CG Sotelo, SP Aubourg, H Rehbein, W Havermeister, B Jorgensen, MK Nielsen. Localisation of formaldehyde production during frozen storage of European hake (*Merluccius merluccius*). *European Food Research and Technology* 213 (1):43–47, 2001.
102. CG Sotelo, C Piñiero, RI Pérez-Martín. Denaturation of fish proteins during frozen storage-role of formaldehyde. *Zeitschrift fuer Lebensmittel-Untersuchung und Forschung* 200 (1):14–23, 1995.
103. Z Sikorski, S Kostuch. Trimethylamine N-oxide demethylase: its occurrence, properties, and role in technological changes in frozen fish. *Food Chemistry* 9 (3):213–222, 1982.
104. ZE Sikorski, S Kostuch, J Kołodiejska. Zur Frage der Protein deaturierung im gefrorenen Fischfleish. *Nahrung* 19:997, 1975.
105. CH Castell, B Smith, WJ Dyer. Effects of formaldehyde on salt extractable proteins of gadoid muscle. *Journal of the Fisheries Research Board of Canada* 30:1205–1213, 1973.
106. H Rehbein. Relevance of trimethylamine oxide demethylase activity and haemoglobin content to formaldehyde production and texture deterioration in frozen stored minced fish muscle. *Journal of the Science of Food and Agriculture* 43 (3):261–276, 1988.
107. I Mackie. The effects of freezing on flesh proteins. *Food Reviews International* 9:575–610, 1993.
108. T Tokunaga. Studies on the development of dimethylamine and formaldehyde in Alaska pollock muscle during frozen storage. I. Torry Research Station, Translation No. 743. *Bulletin of the Hokkaido Regional Fisheries Research Laboratory* 29:108, 1964.
109. NR Draper, H Smith. *Applied Regression Analysis*. New York: John Wiley & Sons, 1966, pp. 1–407.
110. S Srinivasan, YL Xiong. Gelation of beef hear surimi as affected by antioxidants. *Journal of Food Science* 61 (4):707–711, 1996.
111. S Srinivasan, YL Xiong. Sulfhydryls in antioxidant-washed beef heart surimi. *Journal of Muscle Foods* 8 (3):251–263, 1997.
112. B Wang, YL Xiong, S Srinivasan. Chemical stability of antioxidant-washed beef heart surimi during frozen storage. *Journal of Food Science* 62 (5):939–945, 991, 1997.
113. H Davies, P Reece. Fluorescence of fish muscle: causes of change occurring during frozen storage. *Journal of the Science of Food and Agriculture* 33:1143–1151, 1982.
114. C Milo, W Grosch. Changes in the odorants of boiled salmon and cod as affected by the storage of the raw material. *Journal of Agricultural and Food Chemistry* 44 (4):2366–2371, 1996.
115. HHF Refsgaard, PB Brockhoff, B Jensen. Sensory and chemical changes in farmed Atlantic salmon (*Salmo salar*) during frozen storage. *Journal of Agricultural and Food Chemistry* 46:3473–3479, 1998.
116. C Milo, W Grosch. Changes in the odorants of boiled trout (*Salmo fario*) as affected by the storage of the raw material. *Journal of Agricultural and Food Chemistry* 41:2076–2081, 1993.
117. T Ingemansson, P Kaufmann, B Ekstrand. Multivariate evaluation of lipid hydrolysis and oxidation data from light and dark muscle of frozen stored rainbow trout (*Oncorhynchus mykiss*). *Journal of Agricultural and Food Chemistry* 43 (3):2046–2052, 1995.
118. K Miyashita, T Takagi. Study on the oxidative rate and prooxidant activity of free fatty acids. *Journal of the American Oil Chemists' Society* 63:1380–1384, 1986.
119. T Han, J Liston. Correlation between lipid peroxidation and phospholipid hydrolysis in frozen fish muscle. *Journal of Food Science* 53:1917–1918, 1988.
120. J Olley, JA Lovern. Phospholipid hydrolysis in cod flesh stored at various temperatures. *Journal of the Science of Food and Agriculture* 11:644, 1960.
121. J Olley, R Pirie, H Watson. Lipase and phospholipase activity in fish skeletal muscle and its relationship to protein denaturation. *Journal of the Science of Food and Agriculture* 13:501, 1962.
122. TJ Han, J Liston. Lipid peroxidation and phospholipid hydrolysis in fish muscle microsomes and frozen fish. *Journal of Food Science* 52:294–296, 299, 1987.
123. F Mazeaud, E Bilinski. Free fatty acids and the onset of rancidity in rainbow trout (*Salmo gairdneri*) flesh; effect of phospholipase A. *Journal of the Fisheries Research Board of Canada* 33:1297–1302, 1976.
124. RL Shewfelt. Fish muscle lipolysis-a review. *Journal of Food Biochemistry* 5:79–100, 1981.

125. ME Apgar, HO Hultin. Lipid peroxidation in fish muscle microsomes in the frozen state. *Cryobiology* 19:154, 1982.
126. J Pokorny, G Janiček. Reaktion oxydierter Lipide mit mit Eiweisstoffen. 4 Mitt Die Bindung oxydierter Lipide an Casein. *Nahrung* 12:81, 1968.
127. M Karel. Protein–lipid interactions. *Journal of Food Science* 38:756, 1973.
128. M Karel, K Schaich, RB Roy. Interaction of peroxidizing methyl linoleate with some proteins and amino acids. *Journal of Agricultural and Food Chemistry* 23:159, 1975.
129. F Ohta, J Nishimoto. Effect of lipids on insolubilization of protein in frozen fish muscle during storage. *Memoirs of the Faculty of Fisheries of Kagoshima University* 13:45, 1964.
130. KA Al'tufeva, VN Ushkalova. Novyi sposob obrabotki morozhenoi pelyagi. *Rybnoe Khozyaistvo* (Moscow) 11:69, 1971.
131. KA Al'tufeva, VN Ushkalova. Sokhranenie kachestva morozhenoi ryby. *Rybnoe Khozyaistvo* (Moscow) 5:62, 1972.
132. SP Aubourg, I Medina. Influence of storage time and temperature on lipid deterioration during cod (*Gadus morhua*) and haddock (*Melanogrammus aeglefinus*) frozen storage. *Journal of the Science of Food and Agriculture* 79:1943–1948, 1999.
133. JJ Connell. The role of formaldehyde as a protein cross-linking agent acting during the frozen storage of cod. *Journal of the Science of Food and Agriculture* 26 (12):1925–1929, 1975.
134. WM Laird, IM Mackie, T Hattula. Studies of the changes in the proteins of cod-frame minces during frozen storage at −15°C. In: JJ Connell, Staff of Torry Research Station, Eds., *Advances in Fish Science and Technology*. Surrey, UK: Fishing News Books, 1980, pp. 428–434.
135. H Lim, NF Haard. Protein insolubilization in frozen Greenland halibut. *Journal of Food Biochemistry* 8 (3):163–187, 1984.
136. H Rehbein, H Karl. Solubilization of fish muscle proteins with buffers containing sodium dodecyl sulphate. *Zeitschrift fuer Lebensmittel-Untersuchung und Forschung* 180 (5):373–378, 1985.
137. YJ Owusu-Anssah, HO Hultin. Chemical and physical changes in red hake fillets during frozen storage. *Journal of Food Science* 51 (6):1402–1406, 1986.
138. M Tejada, M Careche, P Torrejón, M Del Mazo, MT Solas, ML García, C Barba. Protein extracts and aggregates forming in minced cod (*Gadus Morhua*) during frozen storage. *Journal of the Science of Food and Agriculture* 44 (10):3308–3314, 1996.
139. C Alvarez, A Huidobro, M Tejada, I Vazquez, E de Miguel, IA Gomez de Segura. Consequences of frozen storage for nutritional value of hake. *Food Science and Technology International* 5:493–499, 1999.
140. J Connell. The use of sodium dodecyl sulphate in the study of protein interactions during the storage of cod flesh at −14°C. *Journal of the Science of Food and Agriculture* 16:769–783, 1965.
141. H Buttkus. Accelerated denaturation of myosin in frozen solution. *Journal of Food Science* 35:558–562, 1970.
142. H Buttkus. The sulfhydryl content of rabbit and trout myosins in relation to protein stability. *Canadian Journal of Biochemistry* 49:97–102, 1971.
143. Y Tsuchiya, T Tsuchiya, JJ Matsumoto. The nature of the cross-bridges constituting aggregates of frozen stored carp myosin and actomyosin. In: *Advances in Fish Sciences and Technology*. Jubilee Conference. Aberdeen, UK: Torry Research Station, 1979, pp. 434–438.
144. AD Matthews, GR park, EM Anderson. Evidence for the formation of covalent cross-linked myosin in frozen-stored cod minces. In: JJ Connell, Staff of Torry Research Station, Eds., *Advances in Fish Science and Technology*. Jubilee Conference. Surrey, UK: Fishing News Books, 1980.
145. K Samejima, M Ishioroshi, T Yasui. Relative roles of the head and tail portions of the molecule in heat-induced gelation of myosin. *Journal of Food Science* 46:1412–1418, 1981.
146. K Yamamoto. Electron microscopy of thermal aggregation of myosin. *Journal of Biochemistry* 108:896–898.
147. A Sharp, G Offer. The mechanism of formation of gels from myosin molecules. *Journal of the Science of Food and Agriculture* 58:63–73, 1992.
148. T Sano, SF Noguchi, JJ Matsumoto, T Tsuchiya. Effect of ionic strength on dynamic viscoelastic behavior of myosin during thermal gelation. *Journal of Food Science* 46:1412–1418, 1991.

149. C Boyer, S Joandel, A Ouali, J Culioli. Ionic strength effects on heat-induced gelation of myofibrils and myosin from fast- and slow-twitch rabbit muscles. *Journal of Food Science* 61:1143–1148, 1996.
150. M Careche, ML García, A Herrero, MT Solas, P Carmona. Structural properties of aggregates from frozen hake muscle proteins. *Journal of Food Science* 67 (8):2827–2832, 2002.
151. NF Haard. Biochemical reactions in fish muscle during frozen storage. In: G Bligh, Ed., *Seafood Science and Technology*. London: Fishing News Books, 1990, pp. 176–209.
152. M Tejada, P Torrejón, ML Del Mazo, M Careche. Effect of freezing of formaldehyde on solubility of natural actomyosin isolate from cod (*Gadus morhua*), hake (*Merluccius merluccius*), and blue whiting (*Micromesistius poutassou*). In: JB Luten, T Borrensen, J Oehlesnschläger, Eds., *Developments in Food Science 38. Seafood from Producer to Consumer, Integrated Approach to Quality*. Amsterdam, The Netherlands: Elsevier Science, 1997, pp. 265–280.
153. M Careche, P Torrejón, ML Del Mazo, M Tejada. Study of aggregation of minced hake and blue whiting during frozen storage. In: *Proceedings of the 26th Western European Fish Technolgists' Association Conference*, Gdynia, Poland, 1996.
154. ZE Sikorski, A Kolakowska. Changes in proteins in frozen stored fish. In: ZE Sikorski, BS Pan, R Shahidi, Eds., *Seafood Proteins*. New York: Chapman & Hall, 1994, pp. 99–112.
155. JA Ramírez, MO Martín-Polo, E Bandman. Fish myosin aggregation as affected by freezing and initial physical state. *Journal of Food Science* 65 (4):556–560, 2000.
156. JJ Connell. Changes in amount of myosin extractable from cod flesh during storage at −14°C. *Journal of the Science of Food and Agriculture* 13:607, 1962.
157. JW Park, TC Lanier, JT Keeton, DD Hamann. Use of cryoprotectants to stabilize functional properties of prerigor salted beef during frozen storage. *Journal of Food Science* 52 (3):537–542, 1987.
158. BY Kim, DD Hamann, TC Lanier, MC Wu. Effects of freeze–thaw abuse on the viscosity and gel-forming properties of surimi from two species. *Journal of Food Science* 51:951, 1986.
159. ML Del Mazo, P Torregjon, M Careche, M Tejada. Characteristics of the salt-soluble fraction of hake (*Merluccius merluccius*) fillets stored at −20 and −30°C. *Journal of the Science of Food and Agriculture* 47:1372–1377, 1999.
160. SR Fuselli, ME Almandos, AS Ciarlo, RL Boeri, DH Giannini. The influence of sexual maturity, sex and size on quality aspects of frozen Argentine hake (*Merluccius hubbsi*). *Journal of Aquatic Food Products Technology* 5:81–94, 1996.
161. X Lou, C Wang, YL Xiong, B Wang, G Liu, SD Mims. Physicochemical stability of paddlefish (*Polyodon spathula*) meat under refrigerated and frozen storage. *Journal of Aquatic Food Products Technology* 9:27–39, 2000.
162. N Hedges. Maintaining the quality of frozen fish. In: HA Bremner, Ed., *Safety and Quality Issues in Fish Processing*. Boca Raton, FL: CRC Press LLC, 2000, pp. 379–406.
163. M Tejada, M Careche, P Torrejón, M Del Mazo, MT Solas, ML García, C Barba. Study of protein extracts and aggregates forming in minced cod (*Gadus morhua*) during frozen storage. *Journal of the Science of Food and Agriculture* 44:3308–3314, 1996.
164. JE Kinsella. Relationships between structure and functional properties of food proteins. In: PF Fox, JJ Condon, Eds., *Food Proteins*. London, UK: Applied Science Publishers, 1982, pp. 51–103.
165. SD Kelleher, H Hultin. Lithium chloride as a preferred extractant of fish muscle proteins. *Journal of Food Science* 56:315–317, 1991.
166. WJ Dyer, DJ Fraser. Proteins in fish muscle-13: lipid hydrolysis. *Journal of the Fisheries Research Board of Canada* 16:43, 1959.
167. T Sano. Thermal Gelation of Fish Muscle Proteins. Ph.D. thesis, Sophia University, Tokyo, Japan, 1988.
168. DS Chen, DC Hwang, ST Jiang. Effect of storage temperatures on the formation of disulfides and denaturation of milkfish myosin (*Chanos chanos*). *Journal of Agricultural and Food Chemistry* 37:1228–1231, 1989.
169. M Ito, N Inoue, H Shinana. Denaturation of carp myosin B during slow freezing. *Nippon Suisan Gakkaishi* 57 (2):319–323, 1991.
170. Y Kumazawa, Y Oozaki, S Iwami, I Matsumoto, K Arai. Combined protective effect of inorganic pyrophosphate and sugar on freeze-denaturation of carp myofibrillar protein. *Nippon Suisan Gakkaishi* 56 (1):105–113, 1990.

171. T Ueda, Y Shimizu, W Shimidu. *Bulletin of the Japanese Society of Scientific Fisheries* 28:1005–1009, 1010–1014, 1962. Cited in T Suzuki. Fish and Krill Protein: Processing Technology. London: Applied Science Publishers, p. 260, 1981.
172. K Nilsson, B Ekstrand. Frozen storage and thawing methods affect biochemical and sensory attributes of rainbow trout. *Journal of Food Science* 60:627–630, 635, 1996.
173. HHF Refsgaard, PMB Brockhoff, B Jensen. Free polyunsaturated fatty acids cause taste deterioration of salmon during frozen storage. *Journal of Agricultural and Food Chemistry* 48:3280–3285, 2000.
174. L Ramanathan, NP Das. Studies on the control of lipid oxidation in ground fish by some polyphenolic natural products. *Journal of Agricultural and Food Chemistry* 40 (1):17–21, 1992.
175. J Tillack. The storage characteristics of deep-frozen trout and slice salmon. Archiv Lebensmittelhygiene 26 (2):69–73, 1975.
176. CH Castell. Metal-catalyzed lipid oxidation and changes of proteins in fish. *Journal of the American Oil Chemists' Society* 48 (11):645–649, 1971.
177. A Pearson, J Love, F Shorland. Warmed-over flavor in meat, poultry and fish. *Advances in Food Research* 23:2–61, 1977.
178. G Pigott, B Tucker. Science opens new horizons for marine lipids in human nutrition. *Food Reviews International* 3:105–468, 1987.
179. KS Yoon, CM Lee, LA Hufnagel. Textural and microstructural properties of frozen fish mince as affected by the addition of nonfish proteins and sorbitol. *Food Structure* 10 (3):255–265, 1991.
180. RM Love. Ice-formation in frozen muscle. In: J Hawthorn, EJ Rolfe, Eds., *Low Temperature Biology of Foodstuffs*. New York: Pergamon Press, 1968, pp. 105–124.
181. P Montero, AJ Borderias. Changes in hake muscle collagen during frozen storage due to seasonal effects. *International Journal of Refrigeration* 12 (7):220–223, 1989.
182. M Jahncke, RC Baker, JM Regenstein. Frozen storage of unwashed cod (*Gadus morhua*) frame mince with and without kidney tissue. *Journal of Food Science* 57 (3):575–580, 1992.
183. A Huidobro, C Alvarez, M Tejada. Hake muscle altered by frozen storage as affected by added ingredients. *Journal of Food Science* 63 (4):638–643, 1998.
184. C Rodriguez, T Masoud, MD Heurta. Degradation of trimethylamine oxide for evaluation of quality of frozen fish. *Alimentaria* 288:125–129, 1997.
185. JK Babbitt, DL Crawford, DK Law. Decomposition of trimethylamine oxide and changes in protein extractability during frozen storage of minced and intact hake (*Merluccius productus*) muscle. *Journal of Agricultural and Food Chemistry* 20:1052, 1972.
186. H Rehbein, B Orlick. Comparison of the contribution of formaldehyde and lipid oxidation products to protein denaturation and texture deterioration during frozen storage of minced ice-fish fillet (*Champsocephalus gunnari* and *Pseudochaenichthys georgianus*). *International Journal of Refrigeration* 13:336–341, 1990.
187. PZ Lian, CM Lee, L Hufnagel. Physicochemical properties of frozen red hake (*Urophycis chuss*) mince as affected by cryoprotective ingredients. *Journal of Food Science* 65 (7):1117–1123, 2000.
188. JJ Connell. Changes in the adenosinetriphosphatase activity and sulphydryl groups of cod flesh during frozen storage. *Journal of the Science of Food and Agriculture* 11:245, 1960.
189. JJ Connell. Mechanical properties of fish and fish products. In: AL Copley, G Stainsby, Eds., *Flow Properties of Blood and other Biological Systems*. London: Pergamon, 1960, pp. 316–349.
190. JJ Connell, PF Howgate. Sensory and objective measurements of the quality of frozen stored cod of different initial freshness. *Journal of the Science of Food and Agriculture* 19:342, 1968.
191. WP Cowie, WT Little. The relationship between the toughness of cod stored at −7°C and −14°C, its muscle protein solubility and pH. *Journal of Food Technology* 1:335, 1966.
192. JPH Wessels, CK Simmonds, PD Seaman, LWJ Avery. The effect of storage temperature and certain chemical and physical pretreatments on the storage life of frozen hake mince blocks. Annual Report No 25. Capetown, South Africa: Fishing Industry Research Institute, 1981, pp. 82.
193. L Stodolnki, M Knasiak. Effect of fat content and freezing rate on solubility of myofibrillar proteins and textural properties of minced fish flesh. *Refrigeration Science and Technology* 4:429–435, 1981.
194. DJB Ponte, JP Roozen, W Pilnik. Effects of additives on the stability of frozen stored minced fillets of whiting. I. Various anionic hydrocolloids. *Journal of Food Quality* 8 (1):51–68, 1985.

195. J Toro, T Hwange, J Regenstein. Interaction of frozen storage temperature and food additives with cod fillet and frame mince. In: *Proceedings of the 12th Annual Conference of Tropical and Subtropical Fisheries Technological Society of the Americas*, Gainesville, FL: University of Florida, 1988, pp. 279–293.
196. M Crupkin, CL Montecchia, RE Trucco. Seasonal variations in gonadosomatic index, liver-somatic index, and myosin/actin ratio in actomyosin of mature hake (*Merluccius hubbsi*). *Comparative Biochemistry and Physiology* 89A:7–10, 1988.
197. CL Montecchia, SI Roura, H Roldan O Pérez-Borla, M Crupkin. Biochemical and physicochemical properties of actomyosin from frozen pre- and post-spawned hake. *Journal of Food Science* 62 (3):491–495, 1997.
198. SI Roura, CL Montecchia, H Roldan, O Pérez-Borla, M Crupkin. Ultrastructure of actomyosin in pre- and post-spawning Hake (*Merluccius hubbsi* Marini) during frozen storage. *Journal of Aquatic Food Products Technology* 9 (4):85–94, 2000.
199. RJ Hastings, GW Rodger, R Park, AD Matthews, EM Anderson. Differential scanning calorimetry of fish muscle: the effect of processing and species variation. *Journal of Food Science* 50:503–506, 510, 1985.
200. JF Ang, HO Hultin. Denaturation of cod myosin during freezing after modification with formaldehyde. *Journal of Food Science* 54 (4):814–818, 1989.
201. JJ Connell. The relative stabilities of the skeletal muscle myosins of some animals. *Biochemical Journal* 80:503–509, 1961.
202. RG Poulter, DA Ledward, S Godber, G Hall, B Rowlands. Heat stability of fish muscle proteins. *Journal of Food Technology* 20:203–217, 1985.
203. JR Davies, DA Ledward, RG Bardsley, RG Poulter. Species dependence of fish myosin stability to heat and frozen storage. *International Journal of Food Science and Technology* 29:287–301, 1994.
204. T Misima, T Yokoyama, M Tsuchimoto. The influence of rearing water temperature on the properties of Ca^{2+} and Mg^{2+}-ATPase activity on carp myofibril. *Nippon Suisan Gakkasihi* 56:477–487, 1990.
205. BK Howell, AD Matthews, AP Donelly. Thermal stability of fish myofibrils: a differential scanning calorimetric study. *International Journal of Food Science and Technology* 26:283–295, 1991.
206. O Esturk, JW Park, S Thawornchinsombut. Thermal sensitivity of fish proteins from various species on rheological properties. *Journal of Food Science* 69 (7):E1–E5, 2004.
207. K Nishiya, F Takeda, K Tamoto, O Tanaka, T Fukumi, T Kitabayashit, S Aizawa. Studies on freezing of surimi (fish paste) and its application: Studies on manufacture of frozen surimi for the material of kamaboko and sausage. *Hokkaido Municipal Fish Experiment Station* 18:122–135, 1961.
208. K Nishiya, F Takeda, K Tamoto, O Tanaka, T Kitabayashi. Studies on freezing of surimi (fish paste) and its application IV: On freezing surimi of Atka mackerel meat. *Hokkaido Municipal Fish Experiment Station* 18:391–397, 1961.
209. JJ Matsumoto. Minched fish technology and its potential for developing countries. In: *Proceedings of the Symposium on Fish Utilization Technology and Marketing in the IPFC Region*, Manila, March 8, 1978.
210. S Noguchi. The Control of Denaturation of Fish Muscle Proteins during Frozen Storage. Ph.D. thesis, Sophia University, Tokyo, 1974.
211. JR Herrera, IM Mackie. Cryoprotection of frozen-stored actomyosin of farmed rainbow trout (*Oncorhynchus mykiss*) by some sugars and polyols. *Food Chemistry* 84:91–97, 2004.
212. EEM Santos, JM Regenstein. Effects of vacuum packaging, glazing, and erythorbic acid on the shelf-life of frozen white hake and mackerel. *Journal of Food Science* 55 (1):64–70, 1990.
213. S Jittinandana, PB Kenney, SD Slider. Cryoprotection affects physiochemical attributes of rainbow trout fillet. *Journal of Food Science* 68 (4):1208–1214, 2003.
214. GA MacDonald, TC Lanier. Cryoprotectants for improving frozen food quality. In: MC Erickson, Y-C Hung, Eds., *Quality in Frozen Food*. New York: Chapman & Hall, 1997, pp. 197–232.
215. S Noguchi, JJ Matsumoto. Studies on the control of the denaturation of the fish muscle. I. Preventative effect of Na-glutamate. *Bulletin of the Japanese Society of Scientific Fisheries* 36:1078–1087, 1970.
216. S Noguchi, JJ Matsumoto. Studies on the control of the denaturation of the fish muscle proteins during frozen storage-II. Preventive effect of amino acids and related compounds. *Bulletin of the Japanese Society of Scientific Fisheries* 37:1115–1122, 1971.

217. S Noguchi, JJ Matsumoto. Studies on the control of denaturation of fish muscle proteins during frozen storage-III. Preventive effects of some amino acids, peptides. acetylamino acids and sulfur compounds. *Bulletin of the Japanese Society of Scientific Fisheries* 41:243–249, 1975.
218. S Noguchi, JJ Matsumoto. Studies on the control of denaturation of fish muscle proteins during frozen storage-IV. Preventive effects of carboxylic acids. *Bulletin of the Japanese Society of Scientific Fisheries* 41:329–335, 1975.
219. JJ Matsumoto, S Noguchi. Control of the freezing-denaturation of fish muscle proteins by chemical substances. In: *Proceedings of the XIIIth International Congress of Refrigeration*. Washington, DC, 13 (3):237–241, 1971.
220. J Sych, C Lacroix, LT Adambounou, F Castaigne. Cryoprotective effects of some materials on cod-surimi proteins during frozen storage. *Journal of Food Science* 55 (5):1222–1227, 1990.
221. CM Lee. Surimi process technology. *Food Technology* 38 (12):69–80, 1984.
222. J Grabowska, Z Sikorski. Technological quality of minced fish preserved by freezing and additives. Acta Alimentaria. *Academiae Scientiarum Hungaricae* 2:319, 1973.
223. JF Carpenter, JH Crowe. The mechanism of cryoprotection of proteins by solutes. *Cryobiology* 25:244–255, 1988.
224. SR Payne, D Sandford, A Harris, OA Young. Effects of antifreeze proteins on chilled and frozen meat. *Meat Science* 37 (3):429–438, 1994.
225. SR Payne, OA Young. Effects of pre-slaughter administration of antifreeze proteins on frozen meat quality. *Meat Science* 42 (2):147–155, 1995.
226. A Hunt, JW Park, C Jaundoo. Cryoprotection of Pacific whiting surimi using a less-sweet glucose polymer. In: *Proceedings to the Annual Institute of Food Technologists Meeting*, New Orleans, LA, 2001.
227. A Hunt, JW Park, H Zoerb. Trehalose as functional cryoprotectant for fish proteins. In: *Proceedings to the Annual Institute of Food Technologists Meeting*, Anaheim, CA, 2002.
228. LE Lampila. Functions and uses of phosphates in the seafood industry. *Journal of Aquatic Food Products Technology* 1 (3/4):29–41, 1992.
229. E Tanikawa, M Akiba, A Shitamori. Cold storage of cod fillets treated with polyphosphates. *Food Technology* 17:1425, 1963.
230. A Hunt, JS Kim, JW Park, R Schnee. Effect of various blends of phosphate on fish proteins during frozen storage. In: *Proceedings to the Annual Institute of Food Technologists Meeting*, Las Vegas, NV, 2004.
231. HB Bull, K Bresse. Protein hydration I. Binding sties. *Archives of Biochemistry and Biophysics* 128:488, 1968.
232. PL Privalov. Cold denaturation of proteins. *Critical Reviews in Biochemistry and Molecular Biology* 25:281–305, 1990.
233. K Gekko, T Morikawa. Preferential hydration of bovine serum albumin in polyhydric alcohol–water mixture. *Journal of Biochemistry* 90:39, 1981.
234. TC Lanier, GA MacDonald. Cryoprotection of surimi. In: G Sylvia, MT Morrissey, Eds., *Pacific Whiting-Harvesting, Processing, Marketing, and Quality Assurance*. Oregon Sea Grant. ORESU-W-91–001, 1992, pp. 20–28.
235. H Levine, L Slade. A food polymer science approach to the practice of cryostabilization technology. *Comments on Agricultural and Food Chemistry* 1:315, 1988.
236. H Levine, L Slade. Principles of cryostabilization technology from structure/property relationships of carbohydrate/water systems: a review. *Cryo-Letterrs* 9:21, 1988.
237. M Serpelloni. The food applications of sorbitol, mannitol, and hydrogenated glucose syrups. In: *Proceedings of the International Symposium of Polyols and Polydextrose*, Paris, 1985.
238. M Matsumiya, S Otake. Storage of prepared raw surimi. *Bulletin of the College of Agriculture and Veterinary Medicine of Nihon University*, 40:121, 1983.
239. RE Feeney, Y Yeh. Antifreeze proteins: current status and possible food uses. *Trends in Food Science and Technology* 9:102–106, 1998.
240. G Caple, WL Kerr, TS Burcham, DT Osuga, Y Yeh, RE Feeney. Superadditive effects of fish antifreeze glycoproteins and polyalcohols or surfactants. *Journal of Colloid and Interface Science* 3 (2):299, 1986.

18 Quality and Safety of Frozen Vegetables

Wenceslao Canet Parreño and Maria Dolores Alvarez Torres
Instituto del Frío, CSIC, Madrid, Spain

CONTENTS

I. INTRODUCTION

Looking back over the historical development of quality requirements for processed foods, freezing is undoubtedly the most satisfactory method for the long-term preservation of vegetable produce when properly carried out. The low temperatures commonly prescribed for frozen foods (-18°C) can maintain initial quality and nutritive value practically unchanged, so that frozen and fresh vegetable products differ only in texture [1]. The freezing of vegetables immediately after postharvest guarantees consumers a higher vitamin C content than could be attained by any other form of preservation and distribution. Furthermore, if properly handled before freezing and during distribution, there is no possibility of growth of microbial contaminants between freezing and thawing [2].

The importance of frozen foods is reflected by the constant growth of consumption in all industrialized countries, as confirmed by the latest figures [3]. In 2002, Europe's growing frozen food market passed the 12 million ton, with a monetary value of US$ 46,148.9 million and per capita consumption of 32.3 kg. Vegetables and potato products are the top frozen categories by any count, with consumption standing at 4,707,800 t (value of US$ 10,104.1 million) and per capita consumption of 12.75 kg. Although European frozen food consumption is rising nearly everywhere, growth is not actually strong except in a few countries such as Spain (up by 12%) and Sweden (up by 6.6%). The increase was only 0.9% in Finland and the United Kingdom, 1.3% in Germany, and 1.8% in Italy. The Netherlands reported a drop of 3%. The United States remains the world's largest frozen food market, with per capita consumption of about 62 kg and overall consumption stands at 21,498,825 t (value of US$ 71,089 million) of which 7,937,998 t is vegetables including potato products (value of US$ 12,235 million). Vegetables per capita consumption in the United States is 22.8 kg, but opportunities for major growth are limited due to the development of this market. In Japan, the long running recession affected the frozen food industry in 2002; global consumption fell by 3.6% from 2001 to 2.2 mt (value of ¥705 billion). In contrast, consumption of fresh farm produce (vegetables, potato products, and fruits) rose by 8% from 2001, to 92,090 t (value of ¥24.4 billion) [3].

The importance of freezing and the scientific and technological development of this method of food preservation are also highlighted by the amount of scientific attention centered on this field in recent years. In the 1980s, it was included as a research priority in the framework of food science and technology under the first Intra-European Collaborative Research Programme and the COST 91 Project on the Effects of Thermal Processing on Quality and Nutritive Value of Foods, with a subgroup for collaborative work on the influence of freezing, distribution, and thawing on the quality and nutritive value of foods [4]. More recently, the FAIR Programme CT96-1180 focused on the preservation of quality and safety in frozen foods throughout the distribution chain [2,5]; another 20 projects related to food freezing research have been ongoing or completed in the last few years [6], culminating in interesting scientific findings in the general field of frozen foods and with respect to frozen vegetables in particular.

Before discussing the quality of a food, whether frozen or not and the factors that influence quality, it is essential to clearly define what is meant by quality, given that the term in its broadest sense embraces and is dependent on factors of varied origin and nature. Any food, whether frozen or not, is considered to be of good quality if it meets the following requirements: there must be a total absence of pathogens and compounds toxic to humans (*hygiene and health quality*); it must be easily digestible, with good nutritional value, that is, high concentrations of vitamins, macronutrients, and minerals, and an appropriate caloric content (*nutritional quality*); its sensory attributes such as appearance, flavor, aroma, and texture must be constant, and in the case of frozen products, they must be as close as possible to those of fresh produce (*sensory quality*); and the presentation and mode of preparation must be according to consumers' preferences (*commercial quality*).

The quality of a frozen vegetable depends on a large number of factors including the type, species, and variety of product; ripeness and initial quality; the method of harvesting and the time lapse between harvesting and processing; prefreezing treatments; freezing *per se*; and freezing conditions and packaging. It is this set of factors, known as P–P–P (*Product–Processing–Packaging*), and their optimum way in which they interact that defines quality. This quality can be considerably diminished in the case of adverse storage times and temperatures, that is, T–T–T (*Time–Temperature–Tolerance*) factors.

Given the number of such factors and the possibility of defects in storage conditions throughout the cold chain, one can readily understand the difficulty involved in optimizing the entire process and satisfactorily meet consumer quality expectations in such products. In frozen vegetables, health quality, nutritional quality, and aspects of sensory quality such as color and texture can be objectively assessed and controlled. However, in the case of overall assessment of sensory quality, only the consumer can perceive and process the overall blend of sensations, which denote quality and cause consumers to prefer, accept, or reject a product. The large number of influencing factors makes it difficult first to optimize the final quality of a frozen vegetable, and secondly, to assess and quantify the loss of quality during storage.

Following the general line of this handbook, the present review focuses specifically on how the quality, safety, and nutritional value of frozen vegetables is affected by product-related factors, processing (particularly pretreatments such as blanching), and time during frozen storage.

II. INFLUENCE OF PRODUCT, PROCESSING, AND PACKAGING (P–P–P FACTORS)

The product, processing, and packaging factors briefly referred in the previous section constitute an important field of research inasmuch as these factors considerably influence frozen vegetable quality and stability during storage and distribution.

A. Product

The *raw material* used in the preparation of frozen vegetables is an important conditioning factor affecting both the quality and nutritional value of the final product. A product's suitability for freezing is determined by *agrotechnical practices* and conditions, the *species* and *cultivar* involved, and technological factors such as ripeness and the time elapsing between harvesting and processing. Clearly then, only raw material that is clean, sound, and high nutritional, safe, and having sensory qualities should be selected for freezing.

1. Agrotechnical Practices and Conditions

The factors, including soil structure, fertility, climatic conditions, excessive rainfall or irrigation, temperature, and altitude, exert a pronounced effect on fresh product quality and on product deterioration during processing or cooking before consumption.

Soil structure can affect the success or failure of a culture: a loamy soil allows a high rate of cationic exchange and good water retention, but it is difficult to work (too heavy); and a sandy soil is very soft and has no stagnant water, but there is a low rate of cationic exchange (poor plant nutrition) and low water retention (water stress). The ideal soil mixture would be 20% clay, 40% sand, and 40% slime [7]. The fertility of a soil depends not only on its structure but also on the availability of nutritional elements (N, P, K, microelements, and humus), pH level, and human factors (tillage and fertility maintenance).

Fertilizer composition and fertilizer application conditions also affect the suitability of products for freezing: large amounts of nitrogen fertilizers negatively affect spinach, whose leaves accumulate nitrates that may turn into toxic nitrites. The textural attributes of frozen potato products can be improved by farming practices designed to augment the solid content — the higher the dry matter content of potato, the more suitable it is for processing. Field-grown vegetables tend to retain their texture better after freezing, whereas products grown using forced cultivation methods tend to be unsuitable for freezing [8]. Exposure or sunlight affects photoperiodism (amount of daylight hours) and thermoperiodism (alternation of ambient temperature): depending on the crop orientation, the soil warms up during the day and stores the accumulated warmth during the night, thus preventing freezing, diseases, and accumulation of stagnant water; photosynthesis is improved in north–south oriented fields, where long sunlight exposure and excess of temperature are avoided, thus preventing surface burns and discoloration [9]. If the crop is oriented east–west, the cultivar must have enough foliage to protect the product against sunburn. High winds also impede good growth of crops, as they cause problems during insect pollination; cauliflowers can grow deformed, and peppers, aubergines, beans, and peas can be affected by buffeting. Protective methods include planting the rows in the main direction of the winds, using cultivars with deep roots and taking advantage of natural windbreaks. Low or excessive rainfall or irrigation also affect crop growth. In areas with a short spring and a dry summer, early cultivars are recommended; in areas with excessive rainfall, fast-growing cultivars are necessary, because stagnant water causes mold on peas and harms or kills leaves of leafy crops such as spinach or chard beet. In particular, temperature influences crop quality: low temperatures affect the germination of seeds, with the risk of freezing injury; high temperatures induce reversals of physiological behavior, for example, pigmentation is blocked and water stress affects crop growth. Low temperatures cause internal bractiness and looseness of florets in cauliflowers.

All these agronomic and climatic factors have to be taken into account in planning for correct cultivation and harvesting of industrial crops. Growers have to plan the sowing and transplanting of cultivars, but it is not easy to program the cycle of plant growth. For instance, green beans are simple; they are harvested 50–60 days after sowing. In contrast, broccoli varieties with short life cycles can blossom early at below 10°C. Some spinach varieties react badly with warm spring days, so that it is better to sow later flowering varieties. Pea cultivation is very difficult to organize in countries with steep temperature rises in spring, because the peas harden and lose nutritional and sensory qualities [7].

2. Species and Variety

Vegetable tissues are very sensitive to freeze damage; the physical change of water within the product causes loss of cellular structure, with high drip loss while thawing and loss of shape and texture. This quality deterioration is minimized by cultivar selection and pretreatments before freezing. In terms of species, most vegetable products can be frozen and excellent results have been achieved using asparagus, artichokes, chard beet, green beans, spinach, peas, cauliflower, broccoli, Brussels sprouts, sweet corn, onion, peppers — indeed, almost all vegetables. In particular, potatoes have excellent freezing qualities, and this has led to the appearance of a large number of commercial frozen potato products. In the case of other species such as tomatoes and lettuce, in contrast, flavor and texture are considerably altered by freezing [1].

The cultivar employed is usually of fundamental importance in minimizing such alterations. A new cultivar (cultivated variety) can be obtained by cloning to produce exact duplicates (artichokes), by variety crossing (spinach and potatoes), by hybridization to increase production or disease resistance (maize), and by genetic modification [7]. For many years, there has been extensive research to enlarge the number of available varieties suitable for mechanical harvesting and handling, which afford high yields and optimum levels of quality both as a raw material and as a final frozen product [1]. Kennedy [10] reported that public opinion had turned away from the use of genetic modification of plants for improvement of their agricultural and food quality properties because of concerns over effects on the environment and worries about the effects of ingesting material from these plants. Time will tell whether such worries are well founded. The breeding of plants for frost resistance, nutritional content, and freeze resistance during processing is a desirable goal, which could bring benefits to consumers of frozen vegetables.

3. Technological Factors

In view of the complete cessation of physiological processes (tissue death) brought about by blanching or freezing, fruits and vegetables to be frozen need to be harvested at the exact moment of ripening. Given that the food industry needs continuous supplies of raw material, meaning that it is essential to properly organize the varieties used, planting times, and growing zones; the same applies to mechanical harvesting, prerefrigeration, quick transportation, and handling tailored to each individual product, which may otherwise quickly lose its nutritive value after harvesting.

Now there are harvesting machines for nearly all vegetable crops; with the proper choice of cultivar and uniform ripening, it is possible to minimize waste in mechanical harvesting. Mechanical handling can damage vegetables if they are not tough enough; asparagus, broccoli, cauliflower, mushroom, aubergine, tomato, spinach, and courgette are especially delicate [11]. Low growing plants such as spinach can be damaged during harvesting, and yields are improved by the use of long-stalked cultivars; peas are harvested and shelled simultaneously and should be transported in controlled temperature conditions to avoid loss of quality.

After harvesting, vegetables undergo changes in chemical composition, sensory attributes, and nutritional value: the higher the temperatures and the longer the time elapsing between harvesting and processing, the greater the loss. Thus, in 24 h at 20°C, green beans loses 35% of its initial ascorbic acid, peas lose about 25%, spinach loses 15–55% depending on the season (winter or autumn), and asparagus loses 38%. Low storage temperatures (4°C) reduce these daily losses to 25% in green beans, 6% in peas, 5–25% in spinach, and 6% in asparagus. In comparison, the combined action of blanching and freezing results in average ascorbic acid losses of 20% in flat green beans, 25% in round green beans and peas, and 40% in spinach [12].

Between harvesting and processing, products become dehydrated, undergo wilting to a greater or lesser extent, and lose their shiny, turgid appearance. Asparagus turns yellow and fibrous if not rapidly submerged in an ice-water bath; when damaged, mushrooms undergo intense browning if not processed very quickly; peas give off a considerable amount of respiration heat, accompanied by the development of off-flavors and a substantial rise in tenderometer values [1,8].

Precooling and low temperatures during transportation and storage before processing retard such postharvest losses. Prechilling under pressure is a good method for nearly all vegetables; hydrocooling is especially good for asparagus, carrots, peas, and celery, and vacuum cooling is particularly suitable for leafy vegetables, broccoli, and celery [7,13]. There is a need for basic research into the specific mechanisms producing such losses of quality in individual products. Research is also required into the selection and breeding of varieties resistant to and methods for minimizing such losses when prolonged time lapses between harvesting and processing are unavoidable [14,15].

4. Quality Assessment of Raw Material

Only raw material of high nutritional, safety, and sensory qualities should be selected for freezing. The quality of the raw material can be rarely improved by processing, so at best, it can be only maintained, and this will depend on the degree of excellence of the processing operations. At the factory, the processor has the choice of accepting or rejecting incoming loads of raw vegetables. The decision is made after measuring samples from each consignment against a raw material specification, which should describe precisely what is required and the extent to which the quality may deviate from the standard before it is rejected [16].

The contract between supplier and freezing plant stipulates the quantity and quality expected and the agreed price depending on the quality specifications. Almost all European countries are interested in having official standards for quality of fruits and vegetables; these can be only achieved by studying the quality indices [11]. Such indices or specifications are either established by the industry's own quality control laboratory or produced by public or private institutions. Volume 5A of the FAO/WHO Joint Food Standards Programme (Codex Alimentarius) contains a set of standards for quick-frozen fruits and vegetables [17]. In addition to defining the process and form of presentation for the various products, it deals with other quality factors including definition of defects, sizes of analytical samples and defective units, criteria for acceptance of batches, levels of additives, hygiene, and labeling and packaging. Volume 13 of the Codex also presents analytical and sampling methods [18]. Between 1997 and 2003, Campden and Chorleywood Food Research Association (CCFRA) in conjunction with the U.K. frozen vegetable industry compiled handbooks for many products: The Raw Material Guidelines for Quick Freezing [19], addressing quality standards, detailed sampling plans, and defects — definitions and tolerances; The Specifications for Quick Frozen Fruits and Vegetables [20], including product grades and grade defects — definitions and tolerances — and sensory assessment; other methods used by CCFRA for objective and sensory quality assessment of frozen vegetables are found in the literature [21–24]. There are FDA food defect action levels for vegetables and vegetable products [25] and FDA macroanalytical methods for vegetables and vegetable products [26].

Part IV of this handbook describes methods and techniques used to measure quality by physical (Chapter 24), chemical (Chapter 25), and sensory measurements (Chapter 26) in detail.

5. Nutritional Aspects of Raw Material

Fruits and vegetables are the most important source of vitamins, minerals, sugars, and fibers. The main dietary input from fruit, vegetables, and their derivatives is undoubtedly ascorbic acid — vitamin C; fruit provides practically 33% and vegetables 61%. Vegetables are also the chief sources of provitamin A (43%). They account for 6% of the total caloric content of diet and supply other nutrients including protein (6%), calcium (7%), iron (15%), thiamine (15%), riboflavin (7%), and niacin (14%). They are also the main sources of fiber: fiber is increasing being recognized as beneficial for gastrointestinal functioning and in the prevention of coronary and cerebrovascular disorders [12,27].

The concentrations of nutrients and vitamins vary widely from one vegetable to another. Average vitamin C contents are highest in Brussels sprouts (95 mg/100 g), broccoli (87 mg/100 g), cauliflower (70 mg/100 g), spinach (60 mg/100 g), asparagus (35 mg/100 g), green beans and peas (20–25 mg/100 g), and potatoes (15–40 mg/100 g) — surprisingly, potatoes are one of the main sources of vitamin C in the diet because of the amount consumed. The vegetables with the highest folic acid contents are asparagus (150 μg/100 g), spinach (100 μg/100 g), and Brussels sprouts (70 μg/100 g), whereas green beans, peas, and cauliflower contain 30 μg/100 g.

In the case of vitamin C, adult requirements range from 45 to 80 mg/day. Forty-five milligram per day is the minimum recommended intake; this is believed to be four times the minimum requirement to prevent symptoms of deficiency and enough to maintain reserves at healthy levels. The references show that recommended intakes vary considerably from country to country: Germany recommends 70 mg/day and Russia 125 mg/day [12]. In the case of folic acid, adult requirements range from 200 to 300 μg/day for men and 170 to 300 μg/day for women; in some European countries like the Netherlands, 400 μg/day of supplementary folic acid is recommended for women wishing to become pregnant. The mean folate intake in Europe is very low, and therefore, consumption of folate-rich foods, such as fruits and vegetables, should be encouraged [28].

Vegetable macronutrients and minerals remain stable over the period between harvesting and processing: vitamins are unstable and are affected by heat, light, oxidation, enzymatic action, and solubility; ascorbic acid changes into dehydroascorbic acid and then into 2,3-diketogluconic acid. There is no vitamin C activity in the presence of oxygen. This process starts immediately after harvesting and continues during processing, cold storage, and final cooking. Because ascorbic acid is very unstable, it is considered to be the most representative and important biochemical indicator of the nutritional value of vegetables; if the nutritional value exists we are assured of the presence of all the other vitamins and nutrients. This makes it important to always use cultivars with a high initial concentration of ascorbic acid, because even after losing 30–50% of their ascorbic acid during processing and frozen storage, they will still reach the consumer with nutritional content as good as or better than fresh vegetables purchased at traditional greengrocers.

6. Microbiological and Safety Aspects of Raw Material

There is considerable variation in the number of microorganisms in vegetables from growing fields; they posses a resident microflora that normally subsists on the slight traces of carbohydrates, protein, and inorganic salts, which dissolve in the water exuding from, or condensing on, the epidermis of the host. Vegetables are contaminated by soil, water, dust, and other natural sources and by contact with the soiled surfaces of harvesters and containers during harvesting. Populations of microorganisms vary widely and often depend on the type of vegetable. Aerobic plate counts can be as high as 10^7/g on tubers and other vegetables that are in contact with the soil. Vegetables grown above ground can have comparable populations of bacteria; however, these populations on any given vegetable vary greatly. Cabbage, for example, may have 10^4–10^9 bacteria per gram [29].

A soil-contaminated vegetable will possess the microorganisms that compete best on that particular substrate. The high carbohydrate and low acid content of many vegetables favor lactic acid bacteria, whereas some low-saccharine vegetables have a predominantly aerobic, gram-negative flora. Mold counts of fresh vegetables seldom exceed 10^5/g. Dematiaceous fungi dominate as the preharvest flora, whereas nondematiaceous fungi dominate as postharvest flora. Certain coliform bacteria and enterococci are part of the naturally occurring microflora of plants; the presence of *Escherichia coli* may be due to the use of polluted water for irrigation or washing or due to contaminated surfaces of harvesters and containers.

Populations of microorganisms will normally increase dramatically during the period between harvesting and processing; the degree and rate of this increase will depend on the form of the product and on the environmental conditions. For example, microbes will usually grow faster on cut vegetables; temperature and humidity will also influence the microflora and ultimate populations of bacteria often reach 10^6–10^7 cells per gram before the product appears spoiled. Gram-negative bacteria will usually predominate on stored fresh vegetables; the bacterium of greatest concern is *Erwinia carotovora*, the cause of bacterial soft rot in a large variety of vegetables, but some members of the *Pseudomonas*, *Bacillus*, and *Clostridium* genera are also important spoilage organisms.

The presence of pathogenic microorganisms on raw vegetables does not appear to be a serious problem in developed countries; studies have shown that some vegetables which are widely used in salads may harbour *Salmonella* spp., *Listeria monocytogenes*, *Aeromonas* spp., or *Bacillus cereus*, but these pathogens, although they may be present in fresh products, are generally rare in blanched or frozen vegetables [30]. There is a complete list of pathogens and disease outbreaks on vegetables and vegetable products, adapted from Refs. [31,32], from analysis and evaluation of preventive control measures for the control and reduction or elimination of microbial hazards on fresh and fresh-cut produce.

Rapid methods for microbiological analysis are available [33], but it is effectively impossible to check the microbial quality of all raw materials. It is worth noting that low bacteriological quality (i.e., high but still acceptable bacterial numbers) in a food product at the outset of freezing results in more rapid quality deterioration during storage, thus reducing practical storage life (PSL). Very rapid methods, for example, producing a result in less than 1 h, make it possible to take special precautions against raw materials with excessive numbers of pathogens or microorganisms in general [34].

The absence of contamination from vegetables by pesticides is one of the fundamental requirements that contribute to the nutritional profile of the products. It is imperative for farmers to respect the pesticide-spraying treatment calendar, so that crops do not arrive at the industry in a state potentially harmful to human health. Herbicides, for example, should be used with caution. Residues can remain in the soil even after the end of cultivation and can interfere with successive rotation crops. The FAO/WHO and others [35,36] have issued recommendations for the use of pesticides, particularly for the permissible residue in each product: the maximum residue levels indicate the maximum amount of active molecule in any specific pesticide that is permissible in the product. There is a set of FDA action levels for poisonous or deleterious substances in vegetables and vegetable products [37]. In many countries, these proposals have been incorporated into legislation, and both farmers and industry must comply with these laws as an act of responsibility toward the health of consumers [7]. Part IV of this handbook describes methods and techniques in detail for testing of microorganisms that cause foodborne illnesses and spoilage (Chapter 27).

B. Process

Processing consists of a series of stages from reception of the products at the plant to their final dispatch. The first stage comprises a number of preliminary operations to prepare the product for subsequent freezing. These operations are blow dry cleaning, removal of stones and washing, inspection and selection, classification, peeling, and, depending on the species and variety, chopping and slicing, stoning, and cropping. Hygiene conditions must be absolutely strict, and care must be taken to avoid excessive wastage and mechanical damage. The second stage consists of blanching (heating for a short time to inactivate the enzymatic systems responsible for off-odors and flavors and changes in color during frozen storage), and other prefreezing treatments (use of coadjuvants to improve blanching action, depending on the product), and finally cooling and draining to prevent yield and energy loss during freezing. The third and fourth stages are freezing and frozen storage, respectively. Depending on the type of product, freezing may take place after packaging (e.g., spinach); more commonly, however, products are frozen individually, that is, individual quick-frozen. They are then frozen-stored packaged in small containers weighing anything from 160 g to several kilograms for direct dispatch or bulk stored in polyethylene-lined pallet boxes, which can contain several hundred kilos of product, thus helping to optimize the utilization of storage space [8,15].

1. Main Preparatory Procedures

The purpose of these procedures is to take raw vegetables as received by processors and from them make a product which once processed and frozen is "ready-to-eat" with minimal final preparation by the consumer. Following an initial selection process to meet the quality standards required, the products are cleaned by vibration or air blast to remove unwanted materials (such as leaves, husks, etc.); root vegetables and tubers are brushed to remove earth, stones, and excess dirt.

Washing cleans the product of dirt and impurities (soil and waste matter), of pesticide residues and, in the case of vegetables, of up to 90% of the microbial flora. Different types of washers are used (floating, immersion, rotary, turbulence, and high-pressure sprays) depending on whether the object is a root or tuber (carrot, potatoes, etc.), is leafy (spinach, Swiss chards, etc.), or is a fruit or flower (peas, beans, cauliflower, etc.). Light chlorination by adding gaseous chlorine or sodium hypochlorite enhances the action of water, preventing the formation of sludges of bacterial origin in the equipment and the development of unpleasant odors. Free chlorine contents in the region of 5–10 ppm do not adversely affect product flavor or corrode equipment. Washing accounts for a high proportion of the total water expended in the process (25–35 l/kg of frozen fruit or vegetable) [1].

Peeling, one of the most delicate pretreatments, is performed industrially by abrasion, high-pressure steam, treatment with sodium hydroxide solution, or mechanically. Abrasion is effected by rough, moving surfaces, which remove the outer surface of the product, but it has the drawback of considerable loss of raw material. Steam peeling consists in heating the product to a temperature of up to 80°C and subjecting it to pressures of 392–686 kPa between 30 sec and 3 min. In sodium hydroxide peeling, the product is preheated then immersed in a 10–20% solution at a temperature of 60–90°C between 1 and 5 min depending on the type of the product. The drawback of all these methods is the substantial loss of raw material involved (8–20% in potatoes, depending on their shape and age). Using sodium hydroxide with infrared heating can cut down sodium hydroxide solution consumption by 80%, decrease raw material loss by one third, and reduce water consumption by up to 95%. Abrasion is commonly employed for potatoes and carrots; chemical peeling is used for fruits, especially peaches; mechanical peeling is used for pears and apples [1,38].

After washing and peeling, the product may be subjected to any of the procedures (e.g., sorting, paring, stemming, trimming, cutting, and pulping), depending on the type and variety of the product. There is a wide range of equipment for high-yield performance of these operations [39].

Figure 18.1 depicts the operation of a DiversaCut 2110® Dicer by Urschel Laboratories, Inc. This equipment uniformly dices, strip cuts, and slices a wide variety of vegetables at high production capacities, including potatoes (9.5 × 9.5 mm^2 crinkle strips, 6.4 × 6.4 mm^2 strips, and 9.5 or 6.4 mm dice), onions and carrots (9.5 mm dice and 3.2 × 6.4 × 3.2 mm^3 dice), tomatoes, carrots, turnips, celery, and so on.

All these operations must be carried out with the utmost care and under the most stringent hygiene conditions to prevent contamination of the product and mechanical damage. The varying degrees of complexity and automation of the process according to product type require

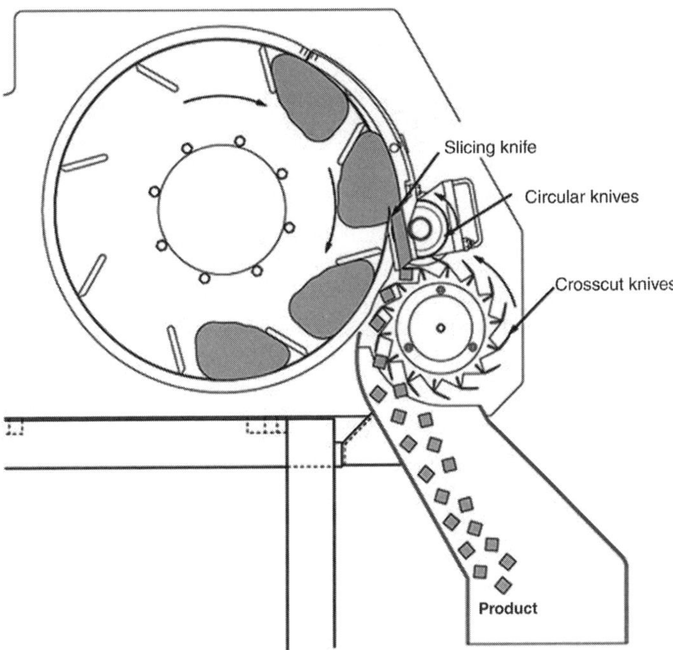

FIGURE 18.1 Diagram showing the operation schema of DiversaCut 2110® Dicer by Urschel Laboratories, Inc. This equipment uniformly dices, strip cuts, and slices a wide variety of vegetables. (From Anonymous. How to cut fruits and vegetable products. Urschel laboratories incorporated. Valparaíso, Indiana, U.S.A. With permission).

a thorough understanding of the mechanical properties of each individual product; further progress is needed in this field to improve automation and to optimize procedures [40].

Depending on the product, after the preparatory procedures and before blanching or immediately after blanching and after cooling and draining, thorough inspections are essential to eliminate unwanted material from the line. If done manually, such inspections either reduce the line output or require a lot of manpower. There are now computerized inspection systems using visible spectrum, infrared or x-ray detectors, or TV images which help to raise output. Nowadays, inspection is faster, more precise, and more economical, and only a small part of the inspection process is manual [41].

2. Blanching

Blanching is a thermal treatment commonly applied in a variety of vegetable preservation treatments and is particularly important in freezing because of its very considerable influence on quality. The product is heated, typically by brief immersion in water at 85–100°C or by steaming at 100°C. The primary objective is to inactivate enzymes responsible for alterations in sensory quality attributes (off-flavors and odors) and in nutritional value (loss of vitamins) during storage. Blanching also affords a series of secondary benefits in that it destroys vegetative cells of microorganisms present on the surface, thus enhancing the effect of washing: it eliminates any remaining insecticide residues, enhances the color of green vegetables, and eliminates off-flavors produced by gases and other volatile substances that may have formed during the period between harvesting and processing. The duration of blanching varies according to the method employed, the type and variety of product, the product size, and the degree of ripening; however, the chief factor affecting processing time is blanching temperature. Oxidases, peroxidases, catalases, and lipoxygenases are destroyed by the heat of blanching, and blanching effectiveness is usually monitored by measuring peroxidase activity in view of its high heat resistance [1].

However, the use of peroxidase as a universal indicator of blanching effectiveness is not in fact a good choice at this time, given that its involvement in flavor and aroma deterioration has not been demonstrated. Moreover, most vegetables contain a number of peroxidase isoenzymes with widely varying heat stabilities, so that complete inactivation requires considerably more heat treatment than is needed to inactivate other enzymes. Sensory analyses of English green peas and green beans have indicated that lipoxygenase is the key enzyme in the development of undesirable odors, and all lipoxygenase activity ceases after about half the heating time required to bring about complete cessation of all peroxidase activity [42]. Total absence of peroxidase activity indicates overblanching, and there is a substantial body of evidence suggesting that the quality of products frozen after blanching is superior if a certain level of peroxidase activity remains at the end of the blanching process. For optimum product quality, it is recommended that blanching be continued only up to the following levels of peroxidase activity: peas 2–6.3% depending on the variety; green beans 0.7–3.2%; cauliflower 2.9–8.2%; and Brussels sprouts 7.5–11.5% [1]. Lipoxygenase activity has been proposed as an indicator of adequate blanching [43]; the recommended ratios between maximal residual lipoxygenase activity and minimal high quality are 10% for peas, 20% for green beans, and 0% for sprouts [44].

a. *Effects of Blanching on Quality*

The heat produced during blanching kills cells and solubilizes pectic substances, causing irreversible alterations in cell structure and in the mechanical properties of plant tissues. Figure 18.2 illustrates the main effects of blanching on a plant cell. Alterations increase the permeability of cytoplasmic membranes, allowing the blanch water to penetrate cells and intercellular spaces and driving out gases and other volatile compounds. At the same time, proteins are denatured and soluble substances, such as vitamins, mineral salts, and sugars, are lost. Chloroplasts and chromoplasts swell and rupture, causing carotenes and chlorophylls to diffuse into the cells and the blanching medium. Starches are similarly affected: they are solubilized and gelatinized,

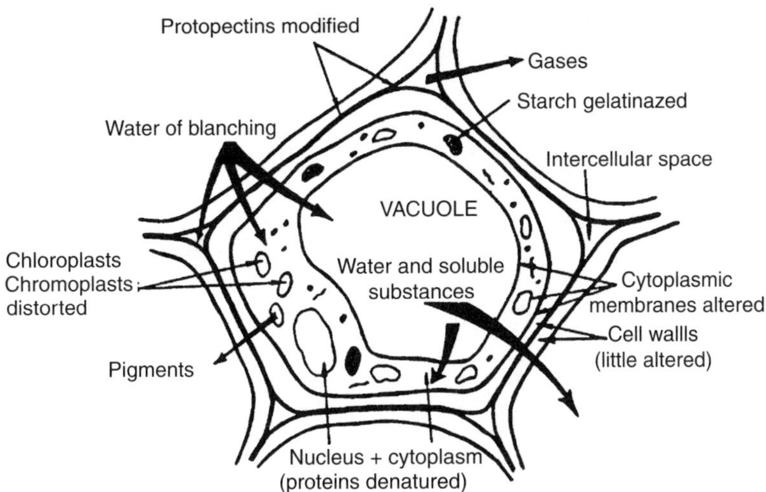

FIGURE 18.2 Diagram showing the main effects of blanching on a generalized plant cell. (From KZ Katsaboxakis. In: P Zeuthen, JC Cheftel, C Eriksson, M Jul, H Leniger, P Linko, G Varela, Eds., *Thermal Processing and Quality of Foods*, London: Elsevier Applied Science, 1984, pp. 559–565. With permission.)

occupying all or part of the cell cytoplasm. The detrimental effects of blanching (chiefly permanent alteration of plant tissue structure, solubilization and destruction of nutrients and vitamins in the blanching medium, and color changes caused by the transformation of chlorophylls into pheophytins) tend to intensify with longer blanching or higher blanching temperatures [44].

In the 1980s, numerous review papers on vegetable blanching were presented [45–53]. The general trend was to test a variety of procedures that would at least partially ameliorate the adverse effects of blanching, either by shortening blanching time or by palliating its detrimental effect on sensory quality (flavor, texture, and color).

Extremely short blanching times (thermal shock) of 10–15 sec on peas and green beans (Table 18.1) produced satisfactory color and flavor values after 1 year of storage at −18°C in spite of unmistakable regeneration of polyphenol oxidase and in some cases of catalase and lipoxygenase; in the case of green beans, texture was superior to that achieved with conventional blanching. These findings conflict with the results of Adams [54], who reported that a tasting panel detected off-flavors after 9 months in storage at −20°C, using blanching times of less than 30 sec for peas and less than 1 min for green beans.

Because the multifoliate structure of Brussels sprouts makes it difficult to achieve deep blanching without overcooking the surface, a preheating treatment at 50°C was proposed so that the accumulated heat from the preheating stage would not damage tissues; this would allow reductions of up to 20% in blanching time. Low-temperature long-time (LTLT) pretreatment (70°C, 10–15 min) followed by cooling and high-temperature short-time (HTST) blanching (97°C) reduced damage to the tissue structure. This stepwise blanching has produced substantial improvements in final product textures of green beans [55,56], potatoes cv. Jaerla [39,57] and cv. Monalisa [58], carrots [59], and peas [60], including after freezing and final preparation [61,62]. Several theories have been presented in the literature reviewed [63] to explain this firming effect in potato: retrogradation of starch; leaching of amylose; stabilization of the middle lamellae and cell walls by activation of the pectin methylesterase (PME) enzyme and by release of calcium from gelatinized starch; and formation of calcium bridges between pectin molecules. Further experimentation is required to elucidate the role of each mechanism, and especially to determine which is the main contributor to the process of firming in different species and varieties. Assays for optimization

TABLE 18.1
Relationship between the Heating Time, Residual Enzyme Activity, and Quality Retention of Frozen Peas and Frozen Green Beans

Heat Treatments in Water (98°C)	Residual Enzyme Activity (%)			Quality Evaluation after 1 yr of Storage				
	Peas and Green beans			Peas		Green Beans		
	Lipoxygenase	Catalase	Peroxidase	Color	Flavor	Color	Flavor	Texture
None	100/100	100/100	100/100	Discolored	Strong off-flavor	Discolored	Strong off-flavor	Good
2.5 sec	80/—	36/—	65/—	Discolored	Off-flavor	Discolored	Off-flavor	Good
5 sec	62/47	28/82	52/74	Discolored	Good	Discolored	Off-flavor	Good
10 sec	6/15	2/41	34/72	Good	Good	Good	Good	Good
15 sec	1/1.1	0.3/34	23/52	Good	Good	—	—	—
20 sec	—/0.2	—/—	—/2.4	—	—	Good	Good	Good
3 min	—/0.13	—/—	0.3/0.8	Good	Good	Good	Good	Soft

Source: From E Steinbuch. In: P Zeuthen, JC Cheftel, C Eriksson, M Jul, H Leniger, P Linko, G Varela, Eds., *Thermal Processing and Quality of Foods*. London: Elsevier Applied Science, 1984, pp. 553–558. With permission.

of stepwise blanching of frozen-thawed potato tissues (cv. Monalisa) by response surface methodology have shown that conditions are optimal within a temperature range of 60–65°C and a time range of 25–35 min; in addition, stationary points presenting maximum PME activity had critical temperature values ≈64°C and time ≈30 min, which were very close to the critical values of some mechanical and textural properties. A high correlation has been found between tissue firmness and increased PME activity caused by blanching in optimal conditions, both before and after freezing and steaming of the product [64,65].

This LTLT pretreatment causes sizeable losses of soluble substances such as vitamins and minerals [66]. Mixed blanching methods consisting of microwaving followed by immersion in boiling water have reduced the duration of blanching in potatoes [67] and Brussels sprouts [68], yielding products in which texture was more homogeneous and acceptable and greater vitamin C retention than in conventional blanching. Microwave-blanched carrots and French beans have been found to present lower residual peroxidase activity, higher retention of ascorbic acid and total carotenoids, and better texture than conventionally blanched (HTST or LTLT) products [69].

Lowering the pH by adding 0.5% citric acid to the blanch water increases the heat sensitivity of the enzyme systems, permitting reductions between 20 and 30% in blanching time for artichokes [48]. Acidification of this kind is not generally practicable, however, as it promotes the transformation of chlorophyll into pheophytin, thereby adversely affecting the color of green vegetables. In contrast, addition of salts such as chlorides and sodium or potassium sulfate does not alter the pH of spinach and Brussels sprouts but substantially reduces he transformation of chlorophyll into pheophytin. This beneficial effect is less pronounced in peas and green beans but continues through freezing and storage [49,50]. Adding 0.5% metabisulfite to the blanch water reduces yellowing in cauliflower; 0.5% sodium bisulfite prevents browning of frozen mushrooms; and calcium chloride or calcium citrates appreciably improve firmness in cauliflower and potatoes. One way of retaining color and improving texture of vegetables through blanching is to blanch in an aqueous Zn solution for 3 min or less: the aqueous blanch solution has a Zn ion concentration of 500 ppm or more [70]. A thorough study of the use of additives in the blanch water to improve vegetable quality or retain soluble substances is required to determine all the possible advantages and drawbacks.

b. Nutritional and Microbiological Aspects of Blanching

There is appreciable loss of ascorbic acid, B-complex vitamins, and folate during blanching and subsequent cooling, particularly when these are carried out in water. Peas blanched for 3 min lose 33% of their initial ascorbic acid content, 20% of their riboflavin, 10% of their niacin, and 5% of their thiamine, even though these vitamins are stable during storage at −18°C or lower. Similarly, losses of total sugars and soluble proteins can also be significant, depending on the blanching method (hot water, steam, or microwave), duration, and temperature; conventionally blanched peas also lose 40% of their minerals, 30% of their sugars, and 20% of their proteins [50]. Vegetables such as peas, sliced beans, and diced carrots can lose half their total vitamin C due to their large surface-to-volume ratio; losses are smaller (about one third of vitamin C) in whole beans, potatoes, and sprouts, where the surface-to-volume ratio is smaller and time appears to be of less importance than the surface area. Reported losses are about the same as when the food is cooked directly from the raw state [71].

Because microwave heating does not involve water and is faster than other methods of heating, it would be expected to cause less damage than hot water and steam. However, although this is generally true, some reports suggest that the three methods differ little or not at all. For example, one report shows 25% loss for green beans in all three methods, whereas another report shows 9% loss in microwave blanching, 14% in hot water, and 18% in steam. Microwave blanching does not inactivate enzymes completely and does not overcook the product, but a mixed method involving a preliminary microwave treatment with a shorter immersion time in boiling water has been found to improve both texture and vitamin retention in potatoes and Brussels sprouts [67,68].

As discussed earlier, vegetables may harbor large numbers of microorganisms at the time of harvest, many of which are removed or destroyed during preparation for freezing. For example, washing removes many of the surface microorganisms — more than 90% of the microorganisms on peas are removed by the first wash; water blanching at 86–96°C destroys the most heat-resistant microorganisms, and only bacterial spores usually survive.

Blanching is the critical control step in the processing of frozen vegetables. Because it destroys most of the contaminating organisms, the microflora of the packaged product reflect recontamination after blanching. The major source of organisms on frozen vegetables is contaminated equipment; the surfaces of conveyors, inspection belts, and filling machines units are difficult to reach for proper cleaning. The degree of difficulty in controlling postblanch contamination also depends on the type of vegetable: with corn, large amounts of starch are released onto equipment surfaces, whereas minimal quantities of soluble solids are leached from green beans and peas. Chopped leafy vegetables usually have higher microbial counts than unchopped products [29].

c. Technological Aspects of Blanching

Blanching and subsequent cooling are major sources of water pollution; although these two operations account for only 5–7 l of water out of a total of 25–35 l/kg of frozen vegetable, they are responsible for 60–70% of the total pollution [46].

During the 1970s and 1980s, there was considerable research and technological development in the field of energy use and environmental protection in a general move to improve the most frequently used water-based methods of blanching and cooling [1,46].

i. Current Technologies

Blanching is usually done in hot water or in steam. Steam blanching at atmospheric pressure takes 30–50% longer than conventional water blanching, yet produces 9–16 times fewer effluents. Improved blanching systems include water recycling, steam blanching methods such as fluidized bed blanching, hydrostatically sealed steam blanching, individual quick blanching, and a spiral vibratory conveyor blanch-cooling system. According to the reviews [1,46], all achieve faster, more uniform blanching with improved color retention, considerably reduced nutrient loss, significantly lower leaching of solubles, energy savings, and better yields. Air cooling systems or systems using water sprays decrease the leaching of solubles and water pollution but entail high product weight loss, and for this reason, they are not competitive with conventional cooling systems using running water. Not all of the improved blanching methods developed have been implemented commercially. Another review [51], comparing water and steam blanchers, considers the cabinet integrated blancher or cooler equipped with a heat exchanger, developed by Odense Cannery Ltd and Cabinplant International A/S, to be the best designed and most advanced solution to energy and pollution problems [72]. This equipment has a heating zone and a cooling zone: a preheating counterflow is produced in the heating zone, where the product is heated to 60°C and then blanched; the product is then counterflow-cooled to 45°C followed by air-cooling in the cooling zone. Water heated to about 80°C in the cooling zone releases its heat to the water in the heating zone via a heat exchanger. This design achieves savings of about 60% of the calculated energy consumption and reduces water consumption by over 90%. The final air-cooling operation has the additional advantage of removing water before the product is conveyed to the freezer.

ii. Emerging Technologies

A report on food blanching process improvement [73] reviews emerging technologies that can be used to reduce energy, processing time, and waste water in the design of blanching equipment. To save energy, steam blanching equipment designs are being improved for more efficient thermal processing by steam seals (steam is confined within the chamber by rotary locks or forced water to prevent evaporation to the atmosphere), insulation (the steam chamber is completely insulated to maximize heat retention), and forced convection (Key Technology has developed a Turbo-Flow^{TM} [74] that uses positive and negative pressure areas within the steam chamber to force steam through the product depth, thus reducing blanching time). In addition, ABCO Industries Ltd. [75] have developed a heat-and-hold process, which transfers only enough steam to heat the surface of the product; the heat is then allowed to spread evenly throughout the product in a separate compartment, raising the center core temperature to the desired blanch level without the addition of more steam. This shortens the blanching time and improves color, flavor, and nutrient retention. Reduced steam also reduces energy costs and effluent volume. Steam recycling (Key Technology's Turbo-Flow) [74] uses forced convection to push the steam through the depth of material. Steam energy that is not absorbed during the initial pass is recirculated back to the product, which reduces waste-water treatment costs and make up water usage. Waste generation can be reduced by 80% through recycling steam.

Intelligent blanching control is difficult because of the natural variability of the unblanched product. Continuous control may be maintained by constantly adjusting the blanching process to characteristics of the product (density, moisture content, thickness, thermal conductivity, and capacitance), and models have been developed to calculate the process variables (steam or water temperature and blanch time). Initial product variability may result in over- or underblanched products, with loss of texture, flavor and nutrients, and excess energy use in the first case and inadequate food preservation for inefficient enzyme inhibition in the second case.

Because of the detrimental effects of blanching on the product, energy costs, and pollution, researchers have also looked at possible alternatives to steam or water that can replace blanching without adversely affecting product quality. In addition to its beneficial effects on texture and vitamin retention as noted earlier [67,68], microwave blanching has the advantage of more uniform volumetric heating, minimal oven temperature gradients, no blanching residual products, lower energy costs, and shorter processing time. High-voltage pulsed electric field, ultraviolet radiation, or ultrahigh pressure treatments could be combined or integrated with steam blanching to improve product quality without the use of chemical preservatives.

As a result of research, onions, leeks, peppers, parsley, and cucumbers can be frozen unblanched with no appreciable loss of quality over relatively short storage periods (6–9 months). Except for these few products, however, blanching remains an essential step in the freezing process, and consequently, further research is needed in this area to determine what blanching treatments, alone or in combination with other procedures, can best counteract the adverse effects of blanching with the final quality of specific frozen vegetable products [1].

3. Freezing

The freezing process consists of lowering the product temperature to $-18°C$ at the thermal center, resulting in crystallization of most of the water and some solutes. Ice crystallization occurs only after a degree of supercooling — that is reduction of the temperature to between -5 and $-9°C$ in a matter of seconds. In the freezing stage, most of the water in the product undergoes a phase change to ice; this change is not complete until the final temperature at the thermal center is at least as low as the storage temperature.

The duration of the freezing process depends on the freezing rate (°C/h). This is defined by the International Institute of Refrigeration 1986 [8] as the difference between initial temperature and final temperature divided by freezing time, freezing time being defined as the time elapsing from the start of the prefreezing stage until the final temperature has been attained. This will be affected by product size (particularly thickness) and shape, as well as by the parameters of the heat transfer process and the temperature of the cooling medium.

Part I of this handbook (Chapter 1–Chapter 8) discusses all the fundamental topics relating to freezing in detail. Here, we look at only the effect of freezing on the quality of vegetables.

a. *Freezing Effects on Structure and Texture*

The prefreezing stages render vegetable tissue membranes more permeable, with a concomitant loss of intracellular pressure; however, it is crystallization that has truly irreversible adverse effects on quality. The ice formation process consists of crystal nucleation and growth: if the temperature of the system drops very quickly to below the nucleation temperature, large numbers of nuclei form. Crystal growth is then more limited, with crystal size inversely proportional to the number of nuclei [75].

The temperature at which vegetable tissues start to freeze is directly dependent on the soluble solids content, particularly sugars, salts, and acids, rather than on the water content. The amount of ice that forms at a given temperature is also related to the initial soluble solids content [76].

When the water has begun to freeze, the crystallization rate and the location of the crystals depend on the rate of heat removal (freezing rate), the tissue structure, and the rate of diffusion of water from the solutions to the surface of the ice crystals. At low freezing rates, few crystallization nuclei form and the cells lose water by extracellular diffusion through the cell membranes; this causes solutes to concentrate, thus impeding the formation of crystallization nuclei inside the cells. Consequently, the extracellular crystals grow relatively large, causing progressive separation of the cells and plasmolysis of the cell protoplasm, which can bring about partial or total collapse depending on the rigidity of the cell structure. In contrast, when the tissues are frozen rapidly to a low enough temperature at the thermal center, numerous tiny ice crystals are distributed evenly inside and outside of the cells. Rapid freezing causes minimal migration of water to the crystallization sites and hence only minor modifications to tissue structure [77]. Nuclear magnetic resonance (NMR) imaging has been used to monitor the formation of ice during food freezing [78]. In this case, formation of ice was seen as a reduction in spatially located NMR signal intensity in vegetables such as potatoes, carrots, or peas. At 350 μm resolution, the ice interface could be seen to advance uniformly and the time taken for the signal to disappear was the same as that required to reach steady-state enthalpy. Magnetic resonance imaging could therefore serve to assess freezing times and the importance of food structure for the freezing process [79].

There has been a considerable amount of research into the effects of ice crystal size and location on the structural and textural qualities of frozen vegetables. These two factors, combined with swelling during freezing, cause more or less irreversible damage to the structure. It has been shown that the freezing rate affects asparagus and green bean tissue structures differently. While comparing freezing times of 1, 5, or $\geq$20 min, structural deterioration was more pronounced in green beans (Figure 18.3), which have a parenchymal structure consisting of larger cells and more vacuoles, than in asparagus (Figure 18.4), in which the structure is mainly protoplasmic with a relatively low level of vacuolation [80,81].

These alterations in tissue structure during freezing cause an irreversible loss of product texture, a topic that has been addressed by researchers because of its growing importance as an attribute of quality [81,82]. The beneficial effect of rapid freezing rates on structure and texture has been reflected in the results of texture analysis by various methods (histological, sensory, imitative, and objective) in studies of green beans [76,83,84], potatoes [40,58,62], carrots, and peas [60–62].

Histological examinations of the beneficial effect of rapid freezing rates have shown that blanching and cooking have adverse effects, in which they mask the different structural alterations caused by rapid or slow freezing [85]. In contrast, other studies using objective methods of texture

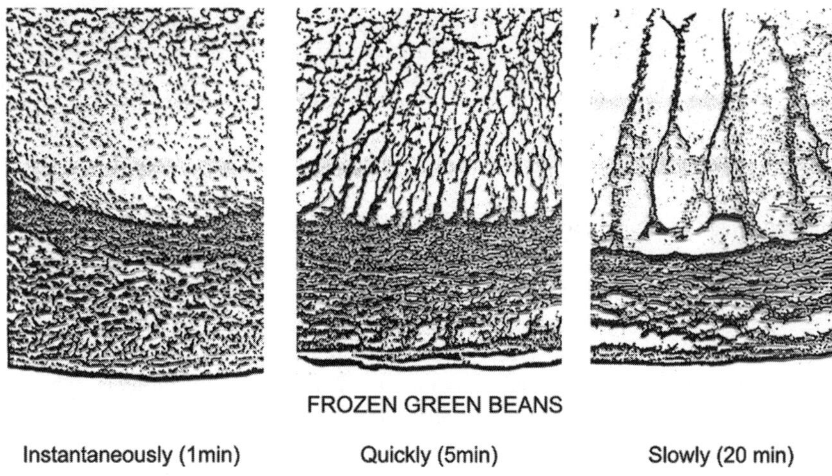

FIGURE 18.3 Sections of green beans (×40) showing the effect of different freezing rate (1, 5, and ≥20 min freezing times), with little, slight, and extensive structural damages, respectively. (From MS Brown. *Journal Science Food Agriculture* 18:77–81, 1967; RM Reeve. *Journal of Texture Studies* 1:247–284, 1970. With permission.)

analysis have detected beneficial effects of rapid freezing rates and stepwise blanching on final texture, even after cooking, in green beans, potatoes, peas, and carrots [40,58,60–62,80–82,86]. Minimum alteration of rheological behavior of slow-thawed potato tissues has been achieved by precooling (3°C/30 min), slow cooling (0.5°C/min) before and after the maximum ice crystallization phase, and quick freezing (2°C/min) [87]. Scanning electron microscopic examination of the tissues has shown varying degrees of mechanical damage to tissue structure and a linear increase in the tissue's mechanical strength caused by precooling [88]. These effects are best studied using the shear test. A comparison of the effect exerted on carrot tissues by slow freezing (2°C/min),

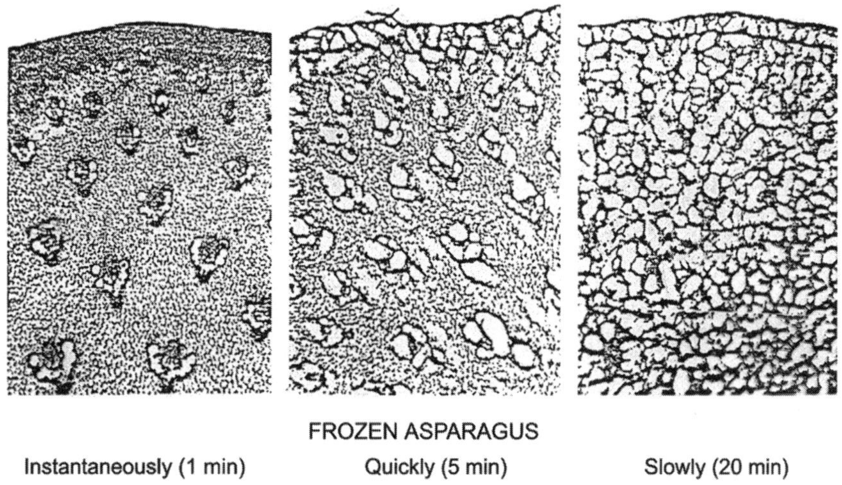

FIGURE 18.4 Transverse sections of asparagus stem (×30) showing the effect of different freezing rate (1, 5, and ≥20 min freezing times), with none, slight, and extensive structural damages, respectively. (From MS Brown. *Journal Science Food Agriculture* 18:77–81, 1967; RM Reeve. *Journal of Texture Studies* 1:247–284, 1970. With permission.)

programmed freezing (slow + quick + slow), and quick freezing (5°C/min) shows that quick freezing produces minimum drip, less structural damage, and good firmness [89]. Drip, cell damage, and softening of carrots tissues have also been prevented by high-voltage-induced electrostatic field thawing [90,91].

b. Importance of the Freezing Rate

It is still widely accepted that there is a close correlation between frozen food quality and rapid freezing rates, although a large body of evidence suggesting that fast freezing rates do not always result in particularly high final frozen food quality. The persistence of this assumed relationship is discussed in detail in the literature. The explanation given by the authors [92] for the persistence of this assumption is that it is readily understandable and has the broadbased support of manufacturers of freezing equipment and associations of frozen food producers, who hold to the view that faster freezing rates yield superior end products.

With general acknowledgment that the vegetable freezing process and frozen vegetables are influenced by other thermal treatments such as blanching, thawing, and cooking and by other factors such as product type, packaging, and storage time, the freezing rate is no longer considered to be the prime factor affecting final quality.

This is not to suggest that the freezing rate does not, in fact, affect quality, because it is true that most fruits and vegetables lose quality when frozen slowly. As noted earlier, green beans, carrots, potatoes, mushrooms, and corn on the cob, all exhibit improved texture and higher water retention when frozen rapidly. Other products, such as peas, are less sensitive to the freezing rate, and only extremely fragile foods such as tomatoes derive some real benefit from ultra rapid freezing (faster than 100 mm/h). The following recommended freezing rates are considered adequate to attain high-quality frozen products [93]: from 5 to 30 mm/h (quick freezing) for packaged frozen foods (e.g., spinach) in air blast or plate freezers; from 50 to 100 mm/h (rapid freezing) for individually frozen products (peas and corn) or for small-sized products (cut green beans, diced carrots, and prefried potatoes) (IQF individual quick freezing in a fluidized bed); and 100 to 1000 mm/h (ultra rapid freezing) by spraying with or immersion in liquid gases (LN_2). The definition of the recommended freezing rate in the earlier cases was defined as the ratio between the minimum distance from the surface to the thermal center and the time elapsed between the surface reaching 0°C and the thermal center reaching 10°C colder than the temperature of initial ice formation at the thermal center, according to IIR [93].

The freezing rates commonly achieved today with commercially available freezing equipment are such that the freezing rate in the freezing process as a whole has ceased to be so important. In practice, other technical and economic factors are considered in selecting a freezing method; these include control and monitoring, processing capacity, energy savings, processing times, capital and operating costs, and also product quality characteristics (texture, color, dehydration, exudates, etc.) [94].

c. Methods and Equipment for Freezing

There have been a number of reviews about freezing methods and equipment and about other physical and engineering aspects of food freezing [1,8,95–99]. Part II (Chapter 9) of this handbook looks in detail at freezing methods and equipment, whereas Chapter 8 deals with the latest innovations in freezing, as reported in a recent review [100]. The three most frequently used methods of freezing vegetables are direct contact (typically plate freezers), air-blast freezing systems, and cryogenic freezing. Plate freezers for vegetables are always horizontal and are used to freeze chopped or sliced products such as spinach in packages of up to 50 mm in height. The product is frozen by contact, with conduction as the main mode of heat transfer. The advantages of this method are economical and minimum weight loss, but freezing rates are moderately slow. Vegetables are preferably presented as individually quick-frozen products, which are obtained using air-blast or cryogenic freezing systems. Air-blast freezing consists of blowing cold air through the product, which can be placed on trays in the case of a batch system (stationary, push-through, or automatic) or in continuous, moving belt systems. Belts may be arranged in line, with one or more belts one above the other or in a rotating spiral if floor space available is limited. Normal operating conditions are between −30 and −40°C, with air velocities of up to 20 m/sec. For proper operation, temperature and air velocity must be adjusted according to product thickness. Precautions must be taken to reduce moisture loss (0.6–2%) and concomitant product dehydration and to maintain surface quality.

In fluidized bed freezers, the cold air is used both for fluidization and freezing. Products entering the fluidized bed are frozen very quickly, with freezing times ranging from 3 min for peas to a maximum of 15 min for strawberries or cauliflower florets. When compared with moving belt freezers, fluidized bed freezing has the advantage of affording true individual quick freezing with lower weight loss, because each product particle is surrounded by a thin layer of frost, which prevents dehydration. Freezing tunnels of this kind are suitable for foods with a uniform shape and a diameter of less than 40 mm and are most commonly used for vegetables such as peas, sliced green beans, sliced carrots, Brussels sprouts, corn, and Lima beans, for fruits such as blueberries, sliced apple, and sliced pineapple, and for prefried potatoes. Throughputs of 2–5 t/h can be achieved.

Tunnel freezers for products that are larger, cut into nonuniform shapes, or fragile, such as cauliflower, Brussels sprouts, and strawberries, combine a first stage based on the fluidized bed principle, in which the product surface (crust freezing zone) is frozen with minimal weight loss, with a second stage in which products are deep frozen (completion freezing zone) on a conveyor belt.

The benefits of individual quick freezing in fluidized bed freezers compared with conventional or moving belt freezing tunnels include greater heat transfer efficiency and hence higher freezing rates, with lower product weight loss and less frequent defrosting. Product handling is also facilitated in that the product circulates freely through the tunnel, with considerable energy savings. The main drawback is that they cannot be used with large or nonuniform products.

More rapid freezing is achieved using cryogenic agents that boil at very low temperatures at atmospheric pressure (e.g., liquid nitrogen, −196°C) in tunnel, cabinet, and spiral freezers. The cryogenic agent is lost to the atmosphere after the vapors have been used to precool and freeze the product. Cryogenic freezing results in rapid crust freezing and reduced weight loss (0.1–0.3%), impermeabilizing the product to oxygen. It causes minimal structural damage, yielding frozen products (particularly fruits and vegetables) with excellent textural qualities. Cryogenic tunnels are relatively short and of simple construction; they thus offer the advantages of low investment outlays and simple operation, making them economical for small, highly seasonal, high-value products despite the high cost of LN_2 [8]. Recent cryogenic equipment designs incorporate continuous monitoring, which helps to reduce gas consumption, thereby lowering operating costs [101]. Sometimes, a good quality product is best ensured by a combination of two different types of freezers. The combination of a cryogenic freezer and a mechanical freezer (a “mixed freezer”) offers the advantages of both techniques. Cryogenic freezing is recommended before mechanical freezing in the case of fragile products, which tend to stick together. Consumption normally ranges from 0.4 to 0.7 l LN_2/kg of product. The line capacity is greater (≈50%) because the product spends less time in the mechanical freezing tunnel, thus reducing weight loss. The line production capacity can be made more flexible (≈10–50%) using the cryogenic freezer after the mechanical freezer; consumption per additional kilogram of frozen product ranges from 0.6 to 0.72 l LN_2 [15,102].

New designs have reduced the problems encountered with the different types of freezing equipment, for example products sticking to conveyor belts; in addition, much attention has been given to hygiene, with food contact surfaces made of stainless steel or high-quality plastic materials and the incorporation of belt washing operations before produce loading [103].

The total freezing cost involves not only the capital expenditure and operating costs but also the cost of product weight losses. A classic comparison of the total freezing costs for the different systems indicated that the most economical were continuous air-blast systems such as belt and fluidized bed freezers for freezing plants, which enjoyed uniformly high production rates, as is the case for frozen vegetables. When freezing volume was not especially high, the operating costs of cryogenic systems dropped, as did their total costs due to the low initial investment required. With the rapid growth of the value-added convenience food market, where the energy costs associated with freezing at air temperatures of -38°C are a very low percentage of the retail price, the trend is likely to be toward the use of lower refrigerant temperatures at the expense of energy costs. This trend is already observable in the increasing use of cryogenic freezers, which have enjoyed considerable growth in the freezing market [104].

d. Nutritional and Microbiological Aspects of Freezing

Freezing itself has no effect on nutrients, and the main losses during the complete freezing process are due to blanching. Subsequent frozen storage and thawing or cooking have variable effects, which are analyzed later on.

The exact mechanisms by which freezing, frozen storage, and thawing kill or damage microbial cells are not fully understood, although a number of studies have been conducted on the nature and sites of such injury. Several factors may be involved, for example, low temperature, extracellular or intracellular ice formation, concentration of solutes, and internal pressure [34]. Of these factors, low temperature and internal pressure seem to be relatively unimportant. The internal pressure in the food may rise to 10^6 MPa or more, especially during very rapid freezing; this pressure is sufficiently high to cause undesirable textural changes in some foods, but not nearly high enough to inactivate microorganisms. Slow freezing encourages growth of a few extracellular ice crystals; the extracellular fluid becomes concentrated, causing dehydration of the cells by forcing water to move out. This makes it difficult for the water molecules to return to their original sites and may injure or kill the microorganisms during and after thawing. It seems that the principal site of bacterial damage during freezing is the membrane, leading to leakage of internal cell material. The cell membrane seems to lose some barrier properties at temperatures below about -15°C. During freezing, cells may be injured as a consequence of dissociation of lipid–proteins; the dissociation may be caused by an increase in the concentration of cell solutes and a resulting increase in ionic strength, by changes in pH, and by physical contact between lipoproteins and the cell wall.

Most studies of the influence of freezing and thawing on microorganism death (or survival) have been model experiments and have concluded that survival of microorganisms depends on the freezing and thawing rates along with numerous other factors.

As shown in Figure 18.5, the combination of rapid freezing and slow thawing may kill more bacteria than that of slow freezing and rapid thawing [34]. In commercial frozen food operations, slow freezing (e.g., plate freezing or freezing in a freezer cabinet) is generally thought to be more damaging than quick or ultrafast freezing such as would occur in a blast freezer or fluidized bed freezer [30]. It is also confirmed that rapid thawing may increase the survival of microorganisms.

Table 18.2 shows the effect of the various operations in the freezing process on the initial bacterial count in peas. Note that how the count drops sharply with washing and blanching, then increases during draining and inspection before placement in the freezer. The freezing stage does not in itself seem to reduce the bacterial count very greatly (736,000 to 560,000 bacteria per gram of peas); only some bacteria are killed, whereas others may be sublethally injured, so that it is essential to reduce all possible risk of recontamination between blanching and freezing [106]. Pathogens that are commonly associated with vegetables (fresh and frozen) are *Salmonella* spp., *Bacillus cereus*, *Listeria monocytogenes*, and *Yersinia enterocolitica*. Those present in fresh produce are generally rare in blanched, frozen vegetables [30]. Frozen vegetables that are boiled or

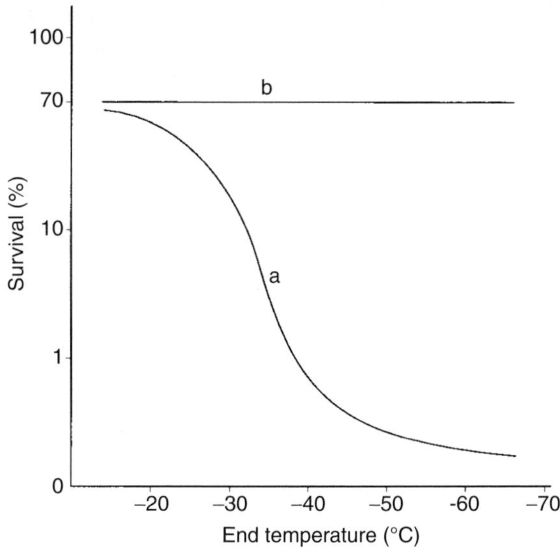

FIGURE 18.5 Survival of bacteria after freezing to different temperatures (a, rapid freezing and slow thawing; b, slow freezing and rapid thawing). (From L Bogh-Sorensen. In: CJ Kennedy, Ed., *Managing Frozen Foods*. Cambridge, England: Woodhead Publishing Ltd., 2000, pp. 5–26; EA Andersen, M Jul, H Riemann. Industriel levnedsmiddelkonservering, Copenhagen, Teknisk Forlag, 1965. With permission.)

cooked without prior thawing have an excellent safety record and have never been cited as the cause of poisoning.

C. Packaging

In addition to the physical and chemical changes causing the progressive deterioration of quality during storage, frozen vegetables can also undergo mechanical (breakage and disaggregation) and photochemical (color and flavor denaturation) alterations. The extent of such alterations in quality is dependent in large measure on the product preparation and the type of packaging

TABLE 18.2
Effect of Processing on Bacterial Count in Peas at Various Stages of Processing

Point of Sampling	Thousands of Bacteria per Gram of Peas
Platform	11346
After washing	1090
After blanching	10
End of flume	239
End of inspection belt	410
Entrance to freezer	736
After freezing	560

Source: From AC Peterson, MF Gunderson. In: DK Tressler, WB van Arsdel, MJ Copley, Eds., *The Freezing Preservation of Foods*, Vol. 2. Westport, Connecticut: Avi Publishing Co., Inc., 1968, pp. 289–326. With permission.

employed [1]. Part V of this handbook deals with several topics connected with packaging of frozen foods, such as materials, packaging machinery, and future developments. At this point, we shall deal with the effect of packaging type on frozen vegetables.

There are numerous articles, books, reviews, and regulations dealing with the mechanical and physical properties of packaging materials [8,107–110] but relatively few dealing with the effect of packaging type on the quality and stability of frozen foods [110,111], and especially frozen vegetables [112,113].

Frozen vegetables have their own special requirements for preparatory treatment and packaging. Certain products, such as cauliflower, Brussels sprouts, and cut green beans, are particularly fragile, needs packages that can withstand the compression and shocks that occur during production. Ultraviolet radiation at a wavelength of 5000 Å can catalyze certain chemical reactions, giving rise to significant denaturation of color in the case of chlorophyll-containing vegetables, so that it is essential to use opaque packaging materials.

Frozen vegetables undergo dehydration during storage, chiefly as a result of fluctuations in storage temperature and the degree of proofness of the packaging to water vapor. Such dehydration is irreversible, giving rise to ice formation inside the package and exerting detrimental effects on quality (alterations in color and flavor, freezer burn, increased risk of oxidation, and structural deterioration). Consequently, packages should ideally be airtight, totally impermeable to water vapor, and effective as thermal insulators to limit possible temperature fluctuations within the product.

The alterations and losses in aroma and flavor, enzymatic browning, and oxidation of ascorbic acid that take place in the presence of oxygen needs the use of packaging materials that are airtight (impervious to oxygen) or permit removal of the oxygen from inside of the package, either by creation of a partial vacuum or by injection of inert gases (N_2 or CO_2).

Most vegetables are frozen using individual quick freezing methods and stored in bulk containers for more or less protracted periods, after which they are repackaged in smaller retail containers. Various factors, primarily economic, have led to the generalized use of polyethylene bags, even though they are ineffective in preventing mechanical damage and dehydration. The use of cardboard coated with paraffin or microcrystalline waves, or plastics such as polyethylene or polypropylene or laminated films and foils impermeable to water vapor and oxygen (polyethylene- or polypropylene-coated cellophane), or aluminum foils laminated with plastic films, helps to prevent mechanical damage and dehydration. These also offer the additional advantage of being able to bear printing or having a transparent window to view the product.

In response to the needs of catering services, the development of new plastic films led to "boil-in-the-bag" packages in which vegetables could be precooked, frozen, and reheated. Such products are typically of high quality, and production is fully automated. The utilization of thermoplastic polyesters (Eastmant TENITE PET Polyester) with a very acceptable level of high-temperature dimensional stability and low-temperature toughness (-80°C to $+200$°C) has made "freezer to oven to table" packaging possible, in which products can be packaged before or after freezing, cooked in the package in conventional or microwave ovens on removal from the freezer, and served. In the case of vegetables, this has permitted the development of precooked frozen products and new forms of preparation and packaging. There is a clear need for research and continuous development of preparation procedures and packagings tailored to the many individual products.

III. INFLUENCE OF STORAGE

Frozen vegetables do not remain stable throughout storage, which frequently lasts for months or even years, resulting in loss of quality; the extent of such loss depends on the storage temperature and product type.

The losses in quality in frozen vegetables stored at -18°C are caused solely by physical and chemical alterations taking place within the product itself; microbial growth is not a factor at the temperatures involved, because very few bacteria can grow below −5°C and no fungi or bacteria have been reported to grow below −12.5°C.

A. Physical Changes during Storage

The physical changes affecting frozen vegetable quality during storage are recrystallization and sublimation phenomena, related to the stability of the ice crystals inside and on the surface of the product. Recrystallization consists of changes in the number, size, shape, and orientation of the ice crystals, which form after initial solidification during freezing. It is the result of the successive melting of small ice crystals on the surface, followed by recrystallization of larger crystals on the surface, so that the crystals become larger in size and fewer in number. The effect of recrystallization during storage and distribution can nullify the benefits derived from rapid freezing [114]. Figure 18.6 shows the variation in ice crystal length in tissue frozen at different freezing times (B) and after 3 months of storage at −20°C (A) [115].

Recrystallization does not occur extensively at low temperatures over average storage periods. Several authors [77,116] have found crystal growth at warm temperatures, but not below −10°C. According to a review [117] of the influence of fluctuating storage temperatures, there is little evidence of any reduction in keeping quality at temperatures below −18°C. A study has been conducted on the effect of temperature fluctuations during frozen storage on the textural quality and the kinetics of softening of potato tissues using a model and mechanical tests (compression, shear, and tension) [58]; the mechanical strength of the frozen tissues decreased when the number of fluctuations increased. It was possible to estimate cumulative loss of texture quality during storage, and this loss was greater at higher storage temperature and greater range of fluctuation. When the number of fluctuations increased by more than four, there was little further deterioration of potato firmness [118,119].

Sublimation of ice at the surface can also occur during storage in improperly packaged food, leading to desiccation, with the water thus extracted accumulating inside the packaging in the form of frost. In addition to causing undesirable weight loss, excessive desiccation can speed up oxidative alterations at the surface of the product, adversely affecting quality. Both recrystallization and surface desiccation are accelerated by fluctuation in storage temperature, although this is less

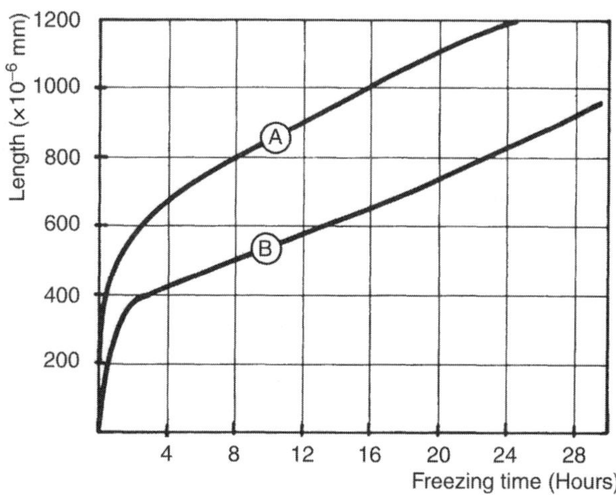

FIGURE 18.6 Average length of ice crystals in frozen tissue after various freezing times (B) and after 3 months in storage at −20°C (A). (From St Astrom. Proceedings Symposium on Frozen and Quick Frozen Food — New Aspects for Agricultural Production and Marketing, FAO, New York, 1977, pp. 149–164. With permission.)

important at low storage temperatures [8]. Except for texture [120], there is no evidence that temperature fluctuations at temperatures below $-18°C$ lead to loss of frozen food quality. The results available as to the effect of recrystallization and sublimation on quality are contradictory, indicating a need for further in-depth studies of these phenomena [121].

B. Chemical Changes during Storage

Despite the low temperatures, a series of enzymatic and nonenzymatic chemical reactions take place in vegetables during frozen storage. Their influence on quality is particularly important, because these reactions are associated with the appearance of off-flavors and odors, changes in color due to breakdown of the chlorophylls and other natural pigments, development of enzymatic browning, and autoxidation of ascorbic acid. Changes in pH during freezing and frozen storage may also be related to alterations in the kinetics of such reactions and to loss of quality [77].

1. Changes Associated with the Appearance of Off-Flavors and Odors

During frozen storage, ethanol (produced by glycolysis) and other volatile compounds may accumulate in the tissues of vegetables that have not been adequately blanched. Such accumulation coincides with the development of off-odors and flavors that can persist even after cooking. Such off-flavors and odors are believed to be at least partially a consequence of the enzymatic oxidation of lipids. Studies on unblanched peas have related the development of off-flavors and odors to the accumulation of volatile substances (hexanal and other aldehydes) produced by oxidation of polyunsaturated fatty acids by lipoxygenase. When unblanched peas were packaged in a nitrogen atmosphere to prevent oxidation reactions during storage, the same off-flavors were found to develop as when the peas were packaged in a normal atmosphere. A wide variety of related substances have been identified by chromatography, but it has not been possible to establish any direct relationship between any individual substance and the characteristic hay and rancid tastes in inadequately blanched vegetables. Blanching for sufficiently long period destroys the oxidases present, and off-flavors and odors do not develop during storage. However, there is still some doubt as to the origin of such flavor and odor alterations, indicating a need for basic research into the chemical and biochemical mechanisms involved [1,45,49,52].

2. Changes Associated with Alterations in Color

Frozen vegetables undergo more or less intense alterations in color during storage, brought about by changes in the natural pigments, such as chlorophylls, anthocyanins, and carotenoids, or by enzymatic browning. The characteristic green color of frozen peas, green beans, and spinach gradually turns to brown during storage at $-18°C$, due to the transformation of chlorophyll a and chlorophyll b into their corresponding pheophytins. Such changes in the initial color occur much more quickly in unblanched vegetables or those stored at insufficiently low temperatures. Another important path of chlorophyll breakdown is by the action of peroxides produced by the lipoxygenase-induced oxidation of polyunsaturated fatty acids in the presence of oxygen. The influence of such alterations of color during storage on final product quality is considerable, because differences in chlorophyll content of as little as 4% in green beans and 1.5% in peas have been detected by 85–95% of taste panellists.

Table 18.3 shows the rate of chlorophyll degradation at various storage temperatures. Note that the rate was higher in the green beans and chopped spinach, in which increased processing led to more rapid breakdown of chlorophyll during storage than in the peas or whole leaf spinach.

Blanching for a very brief time at high temperature is less detrimental to pigments than blanching for a longer time at lower temperatures, which can speed up pigment degradation during storage. Such pigment degradation, and how it is affected by blanching, is to a large extent dependent on the type and variety of vegetable involved and also on the degree of ripeness and the amount

TABLE 18.3
Frozen Storage Time (months) Required for 10% Decrease in Chlorophyll Content of Selected Green Vegetables

	Frozen Storage Time (Months)		
	Storage Temperature		
Product	**−18°C**	**−12°C**	**−7°C**
Peas	43	12	2.5
Whole leaf spinach	30	6	1.6
Chopped spinach	14	3	0.7
Green beans	10	3	0.7

Source: From P Olson, WC Dietrich. In: WB van Arsdel, MJ Copley, RL Olson, Eds., *Quality and Stability of Frozen Foods.* New York: Wiley Interscience, 1969, pp. 117–141. With permission.

of tissue damage sustained during harvesting [77]. Recent studies on color and chlorophyll content of green beans (cv. Bencanta) during frozen storage (250 days at −7, −15, and −30°C) show that chlorophyll content is related to the initial vivid green color but does not give a reliable prediction of color retention in the course of storage. Color coordinates and chlorophyll content do not correlate, which means that chlorophyll content is not a good color index for frozen green beans, and that further research is needed into the relationship between chlorophyll degradation and color changes during frozen storage [123].

The color of blanched and thawed broccoli depends heavily on the pH of the surrounding environment. The rate of color degradation has been found to be linearly related to the concentration of hydrogen ions: the more acidic the pH, the faster the discoloration. Integral analysis of color change, based on a kinetic model that includes all available knowledge of expertise, greatly improves the reliability of the analysis and the understanding of the problem [124].

Anthocyanins are water-soluble pigments responsible for the red colors; under certain conditions, they can be destroyed by enzyme-induced oxidation of the polyphenols, causing color loss during processing and storage. In addition, the oxidation of carotenoids (liposoluble pigments abundant in many vegetables in the form of xanthophylls), carotenes, and acid lycopene isomers is a secondary cause of alterations in color; however, prevention of any substantial degradation of these pigments is worthwhile in view of their role in protecting the chlorophylls against oxidation and in their function as provitamin A. Blanching protects anthocyanins and carotenoids from oxidation by lipoxygenase and by peroxides derived from polyunsaturated fatty acids [1,49].

Alterations in color during the storage of frozen vegetables as a result of enzymatic browning are caused by the oxidation of phenols in the presence of oxygen in such products as cauliflower, potatoes, and mushrooms. The reaction is catalyzed by polyphenol oxidases, giving rise to quinones that condense in the form of brown or reddish-brown compounds with a more or less well-defined chemical composition. The quinones in turn act as oxidants for other substrates such as ascorbic acid, anthocyanins, and so on. Such enzymatic browning can be minimized by thermal inactivation of enzymes, the addition of inhibitors, or the exclusion of oxygen [77].

Blanching is clearly the most appropriate method of preventing enzymatic browning. Enzyme activity can also be controlled in mushrooms by adding citric acid (1.5%) to the blanching water or in cauliflower by adding metabisulfite (2 g/l) to the cooling water after blanching. Adding sequestering agents (e.g., disodium dihydrogen pyrophosphate) to the blanching water prevents the subsequent appearance of undesirable colors in potato products [1,49].

3. Ascorbic Acid Oxidation and Changes in pH

In insufficiently blanched vegetable tissues, ascorbate oxidase catalyzes oxidation of ascorbic acid during storage, and this process is accelerated when oxygen-permeable packaging is employed. The oxidation rate depends on storage temperature and product pH.

Recent research [125] into the effect of storage temperature on the ascorbic acid content of frozen green peas, spinach, green beans, and okra shows in all cases that vitamin C loss is adequately described by an apparent first-order reaction; temperature dependence of vitamin C deterioration is expressed by the Arrhenius equation. Table 18.4 shows the estimated activation energies E_A, the 95% confidence range, the goodness of fit (R^2), and the estimated Q_{10} values (ratio of deterioration for the range -15 to -5°C). This comparison of different green vegetables shows that the type of plant tissue significantly affects the rate of vitamin C loss. Frozen spinach has been found to be the most susceptible to vitamin C degradation; peas and green beans present moderate retention, whereas the loss rate in okra is substantially lower.

Ascorbic acid stability rises as pH drops. In products with a low pH such as strawberries, the ascorbic acid is oxidized at a slower rate than in most green vegetables stored under similar conditions. The pH usually drops after freezing and then undergoes inflection, the intensity and direction of which varies according to the temperature and product. The behavior of peas, green beans, and cauliflower is similar, the pH increasing from 4.5–5 to 6 after 20–30 days of storage. It has been suggested that the decrease in the initial pH postfreezing is due to precipitation of the alkaline phosphates of calcium, magnesium, and sodium. Precipitation of the acid phosphates of potassium and sodium and potassium citrate has been proposed as the cause of the final increase of pH.

Clearly, changes in the pH can play a role in the altered activity of certain enzymes; studies that relate changes in pH to loss of quality during storage would therefore be desirable.

TABLE 18.4
Arrhenius Parameters of Vitamin C Loss (E_A Estimated Activation Energy and k Apparent Reaction Rate), Statistics (R^2), Q_{10} Values (-15 to -5°C), and Shelf Life, at Four Temperatures in the Frozen Storage Range for Frozen Green Peas, Spinach, Green Beans, and Okra

Kinetic Parameters	Green Peas	Spinach	Green Beans	Okra
		Products		
E_A (kJ/mol)	97.9 ± 9.6[a]	112 ± 23.2	101.5	105.9
k_{ref} (1/day)	0.00213	0.00454	0.00223	0.00105
R^2	0.958	0.992	0.967	0.868
Q_{10} (-15 to -5°C)	5.5	7.0	5.8	6.3
Temperature (°C)		**Shelf life (days)[b]**		
−5	24	8	21	40
−10	56	20	50	98
−15	132	55	122	249
−20	325	153	311	660

[a]95% confidence intervals of the kinetic parameters of the Arrhenius model (regression analysis).

[b]Shelf life is based on 50% vitamin C loss.

Source: From MC Giannakourou, PS Taoukis. *Food Chemistry* 83:33–41, 2003. With permission.

C. Combined Effect of Time and Temperature during Storage (T–T–T Factors)

The deterioration of initial quality due to the physical and chemical changes undergone by products during storage is a function of storage temperature and duration. The combined effect of these two factors, time and temperature (T–T), determine product tolerance (T) to frozen storage.

In the Albany TTT Project, a large number of fruits and vegetables were tested at various freezer storage temperatures for various periods of time [126]. Particular attention was paid to the quality and stability of frozen vegetables, with extensive studies of such vegetables as green beans, peas, spinach, and cauliflower [122]. High quality over storage, or high-quality life (HQL), was defined as the storage period during which initial quality was maintained from the time of freezing up to the point where 70% of taste panel members were capable of detecting differences between foods stored at various temperatures and controls stored at −40°C. These keeping times, referred to as the time to a "just noticeable difference" or as "stability time," show an exponential relationship with decreasing temperature [126]. The publications emerging from the TTT Project have provided extremely valuable information on the stability of frozen vegetables, particularly green beans, peas, spinach, and cauliflower [122].

Table 18.5 shows the number of days elapsing in storage at several temperatures before changes are detected in color or flavor. For any given temperature, color is the limiting attribute, as color changes were detected before flavor changes in all the vegetables tested except spinach. The time lapse before a noticeable change in color quality increased at colder temperatures (−4 to −18°C) [122]. HQL indices do not represent finite storage times after which products cannot be consumed, but the length of time during which frozen products retain their initial quality levels [126]; most products remain commercially acceptable even after the end of the stability time. The terms PSL and "acceptability time" [127] are used to designate the storage period during which product quality stays at a level acceptable for consumption or for use in further processing [8]. The collective assessments of scientists, experienced industry representatives, consumers, and taste panellists indicate that for most foods, the PSL is from two to five times longer than the HQL. However in color-sensitive products such as cauliflower, the PSL may be only slightly longer than the HQL [8].

Table 18.6 reflects the PSL or acceptability times for most frozen vegetables [8]. The mean PSLs or acceptability times indicated by the IIR [8] at three different storage temperatures were based on properly processed and packaged products with high initial quality levels. Use of lower quality raw material, inadequate processing or packaging, and fluctuations in the storage temperature can substantially shorten the length of time in storage during which quality remains acceptable. The values in Table 18.6 should therefore not be considered absolute limits to be applied strictly. According to IIR criteria [8], in most circumstances, most vegetables can be categorized as high stability (with a PSL of 15 months at −18°C) or moderate stability (with a PSL of 8–15 months at −18°C); only cut corn and peppers are considered low stability (with a PSL of less than 8 months at −18°C) [1].

Bearing in mind that the combined effects of time and temperature are cumulative and irreversible over the storage period and that the sequence of events does not affect the total cumulative loss in quality, once the relationship between temperature or storage time and a given reduction in quality levels (acceptability times) has been established for a given product, the total reduction in the practical shelf life of that product during storage and distribution can be estimated. Figure 18.7 shows acceptability time curves for peas and cauliflower at various storage temperatures [28].

Using this relationship, if the residence times and temperatures at each stage in the freezing chain are known, the partial percent loss of acceptability and total loss of PSL can be calculated, as shown in Table 18.7. The total loss for peas was 42.51% and for cauliflower 60.47%, in both cases, over a storage period of 257 days; when loss of acceptability reaches 100%, the product

TABLE 18.5
Days in Storage at Various Temperatures Required to Bring about a Perceptible Change in Quality of Some Frozen Vegetables

	Period of Storage (days)							
	Beans		Peas		Cauliflower		Spinach	
Temperature (°C)	Color	Flavor	Color	Flavor	Color	Flavor	Color	Flavor
−18	101	296	202	305	58	291	350	150
−12	28	94	48	90	18	61	70	60
−9.5	15	53	23	49	10	28	35	30
−7	8	30	11	27	6	13	20	20
−4	4	17	5	14	3	6	7	8

Source: From P Olson, WC Dietrich. In: WB van Arsdel, MJ Copley, RL Olson, Eds., *Quality and Stability of Frozen Foods*. New York: Wiley Interscience, 1969, pp. 117–141. With permission.

TABLE 18.6
Vegetables PSL in Months at Several Storage Temperatures

Product	−12°C	−18°C	−24°C
Asparagus (with green spears)	3	12	>24
Green beans	4	15	>24
Beans, lima	—	18	>24
Broccoli	—	15	24
Brussels sprouts	6	15	>24
Carrots	10	18	>24
Cauliflower	4	12	24
Corn-on-the-cob	—	12	18
Cut corn	4	15	>24
Mushrooms (cultivated)	2	8	>24
Green peas	6	24	>24
Red and green peppers	—	6	12
French fried potatoes	9	24	>24
Spinach (chopped)	4	18	>24
Onions	—	10	15
Leeks (blanched)	—	18	—

Source: Adapted from Anonymous, *Recommendations for the Processing and Handling of Frozen Foods*, 3rd Ed., International Institute of Refrigeration, Paris 1986, p. 258. With permission.

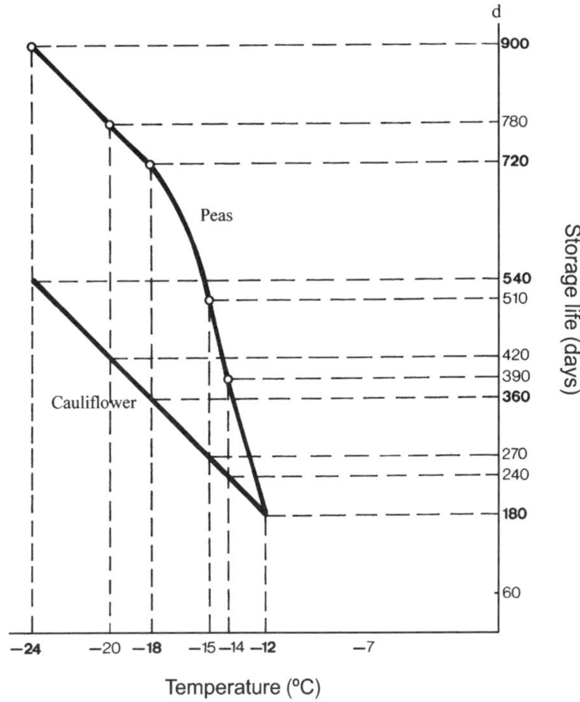

FIGURE 18.7 Acceptability time curves for peas and cauliflower at various storage temperatures. (From JA Muñoz-Delgado. *Fundación Española de la Nutrición, Serie Informes 2*, 1985, pp. 59–88. With permission.)

TABLE 18.7
Loss of PSL at Each Component, or Link, in the Freezer Chain and Total Loss of Acceptability of Frozen Peas and Cauliflower

Stage	Temperature (°C)	Acceptability (days)		Time (days)	Loss per Day		Partial Loss (%)	
		Peas	Cauliflower		Peas	Cauliflower	Peas	Cauliflower
Producer	−24	900	540	30	0.111	0.185	3.33	5.55
Transport	−18	720	360	2	0.138	0.277	0.276	0.554
Wholesaler	−24	900	540	180	0.111	0.185	19.98	33.3
Transport	−15	510	270	1	0.196	0.370	0.196	0.370
Retailer	−12	180	180	30	0.555	0.555	16.55	16.55
Transport	−7	60	60	1/6	1.66	1.66	0.27	0.27
Homefreezer	−18	720	360	14	0.138	0.277	1.932	3.878
Total				257			42.51	60.47

Source: Modified from JA Muñoz-Delgado. *Fundación Española de la Nutrición, Serie Informes 2*, 1985, pp. 59–88. With permission.

is no longer suitable for consumption. Higher losses of acceptability have been reported for peas [127], attaining 70% after storage for 344 days.

The explanation for these relatively high figures is that they are based on old TTT data and thus fail to take account of recent improvements in product selection, processing, and packaging. Nowadays, total losses of PSL over storage and distribution would be lower, as residence times at the various stages of the freezing chain are shorter and more recent acceptability time values are higher. It is not possible to generalize the behavior during storage given the considerable influence of vegetable variety and differences in consumer preferences.

Moreover, differences in the selection of quality attributes, the relative importance attached to each of the various attributes by consumers in different countries, and the differing objective and subjective methods used to measure such attributes all contribute significantly to the wide variation in PSL and HQL results reported in the literature. A T–T–T curve is only valid for a given product with a given raw material quality that has been processed in a given way in a given type of packaging. These P–P–P factors, combined with the taste panel factor, can be as decisive for product quality and stability as storage temperature and time.

There is a clear need to optimize the P–P–P factors and investigate the importance of the T–T–T factors for maintenance of quality and stability in frozen vegetables. Objective and sensory tests to measure quality attributes are also required to determine the acceptability of products as it relates to temperature at each of the various stages of the freezing chain, to be able to assess the true effect and relative importance of each of these stages as regards final quality.

D. Importance of Storage Temperature in the Freezing Chain

The effect of storage temperature on product stability is expressed in T–T–T curves by the Q_{10} factor, which indicates the proportion by which product quality retention time is increased for every 10°C decrease in temperature. In frozen vegetables stored between −15 and −25°C, this factor varies considerably, from 3 to 6; it is clear that temperature is a much more important factor than storage time in determining quality loss. Therefore, it is not important to know the exact amount of time that has elapsed since freezing; dates can mislead consumers as to product quality if the product has not been stored at a sufficiently low and constant temperature. Under certain conditions, losses of quality can be greater and occur faster than those that might otherwise be expected on the basis of the cumulative effect of storage time and temperature. This is true in the case of fluctuations in storage temperature, and therefore, their importance in the freezing chain and the influence of improper handling also need to be addressed by scientists to determine their actual effect on product quality.

In recent years, energy considerations have prompted changes in the recommended minimum storage temperatures (−30°C during production and at wholesalers and −18°C in sales cases and consumers' freezers).

The IIR [8] considers the commonly accepted temperature of −18°C to be the upper limit for storing most vegetables from one season to the next while allowing for a reasonable overlap. As shown in Table 18.6, in practice, the PSL for most vegetables is substantially in excess of 1 year provided that the packaging material used affords adequate protection against moisture migrations and temperature fluctuations. Only mushrooms and asparagus need temperatures of −25°C or colder to attain a storage life of 1 year. When unblanched, aromatic herbs such as parsley, chives, and basil require −30°C, and even then the storage life is less than a year. As can be seen from Table 18.7, producer and wholesale storage temperatures should be as cold as possible (−24°C) and as constant as possible, as this is where most products spend the greater part of their storage lives between freezing and final consumption. Transportation normally entails very short times, and its influence on quality is negligible in most cases (EU legislation accepts tolerances of 3°C up to −15°C during intercity carriage and during distribution) [129]. In contrast, time in retail display cabinets and home freezers is crucial because temperatures there tend to be higher and more subject to fluctuations (EU legislation authorizes tolerances of 6°C up to −12°C in retail display cabinets).

The weakest links in the freezing chain are retail display cabinets and carriage at ambient temperature (commonly when the consumer takes the goods home). Their effects can be avoided by delivering direct from a supermarket cold store to the consumer by refrigerated transport. The Internet direct shopping and home delivery at controlled temperature will improve the adverse effects of the cold chain on frozen vegetables quality.

E. Nutritional and Microbiological Aspects during Storage

Concerning nutrition, as noted in previous sections, care should be taken to prevent vitamin and other nutrient losses during frozen storage. On the scale of the vitamin and macronutrient losses that occur before processing and during blanching, there is practically no decrease in the nutritional value of macronutrients during storage at temperatures below −18°C for periods of up to 1 year. Vitamin C remains fairly stable over a storage period of 1 year if the product temperature is colder than −18°C; losses are associated with temperature and vary according to the product type ($Q_{10} = 6$–20 in vegetables). B-complex vitamins are more stable than vitamin C, except for folic acid, which undergoes comparable moderate losses (20% over 1 year) at −18°C. Losses of carotene are minor (5–20%). The final cooking of frozen vegetables can produce substantial losses: 10–50% of vitamin C; 0–50% of folic acid; 0–25% of vitamin B; 20–40% of vitamin B; and 10–40% of pantothenic acid [130]. Frozen product also compares favorably with chilled raw equivalent purchased at the market, in which vitamin C loss reaches 56% [131]. Recent data [132], included in Table 18.8, provide a direct comparison of the nutritional quality of several cultivars of fresh peas, broccoli, green beans, spinach, and carrots at various stages of distribution and storage, where the same vegetable has been commercially quick-frozen and stored in deep freeze for up to 12 months.

For all vegetables studied, the vitamin C level in the commercially quick-frozen product is (i) equal to or better than in "market fresh" product; (ii) much better than in "market fresh as used" product; or (iii) better than in "supermarket fresh or ambient stored" product, that is, as used (ambient home storage). Vitamin C levels are also higher in quick-frozen product than in (i) all market and supermarket whole green beans; (ii) all market and supermarket spinach; and (iii) all market and supermarket carrots.

As regards folic acid, the recommended mean dietary intake for adults is 200–300 μg/day, increasing to 400 μg/day during pregnancy and lactation. Folate concentrations are especially high in green leafy vegetables (140 μg/100 g), although processing and freezing reduce initial folic acid by 22.5% in peas and by 25% in spinach. This compares favorably with chilled raw equivalent of both products purchased at the market, which presents comparable folic acid loss after 4 days [133].

Experimental data available on table-ready dishes prepared from garden peas indicate that vegetable freezing is less destructive from the nutritional standpoint (vitamin C loss = 61%) than other processing methods such as canning (64%), air-drying (75%), or freeze-drying (65%) [134].

Finally, microorganism growth does not take place at the very cold temperatures employed in frozen storage (−18 to −30°C). During carriage, distribution, and retailing, storage temperatures should be kept under −12°C, below which no bacterial growth occurs and no part of the product should be allowed to warm to above −12°C [8].

IV. CONCLUSIONS

In this chapter on frozen vegetable products and their quality, the diversity of the factors involved is highlighted, and review of the literature indicates the need for research in a number of areas, the most important one being the selection and breeding of high-yield cultivars possessing appropriate

TABLE 18.8
Comparison of Vitamin C in "Fresh" and Frozen Vegetables

	Content of Vitamin C (mg/100 g)							
					Supermarket Fresh as used		Frozen Product	
Product	Garden Fresh (No Storage Sample at Day 0)	Market Fresh (2/3 Days of Ambient Storage)	Market Fresh as Used (Additional 1–4 Days of Ambient Home Storage)	Supermarket Fresh (2/3 Days of Chilled Storage and Transport)	Additional 1–4 Days of Chilled Home Storage	Additional 1–4 Days of Ambient Home Storage	Initial Sampled Day 1 After Processing	Stored 12 Months
Peas	30.9	20.6	12.1	28.8	27.4	17.1	19.6	17.2
	29.6	16.6	11.1	30.4	23.8	17.1	21.8	19.5
	25.6	15.7	9.2	24.7	22.0	16.8	17.3	17.4
Broccoli	77.1	47	34.8	77.8	81.3	50.6	66.1	64.3
	93.1	55.7	40.4	91.5	88.4	61.6	76.7	73.7
Green beans	15.1	7.9	6.9	7.6	6.9	3.7	15.1	17.6
	11.8	7.2	5.1	5.6	4.7	5.1	11.2	N/A
Spinach	31.6	3.2	0	14.3	6.5	0	24.5	16.2
	21.6	2.6	0	12.8	4.2	0	12.6	N/A
Carrots	4.4	3.8	2.9	4.5	3.7	2.8	4.6	N/A
	3.7	3.4	3.0	3.5	3.5	3.2	4.4	N/A

Source: From DJ Favell. *Food Chemistry* 62 (1):59–64, 1998. With permission.

quality attributes for freezing and for mechanical harvesting and processing. In addition, the selection of cultivars with negligible enzyme activity and hence requiring little or no blanching is an important area for future research. Modifications to processes should be the outcome of basic research into the physical, chemical, and biological features of phytosystems at low temperatures aimed at improving our understanding of the behavioral processes that take place during freezing. Another important task is to develop objective methods for measuring product properties, in particular, their mechanical and thermophysical properties, and to apply these to process optimization and the design of control and processing equipment. New T–T–T/P–P–P studies are needed to classify new products according to their stability during storage at different temperatures; such studies should also consider the influence of temperatures under actual freezing chain conditions, taking into account the effect of temperature fluctuations on quality. The development of new generations of frozen vegetable products with higher added values and competitive pricing will depend on creative efforts and technological development arising out of cooperation between scientists and manufacturers. Proper information about and promotion of the safety, the nutritional and sensory qualities and the availability of new vegetable frozen products, combined with the development of the Internet shopping and home delivery services at controlled temperature, will help the frozen vegetable industry to compete better with other food sectors and to satisfy ever-growing consumer expectations regarding quality and health benefits.

REFERENCES

1. W Canet. Quality and stability of frozen vegetables. In: S Thorne, Ed. *Developments in Food Preservation — 5*. London and New York: Elsevier Applied Science Ltd, 1989, pp. 1–50.
2. CJ Kennedy. Introduction. In: CJ Kennedy, Ed., *Managing Frozen Foods*. Cambridge, England: Woodhead Publishing Ltd, 2000, pp. 1–5.
3. JJ Pierce. 2003 Global frozen foods almanac. *Quick Frozen Foods International* 45 (2):107–128, 2003.
4. P Zeuthen. COST 91 — organization, goals and performance. In: P Zeuthen, JC Cheftel, C Eriksson, M Jul, H Leniger, P Linko and G Varela, Eds., *Thermal Processing and Quality of Foods*. London: Elsevier Applied Science, 1984, pp. 1–3.
5. R Gormley. Managing the cold chain for quality and safety. In: *Proceedings of a Flair-Flow Europe Retuert Workshop*, Dublin: The National Food Centre, 1998, pp. 1–53.
6. K Fikiin. Novelties of food freezing research in Europe and beyond. Flair-Flow 4 Synthesis Report SMEs 10, 2003, pp. 1–56.
7. A Maestrelli. Fruit and vegetables: the quality of raw material in relation to freezing. In: CJ Kennedy, Ed., *Managing Frozen Foods*. Cambridge, England: Woodhead Publishing Ltd, 2000, pp. 27–55.
8. Anonymous. *Recommendations for the Processing and Handling of Frozen Foods*, 3rd ed., Paris: International Institute of Refrigeration, 1986, pp. 32, 34, 38, 236, 240, 264, 266.
9. TW Goodwin, M Jamikron. Biosynthesis of carotenes in ripening tomatoes. *Nature* 170:104–105, 1952.
10. CJ Kennedy. Future trends in frozen foods. In: CJ Kennedy, Ed., *Managing Frozen Foods*. Cambridge, England: Woodhead Publishing Ltd, 2000, pp. 263–278.
11. FL Gorini. Comportamento dei frutti nel corso della commercializzazione. *Atti Experimental Institute for Agricultural Product Technology* (*I.V.T.P.A.*) XII:185–210, 1989.
12. W Canet. Estabilidad e importancia de la vitamina C en vegetales congelados. *Alimentación, Equipos y Tecnología* XV (5):75–87, 1996.
13. W Canet. Congelación de vegetales I. *Alimentación, Equipos y Tecnología* IV (6):135–140, 1985.
14. W Canet. Congelación de vegetales II. *Alimentación, Equipos y Tecnología* V (5):77–101, 1986.
15. W Canet, MD Alvarez. Congelación de alimentos vegetales. In: A Madrid Vicente, Ed., *Aplicaciones del frío a los alimentos*. Madrid: AMV y Mundi Prensa, 2000, pp. 201–258.
16. D Arthey. Freezing of vegetables and fruits. In: CP Mallet, Ed., *Frozen Food Technology*. London: Blackie Academic and Professional, 1993, pp. 237–269.
17. Anonymous. *Frutas y hortalizas elaboradas y congeladas rápidamente, Vol. 5A Codex Alimentarius*, 2nd ed., Programa FAO/OMS sobre Normas Alimentarias FAO, Roma, 1995, pp. 301–421.

18. Anonymous. *Métodos de análisis y muestreo, Vol 13 Codex Alimentarius*, 2nd ed., Programa FAO/OMS sobre Normas Alimentarias FAO, Roma, 1995, pp. 11–65.
19. Anonymous. Raw Material Guidelines for Quick Freezing. CCFRA Guidelines, Campden and Chorleywood Food Research Association, Chipping Campden, Gloucestershire, GL55 6LD, U.K., 1995–2003.
20. Anonymous. Specifications for Quick Frozen Fruits and Vegetables. CCFRA Specifications, Campden and Chorleywood Food Research Association, Chipping Campden, Gloucestershire, GL55 6LD, U.K., 1995–2003.
21. V Skegg, MB Springett, JB Adams. Objective methods of quality assessment of frozen vegetables. CCFRA Technical Memorandum 273, Campden and Chorleywood Food Research Association, Chipping Campden, Gloucestershire, GL55 6LD, U.K., April 1981, pp. 1–49.
22. MJ Adams, LV Bedford. QAV — A method for the sensory appraisal of quality of processed vegetable varieties. CCFRA Technical Memorandum 278, Campden and Chorleywood Food Research Association, Chipping Campden, Gloucestershire, GL55 6LD, U.K., July, 1981, pp. 1–33.
23. JB Adams, A Robertson. Instrumental methods of quality assessment. CCFRA Technical Memorandum 307, Campden and Chorleywood Food Research Association, Chipping Campden, Gloucestershire, GL55 6LD, U.K., December, 1982, pp 1–36.
24. JB Adams, A Robertson, MB Springett. Instrumental methods of quality assessment. CCFRA Technical Memorandum 350, Campden and Chorleywood Food Research Association, Chipping Campden, Gloucestershire, GL55 6LD, U.K., July, 1983, pp. 1–85.
25. Anonymous. FDA food defect action levels for vegetables and vegetable products. In: YH Hui, S Ghazala, DM Graham, KD Murrel, WK Nip, Eds., *Handbook of Vegetable Preservation and Processing, Appendix F*. New York: Marcel Dekker Inc, 2004, pp. 649–662.
26. Anonymous. FDA macroanalytical methods for vegetables and vegetable products. In: YH Hui, S Ghazala, DM Graham, KD Murrel, WK Nip, Eds., *Handbook of Vegetable Preservation and Processing, Appendix H*. New York: Marcel Dekker Inc, 2004, pp. 667–683.
27. BK Watt. The nutritive value of frozen foods. In: NW Desrosier, DK Tressler, Eds., *Fundamentals of Food Freezing*. Westport Connecticut: AVI Publishing Company Inc., 1977, pp. 506–536.
28. JK Frans. Folate and health-recommendations. In: *Abstract of 2nd European Symposium on the Health Benefits of Vegetables*. Vlaardingen, Holland: Unilever Nutrition Center, 1997, p. 1.
29. RE Brackett, DF Splittstoesser. Fruits and vegetables. In: C Vanderzant, DF Splittstoesser, Eds., *Compendium of Methods for the Microbiological Examination of Foods*, 3rd ed. Washington: American Public Health Association, 1992, pp. 919–927.
30. GP Archer, Ed. *Introductory Guide to the Microbiological Analysis of Frozen Foods*. EU Concerted Action CT96–1180, 1998, pp. 1–30.
31. Anonymous. Pathogens: vegetables and vegetable products. In: YH Hui, S Ghazala, DM Graham, KD Murrel, WK Nip, Eds., *Handbook of Vegetable Preservation and Processing, Appendix D*. New York: Marcel Dekker Inc, 2004, pp. 585–635.
32. Anonymous. Vegetables, vegetable products and disease outbreaks. In: YH Hui, S Ghazala, DM Graham, KD Murrel, WK Nip, Eds., *Handbook of Vegetable Preservation and Processing, Appendix J*. New York: Marcel Dekker Inc, 2004, pp. 703–720.
33. J Farkas. Rapid detection of microbial contamination in foods using instrumental methods. Flair-Flow 4 Synthesis Report SMEs 9, 2003, pp. 1–27.
34. L Bogh-Sorensen. Maintaining safety in the cold chain. In: CJ Kennedy, Ed., *Managing Frozen Foods*. Cambridge, England: Woodhead Publishing Ltd, 2000, pp. 5–26.
35. Anonymous. Raw Material Guidelines for Pesticide Controls in the Food Chain. CCFRA Guidelines 19, Campden and Chorleywood Food Research Association, Chipping Campden, Gloucestershire, GL55 6LD, U.K., 1998.
36. Anonymous. Considerations of Due Diligence with Respect to Pesticide Residues in Food and Drink. CCFRA Guidelines 2, Campden and Chorleywood Food Research Association, Chipping Campden, Gloucestershire, GL55 6LD, U.K., 1995.
37. Anonymous. FDA action levels for poisonous or deleterious substances in vegetables and vegetable products. In: Y.H. Hui, S Ghazala, DM Graham, KD Murrel, WK Nip, Eds., *Handbook of Vegetable Preservation and Processing, Appendix G*. New York: Marcel Dekker Inc, 2004, pp. 663–666.

38. A Cioubanu, L Niculescu. Fruits and vegetables. In: A Cioubanu, G Lascu, V Bercescu, L Niculescu, Eds., *Cooling Technology in the Food Industry*. Kent, U.K.: Abacus Press, 1976, pp. 377–403.
39. Anonymous. How to cut fruits and vegetable products. Urschel laboratories incorporated. 2503 Calumet Avenue, P.O. Box 2200 Valparaíso, Indiana 46384–2200, U.S.A.
40. W Canet. Estudio de la influencia de los tratamientos térmicos de escaldado, congelación y descongelación en la textura y estructura de patata, Ph.D. thesis, Universidad Politécnica de Madrid, 1980.
41. P Rutledge. Métodos de preparación. In: D Arthey, C Dennis, Eds., Procesado de hortalizas. Zaragoza: Ed., Acribia S.A, 1992, pp. 47–76.
42. DC Williams, MH Lim, AO Chen, RM Pangborn, JR Whitaker. Blanching of vegetables for freezing — which indicator enzyme to choose. *Food Technology* 40 (6):130–140, 1986.
43. E Steinbuch. Heat shock treatment for vegetables to be frozen as an alternative for blanching. In: P Zeuthen, JC Cheftel, C Eriksson, M Jul, H Leniger, P Linko, G Varela, Eds., *Thermal Processing and Quality of Foods*. London: Elsevier Applied Science, 1984, pp. 553–558.
44. KZ Katsaboxakis. The influence of the degree of blanching on the quality of frozen vegetables. In: P Zeuthen, JC Cheftel, C Eriksson, M Jul, H Leniger, P Linko, G Varela, Eds., *Thermal Processing and Quality of Foods*. London: Elsevier Applied Science, 1984, pp. 559–565.
45. R Ulrich. Changes in structure and composition of non-blanched fruits and vegetables, and effects of blanching. *Revue General du Froid* 73 (1):11–19, 1983.
46. J Philippon. Blanching and cooling methods for vegetables before freezing. *Sciences des Aliments* 4 (4):523–550, 1984.
47. W Canet. Congelación de Vegetales III. *Alimentación Equipos y Tecnología* V (5):77–91, 1986.
48. J Philippon, MA Rouet-Mayer. Blanching and quality of frozen vegetables and fruit. Review I. Introduction and enzymatic aspects. *International Journal of Refrigeration* 7 (6):384–388, 1984.
49. J Philippon, MA Rouet-Mayer. Blanching and quality of frozen vegetables and fruit. Review 2. Sensorial aspects. *International Journal of Refrigeration* 8 (1):48–53, 1985.
50. J Philippon, MA Rouet-Mayer. Blanching and quality of frozen vegetables and fruit, Review 3. Nutritional and hygienic aspects, conclusions. *International Journal of Refrigeration* 8 (2): 102–105, 1985.
51. KP Poulsen. Optimization of vegetable blanching. *Food Technology* 40 (6):122–129, 1986.
52. F Pizzocaro, R Ricci, L Zanetti. New aspects of blanching of vegetables before freezing. I Peroxidase and lipoxygenase activity. *Industrie Alimentari* 27 (265):993–998, 1988.
53. JD Selman. The blanching process. In: S Thorne, Ed., *Developments in Food Preservation — 5*. London: Elsevier Applied Science, 1987, pp. 205–250.
54. JB Adams. Heat requirements during blanching of fruits and vegetables for freezing. *Revue General du Froid* 73 (1):21–25, 1983.
55. E. Steinbuch. Improvement of texture of frozen vegetables by stepwise blanching treatments II. *Journal of Food Technology* 12 (4):435–436, 1977.
56. E Steinbuch. Technical note: improvement of texture of frozen vegetables by stepwise blanching treatments. *Journal of Food Technology* 11 (3):313–315, 1976.
57. W Canet, J Espinosa, M Ruiz-Altisent. Effects of the stepwise blanching on the texture of frozen potatoes measured by mechanical tests. *Refrigeration Science and Technology* 4:284–289, 1982.
58. MD Alvarez Torres. Caracterización reológica de tejidos de patata tratados térmicamente. Cinéticas de ablandamiento, Ph.D. thesis, Universidad Politécnica de Madrid, 1996.
59. W Canet, J Espinosa. Influence of the freezing process on the texture of vegetables. Effects of blanching and rate of freezing on the texture of carrots (*Daucus carota* L.). *Revista de Agroquímica y Tecnología de Alimentos* 23 (4):531–540, 1983.
60. W Canet, J Espinosa. Influencia del proceso de congelación en la textura de vegetales. Efecto del escaldado y la velocidad de congelación en la textura de guisante (*Pisum sativum* L.). *Proceeding of the I Congress Nac. Ciencias Hortícolas*, Valencia, España, 1983, 2, pp. 913–922.
61. W Canet, J Espinosa. The effect of blanching and freezing rate on the texture of potatoes, carrots and peas, measured by mechanical tests. In: P Zeuthen, JC Cheftel, C Eriksson, M Jul, H Leniger, P Linko, G Varela, Eds., *Thermal Processing and Quality of Foods*. London: Elsevier Applied Science, 1984, pp. 678–683.
62. W Canet, J Espinosa, M Ruiz-Altisent. Effects of blanching and rate of freezing on the texture of potatoes measured by mechanical tests. *Refrigeration Science and Technology* 4:277–283, 1982.

63. A Andersson, V Gekas, I Lind, F Oliveira, R Ostel. Effect of preheating on potato texture. *Critical Reviews in Food Science and Nutrition* 34 (3):229–251, 1994.
64. MD Alvarez, MJ Morillo, W Canet. Optimisation of freezing process with pressure steaming of potato tissues (cv. Monalisa). *Journal of the Science of Food and Agriculture* 79:1237–1248, 1999.
65. MD Alvarez, W Canet. Optimization of stepwise blanching of frozen-thawed potato tissues (cv. Monalisa). *European Food Research and Technology* 210:102–108, 1999.
66. M Pala. Stepwise blanching and its importance in freezing of vegetables. Refrigeration in the service of man. In: *Proceedings XVIth International Congress of Refrigeration*, Vol. 3, Paris, 1983, pp. 631–638.
67. W Canet, MA Hill. Comparison of several blanching methods on the texture and ascorbic acid content of frozen potatoes. *International Journal of Food Science and Technology* 22 (3):273–277, 1987.
68. W Canet, MJ Gil, R Alique, J Alonso. Efecto de diferentes escaldados en la textura y contenido de ácido ascórbico de coles de Brúselas congeladas. *Revista Agroquímica y Tecnología de Alimentos* 31 (1):46–55, 1991.
69. C Kaur, HC Kapoor. Effect of different blanching methods on the physico-chemical qualities of frozen French beans and carrots. *Journal of Food Science and Technology, India*, 38 (1):65–67, 2001.
70. KK Donato, JW Sabo, JW DeVerna. High-concentration-short-time zinc blanches for colour and texture improvement of thermally processed green vegetables. United States Patent US6004601, 1999.
71. AE Bender. Nutritional aspects of frozen foods. In: CP Mallet, Ed., *Frozen Food Technology*. London: Blackie Academic and Professional, 1993, pp. 123–140.
72. N Hansen, KP Poulsen. Combined blanching/cooling of vegetables. *Internationale Zeitschrift fuer Lebensmittel Technologie und Verfahrenstechnik* 38 (3):175–176, 1987.
73. Anonymous. Food blanching process improvement. *State of the Art Report — Food Manufacturing Coalition for Innovation and Technology Transfer*. RJ Philips and Associates Inc, Food Manufacturing Coalition, Virginia 22066, USA, 1997.
74. Turbo-Flow™ Blancher/Cooker Heat Treatment Principles, Product Documentation, Key Technology.
75. P Holland. Carte blanche to save costs. *Food Processing U.K.* 58 (10):57–58 and 60, 1989.
76. J Gutschmidt. Principles of freezing and low temperature storage. In: J Hawthorn, EJ Rolfe Eds., *Low Temperature Biology of Foodstuffs*. London: Pergamon Press, 1968, pp. 299–318.
77. OR Fennema. Freezing preservation. In: OR Fennema, Ed., *Principles of Food Science, Part II*. New York: Marcel Dekker, Inc., 1976, pp. 173–215.
78. WL Kerr, RJ Kauten, MJ McCarthy, DS Reid. Monitoring the formation of ice during food freezing by magnetic resonance imaging. *Lebensmittel Wissenschaff und Technologie* 31:215–220, 1998.
79. WL Kerr, RJ Kauten, MJ McCarthy, DS Reid. MRI and calorimetric study of freezing rate in potatoes. *Journal of Food Process Engineering* 19:363–384, 1997.
80. MS Brown. Texture of frozen vegetables: effects of freezing rate on green beans. *Journal Science Food Agriculture* 18:77–81, 1967.
81. RM Reeve. Relationships of histological structure to texture of fresh and processed fruits and vegetables. *Journal of Texture Studies* 1:247–284, 1970.
82. MS Brown. Effects of freezing on fruit and vegetable structure. *Food Technology* 30 (5):106–109 and 114, 1976.
83. MS Brown. Texture of frozen fruits and vegetables. *Journal of Texture Studies* 7:391–404, 1977.
84. AA Dos Santos Ferreira. Influência dos tratamentos térmicos de branqueamento, congelaçao e descongelaçao na textura, estructura e cor do feijao verde (*Phaseolus vulgaris* L.). Instituto Nacional de Investigaçao Agrária, Oeiras, 1998.
85. A Manzini, G Crivelli, M Bassi, C Buonocore. Structure of vegetables and modifications due to freezing. Milano, Italy: Istituto Sperimentale per la Valorizzazione Tecnologica dei Prodotti Agricoli ed., 1965, pp. 1–267.
86. M Fuchigami, N Hyakumoto, K Miyazaki. Frozen carrots texture and pectic components as affected by low-temperature-blanching and quick freezing. *Journal of Food Science* 60 (1): 132–136, 1995.

87. MD Alvarez, W Canet, ME Tortosa. Effect of freezing rate and programmed freezing on rheological parameters and tissue structure of potato (cv. Monalisa). *Zeitschrift für Lebensmittel Untersuchung und Forschung A* 204:356–364, 1997.
88. M.D. Alvarez, W. Canet. Effect of pre-cooling and freezing rate on mechanical strength of potato tissues (cv. Monalisa) at freezing temperatures. *Zeitschrift für Lebensmittel Untersuchung und Forschung A* 205:282–289, 1997.
89. M Fuchigami, N Hyakumoto, K Miyazaki. Programmed freezing affects texture, pectic composition and electron microscopic structures of carrots. *Journal of Food Science* 60 (1):137–141, 1995.
90. M Fuchigami, N Hyakumoto, K Miyazaki, T Nomura, J Sasaki. Texture and histological structure of carrots frozen at a programmed rate and thawed in an electrostatic field. *Journal of Food Science* 59 (6):1162–1167, 1994.
91. T Ohtsuki. Process for thawing foodstuffs. European Patent 0409430, 1993.
92. M JUL. Conclusions regarding freezing rates. In: M Jul, Ed., *The Quality of Frozen Foods*. London: Academic Press, Inc., 1984, pp. 33–43.
93. Anonymous. *Recommendations for the Processing and Handling of Frozen Foods*, 2nd ed. Paris: International Institute of Refrigeration, 1972, pp. 82, 36.
94. CH Veerkamp. Engineering parameters for freezing and thawing equipment. In: P Zeuthen, JC Cheftel, C Eriksson, M Jul, H Leniger, P Linko, G Varela, Eds., *Thermal Processing and Quality of Foods*. London: Elsevier Applied Science, 1984, p. 802.
95. SD Holdsworth. Physical and engineering aspects of food freezing. In: S Thorne, Ed., *Developments in Food Preservation — 4*, London: Elsevier Applied Science, 1987, pp. 153–204.
96. PO Persson, G Löndahl. Freezing technology. In: CP Mallet, Ed., *Frozen Food Technology*, London: Blackie Academic and Professional, 1993, pp. 20–58.
97. L Eek. A convenience born of necessity: the growth of the modern food freezing industry. In: WB Blad Ed., *Food Freezing: Today and Tomorrow, Springer Series in Applied Biology*. London: Springer-Verlag, 1991, pp. 143–157.
98. JP Miller. The use of liquid nitrogen in food freezing. The growth of the modern food freezing industry. In: WB Blad, Ed., *Food Freezing: Today and Tomorrow, Springer Series in Applied Biology*. London: Springer-Verlag, 1991, pp. 157–171.
99. JP Miller. Freezer technology. In: CJ Kennedy, Ed., *Managing Frozen Foods*. Cambridge, England: Woodhead Publishing Ltd, 2000, pp. 159–195.
100. B Li, DW Sun. Novel methods for rapid freezing and thawing of foods — a review. *Journal of Food Engineering* 54:175–182, 2002.
101. EMA Willhoft. Contributions to the Symposium on Preparation, Processing and Freezing in Frozen Food Production, Food Engineering Forum, Institute of Mechanical Engineers, London, October, 1986.
102. A Madrid, JM Pastrana, F Santiago. El frío mixto para la congelación de productos alimenticios. In: A Madrid, Ed., *Los Gases en la Alimentación*. Madrid: A Madrid Vicente ediciones, 1991, pp. 125–135.
103. G Campbell-Platt. Recent developments in chilling and freezing. In: A Turner, Ed., *Food Technology International Europe 1987*. London: Sterling Publications Ltd, 1987, pp. 63–67.
104. DW Everington. Contributions to the Symposium on Preparation, Processing and Freezing in Frozen Food Production, Food Engineering Forum, Institute of Mechanical Engineers, London, October, 1986.
105. EA Andersen, M Jul, H Riemann. Industriel levnedsmiddelkonservering, Copenhagen, Teknisk Forlag, 1965.
106. AC Peterson, MF Gunderson. Microbiology of frozen Foods. In: DK Tressler, WB van Arsdel, MJ Copley, Eds., *The Freezing Preservation of Foods*, Vol. 2. Westport, Connecticut: Avi Publishing Co., Inc., 1968, pp. 289–326.
107. T Kadoya. *Food Packaging*. San Diego: Academic Press, 1990.
108. GL Robertson. *Food Packaging: Principles and Practice*, New York: Marcel Dekker, 1993.
109. P Harrison, M Croucher. Packaging of frozen foods. In: CP Mallet, Ed., *Frozen Food Technology*. London: Blackie Academic and Professional, 1993, pp. 59–92.
110. M George. Selecting packaging for frozen food products. In: CJ Kennedy, Ed., *Managing Frozen Foods*. Cambridge, England: Woodhead Publishing Ltd, 2000, pp. 195–212.

111. VM Balasubramanian, MS Chinnan. Role of packaging in quality preservation of foods. In: MC Erickson, YC Hung, Eds., *Quality in Frozen Foods*. New York: Chapman & Hall, 1997.
112. J Philippon. Packaging and maintaining the quality of frozen fruits and vegetables. Literature review. *Revue General du Froid* 71 (3):127–136, 1981.
113. W Canet. Envasado y calidad de vegetales congelados. *Alimentación, equipos y tecnología* VIII (1):158–162, 1989.
114. M Jul. Early investigations. In: M Jul, Ed., *The Quality of Frozen Foods*. London: Academic Press, Inc., 1984, pp. 5–32.
115. St Astrom. How quick should quick freezing be? Proceedings Symposium on Frozen and Quick Frozen Food — New Aspects for Agricultural Production and Marketing, FAO, New York, 1977, pp. 149–164.
116. NE Zaritzky, MC Anon, A Calvelo. Rate of freezing effect on the colour of frozen beef liver. *Meat Science* 7 (4):299–312, 1982.
117. R Ulrich. Variations de temperature et qualité des produits surgeles. *Révue Géneral du Froid* 71 (7/8):371–389, 1981.
118. MD Alvarez, W Canet. Effect of temperature fluctuations during frozen storage on the quality of potato tissues (cv. Monalisa). *Zeitschrift für Lebensmittel Untersuchung und Forschung A* 206:52–57, 1998.
119. MD Alvarez, W Canet. Principal component analysis to study the effect of temperature fluctuations during storage of frozen potato. *European Food Research and Technology* 211:415–421, 2000.
120. MD Alvarez, W Canet. Kinetics of softening of potato tissue by temperature fluctuations in frozen storage. *European Food Research and Technology* 210:273–279, 2000.
121. W Canet. Temperatura de conservación y calidad de vegetales congelados. *Alimentación, Equipos y Tecnología* III (2):145–156, 1988.
122. P Olson, WC Dietrich. Quality and stability of frozen vegetables. In: WB van Arsdel, MJ Copley, RL Olson, Eds., *Quality and Stability of Frozen Foods*. New York: Wiley Interscience, 1969, pp. 117–141.
123. RC Martins, CLM Silva. Modeling colour and chlorophyll losses of frozen green beans (*Phaseolus vulgaris*, L.). *International Journal of Refrigeration* 25 (7):966–974, 2002.
124. LMM Tijskens, SA Barringer, ESA Biekman. Modeling the effect of pH on the colour degradation of blanched broccoli. *Innovative Food Science and Emerging Technologies* 2 (4):315–322, 2001.
125. MC Giannakourou, PS Taoukis. Kinetic modelling of vitamin C loss in frozen green vegetables under variable storage conditions. *Food Chemistry* 83:33–41, 2003.
126. WB van Arsdel, MJ Copley, RL Olson. Introduction. In: WB van Arsdel, MJ Copley, RL Olson, Eds., *Quality and Stability of Frozen Foods*. New York: Wiley Interscience, 1969, pp. 1–19.
127. M Jul. Quality changes during freezer storage. In: M Jul, Ed., *The Quality of Frozen Foods*. London: Academic Press, Inc., 1984, pp. 44–79.
128. JA Muñoz-Delgado. Refrigeración y Congelación de Alimentos Vegetales. *Fundación Española de la Nutrición, Serie Informes 2*, 1985, pp. 59–88.
129. Diario Oficial CEE, 11 Febrero, 1989, DIRECTIVA 89/108.
130. RS Harris, E Karmas. *Nutritional Evolution of Food Processing*. Westport, Connecticut: Avi Publishing Co., Inc., 1975.
131. AE Bender. The nutritional aspects of food processing. In: A Turner, Ed., *Food Technology International Europe 1987*. London: Sterling Publications Ltd, 1987, pp. 273–275.
132. DJ Favell. A comparison of the vitamin C content of fresh and frozen vegetables. *Food Chemistry* 62 (1):59–64, 1998.
133. Anonymous. Programa informativo Verduras y Nutrición, Folatos y Salud, Frudesa Ultracongelados, Valencia, 1998.
134. AE Bender. Nutritional Changes in Food Processing. In: S Thorne, Ed., *Developments in Food Preservation — 4*, London: Elsevier Applied Science, 1987, pp. 1–35.

19 Quality and Safety of Frozen Fruits

Danila Torreggiani and Andrea Maestrelli
Istituto Sperimentale per la Valorizzazione Tecnologica dei Prodotti Agricoli (I.V.T.P.A.), Milano, Italy

CONTENTS

I. INTRODUCTION

The frozen fruit market is now taking its place among other frozen foods. Just as for the fresh market, the demanding consumer is looking for high quality, nutritious, and safe foods. Today, the average person understands the importance of having a varied diet including a large portion of fruit and vegetables as in the so-called "Mediterranean diet."

In the ever-developing field of frozen fruit, there are many diverse aspects to be considered and will be covered in this chapter.

It is well known that there is a deterioration of many fruit quality characteristics caused by severe or unsuitable treatments used for the recovery of the nutritional elements. Temperature reduction is the ideal choice to preserve any biological materials, and among the cooling techniques, freezing is the most valid method.

Fruit freezing is only an apparently simple process, although it really hides some puzzling complexities. The direct, evident, and macroscopic effects of freezing on foods is the phase transition of liquid components, which crystalize and solidify. This fact determines the rupture of cell walls due to the growth of ice crystals during freezing. While thawing, the product will exhibit a loss of cellular integrity, which can manifest itself in increased drainage, shape modification, and less definite structure, so preventing a return to the initial pleasant state [1].

Several factors can infuence the quality of frozen fruit but two of them are universally accepted as being of crucial importance: raw material and pretreatments. It is inevitable that these two factors continuously overlap or they are parallel or dependent on one another.

As far as freezing is concerned, fruits differ from vegetables mainly for the mechanical support organization, which in the fruit group is based on pectins, whereas in the vegetable group on fiber. The pectic system is less resistant than the fiber, being physically, chemically, and enzymatically susceptible to degradation. To emphasize this difference between fruit and vegetable, Table 19.1 shows how cell dimensions and particularly cell wall thickness of the two groups play an important role [2–5].

For the earlier mentioned low-resistence problems of fruit cellular structure, raw material selection for fruit freezing is becoming more and more essential to ensure an extra-high-quality finished product. The exploitation of specific characteristics (intentionally or spontaneously) in the vegetable kingdom to satisfy specific needs leads to the creation of the concept of "varietal or cultivar functionality." A cultivar can be defined as a group of cultivated plants identified by any specific characteristic (morphological, physiological, cellular, chemical, etc.), which on reproduction, whether sexual or not, preserve their original characteristics. The cultivar is the lowest botanic unit.

Some species of fruits and vegetables must undergo thermal treatment before storage at temperatures below zero to reach enzymatic inactivation. Even if this treatment called blanching (a term originating from the French cuisine, "blanchiment," a word for not complete cooking [6]) is normally carried out for a short time and followed by rapid water chilling; its detrimental effect on the fruit structure cannot be avoided because of the noticeable fragility of fruit. Apart from blanching, there are many other pretreatments that are used for specific purposes, not only for enzymatic inactivation but also for protecting or enhancing the structure resistance, color maintenance, and vitamin retention during frozen storage. These pretreatments will be looked into further on in this chapter.

II. FACTORS INFLUENCING FRUIT CULTIVAR SELECTION FOR FREEZING

A. Agronomical Factors

The cultivation of fruit intended for processing, particularly for freezing, has been modified over the years. In southern Europe, farms used to consist of fields of just a few hectares; the larger country zones were destined for pastures and vast private estates. In recent years, the dimension

TABLE 19.1
Average Size of Cells (μm) of Some Fruits and Vegetables

Product	Cell Dimension (μm)			References
	Length	Width	Wall Thickness	
Strawberry (cv Sparkle)	161–385	53–122	1.9	[2]
Clingstone Peach (cv Vivian)	64–80	45–60	4–7	[3]
Apple (cv Red delicious)	100	60–80	—	[4]
Carrot (cv N.K.)	50–60	50–60	1.5–1.9	[5]

of the fields has been noticeably increased, especially in the Mediterranean area, rather than in northern Europe. This new change in dimension has the advantage of introducing mechanical tillage and harvesting and helping form a link between production and industry.

Industry normally uses "on line" equipment with a known capacity. In contrast, the producers have to provide a regular crop to industry, without minimum or maximum peaks. Only the background of the cultural cycles of the varieties and parallely the equipment capacity can ensure this regularity. Therefore, it is crucial to achieve an integrated field-industry system.

The cultivar concept is the solution to this problem. The farmer is required to supply industry with cultivars that mature at different periods spread over a period of time, meaning that as one variety finishes production, there is another one ready to take its place. Obviously, the nutritional and morphological characteristics of cultivar have to be similar, to supply industry with qualitatively homogeneous stock.

For this reason, the breeding of the specific varieties is imperative to obtain an exact type of behavior from the series of cultivar used, especially in the face of agronomical but principally climatic conditions. Such precision means that it is possible to create a ripeness calendar where the variability can be limited to a range of 2–3 days, at maximum. Figure 19.1 shows an example of a *ripeness calendar* for some clingstone peaches destined for Central Italy [7]. In this calendar, "0" is the reference cultivar and the negative or positive numbers indicate ripeness of early or late cultivar. For the clingstone, the reference cv is Redhaven, which is harvested from 5th to 10th of July.

Soil has always played an important role but looking into the future, it can be considered as a simple substrate of the plant as it is possible to correct difficult environmental conditions using specific protective tunnels (plastic and so on; fixed or temporary), which more or less give individual nutritional requirements and modify the amount of different macro-elements according to the needs of the plants (*fertigation*). A fertilizer already exists in a waxy capsule form, which regulates the amount of nutrients given to the soil depending on external temperature; the hotter the weather, the more nutrients are given out, while the cooler the weather, the fewer the nutrients. Yet, these modern concepts should not go against the traditional idea of the irreplaceable link between the soil and the product.

Through this unusual "blood supply" system, it is also easy to distribute water and pesticide, when necessary (drip irrigation). *Soil-less* strawberry or other berry fruit are commonly produced in tunnel and the fruit is grown on suspended peat blocks at human eye level to facilitate picking, and so on. At the moment, these agricultural models are expensive and not exactly suitable for industry [8]. In extensive agriculture, it is difficult to put these protective models into practice.

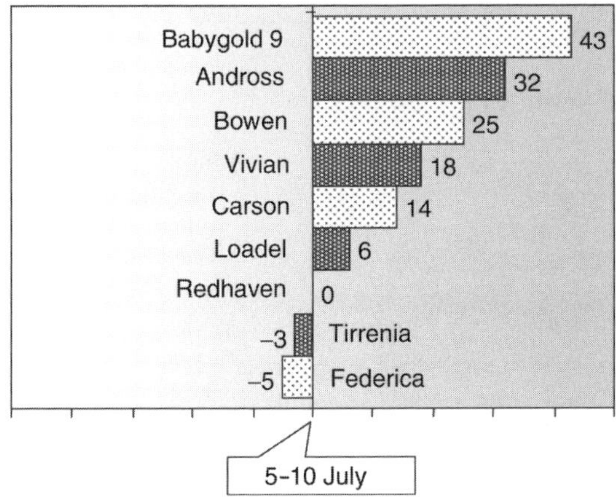

FIGURE 19.1 Ripeness calendar in Central Italy of different clingstone peach cultivars. (From E Bellini, V Nencetti, L Conte, A Liverani, O Insero. Liste varietali fruttiferi. Pesco (pesche, nettarine, percoche). *Terra e Vita* 52 (18):51–70, 2001).

There is no way in which a farmer can protect his production from severe weather conditions, particularly under excess rain and hard frost.

An interesting example can be found in Spain, where 10,000 ha are occupied by only one cultivar, called Camarosa, which is suitable for retail and recently also for freezing. In 2001 and 2002, unseasonal rain and cold during the last phase of ripening seriously damaged most of the crops. The price of the crop thus fell and the product could not be competitive with the investment made. Most of the crop was eventually left in the field, creating difficult sanitary conditions for the following year [9]. Camarosa has an excellent market value, but it would be better to include it with other varieties.

There is also a market for small berry fruits which are important not only for retail and for their very interesting nutritional characteristics but also for industry. In Central Europe, these berries are grown in hilly and mountain areas (in some cases, up to an altitude of 1500 m) [10]. In such conditions, it is not rare to have very cold summer nights (below 0°C, near the harvesting time).

However the "abnormal" weather conditions should be recognized by the very highly specialized farmers. In some cases, the success of the berries production depends only on 0.5°C difference in temperature, when a winning cultivar can survive and give a good yield. Table 19.2 gives an example of the harvesting period of some cv of raspberry and blueberry [11].

Table 19.3 is entirely dedicated to strawberry showing the different behavior of some quality and nutritional parameters of two very well known varieties, with respect to the place of cultivation, showing another simple and clear example of how the soil and weather conditions are linked. The earlier mentioned conditions indicate that in reality, there is only a slight line dividing the agronomical factors from the climatic conditions, both of which should be considered together.

B. Technological Factors

The suitability for mechanical harvesting has at last been accepted as one of the most important technological characteristics of strawberry destined to freezing. However, not all small fruits can be harvested mechanically. Wild strawberry and raspberry are too delicate and must be picked manually. In contrast, blueberry, blackberry, and redcurrant are already harvested automatically. For strawberry, the best way to pick fruit seems to be based on a rake moving slowly through the rows of strawberry [12,13], meaning that the stalk of the fruit can easily be separated. Destalking characteristic is a genetic factor influenced by the cultivar. Figure 19.2 illustrates the force needed to destalk the fruit of three varieties. As it can be seen, all varieties are in a similar force

TABLE 19.2
Harvesting Calendar of Some Berry Cultivars

	Cultivar			
Fruit	**Early Ripening**	**Standard**	**Late Ripening**	**Reflowering**
Raspberry	Malling Exploit	Waki	Malling Admiral	Héritage (end September)
	Glen Moy	Gradina	Malling Léo	
	Malling Promise	Puyallup	Meeker	Baron de Wavre (September)
	Williamette	Delmes	Radbound	
	Lloyd George	Haïda	Schoeneman	Zéva reflowering
	Méco (very early)	Zéva 2	Rose de Cote-d'Or	
Blueberry	Early Blue	Goldtraube	Jersey	

Source: Adapted from A Maestrelli, JM Chourot. Sélection des cultivars en relation avec la transformation. In: G Albagnac, P Varoquaux, JC Montigaud, Eds., *Technologies de Transformation des Fruits*, Paris (France): Lavoisier, 2002, pp. 41–77. With permission.

TABLE 19.3
Influence of Origin and Harvesting Season on Quality and Nutritional Characteristics of Camarosa and Senga Sengana Cultivars

	Quality Parameters						
Cultivar	**Origin**	**Year**	**°Bx**	**Acidity (meq/100 g Fruit Weight)**	**pH**	**Weight (g)**	**Penetration Force (g)**
Camarosa	Cesena, Italy	1996	8.20	12.30	3.38	20.3	700
		1997	7.20	11.60	3.39	13.6	—
		1998	8.40	11.80	2.87	19.4	336
	Palermo, Italy	1998	7.90	9.60	3.50	28.0	415
		1999	7.36	—	—	25.0	461
	Spain	1999	8.00	—	—	26.0	—
Senga Sengana	Cesena, Italy	1998	5.18	9.63	2.84	21.3	125
	Cuneo, Italy	1999	6.20	10.50	3.44	10.0	245
	Poland	2000	10.50	13.90	3.30	25.0	—

range, but what really happens is better explained in the pictures (Maestrelli, personal communication). The 734 (cv Senga Sengana) shows a perfect destalking without defects, whereas only some of strawberry of 745 are destalked. The calix still remains on some of the fruits, and in others, the stalk is broken far from the fruit. The worst that can happen is what can occur to

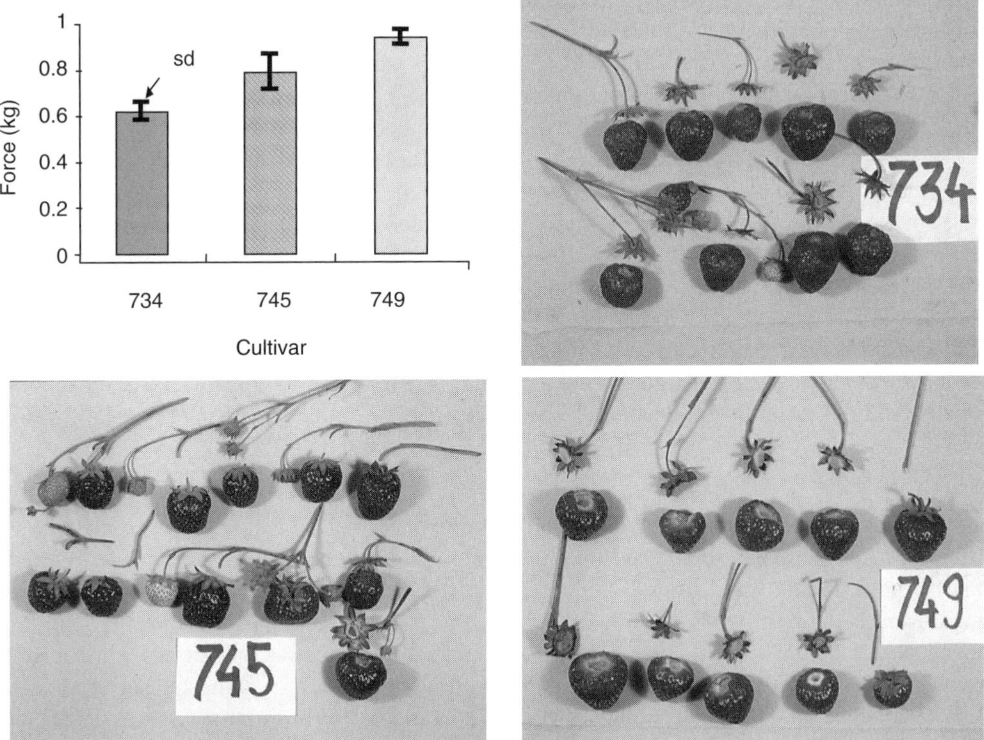

FIGURE 19.2 Destalking force (kg) of three strawberry cultivars (734, Senga Sengana; 745, Mimek × Sella; 749, 89.209.9 × 87.38.1). SD, standard deviation.

strawberry 749, for which the stalk breaks and calix remains, but more seriously a portion of pulp is taken away by the destalking action. This is an example of physical unsuitability of sample under selection, which can also reflect on the nutritional value. Danger of oxidation, mold pollution, and drip loss (also before freezing), and lack of natural resistance, which can all occur around the injured zone, all contribute to the erosion of the nutritional value.

Some agronomic practices already exist which are oriented toward the improvement of the technological functionality of fruit intended for processing and particularly freezing, although these practices are always connnected to breeding programs or cultivar selection.

A good example is the grafting technique for some fruits, normally used on fruit trees and flowers. The main purpose of tree grafting is to defend them from the attacks of soilborne pathogens throughout the biological life (3, 5, and 10 years). The rootstock and the graft often came from different cultivar or species. Another reason is to allow the plant to stay the 2- or 3-year cultivation cycle in the same soil, without showing any signs of illness or "tiredness" without rotation. The principle of crop rotation is normally used as a protective measure, but in this way, it can become less rigid. The fight against mushroom illness (*Fusarium* sp.) of Cantalup melon has been going on for many years by grafting cultivar on specific rootstock. Supermarket is a high-quality cultivar melon with good nutritional and sensory values but it is easily attacked by mushrooms. This melon is well protected when grafted onto Vector, a cv which produces mediocre quality fruit [14].

C. Sensory Factors

Sensory quality of fruit is formed by a combination of different and very important "subunits," such as appearance, odor, texture, taste, flavor, and so on. An example of what can happen, if only one subunit is favored is given by a case observed in the United States [15], where cherry growers chose firmness and color attributes of the fruit over sugar and taste. Early harvested cherry (cv Brooks and Tulare) completely lacked taste and the consumers refused to buy them.

This example is obviously referring to fresh product. It is very easy to imagine what could happen to pretreated and processed products. Spreading out the branches of technological steps can easily hide raw material that is of bad quality or where the process is carried out incorrectly, so multiplying the gray zone of the errors.

The performances of processed fruit depend on suitability of cultivar and type of processing. This is confirmed by what was observed in processed melons [16]. Two cultivar of muskmelon (Rony and Mirado) were pretreated by osmotic dehydration and air dehydration before freezing and 4 months storage ($-20°C$). Among the different quality parameters, the most important for muskmelon is the flavor, and Rony and Mirado raw varieties have the same strong flavor and overall acceptance. Freezing caused an unexpected behavior: while the sensory evaluation of Rony showed a high standard, that of Mirado decreased noticeably, indicating that not only is cultivar a crucial point but also processing, which can influence the quality. In particular, a "simple" unit operation like freezing that normally does not affect flavor, becomes heavily detrimental for that of melon (Table 19.4) [16].

D. Nutritional Factors

Fruits have always been considered as important components of a healthy diet, especially when refering to macronutrients: their role in the prevention of vitamin deficiency; adequate folic acid intake; a quick energy boost from easily digestable sugars and their important source of dietary fiber. However, in recent years, the role of phytochemical activity of fruits has come to light. Phytochemicals are complex plant chemicals which could act as protective factors in a wide range of chronic disorders. Once consumed, phytochemicals are subjected to enzymatic and bacterial degradation to produce an even wider range of products available for absorption. As previously discussed, regarding the link between agronomical and climatic factors, there exists a strong connection between sensory and nutritional factors. The classification of nutritional compounds

TABLE 19.4
Flavor and Overall Acceptance Scores of Melon Spheres Cultivars Rony and Mirado, not Pretreated (Raw) and Pretreated (Air Dehydrated to 50% Weight Reduction, AD; Osmo-dehydrated for 60 min, at 25°C at Atmospheric Pressure in 60% (wt/wt) Sucrose Solution, DIS), after 0 (T0) and 4 (T4) Months of Frozen Storage at −20°C

Storage (Months)	Cultivar	Raw	AD	DIS
		Flavor		
0	Rony	50.75^{Bb}	34.83^{Aa}	40.17^{ABb}
	Mirado	37.67^{Ba}	34.42^{ABa}	36.42^{Ba}
4	Rony	40.75^{Bb}	33.17^{Aa}	38.42^{Ba}
	Mirado	37.75^{Ba}	33.50^{Aa}	36.33^{Ba}
		Overall acceptance		
0	Rony	53.58^{Bb}	33.58^{Aa}	40.25^{Bb}
	Mirado	39.25^{Ba}	35.33^{ABa}	37.42^{Ba}
4	Rony	44.17^{Bb}	30.08^{Aa}	36.67^{ABa}
	Mirado	39.67^{Aa}	30.17^{Aa}	35.42^{Aa}

Note: Different letters (uppercase: among the different pretreatments for a single cultivar; lowercase: between the two cultivars for the same pretreatment) indicate significant difference ($P \leq 0.05$).

Source: Adapted from A Maestrelli, R Lo Scalzo, D Lupi, G Bertolo, D Torreggiani. *Journal of Food Engineering* 49: 255–260, 2001. With permission.

in "macronutrients" and "phytochemicals" can help us to describe the crucial influence of cultivar and agronomical and processing factors on maintaining high nutritional levels.

As for macronutrients, among the 2000 varieties of strawberry known around the world, the balance of the nutrients significantly change and contribute to establishing the real qualitative difference. The ratio between sugar and acid content has always been considered as both a technological and nutritional index. It identifies very important characteristics such as maturity stage, harvesting index, sensory acceptance (described by the correct harmony between two sensory parameters: sweet and sour), and balanced nutritional intake. In Table 19.5, sugar/acid ratio of five strawberry varieties harvested in 1999 is shown (Testoni and Lovati, personal communication).

TABLE 19.5
Average Content (mg/100 g fr. wt.) of Acids and Sugars and Total Sugars/Total Acids Ratio of Some Strawberry Cultivars

	Content (mg/100 g)							
	Acids				Sugars			Total Sugars/Total
Cultivar	Qui	Mal	Cit	Asc	Suc	Glu	Fru	Acids Ratio
Andana	227	472	792	39	830	1600	1770	2.74
Cigaline	334	382	880	56	410	2430	2700	3.35
Red Chief	262	597	1135	41	630	2060	2340	2.48
Cigoulette	195	228	977	43	300	1620	1880	2.63
Kimberly	173	731	638	37	390	1300	1550	2.05

Note: Qui, quinic; Mal, malic; Cit, citric; Asc, ascorbic; Suc, sucrose; Glu, glucose; Fru, fructose.

A further interesting example is the importance of the sugar/acid ratio in oranges, which indicates the suitability of both the cultivar and the quality level for juice production. To obtain high quality red orange juice, the ratio °Bx/acidity should not be lower than 8.0, independently from the cultivars (Moro, Tarocco, and Sanguinello) [17].

As for phytochemicals, an important step toward this new research area has been the screening of fruit cultivar to identify the variety with the highest and most stable phytochemical activity. Until now, cultivar selection has been made by industry with the specific goal of extracting from the fruit single nutrients such as vitamins, antioxidant compounds, and so on. Today, it is understood that a daily consumption of fruits is far more important to have a regular intake of protective factors. Indeed, it has been clearly demonstrated that the *in vivo* activity of whole consumed fruit is higher than that of single extracted molecules [18]. An example fruit showing the variability of antioxidant capacity related to variety is presented in Table 19.6 [19].

III. THE RANGE OF PRETREATMENT TECHNIQUES AND THEIR IMPACT ON QUALITY

A. Blanching

Some fruit species need a treatment of enzymatic inactivation before freezing and low-temperature storage, which is much more important for horticultural products than for fruit.

Table 19.7 shows advantages and disadvantages of immersion blanching as a prefreeze treatment [20]. Enzymatic inactivation (blanching) is carried out normally using heat, particularly boiling water or steam, where the heat action must be intense and quick, being more efficient if the product is divided into small pieces such as cubes, spheres, slices, disks, and halves. The length of time and intensity of the thermal effect must be accurately calculated and, when terminated, has to be immediately stopped often by spraying with or immersing in cold water. The treatment time

TABLE 19.6
Antioxidant Activity, Anthocyanin, and Phenolic Contents of Acetonitrile Extracts of Berries from Different Commercially Available Cultivars of *Vaccinium* Species

Cultivar (State)	ORACROO[a] (μmol TE/g)	Anthocyanin[b] (mg/100 g)	Phenolics[c] (mg/100 g)
	***Vaccinium corymbosum* L. (Northern Highbush)**		
Bluecrop (MI)	17.0 ± 1.0 (70.5)	93.1 ± 1.6	189.8 ± 10.9
Jersey (MI)	20.8 ± 0.6 (63.2)	100.1 ± 2.3	206.2 ± 4.1
Jersey (NJ)	21.4 ± 0.4 (91.9)	116.6 ± 1.1	221.3 ± 4.3
Rubel (MI)	37.1 ± 0.5 (182.8)	235.4 ± 6.1	390.5 ± 6.5
	***Vaccinium corymbosum* L. (Southern Highbush)**		
O'Neal (NC)	16.8 ± 1.9 (105.0)	92.6 ± 4.6	227.3 ± 6.9

Note: MI, Michigan; NJ, New Jersey; NC, North Carolina.

[a]Oxygen radical absorbance capacity expressed as micromole Trolox equivalents per gram of fresh fruit. Data in parentheses expressed per gram of dry matter.

[b]Concentration based on cyanidin-3-glucoside as standard expressed per gram of fresh weight.

[c]Concentration based on gallic acid as standard expressed per gram of fresh weight.

Source: Adapted from RL Prior, G Cao, A Martin, E Sofic, J Mc Ewen, C O'Brien, N Lischner, M Ehlenfeldt, W Kalt, G Krewer, CM Mainland. *Journal of Agricultural and Food Chemistry* 46:2686–2693, 1998. With permission.

TABLE 19.7
Advantages and Disadvantages of Immersion Blanching as Pretreatment

Advantages	Disadvantages
Inactivating enzymes responsible for browning and off-colors	Leading to the death of cellular tissue with high possible deterioration of texture
Partially destroying microorganisms	Higher sensitivity to microbial growth afterwards
Decreasing pesticide	Loss of food's own solutes resulting in pollution of the blancing bath and decrease of nutritional value
Specific applications	Absorption of water by food and modification of yield
Reducing volume of pieces increasing storage capacity	High water consumption
Leading to exclusion of air from tissues	
Increasing extraction of antioxidants in juice production (blueberry and orange)	

Source: Adapted from D Torreggiani, T Lucas, AL Raoult-Wack. In: CJ Kennedy, Ed., *Managing Frozen Foods*, Cambridge (England): Woodhead Publishing Limited and CRC Press LLC, 2000, pp. 57–80. With permission.

depends on the kind and the vitality of the enzymes and on the type of raw material and its maturity stage.

A "dualistic" effect acts in botanical field as the fruit "system" is made up of, on one hand, oxidation catalyzer compounds, and on the other hand antagonist antioxidant compounds. The more the technological process is correctly targeted (including blanching), the more the integrity of the protective compounds (antioxidants) is expected.

Lipids are the most susceptible substances to oxidation in the food system and many heterogeneous factors (chemical, physical, and biological) can create a lot of problems when lipid oxidation must be controlled. Even blanching as a specific protective pretreatment influences oxidative reaction in a contradictory way.

The most important oxidative catalyzers present in vegetable cells are enzymatic and nonenzymatic, and they are active, preferably, against the unsaturated fatty acids. The lypoxygenase enzyme favors the attachment of oxygen on fatty acids [21,22]. This enzyme's activity can provoke at least three undesired effects: destruction of essential fatty acids, production of free radicals (reactive against vitamins and proteins), and development of off-flavours. These effects justify the need to completely inactivate the lipoxygenase before freezing.

The hemoproteins are other catalytic agents, which are important enzymes in the vegetable world (such as peroxidase and catalase). However they react with fatty acids because of their physical structure; this reaction occurs through the intervention of the iron atom inside their porphyrin ring (heme group), which is hidden within the hydrophobic "fessure" of the proteic area. During blanching, heating denaturates hemoproteins, facilitates the exposure of the heme group, and reveals the porphyrin ring. In this way, the fatty acids can come into contact with the iron atoms (specific mineral, catalytic agent), allowing it to express its catalytic ability. This can be called as nonenzymatic activity.

For this reason, enzymes like peroxidase and catalase, which have no specific action on fatty acids, can aid the lipids oxidation through the nonenzymatic activity of their prosthetic group [23,24]. It would seem that the lipoxygenase inactivation may not be enough to control or stop the oxidation of fatty acids. The enzymatic role of vegetables hemoprotein is still not clear, whereas the peroxidase role is very clear as an indicator of the inactivation level of the enzymes during blanching. The technological importance of peroxidase comes from its thermal resistance, which is higher than that of other enzymes [25].

Nevertheless, some factors should be considered to indicate the lipoxygenase as "key enzyme" suitable for evaluating the blanching effect. Among these factors are (1) technological trend limiting the thermal treatment intensity and (2) a better understanding of the biochemical phenomena involved in the quality loss of processed fruit. These factors show that there is a delicate equilibrium inside different enzymatic activities and the nonenzymatic activities.

It is very difficult to foresee the potential lipid oxidation by enzymes and hemoproteins. The forecast is more and more approximate, as the enzymatic activities are contrasted by natural antioxidant activities. Most fruits contain antioxidant compounds such as flavonoids, anthocyanins, tocopherols, carotenoids, and ascorbic acid, acting as reducing factors or free radical chelators. These compounds are present, for example, in aromatic herbs, berries, and citrus fruit.

In this biological world, there exists the speed in which the activity of the two antagonistic agents decay. As the speed can be different for both agents, there is a shifting from the left or from the right of the following reaction:

$$\text{Antioxidant activity} \rightleftarrows \text{Pro-oxidant activity}$$

Therefore, the ideal fruit thermal treatment can be achieved by (1) knowing the parameters of the lipoxygenase thermal inactivation and simultaneously controlling the nonenzymatic oxidant activity of lipids and (2) studying the blanching effect on natural antioxidant activity.

The blanching influence on antioxidant activity in fruit can be explained with two examples: blueberry and orange.

The addition of a blanching step of the fruits in juice processing proved to be a very important factor when evaluating processed blueberry products for their possible health benefits. In fact, the inactivation of polyphenol oxidase through steam blanching significantly increased the anthocyanin and cinnamate recovery when blueberries were pressed into juice [26]. Besides having a higher content in phenolic compounds, juices obtained from blanched blueberry were more blue due to the positive effect of the thermal treatment on the extraction of the most soluble anthocyanin pigments, which are also the most intense blue. The higher recovery of phenolic compounds led to a significant increase of the radical-scavenging activity of the juice.

Blanching can be even advantageous for blood oranges, in fact, the antiradical activity of cv Moro blood orange juice obtained from blanched orange segments was significantly higher when compared with that of the nontreated one [27]. This phenomenon could be linked to a better extraction of compounds with antioxidant and radical-scavenging activity, such as free and bound hydroxycinnamic acids and anthocyanins, as already observed in blueberry.

During frozen storage of blanched products, the cryolability of the catalytic agents, which is in itself favorable, does not occur with any regularity. As nonenzymatic activity in some blanched and frozen vegetables dramatically increases over some weeks of storage at $-20°C$ and then diminishes [28], it would be also interesting to follow this pattern in fruit. The revealing exposition of the heme group, probably arising during blanching, can continue through frozen storage, caused by the denaturating effect of low temperature on proteins. Although all this happens, the inactivated lipoxygenase tends to naturally disappear.

Blanching pretreatment could be applied correctly by taking into account the great number of variables (species, cultivar, storage stage, way of cutting, handling before pretreatment, time and temperature of treatment, etc.).

B. Partial Dehydration and Formulation Techniques

Currently, there is renewed interest in implementing partial dehydration and formulation stages before freezing. The reason for this is the versatility of these techniques, which make it possible to reduce water content, improve quality, and develop new products.

Partial removal of water from the fruit leads to (1) concentration of cytoplasmatic components within the cells; (2) depression of the freezing point; (3) increase of supercooling and microcrystallization; and (4) lower ratio of ice crystals to unfrozen phase. There is consequently a reduction of the detrimental phenomena of loss of cellular structure and drip loss caused by freezing in fruit tissues [29]. Simple it may be, yet it is the choice of the dehydration method that is the key point because of the great variability of fruit tissue structure linked to species and cultivar.

The dehydration methods that can be used include air dehydration, osmotic dehydration (also defined as dewatering impregnation soaking in concentrated solutions, DIS) and a combination of both.

Partial dehydration is generally achieved by air drying. The resulting process is termed dehydrofreezing. The advantages over conventional freezing include (1) energy saving, as the water load to the freezer is reduced, leading to a reduction in transport, storage, and wrapping costs and (2) better quality and stability as well as better thawing behavior (lower drip loss). When using partial air dehydration, food ingredients of high water activity are generally obtained, as water removal is limited to 50–60% of the original content.

Conventional air drying can be substituted by (or combined with) DIS, mainly for fruits, whose color can be affected by heat modification under any form of air dehydration, as DIS is effective at room temperature and operates away from oxygen. This process involves placing the solid food (whole or in pieces) into solutions of high sugar or salt concentration [30–32]. Soaking gives rise to at least two major countercurrent flows: an important outflow of water from the food into the concentrated solution and a simultaneous transfer of solute from the concentrated solution into the food. However, the main unique feature of DIS compared with other dehydration processes is the penetration of solutes into the food material, making it more suitable for freezing process [33]. As a result, it is possible to adapt further the nutritional properties of the dehydrofrozen fruit and thereby formulate new fruit products suitable for various industrial uses. This can be done by (1) adjusting the physicochemical composition of food by reducing water content or adding water activity lowering agents; (2) incorporating ingredients or additives with antioxidant or other preservative properties (spices, sugars, ascorbic acid, etc.) into the food before freezing; (3) adding solutes of nutritional or sensory interest; and (4) providing a larger range of food consistency.

There is vast literature that indicates the usefulness of partial water removal before freezing for numerous species of fruits [31]. Keeping the fruit firmness and structural integrity is important not only for fruit quality but also for preventing the loss of nutrients, which is manifested through exudate loss at thawing. The quality characteristics of frozen fruit, which could be improved through the application of a dehydration step, are numerous and will be analyzed here.

1. Texture

Moisture reduction has been proved to be useful even to improve quality of a delicate tissue such as that of strawberry. The structural collapse after thawing–rehydration of strawberry slices is reduced by adopting partial removal of water through air dehydration, DIS, or their combination [34]. A reduction in moisture content of at least 60% is needed to improve the texture characteristics of thawed–rehydrated fruits, irrespective of the dehydration method used.

These findings were confirmed by the results of microscopic analysis performed on predehydrated and freeze–thawed strawberry slices [35]. The results clearly showed that the freezing damage is reduced due to the decrease in moisture content. Predehydrated strawberry slices retain the tissue structure after thawing, whereas the untreated ones show a definite continuity loss and thinning of cell wall.

Even though osmotic treatments have been proved to be a useful tool in fruit and vegetable cryoprotection, the changes in mechanical properties caused by the process itself have to be taken into account [36]. Different factors contribute to mechanical properties of plant tissue: cell turgor, which is one of the most important ones, cell bonding force through middle lamella, cell wall resistance to compression or tensile forces, density of cell packaging that defines the free spaces with gas or liquid, and some other factors, which are also common to other products such as sample size and shape, temperature, and strain rate [37].

As an example, treatment using DIS alone for 4 h at atmospheric pressure is needed to obtain texture improvement of strawberry slices, but at the same time, structural modifications in the fruit tissue occur [38]. Light photomicrographs of DIS-treated strawberry tissues show that there is a deterioration in the cell links and that the cell walls already lose their shape after 4 h of DIS treatment (Figure 19.3). Considering the good agreement obtained between structural and texture changes for strawberry, the DIS pretreatment has to be shorter than 2 h and combined with air dehydration if texture improvements must be obtained. The combined dehydration is proposed because through the incorporation of sugars, it could be possible to improve color, flavor, and vitamin retention, during frozen storage.

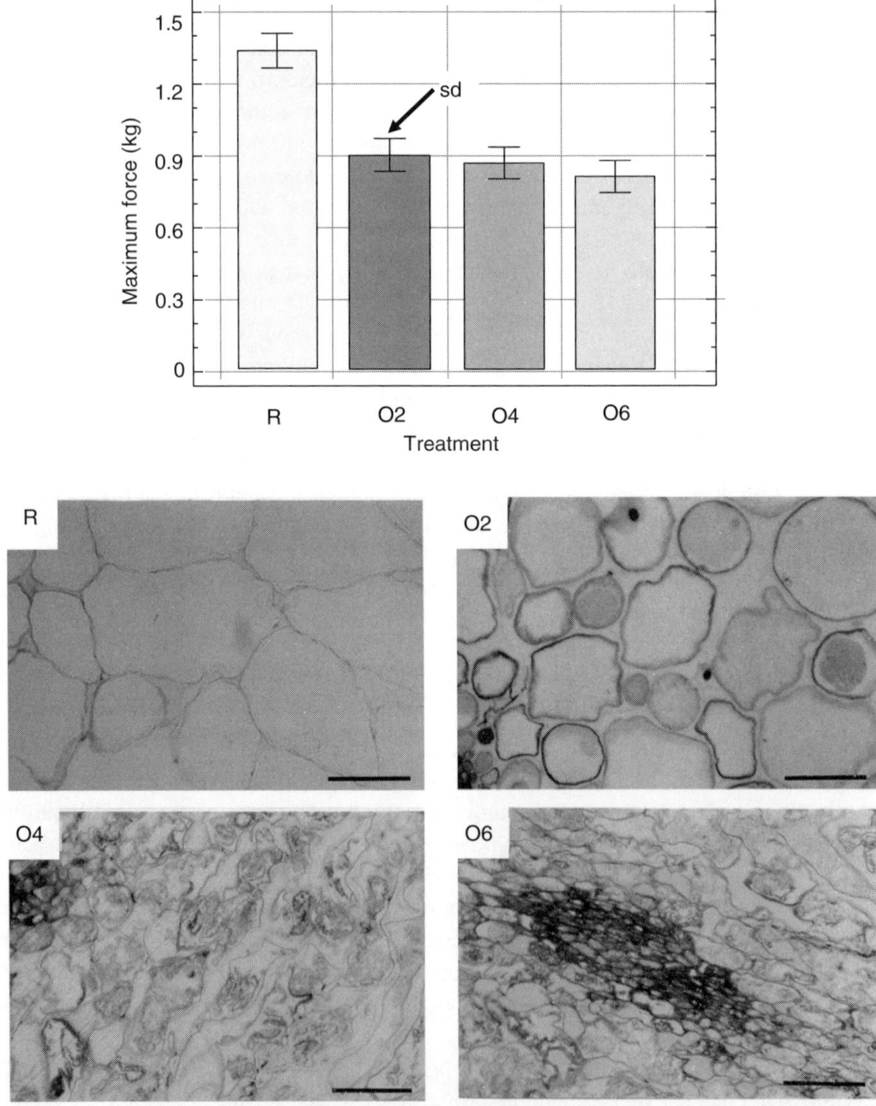

FIGURE 19.3 Photomicrographs and texture values, expressed as shear press cell maximum force (kg) on dry basis, of strawberry slices before (R) and after osmotic dehydration for 120 (O2), 240 (O4), and 360 min (O6) in 60% (wt/wt) sucrose solution, at 25°C at atmospheric pressure. Bars: 100 μm. SD, standard deviation.

Referring to the DIS pretreatment, it should be kept in mind that the earlier mentioned results refer to a specific cultivar, that is, the cultivar Chandler. Therefore, if other cultivars were used, different results might be obtained. In fact, as shown in Figure 19.4, the cultivar influenced the solid–liquid exchanges during DIS treatment applied for 240 min in a 50% (wt/wt) glucose syrup at atmospheric pressure, showing the great importance of the tissue structure, size, and architecture of the intercellular spaces [39]. Furthermore textural properties of fruits are closely linked to cellular structure and pectic composition, and solid–liquid exchanges can influence the texture characteristics of the end product. DIS causes a slight decrease in texture values of strawberry correlated with a decrease in the soluble oxalate and residual pectin (protopectin) fractions, which are correlated with fruit firmness.

The analysis of how differently soluble pectin fractions of strawberry slices are modified by air dehydration or combined osmo-air dehydration applied before freezing, and freezing itself, indicated that protopectin (residual insoluble pectin fraction) content significantly decreases during air dehydration, with the osmotic step reducing the loss [40]. Freezing causes a significant reduction of protopectin content, which is the biggest effect occurring in strawberry that is not predehydrated before freezing. The different losses of protopectins in different predehydrated fruits could explain the differences in texture observed in freeze–thawed fruits. Osmotic treatments using selective solutes can also allow cryoprotection of the cell during freeze–thawing [41].

Another interesting treatment is vacuum infusion with cryoprotectants (sugars from concentrated grape must) and cryostabilizers (HM pectin), which was applied to reduce ice crystal damage in frozen apple cylinders and to improve the fruit resistance to freezing damage through a notable reduction of freezable water [42,43]. Addition of cryoprotectants and cryostabilizers in the formulation changed the glass-transition temperature ($T_{g'}$) of the maximally cryoconcentrated food liquid phase and the freezable water content of strawberry impregnated under vacuum or at atmospheric pressure, with sucrose and sorbitol acqueous solutions with or without the addition of ascorbic acid [44]. The analysis of the product microstructure by light transmission electron microscopy showed that tissues subjected to vacuum had higher cellular tissue integrity and the ascorbic acid addition preserved the cellular tissue better in all the samples.

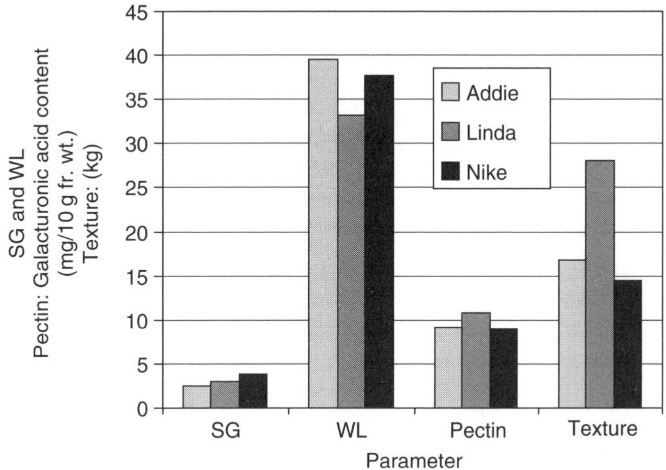

FIGURE 19.4 Solid gain (SG) and water loss (WL) expressed on 100 g of initial fresh fruit, after DIS treatment of three cultivars of strawberry slices. On the same cultivars, after DIS treatment, insoluble pectin fraction (mg galacturonic acid/10 g fresh weight) and texture values expressed as shear press cell maximum force (kg). (Adapted from E Forni, A Sormani, D Torreggiani. In: S Porretta, Ed., *Ricerche e innovazioni nell'industria alimentare*, 4th CISETA (Congresso Italiano di Scienza e Tecnologia degli Alimenti). Pinerolo, Italy: Chiriotti Editori, 2000, Vol. 4, pp. 750–762. With permission.)

Partial dehydration before freezing could even enhance the resistance of texture of frozen strawberry slices and apricot cubes to a thermal treatment [45,46]. For fruit to be incorporated as a food ingredient, for example, in yogurt, a heat treatment should be applied, but this causes texture damage and so does freezing. As shown in Figure 19.5 for strawberry, the most delicate but most requested fruit for yogurt, the fruit undergoes several thermal shocks (two cooling and two heating steps) during processing. The texture measurement data, obtained with a dynamometer, emphasize that the major rheological damage occurs at freezing. Even a drastic treatment such as pasteurization does not determine any further texture decrease. To reach a texture improvement after the proposed heat treatment, a moisture reduction of at least 50% before freezing is needed for both strawberry and apricot, irrespective of the dehydration method used. This percentage of moisture reduction is what is required to reduce the freezing damage of the fruits at thawing [34], so confirming that the freezing step is the most crucial point in the production process of thermally stabilized strawberry and apricot ingredients. If the freezing damage is limited, then the fruit texture can be improved even after heat treatment.

2. Pigments, Color, and Nutrients

Together with a texture improvement, during frozen storage, the penetration of solutes combined with a dehydration effect due to the DIS pretreatment could modify the fruit composition and improve retention of pigment, color, and nutrients such as vitamin and antioxidant substances.

According to the kinetic interpretation based on the glass-transition concept, physical and chemical stability is related to the viscosity and molecular mobility of the unfrozen phase, which in turn depends on the glass-transition temperature [47,48]. When the temperature is at or

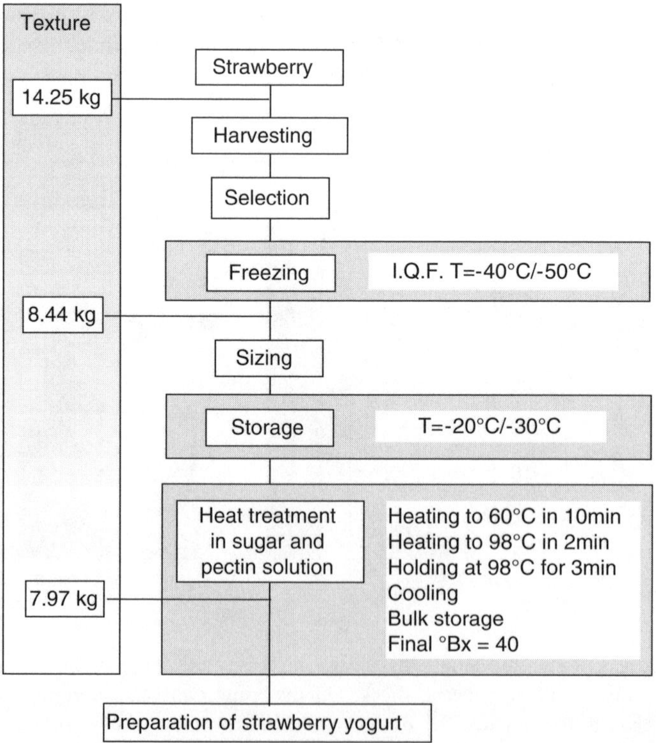

FIGURE 19.5 Production flow sheet of strawberry ingredients destined for yogurt. Texture values expressed as shear press cell maximum force (kg).

below $T_{g'}$, diffusion controlled changes occur at very slow rates, that is, the stability if based on diffusion-controlled events is excellent. Yet, it should be remembered that many chemical changes are not diffusion controlled. The rates of presumed diffusion controlled reactions are considered proportional to the difference between $T_{g'}$, which is also called mobility temperature [49] and the temperature of study. Manipulation of mobility temperatures through composition could therefore influence reaction rates. Therefore, if through DIS the fruit formulation can be modified and thereby an increase in glass-transition temperature could be obtained, then there could also be an increase in the storage stability.

Although the kinetic interpretation based on the glass-transition temperature holds for chlorophyll and vitamin C stabilization in kiwi fruit, for anthocyanin in strawberry and blueberry, a simple relationship does not exist between the loss of these antioxidant phytonutrients and the amplitude of the difference between the storage temperature and the glass-transition temperature of the maximally freeze-concentrated phase. The incorporation by DIS of different sugars into kiwifruit slices modified their low-temperature phase transitions and significantly increased chlorophyll and vitamin C stability during frozen storage at $-10°C$ [50]. Kiwifruit pretreated in maltose and thus having the highest $T_{g'}$ values showed the highest vitamin C retention (Figure 19.6).

Strawberry and blueberry juices, added with different sugars and used as a model, show anthocyanin retention significantly higher than that observed in the juices frozen without the addition of sugar, but no differences were observed among the juices added with different sugars, thus having different glass-transition temperatures [51,52]. The sorbitol-added strawberry and blueberry juices, which have the lowest glass transition temperature, show the same anthocyanin retention as the sucrose and maltose added juices (Figure 19.7). Other factors such as the pH of the unfrozen phase and the specific chemical nature of sorbitol could have influenced the anthocyanin degradation.

These results on model systems such as juices were confirmed on strawberry halves osmodehydrated in different sugars [53,54]. Sugar incorporation improved the health benefit of fruit, increasing the stability of the antioxidant activity linked to anthocyanin pigments.

An osmotic step could also improve the stability of vitamin C and color during air drying and frozen storage of osmodehydrofrozen apricot cubes by the modification of sugar composition

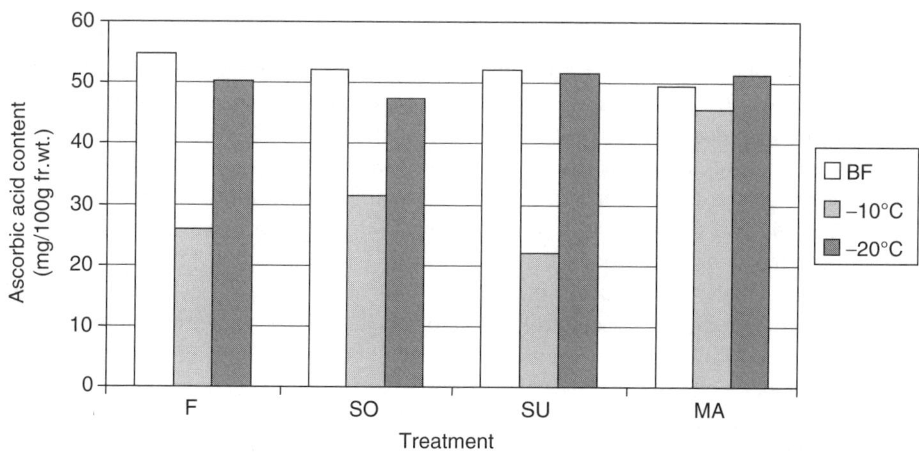

FIGURE 19.6 Ascorbic acid content of kiwi fruit slices, not pretreated (F) and pretreated for 120 min, at 25°C at atmospheric pressure, in 65% (wt/wt) sorbitol (SO), sucrose (SU), and maltose (MA) solution, after 9 months of frozen storage at $-10°C$ and $-20°C$; BF, content before freezing. (Adapted from D Torreggiani, G Bertolo. *Journal of Food Engineering* 49:247–253, 2001. With permission.)

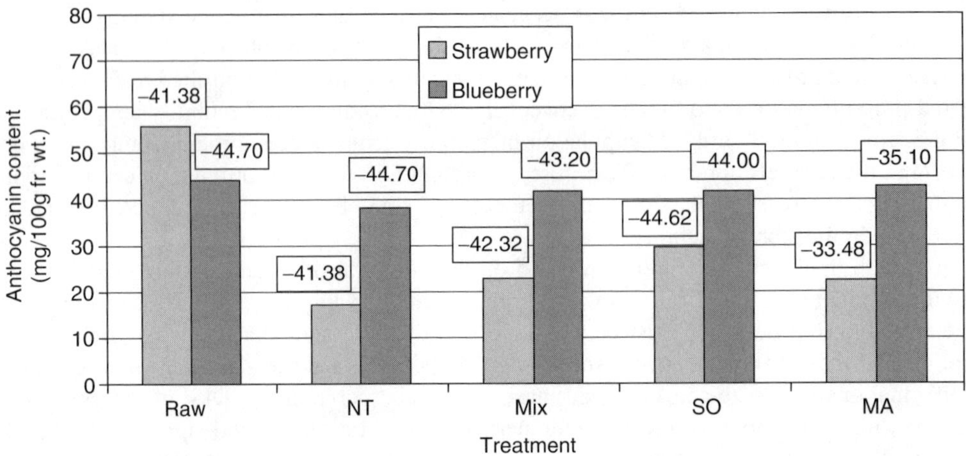

FIGURE 19.7 Anthocyanin content of strawberry and blueberry juices, not added (NT) and added with 20% (wt/wt) glucose–fructose mixture (Mix), sorbitol (SO), and maltose (MA), after 6 months of storage at $-10°C$. Raw, content of the fresh juice. Tg′ values (°C) are reported at the top of each bar.

[55,56]. The higher the sugar enrichment, the higher the protective effect on vitamin C during air drying at 65°C, with maltose being the most effective carbohydrate.

3. Aroma Compounds

Aroma is one of the major determinants of fruit quality, however, it can be easily modified or even greatly reduced during processing. One of the most interesting fruits displaying this property is muskmelon, whose aroma is a very complex mixture and is highly influenced by both the cultivar and harvest time. The retention or loss during DIS and air dehydration applied before freezing was investigated on muskmelon spheres to obtain high-quality innovative frozen products [16,57]. The results ascertained the crucial importance of the cultivar, which had a great influence on the quality characteristics of the end products. Among the pretreatments, air dehydration caused a significant increase of alcohols, whereas these "negative" aroma compounds responsible for the fermented taste were stable in the DIS-treated fruits (Figure 19.8). Furthermore DIS prevented the increase of alcohols during the freezing process. This finding could explain the higher sensory acceptability of the fruit pre-DIS when compared with those pre-air-dehydrated. These considerations are also dealt with in the section on "sensory factors" (Section IIC) to emphasize the link between cultivar and pretreatment (Table 19.4).

The effect of freezing and frozen storage together with that of osmotic process conditions was also analyzed on the volatile fraction of strawberries [58]. Treatments with 65% (wt/wt) sucrose solutions showed the same behavior as that observed by Di Cesare et al. [59] and Escriche et al. [60,61]: there was an increase in some ethylesters and furaneol but a decrease in isobutylester and hexanal, with the changes being slightly lower in pulsed vacuum osmotic treatments (PVOD). Freezing and frozen storage implied losses in all components, although in predehydrated strawberries, the concentration of some esters (and furaneol) remained greater than in the fresh ones due to the formation of these esters promoted during the osmotic step.

Osmotic dehydration also caused changes in the volatile profile of kiwi fruit, depending on the treatment conditions applied [62]. With osmotic dehydration, the concentration of the ester fraction increased, whereas aldehydes and alcohols decreased. After 1 month of frozen storage of kiwi fruit slices, a severe reduction of all compounds (esters, aldehydes, and alcohols) occurred, which resulted in very small differences in the volatile profile of fruit directly frozen and

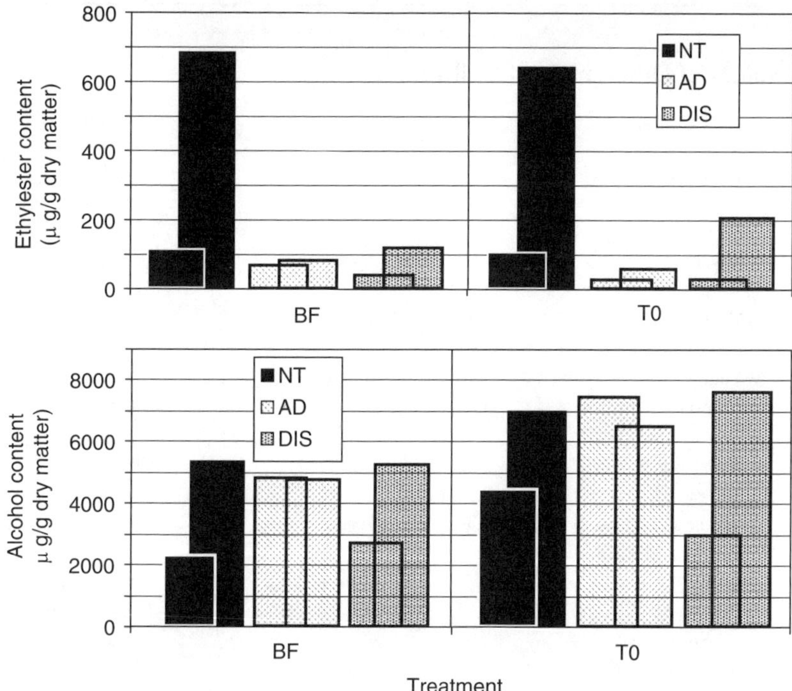

FIGURE 19.8 Ethylester and alcohol content of melon spheres (left bars of each twin set refer to cv Rony and right bars to cv Mirado) not pretreated (NT) and pretreated (air dehydrated to 50% weight reduction, AD; osmodehydrated for 60 min, at 25°C and at atmospheric pressure in 60% (wt/wt) sucrose solution, DIS), before freezing (BF), and immediately after freezing (T0). (Adapted from A Maestrelli, R Lo Scalzo, D Lupi, G Bertolo, D Torreggiani. *Journal of Food Engineering* 49:255–260, 2001. With permission.)

previously dehydrated in different conditions. The sensory impact of these differences needs to be analyzed.

IV. SAFETY ASPECTS OF FROZEN FRUITS

Frozen fruits have an excellent safety record and there have never been any cases of food poisoning, because microorganisms do not grow when the temperature is -10°C or lower. However, freezing does not eliminate microorganisms or microbial toxins present in the food product before freezing, yet in most fruit bacterial growth is inhibited by the pH ≤ 4.5 (Table 19.8). There are few exceptions which have higher pH values, thereby requiring particular attention as poisoning and spoilage microorganisms can proliferate [63].

Besides this important factor along the food chain there are inevitably some levels of contamination that have been identified by the well-known system called Hazard Analysis Critical Control Points (HACCP), which includes the following steps: (1) identification and evaluation of severity of hazards plus risks from raw material, processing steps, any packaging, storage conditions up to final use of the product; (2) determination of critical control points (CCPs); (3) specifications of limits of physical (e.g., temperature), chemical (e.g., pH), or biological (sensorial/microbiological) nature that indicate if a certain CCP is under control; (4) monitoring each CCP through visual inspection test measurements to check whether it is under control; (5) corrective intervention to be put into action when a particular CCP is not under control; (6) verification plus documentation of the system.

TABLE 19.8
Average pH Values of Different Fruit Cultivars

Fruit	Cultivar	pH	Fruit	Cultivar	pH
Apricot	Tonda di Costigliole	3.51	Raspberry	Zeva I	3.38
	S.Castrese	3.43		Williamette	3.38
	Boccuccia	3.41		Lloyd George	3.40
Orange	Navel	4.02	Apple	Delicious	3.90
	Tarocco	3.57		Golden Delicious	3.42
	Moro	3.51	Melon	Casaba	5.89
Banana	Red	4.67		Honey Dew	6.34
	Yellow	5.15	Blueberry	Berkeley	3.27
Cherry	Vittoria	3.84		Bluehaven	3.09
	Durone nero	3.72		Jersey	3.12
	Royal Ann	3.82	Blackberry	Black Satin	3.35
Sour cherry	Mailot	3.50		Thornfree	3.70
	Marasca di Chieti	3.40	Pear	Williams	3.70
	Schateen Morelle	3.40		Bartlett	4.05
Strawberry	Tethis	3.57	Clingstone peach	Andross	3.80
	Chandler	3.55		Baby Gold 6	3.55
	Camarosa	3.38		Vivian	3.77
	Senga Sengana	3.30	Plum	Blue	3.10
Kiwi	Abbot	3.20		Green Gage	3.95
	Hayward	3.20		Damson	3.00

This system not only addresses microbiological safety, which is its primary objective, but it should also address other quality parameters and safety aspects that are not microbial [64]. If HACCP is applied correctly to the entire chain, there are many benefits to both consumer and industry. Emphasis is moved from retrospective quality control to preventive quality assurance.

A. Raw Materials

Just as for chilled foods, fruit safety may be compromised in several ways, apart from microbial contamination. An important consideration is the time and temperature between harvest and receipt of fruits into the factory: short time and low temperature would be the most suitable. Raw fruits may contain pesticides, herbicides, heavy metals, and so on, that are above permissible legal limits, which is considered totally unacceptable by health authorities and increasingly by the consumers. Some large industries and supermarkets already consider this as a CCP, together with the presence of foreign matter. Some of the latter are esthetically unacceptable, but others such as metallic pieces and glass or wood splinters are dangerous. Apart from the metal detector for iron, the only way to detect and eliminate such objects is by visual inspection.

B. Processing before Freezing

Microbial growth is highly dependent on time and temperature, therefore, the time that each lot of fruit spends in each part of the plant, including chill rooms, must be recorded according to specifications and should be included in HACCP.

Normally, fruit is sorted, graded, and washed. An accurate washing must be carried out, particularly for fruit such as strawberry, which grows near or just above the soil. Sometimes, the fruit is peeled and sliced; these operations must be carried out in a separate room from the preliminary sorting, and so on.

As the low pH protects fruit from the growth of pathogens, blanching is not normally necessary and it is only used when enzyme inactivation is required (Section IIIA). If fruit undergoes the blanching process, then it must be considered as a CCP.

C. Freezing Process

During the freezing process, most of the microorganisms in fruit are inactivated, thus the process itself should not present a risk. However, to inhibit prefreezing growth, the time until the initiation of freezing may be critical and may be a CCP. If fruit is kept for too long above chill temperatures before being transferred to the freezing equipment, a pronounced microbial growth or toxin formation may take place.

Another critical point is when the fruit is overloaded in the freezer tunnels, even though there is an appropriate air circulation and air temperature is under $-30°C$. The inner parts of the mass of the fruit take much longer to freeze, resulting in microbial or enzymatic risk of deterioration before the freezing process is completed.

D. Frozen Storage

Microbial growth does not take place during frozen storage; therefore, it is not a food safety problem. For fruit, it is common practice to use bulk packaging immediately after harvesting and freezing. Retail packaging is carried out at intervals and it may be regarded as a CCP to ensure that the fruit does not thaw at any time during repackaging.

E. Thawing

Frozen fruits must be thawed and heat-treated before consumption. They require no special precautions, especially if prepared directly from the frozen state. Thawed fruits should be treated as carefully as unfrozen fruits to prevent cross-contamination from one foodstuff to another. Because, when thawing, there will normally be drip loss, cleaning and disinfecting of food contact surfaces, dishes, knives, and so on should adequately reduce the microbial population. If fruits are not utilized frozen, the thawing process may be as important as the freezing process.

F. Osmotic Pretreatments

When partial removal of water is applied before freezing using osmotic dehydration, the microbial contamination of the solution in contact with the fruit and its sanitation have to be taken into account and be considered as a CCP.

1. Microbial Contamination of Osmotic Solution

Different sources of contamination can affect the microbial stability of the solutions, although the water activity values, ranging around $a_w = 0.90–0.95$, should be able to limit the growth of nonosmotolerant bacteria and yeasts [65,66].

During the processing of fruit with a pH ≤ 4.5 yeasts, molds, and lactic bacteria are the most frequent microorganisms released from the product into the solution. In this situation, pathogenic bacteria are not able to grow. Depending on the environmental process conditions, the microbial load after several osmotic cycles can range from 2×10^2 CFU ml^{-1} [65] to high levels of yeast and fungi only after the 15th cycle [65] and 10^5 CFU ml^{-1} after 8 h of continuous treatment [67].

The use of ozone has been proposed for hygenic management of the osmotic solutions in continuous systems [66]; the spoilage is minimized by destroying bacteria and off-odors in several treatments [68].

2. Sanitation of the Osmotic Solution

If the osmotic process is carried out in a nonsterile environment and the restoring of the concentration takes place by evaporation at low temperature or by nonevaporative processes, the sanitation of the solution comes out as a priority, so as to maintain the microbial load at low level. The necessity of microbial assessment during and at the end of the process allows the optimization of the heat treatment. Microbial targets have to be individualized to assure the safety and to save the solution from over heating.

Plate heat exchangers can be used even if the solution is of high viscosity [67]. Nevertheless, an adequate pumping apparatus must be implemented to assure the desired solution flow. One of the main problems related to heat treatment is nonenzymatic browning, such as caramelization and Maillard reactions, since some amino acids or proteins can be extracted from the fruit. The susceptibility to thermal degradation mainly depends on the presence of reducing sugars used as osmotic agents and on the pH of the solution. For this reason, the use of corn syrup instead of mono- or disaccharides solutions has been proposed [32].

The use of high molecular weight saccharides while preventing nonenzymatic browning, reduces the rate of heat inactivation of microorganisms. Yeasts can grow at relatively low a_w, high sugar concentration, and low pH; they also show increased heat resistance in dry foods [69]. Furthermore, the heating medium, besides increasing heat resistance, also influences the extent of sublethal injuries and repair of thermally stressed cells [70]. The rate of heat inactivation varies in function of the type of sugar and decreases with decreasing a_w in glucose and fructose but not in sucrose solutions. Moreover, at any a_w, the order of susceptibility to inactivation of yeast cells is consistently fructose, glucose, and sucrose. In fructose, a large proportion of the survivors exhibit sublethal injury. Depending on the raw fruit characteristics, the right balance between high and low-molecular-weight saccharides in the osmotic solution should be defined. In this way, both avoidance of enzymatic browning and microbial inactivation can be achieved.

V. CONCLUSIONS

The most important factors on which quality and safety of frozen fruits depend are raw material and pretreatments, whose interaction is essential for extra-high-quality final products.

Through a correct cultivar selection for freezing, it is possible to enhance the nutritional and sensory quality of frozen fruits and to improve suitability of fruits to further processing with a good compatibility with environmental aspects. Yet, this selection is made very difficult by the great variability among cultivars and environments where the cultivar is grown.

Future perspectives of fruit cultivar selection should include the creation of a multidisciplinary research sector that is involved in selection, cultivation, and quality assessment of new or old fruit varieties ideally suitable for freezing.

Within this sector, it would be very interesting to understand whether known and newly available cultivars could be ideal for organic cultivation, and consequently, to evaluate the suitability to freezing of the cultivars grown organically. The gap that exists between fruit production and processing is even wider when considering organic cultivation and hardly any research has been made concerning freezing.

In parallel, and strictly connected with cultivar selection, there is the application of dehydration pretreatments. Just as cultivar selection, combined air-osmotic dehydration pretreatments are aimed at the improvement of the suitability of fruit to freezing, reducing structural damage. Furthermore through the addition of solutes of nutritional and sensory interest, it is possible to improve quality and even increase stability of the pigments, vitamins, and aroma compounds during frozen storage. Such an advantageous step requires small investment as simple equipment is required, and it is a low-cost process which needs reduced energy input over traditional drying processes (40–50% energy cost reduction). It can even be considered as a minimal processing, as thermal stress is minimized it is compatible with environmental aspects of food processing. Pretreatments could also be a tool to develop new products and to prepare fruit ingredients with functional properties tailored for specific food systems, so opening up a tremendous market potential for high-quality frozen fruit products and ingredients, and an increased variety of value-added frozen fruits available to the consumer.

The weakness of the proposed combined system pretreatments–freezing is the difficulty to define a "general" predictive processing model, owing to the great variability of plant materials (species, cultivar, maturity stage, etc.). There is also a lack of adequate responses to problems related to the management of the osmotic solution (reconcentration, reuse, microbial contamination, reutilization and discharge of the spent solution, etc.). Future in-depth studies should be made on how to manage the solution and how to maintain at low level the microbial load of fruits destined for freezing. To avoid sanitary problems, a control system such as HACCP has to be implemented, together with the reduction of the volume of solution involved in processing and the reduction of the amount of discharged spent solutions.

The effective monitoring of product's safety and quality is based on the fact that the microbiology of frozen fruits is not very different from that of unfrozen (chilled) ones, but frozen fruits have the distinct advantage that microorganisms cannot grow at such low temperatures. However, for frozen fruit safety, HACCP system should be used to identify and monitor the critical points. The actual freezing process should be considered as the particular control point and there should be as little delay as possible in initiating the process and ensuring a rapid freezing time. Even the thawing process should be carefully monitored, to ensure that there is a minimum of microbial growth during and after the process. Yet, further studies are necessary in this field also regarding the quality and shelf life of thawed fruits as there is some doubt about the opinion that microbial growth is similar in thawed and unfrozen foods.

REFERENCES

1. OR Fennema, Chilled and frozen biological materials: the lesson of roses. In: *Proceedings of 19th International Congress of Refrigeration*, The Hague, The Netherlands, 1995, Vol. 1, pp. 13–20.
2. AS Sczesniak, BJ Smith. Observation on strawberry texture. A three-pronged approach. *Journal Texture Studies* 1:41–44, 1969.
3. VX Nguyen, PA Phan. Exsudation de peches congelèes, son importance. *Revue general du Froid* 5:479–486, 1973.
4. C Sterling. Effect of low temperature on structure and firmness of apple tissue. *Journal of Food Science* 33 (6):577–580, 1968.
5. G Van Hulle, O Fennema, WD Powrie. A comparison of methods for the microscopic examination of frozen tissue. *Journal of Food Science* 30 (4): 601–603, 1965.
6. E Pauli. *Technologie Culinaire*. 1st ed. Lucerne (Switzerland): Union Helvetia, 1976, pp. 209–210.
7. E Bellini, V Nencetti, L Conte, A Liverani, O Insero. Liste varietali fruttiferi. Pesco (pesche, nettarine, percoche). *Terra e Vita* 52 (18):51–70, 2001.
8. LF D'Antonio, R Fiori, G Baruzzi, W Faedi. La qualità delle fragole in tre sistemi di coltivazione. *Frutticoltura* 12: 69–76, 2000.
9. Anonimous. Fragola — ultimo rapporto. *Weekly Report of CSO (Centro Servizi Ortofrutticoli) Week* 28:1–6, 2001.
10. G Bonous. *Piccoli frutti*. Bologna: Edizioni Agricole, 1996, pp. 434–442.
11. A Maestrelli, JM Chourot. Sélection des cultivars en relation avec la transformation. In: G Albagnac, P Varoquaux, JC Montigaud, Eds., *Technologies de Transformation des Fruits*, Paris (France): Lavoisier, 2002, pp. 41–77.
12. A Dale, EJ Hanson, DE Yarborough, RJ McNicol, EJ Stang, R Brennan, JR Morris, G.B. Hergert. Mechanical harvesting of berry crops. *Horticultural Review* 16:255–382, 1994.

13. P Desmarest, D Derchue. Mechanical harvest of strawberry in France. *Acta Horticulturae* 348: 227–233, 1993.
14. O Temperini, P Crinò, R Campanelli, C Piccioni, F Saccardo. Influenza di cultivar e portainnesti sulla qualità e la produttività del melone. *Informatore Agrario* 44:14–16, 1999.
15. M Arax. Tasteless cherries, big bucks. *Los Angeles Time*, June 13:A25, 2001.
16. A Maestrelli, R Lo Scalzo, D Lupi, G Bertolo, D Torreggiani. Partial removal of water before freezing: cultivar and pre-treatments as quality factors of frozen muskmelon (*cucumis melo*, cv reticulatus naud). *Journal of Food Engineering* 49:255–260, 2001.
17. R Bazzarini, B Frullanti, C Zoni, A Zanotti, G Saccani, A Trifirò, S Gherardi, E Pastorino. Caratteristiche analitiche del succo di arancia rossa non da concentrato. *Industria Conserve* 76:195–200, 2001.
18. A Ghiselli, A D'Amicis, A Giocosa. The antioxidant potential of the Mediterranean diet. *European Journal of Cancer Prevention* 6:9–15, 1997.
19. RL Prior, G Cao, A Martin, E Sofic, J McEwen, C O'Brien, N Lischner, M Ehlenfeldt, W Kalt, G Krewer, CM Mainland. Antioxidant capacity as influenced by total phenolic and anthocyanin content, maturity, and variety of *Vaccinium* species. *Journal of Agricultural and Food Chemistry* 46:2686–2693, 1998.
20. D Torreggiani, T Lucas, AL Raoult-Wack. The pre-treatment of fruits and vegetables. In: CJ Kennedy, Ed., *Managing Frozen Foods*, Cambridge (England): Woodhead Publishing Limited and CRC Press LLC, 2000, pp. 57–80.
21. NAM Eskin, S Grossman, A Pinsky. Biochemistry of lipoxygenase in relation to food quality. *Critical Review in Food Science and Nutrition* 15:1–40, 1977.
22. J Kanner, JB German, JE Kinsella. Initiation of lipid peroxidation in biological systems. *Critical Review in Food Science and Nutrition* 25:317–364, 1987.
23. CE Eriksson, P Olsson, S Svensson. Denaturated hemoproteins as catalysts in lipid oxidation. *Journal of American Oil Chemistry Society* 48:442–447, 1971.
24. CE Eriksson, K Vallentin. Thermal inactivation of peroxidase as a lipid oxidation catalyst. *Journal of American Oil Chemistry Society* 50:264–268, 1973.
25. T Richardson, DB Hyslop. Enzymes. In: OR Fennema, Ed., *Food Chemistry*, New York: Marcel Dekker Inc., 1985, pp. 1451–1473.
26. M Rossi, E Giussani, R Morelli, R Lo Scalzo, RC Nani, D Torreggiani. Effect of fruit blanching on phenolics and radical scavenging activity of highbush blueberry juice. *Food Research International* 36:999–1005, 2003.
27. R Lo Scalzo, T Jannoccari, C Summa, R Morelli, P Rapisarda. Effect of thermal treatments on antioxidant and antiradical activity of blood orange juice. In: *Proceedings of International Society of Citrusculture*, Agadir, Marocco, 2004, Paper 171.
28. L Panzeri. Alcuni aspetti dell'inattivazione termica precongelamento nello spinacio. Attività ossidasiche e antiossidanti in relazione alla stabilità dei lipidi durante la conservazione allo stato congelato. Master Thesis, Università degli Studi, Milano, Italy, 1988.
29. CC Huxsoll. Reducing the refrigeration load by partial concentration of food prior to freezing. *Food Technology* 5:98–102, 1982.
30. AL Raoult-Wack. Recent advances in the osmotic dehydration of foods. *Trends in Food Science and Technology* 5 (8):255–260, 1994.
31. D Torreggiani. Technological aspects of osmotic dehydration in foods. In: GV Barbosa-Cànovas, J Welti-Chanes, Eds., *Food Preservation by Moisture Control: Fundamentals and Applications*, Lancaster, PA (USA): Technomic Publishing Co. Inc., 1995, pp. 281–304.
32. HN Lazarides, P Fito, A Chiralt, V Gekas, A Lenart. Advances in osmotic dehydration. In: ARF Oliveira, JC Oliveira, Eds., *Processing Foods, Quality Optimisation and Process Assessment*, London: CRC Press, 1999, pp 175–199.
33. D Torreggiani, G Bertolo. Osmotic pre-treatments in fruit processing: chemical, physical and structural effects. *Journal of Food Engineering* 49:247–253, 2001.
34. A Maestrelli, G Giallonardo, E Forni, D Torreggiani. Dehydrofreezing of sliced strawberries: a combined technique for improving texture. In: R Jowitt, Ed., *Engineering and Food, ICEF 7*. Sheffield (UK): Sheffield Academic Press, 1997, Part 2, pp. F37–F40.

35. A Sormani, D Maffi, G Bertolo, D Torreggiani. Textural and structural changes of dehydrofreeze-thawed strawberry slices: effects of different dehydration pre-treatments. *Food Science and Technology International* 5 (6):479–485, 1999.
36. A Chiralt, N Martìnez-Navarrete, J Martìnez-Monzò, P Talens, G Morata, A Ayala, P Fito. Changes in mechanical properties throughout osmotic processes: cryoprotectant effect. *Journal of Food Engineering* 49:129–135, 2001.
37. JFV Vincent. Texture of plants. In: HF Linskens, JF Jackson, Eds., *Vegetables and Vegetable Products*, Berlin, Germany: Springer, 1994, pp. 57–72.
38. A Brambilla, D Maffi, G Bertolo, D Torreggiani. Effect of osmotic dehydration time on strawberry tissue structure. *Book of Abstracts of Eight International Congress on Engineering and Food ICEF 8*, Puebla, Mexico, 2000, pp. 211.
39. E Forni, A Sormani, D Torreggiani. Influenza del pre-trattamento osmotico e della varietà sulla qualità di fragole congelate. In: S Porretta, Ed., *Ricerche e innovazioni nell'industria alimentare*, 4th CISETA (Congresso Italiano di Scienza e Tecnologia degli Alimenti). Pinerolo, Italy: Chiriotti Editori, 2000, Vol. 4, pp. 750–762.
40. R Lo Scalzo, H Brimar, A Avitabile Leva, D Torreggiani, A Maestrelli. Texture and pectin composition changes of dehydrofreeze-thawed strawberry slices: influence of different dehydration pre-treatments. Proceedings of the 21st International Congress of Refrigeration, Washington DC, 2003, Paper ICR0553.
41. NB Tregunno, HD Goff. Osmodehydrofreezing of apples: structural and textural effects. *Food Research International* 29:471–479, 1996.
42. J Martìnez-Monzò, N Martìnez-Navarrete, A Chiralt, P Fito. Osmotic dehydration of apple as affected by vacuum impregnation with HM pectin. In: CB Akritis, D Marinos-Kouris, GD Saravacos, Eds., *Drying '98*. Thessaloniki, Greece: Ziti Editions, 1998, Vol. A, pp. 836–843.
43. J Martìnez-Monzò, N Martìnez-Navarrete, A Chiralt, P Fito. Mechanical and structural changes in apple (var. Granny Smith) due to vacuum impregnation with cryoprotectants. *Journal of Food Science* 63:499–503, 1998.
44. SL Vidales, SM Alzamora, G Bertolo, D Torreggiani. Thermal and structural changes of frozen strawberries using vacuum and atmospheric impregnation with cryoprotectans. In: J Welti-Chanes, GV Barbosa-Cànovas, JM Aguilera, Eds., *Proceedings of International Congress on Engineering and Food, ICEF 8*, Lancaster, PA, U.S.A.: Technomic Publisher Co. Inc., 2001, Vol. 1, pp. 803–807.
45. D Torreggiani, B Rondo Brovetto, A Maestrelli, G Bertolo. High quality strawberry ingredients by partial dehydration before freezing. In: *Proceedings of the 20th International Congress of Refrigeration*, Sidney, 1999, Vol. IV, pp. 2100–2105.
46. D Torreggiani, G Giallonardo, B Rondo Brovetto, G Bertolo. Impiego della deidrocongelazione per il miglioramento qualitativo di semilavorati di albicocca in pezzi per yogurt. In: S Porretta, Ed., *Ricerche e innovazioni nell'industria alimentare*, 4th CISETA (Congresso Italiano di Scienza e Tecnologia degli Alimenti). Pinerolo, Italy: Chiriotti Editori, 2000, Vol. 4, pp. 700–708.
47. L Slade, H Levine. Beyond water activity: recent advances based on an alternative approach to the assessment of food quality and safety. *Critical Review of Food Science and Nutrition* 30:115–360, 1991.
48. M Karel, MP Buera, Y Roos. Effects of glass transitions on processing and storage. In: JMV Blanshard, PJ Lillford, Eds., *Glassy State in Foods*, Loughborough, U.K.: Nottingham University Press, 1993, pp. 13–85.
49. DS Reid. Factors which influence the freezing process: an examination of new insights. In: *Proceedings of the 20th International Congress of Refrigeration*, Sidney, 1999, Vol. IV, pp. 735–744.
50. D Torreggiani, E Forni, L Pelliccioni. Modificazione della temperatura di transizione vetrosa mediante disidratazione osmotica e stabilità al congelamento del colore di kiwi. In: S Porretta, Ed., *Ricerche e innovazioni nell'industria alimentare*, 1st CISETA (Congresso Italiano di Scienza e Tecnologia degli Alimenti). Pinerolo, Italy: Chiriotti Editori, 1994, Vol. 1, pp. 621–630.
51. D Torreggiani, E Forni, I Guercilena, A Maestrelli, G Bertolo, GP Archer, CJ Kennedy, S Bone, G Blond, E Contreras-Lopez, D Champion. Modification of glass transition temperature through carbohydrates additions: effect upon colour and anthocyanin pigment stability in frozen strawberry juices. *Food Research International* 32:441–446, 1999.

52. A Rizzolo, CR Nani, D Viscardi, G Bertolo, D Torreggiani. Modification of glass transition temperature through carbohydrates addition and anthocyanin and soluble phenol stability of frozen blueberry juices. *Journal of Food Engineering* 56:229–231, 2003.
53. D Torreggiani, E Forni, A Maestrelli, G Bertolo, A Genna. Modification of glass transition temperature by osmotic dehydration and color stability of strawberry during frozen storage. In: *Proceedings of the 19th International Congress of Refrigeration*, The Hague, The Netherlands, 1995, Vol. 1, pp. 315–321.
54. E Forni, A Genna, D Torreggiani. Modificazione della temperatura di transizione vetrosa mediante disidratazione osmotica e stabilità al congelamento del colore delle fragile. In: S Porretta, Ed., *Ricerche e innovazioni nell'industria alimentare*, 3rd CISETA (Congresso Italiano di Scienza e Tecnologia degli Alimenti). Pinerolo, Italy: Chiriotti Editori, 1998, Vol. 3, pp. 123–130.
55. E Forni, A Sormani, S Scalise, D Torreggiani. The influence of sugar composition on the colour stability of osmodehydrofrozen intermediate moisture apricots. *Food Research International* 30:87–94, 1997.
56. G Camacho, G Bertolo, D Torreggiani. Stabilità del colore di albicocche disidratate: influenza del pre-trattamento osmotico in sciroppi zuccherini diversi. In: S Porretta, Ed., *Ricerche e innovazioni nell'industria alimentare*, 3rd CISETA (Congresso Italiano di Scienza e Tecnologia degli Alimenti). Pinerolo, Italy: Chiriotti Editori, 1998, Vol. 3, pp. 553–561.
57. R Lo Scalzo, C Papadimitriu, G Bertolo, A Maestrelli, D Torreggiani. Influence of cultivar and osmotic dehydration time on aroma profiles of muskmelon (*cucumis melo*, cv reticulatus naud) spheres. *Journal of Food Engineering* 49:261–264, 2001.
58. P Talens, L Escriche, N Martìnez-Navarrete, A Chiralt. Study on the influence of osmotic dehydration and freezing on the volatile profile of strawberries. *Journal of Food Science* 67 (5):1648–1653, 2002.
59. LF Di Cesare, D Torreggiani, G Bertolo. Preliminary study of volatile composition of strawberry slices air dried with or without an osmotic pre-treatment. *Proceedings of the 5th Plenary Meeting of Concerted Action FAIR-CT96-1118, Improvement of Overall Food Quality by Application of Osmotic Treatments in Conventional and New Process*, Valencia (Spain), 1999, pp. 39–44.
60. L Escriche, A Chiralt, J Moreno, JA Serra. Influence of blanching–osmotic dehydration treatments on volatile fraction of strawberry. *Journal of Food Science* 65 (7):1107–1111, 2000.
61. L Escriche, A Chiralt, J Moreno, JA Serra. Influence of osmotic treatment on the volatile profile of strawberry (*Fragaria* × *ananassa* var. Chandler). In: J Welti-Chanes, GV Barbosa-Cànovas, JM Aguilera, Eds., *Proceedings of International Congress on Engineering and Food, ICEF 8*. Lancaster, PA, USA: Technomic Publisher Co. Inc., 2001, Vol. 1, pp. 151–155.
62. P Talens, L Escriche, N Martìnez-Navarrete, A Chiralt. Influence of osmotic dehydration and freezing on the volatile profile of kiwi fruit. *Food Research International* 36 (6):635–642, 2003.
63. L Leistner. Update on hurdle technology. In: J Welti-Chanes, GV Barbosa-Cànovas, JM Aguilera, Eds., *Engineering and Food for the 21st Century*. Boca Raton, FL, USA: CRC Press, 2002, pp. 615–629.
64. L Bogh-Sorensen. Maintaining safety in the cold chain, in *Managing frozen foods*, CJ Kennedy, Ed., Cambridge (England): Woodhead Publishing Limited and CRC Press LLC, 2000, pp. 5–26.
65. A Valdez-Fragoso, J Welti-Chanes, F Giroux. Properties of sucrose solution reused in osmotic dehydration of apples. *Drying Technology* 16 (7):1429–1445, 1998.
66. M Dalla Rosa, F Giroux. Osmotic treatments (OT) and problems related to the solution management. *Journal of Food Engineering* 49:223–236, 2001.
67. M Dalla Rosa, F Bressa, D Mastrocola, P Pittia. Use of osmotic treatments to improve the quality of high-moisture minimally processed fruits. In: A Lenart, PP Lewicki, Eds., *Osmotic Dehydration of Fruits and Vegetables*, Warsaw, Poland: Warsaw Agriculture University Press, 1995, pp. 69–87.
68. BW Sheldon, AL Brown. Efficacy of ozone as a disinfectant for poultry carcasses and chill water. *Journal of Food Science* 51:305–309, 1986.
69. L Leistner, LGM Gorris. Food preservation by hurdle technology. *Trends in Food Science and Technology* 6:41–46, 1995.
70. D Torreggiani, RT Toledo. Influence of sugars on heat inactivation, injury and repair of *Saccharomyces cerevisiae*. *Journal of Food Science* 51:211–215, 1986.

20 Quality and Safety of Frozen Dairy Products

H. Douglas Goff
University of Guelph, Guelph, Ontario, Canada

CONTENTS

I. INTRODUCTION

Frozen dairy products can be divided into two categories: (i) products frozen for increasing their shelf-life and thawed before consumption or further processing and (ii) products in which the freezing process is responsible for the development of the desired structure and texture and which are consumed in the frozen state. Unlike most other frozen food commodities, the majority of frozen dairy products fall into the latter category, namely, ice cream and related frozen, aerated dairy desserts such as ice milk, sherbet, and frozen yogurt. The amount of dairy products that fall into the first category, those that need thawing before further processing or consumption, is very small relative to the frozen dairy dessert industry. Therefore, a major portion of this chapter will overview the technology of ice cream and related products, with a focus on quality and safety. Following a review of ice cream freezing, the freezing of other dairy products will be covered.

The major components of normal bovine milk include milkfat (~4%) in the form of an emulsion and the milk solids-not-fat (SNF) fraction, consisting of casein proteins (~2.6%) suspended in the form of micelles, whey or serum proteins (~0.45%), dissolved lactose (~4.8%), and mineral salts (~0.7%) [1]. The first section of this chapter will cover the major effects that freezing has on the two main categories of milk components, the milkfat emulsion, and the milk SNF. Most dairy products are based on altering the composition by physically separating one of the components, for example, concentrating the fat phase to produce creams or butter or concentrating the protein phase to produce cheese or by removal of water, for example, the production of condensed or dried milk products. The amount of water in the various products varies considerably. Ice cream and related products are usually manufactured by combining several dairy ingredients, together with sweetening and flavoring agents. In addition to ice cream, some fluid products, cheeses, and butter can also be frozen, although few of these products are distributed directly to the consumer in the frozen state.

II. INFLUENCE OF FREEZING ON MILK COMPONENTS

A. Fat Fraction

The fat fraction in milk and most other dairy products exists in the form of an emulsion with droplet sizes ranging from 0.5 to 5.0 μm [1]. The fat globules are coated with a protein and phospholipid membrane after secretion from the mammary gland. Homogenization greatly increases the number and surface area of the fat globules present, and milk proteins quickly adsorb homogenized fat globules to reduce their interfacial tension. During freezing, the physical defect of greatest concern related to the fat phase is the loss of the emulsified state and the separation of the fat phase. Coalescence of the fat during static freezing is caused by mechanical damage to the fat globule membranes by the expanding ice crystals, and the degree is closely related to the extent of freezing [2]. Homogenization before freezing greatly overcomes the de-emulsification of the fat globules in milk, concentrated milk, or low-fat cream because of their smaller size and enhanced surface layers, making them less prone to rupture, although some fat coalescence can still be evident, dependent largely on the rates of freezing and thawing. The addition of sugar also helps to stabilize frozen milkfat emulsions, as it creates an unfrozen phase that limits the close approach of emulsions droplets.

B. Milk SNF Fraction

Freezing and frozen storage can have a large effect on the proteins of milk, causing the casein micelles to lose their stability and precipitate on thawing [3]. This manifests itself either as thickened product or as flocs of casein evident either on the sides of thawed glass or plastic containers or as a precipitate at the bottom. In fact, commercial separation of casein from concentrated milk serum may be accomplished by freezing, storage at −10°C, and thawing to produce a product that has been referred to as cryo-casein [4]. The flocculation of casein from frozen milk is initially reversible with heat and agitation, but becomes irreversible with continued storage. Even minor amounts of casein flocculation can lead to the perception of a chalky texture upon consumption. Factors contributing to its instability include a high degree of concentration of the milk, preheating or pasteurization temperatures in excess of 77°C, storage temperatures above −23°C and especially higher than −18°C, cooling and holding the product under refrigeration between concentrating and freezing, and lengthening storage periods [5]. These factors are all related to the state of the lactose and serum proteins in the product [6,7]. Slow freezing has also been reported to result in greater protein stability than fast freezing, possibly related to the effect of rapid freezing in promoting lactose nucleation [5].

The stability of casein in frozen milk depends heavily on the state of the lactose [6]. Owing to the freeze-concentration process, very little liquid water remains as a solvent for the lactose and mineral salts. A tenfold concentration of the salts may occur at $-7.5°C$, and up to 30-fold increase may occur at lower temperatures, especially in the presence of lactose crystallization. Considerable precipitation of soluble $CaHPO_4$ and $Ca(H_2PO_4)_2$ as colloidal $Ca_3(PO_4)_2$ leads to the release of H^+, resulting in a decrease in the pH from 6.7 to ~6.0 and an increase in Ca^{2+}, eventually leading to solubilization and precipitation of the casein micelle [4]. When dissolved, the lactose maintains a lowered freezing point, limits the concentration of the dissolved salts in the unfrozen phase, and contributes to high viscosity in the unfrozen phase [5,8]. Studies have shown that casein remains relatively stable under these conditions [5]. However, the crystallization of the lactose leads to an increased freezing point of the unfrozen phase, which in turn leads to a further crystallization of water at constant temperatures. Consequently, further concentration of the salts, increased acidity, a change in the balance of colloidal calcium phosphate, and resultant flocculation of the casein occurs. As little as 40% lactose crystallization may be sufficient for casein instability [8]. Dialysis of milk before freezing has led to enhanced protein stability, demonstrating the effects of milk salts on this process. *Ultrafiltration* of milk has also been demonstrated to increase the storage stability by three times, up to an optimum concentration level beyond which removal of soluble phosphate accounted for decreased stability [5]. Lactose does not crystallize readily from freeze-concentrated solutions and remains largely in the supersaturated state, provided no nucleation has occurred before freezing [8]. At high solution viscosities resulting from high degrees of supersaturation and low temperatures, an amorphous solid (*glass*) may easily form in the unfrozen phase, rendering greater stability to the product and demonstrating the importance of low storage temperatures [8].

C. Freezing Point Depression

Freezing point is a colligative property that is determined by the molarity of solutes rather than by the percentage by weight or volume. The ideal molal depression constant for water as defined by Raoult's law is 1.86 for dilute solutions (i.e., each mole of solute will decrease the freezing point of water by 1.86°C) [9]. Freezing point can therefore be used to estimate the molecular weight of pure solutes or the average molecular weight of mixed solutes. The freezing point of milk is usually very constant as milk is in osmotic equilibrium with blood when it is synthesized [10]. Lactose accounts for about 55% of freezing point depression of whole or skim milk, chloride accounts for about 25%, and the remaining 20% is due to other soluble components including calcium, potassium, magnesium, lactates, phosphates, and citrates [10]. In the dairy industry, the invariability of the freezing point of raw milk is an important quality control parameter as it is used to determine unintentionally or intentionally the added water in milk [11].

The freezing point of milk is usually in the range of -0.512 to $-0.550°C$ with an average of about $-0.522°C$ [9]. Freezing points of other dairy products depend on their water content and dissolved lactose and other solids in the water phase (Table 20.1). Fats, proteins, and colloids generally have a negligible effect on the freezing point of water or solutions in which they are dispersed [9,10]. Thus, the freezing points of cream and skim milk are identical with those of the milk from which they were separated. Soured or fermented milk is unsuitable for added water testing because the freezing point is lowered by lactic acid and increased concentrations of soluble minerals. Several reports suggest that heat treatment of milk, including ultra-high temperature and retort sterilization, produces little permanent effect on freezing points [9]. If the freezing point of unwatered milk is known, the relationship between added water and freezing point depression is given subsequently. If the actual freezing point of the unwatered milk is not known, a reference value can be used:

$$W = \frac{(C - D)(100 - S)}{C} \tag{20.1}$$

TABLE 20.1
The Initial Freezing Point of Various Dairy Products

Product	Freezing Point (°C)
Condensed milks	
Skim milk (64% moisture)	− 3.1
Sweetened condensed whole milk (20% SNF, 42% sugar, 30% moisture)	− 15
Cheese	
Cottage (79% moisture)	− 1.2
Cheddar (34% moisture)	− 13
Processed cheddar (39% moisture)	− 7
Brick (42% moisture)	− 9
Swiss (40% moisture)	− 10
Butter	
Unsalted (20% moisture)	0
Salted (18% moisture)	− 9 to −20

Source: From PG Keeney, M Kroger. In: BH Webb, AH Johnson, JA Alford, Eds., *Fundamentals of Dairy Chemistry*, 2nd ed. Westport, CT: AVI Publishing Co. Inc., 1974, p. 873. With permission.

where W is the percent (w/w) extraneous water in the suspect milk, C the actual or reference freezing point of genuine milk, D the freezing point of suspect milk, and S the percent (w/w) total solids in the suspect milk.

III. ICE CREAM AND FROZEN DAIRY DESSERTS

The range of compositional variables found in most ice cream mix formulations is shown in Table 20.2. The composition of frozen dairy dessert products in most countries is standardized

TABLE 20.2
A Typical Compositional Range for the Components Used in Various Ice Cream Mix Formulations

Component	Range (%)
Milkfat	
Premium ice cream	14–16
Ice cream	10–14
Light ice cream	6–10
Lowfat ice cream	2–6
Milk SNF	9–12
Sucrose	10–16
Corn syrup solids	0–6
Stabilizers/emuslifiers	0–0.5
Total solids (100 − water content)	
Premium ice cream	40–45
Ice cream	36–40
Light ice cream	32–36
Lowfat ice cream	28–32

and regulated. These products are all consumed in the frozen state and rely on a concomitant freezing and whipping process to establish their structure and texture. The manufacturing process for most of these products is similar and involves the preparation of a liquid mix; whipping and freezing this mix dynamically under high shear to a soft, semifrozen slurry; incorporation of flavoring ingredients to this partially frozen mix; packaging the product, and further freezing (hardening) of the product under static, quiescent conditions. Swept (scraped)-surface freezers are used for the first freezing step, whereas forced convection freezers, such as air-blast tunnels or rooms, or plate-type conduction freezers are used for the second freezing step. Although this chapter overviews the fundamentals of processing and ingredients, the reader is referred to several recent comprehensive reviews for further details: References [13–18] and references therein.

In frozen aerated dairy desserts, ice, air, and fat each occupy distinct but interrelated phases (Figure 20.1) and together establish the structure and resulting texture. The ice phase is of critical importance to the quality and shelf life of frozen products. The objective of ice cream manufacturers is to produce ice crystals that are below, or at least not significantly above, the threshold of sensory detection at the time of consumption, i.e., 40–50 μm [14]. Consequently, the freezing steps of the manufacturing process and the temperature profile throughout the distribution system are the critical factors in meeting this objective.

A. Ingredients

1. Fat

The fat content is an indicator of the perceived quality and value of the ice cream. The fat component of the mix increases the richness of flavor of ice cream, produces a characteristic smooth texture by lubricating the palate, helps to give body, and aids in producing desirable melting properties [14]. Milkfat, from cream, sweet (unsalted) butter, frozen cream, condensed milk blends, or

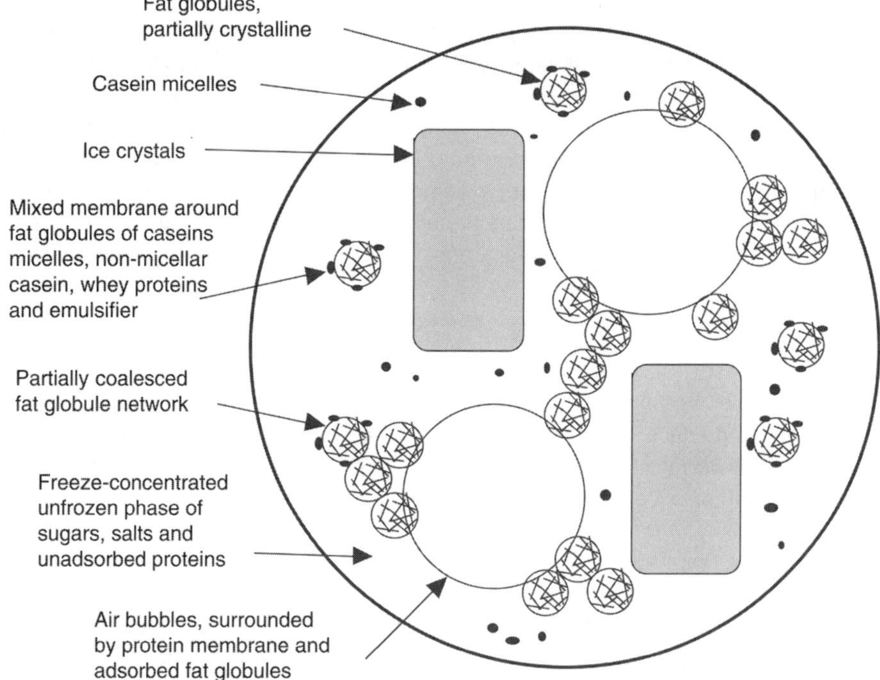

FIGURE 20.1 Schematic diagram of the structure of ice cream.

whey cream, is the principal or only fat source for dairy ice cream formulations. Vegetable fats can also be used as fat sources in nondairy ice cream. Blends of oils are often used in ice cream manufacture, selected to take into account physical characteristics, flavor, availability, and cost. Palm kernel oil, coconut oil, palm oil, sunflower oil, peanut oil, fractions thereof, and their hydrogenated counterparts are all used to some extent [16]. Regardless of the source, it is important that a considerable quantity of the fat be crystallized at the time of whipping or freezing. During freezing of ice cream, the fat emulsion that exists in the mix will partially coalesce or destabilize as a result of the presence of crystals of fat, emulsifier action (Section III.A.5), air incorporation, ice crystallization, and high shear in the freezer (see also Section III.C) [14,19,20].

2. Milk SNF

The SNF or serum solids contain the lactose, caseins, whey proteins, minerals (ash), vitamins, acids, enzymes, and gases of the milk or milk products from which they were derived. The proteins are essential for their functional contributions of emulsification, aeration and water holding capacity or viscosity enhancement. An excess of lactose, however, may lead to problems due to excessive freezing point depression or lactose crystallization, leading to a textural defect. Thus, SNF sources should be chosen to optimize protein functionality but limit lactose content. Traditionally, the best sources of milk SNF for high-quality products have been fresh concentrated skimmed milk or spray dried low-heat skim milk powder. Others include those containing whole milk protein (e.g., condensed or sweetened condensed whole milk, milk protein concentrates, and dry or condensed buttermilk), those containing casein (e.g., sodium caseinate), or those containing whey proteins (e.g., dried or condensed whey, whey protein concentrate, and whey protein isolate) [14].

3. Sweeteners

Sweeteners improve the texture and palatability of ice cream and enhance flavors. Their ability to lower the freezing point of a solution imparts a measure of control over the temperature–hardness relationship (Section III.C) [14]. The most common sweetening agent is sucrose, alone or in combination with other sugars. In determining the appropriate blend of sweeteners to use, the sweetness, freezing point depression, and contribution to total solids all have to be considered. Sucrose and lactose are most commonly present in ice cream in the supersaturated or glassy state, with few crystals being present [15]. In many ice cream formulations, sweeteners derived from corn syrup are substituted for all or a portion of the sucrose. The use of corn starch hydrolysis products (corn syrups or glucose solids) in ice cream is generally perceived to provide greater smoothness by contributing to a firmer and more chewy body, to provide better meltdown characteristics, to reduce heat shock potential, which improves the shelf life of the finished product, and to provide an economical source of solids.

4. Stabilizers

Ice cream stabilizers are a group of hydrocolloid ingredients (usually polysaccharides) used in ice cream formulations to produce smoothness in body and texture and retard or reduce the growth of ice and lactose crystals during storage, especially during periods of temperature fluctuation, known as heat shock. Thus, by physical means, they effectively increase the shelf life of ice cream. They also increase the viscosity of the mix, aid in suspension of flavoring particles in the semifrozen ice cream, produce a stable foam with easy cut-off and stiffness at the barrel freezer for packaging, slow down moisture migration from the product to the package or the air in frozen product, and help to prevent shrinkage of the product volume during storage [14]. Stabilizers commonly used include: locust bean (carob) gum, guar gum, carboxymethyl cellulose, sodium alginate, xanthan, and gelatin. Each stabilizer has its own characteristics and often two or more of these stabilizers are used in combination to lend synergistic properties to each other and improve their overall effectiveness.

Carrageenan is a secondary hydrocolloid used to prevent serum separation in the mix, which is usually promoted by one of the other stabilizers [14]. Hence, it is found in most stabilizer blends.

5. Emulsifiers

Small-molecule surfactants (*emulsifiers*) are usually integrated with the stabilizers in proprietary blends, but their function and action is very different from that of the stabilizers. They are used to improve the aeration properties of the mix; produce an ice cream at extrusion with good shape retention properties (referred to as "dryness") for facilitating molding, fancy extrusion, and novelty product manufacture; produce a smoother body and texture in the finished product; and to produce a product with good shape retention properties during melting [14]. Their mechanism of action can be summarized as follows: they lower the fat or water interfacial tension in the mix, resulting in protein displacement from the fat globule surface, which in turn reduces the stability of the fat globule to partial coalescence, which occurs during the whipping and freezing process, leading to the formation of an aggregated fat structure in the frozen product which contributes greatly to texture and melt-down properties [14,19,20]. Their interaction with proteins and role in structure formation will be described in Section III.C. Emulsifiers used in ice cream manufacture are of two main types: mono- and diglycerides and sorbitan esters. Of the latter, polysorbate 80 is a very strong promoter of fat destabilization in ice cream [14] and is used in many commercial stabilizer and emulsifier blends.

B. Processing

1. Mix Manufacture

Ice cream processing operations can be divided into two distinct stages, mix manufacture and freezing operations (Figure 20.2). The manufacture of ice cream mix involves the following unit

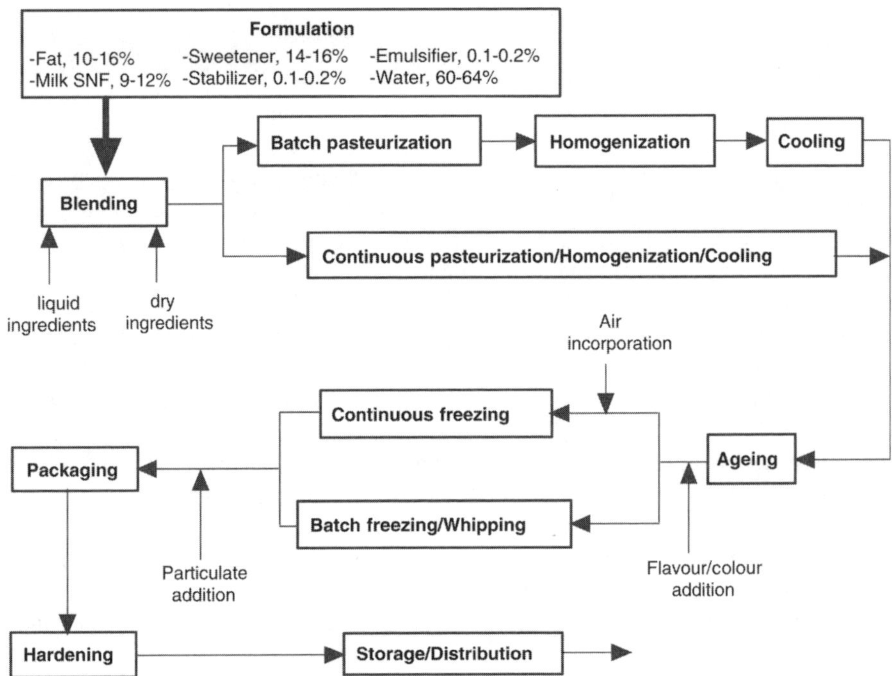

FIGURE 20.2 Processing flow chart for the manufacture of ice cream.

operations: combination and blending of ingredients, batch or continuous pasteurization, homogenization, and aging [13]. Ingredients are chosen to supply the desired components, for example, cream or butter to supply fat, on the basis of availability, ease of handling, and desired quality and cost. An algebraic solution of the formulation is required, as many of the ingredients supply more than one component [14,15,18].

Pasteurization is designed to kill pathogenic bacteria. In addition, it serves a useful role in reducing the total bacterial load and in solubilization of some of the components (proteins and stabilizers). Both batch (>~69°C for ~30 min) and continuous (high temperature–short time, >~80°C for ~15–25 sec) systems are in common use. Provided that mix has been properly pasteurized and no postpasteurization contamination has occurred, ice cream should be safe for human consumption (Section III.E).

Homogenization is responsible for the formation of the fat emulsion by forcing the hot mix through a small orifice under pressures of 15.5 to 18.9 MPa (2000 to 3000 psi gage), depending on the composition of the mix. A large increase in the surface area of the fat globules is responsible partly for the formation of the fat globule membrane, comprised of adsorbed materials that lower the interfacial free energy of the fat globules [19,20]. With single stage homogenizers, fat globules tend to cluster as bare fat surfaces come together or adsorbed molecules are shared. Therefore, a second homogenizing valve is frequently placed immediately after the first with applied back pressures of 3.4 MPa (500 psi gage), allowing more time for surface adsorption to occur.

An aging time of 4 h or greater is recommended following mix processing prior to freezing to produce a smoother texture and better quality product. The temperature of the mix should be maintained as low as possible without freezing (≤4°C). Aging allows for hydration of milk proteins and stabilizers (some increase in viscosity occurs during the aging period), crystallization of the fat globules, and membrane rearrangement. The appropriate ratio of solid–liquid fat must be attained at this stage, a function of temperature and the triglyceride composition of the fat used, as a partially crystalline emulsion is needed for partial coalescence during the whipping and freezing steps [18]. Emulsifiers generally displace milk proteins from the fat surface during the aging period [19,20]; this is discussed in Section III.C.

2. Ice Cream Freezing

Ice cream freezing also consists of two distinct stages: passing mix through a swept (scraped)-surface heat exchanger under high shear conditions to promote extensive ice crystallization and air incorporation and freezing the packaged ice cream under conditions that promote rapid freezing and small ice crystal sizes [14,15]. The concomitant freezing and whipping process is one of the most important unit operations for the development of quality, palatability, and yield of the finished product. Flavoring and coloring can be added as desired to the mix before passing through the barrel freezer, and particulate flavoring ingredients, such as nuts, fruits, candy pieces, or ripple sauces, can be added to the semifrozen product at the exit from the barrel freezer before packaging and hardening.

Continuous freezers dominate the ice cream industry [14,17]. In this type of process, mix is drawn from the flavoring tank into a swept surface heat exchanger, which is jacketed with a liquid, boiling refrigerant (e.g., ammonia). Following the incorporation of air into the mix, the water in the mix is partially frozen as the mix and air combination passes through the barrel of the scraped-surface heat exchanger. Rotating knife blades scrape the ice layer off the surface and dashers keep the product agitated, which incorporates the air phase as tiny bubbles and maintains discrete ice crystals as they grow in the bulk liquid. Residence time for mix through the annulus of the freezer varies from 0.4 to 2 min; freezing rates can vary from 5 to 27°C min^{-1}, and draw temperatures of −6°C can easily be achieved [14,15,17].

Incorporation of air into ice cream, termed the *overrun*, is a necessity to produce desirable body and texture. Overrun is calculated as follows [14]:

$$\%\text{Overrun} = \frac{\text{Volume of ice cream produced} - \text{Volume of mix used}}{\text{Volume of mix used}} \times 100\% \qquad (20.2)$$

As a guide, maximum overrun should be 2.5–3 times the total solids content to avoid possible defects in the finished ice cream [14]. There are three types of air incorporation systems used in continuous freezers. In older systems, the pump configuration resulted in a vacuum either at the pump itself or on the mix line entering the pump. Air was then incorporated through a spring-loaded, controllable needle valve. Newer types of freezers utilize compressed air, which is injected into the mix. This type of air handling system allows for filtration of the air before entering into the mix and for better control of overrun. Additionally, preaeration systems in which the mix is whipped independently before entering into the barrel freezer can produce much smaller air bubble sizes, which may improve body and texture of the ice cream [14,17].

It is extremely important to maintain freezers in excellent condition to obtain rapid freezing by keeping the heat exchange surfaces free of oil or debris on the refrigerant side and by keeping the blades sharp and straight on the mix side. Moving the ice cream through the ingredient feeders, filling pipelines, and packaging machines should be done rapidly, so that the tiny preformed crystals do not have a chance to melt. Measures such as precooling ingredients before addition to the fruit feeder, insulating the pipelines to the filling machines, and keeping pipelines short are essential to small, uniform ice crystal size distributions.

Batch freezing processes differ slightly from the continuous systems as described. The barrel of a batch swept surface heat exchanger is jacketed with refrigerant and contains a set of dashers and scraper blades inside the barrel. It is filled to about one half volume with the liquid mix. Barrel volumes usually range from 2 to 12 l. The freezing unit and agitators are then activated and the product remains in the barrel for sufficient time to achieve the desired degree of overrun and stiffness. Whipping increases with time and cannot exceed the amount that will fill the barrel with product (i.e., 100% overrun when starting half full). Batch freezers are used in smaller operations where it is desirable to run individual flavored mixes on a small scale or to retain an element of the "homemade"-style manufacturing process. They are also operated in a semicontinuous mode for the production of soft-serve type desserts. A hopper containing the mix feeds the barrel as product is removed.

Ice cream following dynamic freezing, ingredient addition, and packaging is immediately transferred to a hardening chamber (−30°C or colder, either forced convection or plate-type conduction freezers) where the majority of the remaining water freezes [15,21]. Rapid hardening is necessary for product quality, as it helps to maintain the small ice crystal size distribution that was created in the scraped-surface freezer. When hardening is slow, there is a great opportunity for small ice nuclei formed to recrystallize, resulting in larger ice crystals and coarser product. Many factors need to be considered during the hardening process. The main factors affecting heat transfer are the temperature difference between the product and the freezing medium, the area of product being exposed to the freezing medium, and the heat transfer coefficient for the particular operation [15]. The temperature of the ice cream when placed in the hardening room should be as cold as possible [15]. Draw temperatures from the barrel freezer are limited by the necessity of flowability for packaging the product. The addition of ingredients and the packaging operation should not increase the temperature of the ice cream as it is drawn from the barrel freezer any more than necessary. The temperature of the hardening chamber is also critical for rapid freezing and smooth product. The surface area of the ice cream also needs to be considered and is especially important when packaging in large packages or in shrink-wrapping product bundles. Palletizing or stacking of product should not interfere with rapid air circulation and fast freezing. Convective heat transfer coefficients are greatly increased through the use of forced convection systems.

The evaporator should be free of frost from the outside and oil from the inside of the coils as these act to reduce the heat transfer coefficients. Following rapid hardening, ice cream storage should occur at low, constant temperatures, usually at −25°C [14,15,17].

C. Structural Changes Occurring during Freezing

The concomitant whipping and freezing process occurring in the scraped-surface freezer accounts for the formation of the freeze-concentrated unfrozen, continuous phase, and all of the discrete phases of ice cream structure: the ice crystals, the air bubbles, and the partially coalesced fat globule structure (Figure 20.1) [14,19,20]. The crystallization of water to ice involves two major steps: nucleation and crystal growth. Nucleation occurs at the wall of the heat exchanger during startup. After startup, the continual scraping action of the blades act as a seeding mechanism, by providing a source of tiny crystals into the bulk where they grow. Rapid heat removal, which results from low temperatures (large ΔT) in the freezing medium, and rapid agitation, which are both present in the barrel freezer, create numerous, tiny ice crystals. Further temperature reduction during hardening accounts for continued growth of the preformed crystals [15].

The dissolved sugars, lactose, and salts, result in an initial freezing temperature in the mix of about −2.5°C, depending on their concentration. From Raoult's law, the freezing point is a function of both the concentration and the molecular weight of the solutes, as it is predicted by the number of solutes in solution. As water freezes out of solution in the form of pure ice crystals, solutes are excluded from the growing ice and concentrate in an ever-decreasing amount of water. Water and its dissolved components are referred to as the serum or matrix of the mix. Because the freezing point of the serum is a function of the concentration of dissolved solids, the formation of more ice concentrates the serum and results in an ever-decreasing freezing temperature for the remaining serum. Thus, at temperatures of several degrees below the initial freezing temperature, there is always an unfrozen phase present (Figure 20.3). Ice cream hardness is a function of temperature due to its effect on this conversion from unfrozen water to ice and further concentration of the serum phase surrounding the ice crystals, which helps to give ice cream its ability to be scooped and chewed at freezer temperatures. Freezing curves for ice cream based on composition can be readily calculated [14,21–23].

As freeze concentration progresses, crystallization of solutes at their maximum solubility concentration (eutectic temperature) is unlikely due to the high viscosity and low temperature of the freeze-concentrated, unfrozen solution in the ice cream. This unfrozen phase is then capable of forming a glass, a noncrystalline solid with a high viscosity (10^{13}–10^{15} Pa sec), at less than −30°C [8,15]. In the glassy state, the serum phase exists as an unreactive amorphous solid, and consequently, no changes that rely on diffusion, such as ice recrystallization, can occur.

The dynamic whipping and freezing process is also responsible for the formation of a fat network or structure in the product [19,20,24]. Ice cream is both an emulsion and a foam. The milkfat exists in tiny globules that have been formed by the homogenizer. There are many proteins that act as emulsifiers and give the fat emulsion its needed stability. The emulsifiers discussed in Section III.A.5 actually reduce the stability of this fat emulsion, because they replace proteins on the fat surface. When the mix is subjected to the whipping action of the scraped-surface freezer, the fat emulsion begins to partially coalesce and the fat globules begin to flocculate (Figure 20.1). The air bubbles that are being beaten into the mix are stabilized by this partially coalesced fat [24]. This process is similar to that which occurs during the whipping of heavy cream as the liquid is converted to a semisolid with desirable stand-up qualities and mouthfeel. In ice cream, the emulsion destabilization phenomenon also results in desirable textural qualities both at the time of draw from the barrel freezer and during consumption, resulting in a smoother, creamier product with a slower meltdown [19,20].

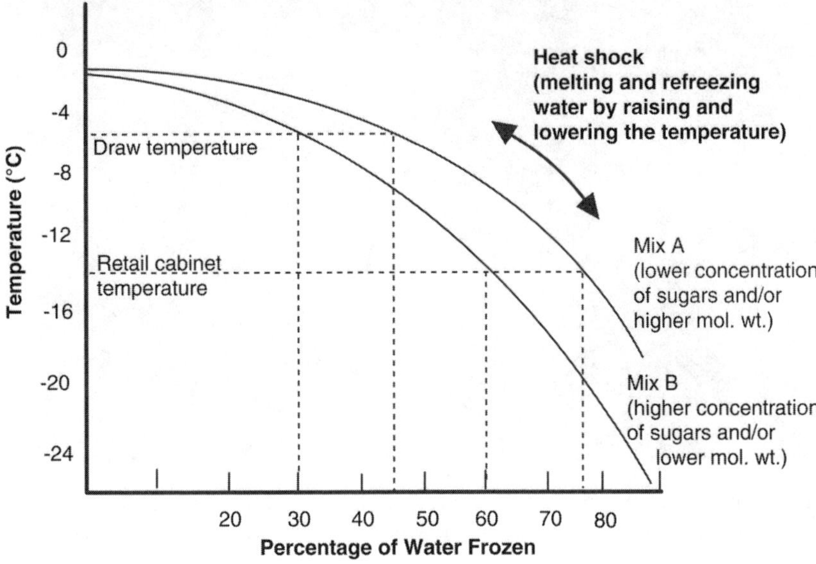

FIGURE 20.3 Ice cream freezing curve showing the relationship between temperature and frozen water for two mixes that vary in solute concentration. There is more water frozen at both draw temperature and retail cabinet temperature for the mix with lower concentration and higher molecular weight of solutes, when compared with that with higher concentration and lower molecular weight of solutes.

D. Factors Affecting Quality

The main factors affecting ice cream quality can be divided into two categories: compositional factors and processing factors. The reader is referred to an excellent review on ice cream defects by Bodyfelt et al. [25] for further details. Defects in ice cream quality associated with compositional factors are related to flavor, body and texture, meltdown, color, and appearance. Dairy ingredients may contribute to rancid or oxidized flavors associated with fat content, stale or old ingredient flavors associated especially with the use of dried milk ingredients, salty flavors associated with the use of high levels of whey, or sandiness, a textural defect associated with excessive lactose crystallization. The sweetening system may also contribute either to a lack of or too much sweetness or to a syrupy flavor from the corn sweeteners. Stabilizers can produce gummy types of defects if used in excess. The flavoring system may likewise produce many flavor defects, associated with either too high or not enough flavor or to an unnatural type of flavor [14,25].

During manufacture, a great deal of emphasis is placed on producing an optimal size distribution of ice crystals, which is as small and numerous as possible, to produce optimal texture. However, ice crystals are relatively unstable and undergo morphological changes in number, size, and shape during frozen storage. This is known collectively as recrystallization and leads to a coarse, icy texture and the defect of iciness (Figure 20.4). Recrystallization is probably the most important change, producing quality losses and limitations in shelf life [14,15]. It also probably accounts for countless lost sales through customer dissatisfaction with quality. Some recrystallization, due to a process known as Ostwald ripening [14], occurs naturally at constant temperatures. This is related to the differences in vapor pressures and relative free energy between small and larger ice crystals, resulting in a driving force for a reduction in small crystals and the growth of large crystals to minimize free energy [14,15]. However, the majority of recrystallization is stimulated by fluctuating temperatures (known as heat shock) and can be minimized by maintaining a low and constant storage temperature [14,15].

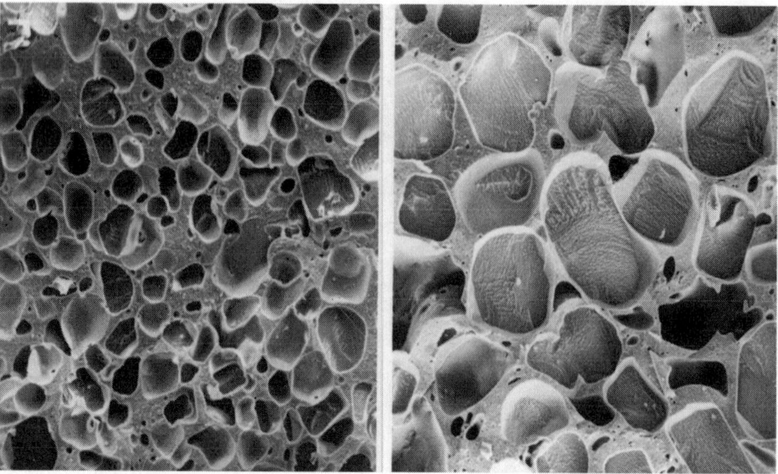

FIGURE 20.4 Scanning electron micrographs of ice crystals in ice cream showing the effects of ice recrystallization due to heat shock. The image on the left is from freshly hardened ice cream, whereas that on the right is after heat shock. Bar, 100 μm.

Heat shock occurs readily in ice cream products. If the temperature during the frozen storage increases, some of the ice crystals melt, particularly the smaller ones as they have the highest free energy and lowest melting point. Consequently, the amount of unfrozen water in the serum phase increases. Conversely, as temperatures decrease, water does not renucleate rather refreezes on the surface of larger crystals, resulting in the total number of crystals diminishing and the mean crystal size increasing. Temperature fluctuations are common during frozen storage, due to the cyclic nature of refrigeration systems and the need for automatic defrost. However, mishandling of product is probably the biggest culprit [14]. If one tracks the temperature history of ice cream during distribution, retailing, and, finally, consumption, one would find a great number of temperature fluctuations. Each time the temperature changes, the ice to serum content changes, and the smaller ice crystals decrease in numbers or disappear, whereas the larger ones grow even larger, leading to a change in the mean crystal size distribution.

Processors have recognized for a long time how to prevent iciness: formulate the ice cream properly to begin with, freeze the ice cream quickly in a well-maintained barrel freezer, harden the ice cream rapidly, and avoid temperature fluctuations during storage and distribution to the extent possible [14,15]. Proper formulation with stabilizers designed to combat heat shock is an essential defense against the inevitable growth of ice crystals. As was discussed previously in Section III.A.4, the most important function of the polysaccharide stabilizer is to limit the growth of ice crystals during storage. Mixes with low total solids are also more difficult to effectively stabilize, as the increased content of water leads to more ice at any given temperature. In addition, high concentrations of sucrose or lactose change the ratio of water to ice and lead to increased problems of recrystallization. Education of people involved in the retailing sector and consumers regarding the causes of iciness and preventative action to maintain a smooth-textured ice cream is also recognized as good preventative action in minimizing heat shock.

E. Factors Affecting Safety

Food safety should be the first consideration of any food manufacturer in all processing issues. Hazard Analysis Critical Control Point (HACCP) programs, as part of an overall total quality management program, have proved to be effective in ensuring product safety and are widely implemented in ice cream manufacturing facilities [14]. Pasteurization is the biological control point in the ice cream processing system. Provided that mix has been properly pasteurized and no postpasteurization contamination has occurred, ice cream should be safe for human consumption. Pathogens are continually tested, in the presence of various ice cream mix components, to ensure thermal destruction by pasteurization, for example, *Listeria monocytogenes*, in which it was concluded that pasteurization guidelines are adequate to ensure inactivation [26].

Examples of postpasteurization contamination include contamination of raw mix with pasteurized mix, improperly cleaned and sanitized equipment or utensils, contamination from the processing environment, and contamination by personnel handling the process. Effective cleaning is required on a regular basis to ensure low standard plate counts and coliform counts. Problem areas include raw product blending equipment, storage tanks, and packaging machines [27]. All incidents of pathogenic contamination of ice cream from modern ice cream factories in the last two decades have resulted from either improper pasteurization or postpasteurization contamination. For example, in a major outbreak of *Salmonella enteritidis* associated with ice cream infecting 224,000 individuals in the United States, the cause was found to be transportation of pasteurized mix that was not subsequently repasteurized in tanker vehicles which had formerly been used to transport nonpasteurized liquid egg products and improperly cleaned [28,29]. Such transportation is no longer permitted.

It should also be recognized, however, that the incorporation of flavoring ingredients after the dynamic freezing step poses another source of potential microbiological contamination. The use of precooked or pasteurized ingredients can control this. Ingredient additions should be routinely checked, by the supplier and the processor, for contamination of the relevant potential chemical and microbiological hazards before use.

IV. FREEZING OF OTHER DAIRY PRODUCTS

Dairy products vary considerably in their moisture content, from 87 to 91% for whole and skim milk to 3% for milk powders (Table 20.1). Obviously, products with higher water content will involve significantly more physical alterations during freezing than those with lower water contents. Although milk can be frozen for preservation or extended shelf life, its high water content makes this process somewhat economically unattractive and suggests the need for concentration before freezing [5]. The flavor of dairy products after freezing and thawing generally are comparable to their fresh counterparts under normal circumstances, with lipid oxidation being the greatest concern. However, the physical effects of freezing may be quite noticeable upon thawing [4], as discussed in Section II. Although flavor deterioration, if present, cannot be rectified, physical changes that may have occurred during storage may be reversible, if the thawed product is to be pasteurized and homogenized for further processing by heating or homogenization into dairy or other food products.

A. Fluid Milk and Condensed Milks

Very little commercial freezing of fluid milk occurs commercially, primarily due to the widespread and constant supply of fresh milk on a year-round basis and the unfavorable economics of preservation of milk in this form. There are a few technological problems associated with the freezing of milk, namely, protein coagulation, fat separation, and fat autoxidation, but these can be overcome if commercial interest develops in application of the process. During freezing of homogenized milk, a considerable degree of phase separation usually occurs, as the solids and the unfrozen phase drain or migrate away from the frozen ice, which may produce a distinct layering of the product during quiescent thawing. However, the phases are usually redispersed quickly with agitation. Storage at -18°C greatly improves textural quality of frozen fluid or condensed milk products when compared with storage at -10°C [5]. The microbiological implications of milk freezing have also been studied. El-Kest and Marth [30,31] examined the influence of freezing and frozen storage of milk on the survival of *Listeria monocytogenes* and observed that death and injury after 4 weeks of storage at −18°C ranged from 11 to 67%, depending on the strain. Cells were protected from death and injury by the presence of milkfat, casein, and lactose. Therefore, survival rates were higher in milk than in buffer solutions.

Milk could be sold in the frozen, concentrated form (evaporated or condensed), in a similar fashion to the widespread distribution of orange juice, if it was not for the instability of the casein micelle system, as discussed in Section II.B [7]. The stability of these products can be enhanced by removing calcium ions through dialysis or ultrafiltration of milk or by adding calcium ion-complexing chemicals, such as hexametaphosphate or sodium polyphosphate to form more soluble casein aggregates and submicelles [5,7]. Lonergan et al. [32] examined the use of electrodialysis at concentrations up to 3.3 times to remove calcium and thereby dissociate micellar casein before freezing of skim milk. Removal of 40% of the total calcium resulted in protein stability greater than 17 weeks at −8°C when compared with less than 1 week for the control sample. A pasteurization treatment just before freezing, presumably to solubilize all lactose crystals, or removal of calcium to 70%, extended storage stability up to 1 year at −8°C. Other techniques to stabilize the casein focus on the crystallization of lactose, as discussed in Section II.B. These include the crystallization and removal of lactose from concentrated products before freezing, the addition of hydrocolloid stabilizers to increase viscosity and suppress crystallization of lactose, the addition of sugar to suppress lactose crystallization, the use of heat treatments postconcentration to dissolve lactose nuclei, and the enzymatic hydrolysis of lactose by β-galactosidase (lactase), which results in a slightly sweeter flavor in the reconstituted milk [7,8].

B. Cream and Butter

The freezing of cream in bulk containers increases its shelf life greatly, but can also lead to gross separation of fat and serum solids upon thawing. Cream processed in this manner is suitable for reprocessing into cream soups, recombined milk, butter, or ice cream, where pasteurization and homogenization will restore the original fat emulsion or churning will continue to induce fat (butter) separation [33]. Bulk frozen cream of desirable quality can be prepared with fresh, low acid cream, with desirable flavor, and 40–60% fat. Manufacturing steps include addition of sugar at a rate of 10% before pasteurization, if desired, pasteurizing at 88°C for 5 min, and freezing the cream in stainless steel cans or plastic containers to −23°C or lower. Cream for further processing can be stored for periods of 6–10 months at temperatures below −18°C if handled properly [12,33]. The addition of sugar to cream considerably retards fat coalescence, primarily due to the lowering of the freezing point by the dissolved sugar. Usually 5 to 20% of the weight of the cream is added. Frozen sweetened cream melts much more readily and is therefore easier to handle than unsweetened cream [12]. The same result can be accomplished by increasing the milk SNF content or by rapid freezing through the use of low temperatures, thin films, or small containers. Rapid freezing will not damage the fat emulsion to the same extent as slower freezing, even after long periods of frozen storage.

Rotary drum freezing provides a means for the continuous production of rapidly frozen, "chipped," or flaked cream of high quality [33]. The packaged flakes have a lower density than bulk frozen cream and may be convenient to use in further processing applications. Storage temperatures of −18°C or lower are required for such a product. Consumer products of this nature (e.g., whipping cream) can also be prepared utilizing rapid freezing techniques. These products could be distributed in bags, similar to frozen vegetables. Additionally, packaged cream in small, consumer packages can be frozen rapidly using cryogenic techniques (e.g., liquid nitrogen tunnels) to produce a high-quality product upon thawing [33]. It is essential that the fat emulsion is very stable before freezing to prevent fat coalescence upon thawing. Emulsifiers and stabilizers may also assist in the freeze–thaw stability of cream, if rapid freezing is not possible. Low, constant storage temperatures are equally as important for the maintenance of quality. Temperature cycling may lead to the formation of large crystals with resultant damage to the fat globules [33].

Freezing has no observable effect of the characteristics or quality of butter. The water content of butter is less than 20%, which is evenly distributed in the form of tiny droplets throughout the product. The freezing point of the water phase is determined primarily by the addition of salt and the degree of residual lactose and dissolved milk salts after washing. Temperatures of −20°C for a few months or −30°C for storage periods of 1 year or more are normally employed during the storage of butter to minimize flavor deterioration. When butter is to be frozen, freezing immediately after processing has shown to produce better quality product than freezing after chilled storage (4°C) for several days. Butter is notorious for absorption of flavors from its storage environment. Therefore, butter should be properly wrapped to prevent penetration of air and off-odors [12].

C. Cheese

Interest in the freezing of cheese has focused on two distinct processes, the freezing of curd for further processing and the freezing of fresh or aged cheese for extended storage. Le Jaouen [34] has reviewed the process of freezing cheese curd for deferred utilization. The following discussion is taken from this report. It was considered most desirable to freeze the curd rapidly in thin layers or flat blocks less than 100 mm thick and wrapped tightly to avoid oxygen exposure. Constant storage temperatures at less than −20°C are needed before utilization of the curd. Thawing can be accomplished at 4°C for 24–48 h before warming the curd to 22°C for further processing, molding, inoculation, ripening, etc. Flavor and texture defects of products manufactured from frozen curd, particularly lipid oxidation and crumbly, grainy textures, may readily occur. The lactic acid microflora may be reduced to about one tenth of the original numbers after prolonged storage of curd, and yeasts and coliforms may completely disappear, whereas other microorganisms may undergo relatively little change in their numbers. These changes in microflora may have an effect on the ripening process after curd utilization. Salt contributes greatly to the formation of oxidized flavor. Only unsalted curd can be stored for deferred utilization. During ripening of cheeses made from frozen curd, a distinctly higher noncasein nitrogen is found, presumably due to the increased susceptibility of the casein fraction to degradation. A lack of characteristic flavor may also result from a loss of carbonyl compounds during frozen storage of the curd. Similarly, dry uncreamed cottage cheese curd is sometimes frozen for extended storage, before creaming, packaging, and retail distribution. Such curd should be less than 80% moisture, salted lightly, fast frozen to −30°C or lower in air-tight packaging, and stored at −23°C or lower for periods no longer than 3–6 months.

During the freezing of cheese, the flavor usually remains good, unless lipid oxidation occurs. However, the body and texture become more crumbly and mealy after thawing, especially after extensive freezing. An increase in the unordered structure of the protein in frozen cheese, particularly in slowly frozen samples, consistent with greater damage to the microstructure observed by scanning electron microscopy and greater proteolysis, has been reported [35]. Cheese curds made from frozen milk concentrate show lower breaking stress and lower syneresis from a change in curd protein structure [36]. Studies on frozen mozzarella cheese suggest similar change in protein structure as evident by an alteration in the stretch and melting properties of shredded or bulk mozzarella after thawing [37–39].

V. CONCLUSIONS

The freezing of dairy products presents an interesting study in contrast to the freezing of many other commodity groups. First, milk is readily available both seasonally and geographically, making its long-term preservation less critical than that of seasonal fruits and vegetables. Secondly, many dairy products from a structural viewpoint are carbohydrate solutions, protein suspensions, and fat emulsions and thus present very different physical challenges than freezing of tissue-based systems such as plant and animal products. Thirdly, the vast majority of frozen dairy products, namely, the dessert category, are eaten in the frozen form, unlike other categories of frozen foods. The dairy industry has made a tremendous success of the frozen dessert category but has used freezing to a far less extent in other dairy products. Thus other frozen dairy products generally have only little commercial significance. One must also distinguish the freezing of dairy products for further processing and the freezing of dairy products for the consumer market. In the latter category, the greatest opportunities may lie in two areas, the frozen pelleted or chipped cream product described earlier and the distribution of frozen concentrated milk in sizes suitable for easy reconstitution, analogous to the tremendous success in the frozen orange juice market. Although there is little use of these products in the consumer market now, evidence of commercial interest in these and other frozen dairy products exists.

REFERENCES

1. PF Fox, PLH McSweeney. *Dairy Chemistry and Biochemistry*. New York: Blackie Academic, 1998.
2. P Walstra, TJ Geurts, A Noomen, A Jellema, MAJS van Boekel. *Dairy Technology*. New York: Marcel Dekker, Inc., 1999, pp. 322–323.
3. P Walstra. On the stability of casein micelles. *Journal of Dairy Science* 73: 1965–1979, 1990.
4. PF Fox. Milk proteins: general and historical aspects. In: PF Fox, PLH McSweeney, Eds., *Advanced Dairy Chemistry, 1. Proteins*, 3rd ed. New York: Kluwer Academic, 2003, p. 8.
5. CV Morr, RL Richter. Chemistry of processing. In: Wong, NP, Jenness, M Keeney, EH Marth, Eds., *Fundamentals of Dairy Chemistry*. 3rd ed. New York: van Nostrand Reinhold, 1988, pp. 755–756.
6. L Tumerman, H From, KW Cornely. The effect of lactose crystallization on protein stability in frozen concentrated milk. *Journal of dairy Science* 37:830–839, 1954.
7. RR Mahoney. Lactose: enzymatic modification. In: *Advanced Dairy Chemistry. 3. Lactose, Water, Salts and Vitamins*, 2nd ed. New York: Chapman and Hall, 1997, pp. 109–110.
8. YH Roos. Water in milk products. In: *Advanced Dairy Chemistry. 3. Lactose, Water, Salts and Vitamins*. 2nd ed. New York: Chapman and Hall, 1997, p. 331.
9. H Singh, OJ McCarthy, JA Lucey. Physico-chemical properties of milk. In: *Advanced Dairy Chemistry. 3. Lactose, Water, Salts and Vitamins*. 2nd ed. New York: Chapman and Hall, 1997, pp. 495–496.
10. JW Sherbon. Physical properties of milk. In: Wong, NP, Jenness, M Keeney, EH Marth, Eds., *Fundamentals of Dairy Chemistry*. 3rd ed. New York: van Nostrand Reinhold, 1988, p. 409.
11. F Harding. Measurement of extraneous water by the freezing point test. *Brussels: International Dairy Federation Bulletin*. Vol. 154, 1983, p. 1.
12. PG Keeney, M Kroger. Frozen dairy products. In: BH Webb, AH Johnson, JA Alford, Eds., *Fundamentals of Dairy Chemistry*. 2nd ed. Westport, CT: AVI Publishing Co. Inc., 1974, p. 873.
13. HD Goff, BW Tharp, Eds. Ice Cream: Proceedings of the 2003 IDF Ice Cream Symposium, Thessaloniki, Greece. Brussels: International Dairy Federation, 2004.
14. RT Marshall, HD Goff, RW Hartel. *Ice Cream*. 6th ed. New York: Kluwer Academic/Plenum Publishers, 2003.
15. HD Goff, RW Hartel. Ice cream and frozen desserts. In: YH Hui, Ed., *Handbook of Food Freezing Technology*, New York: Marcel Dekker, Inc., 2003, pp. 499–570.
16. KG Berger. Ice cream. In: SE Friberg and K Larsson, Eds., *Food Emulsions*. 3rd ed. New York: Marcel Dekker, Inc., 1997, pp. 413–490.
17. HL Mitten, JM Neirinckx. Developments in frozen products manufacture. In: RK Robinson, Ed., *Modern Dairy Technology, Vol. 2, Advances in Milk Products*. 2nd ed. New York: Elsevier Applied Science Publishers, 1993, pp. 281–329.
18. R Jimenez-Flores, NJ Klipfel, J Tobias. Ice cream and frozen desserts. In: YH Hui, Ed. *Dairy Science and Technology Handbook. Vol. 2. Product Manufacturing*. New York: VCH Publishers, Inc. 1993, pp. 57–159.

19. HD Goff. Formation and stabilization of structure in ice cream and related products. *Current Opinion in Colloid and Interface Science* 7:432–437, 2002.
20. HD Goff. Colloidal aspects of ice cream — a review. *International Dairy Journal* 7:363–373, 1997.
21. T Livney, E Verespej, HD Goff. On the calculation of ice cream freezing curves. *Milchwissenschaft* 58:640–642, 2003.
22. R Bradley, K Smith. Finding the freezing point of frozen desserts. *Dairy Record* 84 (6):114–115, 1983.
23. R Bradley. Plotting freezing curves for frozen desserts. *Dairy Record* 85 (7):86–87, 1984.
24. HD Goff, E Verespej, AK Smith. A study of fat and air structures in ice cream. *Internatational Dairy Journal* 9:817–829, 1999.
25. FW Bodyfelt, J Tobias, GM Trout. *The Sensory Evaluation of Dairy Products.* New York: van Nostrand Reinhold, 1988, p. 166.
26. VH Holsinger, PW Smith, JL Smith, SA Palumbo. Thermal destruction of *Listeria monocytogenes* in ice cream. *Journal of Food Protection* 55:234–237, 1992.
27. S Holm, RB Toma, W Reiboldt, C Newcomer, M Calicchia. Cleaning frequency and the microbial load in ice cream. *International Journal of Food Sciences and Nutrition* 53:337–342, 2002.
28. TW Hennessy, CW Hedberg, L Slutsker, KE White, JM Besser-Wiek, ME Moen, J Feldman, WW Coleman, LM Edmonson, KL MacDonald, MT Osterholm. A national outbreak of *Salmonella enteritidis* infections from ice cream. *New England Journal of Medicine* 334:1281–1286, 1996.
29. WL Oemichen. The Schwan's *Salmonella enteritidis* experience. *Journal of the Association of Food and Drug Officials* 59 (3):48–68, 1995.
30. SE El-Kest, EH Marth. Injury and death of frozen *Listeria monocytogenes* as affected by glycerol and milk components. *Journal of Dairy Science* 74:1201–1208, 1991.
31. SE El-Kest, EH Marth. Strains and suspending menstrua as factors affecting death and injury of *Listeria monocytogenes* during freezing and frozen storage. *Journal of Dairy Science* 74:1209–1213, 1991.
32. DA Lonergan, O Fennema, CH Amundson. Use of electrodialysis to improve the protein stability of frozen skim milks and milk concentrates. *Journal of Food Science* 47:1429–1434, 1443, 1982.
33. C Towler. Developments in cream separation and processing. In: RK Robinson, Ed. *Modern Dairy Technology, Vol. 1. Advances in Milk Processing*, 2nd ed. New York: Elsevier Applied Science Publishers, 1993, pp. 61–105.
34. JC Le Jaouen. Curd carry-over In: A Eck and JC Gillis, Eds., *Cheesemaking*, 2nd ed. New York: Lavoisier Publishing Inc., 2000, pp. 341–349.
35. J Fontecha, J Bellanato, M Juarez. Infrared and Raman spectroscopic study of casein in cheese: effect of freezing and frozen storage. *Journal of Dairy Science* 76:3303–3309, 1993.
36. CW Lin, CH Hsieh, HP Su. Breaking stress and syneresis of rennin curds from reconstituted skim milk frozen concentrate. *Journal of Food Science* 59:952–955, 1994.
37. YH Kim, TH Lee, JH Yu. The changes of free oil contents in Mozzarella cheese during refrigerated and frozen storage periods and for thawing conditions. *Korean Journal of Dairy Science* 19:297–302, 1997.
38. NC Bertola, AN Califano, AE Bevilacqua, NE Zaritzky. Textural changes and proteolysis of low moisture Mozzarella cheese frozen under various conditions. *Lebensmittel Wissenschaft und Technologie* 29:470–474, 1996.
39. CJ Oberg, RK Merrill, RJ Brown, GH Richardson. Effects of freezing, thawing, and shredding on low moisture, part-skim mozzerella cheese. *Journal of Dairy Science* 75:1161–1166, 1992.

21 Quality and Safety of Frozen Ready Meals

Philip G. Creed
Bournemouth University, Poole, Dorset, UK

CONTENTS

I. INTRODUCTION

Freezing and frozen storage have long been the important method of refrigeration to ensure the safety and quality of meat, poultry, fish, vegetable, and fruit products as described in other chapters.

However, more recently the emphasis has changed from simply preserving these raw materials as the basis for further processing to adding value by manufacturing prepared foods. One such example, the frozen ready meal, is consumed often with little or no further processing except reheating. Therefore, the responsibility for building into the frozen ready meal the attributes of sensory and nutritional quality, microbiological safety, convenience of handling, and reheating demanded by the consumer, lie almost wholly in the hands of manufacturers and food retailers, making use of their quality management processes. Satisfying consumers' demands for high-quality frozen ready meals thus brings together many of the challenges facing the frozen food industry. These include selecting the most appropriate manufacturing treatments and raw materials for the wide range of often complex recipe dishes, maintaining the quality of these food materials up to the point of consumption, while taking into account the consumers' sometimes unreliable role in the final part of the chain, storage and reheating.

II. DEVELOPING AND MARKETING FROZEN READY MEALS

A. Brief History of Frozen Ready Meals

The work of Clarence Birdseye, in pioneering the development of frozen foods is well known and documented [1,2]. These foods were meal components rather than complete meals. The first complete frozen meals, consisting of an entrée with two vegetables on a paperboard tray treated with Bakelite resin, were devised by William Maxson in 1944 for in-flight feeding of troops going abroad and later for general sale to the public with a wider range of meals [3–5]. A contemporary review describes the meals available, how they were frozen in stages to avoid quality problems such as soggy bread, the incomplete cooking before freezing to allow completion during the defrosting or reheating step, and the special hot-air-blast equipment invented to defrost from 6 to 120 dinners at a time [6]. Later, in 1953, frozen ready meals on aluminum foil trays were first marketed as 'TV dinners' by the Swanson Company in the United States [3]: this was the start of the increasing popularity of frozen ready meals over the next 50 years until more recently being rivaled by the chilled ready meal.

The reasons for the increasing use and range of ready meals can be summarized as the increased desire for convenience coupled with less time for food preparation because of more women working full time and the increased availability and relative cheapness of freezers to store ready meals and microwave ovens to speed up the time to make them ready for consumption [7]. Other factors, which can also play a part, are the decreasing family household size because of changes in family structure, increasing levels of disposable income, wider travel abroad, and experience of different cuisines [8]. The use of freezers has also changed from the 1960s and 1970s when large chest freezers were mainly used in rural areas for storing home-grown produce and taking advantage of bulk buying. In the beginning of the 21st century, the trend is towards using smaller freezers for storing manufactured convenience products such as ready meals. These products usually require a microwave oven and the unseen network of "manufacturers, frozen food producers, global transport systems, and agricultural practices," which provide the infrastructure for a new way of consumption [9].

However, it should also be remembered that frozen ready meals also have a strong market outside the home in commercial restaurants and in the institutional food service sector. Many restaurant and fast food chains rely on reheating and assembling frozen meal components to provide the speedy service and meal quality required by their customers. In institutional settings where costs are always under scrutiny, such as hospitals, care homes for the elderly, school meals, and for providing meals to older people in the home, the use of frozen ready meals can offer a solution which can, with good management, ensure adequate nutritional levels, a high level of safety and a wide menu range.

B. The Market for Frozen Ready Meals

1. World Market

Frozen ready meals can be defined as complete meals which need no additional ingredients as opposed to part meals also known as "meal centers" such as fish portions, fish fingers, and pies, to which consumers can add vegetables, pasta, or rice [9]. Figure 21.1 illustrates recent growth in world sales of frozen ready meals with North America and Western Europe forming the largest part of the total volume and the volume growth rate over this period averaging 5.2% per annum [10]. Western Europe and North America form the largest part of the market growing at around 5% per annum between 1998 and 2003 but Latin America and Eastern Europe have been growing at 10.5 and 8% per annum, respectively, over the same period. In 2003, the average retail selling price per tonne of frozen ready meals was $7108 for the world but varied from $3736 for Latin America and $4130 for the Africa and Middle East region, up to $8136 for North America and $8980 for the Asia and Pacific region. This compares with an average retail selling price per tonne of chilled ready meals of $8706 for the world [11]. However, these figures are subject to some error owing to exchange rate fluctuations and the market forces dictating price levels in different regions. There are also wide variations in preference as shown by comparing the tonnage sold of frozen and chilled ready meals. For comparison, Figure 21.2 shows recent growth in world sales of chilled ready meals, which over the period shown averaged 4.7% per annum, and emphasizes these variations in preference: for example, in 2003, the tonnage of frozen ready meals sold in the North American region was 11 times that of chilled ready meals, in Western Europe it was about equal with frozen only 2% higher than the chilled meals, but in Asia or Pacific region, the tonnage of frozen ready meals was only 16% that of chilled ready meals mainly because of the high popularity of chilled meals in Japan.

2. UK Market

In the UK, sales of the main competitor to frozen ready meals, chilled ready meals, have grown rapidly since 1999 as shown in Figure 21.3 [13]. Most chilled ready meals sold consist of supermarket-owned brands (private label) where the rapid turnover overcomes problems of their short shelf life. In contrast, frozen ready meals tend to be branded products sold more often in the smaller independent stores where turnover can be much lower and thus a longer shelf life is

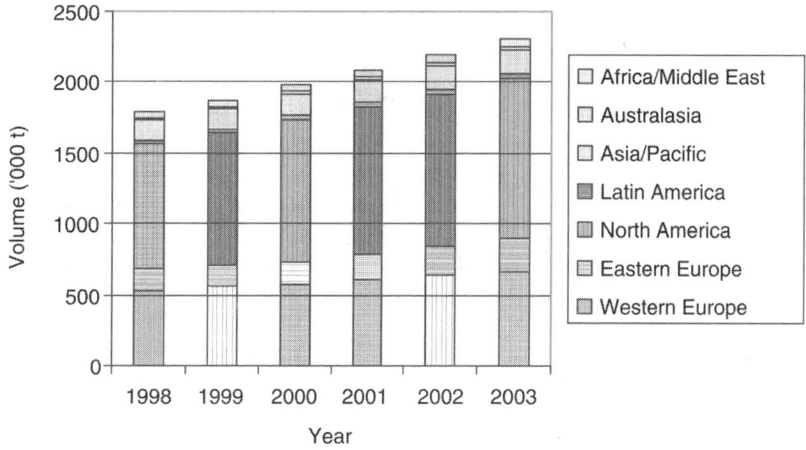

FIGURE 21.1 Volume sales of frozen ready meals (in thousands of tonnes) from 1998 to 2003. *Source*: Anonymous. Frozen Ready Meals Statistics, Global Market Information Database, Euromonitor International, London, 2004.

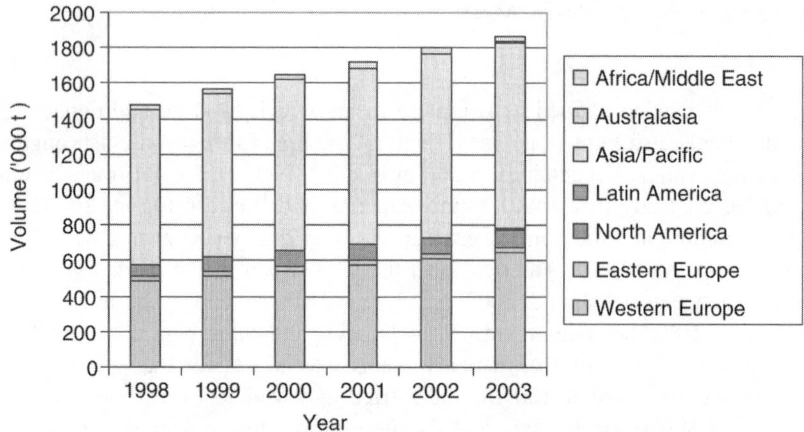

FIGURE 21.2 Volume sales of chilled ready meals (in thousands of tonnes) from 1998 to 2003. *Source*: Anonymous. Chilled Ready Meals Statistics, Global Market Information Database, Euromonitor International, London, 2004.

required [7]. Over the period from 1999 to 2004, UK sales of frozen ready meals in real terms have remained static while sales of chilled ready meals have increased at an average annual rate of 13% to more than double the value of frozen meals. This has prompted many manufacturers of frozen ready meals to launch new products: some linked to popular dieting programs, others with reduced-fat healthy images, recipes with higher quality ingredients, and improvements in packaging [13].

3. Range of Meals Available

The range of frozen ready meals can be divided into "traditional" meals which in the UK include roast dinners, chicken breast with sauce, "international" meals such as many types of curry with rice and "healthy" versions of mainly "traditional" meals. Table 21.1 shows some examples of complete frozen ready meals available from various UK manufacturers. In general, these complete meals

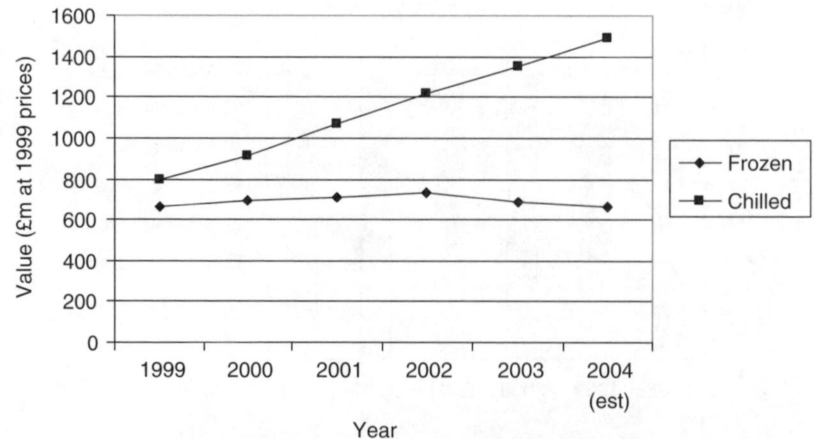

FIGURE 21.3 UK retail sales of frozen and chilled ready meals (in £m at 1999 prices) from 1999 to 2004 (estimated). *Source*: Anonymous. Chilled Ready Meals — UK, Mintel International Group Ltd., London, May 2004.

TABLE 21.1
Examples of the Composition of Frozen Ready Meals per Pack Available in UK (Information from Packs and Websites)

	Weight (g)	Energy (kJ/kcal)	Fat (g)	Salt (g)	Summary of Contents	Source[a]
			Traditional Meals			
Beef stew and dumplings	320	1192/285	7.4	1.8	20% beef, 80% gravy, vegetables and dumplings.	1
Beef stew and dumplings	400	1628/389	13	4	12% beef, 12% vegetables (onions, carrots), 14% dumplings, 62% gravy	4
Chicken breast in a tomato or white wine sauce	365	1360/325	9.6	1.8	27% chicken, 29% sauce, 44% vegetables (potato, onion, mushrooms, rice, broccoli, carrots)	1
Beef in an ale gravy	340	1213/290	7.1	1.3	25% beef, 28% sauce, 47% vegetables (potato, carrots, peas)	1
Roast beef dinner	340	1674/400	13	1	26% beef, 19% gravy and Yorkshire pudding, 55% vegetables (potatoes, carrots, peas)	1
Roast pork dinner	340	1423/340	8.6	1.8	22% pork, 26% gravy and stuffing, 52% vegetables (potatoes, carrots, peas)	1
Salmon fillet in a dill sauce	340	1464/350	16	1	18.5% salmon, 33.5% sauce, 48% vegetables (potato, carrots, green beans)	1
			Healthy Meals			
Lean cuisine Beef oriental	350	1757/420	9	2.5	11% beef, 48% sweet and sour sauce, 38% rice, 3% pineapple or waterchestnuts	2
Lean cuisine Chicken sweet and sour	350	1632/390	9	2	10% chicken, 53% sweet and sour sauce and vegetables, 37% noodles	2
Lean cuisine Chicken tikka masala	350	1799/430	12	2.5	10% chicken, 52% curry sauce and vegetables, 38% rice	2
Lean cuisine Glaze chicken	350	1130/270	6.5	1.5	10% chicken, 66% tarragon sauce and vegetables, 24% rice	2
Good for you chicken and mushroom	400	1151/275	7	1.5	13% chicken, 47% vegetables (potatoes, peas, carrots, mushrooms), 40% sauce	4
Eat smart lasagne	400	1611/385	8	1.8	16% pasta, 10% beef, 74% sauce and cheese	5
Be good to yourself lasagne	400	1335/319	7.9	1.9	13% pasta, 9% beef, 10% tomato, 68% sauce and cheese	6
Healthy living beef lasagne	400	1724/412	10	1.9	16% pasta, 12% beef, 72% sauce and cheese	7
Good for you beef lasagne	350	1602/383	7	3.6	11% pasta, 26% beef, 63% sauce and cheese	4
Lean cuisine tuna pasta	350	1590/380	9.5	2.8	6% tuna, 56% sauce and vegetables, 38% pasta	2
Lean cuisine vegetable curry	350	1590/380	8.5	2	26% vegetables, 36% curry sauce, 38% rice	2

(Table continued)

TABLE 21.1 *Continued*

	Weight (g)	Energy (kJ/kcal)	Fat (g)	Salt (g)	Summary of Contents	Source[a]
			International Meals			
Lasagne	400	1904/455	18.4	3	10% beef, 5.5% cheese, 84.5% pasta, tomatoes and sauce	5
Beef lasagne	400	1874/448	16.4	4	13% pasta, 13% beef, 11% tomato, 63% sauce and cheese	6
Beef lasagne	400	2293/548	24.8	4.1	16% pasta, 11% beef, 73% sauce and cheese	7
Beef lasagne	400	1582/378	14	4.1	13% pasta, 16% beef, 73% sauce and cheese	4
Spaghetti bolognese	350	1590/380	6.5	1.8	48% pasta, 8% beef, 44% sauce and vegetables	2
Spaghetti bolognese	340	1778/425	14	1.5	50% pasta, 13% beef, 37% sauce and vegetables	1
Chicken jalfrezi with rice	400	2326/556	18.4	1.9	31% rice, 20% chicken, 6% peppers, 43% sauce and vegetables	3
Chicken jalfrezi with rice	375	1966/470	7.1	0.8	50% rice, 14% chicken, 36% sauce and vegetables	1
Chicken tikka masala and rice	400	2427/580	20	4	36% rice, 20% chicken, 44% sauce	3
Chicken tikka masala	400	2008/480	13	1.8	50% rice, 14% chicken, 36% sauce and vegetables	1
Chicken korma	400	2694/644	29.6	3	36% rice, 20% chicken, 44% sauce	3

[a]1: Birdseye [14]; 2: Findus [15]; 3: Patak's [16]; 4: Asda Stores; 5: Safeway; 6: Sainsbury's; 7: Tesco Stores.

consist of meat, fish, poultry, or pasta with sauce and vegetables, the sauce often being the greatest component by weight, mostly varying from 30 to 48% or up to 73% for beef lasagne (Table 21.1). "Healthy" frozen meals tend to include a smaller proportion of the protein ingredient to stay around 1674 kJ (400 kcal) per serving even though most "traditional" meals are also around or below this value but they contain about half the fat content of equivalent "non-healthy" versions. Similarly, "international" meals produced by different manufacturers also vary; for example, the ratio of rice to chicken varies from 1.8:1 to 3.6:1 for Chicken tikka masala (Table 21.1).

4. Consumer Acceptability of Frozen Ready Meals

The reason for the increasing popularity of chilled over frozen ready meals can be traced back to the long-held belief that frozen foods were intrinsically inferior to those not frozen. In the 1930s the description "frozen" was often synonymous with "spoiled" and manufacturers realized the need to have control over the whole food chain from manufacture to the retailer's freezer display cabinets to ensure that their products did not deteriorate through mishandling and temperature abuse during distribution [17]. Other descriptions such as "quick frozen" and "frosted" were used to differentiate products on the market from "frozen" [18]. In the UK, manufacturers have often had to use promotional offers to obtain an estimated 40–50% of their sales which further prolongs the consumers' perception that low prices must imply low quality [8]. In the United States, in order to maintain their market share, manufacturers have launched better quality frozen meals branded with nationwide restaurant chains to take advantage of the associated high-quality image [19].

When consumers were questioned about their perception of the acceptability of meals prepared using different methods, any meal thought not to be prepared "traditionally" from "fresh" ingredients, for example, using chilled, frozen, sous-vide, dehydrated items, or a mixture of these items, was considered less acceptable [20]. The method closest to reality, where meals have been prepared from ingredients processed by a mixture of methods, was thought to be only an average of 51% acceptable, as "freshly" prepared, with meals produced by cook-freeze and cook-chill systems slightly more acceptable at average levels of 65 and 56%, respectively. There was also a trend for acceptability to decline with increasing age group with, for example, those aged 65 and over considering cook freeze to be 46% as acceptable as "freshly" prepared compared to 73% for the 20–24 age group. The corresponding figures for the acceptability of cook-chill meals were 29 and 77%.

Thus, it can be seen that frozen meals are considered more acceptable than those prepared using other nontraditional processes by older consumers who are likely to use them if confined largely to the home [21–23]. Other factors considered important for this group are ease of opening of the frozen meal packs, fortification with calcium and vitamins and ease of ordering and delivery [24]. In the United States, frozen meals have been prepared commercially for those older people who need to control blood pressure, cholesterol, and sugar [25]. More recently, surveys have found that the percentage of those agreeing with the statement that "frozen foods are as good for you as fresh foods" had increased from 27.1% in 1997 to 32.8% in 2003, while those disagreeing had decreased from 42.4% in 1997 to 37.9% in 2003 [8]. Thus it seems that the consumer's prejudice against frozen meals may slowly be declining.

C. The Complexities of Frozen Ready Meals

Unlike other chapters of this handbook which focus on many aspects of freezing particular food commodities, this chapter aims to review those same aspects for the frozen ready meal, a multicomponent product where the ingredients are cooked and then frozen — a sequence of two treatments. During manufacture, these components can be treated separately if necessary before final assembly and freezing but in the hands of the consumer or end user they will all have to be reheated together. This raises the question of how precooked food products with differing physical and thermophysical properties and differing reactions to the freezing, frozen storage, and reheating processes, can ultimately be brought to a consistent level of quality on the plate for the consumer. At the same time, the reheated ready meal must, of course, be microbiologically safe, be appetizing in terms of its visual and sensory qualities and also provide a significant contribution to the recommended levels and balance of nutrients in the consumer's diet. This will inevitably mean that compromises must sometimes be made during the product development and manufacturing procedures when choosing ingredients, pretreatments for meal components, and packaging.

III. METHODS OF MANUFACTURE

A. Pretreatment of Meal Components

As discussed in other chapters, freezing has some significant effects on the properties of food materials when thawed to their original state to enable further processing. It is not always clear how freezing affects these same materials when they have been cooked so that they can regain their original properties after reheating. For frozen ready meals, these effects would be particularly relevant to the stability of the sauces which accompany most ready meals, to the quality of vegetables and to minimize the risk of off-flavors developing during frozen storage.

1. Preparing and Modifying Sauces

Sauces serve the culinary purpose of lubricating the foods as they are chewed and swallowed and of carrying flavors to enhance the sensory quality of the food itself. The French chef, Escoffier, lists almost 200 different versions mostly derived from three basic "mother" sauces [26,27].

Manufacturing sauce for ready meal production in a similar way as a chef would prepare is not practicable, and consequently large-scale production demands using various types of blenders and heat exchange equipment for cooking and cooling [28,29]. The main types of sauces used for frozen ready meals are brown sauce (a light sauce with low solid content), cream sauce (a rich sauce with dairy ingredients), and tomato sauce (with a high solid content which needs less thickening agent). Brown sauce is the simplest to make while cream sauces need emulsification using homogenization to give the desired appearance. A typical process would start with the thorough mixing of the dry ingredients, which might serve for several batches of sauce. For each batch, the dry materials would be prewetted with a portion of the liquid ingredients and blended into a slurry before being added to the cooking vessel where the remaining ingredients such as stock (beef, chicken, or fish), tomato solids, and seasoning mixes are added depending on the recipe. The cooking vessels, heated by steam jackets with continuous mixing, bring the ingredients to 88°C which then simmer for 10 min to ensure that the thickening agents are fully gelatinized. When the sauce has reached the required viscosity, it must then be cooled rapidly to avoid microbiological growth and flavor deterioration [28]. This can be achieved by using the jacket to circulate cooling water or else by pumping the sauce through heat exchangers. For some sauces, particulate foods such as diced mushrooms, onion, carrots, peppers, celery, or cooked ground meat can be added to the cooled sauce before assembly with the other ready meal ingredients.

The inclusion of a thickening agent such as wheat flour in the dry ingredients is designed to produce the desired viscosity and mouthfeel. Freezing tends to cause sauces to separate when thawed — a phenomenon known as syneresis – because of starch retrogradation caused by gelation of the amylose fraction and a slow recrystallization of short amylopectin segments, leading to a reduction in the water-holding capacity [30,31]. Therefore, modification of recipes is necessary to provide freeze- and thaw-stable sauces by using other starches such as tapioca with lower amylose content, varieties of cereal such as waxy maize, which have mainly amylopectin, or by using chemically modified starches to form crosslinks, the most common method. The importance of using sauces as a protective coating around the cooked protein components of frozen foods to avoid oxidation and dehydration has been known since the frozen ready meal was first devised [4,31].

2. Blanching Vegetables

As with the production of all convenience foods such as frozen ready meals, any vegetables used need to be washed, cleaned, peeled, diced, or sliced as required before blanching [32]. The blanching operation is necessary before freezing to destroy enzymes, which if still present, would cause deterioration of the texture, color, flavor, and nutritional quality of the vegetables during frozen storage. This operation is usually accomplished by immersion in water between 70° and 105°C for 1–10 min depending on the size of the vegetable pieces. It is then essential to cool the vegetables to prevent any further internal cooking, which would adversely affect the sensory and nutritional qualities [28]. This can be done using air-blast cooling as using chilled water would lead to a deterioration through leaching of flavor and nutritional components. The preparation steps for vegetables produce large amounts of waste and are also a source of contamination, thus many frozen ready meal manufacturers buy in prefrozen individually quick frozen IQF vegetables of the required size and shape from specialist suppliers to assemble with the other precooked meal components [33]. As vegetables require relatively little cooking to suit the taste of consumers today, the final reheating step of a frozen meal would often be sufficient to complete the cooking of these prefrozen vegetables.

3. Marinating, Coloring, and Cooking Meat

Marinating is often used to ensure that the main focus of the frozen ready meat for the consumer, the portion of meat, poultry or fish, has desirable sensory qualities such as moistness, flavor, and tenderness as well as improving the cooking yield. The marinade consists of a solution of salt, vinegar, oil, citric acid, spices, and seasonings in which meat portions can be soaked or can be injected directly into the meat [28,33].

Frying is often used to "seal" meat and to produce the desirable brown color on the meat itself or on any coating of batter or breadcrumbs. Continuous fryers take the meat pieces through the hot oil between two conveyors, which ensure total immersion [28,34]. Oven-cooking is also often used for this most important part of a frozen ready meal. Batch or continuous ovens transferring heat into the meat portions by conduction or convection of hot air can achieve consistent results before cooling and size reduction of the meat pieces as necessary before assembly. The sous-vide process can also be used for cooking meat: this entails vacuum sealing the meat portions in plastic pouches, cooking in hot water vats for up to 4 h followed by rapid cooling at 1°C, surface drying of the pouches, and flash-freezing for incorporation with the final meal [28].

4. Modifying Other Ingredients

The other main components of a typical frozen ready meal will be rice, pasta, or potatoes, which are cooked and cooled before being assembled with the other meal components. The particular type and size of pasta has to be chosen carefully to avoid thin versions, which cook very quickly and become soft and to select types, which complement the sauce [28]. Coating the pasta with vegetable oil can also reduce the ingress of water after freezing. It is essential to rinse rice and pasta after cooking to remove the soluble carbohydrates on the food surface and any small broken pieces. Like vegetables, pasta needs to be cooked "al dente" so that on reheating, the texture will still be acceptable for the consumer [35]. Thus, it is necessary for the cooked and rinsed product to be cooled quickly to inhibit any further cooking [28]. High quality mashed potato can be produced using granules and flakes blended with milk solids, seasonings, butter, or margarine and boiling water in a mixer with scraper blades. After the mixture is smooth, it can then be pumped out and chilled ready for assembly into the meal.

Other developments include methods for creating battered frozen products, which avoid the prefrying step and decrease the absorption of oil. The batter can be formed by immersing the coated product in water at 75°C to coagulate the methylcellulose added in the batter, followed by flash heating in a microwave before freezing, storage, and the final frying stage [36].

Freezing and thawing cooked eggs also results in syneresis, leading to a loss of water and a porous texture that can be reduced by adding a sodium caseinate solution [37].

B. Manufacturing Methods

1. Cook–Assemble–Freeze

The manufacturing process for frozen ready meals is essentially an assembly process where the ingredients, pretreated as necessary, are cooked, cooled as discussed earlier, and then assembled into ovenproof trays manually or by depositors. This is then followed by sealing, packing into cardboard sleeves or boxes, and then freezing after which the individual packs can be placed in cartons and palletized for storage and distribution. Some meals might be suitable for depositing into bags, which are frozen and eventually reheated by immersion in boiling water.

It is essential that the area where the chilled or frozen cooked components are assembled before freezing is treated as a "high-risk" area, which is physically separated and uses different personnel from "low-risk" areas where the uncooked components have been handled [33].

2. Frozen Sous-Vide

The sous-vide process for the production of high-quality chilled meals has developed extensively over the last 25 years [38]. The initial fears that the process would increase the risk of food poisoning during chilled storage because of the anaerobic conditions in the vacuum-sealed packs appear to have been counteracted by more stringent quality management systems. However, several manufacturers have decided to freeze the packs and distribute them in the frozen state to assure their customers that the product was safe [39]. This has prompted research to investigate the effect of freezing on many aspects of quality [40].

IV. QUALITY AND SAFETY ISSUES

Quality and safety for frozen ready meals can be divided into those aspects affecting the eating quality in terms of flavor, texture, aroma, color, and appearance, those affecting the microbiological safety and those affecting the hidden nutritional qualities.

As mentioned earlier in this chapter, frozen ready meals are complex multicomponent products with a wide variety of ingredients. The quality of each of these ingredients can be affected by storage time and storage temperature to different extents based on the kinetics of reactions, causing the deterioration of pigments, vitamins, fats, and changes in enzyme activity. Figure 21.4 illustrates the overall effect of storage temperature over time on quality [41]. Figure 21.5 shows how different sensory characteristics can be affected differentially by storage temperature. Figure 21.6 shows how different products with varying sensitivity to storage temperature because of different ingredients will exhibit a varying shelf-life. Therefore, this means that where a hypothetical ready meal's shelf life at −25°C is limited by the rate of lipid oxidation in the meat component, at −15°C the limiting factor may be the deterioration owing to enzyme activity in the vegetable component, and at −5°C this may change to the unacceptable dehydration of a mashed potato component because of moisture migration [42]. This changing pattern of deterioration with storage temperature will influence how frozen ready meals are handled during storage and distribution. As frozen ready meals are the result of a long process of product development by manufacturers, unlike particular food groups and commodities, detailed information on the complex series of interactions affecting quality and safety is sparse because of the manufacturers' need for commercial confidentiality.

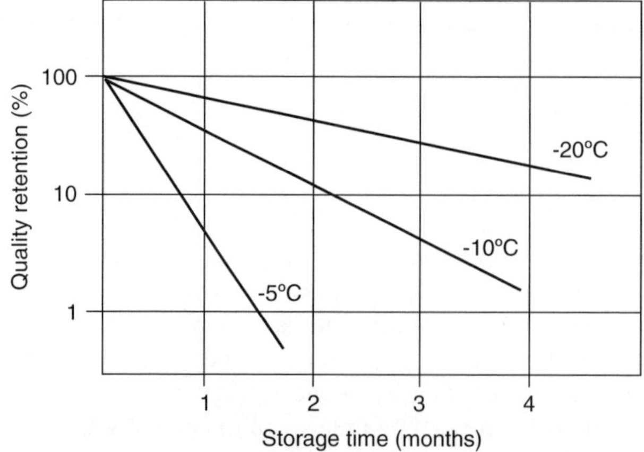

FIGURE 21.4 The influence of storage temperature and time on quality retention. Adapted from Heldman and Hartel, Principles of Food Processing, Chapman & Hall, USA, 1997.

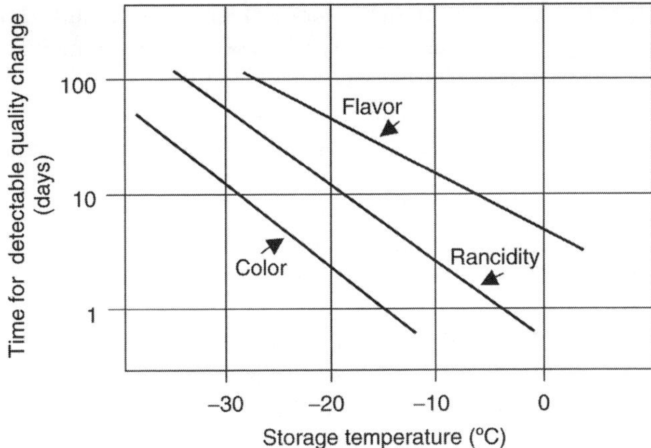

FIGURE 21.5 The influence of storage temperature on time to a detectable change in quality (color, flavor, rancidity). Adapted from Heldman and Hartel, Principles of Food Processing, Chapman & Hall, USA, 1997.

A. Factors Affecting the Eating Quality of Frozen Meals

Most published information can be divided into: information derived from measuring the content of substances indicating the development of off-flavors because of rancidity development such as TBA (thiobarbituric acid), free fatty acid, and peroxide values; instrumental measurements of texture, color, and drip loss; and use of sensory analysis for judging and maintaining quality standards during production and for predicting consumer reactions.

1. Rancidity Development

Rancidity develops through lipid oxidation caused by a series of complex chemical reactions involving free radicals [43,44]. Temperature changes, such as those caused by fluctuating storage conditions or temperature abuse, reducing water activity as ice forms and allowing large surface areas to encourage oxygen solubility, are the main factors increasing lipid oxidation in frozen foods. In frozen ready meals, rancidity is most likely to develop in the fat components of meat and sauces

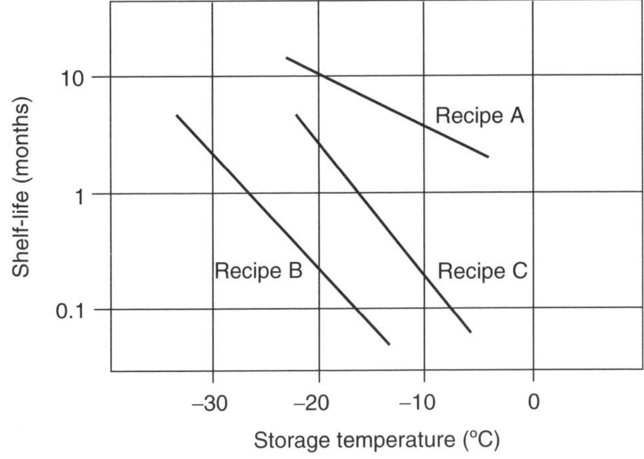

FIGURE 21.6 The influence of storage temperature on shelf-life for different frozen ready meals. Adapted from Heldman and Hartel, Principles of Food Processing, Chapman & Hall, USA, 1997.

[33]. Most chemical reactions speed up initially during freezing as the unfrozen matrix becomes more concentrated but then slow down as viscosity increases to stop the mobility of the reactants. However, this does not apply to the oxidation of unsaturated fatty acids where removal of water allows the free radicals to react and thereby, increase the development of rancidity. Lipid oxidation leads to flavor deterioration although it is not always possible to correlate chemical measurements of oxidation with taste panel results [43]. Deterioration of color can also result from pigment degradation in meat and vegetables because of lipid oxidation.

TBA value is used to measure secondary oxidation products, in particular malonaldehyde. Experiments in a hospital cook-freeze unit found that pork which was cooked and then allowed to cool overnight before slicing and freezing, resulted in TBA values of 2.7–3.5 mg malonaldehyde per kg meat, levels with detectable rancidity attributes [45]. Work on cooked mutton meat [46] and cooked water buffalo meat [47] showed that TBA values increased over 90 days of storage at −10°C but did not reach the threshold value of 1–2 mg malonaldehyde per kg meat associated with rancidity development. Frozen chicken meals with sauce have shown a smaller increase in free fatty acids, the result of hydrolytic rancidity, over 13 days of storage than those without sauce illustrating the protective role of sauces in reducing rancidity development [48].

As part of an extensive project on freeze-chilling (thawing frozen cooked products and then treating as chilled), steamed salmon was blast frozen at −35°C for 2.5 h, stored at −25°C for up to 32 weeks and then assessed by measuring free fatty acid and peroxide values but no effect was found between treatments [49]. No significant differences in the low values measured were found between the conventional freezing and freeze-chilling treatments.

2. Texture, Color, and Drip Loss Measurement

Sauce again shows its usefulness in frozen ready meals by covering meat components and thereby, reducing dehydration, the main cause of deterioration in texture during frozen storage [33]. In the project on freeze-chilling cited earlier, instant mashed potato, steamed salmon, and steamed broccoli blast-frozen and stored at −25°C for up to 32 weeks, were assessed for texture (Kramer shear press), color (Hunterlab), and centrifugal drip loss [49]. There were significant differences between these freezing treatments and freshly cooked products but over the frozen storage period for the conventionally frozen products, no significant changes occurred in mashed potato; in steamed salmon, redness declined ($p < 0.05$) and softness increased significantly ($p < 0.01$); and in steamed broccoli, shear values changed little by the end of storage.

Further work was done on the effect of different potato varieties on the texture, color, and drip loss of mashed potato showing that Rooster and Golden Wonder varieties were more suitable for frozen mashed potato and Maris Piper better for frozen potato wedges [50]. The length of time in frozen storage at −25°C up to 12 months had no effect on drip loss although values were, as expected, significantly higher compared with fresh or chilled mashed potato ($p < 0.001$). The color as measured by L/b values showing the whiteness to yellow ratio, increased during storage ($p < 0.001$) to become brighter. The work was extended to cooked green beans and carrots again assessing the same attributes for storage times at −25°C for up to 12 months [51]. In this work, differences were found between freeze-chilled and other treatments but for conventionally frozen beans and carrots, no significant differences were found for texture, drip loss, and color.

The effect of freezing on foods prepared using the sous-vide method on texture (shear values) has found that freezing softened the texture of carrots, broccoli, pasta shells, and potato slices but hardened salmon, cod, and rice compared with the normal chilled sous-vide product [40]. Drip loss increased from 2% for chilled sous-vide to 2.5% for frozen sous-vide for carrots, from 8.4 to 18.5% for broccoli, from 14.1 to 16.3% for salmon, and from 1.4 to 18.9% for potato slices but decreased from 22.6 to 20.1% for cod, from 1.92 to 0.9% for rice, and from 1.07 to 0.58% for pasta shells.

3. Acceptability and Sensory Attributes

The project on freeze-chilling previously cited assessed the sensory quality of instant mashed potato, steamed salmon and steamed broccoli blast frozen at −35°C for 2.5 h and stored at −25°C for up to 32 weeks [49]. The assessment using 20 assessors showed no differences in acceptability between treatments. The work on frozen mashed potato stored at −25°C for up to 12 months revealed no effect of storage time or potato variety on sensory score but the conventionally frozen mashed potato was significantly ($p < 0.01$) more acceptable than the freeze-chilled product [50]. Other work has also studied the effect of 19 potato varieties on the acceptability of blanched, cooked, frozen-peeled and, quartered potatoes [52]. No variety performed consistently better but fresh potatoes were more acceptable than stored and cooking at less than 100°C produced a less acceptable result. The sensory analysis of cooked green beans and carrots also showed no significant effect on acceptability for storage times at −25°C for up to 12 months [51].

Chicken Velouté prepared using cook-chill, cook-freeze, and sous-vide systems was compared with freshly prepared samples after 6 days of storage at 4 or −14°C [53]. No significant differences were found between samples for aroma, appearance, flavor, and tenderness. Frozen sous-vide carrots compared well with steamed carrots in sensory tests [54].

Ideally, frozen ready meals will be reheated as required for almost immediate consumption but in many food service and institutional settings, consumption may be delayed for logistical reasons. During this period of warmholding, visual and sensory deterioration can take place, often within 20 min: mashed potatoes will become "waxy", baked potatoes become soggy, meat will dehydrate, and vegetables will become discolored [55].

B. Effect of Speed of Freezing on Eating Quality

As freezing has become a significant method of preservation, the influence of the speed of freezing on quality has been an area of debate. In the early days of using this technology, freezing was slow because of the low efficiency of the refrigeration equipment then available producing relatively high storage temperatures of −10°C. Often cold stores were used for freezing rather than purpose-designed blast freezers, therefore food could take days to freeze completely, leading to the growth of large ice crystals which disrupted food structure more than faster freezing which produced smaller ice crystals and hence less damage. The efficiency of freezing is now much greater so that modern air blast freezers and liquid nitrogen freezers can provide rapid freezing as discussed in other chapters.

A survey of frozen ready meals for institutional use in Italy showed no differences in quality between air-blast and liquid nitrogen methods [56]. Similar results were found for the effects on the quality total viable count (TVC), TBA value, color, hedonic) of three prepared meals studied in Taiwan [57]. No differences were found for the lemon chicken, pineapple chicken and Kong Pao chicken dishes except for the better color of carrots in pineapple chicken frozen with liquid nitrogen.

C. Effect of Fluctuating Storage Conditions on Eating Quality

The main modes of deterioration in frozen convenience foods during storage have been suggested as rancidity in meat portions, weeping and curdling of sauces, discoloration, and package ice [58].

As fully explained in other chapters, storage temperature for frozen foods has a significant effect on the quality. These temperatures are usually defined at given levels but in the commercial reality, foods can be subjected to fluctuations in storage temperature, which may reduce the shelf-life considerably. Work in this area has compared several foods stored for 8 months at a constant −30°C as a control; at −60°C (superfreezing), and with mild abuse — subjected to an initial 3-week period where each week there was a 48-h period at −10°C instead of −30°C then followed by storage at −30°C for the remaining time up to 8 months [59]. Among the foods assessed by sensory analysis, TVC, water content, texture (instrumental), water-holding capacity, free fatty acids, and peroxide value, was pork stewed in gravy. The peroxide values for the stewed pork increased significantly ($p < 0.001$) over 8 months from 1.5 to 2 mequiv./kg fat for −60°C storage, to 3.5 mequiv./kg fat for −30°C storage and to 3.8 mequiv./kg fat for the temperature-abused samples. The free fatty acid values also increased significantly ($p < 0.001$) over 8 months from 0.4 to 0.5% oleic acid for −60°C storage, to 1.0% oleic acid for −30°C storage and to 1.7% oleic acid for the temperature-abused samples. Sensory analysis consistently showed the temperature-abused pork as the least preferred. It was concluded that the increased cost of storage at −60°C for sensitive products, especially those with high fat levels, would have to be balanced against the gain in quality and that the fluctuating temperatures promoted more rapid development of rancidity.

The storage life of frozen foods is often defined in terms of whether a taste panel can detect statistically significant changes in acceptability or particular aromas or flavors at a given confidence level — the high-quality life (HQL) [60]. In practical terms, the storage life for the consumer may be much longer than this. Data collected from several sources on HQL for a range of frozen ready meals offered times varying from 4 to 18 months at −10°C, from 3 to 26 months at −20°C, and from 6 to 18 months at −30°C [61]. Therefore, no pattern was apparent to guide the consumer or the manufacturer. This again, illustrates the general principles of the differential effects of storage time and temperature on HQL or shelf-life of frozen ready meals shown in Figure 21.4 to Figure 21.6.

D. Factors Affecting Safety of Frozen Meals

Over the years, frozen foods have had a good reputation for safety with few related illnesses [62]. Although many microorganisms are preserved very well by freezing, many are damaged. Consequently, future research might aim to design food systems where pathogens, if present, would not survive the freezing process.

1. During Processing and Storage

As emphasized earlier, when discussing the assembly of ready meals before freezing, it is essential to treat this production area as "high risk". Guidelines on Good Manufacturing Practice (GMP) enable the manufacturer to plan the process steps [63] and procedures to monitor the effectiveness of heat treatments using hazard analysis critical control point (HACCP) methods as discussed in other chapters. The design of the high-risk area aims to eliminate the possibility of cross-contamination by physical separation and controlling the flow of food materials and personnel through the production environment. Combining these steps with a high level of personnel training and hygiene will minimize the risk of food safety problems occurring.

For cooking components of frozen ready meals, the pasteurization heat treatment of 70°C for 2 min at the center of the product to destroy pathogens followed by immediate cooling to chill temperatures before assembly and freezing will avoid food safety problems. However, this has to be done in an environment where the temperature is controlled at around 10°C to prevent cooked-chilled food from warming up. This can have an adverse effect on staff working conditions so localized cooling, where the food components' immediate environment is controlled at chill temperatures rather than the whole production environment, has been suggested as being a safer and more energy-saving method of production [64].

The production of safe food has to start with the use of high-quality raw materials as free as practicable from microbiological and other sources of contamination. Buying in frozen or chilled meal components such as prepared vegetables eliminates the risk of contamination from the washing, peeling, and slicing or dicing procedures: similarly, meat prepared in the same way will offer advantages.

Microbiological surveys can offer case studies of how well frozen ready meal manufacturers are succeeding in using safe production techniques. For example, in Spain, a survey of meat, fish, pasta, and vegetable frozen ready meals found that significant numbers of samples exceeded limits for *Escherichia coli* and *Staphylococcus aureus*. Overall, 112 out of 353 samples were considered unfit for sale [65]. A later survey of the microbiological safety of three frozen ready meals, York ham flamenco, chicken croquettes, and hake fish fingers, found that counts of bacteria were higher in the final products than in the raw materials especially the ham and chicken dishes [66]. It was concluded that the heat treatment was not adequate to overcome the poor quality of the raw materials where *S. aureus* and *Clostridium perfringens* were detected. These surveys emphasize the need for only high-quality raw materials to be used for the manufacture of frozen ready meals and for strict control during the manufacturing process.

The project on freeze-chilling previously cited measured the TVC of instant mashed potato, steamed salmon, and steamed broccoli blast frozen at $-35°C$ for 2.5 h and stored at $-25°C$ for up to 32 weeks [49]. The assessment showed no significant changes in TVC for mashed potato and steamed broccoli between different treatments but for steamed salmon the TVC changed from 1 to 1.5 $\log_{10}$ cfu/g during storage. The work was extended to cooked green beans and carrots in frozen storage at $-25°C$ for up to 12 months where TVC declined from 2.9 to 2.2 $\log_{10}$ cfu/g for cooked carrots and significantly from 5.5 to 1.2 $\log_{10}$ cfu/g for green beans ($p < 0.05$) [51].

Microbiological standards are available for determining the quality of frozen ready meals [67]. These specify the number of packs required for testing, the maximum number of defective packs allowed, and the microbiological limits for different organisms which separate "good" quality from "defective" in a two-class plan or from "marginally acceptable" in a three-class plan.

2. During Thawing and Reheating

After manufacture, frozen storage, and distribution, the responsibility for safety of ready meals effectively moves into the hands of the consumer or into the hands of catering staff for institutional feeding. In the case of consumers, cooking instructions form the only viable method of ensuring that frozen meals are reheated sufficiently. For foodservice operations in hospital foodservice, meal-on-wheels, and so on, education and training of the foodservice employees or volunteers is essential to maintain the quality of frozen ready meals assured during the high-quality processes to be expected from well-run frozen food manufacturing organizations.

During the development of frozen ready meals, as well as investigating alternative ingredients, problems of scaling up to large-scale production, and so on, great efforts should be put into providing instructions which will enable the consumer to reheat the meals while maintaining the maximum quality. It is often this final step of reheating where eating quality can be reduced but more importantly, where insufficient heating could result in an increased risk of not destroying any pathogen bacteria, which if present, could lead to food poisoning [68].

The prediction of reheating times and temperatures using microwave or conventional ovens is possible using computer programs with data on the thermophysical properties of the food, the reheating conditions, and the growth behavior of the pathogenic bacteria in question [69–71]. The complexity of the prediction is increased by the range of different food products, their layout in the pack, variations in thickness, product density, moisture content, and so on.

When microwave ovens became more popular as the market in chilled and frozen ready meals grew, attention focused on their safe use by the consumer. Concern was raised that microwave ovens had no consistent method of power settings unlike conventional gas and electric ovens [72]. This made it very difficult for food manufacturers to provide meaningful instructions to consumers on how to set the controls on their microwave ovens for thawing or reheating frozen meals. A survey showed that when following the instructions provided on the packs of frozen ready meals, many microwave ovens were failing to heat the center of these meals to 70°C for 2 min or equivalent, that is, a pasteurization treatment, which is necessary for microbiological safety [73]. A later report reviewed progress in overcoming this problem using beef lasagne meals and showed that the situation had changed little [74]. This led to a voluntary scheme where the power rating of microwaves could be taken into account on the frozen meal reheating instructions. More detailed instructions were suggested to allow longer heating times, use resting periods for temperature equalization, and retaining the pierced film lid during reheating.

The layout of meal components also has a significant effect on the heating characteristics of chilled, frozen, and ambient stable ready meals [75]. It was found that foil plates provided a more consistent temperature distribution in the food during reheating but increased the reheating time; that using a cover reduced reheating time and weight loss and that each microwave oven design gave different heating capabilities.

A solution put forward by one manufacturer is to have codes printed on packs of frozen or chilled food which when entered into a particular microwave via a keypad will provide the information necessary to adjust itself for that particular food and give optimum results in terms of cooking time, power, and so on [76].

Dual ovenable containers for frozen ready meals have to withstand temperatures from $-40°C$ during storage to reheating at up to 220°C in microwave as well as conventional gas or electric ovens [77]. They must also be able to protect the food and facilitate cooking or reheating. Suitable materials are aluminum foil trays, paperboard trays, and heatproof plastic containers that can have anti-stick surfaces or be suitable for direct serving to the table. Several types of ovenable board used in these frozen ready meals containers have been tested [78]. Results showed that migration of chemical components from the plastic coated boards into a simulated food at 160° and 230°C was not above the overall acceptable limits.

A study of many of the cooking instructions of the meals shown in Table 21.1 stress that they are for guidance only and that the consumer should make sure that the food is "piping hot" before consumption. This vague and undefined phrase could be seen acting essentially as a "get-out clause." Perhaps, persuading users of frozen ready meals to invest in an electronic thermometer might be a more sensible method to check the end temperature of reheated meals.

E. Factors Affecting Nutritional Quality of Frozen Meals

The effect of frozen storage on the nutritional quality of individual types of frozen foods is well documented in other chapters. A common assumption is that fresh home-prepared foods will have a higher nutritional value than manufactured foods such as frozen ready meals [33]. However, in many cases this is untrue because of the rapid and highly organized methods of harvesting and freezing, for example, peas. This contrasts with the time which food can spend in the chain of producers, wholesalers, and retailers before the consumer is able to purchase and prepare them, during which the nutritional content can decrease significantly.

Different systems used for preparing frozen ready meals can have varying effects on the nutritional content of the meals. A survey of the vitamin C content of potato processed through different meals-on-wheels delivery systems (cook-chill, cook-hold in warm or insulated containers) found that industrially cooked and frozen potatoes provided the highest retention (70%) but lowest sensory quality because of poor texture [79].

Surveys of cook-freeze systems in hospitals produced a large amount of data comparing vitamin C content of frozen vegetables just freshly cooked to those at the point of service [45]. For cabbage, the percentage retention varied from 51 to only 9% (11–1 mg/100 g), for peas from 45 to 37% (5–3 mg/100 g), and for new potatoes from 70 to 80% (12–9 mg/100 g). In other work, the use of frozen vegetables in hospital feeding was simulated using commercially frozen and chilled broccoli, peas, boiled potatoes, and mashed potatoes [80]. The products were reheated in forced convection ovens directly from the frozen state or after 3 or 5 days of chilled storage. The percentage retention of vitamin C compared to the initial level just before reheating was 27.2% for frozen broccoli (5.2 mg/100 g), 53.3% for frozen peas (5.5 mg/100 g), 82.6% for frozen boiled potatoes (9.8 mg/100 g) and 46.6% for frozen mashed potatoes (0.7 mg/100 g).

In the project on freeze-chilling previously cited, the vitamin C content of instant mashed potato and steamed broccoli, blast-frozen at −35°C for 2.5 h and stored at −25°C for up to 32 weeks was measured [49]. Vitamin C content declined significantly over the period: from 3 to 1.2 mg/100 g for mashed potato ($p < 0.05$) and for steamed broccoli from 37 to 25 mg/100 g ($p < 0.01$). Further work on the effect of freeze-chilling on vitamin C content of frozen mashed potato frozen at −35°C for 2.5 h and then stored at −25°C for up to 12 months, showed that freeze-chilling led to a significantly lower value than conventionally frozen mashed potato: 0.64 compared with 1.68 mg/100 g but storage time had no effect [50].

The effect of freezing on foods prepared using the sous-vide method on β-carotene in carrots, vitamin C in broccoli, and thiamine in cod and salmon found that β-carotene values and vitamin C levels did not change through the freezing step [40]. However, thiamine levels in salmon decreased from 0.23 to 0.191 mg/100 g and in cod from 0.084 to 0.064 mg/100 g. The vitamin C content of the spinach component of Chicken Velouté prepared using cook-chill, sous-vide (storage at 4°C) or cook-freeze (storage at −14°C) systems was compared with freshly prepared samples after 6 days [53]. The content decreased over the whole 6 days for cook-chill and over the last 3 days for sous-vide and cook-freeze.

The effects on the fatty acid composition of sardine fillets of using three different methods for cooking (frying, oven-baking, grilling), followed by frozen storage at −20° for 4 months and two reheating methods (oven, microwave), showed that fried fish reheated by microwave were more dehydrated, while oven baking changed the fatty acid profile least [81]. Overall, the sequence of cooking, freezing, and reheating led to thermal oxidation, which would reduce the level and therefore, the positive health effect of docosahexaenoic acid, an essential omega-3 fatty acid in the sardine and other oily fish.

A survey of the trans-fatty acid (TFA) content (linked to coronary heart disease) of various convenience foods found that frozen prepared meals were not major sources of TFA, providing less than 1 g/100 g meat or serving, much less than dehydrated convenience foods [82]. Work on another lipid component related to heart disease, oxysterols, an oxidation product of cholesterol found that its quantity doubled in frozen ready meals over 3 months of storage at −20°C [83].

Surveys of commercially manufactured frozen ready meals in Germany [84] and Netherlands [85] both concluded that the meals contained too much fat and protein and too little carbohydrate compared with the optimal nutritional content.

The salt content of ready meals has recently been surveyed showing that many meals provide in a single serving up to 98% of the recommended daily intake of 6 g [86]. Some meals aimed at children provide more than 40% of the 5 g intake recommended for children. Table 21.1 shows some examples of the amount of salt included in some UK frozen ready meals. In most cases, the figure given on packs is for sodium, which has to be multiplied by about 2.5 to give the actual amount of salt.

Ideally, frozen ready meals will be reheated as required for almost immediate consumption but in many food service and institutional settings, consumption may be delayed for logistical reasons. During this period of warm-holding, nutritional deterioration can take place as well as the visual deterioration mentioned earlier.

V. CONCLUSIONS

The frozen ready meal has undoubtedly developed since being fed to troops heading overseas in 1944. Most aspects of understanding its quality attributes have now been the subject of research encouraged by the drive of manufacturers to develop new products and recipes and to exploit the consumer's desire for convenience in partnership with the rise in the use of microwave ovens.

The manufacture of frozen ready meals has advanced as pretreatment procedures for sauces, meat, vegetables, and carbohydrate components have become better understood and hence can avoid problems of deterioration during frozen storage and reheating. The main concerns for manufacturers relate to the nutritional composition of frozen ready meals, especially the level of fat and salt. These are of increasing concern to consumers and to those responsible for promoting and encouraging healthier diets in the population so these topics will have to be addressed.

Retaining the qualities put into frozen ready meals by manufacturers so that they are still there to be enjoyed and appreciated by the consumer still relies very much on education and training. Consumers are becoming more familiar with the use of microwave ovens for providing quick and convenient meals but in the food service and institutional feeding sectors, catering systems including cook-freeze still seem to have a poor reputation for food quality. In this case, training staff to recognize the limitations imposed by the systems and to appreciate their role in assuring the quality of the food produced by the system would go a long way in satisfying the consumer's needs.

REFERENCES

1. S Shephard. *Pickled, Potted and Canned — The Story of Food Preserving*. London: Headline, 2000, pp. 298–305.
2. Anonymous. *About Birdseye — Our history*. Birdseye Ltd., Freepost UK, 2004.
3. S Martin. The rise of prepared and precooked frozen foods. In: DK Tressler, WB Van Arsdel, MJ Copley, Eds., *The Freezing Preservation of Foods*. Vol. 4. USA: AVI Publishing Co Inc., 1968, pp. 1–30.
4. DK Tressler. Complete meals. In: DK Tressler, WB Van Arsdel, MJ Copley, Eds., *The Freezing Preservation of Foods*. Vol. 4. USA: AVI Publishing Co Inc., 1968, pp. 512–521.
5. W Woloson. Frozen Entrées. *St James Encyclopaedia of Popular Culture*. USA: St James Press, 1999.
6. L Ross. *Defrosted Dinners*. The New Yorker, 4th August, 1945.
7. PG Creed. Chilling and freezing of prepared consumer foods. In: D-W Sun, Ed., *Advances in Food Refrigeration*. U.K.: Leatherhead Publishing, 2001, pp. 438–471.
8. Anonymous. *Frozen Ready Meals — UK*. London: Mintel International Group Ltd., March 2004.
9. E Shove, D Southerton. Defrosting the freezer: from novelty to convenience — a narrative of normalization. *Journal of Material Culture* 5 (3):301–319, 2000.
10. Anonymous. *Frozen Ready Meals Statistics, Global Market Information Database*. London: Euromonitor International, 2004.
11. Anonymous. *Chilled Ready Meals Statistics, Global Market Information Database*. London: Euromonitor International, 2004.
12. Anonymous. *The World Market for Frozen Food — Frozen Ready Meals*. London: Euromonitor International, January 2004.
13. Anonymous. *Chilled Ready Meals — UK*. London: Mintel International Group Ltd., May 2004.
14. Anonymous. *Our Food*, Birdseye Ltd., Freepost NATE 139, Milton Keynes, MK9 1BR, UK, 2004.
15. Anonymous. *Products*, Findus Ltd., PO Box 188, Newcastle upon Tyne, NE12 8WP, UK, 2004.
16. Anonymous. *Products*, Patak's Foods Ltd., Kiribati Way, Leigh, WN7 5RS, UK, 2004.
17. E Jones. Marketing frozen foods. In: MC Erickson, YC Hung, Eds., *Quality in Frozen Food*. New York: Chapman & Hall, 1997, pp. 426–441.
18. H Symons. Frozen foods. In: CMD Man, AA Jones, Eds., *Shelf Life Evaluation of Foods*. London: Blackie Academic & Professional, 1994, pp. 296–316.
19. T Miner. Restaurant Branded Products: The Future of HMR? — home meal replacement. *Prepared Foods* October, 79, 2001.
20. PG Creed. The potential of foodservice systems for satisfying consumer needs. *Innovative Food Science and Emerging Technologies* 1:219–227, 2001.
21. J Tak, MB Gregoire, S Hearne-Morcos. Commercial frozen meals: a cost-effective alternative for home-delivery in feeding programs for the elderly? *Journal of Nutrition for the Elderly* 12 (3): 15–25, 1993.

22. MB Gregoire, N Nyland, S Morcos. Use of frozen meals by and food preferences of various age groups of adults. *Journal of Nutrition for the Elderly* 13 (2):23–27, 1993.
23. C Leighton, C Seaman. Food retailing: an opportunity for meeting elderly consumers' needs. *Nutrition and Food Science* 97 (4/5):4i–4iv, 1997.
24. F Katz. 'How nutritious?' meets 'How convenient?' *Food Technology* 53 (10):44, 47, 48, 50, 1999.
25. PK Yen. Ready-to-eat meals. *Geriatric Nursing* 18:182–183, 1997.
26. GA Escoffier. *A Guide to Modern Cookery*, 2nd ed., London: Heinemann, 1957.
27. H McGee. *On Food and Cooking*. London: Allen and Unwin, 1986, pp. 327–366.
28. ML Shaevel. Manufacturing of frozen prepared meals. In: CP Mallett, Ed., *Frozen Food Technology*. London: Blackie Academic & Professional, 1993, pp. 270–302.
29. E Vogelaers. Recipe development. In: *Proceedings of the Second European Symposium on Sous Vide*. Belgium: Leuven, 1996, pp. 157–172.
30. MA Hill. The effect of refrigeration on the quality of some prepared foods, In: S Thorne, Ed., *Developments in Food Preservation* — 4. London: Elsevier Applied Science, 1987, pp. 123–152.
31. NE Zaritzky. Factors affecting the stability of frozen foods. In: CJ Kennedy, Ed., *Managing Frozen Foods*. Boca Raton: CRC Press, 2000, pp. 111–135.
32. MP Cano. Vegetables. In: LE Jeremiah, Ed., *Freezing Effects on Food Quality*. New York: Marcel Dekker, 1996, pp. 247–298.
33. CJ Kennedy. Freezing processed foods. In: CJ Kennedy, Ed., *Managing Frozen Foods*. Boca Raton: CRC Press, 2000, pp. 137–158.
34. RG Moreira, ME Castell-Perez, MA Barrufet. *Deep-Fat Frying — Fundamentals and Applications*. U.S.A.: Aspen Publishers Inc., 1999.
35. L Kobs. Frozen pasta and rice dishes. *Food Product Design*, 10 (8):124–126, 129–130, 133–134, 137–143, 2000.
36. SM Fiszman, A Salvador. Recent developments in coating batters. *Trends in Food Science and Technology* 14:399–407, 2003.
37. PL Dawson. Effects of freezing, frozen storage, and thawing on eggs and egg products. In: LE Jeremiah, Ed., *Freezing Effects on Food Quality*. New York: Marcel Dekker, 1996, pp. 337–366.
38. PG Creed, WG Reeve. Principles and applications of sous vide processed foods. In: S Ghazala, Ed., *Sous Vide and Cook Chill Processing for the Food Industry*. U.S.A.: Aspen Publishers Inc., 1998, pp. 25–56.
39. MS Wexler. The new high-tech cooking. *Hotels* 26 (4):64–66, 68, 1992.
40. FS Tansey, TR Gormley, P Bourke, D O'Beirne, JC Oliveira. Texture, quality and safety of sous vide/ frozen foods. In: *Proceedings of the Fourth International Conference on Culinary Arts and Sciences*. Sweden: Örebro, 2003, pp. 199–207.
41. DR Heldman, RW Hartel. *Principles of Food Processing*. U.S.A.: Chapman & Hall, 1997.
42. CJ Kennedy. Future trends in frozen foods. In: CJ Kennedy, Ed., *Managing Frozen Foods*. Boca Raton: CRC Press, 2000, pp. 263–278.
43. MC Erickson. Lipid oxidation: flavor and nutritional quality deterioration in frozen foods. In: MC Erickson, YC Hung, Eds., *Quality in Frozen Food*. New York: Chapman & Hall, 1997, pp. 141–173.
44. J Kristott. Fats and oils. In: D Kilcast, P Subramaniam, Eds., *The Stability and Shelf-Life of Food*. Boca Raton: CRC Press, 2000, pp. 279–309.
45. M Turner, J Mottishaw, R Zacharias, A Bognar. Sensory quality and nutritive value of meals prepared from fresh and pre-processed components. In: P Zeuthen, JC Cheftel, C Eriksson, M Jul, H Leniger, P Linko, G Varela, G Vos, Eds., *Thermal Processing and Quality of Foods*. London: Elsevier Applied Science, 1984, pp. 371–402.
46. BN Kowale, VK Rao, NP Babu, N Sharma, GS Bisht. Lipid oxidation and cholesterol oxidation in mutton during cooking and storage. *Meat Science* 43:195–202, 1996.
47. VK Rao, BN Kowale, NP Babu, GS Bisht. Effect of cooking and storage on lipid oxidation and development of cholesterol oxidation products in water buffalo meat. *Meat Science* 43:179–185, 1996.
48. AS Ratushnyi, LL Bushkova. Variations in lipids contents of hen meat in frozen meals during storage (in Russian). *Isvestiya Vysshikh Uchebnykh Zavedenii–Pishchevaya Tekhnologiya* (2):40–43, 1980.
49. E O'Leary, TR Gormley, F Butler, N Shilton. The effect of freeze-chilling on the quality of ready-meal components. *Lebensmittel-Wissenschaft und Technologie* 33:217–224, 2000.

50. GA Redmond, TR Gormley, F Butler. The effect of short- and long-term freeze-chilling on the quality of mashed potato. *Innovative Food Science and Emerging Technologies* 4:85–97, 2003.
51. GA Redmond, TR Gormley, F Butler. The effect of short- and long-term freeze-chilling on the quality of cooked green beans and carrots. *Innovative Food Science and Emerging Technologies* 5:65–72, 2004.
52. NU Haase, B Putz. Freeze suitability of potatoes — a challenge to technology (in German). *Luft und Kälte-Technik* 32 (32):310–313, 1996.
53. DB Smith, L Fullum-Bouchard. Comparative nutritional, sensory and microbiological quality of a cooked chicken menu item produced and stored by cook/chill, cook/freeze and sous vide cook/chill methods. In: *Proceedings of Canadian Dietetic Association Annual Conference*, Ottawa, Canada, 1990, 6 pp.
54. FS Tansey, TR Gormley. Sous vide/freezing of ready meals. *Farm Food* 12 (1):18–22, 2002.
55. EW Holynski, JN Auckland, G Glew. A review of the literature concerning warmholding of foods in catering. In: P Zeuthen, JC Cheftel, C Eriksson, M Jul, H Leniger, P Linko, G Varela, G Vos, Eds., *Thermal Processing and Quality of Foods*. London: Elsevier Applied Science, 1984, pp. 403–424.
56. J Koscher, A Bart, JP Dezavelle, R Rosset, F Lebert, P Liger, E Lecrivain, S Dunas. Experimental preparation of frozen cooked meals for institutional catering (in Italian). *Freddo* 35 (4):251–255, 1981.
57. JT Lin, SH Chen, HH Kuo. Research and development of Chinese-style frozen prepared dishes and evaluation of their quality (in Chinese). *Food Science — Taiwan* 24:203–219, 1997.
58. B Fu, TP Labuza. Shelf-life testing: Procedures and prediction methods. In: MC Erickson, YC Hung, Eds., *Quality in Frozen Food*. New York: Chapman & Hall, 1997, pp. 377–415.
59. R Gormley, T Walshe, K Hussey, F Butler. The effect of fluctuating vs. constant frozen storage temperature regimes on some quality parameters of selected food products. *Lebensmittel-Wissenschaft und Technologie* 35:190–200, 2002.
60. PJ Fellows. *Food Processing Technology — Principles and Practice*, 2nd ed., Cambridge: Woodhead Publishing, 2000.
61. WEL Spiess. The shelf life of deep frozen food products — Confidential draft for Codex Alimentarius, 1980 cited in M Jul, Ed. *The Quality of Frozen Foods*. London: Academic Press, 1984.
62. DL Archer. Freezing: an underutilized food safety technology? *International Journal of Food Microbiology* 90:127–138, 2004.
63. Anonymous. *Food and Drink Good Manufacturing Practice — A Guide to its Responsible Management*, 4th ed., IFST (Institute of Food Science and Technology), London, 1998.
64. D Burfoot, K Brown, Y Xu, SV Reavell, K Hall. Localised air delivery systems in the food industry. *Trends in Food Science and Technology* 11:410–418, 2000.
65. A Fabrega-Fernandez, ML Forcadell-Berenguer. Microbiological study of frozen ready meals for bulk sale, (in Spanish). *Alimentaria* (229):61–68, 1992.
66. MG Cordoba. Microbiological quality of frozen ready meals, (in Spanish). *Alimentaria* (296):85–88, 1998.
67. DA Shapton, NF Shapton. *Principles and Practices for Safe Processing of Foods*. Oxford: Butterworth-Heinemann, 1991, pp. 377–444.
68. H Kolb. Defective cooking instructions impair results. Main report on the 1998 German Agricultural Society quality tests on frozen food (in German). *Fleischwirtschaft* 79 (4):42–44, 1999.
69. P Nesvadba, M Houska, W Wolf, V Gekas, D Jarvis, PA Sadd, AI Johns. Database of physical properties of agro-food materials. *Journal of Food Engineering* 61:497–503, 2004.
70. P Verboven, AK Datta, NT Anh, N Scheerlinck, BM Nicolaï. Computation of airflow effects on heat and mass transfer in a microwave oven. *Journal of Food Engineering* 59:181–190, 2003.
71. T Martens, M Schellekens, B Nicolaï, J De Baerdemaeker. Computer aided process design for minimally processed foods. In: *Proceedings of the ACoFoP 3 Symposium — Automatic Control of Food and Biological Processes*. Paris, France, 1994, pp. 664–652.
72. M Hill. The effect of microwave processing on the chemical, physical and organoleptic properties of some foods. In: S Thorne, Ed., *Developments in Food Preservation — 1*. London: Applied Science Publishers, 1981, pp. 121–151.

73. JR Bows, PS Richardson. The influence of the thermal, electrical, and physical properties on the quality of foods heated by microwaves: II, Technical Memorandum No. 573, Campden Food and Drink Research Association, UK: Chipping Campden, 1990.
74. RM George, SA Burnett, PS Richardson. The influence of the thermal, electrical, and physical properties on the quality of foods heated by microwaves: final report. Technical Memorandum No. 624, Campden Food and Drink Research Association, UK: Chipping Campden, 1991.
75. RM George, DG Evans, GI Hooper, GM Campbell, PAP Dobie. Assessment and improvement of the manufacturers' reheating instructions for microwaveable lasagne based upon the voluntary UK new microwave labelling scheme, Microwave Science Series — 14th Report, London: MAFF Publications, 1995.
76. P Berezai. *Ready Meal Trends*, World of Ingredients, (December): 32–33, 2001.
77. VM Balasubramaniam, MS Chinnan. Role of packaging in quality preservation of frozen foods. In: MC Erickson, YC Hung, Eds., *Quality in Frozen Food*. New York: Chapman & Hall, 1997, pp. 296–309.
78. B Aurela, M Vuorimaa, H Lindell. Migration from ovenable boards at high temperatures. *Nordic Pulp and Paper Research Journal* 15:150–154, 2000.
79. BE Mikkelsen. The quality of potatoes in four different meals-on-wheels systems. *Journal of Foodservice Systems* 3:241–256, 1985.
80. A West. Meals assembly system and nutritional implications. In: *Proceedings of the Third International Conference on Culinary Arts and Sciences*. Egypt: Cairo, 2001, pp. 175–184.
81. MT García-Arias, E Álvarez-Pontes, MC García-Linares, MC García-Fernández, FJ Sánchez-Muniz. Cooking-freezing-reheating (CFR) of sardine (*Sardina pilchardus*) fillets. Effect of different cooking and reheating procedures on the proximate and fatty acid composition. *Food Chemistry* 83:349–356, 2003.
82. M Henninger, F Ulberth. Trans-fatty acid content of convenience food (in German). *Zeitschrift für Ernährungswissenschaft* 36:161–168, 1997.
83. JE Pie, K Spahis, C Seillan. Cholesterol oxidation in meat-products during cooking and frozen storage. *Journal of Agricultural and Food Chemistry* 39 (2):250–254, 1991.
84. CC Metges, B Kastel, G Wolfram. Meals-on-wheels for the elderly in Munich — Evaluation of frozen lunch meals and characteristics of recipients (in German). *Ernährungs-Umschau* 41 (5):191–194, 1994.
85. JPH Linssen, JL Cozijnsen, AA van den Driessche. Composition of various deep frozen meals (in Dutch). *Voedingsmiddelen Technologie* 22 (5):36–37, 1989.
86. Anonymous. Ready meal salt levels revealed, Food Standards Agency, London, 2003.

22 Quality and Safety of Frozen Bakery Products

Virginia Giannou and Constantina Tzia
National Technical University of Athens, Athens, Greece

Alain Le Bail
ENITIAA, Ecole des Mines de Nantes, Univ. Nantes, Nantes, France

CONTENTS

I. INTRODUCTION

Bread and cereal products are considered as some of the oldest components of the human diet. With the ages, enormous changes have been accomplished in their formulation, characteristics, and methods of preparation. Nowadays, bakery products present a great variety, capable of satisfying the needs and demands of almost every consumer. However, they have a common attribute. The shelf-life of these products is limited and their flavor, aroma, and textural characteristics are degraded rapidly after baking. To overcome these problems and extend the product shelf-life, several preservation methods have occasionally been proposed, the most important of which is freezing.

The application of low temperatures renders the production of bakery products particularly flexible and effective. Frozen bakery products and products made from frozen dough can remain unadulterated for weeks or months provided that they are stored under suitable conditions. They can be stored, thawed, proofed, and baked in quantities proportional to daily demand, even from minimally skilled personnel at in-store bakeries, restaurants, institutions, and supermarkets, with limited requirement in equipment, providing consumers with freshly baked products any time of the day [1–5].

Nevertheless, the use of freezing might result in some limitations or in specific quality problems of the final products such as increased proof times, decreased inflation ability, and variable textural properties. These problems render generation of research and quest for solution more compulsory than ever before because consumers and buyers are becoming increasingly aware of the importance of high-quality and safe food products. The modern trend in the breadmaking industry is to keep processing as much as possible at the industrial level to produce products which require minimal knowhow for final preparation before consumption. The main steps of freezing in the breadmaking will be presented in this chapter. The quality-related problems and the technological methods through which frozen bakery products can maintain their performance during prolonged frozen storage and increase their appeal will also be discussed.

II. MANUFACTURE OF BAKERY PRODUCTS

A. Western European Bakery Market

The western European bread industry produces 25 million tons of bread per year and 5 million tons of biscuits, cakes, and pastries. Germany is by far the largest producer of bread within the European market, followed by the U.K., France, Italy, and Spain, whereas the average consumption of bread in Europe at the present is 65 kg per capita per year [6].

However, bread demands and eating habits may extremely vary across Europe. More specifically, in countries such as Germany and France, bread is considered as a major daily meal component, whereas in the U.K. it is an accompaniment to other foods. In addition, some bread types are more favorable than others in different countries. Germany, Norway, Belgium, and the Netherlands mainly consume brown, wholemeal, and mixed wheat/rye bread, whereas in France white bread, mainly baguette, dominates the market [6].

B. Conventional Bakery Products

Bakery products are considered to be important for a balanced and nutritious diet because they are rich in carbohydrates, which offer the essential energy for human body functions. They are also supplied with several important B complex vitamins, vitamin E, minerals (calcium, iron), and dietary fiber, especially the wholegrain and wholemeal ones. Although bread is a carbohydrate-based food, it is also considered a good source of protein in the diet. Moreover, most bakery products are also low-fat.

Dough products can be categorized according to the ingredients used in the baking formula into lean, normal, sweet, and dietetic products. Lean dough bakery products consist of flour, water, yeast, and salt. Their main characteristic is total lack of lipids or shortening. Normal dough bakery products may additionally include sugar, lipids, or milk in small quantities. Sweet dough formula includes relatively large quantities of ingredients such as sugar, lipids, or milk and it can also include eggs, spices, or aromatic compounds. Dietetic products may be low-sugar, sugar-free, low-cholesterol, diet, fiber-enriched, and so on, and their formulation varies according to product characteristics. There are also several ethnic bakery products, which present an enormous variety in ingredients and production methods according to every country's history, tradition, or religion.

Breadmaking and especially dough formation is one of the most complex and impressive functions in food preparation. The production of bakery products begins by mixing of raw material (mainly flour, yeast, and salt) with water and occasionally with various other ingredients (sugars, shortening, oxidizing agents, etc.). This results in a series of complex changes and interactions between those diverse components and finally in the formation of the gluten network and the development of a cohesive and viscoelastic dough [5,7,8].

Bakery products that contain yeast as a leavening agent can be produced according to the following methods [1,9–11].

1. Straight Dough Method

This is the simplest fermentation method and no preferments are involved. It is a single-step process in which all the ingredients of the formula are mixed into a single batch until an optimally developed dough occurs. The dough is then allowed to ferment. The sequence in which ingredients are accumulated into the dough may differ due to different equipment or manufacturer preference. Sometimes, salt is added after the dough is partially mixed because it tends to interfere with the gluten development in the dough. Retarding salt incorporation in the mixer also assists yeast dissemination and limits the osmotic stress.

2. Sponge and Dough Method

In the sponge and dough method, fermentation is mostly carried out through a preferment called sponge because when it is fully fermented and conditioned it resembles to the sponge used for cleaning. Part of the formula flour (normally 50–70%) is subjected to the physical, chemical, and biological actions of fermenting yeast until a fairly firm yeast culture is obtained. Sponge fermentation times may vary considerably, as may the composition of the sponge. After fermentation, the sponge is mixed with the rest of the formula ingredients for optimum dough development. In some cases, the sponge component may be replaced with a flour brew in which the proportion of liquid is much higher than that used in a sponge.

3. Sourdough (Levain) Method

Sour dough fermentation includes the development of a typical microflora, which consists of hetero-fermentative lactic acid bacteria (lactobacilli) and wild yeasts. Lactic acid bacteria lower the pH by producing lactic and acetic acids, which modify products flavor toward an acid character. The yeasts are generally responsible for the leavening action via carbon dioxide production. This microflora usually must undergo two or three builds to be fully conditioned and incorporated into the dough as the only leavening agent.

4. Mechanical Dough Development

The common attribute of all mechanical dough development methods is that there is no fermentation period, when dough is largely, if not entirely developed in the mixing machine. The physicochemical changes, which normally occur during bulk fermentation periods, are achieved in the mixer through the addition of improvers, extra water, and a specifically planned level of mechanical energy. Several systems have been developed and some of the most popular are: the continuous system in the U.S., the Chorleywood process in the U.K., and the Brimec process in Australia.

5. Rapid Processing

Rapid dough processing includes a multitude of slightly different breadmaking methods, which may have evolved based on different combinations of active ingredients and processing techniques. The common characteristic of all those breadmaking processes is the inclusion of improvers in the formula to assist dough development and reduce the fermentation period to less than 1 h.

C. Ingredients Used in Breadmaking

1. Flour

Flour is the most important ingredient in essentially all bakery products as it is the determinant for the rheological properties of both doughs and batters. It can derive from wheat, rye, barley, corn, oats, amaranth, millet, and so on and consists of proteins, starch and nonstarch polysaccharides, fibers, lipids, water and small amounts of vitamins, minerals, and enzymes [12]. Wheat flour is the most common flour used in breadmaking because of its unique ability to provide a light, palatable, well-risen loaf of bread when processed into fermented dough [13]. The two classes of proteins in wheat flour are prolamins called gliadin and glutelins called glutenin. When these proteins are wetted separately, they present totally different behavior; gliadin forms a viscous, sticky, and inelastic liquid, whereas glutenin forms a more elastic and tenacious rubbery material [14]. On the contrary, when their mixture, called gluten, is wetted, as during the preparation of dough, they form a cohesive and elastic three-dimensional network stabilized by thiol–disulfide exchange reactions among gluten proteins, which provide wheat with its functional properties [15].

The commercial value of wheat flour depends on a number of factors such as hardness, gluten strength, protein content, ash content, color, moisture content, and level of enzyme activity. Different kinds of bakery products require flours with different properties. For example, flour that is good for bread production may not be satisfactory for producing cookies. Yeast-leavened products (bread, rolls, etc.) usually require flours with relatively high protein content (>10.5%), high water absorption, and moderate levels of enzyme activity, which can produce doughs with a good balance of elastic and viscous properties. In reverse, chemically leavened products (cookies, cakes, crackers, etc.) require wheat flours with low ash, relatively low protein content (8–10%), and low contents of damaged starch. However, even relatively poor-quality wheat can produce bread that is significantly more palatable than that made with flour from other cereal grains [13].

2. Water

Water is a unique compound, which has the ability to impart to the materials dissolved in it unusual and often unexpected properties. Water is needed for the formation of dough and the development of its rheological properties while it is responsible for dough fluidity and acts as a plasticizer and a reaction medium. Water is essential for the dissolution of salt or sucrose and for the hydrolysis of sugars or starch. It assists yeast cells dispersion and food transportation to them through cell membranes. It is important for starch gelatinization during baking and contributes to ovenspring through vaporization. Finally, it can activate flour enzymes and provoke the formation or alteration of bonds between flour macromolecules [16,17].

3. Yeast

The most common yeast used in breadmaking is *Saccharomyces cerevisiae*, but sometimes other strains can be used as well (e.g., wild yeasts and *S. exiguus* in sour dough processing). Yeast cells metabolize fermentable sugars (glucose, fructose, dextrose, and maltose), under anaerobic conditions, into alcohol and carbon dioxide, which acts as a leavening agent and enhances dough structure by inflating air cells formed during mixing within the three-dimensional gluten matrix. Yeast also supports aromatic compounds production and forms carbonic acid, which lowers the pH of the dough and enhances the flavor characteristics of finished baked products. Active cells of yeast are available in compressed or in dried form. The compressed type contains 15.5% protein, about 12–14.5% carbohydrates, and approximately 70% moisture and so it is highly perishable unless it is refrigerated. Active dry yeast is produced by extruding compressed yeast in fine strands, which are dried to low-moisture content. Instant yeast is made from more active strains of yeast and dried faster to a lower moisture level [12].

4. Salt

Salt is generally used at levels of about 1–2%, based on the flour weight, and is considered a fundamental ingredient for the production of many bakery products. Apart from enhancing the flavor and appeal of final products, it has several other functions in breadmaking. It affects the rheological properties of dough, as it improves both cohesiveness and elasticity, and strengthens the gluten network. Salt influences yeast performance, favors the action of amylases, and therefore induces the production of maltose as yeast food, and inhibits the action of flour proteases, which otherwise would depolymerize gluten proteins [12,18]. Finally, it is believed to facilitate the development of crust color and influence the shelf-life of bakery products due to its hydroscopic properties [10].

5. Sugars

Sugars are normally incorporated in the formula of the most bakery products as the initial source of fermentable carbohydrates for the yeast. This is important because flour enzymes are not capable of producing sufficient amounts of sugars to maintain fermentation and gas production during the early stages of fermentation. Sugars are commonly added in small amounts and in the form of cane sugar or beet sugar or of various hydrolysates of corn starch (corn syrup, dextrose, etc.). Sugars can also affect the textural characteristics, the taste, and the appearance of bakery products mainly by improving crust color. They also act as antiplasticizers retarding pasting of native starch or function as antistaling ingredients inhibiting starch recrystallization [19]. Sometimes, several other sweeteners can be used in breadmaking for special purposes, such as to enhance flavor and avoid health or diet aggravation [16].

6. Lipids

Lipids can be used in breadmaking either in the form of fats or oils and are usually referred to as "shortening" because they "shorten" (tenderize) the texture of the finished product. They are an optional ingredient in bread but when incorporated in the formula, they can improve dough handling, viscoelastic properties, gas retention ability, and ovenspring by interacting with gluten proteins during dough mixing [20]. They also affect the mouthfeel, flavor, crumb appearance, and crust texture of the final products [21,22]. Lipids improve product quality by enhancing softness/tenderness, moistness/lubricity, and due to their antistaling properties they extend shelf-life. Both endogenous lipids and added fats are known to play an important role during bread-making and staling of bread [23,24].

7. Other Ingredients

Apart from the ingredients mentioned above, several others can be incorporated in the dough formula, usually in small proportions or in special bakery products. These can either be milk or egg products, flour additives (improvers, maturing agents, enzymes, or vital wheat gluten), dried fruits, spices and herbs, dough conditioners/strengtheners, oxidizing agents, preservatives, and so on [11].

Traditional breadmaking initially involves mixing of raw materials in horizontal, planetary, vertical, or continuous mixers. Then molding or proofing occurs. Dough make-up is accomplished with appropriate dividers, which scale the bulk dough into units of predetermined weight, rounders, which impart a spherical form to the dough pieces and seal their raw cut surfaces with a fine skin to preserve excessive loss of carbon dioxide, and molders which sheet and mold dough pieces into the final loaf form. Proofing (initial, intermediate, and final) usually takes place in controlled climate rooms of sectional construction, which should maintain uniform humidity and temperature, and minimize physical shock to dough pieces. Finally baking is conducted in deck, reel, rack, band, traveling tray, tunnel, or conveyorized ovens [11,25].

D. Sensory Characteristics of Bakery Products

Bakery products, especially when they are freshly baked, present extremely attractive sensory characteristics, which are attributed to dough ingredients, processing methods, and duration of fermentation and of course baking. Most of the ingredients incorporated in fermented products mixture generate the formation of flavor compounds within the dough. Flour lends a fairly bland flavor with most of its contribution coming from the germ (embryo) oils and bran particles. Therefore, wholemeal, wholewheat, bran, and germ-enriched white flours are expected to provide bread with more enhanced flavor than white flours. The addition of salt or other ingredients such as fat, sugar, or milk also generally imparts special flavor to bakery products, which strongly depends on their proportion in the dough formula [26].

When bread is prepared from a naturally fermented sponge or sourdough, it exhibits a light cream-colored crumb with grayish tones, a very definite and distinctive odor and taste, a sharp acetic acid flavor, and a wholesome rustic flavor and aroma. It is pleasant to chew and has especially attractive eating qualities. Bread leavened with baker's yeast presents a golden crust, a creamy white crumb, and an attractive aroma or flavor, which derives from the combination of wheat flour with alcoholic fermentation and caramelization of sugars in the crust and provides particularly pleasant eating properties and wonderful taste. The level of yeast used in the recipe also makes its own unique contribution to bread flavor.

However, the most important contribution to bread flavor hails from baking. During this processing step, several flavor compounds are suppressed, modified, or formed. Numerous changes and reactions occur, the most important of which is the nonenzymatic "Maillard browning." This involves the reaction of reducing sugars with amino groups during baking and is responsible for the formation of a highly flavored dark, mostly brown crust on the outer surfaces of the dough which is very important to the flavor perception of most baked foods [1,10,26].

III. FREEZING OF BAKERY PRODUCTS

A. Frozen Bakery Products

During the last decades, breadmaking has been slowly but significantly captured by the industry. The use of refrigeration or freezing in the food industry, which started in the late 19th century, is becoming increasingly popular to the breadmaking industry as it provides bakery products with extended shelf-life, postpones the proofing-baking phase, and allows the benefits of producing freshly baked products while saving on equipment and labor costs [27].

The first kind of product developed in the 1970s was the frozen fully baked bread. Earlier in the 1960s, the frozen part baked bread called "brown and serve" was already proposed. This product was the main frozen bakery product in France. Frozen fully baked products had a limited success due to crust-flaking problems. At that time, yeasted frozen dough was introduced, which has now become the leading product in terms of the market share of frozen bakery products. This technology consists of preparing a dough and freezing it before fermentation starts or after limited fermentation prior to freezing. The final transformation of frozen dough is a three-step process requiring thawing, fermentation, and finally baking. Approximately a decade later, frozen partially baked bread (or frozen part baked bread) developed significantly in the industry. At industrial level, this technology consists of preparing bread with partial baking usually done at moderate temperature and bread being thereafter chilled and frozen. This frozen product can be placed directly into the oven and perform thawing–baking in a single unit operation. One could mention a French patent related to this type of product [28]. In the end of the 1990s, the idea of producing fermented frozen products came up and this technology, which was considered as a rather ascending technique until few years ago, seems to attract more the breadmaking industry. These products are also called "frozen ready to bake."

Nowadays, frozen bakery products occupy an important share of the market. Researchers believe that in 2006, 17% of "fresh bread" will be done from frozen products (13% in 2001). In Europe, the overall consumption of bread and viennoiserie increases by 1% per year; at the same time, the production of frozen bread and viennoiseries should increase by 7% by 2006 [29,30].

This expansion of the frozen bakery products in Europe is driven by two patterns:

- The research for convenient products that can be quickly prepared and proposed "as fresh" to the consumer.
- The consumers demand for a large variety of bakery products that is unprofitable to be prepared by retailers.

The market of frozen bakery products is therefore expected to increase in the coming years.

B. Refrigeration Applied to Bread Dough

Frozen bakery products can be mainly divided into two categories: frozen yeasted dough and part baked products. These are two complementary products although they produce two different qualities of bread. Frozen bread dough products are especially formulated to survive freezing and thawing. They present quality similar to conventional bread but require a minimum preparation of 2–3 h. They are normally allowed to thaw and rise (proof) at temperatures slightly above ambient to provide an expanded open grain dough structure and then baked to produce a suitable finished product. The time required for thawed dough proofing is usually determined as "slack time" in the baking industry.

On the contrary, frozen partially baked or part-baked products exhibit shorter preparation time, as they can be ready in less than 20 min, but give a bread with slightly lower sensorial quality. The freshly made dough is allowed to rise and then is partially baked, usually at milder temperature than in the case of conventional breadmaking (i.e., 180°C vs. around 230°C for a French baguette).

Baking must be interrupted before Maillard reactions take place; a sufficient baking is required to achieve a rigid product center at the end of the postbaking chilling. Afterward, the product is frozen and then distributed. Thawing is sometimes recommended before final baking, which mainly consists of reheating the product for a short baking time; it is recommended to bake products for up to two thirds of the time required for full baking, until the color change of the crust due to Maillard reactions is achieved [31].

Retailers using refrigerated bakery products are very often combining the use of frozen dough to cover customary needs and part-baked products to deal with increased consumer demand during peak periods.

Frozen part-baked bread is dragging the innovation and its market share is continuously growing in Europe due to its convenience and the reduced requirements in equipment and labor as it requires a very limited know-how for the final transformation before retailing [32]. It also allows the production of more elaborated products at industrial level, whereas frozen dough is usually applied for the mass production of conventional products.

In 2002, the industrial production of frozen bakery products in France was 65% for yeasted frozen dough and 35% for frozen part-baked, whereas 40% of the frozen part-baked breads is exported abroad.

Figure 22.1 and Figure 22.2 illustrate the process flow diagrams for the production of frozen dough and part baked products as well as the modifications from conventional breadmaking for yeasted and fermented bakery products, respectively.

C. Refrigeration Applied to Rich Dough

Sandwich bread, pizza dough, puffing dough (such as those used for croissant), brioche, and other similar type of dough are referred to as rich dough products. The use of refrigeration for this type of dough products was introduced long after its use in bread. Indeed, the shelf-life of these products after baking is not as limited as it is for lean dough in which the interval before staling occurs is very

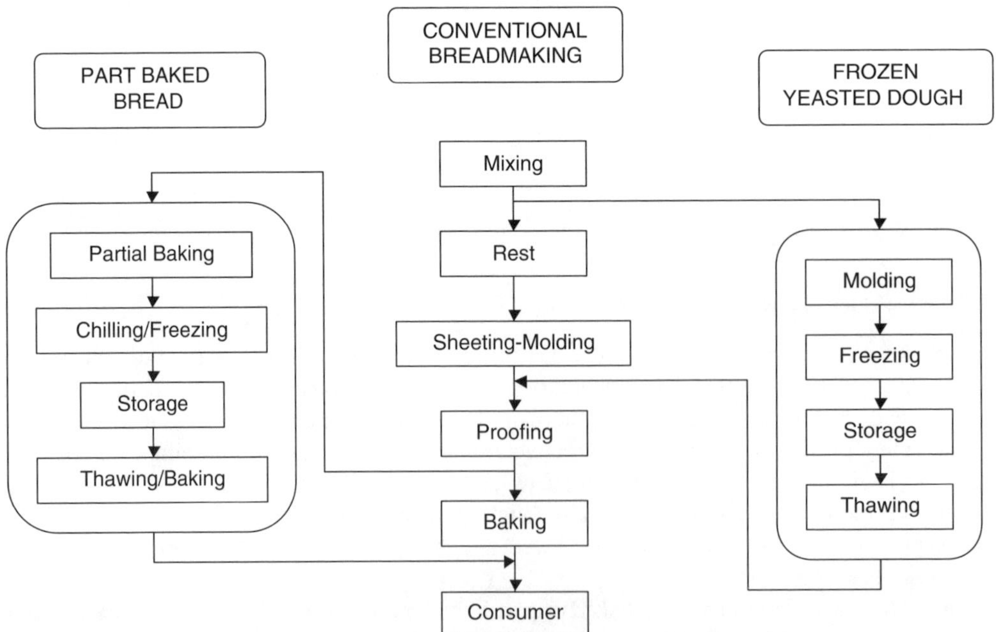

FIGURE 22.1 Flowsheet for the application of freezing in breadmaking No. 1 frozen yeasted dough and part baked bread.

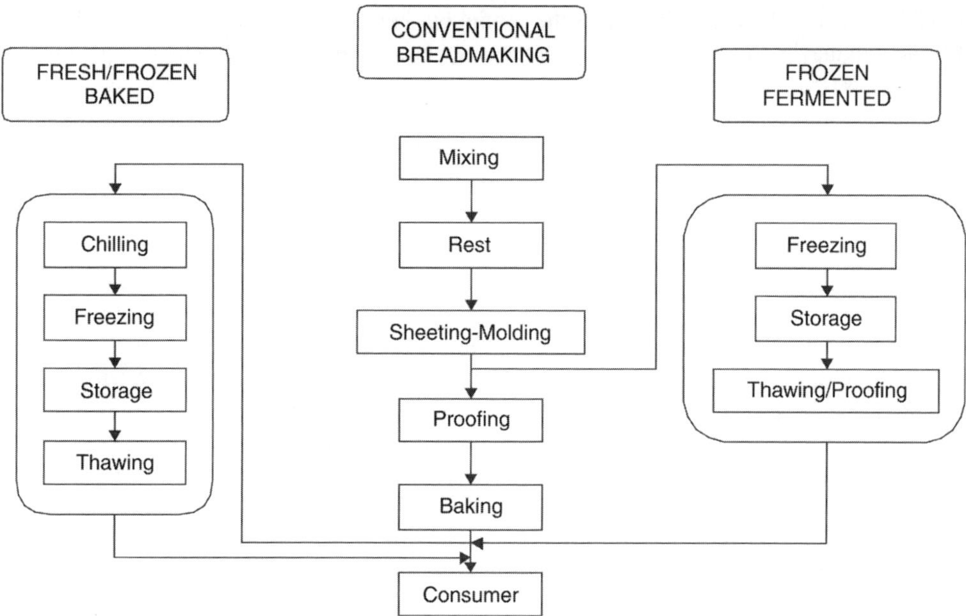

FIGURE 22.2 Flowsheet for the application of freezing in breadmaking No. 2 frozen fermented and baked bread.

short (i.e., 24 h or less for a French baguette). There is thus an important issue regarding the formulation of lean bread and especially the amount and type of lipids (in a broad spectra of consideration) that is introduced in the mixture; indeed, endogenous lipids from flour and added lipids strongly affect the staling phenomena. In France, for example, the use of shortening or emulsifiers is not allowed to produce "French bread." Moreover, this type of product is very often prepared by using some chemical leavening agents (baking powder like sodium bicarbonate) or a mixture of yeast and these agents to obtain the desired volume intake during baking. The freezing of rich dough is also very often associated with other functionalities such as freezing of pizza (i.e., to preserve the toppings), freezing of cakes and tarts (to preserve the filling), and freezing of croissant and puffing pastry (to reduce the manual work needed to produce this type of product). It is thus difficult to clearly identify technologies such as those proposed for lean dough breadmaking. The industrial making of puffing pastry is maybe the easiest technology to identify; two alternatives can be found in the existing products namely nonfermented frozen and partially fermented and frozen. Frozen partially baked puffing pastry is not so common in the industry.

D. Freezing Technology — Equipment

Apart from the equipment used in conventional breadmaking such as mixers, dividers, molders, and so on, in frozen bakery products, a freezing system is additionally required. The two basic freezing systems that are available commercially are the cryogenic method that uses liquid nitrogen or carbon dioxide as the cooling agent, and the more commonly used mechanical refrigeration, which relies on air blast.

1. Mechanical Refrigeration

Blast freezers usually consist of an enclosed insulated area with a plurality of product-carrying trolleys, which is properly arranged to convey and expose bakery products to the cold air currents, a mechanical refrigeration system, and blowers for distributing cold air throughout the unit. An advanced model of this type of freezers is the continuous belt freezer, which consists of a long metal belt located in the freezing room. Belt-type freezers can carry the product under a large number of cold air outlets or automatically position product on moving trays, which travel through the freezer compartment. Spiral conveyor freezers provide extremely long product exposure paths and are designed to minimize product weight loss and assure gentle handling during the freezing process. The operational cost for these freezers is not necessarily less, but they usually fit better into plant layouts and therefore are more popular. Furthermore, they can also be used for ambient cooling, proofing, and setting.

2. Cryogenic Freezing

The second major category of food freezing equipment is the cryogenic units, which operate with liquid nitrogen or carbon dioxide. They consist of a conveying system, usually a wiremesh belt, operating inside an insulated tunnel and a recirculating system, which moves cold gas at high velocity over the products. Products are usually precooled with nitrogen gas and then either sprayed with liquid nitrogen or completely frozen by the gas. The freezing temperatures achieved in a cryogenic unit are lower than in air blast system since the boiling point of liquid nitrogen, for example, is -196°C. Cryogenic freezing rates are consequently very fast compared with the mechanical ones. However, the continuous expenditure of the cooling agent considerably raises the operating cost [3,5,33].

IV. TECHNOLOGICAL PROBLEMS, SOLUTIONS, AND REQUIREMENTS

A. Problems Associated with Frozen Dough Products Freezing

During food freezing, a number of serious physical changes occur such as the uneven growth of ice crystals within products or moisture migration due to water vapor pressure variance. This results in the accumulation of moisture particularly at the surface of the products and can be detrimental to their textural or sensory characteristics [31].

The quality of the bread made out of frozen dough in specific is influenced by dough formulation as well as by processing parameters such as dough mixing time, freezing rate, frozen storage temperature, storage duration, and thawing rate [34–40]. It appears that these factors may act either independently or synergistically to reduce yeast activity, which results in reduced CO_2 production or weakening or damage to the gluten network and entails in poor retention of CO_2 and poor baking performance [27,41–44]. The main consequences of these phenomena include longer proof or fermentation times, increased extensibility, decreased loaf volume, textural characteristics deterioration, and variable performance [45,46].

The rheological characteristics of frozen dough bread have been studied extensively. The presence of dead yeast cells in the dough has been implicated in poor bread quality but some researchers [47] did not observe significant modification in the rheology of the dough with or without yeast. Others [48] showed that the gluten structure in frozen dough could be damaged by the formation of ice crystals. Researchers [46,48] who worked with extensigraph observed the strengthening of dough (increase of extensigraph resistance and decrease of extensibility) submitted to freeze–thaw cycles because of the reducing substances leached out from yeast cells (mainly glutathione), which cause depolymerization by cleaving disulfide bonds and subsequently weaken the gluten matrix, or the redistribution of water caused by a change in water-binding capacity of flour constituents. The opposite effect was observed by others [35]. The use of different oxidants may explain this result (potassium bromate for [46] vs. ascorbic acid for [35]) [27,42,45].

B. Problems Associated with Partially Baked Products

There is a quite limited amount of literature concerning this product. One of the main problems concerning the quality of the crust is that the undergoing intense heating and cooling phases

FIGURE 22.3 Crust flaking of a frozen part baked baguette.

result in a risk of excessive surface dehydration. In some extreme cases, crust flaking might occur (Figure 22.3). Crust is the result of successive dehydration of the surface area of the dough during proofing and baking. Even though most of the literature recommends the use of moist air during proofing, there is no clear evidence about this allegation. Neither the effect of post-baking chilling nor the effect of the freezing conditions has been studied. One evidence is that crust flaking is visible at the end of the freezing process [49]. Poor storage conditions may magnify the problem but cannot be considered as solely responsible for the flaking phenomena.

C. Solutions Proposed for Confrontation of Problems

Different ways to minimize the effect of freezing on doughs and prevent loss of dough quality are suggested in the literature: maintaining yeast viability during freezing and thawing, improving parts of the breadmaking process, or using suitable ingredients, additives, and cryoprotectants for frozen doughs. All these parameters are individually developed and discussed subsequently.

1. Freezing Effect on Yeast Performance — Requirements and Suggestions

Yeast cells in bulk are regarded cryoresistant and their ability to produce CO_2 is not affected considerably by successive freeze–thaw cycles. However, when the cells are dispersed in a dough, and especially when unfavorable processes such as freezing intervene, this resistance is seriously restricted [50]. The loss of cell viability in the dough during freezing has been attributed to intracellular freezing and increased internal solute concentrations, which may result in pH lowering, dehydration, ionic toxicity, damage to essential membrane processes, impairment of cytoskeletal elements, and decreased glycolytic enzymes activity [51].

Yeast survival and gassing power are strongly affected by freezing rate, frozen storage temperatures, and duration of frozen storage [52]. From previous studies, it appears that a slow freezing rate is preferable to preserve yeast activity [53]. Yeast strain, age of cells, protein content, as well as nature and concentration of cryoprotectants (e.g., trehalose) also influence the yeast activity [54–56]. In addition, processing conditions such as fermentation prior to freezing may reduce yeast cryoresistance [57,58].

To minimize the freezing effect on product stability, several suggestions have been proposed. Some researchers support that dry yeast may be superior to compressed yeast in preserving the shelf-life of frozen dough as it presents longer lag period and consequently more restrained fermentation before freezing, providing a more stable dough. However, reports also show that doughs made with dried yeast exhibit slightly longer proof times and could contain more broken cells that might release glutathion, which is known to affect the gluten network [46]. Another approach for the maintenance of yeast viability is the commercial production of new yeast strains that are more resistant to freeze damage [59]. Finally, it is suggested that yeast content in the dough formula should normally be higher than in conventional breadmaking to overcome the prospective loss of activity during freezing and storage and any inadequacy in proofing conditions.

2. Processing Parameters — Requirements and Suggestions

The poor baking performance of frozen dough can be overcome to a great extent through the use of appropriate processing conditions, which aim at the restraint of yeast damage and the enhancement of gluten network ability to retain gas [42]. Mixing duration, dough temperature, and resting after mixing are very important parameters. If dough is undermixed, starch and proteins are unevenly distributed, and when it is overmixed gluten proteins become stressed and partially depolymerized [21]. To minimize yeast activity before freezing, dough temperature after mixing should be slightly lower than conventional (usually between 24 and 26°C) breadmaking, and range between 19 and 22°C [60]. Several researchers suggest that in frozen bakery products, dough resting after mixing should be completely avoided to minimize fermentation before freezing, whereas others consider short rest times (8–10 min) to be beneficial [4].

The influence of sheeting and molding conditions on the stability of frozen dough was not found to be very significant. However, as far as dough shape is concerned, it is believed that round-shaped dough pieces produce less satisfactory bread than slabs and cylinders [40]. Packaging is also very important as it performs a number of functions: it contains, protects, identifies, and merchandizes food products. It should provide an effective barrier to contamination and variable moisture conditions, compressive strength to withstand stresses, and perform satisfactorily during storage and transport. The packaging materials and their shapes may vary according to product specifications but the most popular materials applied to frozen bakery product are plastic (films, membranes, etc.) and aluminum [61]. Films used for frozen dough products should present good oxygen and moisture barrier characteristics, physical strength against brittleness and breakage at low temperature, stiffness to work on automatic machinery, and good heat sealability [4,62].

As far as freezing is concerned, reports show that slow rates ($<2°C/min$) provide higher yeast survival levels and bread scores [63]. It has also been shown that slow freezing at $-20°C$ is better than freezing at $-40°C$ [35]. At relatively slow freezing rates, ice is formed outside the yeast cell and can lead to a relative increase in external solute concentration and hence hyperosmotic conditions that cause efflux of intracellular water and relative dehydration. On the contrary, high freezing rates result in the formation of small ice crystals inside yeast cells. This can eventually lead to the rupture of cell membranes if recrystallization into larger crystals occurs especially during prolonged frozen storage and slow thawing at low temperatures [52]. However, studies on the influence of both freezing rate and dough geometry show that a high freezing rate and a low yeast activity were observed at the core of cylindrical dough pieces whereas lower freezing rate and a higher yeast activity were noticed at the surface of these samples [64,65]. These observations on the freezing rate are in agreement with the heat transfer theory in cylindrical geometry, which demonstrates that the velocity of the freezing front is the lowest at the surface and in the bulk zone and tends to become infinite at the center [66].

Another factor that should be considered is the influence of storage time and conditions on the gluten structure, which appears to be disrupted during extended storage resulting in dough weakening, loss of gas retention ability, and deterioration of product quality [62]. Temperature fluctuations during storage were shown to be as important as the storage duration [40,67].

Thawing should preferably be performed under stepwise temperature increase to avoid the formation of an excessive temperature differential between dough surface and surrounding air, which can cause condensation in the crust and generate crust spotting and blistering. Rapid thawing also results in temperature rising only to the outer surface of the dough, which becomes ready for proofing, whereas the center of the dough still remains frozen [60]. This can lead to unconformable gas cell structure and significantly affects frozen storage stability because when dough comprises a large number of small bubbles with narrow size distribution and thick walls, it appears to be more stable than a dough that contains bubbles with less uniform size distribution [21].

3. Raw Material — Requirements and Suggestions

The ability of dough to withstand harsh freezing and thawing conditions significantly depends on flour type and protein quality, which are important variables for the stability of frozen dough. A medium to strong flour is recommended for frozen doughs to maintain their ovenspring potential even after losing some intrinsic strength over storage period [46,52]. Doughs made from strong flours are generally more resistant to freeze damage, and hard red spring (HRS) wheat is preferred for frozen dough because of its superior gluten strength [53]. However, flour protein content is found to be less important than flour protein strength for optimum frozen dough performance [42].

Dough rheology and consequently dough machinability are strongly affected by water content. It is recommended to incorporate less water in the formula of frozen bakery products to minimize free water in the dough. This is important because free water is responsible for water migration and ice crystal formation, which can be detrimental for yeast cells and gluten proteins and may result in reduced gas retention [42,63]. The use of chilled water is also preferable as it retards yeast activity and accelerates freezing of dough pieces.

Reports indicate that higher levels of sugars should be used as well because due to their hydroscopic properties they can reduce the amount of free water in frozen dough products and therefore constrain its undesirable effects on dough stability. Besides, sugars are responsible for the osmotic stress of yeast cells during freezing and therefore, specific yeast strains with increased "osmotolerance" are proposed by yeast companies [12]. Salt addition is also important, as it has been found to retard the production of carbon dioxide by the yeast thus delaying dough fermentation. Finally, shortening is considered to improve dough processing and freezing tolerance. More preferably, saturated or partially saturated shortening should be used in frozen bakery products [4].

4. Additives — Requirements and Suggestions

Additives are used in breadmaking to facilitate processing, confront raw material variations, ensure stability in quality characteristics, and sustain freshness [43]. Especially when incorporated in the frozen dough formula, they are able to counteract with several of the changes occurring during freezing, frozen storage, and thawing. More specifically, they can decrease final proof time, improve dough rheological properties, and increase loaf volume and bread softness [27]. The most frequently used additives in frozen dough products are: oxidative substances (L-ascorbic acid, azidocarbonamide, and potassium bromate), emulsifiers (monoglyceride, sodium or calcium stearoyl-2-lactylate, diacetyl tartaric acid esters of monoglycerides, etc.), and enzymes [68].

The use of optimum levels of oxidant ingredients, whether from natural or chemical origin, exerts an improving effect on dough rheology and handling and on the overall quality of the finished product. During mixing, oxidizing agents convert sulfhydryl (SH) groups of the gluten protein to disulfide (SS) linkages between adjacent molecules, building up the gluten matrix and providing a stronger dough [20,59]. An oxidant also exhibits its improving effect by increasing the loaf volume during the first few minutes of the baking process.

Ascorbic acid is probably the most popular oxidizing agent used in frozen bakery products. It is reported to significantly reduce dough stickiness, decrease extensibility, and increase its elasticity by inducing intermolecular interactions between dough protein molecules, and consequently increase ovenrise and bread score [27,69,70]. The use of potassium bromate is prohibited in most countries even though it shows better performance than ascorbic acid.

Emulsifiers may also be used in the formulation of yeasted frozen dough products. They are commonly added to commercial bread products to improve bread quality and dough handling characteristics and usually result in a foamy crumb with fine and uniformly dispersed cells. Such a foamy crumb is easy to recognize and is more or less accepted by the consumer. The effect of emulsifiers has been extensively studied and numerous references can be found in the literature on this aspect. They are reported to improve mixing tolerance, gas retention, and dough resistance, to increase the loaf volume of the final product and to endow it with resilient texture, fine grain as well as slicing properties [27,43].

Monoglyceride (MG) was the first industrial emulsifier; since then other emulsifiers has been developed, some of them requiring the combined presence of "conventional" emulsifiers such as MG. One could mention propylene glycol monostearate (PGMS), which is an additive typically used in the United States and enhances the efficiency of MG. Polyglycerol esters of fatty acids (PGEs, HLB 3-13) has been patented by "PURATOS" and must be used with MG as well. Diacetyl tartaric acid esters of monoglycerides (DATEMs), acetic acid esters of monoglycerides (AMGs), and lactic acid esters of monoglycerides (LMGs) can be used alone or with other emulsifiers. These emulsifiers can be used as crumb softeners or antistaling agents as they interact with amylose by forming complexes that are known to delay the retrogradation, resulting in the staling of bread. They also interact with lipids and shortening, resulting in a foamy structure with refined cells (i.e., sandwich bread) [71]. Sodium stearoyl lactylate (SSL) has also been shown to be effective in maintaining both volume and crumb softness and provide longer shelf-life stability in fresh and frozen dough products subjected to extended storage [27,46].

Enzymes such as amylases, proteases, hemicellulases, lipases, and oxidases can be naturally present in foods such as wheat and soy products (as long as they are not removed or inactivated) or incorporated in them as additives. They have been shown to influence the entire breadmaking process by decolorizing (bleaching) dough, improving dough volume and texture, and maintaining shelf-life by extending products freshness during storage [68,72]. Their effects on frozen dough products can be seriously enhanced when used in combination with selected oxidants such as ascorbic acid [73]. Transglutaminase is also reported to significantly improve the cohesion of the gluten network and seems to be well adapted in the puffing pastry [74].

The supplementation of vital gluten to relatively weak doughs improves the mixing tolerance and stability of dough, resulting in increased loaf volume, improved crumb texture and softness, and prolonged shelf-life of bakery products. As gluten is a protein, it also enhances nutritional value of the products [2,45]. The embedment of pentosans is particularly important as well in breadmaking because of their physical properties. They improve dough machinability, increase bread volume, and decrease the rate of starch recrystallization, and therefore reduce staling rate of baked bread [72,75].

Finally, the incorporation of cryoprotective materials in frozen bakery products is increasingly becoming appealing. Cryoprotectants are substances, which have the ability to protect protein cells from chemical changes and loss of functionality during freezing or thawing and consequently improve quality and extend the shelf-life of frozen foods. Cryoprotective agents may include sugars, amino acids, polyols, methyl amines, carbohydrates, some proteins, and inorganic salts such as potassium phosphate and ammonium sulfate [70,76].

V. SAFETY AND QUALITY ISSUES

A. Safety Considerations

Before making any comments and considerations on the quality characteristics of food products, it is necessary to ensure their safety as it is considered a prerequisite. The food safety system that is currently applied by many regulatory agencies to ensure that all operations of a manufacturing process are controlled so as to preclude potential health hazards is hazard analysis critical control points (HACCP). The HACCP system can be used to identify the critical control points (CCPs) in the food production process, which may contribute to a hazardous situation whether it is contaminants, pathogenic microorganisms, or foreign materials, respectively, and involves a systematic study of the raw materials and ingredients, the special conditions of the manufacturing process, handling, storage, packaging and distribution of food products, and consumer use [77].

Frozen bakery products, in general, are not considered as high-risk food products because baking at relatively high temperatures is involved in their preparation. However, raw material should meet certain specifications before being incorporated in the dough formula.

The possible hazards in flour, from the microbiological standpoint, may be molds and yeasts (species of the genera Penicillium, Aspergillus, Rhizopus, Eurotium, Torula, Fusarium, and Cladosporium) and bacteria (mainly of the Bacillus species, Pseudomonas, Streptococcus, Achromobacter, Flavobacterium, Micrococcus, and Alcaligenes). They may also be mycotoxins (Alternaria toxins, Aflatoxins, Citrinin, Cyclopiazonic acid, Achratoxin, Viomellein, and Xanthomegnin), traces of insecticides, pesticides or parasiticides, pieces of rocks, wood, and insects. Careful handling and storage of flour is very important to avoid spore contamination of bakery products from the environment, equipment surfaces, and other raw materials and additives, which may result in primary mycotoxin contamination [78]. Improper handling of products may enable their existence after baking or their recontamination from the bakery dust consisting of flour particles [2,79].

Yeast should be suitable for foods applications and potable water must be used for both dough formation and washing of equipment. Sugars may contain thermophilic spores, insecticides, pesticides, parasiticides, dirt, or foreign materials. The last two can also appear in salts used for dough preparation. Lipolytic bacteria (Pseudomonas, Flavobacterium, Micrococci, Zymomonas, and Bacillus) and spoilage yeast and molds (Trichoderma viride and harzianum, Rhizopus, Aspergillus, Gladosporium, Paecilomyces, Penicillium, Geotrichium candidum, Candida lipolytica, and Alternaria) may be present in lipids.

B. Quality Considerations

Product quality is usually affected by many mutually associated external and internal conditions, which determine their acceptability and merchantability. Therefore, the implementation of a quality assurance system is very important to identify the control points (CPs) of the productive procedure, which are determinant for final product quality. The most important quality parameters in foods are: appearance, taste, flavor, texture, and nutrition. Especially for frozen bakery products, these can be further analyzed as follows.

Appearance basically comprises product color, shape, size, and gloss. Bakery products should present adequate volume and symmetrical expansion, appealing, and uniform crust and crumb appearance. They must be proportioned according to product specification and be appropriately shaped with a well-rounded, smooth top, and without excessive cracks bulges or streaks. Crust should have an even and pleasant brownish color and proper thickness while crumb bubbles should preferably have similar size and be uniformly distributed. Nevertheless, in some products such as French baguette, nonuniformly distributed cells are rather eligible by the consumers.

Flavor, which is the complex result of the taste and odor, is the response of the receptors in the oral and nasal cavities to chemical stimuli. Bakery products should present a pleasing, wheaty, and sweet taste without off-flavors and fine roasty aroma with a mild yeast overtone. Sourdough and sponge dough products usually exhibit a more acid aroma, which is considered very appealing by some consumers.

Texture is primarily the response of the tactile senses to physical stimuli that result from contact between some part of the human body and the food. However, it can be further evaluated by kinesthetics (sense of movement and position) and sometimes sight and sound, which is associated with crisp or crackly characteristics. Bakery products should exhibit soft, tender, smooth, and slightly moist mouthfeel with fine grain. Crumb should also be satisfactorily elastic and cohesive and present decreased adhesiveness.

TABLE 22.1
Quality Defects in Frozen Bakery Products

Cause	Defect
Raw Materials	
Flour	
Very weak	Coarse crumb
	Small volume
	Inadequate shape
	Unsatisfactory crumb color
Very strong	Small volume
	Inadequate shape
	Unsatisfactory crumb color
	Leachy crumb
Yeast	
Smaller amount	Small volume
	Excessively dark crust color
	Leachy crumb
Larger amount	Small volume
	Unsatisfactory crust/crumb color
	Inadequate shape
	Excessive crumb brittleness
	Skinning
	Fissures on crust
	White spots or blisters
Salt	
Smaller amount	Poor crust/crumb color
	Excessive crumb brittleness
	Excessive volume
Larger amount	Leachy crumb
	Excessively dark crust color
	Small volume
Water	
Smaller amount	Small volume
	Inadequate shape
	Crust–crumb separation
	Excessive crumb brittleness
	Leachy crumb
	Unsatisfactory crumb color
Larger amount	Coarse crumb
	Unsatisfactory crumb color
	Small volume
	Inadequate shape
Sugars (larger amount)	Dark crust color
Processing	
Mixing	
Restricted	Crumb fissures
	Leachy crumb
	Unsatisfactory crumb color
Excessive	Leachy crumb
	Excessive crumb brittleness

(*Table continued*)

TABLE 22.1 *Continued*

Cause	Defect
Molding (inadequate)	Leachy crumb Large blisters either on the surface causing shape distortion or under the top crust
Freezing (very slow)	Crust fissures Uneven or open cell structure
Storage (very long)	White spots and blisters Patches of uneven color on the side and bottom crusts of breads
Proofing/thawing	
Very slow	White spots and blisters Small volume Inadequate shape
At higher temperatures	Areas of dense crumb Excessive volume Ragged crust breaks Overlapping top Thick and hard crust
Baking	
At lower temperatures	Coarse/leachy crumb Unsatisfactory crumb/crust color Excessive dough volume
At higher temperatures	Coarse/leachy crumb Small volume Excessively dark crust color Inadequate shape

Finally, nutrition concerns products content of major (carbohydrates, fat, protein) and minor nutrients (minerals, vitamins, fiber). This quality parameter, however, is not very obvious for most consumers as it cannot be perceived by human senses and relies both on the quality characteristics of raw material used in breadmaking and product processing and handling conditions.

Other factors, which may also be important even though they are not considered as food quality characteristics are: product availability, cost, convenience, and packaging [68,80].

C. Quality and Safety Controls

The most common quality defects, which appear in frozen bakery products, the cause of their-existence as well as the controls that should be applied to maintain the safety and quality characteristics of the final product, are presented in Table 22.1 and Table 22.2 [81].

By studying Table 22.1 and Table 22.2 and following the recommendations listed in Table 22.2, one can efficiently control the quality and safety of frozen bakery products.

TABLE 22.2
Controls for Safety and Quality Maintenance

Process Stage	Controls
Raw Material	*Determination of specification-inspection of suppliers*
Flour (white/wholewheat) [78,79,82–85]	Molds-Bacteria-Salmonellae control
	Foreign materials–insects–rodents control
	Low a_w
	Storage at dry/low temperature places
	Good quality flour (from hard wheat, low in damaged starch-enzyme activity, farinograph)
Yeast	Storage in dry and cool places
	Minimum fermentation prior freezing
	Control of viability and gassing power
	Control of freezing and thawing condition
Water	Disinfectants control
	Water distribution system control
Sugar [85]	Suitable packaging
	Storage under appropriate temperature and relative humidity
	Foreign materials-insects-rodents control
Salt	Suitable packaging
	Storage under appropriate relative humidity
	Foreign materials control
Shortening [86]	Storage conditions control
	Spoilage control
	Melting point control (if in solid state)
Production	*GMP-sanitation and cleaning programs*
Mixing	Adequate cleaning of mixer and moving parts
	Proper weight of ingredients-no overloading
	Temperature control
	According to farinogragh, extensigraph, mixograph
Molding	Clean environment (no contamination)
	Gentle handling of dough
	Short preparation time (no fermentation)
Packaging [84]	Superior moisture barrier materials
	Adequate storage conditions
Freezing [84]	Freezing rate control
	Storage temperature control
Thawing	Temperature control
	Relative humidity control
Baking [84]	Time–temperature–oven humidity control
	Sensory evaluation

VI. CONCLUSIONS

Refrigeration applied to the breadmaking process was almost totally ignored by the industry until the 1960s. Since then, the industry has seriously evolved, and even though local traditional bakeries remain the ascendant distributors in most countries, the breadmaking industry is slowly increasing its market share. Complex products, specialty breads, and "ethnic" breads demand is growing and freezing offers a well appropriate mean of providing products with great variety, affordable cost, and satisfactory quality level. Partially baked products are growing quickly because of their convenience. Their mixing–proofing–baking process is very similar to conventional processing, and freezing, which is just used to extend the shelf-life, provides flexibility and permits the retailing of freshly baked products all day long. Even though bakery products are not considered as high-risk products from sanitary point of view, the risk of impairing human health is not negligible. Chemicals, pesticides, mycotoxins, or specific improvers can become a problem. Allergy to specific ingredients such as gluten is also a matter of concern.

REFERENCES

1. PS Cauvain. Improving the control of staling in frozen bakery products. *Trends in Food Science and Technology* 9:56–61, 1998.
2. CA Stear. *Handbook of Breadmaking Technology*. London: Elsevier Applied Science, 1990, pp. 45–53, 322–326, 689–690.
3. SA Matz. *Bakery Technology — Packaging, Nutrition, Product Development, Quality Assurance*. London: Elsevier Science Publishers, 1989, pp. 103–106, 119–129.
4. K Kulp, K Lorenz, J Brümmer. *Frozen and Refrigerated Doughs and Batters*. Minnesota: American Association of Cereal Chemists, 1995, pp. 1–3, 93, 148.
5. EJ Pyler. Baking Science and Technology, Vol. II, 3rd ed. Kansas City: Sosland Publishing Company, 1988, pp. 589–697, 1071–1072, 1107–1123, 1257–1261.
6. P Hy. Baking new strategies. *Food Ingredients and Analysis International* 3–4:27–28, 30, 1998.
7. CR Hoseney, ED Rogers. The formation and properties of wheat flour doughs. *Critical Reviews in Food Science and Nutrition* 29:73–93, 1990.
8. MA Rao, JF Steffe. *Viscoelastic Properties of Foods*. London: Elsevier Applied Science, 1992, pp. 77–83.
9. M Gobetti. The sourdough microflora: interactions of lactic acid bacteria and yeasts. *Trends in Food Science and Technology* 9:267–274, 1998.
10. R Calval, R Wirtz, JJ MacGuire. *The Taste of Bread*. Maryland: Aspen Publishers, 2001, pp. 19, 31–32, 39, 190–192.
11. W Doerry. *Baking Technology*. Vol. I. *Breadmaking*. Kansas City: The American Institute of Baking, 1995, pp. 5–17, 20–51; Vol. II. *Controlled Baking*, pp. 108–126.
12. H Charley, C Weaver. *Foods—A Scientific Approach*. New Jersey: Prentice-Hall, 1998, pp. 180–183, 202–205, 207–208, 223–226.
13. W Bushuk, VF Rasper. *Wheat—Production, Properties and Quality*. London: Chapman & Hall, 1994, pp. 25–27.
14. H Singh, F MacRitchie. Application of polymer science to properties of gluten. *Journal of Cereal Science* 33:231–243, 2001.
15. MJ DeMan. *Principles of Food Chemistry*. London: Chapman & Hall, 1990, pp. 281–282, 320–322.
16. SA Matz. *Technology of the Materials of Baking*. Texas: Pan-Tech International, 1989, pp. 96, 142–144.
17. JM Gil, JM Callejo, G Rodríquez. Effect of water content and storage time on white pan bread quality: instrument evaluation. *Lebensmittel Wissenschaft und Technologie* 205:268–273, 1997.
18. BJB Wood. *Microbiology of Fermented Foods*. Vol. I, 2nd Ed. London: Chapman & Hall, 1998, 176 pp.
19. H Faridi, JM Faubion. *Dough Rheology and Baked Product Texture*. New York: Avi Books, 1990, pp. 252–256, 372–374.
20. H Demiralp, S Celik, H Köksel. Effects of oxidizing agents and defatting on the electrophoretic patterns of flour proteins during dough mixing. *European Food Research and Technology* 211:322–325, 2000.
21. K Autio, T Laurikainen. Relationship between flour/dough microstructure and dough handling and baking properties. *Trends in Food Science and Technology* 8:181–185, 1997.
22. CE Stauffer. *Fats and Oils*. Minnesota: Eagan Press, 1999, pp. 61–66.
23. C Collar, E Armero, J Martínez. Lipid binding of formula bread doughs: relationships with dough and bread technological performance. *Lebensmittel Wissenschaft und Technologie* 207:110–121, 1998.

24. CE Stauffer. Fats and oils in bakery products. *Cereal Foods World* 43:120–126, 1998.
25. SA Matz. *Equipment for Bakers.* Texas: Pan-Tech International, 1988, pp. 89–109, 125–130, 141, 333–339.
26. P Schieberle. Intense aroma compounds — useful tools to monitor the influence of processing and storage on bread aroma. *Advances in Food Science* 18:237–244, 1996.
27. S Kenny, K Wehlre, T Dennehy, KE Arendt. Correlations between empirical and fundamental rheology measurements and baking performance of frozen bread dough. *Cereal Chemistry* 76 (3):421–425, 1999.
28. L LeDuff. Pain Français précuit congelé et son procédé de fabrication. French Patent. Patent number FR 2 589 044-A1, 1985.
29. Anon. Le Prêt à cuire, une révolution prometteuse. *Filière Gourmande* 93:24–26, 2002.
30. FJ Aubry. Table ronde pain et viénnoiserie; Du pain, du vin et de la diversité. *Le monde du Surgelé* 87:16–27, 2003.
31. KH Kraklow, RC Kandler. Frozen microwaveable bakery products. PCT International Application number WO03092388, 2003, 61 pp.
32. LG Carr, CC Tadini. Influence of yeast and vegetable shortening on physical and textural parameters of frozen part baked French bread. *Lebensmittel Wissenschaft und Technologie* 36 (6):609–614, 2003.
33. EA El-Hady, SK El-Samahy, W Seibel, JM Brummer. Changes in gas production and retention in non-prefermented frozen wheat doughs. *Lebensmittel Wissenschaft und Technologie* 73 (4):472–477, 1996.
34. P Mazur, JJ Schmidt. Interactions of cooling velocity, temperature and warming velocity on the survival of frozen and thawed yeast. *Cryobiology* 5 (1):1–17, 1968.
35. Y Inoue, W Bushuk. Studies on frozen doughs. I. Effects of frozen storage and freeze–thaw cycles on baking and rheological properties. *Cereal Chemistry* 68 (6):627–631, 1991.
36. Y Inoue, W Bushuk. Studies on frozen doughs. II. Flour quality requirements for bread production from frozen dough. *Cereal Chemistry* 69 (4):423–428, 1992.
37. O Neyreneuf, B Delpuech. Freezing experiments on yeasted dough slabs: effects of cryogenic temperatures on the baking performance. *Cereal Chemistry* 70 (1):109–111, 1993.
38. Y Inoue, DH Sapirstein, S Takayanagi, W Bushuk. Studies on frozen doughs. III. Some factors involved in dough weakening during frozen storage and thaw–freeze cycles. *Cereal Chemistry* 71 (2):118–121, 1994.
39. Y Inoue, DH Sapirstein, W Bushuk. Studies on frozen doughs. IV. Effect of shortening systems on baking and rheological properties. *Cereal Chemistry* 72 (2):221–225, 1995.
40. M Havet, M Mankai, A LeBail. Influence of the freezing condition on the baking performances of French frozen dough. *Journal of Food Engineering* 45 (3):139–145, 2000.
41. J Rasanen, T Laurikainen, K Autio. Fermentation stability and pore size distribution of frozen prefermented lean wheat doughs. *Cereal Chemistry* 74 (1):56–62, 1997.
42. CE Perron, OM Lukow, W Bushuk, F Townley-Smith. The blending potential of diverse wheat cultivars in a frozen dough system. *Cereal Foods World* 44 (9):667–672, 1999.
43. PD Ribotta, AE León, MC Añón. Effect of freezing and frozen storage on the gelatinization and retrogradation of amylopectin in dough baked in a differential scanning calorimeter. *Food Research International* 36 (4):357–363, 2003.
44. PD Ribotta, TG Pérez, AE León, MC Añón. Effect of emulsifier and guar gum on micro structural, rheological and baking performance of frozen bread dough. *Food Hydrocolloids* 18 (2):305–313, 2004.
45. MJ Wolt, BL D'Appolonia. Factors involved in the stability of frozen dough. I. The influence of yeast reducing compounds on frozen dough stability. *Cereal Chemistry* 61 (3):213–221, 1984.
46. PD Ribotta, AE León, MC Añón. Effect of freezing and frozen storage of doughs on bread quality. *Journal of Agricultural and Food Chemistry* 49 (2):913–918, 2001.
47. K Autio, E Sinda. Frozen doughs: rheological changes and yeast viability. *Cereal Chemistry* 69 (4): 409–413, 1992.
48. EK Varriano-Marston, KH Hsu, J Mahdi. Rheological and structural changes in frozen dough. *The Bakers Digest* 54 (1):32–41, 1980.
49. A LeBail, JY Monteau, F Margerie, T Lucas, A Chargelegue, Y Reverdy. Impact of selected process parameters on crust flaking of part baked bread. *Journal of Food Engineering* 69:503–509, 2005.

50. O Neyreneuf, JB Van Der Plaat. Preparation of frozen French bread dough with improved stability. *Cereal Chemistry* 68 (1):60–66, 1991.
51. DK Myers, PV Attfield. Intracellular concentration of exogenous glycerol in *Saccharomyces cerevisiae* provides for improved leavening of frozen sweet doughs. *Food Microbiology* 16:45–51, 1999.
52. W Lu, LA Grant. Role of flour fractions in breadmaking quality of frozen dough. *Cereal Chemistry* 76 (5):663–667, 1999.
53. M Bhattacharya, TM Langstaff, A Berzonsky. Effect of frozen storage and freeze–thaw cycles on the rheological and baking properties of frozen doughs. *Food Research International* 36:365–372, 2003.
54. Y Oda, K Uno, S Otha. Selection of yeasts for breadmaking by the frozen-dough method. *Applied Environmental Microbiology* 11:339–356, 1986.
55. Y Oda, K Tonomura. Applicability of the yeast Torulaspora pretoriensis YK-1 to breadmaking by the frozen dough method. *Journal of Japanese Society of Food Science and Technology* (*Nippon Shokuhin Kogyo Gakkaishi*) 41 (3):214–217, 1994.
56. L Meric, GS Lambert, O Neyreneuf, MD Richard. Cryoresistance of baker's yeast *Saccharomyces cerevisiae* in frozen dough: contribution of cellular trehalose. *Cereal Chemistry* 72 (6):609–615, 1995.
57. L Kline, FT Sugihara. Frozen bread doughs prepared by the straight dough method. *The Bakers Digest* 42 (5):45–50, 1968.
58. V Giannou, V Kessoglou, C Tzia. Quality and safety characteristics of bread made from frozen dough. *Trends in Food Science and Technology* 14 (3):99–108, 2003.
59. EA El-Hady, SK El-Samahy, JM Brummer. Effect of oxidants, sodium-stearoyl-2-lactylate and their mixtures on rheological and baking properties of nonprefermented frozen doughs. *Lebensmittel Wissenschaft und Technologie* 32:446–454, 1999.
60. S Kenny, H Grau, EK Arendt. Use of response surface methodology to investigate the effects of processing conditions on frozen dough quality and stability. *European Food Research and Technology* 213:323–328, 2001.
61. CP Mallett. *Frozen Food Technology*. London: Chapman & Hall, 1993, pp. 59–60.
62. LJ Nemeth, FG Paulley, KR Preston. Effect of ingredients and processing conditions on the frozen bread quality of a Canada Western Red Spring wheat flour during prolonged storage. *Food Research International* 29 (7):609–616, 1996.
63. P Gélinas, I Deaudelin, M Grenier. Frozen dough: effects of dough shape, water content, and sheeting-molding conditions. *Cereal Foods World* 40 (3):124–126, 1995.
64. A LeBail, M Havet, M Pasco. Influence of the freezing rate and of storage duration on the gassing power of frozen bread dough. In: *Proceedings of the Symposium of the International Institute of Refrigeration*. Nantes, France: International Institute of Refrigeration, 1998.
65. M Havet, A LeBail. Frozen bread dough: impact of the freezing rate and the storage duration on gassing power. In: *Proceedings of the Workshop on Process Engineering of Cereals*. France: Montpellier, October 8, 1999.
66. R Plank. *Hanbuch der kältetechnik*. Band X. *Die anwendung der kälte en der lebensmittelindustrie*. Berlin: Springer, 1941.
67. A LeBail, C Grinand, S LeCleach, S Martinez, E Quilin. Influence of storage conditions on frozen French bread dough. *Journal of Food Engineering* 39 (3):289–291, 1999.
68. B Hozova, J Jancovicova, L Dodok, V Buchtova, L Staruch. Use of transglutaminase for improvement of quality of pastry produced by frozen-dough technology. *Czech Journal of Food Science* 20 (6): 215–222, 2002.
69. M Nakamura, T Kurata. Effect of L-ascorbic acid and superoxide anion radical on the rheological properties of wheat flour-water dough. *Cereal Chemistry* 74 (5):651–655, 1997.
70. J Rouillé, A LeBail, P Courcoux. Influence of formulation and mixing conditions on breadmaking qualities of French frozen dough. *Journal of Food Engineering* 43:197–203, 2000.
71. L Stampfli, B Nersten, EL Molteberg. Effects of emulsifiers on farinograph and extensograph measurements. *Food Chemistry* 57 (4):523–530, 1996.
72. ME Bárcenas, M Haros, CM Rosell. An approach to studying the effect of different bread improvers on the staling of pre-baked frozen bread. *European Food Research and Technology* 218 (1):56–61, 2003.
73. F Faisy, O Neyreneuf. Performance d'une association enzymatique "Glucose oxydase – Hémicellulases" pour remplacer l'acide ascorbique en panification, Industrie des céréales. (Avril-Mai-Juin):4–12, 1996.

74. JA Gerrard, MP Newberry, M Ross, AJ Wilson, SE Fayle, S Kavale. Pastry lift and croissant volume as affected by microbial transglutaminase. *Journal of Food Science* 65 (2):312–314, 2000.
75. TJ Laaksonen, T Kuuva, K Jouppila, YH Roos. Effects of arabinoxylans on thermal behavior of frozen wheat doughs as measured by DSC, DMA, and DEA. *Journal of Food Science* 67 (1):223–230, 2002.
76. CJ Kennedy. *Managing Frozen Foods.* Cambridge: Woodhead Publishing Limited and CRC Press LLC, 2000, 105 pp.
77. C Tzia, A Tsiapouris. *HACCP in the Food Industries.* Athens, Greece: Papasotiriou, 1996, pp. 17–18.
78. PC Bailey, A Von Holy. Bacillus spore contamination associated with commercial bread manufacture. *Food Microbiology* 10:287–294, 1993.
79. M Weidenbörner, C Wieczorek, S Appel, B Kunz. Whole wheat and white wheat flour — the mycobiota and potential mycotoxins. *Food Microbiology* 17:103–107, 2000.
80. MC Bourne. *Food Texture and Viscosity: Concept and Measurements*, 2nd ed., San Diego: Academic Press, 2002, pp. 1–6.
81. SP Cauvain, LS Young. *Technology of Breadmaking.* Maryland: Aspen Publishers, 1999, pp. 172–175.
82. JD Legan. Mould spoilage of bread: the problem and some solutions. *International Biodeterioration and Biodegradation* 32:33–53, 1993.
83. WO Ellis, AK Obubuafo, A Ofosu-Okyere, EK Marfo, K Osei-Agyemang, JK Odame-Darkwah. A survey of bread defects in Ghana. *Food Control* 8 (2):77–82, 1997.
84. SJ Forsythe, PR Hayes. *Food Hygiene, Microbiology and HACCP.* Maryland: Aspen Publications, 1998, pp. 39–41, 123–124, 133–136, 239–240, 309.
85. BM Lund, TC Baird-Parker, GW Gould. *The Microbiological Safety and Quality of Food*, Vol. I. Maryland: Aspen Publications, 2000, pp. 766–768, 945–948.
86. S Delamarre, CA Batt. The microbiology and historical safety of margarine. *Food Microbiology* 16:327–333, 1999.

23 Quality and Safety of Frozen Eggs and Egg Products

Lih-Shiuh Lai
National Chung Hsing University, Taichung, Taiwan

CONTENTS

I. INTRODUCTION

Eggs are widely used in the preparation of various processed foods because of their versatile functions such as foaming properties, heat-induced gelation of egg albumen, emulsifying properties of egg yolk, and so on. To extend the shelf life of liquid eggs, they are commonly pasteurized, dehydrated, or frozen. The adequacy of performing functionalities of eggs after pasteurization, dehydration, or freezing determines the value of eggs in food products. In this chapter, issues related to improving functional performance of eggs after pasteurization and freezing would be reviewed. Generally, top quality frozen egg products should be prepared with appropriately stored fresh shell eggs with low bacteria count, followed by breaking operations with strict sanitation, temperature control, and rapid freezing operations, together with application of cryoprotectants.

II. SELECTION OF RAW MATERIALS FOR THE PROCESSING OF FROZEN EGGS AND EGG PRODUCTS

Eggs are generally considered as tasty, wholesome, and nutritious food. Their protein value is high, and their calories and fat content are in moderation. In addition, eggs are easy to digest. Generally, they contain approximately 75% water, 12% protein, 10% lipid, all necessary vitamins (except vitamin C), and minerals [1–3]. Egg protein is known to be a nutritionally complete protein with an unenviable balance of amino acids. The protein value of whole egg proteins is considered to be 100. According to the World Health Organization, egg protein has the highest true digestibility among major food proteins. Because of its high quality, egg protein is used as a standard for measuring the nutritional quality of other food proteins. One egg contributes the same dietary requirements of protein as 35 g of meat. The protein content of two eggs is about 12 g, which corresponds to 30% of the dietary allowance recommended by the National Research Council in the United States [2]. Most of the egg lipids are contained in the yolk. Egg yolk contains triglycerides, phospholipids, and sterols. The fatty acids in eggs are more unsaturated than those of most animal lipids. One egg may supply almost 12% vitamin A, more than 6% vitamin D, 9% riboflavin, and 8% pantothenic acid of the recommended daily allowance in the United States [2]. The high nutritional value, low caloric content, blandness, and ease of digestibility make eggs quite popular. However, in the last two decades, these positive attributes of egg has suffered a blow because of the issues of cholesterol, food safety, allergies induced by eggs, mainly in children, as well as lack of convenience in preparation [1–3]. These issues have been addressed to a certain extent to fit in the consumers' new lifestyles and preferences.

The functionality of eggs, such as coagulation, foaming, emulsifying, and contributing nutrients, make them widely used in cooking and in the preparation of various processed foods. In addition, eggs serve as color and flavor ingredients, and in some instances, they are used to control the growth of sugar crystals. The adequacy of performing these functions determines the value of eggs in food products [4,5]. To be manufactured into various processed foods, shell eggs are first broken into liquid products, including liquid whole egg, liquid egg whites, and liquid egg yolk. Liquid egg products are pasteurized to eliminate *Salmonella.* Unfortunately, many spoilage bacteria and spores remain viable after pasteurization. These organisms multiply rapidly to about 10 million cells per gram, by which time the egg product starts to deteriorate [6]. Furthermore, microorganisms release enzymes into the media in which they live. Some of these enzymes are heat-resistant, thus they will survive pasteurization. Refrigeration slows down this process and increases the shelf life of the product. Therefore, rapid breaking and pasteurization of the eggs immediately after being laid, as well as rapid cooling in pre- and postpasteurization periods dramatically reduces microbial and enzymatic risks.

In addition to the microbial and enzymatic risks, the pH value of very fresh egg albumen is generally 7.6–7.9. Upon storing eggs at 25°C for 6 days, the pH rises to 9.2–9.5, possibly due to carbon dioxide release from the eggs. A fall in the freshness of eggs is generally accompanied by a decrease in the viscosity and gel strength of egg albumen as well as a decrease in the foam stability of albumen and an increase in the temperature for heat-induced gelation during the processing applications [7]. Lowering storage temperature and shell oiling with light, food-grade mineral oils could slow down the escape of carbon dioxide and moisture of shell eggs and prevent the shrinkage and thinning of the egg white [8]. Therefore, top quality frozen egg products should be prepared with appropriately stored fresh shell eggs with low bacteria count, followed by breaking operations with strict sanitation, temperature control, and rapid cooling during pre- and post-pasteurization process to minimize microbial hazards. Further-processed liquid eggs are then used to prepare chilled, frozen, or dried products.

A. Before Breaking

As mentioned earlier, top quality frozen egg products should be prepared with appropriately stored fresh shell eggs with low bacteria count. Candling is the most commonly used method to measure the freshness of an egg before breaking [8]. When an egg is candled, the yolk creates a definite shadow. For fresh eggs, the shadow is light, as the thick albumen tends to keep the yolk centralized within the shell. However as the egg ages, the albumen becomes thinner, allowing the yolk to approach the shell during candling with rapid rotation of the egg, therefore, creating a darker shadow. Yolk shadows could be affected by the color of the shells and yolks.

B. After Breaking

1. Albumen Quality

After the egg is broken, the most widely used measurement of albumen quality is the Haugh unit [8], which is a measure of the height of the albumen after correcting the reading for differences in egg weight as follows:

$$\mathrm{HU} = \frac{100\ \log H - \{\sqrt{G}[30(W \times 1000)^{0.37} - 100] + 1.9\}}{100} \tag{23.1}$$

where HU is the Haugh unit, H the thick albumen height (mm), G a dimensional constant of 32.2, and W the weight of egg (kg). Generally, the HU values of albumen range from a high of above 100 to a low of less than 20. The higher the HU value, the thicker the albumen and the better the albumen quality.

2. Yolk Quality

Yolk quality is determined by the shape and the color of yolks. The shape of yolk depends on the strength of the vitelline membrane and the chalaziferous albumen layer surrounding the yolk. In a freshly laid egg, the yolk is nearly spherical, and when the egg is broken out onto a flat surface, the yolk stands high with only a little change in shape. After oviposition, the vitelline membrane and the chalaziferous albumen layer surrounding the yolk gradually undergo physical and chemical changes, which decrease their ability to maintain the yolk's spherical shape. A general flattening of the yolk upon breaking therefore results. For yolk color, processors of liquid, frozen, and dried egg products generally desire a darker yolk than do users of table eggs, as these products are usually used for mayonnaise, doughnuts, noodles, pasta, and other foods that depend on eggs for their yellowish color [8].

III. PRETREATMENT BEFORE FREEZING TO MINIMIZE PRODUCT CHANGES

A. Microbial Aspects: Pasteurization

The egg breaking process transforms shell eggs into liquid products, including liquid whole egg, liquid egg whites, and liquid egg yolk. The Egg Products Inspection Act of 1970 [9] led to regulations requiring that all egg products, including liquid egg, be rendered free from *Salmonella* by the application of appropriate pasteurization process. Furthermore, the USDA [10] requires that all prepasteurized liquid egg products be refrigerated to holding temperature between 4.4 and 21.1°C (40 and 70°F) within 2 h from the time when the eggs are broken, as shown in Table 23.1. Minimum pasteurization temperature and holding times for pasteurizing various types of liquid egg products, as specified by the USDA [10], are listed in Table 23.2. It requires that liquid whole egg be heated to a least 60°C (140°F) and be held for no less than 3.5 min for the average particle. Pasteurization specifications could vary, as different time–temperature combinations will provide the same pasteurization effect. In addition, pH values also affect the successful pasteurization of liquid eggs. For example, higher pH generally requires lower pasteurization temperature, as the alkaline pH of 9.0 is the most effective for destroying *Salmonella*. This is partly, why *Salmonella* is more heat-resistant in yolk than in whole egg. Yolk has lower pH and higher solid content and hence must be pasteurized under a higher temperature than whole egg

TABLE 23.1
Maximum Holding Temperature Allowed for Liquid Egg Products within 2 h from the Time the Eggs are Broken

Products	Time to be Held before Next Operation	Maximum Holding Temperature Allowed °C	°F
Liquid egg whites (not to be stabilized)			
Unpasteurized	8 h or less	12.8	55
Unpasteurized	More than 8 h	7.2	45
Pasteurized	—	7.2	45
Liquid egg whites (to be stabilized)			
Unpasteurized	8 h or less	21.1	70
Unpasteurized	More than 8 h	12.8	55
Pasteurized	—	12.8	55
Products with 10% or more salt added			
Unpasteurized	30 h or less	18.3	65
Unpasteurized	More than 30 h	7.2	45
Pasteurized	—	18.3	65
All other products			
Unpasteurized	8 h or less	7.2	45
Unpasteurized	More than 8 h	4.4	40
Pasteurized	—	7.2	45

Source: Anonymous. *Regulations Governing the Inspection of Eggs and Egg Products, 7CFR Part 59*, Department of Agriculture, Washington, DC, USA, 1991. With permission.

or egg white. In contrast, egg white is more sensitive to higher temperature than whole egg or egg yolk due to the possibility of coagulation of protein. The maximum stability of most egg white proteins occurs at near neutral pH, except conalbumin. Addition of lactic acid, which adjusts the pH of albumen to 7.0, allows the albumen to withstand temperatures of 60.5–61.7°C (141–143°F). Aluminum sulfate is generally added together with lactic acid to protect conalbumin, although this process generally results in products with lower whipping ability. Cotterill et al. [11] reported the thermal destruction curves for a wide range of egg products. Table 23.3 lists' minimum requirements for pasteurization time–temperature combinations of liquid egg products in various countries [6,12,13].

Pasteurization is generally done using a high temperature–short time process (HIST) equipped with a plate heat exchanger. A more advanced technology known as UHT is currently available in the U.S. and other countries [14]. This technique using modified milk ultrapasteurization technology allows the ultrapasteurized liquid eggs to be processed at temperature above 64°C (147°F) in a very short time. However, as the ultraheat treatment of liquid eggs is conducted at a much lower temperature and longer time than that of UHT milk (e.g., 135°C for 2–5 sec), the UHT egg products with a shelf life of 60 days or longer need to be stored under refrigeration. Pasteurization method without heat is also available. For example, using high-energy radiation, particularly gamma rays, to pasteurize frozen egg products results in destruction of bacteria, in addition to the benefits of eliminating the costs of thawing, heat pasteurization, and refreezing. In a study of the effect of gamma irradiation on the physicochemical and functional properties of frozen liquid egg products, Ma et al. [15] found that the apparent viscosity of frozen egg yolk was significantly decreased by radiation at pasteurization dosages of 1–4 kGy. The functional properties of the egg whites,

TABLE 23.2
Pasteurization Requirements for Various Liquid Egg Products by USDA

Products	Minimum Holding Temperature °C	°F	Minimum Holding Time[a] (min)
Albumen (without use of chemicals)	56.7	134	3.5
	55.6	132	6.2
Whole egg	60.0	140	3.5
Whole egg blends	61.1	142	3.5
(less than 2% added nonegg ingredients)	60.0	140	6.2
Fortified whole egg and blends	62.2	144	3.5
(24–38% egg solids, 2–12% added non-egg ingredients)	61.1	142	6.2
Salted whole egg	63.3	146	3.5
(with 2% or more salt added)	62.2	144	6.2
Sugared whole egg	61.1	142	3.5
(2–12% sugar added)	60.0	140	6.2
Plain yolk	61.1	142	3.5
	60.0	140	6.2
Sugared yolk (2% or more sugar added)	63.3	146	3.5
	62.2	144	6.2
Salted yolk (2–12% salt added)	63.3	146	3.5
	62.2	144	6.2

[a] For the average particles.

Source: Anonymous. *Regulations Governing the Inspection of Eggs and Egg Products, 7CFR Part 59*, Department of Agriculture, Washington, DC, USA, 1991. With permission.

including foaming, emulsifying, and gelling, were generally not significantly affected or slightly decreased by radiation. Angel food cakes prepared with irradiated frozen egg white had increased cake volume. Mayonnaise prepared with irradiated frozen egg yolk had increased stiffness and stability. However, this method has not been generally accepted by the egg industry yet [13].

Some spoilage bacteria and spores unfortunately remain viable after pasteurization. Furthermore, microorganisms release enzymes into the media in which they live. Some of these enzymes are heat-resistant, thus they survive pasteurization. Therefore, rapid breaking and pasteurization of the eggs immediately after being laid, as well as rapid cooling in pre- and postpasteurization periods dramatically reduce microbial and enzymatic risks [6].

B. Rheological Aspects: Minimizing Gelation Reaction

Freezing is an effective way to preserve the quality and nutritive value of many foods. As shown in Table 23.4 [16–22], approximately 6–9% of the total liquid egg production is frozen. Frozen egg and egg products are widely used in the food industry as ingredients for other food products, and in quantity food preparation, such as restaurants, hotels, and institutions, due to its longer shelf life and high quality if processed appropriately. Appropriate freezing process generally causes only minor changes in raw egg white. However, freezing egg yolk to below −6°C (21°F) causes an irreversible change in its fluid texture called gelation [23]. The thawed yolks would not return to their original smooth texture, but instead show a higher viscosity with a separation of water as well as a lumpy and gummy texture, which would not mix well with other ingredients after thawing. Yolk gelation is an undesirable process because it reduces functionality [23,24]. Chen and Chang [25] reported

TABLE 23.3
Minimum Requirement for Pasteurization of Liquid Egg Products in Various Countries

Country	Product	Temperature °C	Temperature °F	Time (min)	Reference
Australia	Whole egg	62.5	144.5	2.5	[6]
China (PRC)	Whole egg	63.3	146	2.5	[6]
Denmark	Whole egg	65–69	149–156.5	1.5–3.0	[6]
	Whole egg	68.0	154.4	4.5	[12]
	Egg white	61.0	141.8	3.0	[12]
	Egg yolk	68.0	154.4	4.5	[12]
France	Whole egg	58.0	136.4	4.0	[12]
	Egg white	55.5	132	3.5	[12]
	Egg yolk	62.5	144.5	4.0	[12]
Germany	Whole egg	65.5	150	5.0	[12]
	Egg white	56.0	132.8	8.0	[12]
	Egg yolk	58.0	136.4	3.5	[12]
Japan	Whole egg	60	140	3.5	[12]
	Egg white	55–56	131–132.8	3.5	[12]
	Egg yolk	60	140	3.5	[12]
Poland	Whole egg	66.1–67.8	151–154	3.0	[6]
Taiwan	Whole egg	60	140	10	Industry info
	Egg white	56	132.8	10	Industry info
	Egg yolk	63	145.4	20	Industry info
U.K.	Whole egg	64.4	148	2.5	[6]
U.S.A.	Whole egg	60	140	3.5	[6]
	Egg white	56.7	134	3.5	[6]
	Egg yolk	61.1	142	3.5	[6]

that the unpasteurized yolk quality was acceptable after frozen storage at −3 to −6°C for 4 weeks. However, the emulsifying capacity and emulsifying stability of yolk were significantly reduced when frozen-stored at −18°C. The sponge cake made from −18°C frozen yolk showed lowered cake volume and poorer quality when compared with those made from −3 to −6°C frozen yolk.

The mechanism of egg yolk gelation is still not completely understood. As egg yolk is a dispersion of particles in clear plasma, constituents in both plasma and in granules should contribute to gelation during freezing. It has been generally accepted that the protein and phospholipid moieties of the low-density lipoprotein (LDL) may participate in the formation of a LDL–water–sodium chloride complex. Yolk gelation seems to be caused by denaturation of the LDL largely presented in egg yolk. The breakdown of the water shell surrounding the molecules during freezing could further promote rearrangement and aggregation of yolk lipoproteins [26,27]. As constituents in both plasma and in granules contribute to the gelation problem during freezing, and the fatty acid compositions of yolk lipids are influenced by the types of fat diets for hens and the genotypes of hens, it might be expected that factors such as breed, diet, and age of the hen may impact the compositions of LDL particles in such a way that would influence the propensity for yolk gelation during freezing [28]. Other factors, such as freezing and thawing rates, storage temperatures, additives, homogenization, and so on, which influence the

TABLE 23.4
Liquid Egg and Frozen Egg Productions in the United States

Year	Edible Liquid from Shell Eggs Broken (1000 lb)	Frozen Eggs (1000 lb)					Frozen to Liquid Production Ratio (%)
		Whites	Yolks	Whole and Mixed	Unclassified	Total	
1996	1,886,003	15,998	21,807	98,814	18,860	155,479	8.24
1997	2,009,295	12,617	12,950	82,374	9,834	117,775	5.86
1998	2,064,563	11,326	12,019	84,829	17,095	125,269	6.07
1999	2,213,090	8,397	8,571	87,965	13,422	118,355	5.35
2000	2,300,156	38,834	8,716	116,382	13,570	177,502	7.72
2001	2,319,322	34,962	9,001	124,809	19,526	188,298	8.12
2002	2,379,668	22,614	9,633	76,634	41,015	149,896	6.30
2003	2,334,058	30,964	12,632	119,846	40,426	203,868	8.73

Source: Anonymous. Dairy and poultry statistics. Agricultural Statistics 1998–2004. Department of Agriculture, Washington, DC, USA, 1998.

rate and size of ice crystal formation, dehydration of proteins, and concentration effects due to increased salt concentrations or ionic strengths, are also important [6,26,29,30]. In a study of the viscoelastic properties of frozen or thawed egg yolk, Telis and Kieckbusch [31] concluded that the gel formed during freezing was based on physical aggregation rather than chemical binding, with a nonhomogeneous structure. Ice crystals formation and associated freeze concentration of the unfrozen phase were hypothesized to be fundamental causes of gelation, because undisturbed supercooled samples did not show notable changes in complex modulus, G^*; but considerable increase in G^* was observed for yolks that were disturbed and became frozen at the same temperature for the same time.

A variety of approaches have been developed to minimize the gelation of frozen egg products. This section looks at a range of developments in frozen egg products, many of which exploit the benefits of addition of cryoprotectant and rapid freezing on product quality.

1. Addition of Cryoprotectant

Cryoprotectants are compounds that improve the quality and extend the shelf life of frozen foods. A wide variety of cryoprotective compounds are available, including sugars (e.g., sucrose, galactose, glucose, and fructose), amino acids, polyols, methyl amines, carbohydrates, some proteins, enzymes, and even inorganic salts, such as potassium phosphate and ammonium sulfate [32–35].

For frozen egg products, sodium chloride and sucrose at a level of 10% are commonly added to the yolk to prevent gelation. Syrup, glycerin, phosphates, and other sugars can also be used. The slated yolk is then used for mayonnaise and salad dressing. In contrast, the sugared yolk is used for bakery products and ice cream. Sato and Aoki [36] observed that LDL gelation was inhibited by the addition of salts when frozen at higher than eutectic temperature of coexisting salts. Such inhibitory effects are probably attributed to solvation of the adsorbed layers or the formation of a complex between LDL, water, and sodium chloride, which stabilize LDL particles during freezing [27,37]. Jaax and Travnicek [38] reported that sodium chloride increases both emulsifying capacity and viscosity; however fructose reduces emulsifying capacity of the thawed yolk. On the basis of dynamic rheological measurements, Telis and Kieckbusch [39] found that sucrose, glycerol, and magnesium chloride could prevent egg yolk gelation at concentration of 2% and higher. These additives showed improved cryoprotectant effects as their concentration was increased. Sodium chloride at 2% also prevented gelation, but at 10%, it caused a considerable increase in the viscosity of unfrozen yolk. Instead of preventing yolk gelation, calcium chloride showed an opposite effect. It could even promote protein coagulation before freezing. Egg yolk with 2% calcium chloride was found to be gelled completely after 36 h at −24°C. Potassium chloride in the range of 2–10% had an effect similar to that of sodium chloride before freezing. However, yolk with 2% potassium chloride showed very elastic behavior after 36 h at −24°C. Ibarz and Segales [40] studied the steady-shear rheological behavior of salted yolk frozen stored at −20°C, and found that the shear thinning flow behavior of frozen salted yolk could be described by the power-law equation as follows:

$$\tau = K\dot{\gamma}^n \tag{23.2}$$

where τ is the shear stress (Pa), $\dot{\gamma}$ the shear rate (sec^{-1}), K the consistency index (Pa s^n), and n the flow behavior index (dimensionless); the power law parameters, K and n, could be obtained using linear regression analysis of the log (shear stress) and log (shear rate) data. The flow behavior index of frozen salted yolk was found to be decreased, but consistency index increased with increasing salt concentrations ranging from 2 to 14% [40]. In a study of the effect of antifreezing agent addition (10% glycerol), pasteurization treatment (61°C, 3.5 min for yolk, 60°C, 3.5 min for whole egg), and frozen storage (0–60 days) on the rheological properties and functional properties of freeze–thawed yolk and whole egg, Chou et al. [29,30] reported that 10% glycerol addition could significantly improve the gelation problems associated with frozen yolk and whole egg, as indicated by the lesser extent of changes in rheological behavior.

2. Processing

Huang and Yang [41] reported that the texture and stability modification of frozen–thawed egg white gels are affected by the adjustment of pH values of egg white in conjunction with sodium chloride or sucrose addition. At pH 9, sodium chloride or sucrose addition could significantly reduce the toughness of egg white gels made from frozen–thawed egg white. However, at pH 7, the toughness of egg white gels made from frozen–thawed egg white was significantly lowered and sodium chloride or sucrose addition did not modify the toughness of egg white gels significantly. Lopez et al. [42] reported that treatment of yolk with proteolytic enzymes (papain, trypsin, or rhizome) inhibited gelation. Only papain did not seriously affect its organoleptic properties. Feeney et al. [43] reported that gelation was reduced by incubation with crotoxin (lecithinase A) before or after freezing. Haard [28] also reported that treatment of egg yolk with proteolytic enzymes such as papain, trypsin also prevents gel formation, but the product has reduced emulsifying capacity. However, frozen egg yolk products modified with a natural enzyme designed for sauces and dressings are commercially available [44]. Mechanical treatment, such as homogenization, colloid milling, or excessive mixing, also reduces the viscosity of frozen yolk. Low levels of salt, sugar, and skim milk combined with homogenization stabilizes frozen eggs used for scrambling [45].

Fast freezing and thawing result in less gelation for egg yolks than slow freezing and thawing, possibly due to the decreased damage to protein structures by formation of smaller ice crystals, less dehydration of proteins, and less concentration effects by less increase in ionic strength and salt concentrations [23,24,46]. There is evidence that some lipoprotein complexes are altered by freezing; however, when the protein moiety is hydrolyzed, gelation could be inhibited or significantly retarded [26]. Fast freezing and thawing also result in less damage to whole eggs. Using a differential scanning calorimetry, Wooton et al. [47] reported that the loss of denaturation enthalpy was increased by slower freezing rates, higher thawing temperature, higher storage temperature, and longer storage time. Conalbumin suffered greater losses, and ovalbumin had smaller losses than egg white itself. Viscosity and foam instability of egg white was reduced by slower freezing rates, higher thawing temperatures, increased storage times, and lower storage temperatures [26]. Moreover, the magnitude of protein changes and resulting functionality alterations due to freezing are less pronounced for whole eggs when compared with yolks.

IV. FREEZING AND PACKAGING OF EGG PRODUCTS

Freezing of liquid egg products is a time–temperature related process. As mentioned earlier, rapid freezing is generally beneficial to product quality and also reduces processing costs. Air-blast freezers with various designs using cold air as the medium to remove heat from egg products are the most commonly used freezing systems. The cooling rate and the efficiency of the process depend on the contact between the air and the product. Nonami and Akasawa [48] studied the quality of whole liquid egg packaged in unit tin container and frozen at -5 to $-20°C$. It was found that the solid content and the apparent viscosity of tinned frozen egg were lower for samples in the outside than in the center, particularly evident for those frozen at $-5°C$. Such results also implied the benefits of fast freezing. Efstathiou [49] developed a process for making the frozen concentrated whole egg. This process involves two passes of a film of liquid whole egg (3 mm) over the surface of a plate evaporator for preheating (51.7°C or 125°F) and concentration (54.4°C or 130°F for 8–10 sec) at a time and temperature effective for removing water (33–49% solids) but not for coagulating and freezing the concentrated liquid whole egg at -23.3 to $-28.9°C$ (-10 to $-20°F$) using the known equipment.

Individual quick freezing is a relatively more advanced design. With this method, egg whites or liquid whole eggs could first be concentrated using the known techniques such as vacuum evaporation. Concentrates are then forced through a nozzle to create droplets, followed by falling droplets into liquid nitrogen bath where they are immediately frozen, forming small pellets. The free flowing pellets are easy to handle and are thawed rapidly upon heating or mixing with other ingredients. The similar functional properties to standard pasteurized products are closely related to the cryogenic freezing process, with rapid freezing eliminating the risk of visible gel formation. This product is widely used for manufacturing commercial scrambled eggs and omelets [6,50].

The majority of frozen egg products by volume are for further-processing markets. The 13.6-kg (30 lb) containers, including cans, cartons, and plastic bags, are the standard commercial packages for frozen egg products. One of the major drawbacks of frozen egg products is the lengthy thawing time, which reduces the flexibility of work scheduling, especially in emergencies or when additional eggs are needed [6,26]. Smaller units in 2.3 and 4.5 kg (5 and 10 lb) and 3.785×10^{-3} m^3 (1 gal) sizes are also available to overcome the problems associated with lengthy thawing time. The compositions of some frozen products are shown in Table 23.5 [51].

V. QUALITY OF FROZEN EGGS AND FROZEN EGG-RELATED PRODUCTS

A. Microbial Aspects

Normally less than 1% of the bacteria in raw egg products survive pasteurization. The principal genera found in pasteurized egg products are *Alcaligenes, Bacillus, Proteus, Escherichia, Flavobacterium*, and Gram-positive cocci. However, the last three genera are not found after freezing [13]. The approximate total number of bacteria (psychrophiles, thermophiles, and anaerobes) found in 44 pasteurized, frozen commercial egg products are reported in the study of Shafi et al. [52]. *Bacillus* was the predominate genus present in these products.

B. Functional Aspects

Denaturation and performance impairment brought by pasteurization on egg products before freezing or after freeze–thawing is a function of time and temperature [13]. It has been reported consistently that pasteurizing egg white in the range of 54–60°C (129–140°F) damages the foaming power of egg whites. However, for yolk-contaminated egg white, heating improves its foaming properties. The extent of improvement varies with pH [13]. Whole eggs pasteurized under commercial conditions in the United States show a small (about 5%) reduction in volume and functional properties, but performed satisfactorily in commercial baking test [13].

Freezing causes major textural changes and reductions in microbial counts in some egg products [46]. However, functional properties of frozen stored liquid egg white may be only slightly affected, such as some thinning of thick white, possibly due to the denaturation of albumen [46]. Upon freezing and storing raw egg yolk below −6°C (21.1°F), the viscosity increases and gelation occurs. Generally, this gelation has been considered irreversible, although Palmer et al. [53] observed that heating thawed yolk at 45–55°C (113–131°F) for 1 h partially reversed this gelation. The functional properties of plain egg yolk are little affected by freezing [26,29,30,46]. However, pasteurized frozen whole egg generally has more separation of a watery portion after thawing when compared with unpasteurized frozen whole egg. In addition, the combination of pasteurization and freezing reduces the viscosity of the product when thawed and increases the heating time required for the preparation of a sponge cake, but improves the foam stability [13]. Significant damage to functional properties of frozen whole egg may occur when the egg is pasteurized above 63°C (145.4°F) for 3.5 min or 74°C (165.2°F) for 2–3 sec [13]. In a study of the rheological and functional properties of freeze–thawed whole egg and egg yolk, Chou et al. [29,30] found that the emulsifying capacity of frozen yolk and frozen whole egg were not significantly affected by frozen

TABLE 23.5
Nutritive Values Per 100g of Edible Portions of Various Eggs and Egg Products

Nutrient	Whole Egg		Egg White		Egg Yolk		Salted Frozen Yolk	Sugared Frozen Yolk	Frozen Scrambled Egg Mixtures
	Fresh	Frozen	Fresh	Frozen	Fresh	Frozen			
Proximate									
Energy (cal)	147	148	52	47	322	303	274	307	131
Water (g)	75.84	75.85	87.57	88.55	52.31	56.20	50.80	51.25	72.70
Protein (g)	12.58	11.95	10.90	9.80	15.86	15.50	14.00	13.80	13.10
Total fat (g)	9.94	10.20	0.17	0.00	26.54	25.60	23.00	22.75	5.60
Saturated fatty acid (g)	3.099	3.147	0.000	0.000	9.551	7.820	7.028	6.970	1.052
Monounsaturated fatty acid (g)	3.810	3.886	0.000	0.000	11.738	9.747	8.849	8.614	2.339
Polyunsaturated fatty acid (g)	1.364	1.412	0.000	0.000	4.204	3.628	3.150	3.244	1.778
Cholesterol (mg)	423	432	0	0	1234	1075	955	959	65
Carbohydrate (g)	0.77	1.05	0.73	1.05	3.59	1.15	1.60	10.80	7.50
Ash (g)	0.86	0.95	0.63	0.60	1.71	1.55	10.60	1.40	1.10
Fiber (g)	0.0	0.0	0.0	0.0	0.0	0.0	0.0	0.0	0.0
Minerals									
Ca (mg)	53	59	7	7	129	138	114	123	17
Fe (mg)	1.83	1.85	0.08	0.05	2.73	3.34	3.75	3.14	0.23
Mg (mg)	12	11	11	10	5	9	10	10	10
P (mg)	191	202	15	13	390	417	431	384	30
K (mg)	134	130	163	136	109	118	117	103	147
Na (mg)	140	133	166	158	48	67	3780	67	162
Zn (mg)	1.11	1.38	0.03	0.02	2.30	2.88	2.84	2.81	0.14
Cu (mg)	0.102	0.053	0.023	0.012	0.077	0.024	0.109	0.012	0.030
Mn (mg)	0.038	0.034	0.011	0.007	0.055	0.062	0.062	0.059	—[a]
Se (μg)	31.7	30.8	20.0	17.6	56.0	41.8	37.7	37.7	22.9

(*Table continued*)

TABLE 23.5 *Continued*

Nutrient	Whole Egg		Egg White		Egg Yolk		Salted Frozen Yolk	Sugared Frozen Yolk	Frozen Scrambled Egg Mixtures
	Fresh	Frozen	Fresh	Frozen	Fresh	Frozen			
Vitamins									
Vitamin C (mg)	0.0	0.0	0.0	0.0	0.0	0.0	0.0	0.0	0.0
Thiamin (mg)	0.069	0.060	0.004	0.005	0.176	0.155	0.130	0.135	0.010
Riboflavin (mg)	0.478	0.460	0.439	0.400	0.528	0.520	0.430	0.530	0.300
Niacin (mg)	0.070	0.075	0.105	0.100	0.024	0.045	0.040	0.023	0.090
Pantothenic acid (mg)	1.438	1.480	0.190	0.155	2.990	3.530	3.230	3.200	—[a]
Vitamin B_6 (mg)	0.143	0.162	0.005	0.004	0.350	0.345	0.261	0.284	0.010
Folate (μg)	47	73	4	3	146	116	107	139	17
Vitamin B_{12} (μg)	1.29	1.07	0.09	0.06	1.95	1.82	2.52	1.77	0.17
Vitamin A (IU)	487	525	0	0	1442	1609	1190	1315	410
Retinol (μg)	139	158	0	0	371	433	357	395	0
Vitamin E (mg)	0.97	—[a]	0.00	—[a]	2.58	2.49	—[a]	—[a]	0.84
Vitamin D (IU)	34.548	—[a]	0.000	—[a]	107.423	—[a]	—[a]	—[a]	—[a]
Vitamin K(μg)	0.3	—[a]	0.0	—[a]	0.7	0.7	—[a]	—[a]	1.8

[a]Data unavailable.

Source: Anonymous. *National Nutrient Database for Standard Reference, Release 16–1.* Agricultural Research Service, Department of Agriculture, Washington, DC, USA, 2004. With permission.

storage at $-20°C$ or by pasteurization treatments at 60°C, 3.5 min for whole egg, and 61°C, 3.5 min for yolk. However, addition of 10% glycerol significantly increased the emulsifying capacity of frozen whole egg, but not that of frozen yolk.

C. Product Performance Aspects

The majority of frozen egg products by volume are for further-processing markets. Omelets and scrambled eggs are major items. A wide range of ingredients have been added to whole egg to prepare commercial frozen scrambled egg mixes. The most common added components include nonfat dry milk, whey, vegetable oil, water (if dry ingredients are used), gums (e.g., carboxyl methyl cellulose or xanthan gums), organic acids or other chelators (e.g., citric acid, lactic acid, or phosphates), salt, and egg white. When they are held at serving temperatures (at or above 60°C or 140°F) or for long periods of time, a fluid may separate (syneresis) and a green discoloration due to the formation of the iron–sulfur compound may occur. The problem of greening can be easily prevented by the addition of ingredients to lower the pH (e.g., lemon juice) and to chelate the iron (e.g., citric acid, lactic acid, or EDTA) [26].

Other product lines using frozen egg products include pancakes, crepes, waffles, French toast, egg tofu, and sauces, and so on [26,54–56]. Schell and Schell [46] successfully developed a method of preparing frozen egg butter sauces without loss of texture or appearance. Manderfeld et al. [55] developed a method for manufacturing egg patties that could maintain their shape when thawed. Feiser and Cotterill [57,58] reviewed the performance of cooked–frozen–thawed–reheated egg products, including scrambled egg, quiches, soufflés, meringues, omelets, and egg rolls. The increase in pH due to the loss of carbon dioxide is commonly encountered in these products. Greening is therefore likely if the pH of the cooked products exceeds 8.2. Gossett and Baker [59] reported the optimum conditions for the prevention of greening discoloration in cooked liquid whole egg by the addition of the chelator. O'Brien et al. [60] reported that by addition of 0.1% xanthan gum, omelets that has been steamed for 5 min and then cryogenically frozen using liquid carbon dioxide ($-79°C$) or liquid nitrogen ($-196°C$) could minimize the moisture loss and shear force of frozen omelets, and provide satisfactory organoleptic properties. Chang and Ho [61] studied the qualities of liquid whole eggs with various cryoprotectants such as xanthan gum, sodium carboxymethyl cellulose, polyphosphate, or xanthan gum in conjunction with tapioca starch, after heating at 85°C for 20 min, followed by freezing at -25, -35, and $-45°C$ freezer or at $-196°C$ liquid nitrogen to lower the internal temperature to $-20°C$. After frozen storage of up to 3 months, samples were thawed for drip loss, color, hardness, and microstructure determination. It was found that the drip loss of frozen-thawed whole egg coagulates increased with increasing freezing temperature and frozen storage time. The drip loss of frozen–thawed whole egg coagulates can be significantly lowered by the addition of cryoprotectants as mentioned earlier. L values of the frozen–thawed whole egg coagulates decrease and a/b ratios increase as a result of freezing and frozen storage. Quick freezing at $-196°C$ with the addition of cryoprotectants, particularly 0.2% polyphosphate, lowers the damage to microstructure caused by ice crystal formation.

Wang et al. [62] found that the quality of frozen brined duck yolk was significantly affected by the brining process and freezing temperature. Generally, the shear value of frozen brined duck yolk is the highest when frozen for 2 months, and then decreases after 6-month frozen storage. The thiobarbituric acid and volatile basic nitrogen values of frozen brined duck yolk are within acceptable level, and the microstructure becomes smoother after 6-month frozen storage.

VI. CONCLUSIONS

Eggs have been generally considered as nutritious food. The nutritive values and various functionalities of eggs make them quite useful for food industry. To be manufactured into various processed foods, shell eggs are generally broken into liquid products, pasteurized to eliminate *Salmonella*, and frozen for longer shelf life. Appropriate freezing process generally causes only minor changes in raw egg white. However, freezing egg yolk to below $-6°C$ (21°F) causes a irreversible change in its fluid texture called gelation. A variety of approaches have been developed to minimize the gelation problems associated with frozen egg products, many of which exploit the benefits of addition of cryoprotectant and rapid freezing on product quality. Functional properties of the resulting frozen egg products may therefore be only slightly or little affected by freezing.

NOMENCLATURE

HU	Haugh units (dimensionless)
H	thick albumen height (mm)
G	constant of 32.2 (dimensionless)
W	weight of egg (kg)
τ	shear stress (Pa)
$\dot{\gamma}$	shear rate (sec^{-1})
K	consistency index (Pa sec^n)
n	flow behavior index (dimensionless)

REFERENCES

1. BA Watkins. The nutritive value of egg. In: WJ Stadelman, OJ Cotterill, Eds., *Egg Science and Technology*. 4th ed. New York: The Haworth Press, Inc., 1995, pp. 177–194.
2. MA Gutierrez, H Takahashi, LR Juneja. Nutritive evaluation of hen eggs. In: T Yamamoto, LR Juneja, H Hatta, M Kim, Eds., *Hen Eggs, Their Basic and Applied Science*. Boca Raton: CRC Press Inc., 1997, pp. 25–35.
3. G Zeidler. Shell eggs and their nutritive value. In: DD Bell, WD Weaver Jr, Eds., *Commercial Chicken Meat and Egg Production*. 5th ed. Massachusetts: Kluwer Academic Publishers, 2002, pp. 1109–1128
4. SC Yang, RE Baldwin. Functional properties of eggs in foods. In: WJ Stadelman, OJ Cotterill, Eds., *Egg Science and Technology*, 4th ed. New York: The Haworth Press, Inc., 1995, pp. 405–463.
5. G Zeidler. Quality and functionality of egg products. In: DD Bell, WD Weaver Jr, Eds., *Commercial Chicken Meat and Egg Production*. 5th ed. Massachusetts: Kluwer Academic Publishers, 2002, pp. 1219–1228.
6. G Zeidler. Further-processing eggs and egg products. In: DD Bell, WD Weaver Jr, Eds., *Commercial Chicken Meat and Egg Production*. 5th ed. Massachusetts: Kluwer Academic Publishers, 2002, pp. 1163–1197.
7. H Hatta, T Hagi, K Hirano. Chemical and physicochemical properties of hen eggs and their application in foods. In: T Yamamoto, LR Juneja, H Hatta, M Kim, Eds., *Hen Eggs, Their Basic and Applied Science*. Boca Raton: CRC Press Inc., 1997, pp. 117–133.
8. G Zeidler. Shell egg quality and preservation. In: DD Bell, WD Weaver Jr, Eds., *Commercial Chicken Meat and Egg Production*. 5th ed. Massachusetts: Kluwer Academic Publishers, 2002, pp. 1199–1218.
9. Anonymous. Egg products inspection act 1970. *Food and Drug Authority*, Rockville, MD, USA, 1971.
10. Anonymous. *Regulations Governing the Inspection of Eggs and Egg Products, 7CFR Part 59*. Department of Agriculture, Washington, DC, USA, 1991.
11. OJ Cotterill, J Glauert, GF Krause. Thermal destruction curves for *Salmonella* oranienburg in egg products. *Poultry Science* 52:568–577, 1973.
12. M Kobayashi, MA Gutierrez, H Hatta. Microbiology of eggs. In: T Yamamoto, LR Juneja, H Hatta, M Kim, Eds., *Hen Eggs, Their Basic and Applied Science*. Boca Raton: CRC Press Inc., 1997, pp. 179–191.
13. FE Cunningham. Egg-product pasteurization. In: WJ Stadelman, OJ Cotterill, Ed., *Egg Science and Technology*. 4th ed. New York: The Haworth Press, Inc., 1995, pp. 289–321.
14. KR Swartzel, HR Jr Ball. Method for pasteurizing liquid whole egg products. US Patent: US5019407, 1991.

15. CY Ma, VR Harwalkar, LM Poste, MR Sahasrabudhe. Effect of gamma irradiation on the physicochemical and functional properties of frozen liquid egg products. *Food Research International* 26 (4):247–254, 1993.
16. Anonymous. Dairy and poultry statistics. *Agricultural Statistics 1998*. National Agricultural Statistics Service, Department of Agriculture, Washington, DC, USA, 1998.
17. Anonymous. Dairy and poultry statistics. *Agricultural Statistics 1999*. National Agricultural Statistics Service, Department of Agriculture, Washington, DC, USA, 1999.
18. Anonymous. Dairy and poultry statistics. *Agricultural Statistics 2000*. National Agricultural Statistics Service, Department of Agriculture, Washington, DC, USA, 2000.
19. Anonymous. Dairy and poultry statistics. *Agricultural Statistics 2001*. National Agricultural Statistics Service, Department of Agriculture, Washington, DC, USA, 2001.
20. Anonymous. Dairy and poultry statistics. *Agricultural Statistics 2002*. National Agricultural Statistics Service, Department of Agriculture, Washington, DC, USA, 2002.
21. Anonymous. Dairy and poultry statistics. *Agricultural Statistics 2003*. National Agricultural Statistics Service, Department of Agriculture, Washington, DC, USA, 2003.
22. Anonymous. Dairy and poultry statistics. *Agricultural Statistics 2004*. National Agricultural Statistics Service, Department of Agriculture, Washington, DC, USA, 2004.
23. WD Powrie, H Little, A Lopez. Gelation of egg yolk. *Journal of Food Science* 28:38–46, 1963.
24. WD Powrie. Cryopreservation of egg yolk. In: OR Fennema, Ed., *Low Temperature Preservation of Food and Living Matter*. New York: Marcel Dekker Inc., 1973, pp. 264–281.
25. YC Chen, HS Chang. Effects of freezing storage temperature on quality of hen yolk. *Journal of the Chinese Society of Animal Science* 27 (3):421–431, 1998.
26. OJ Cotterill. Freezing egg products. In: WJ Stadelman, OJ Cotterill, Eds., Egg Science and Technology. 4th ed. New York: The Haworth Press, Inc., 1995, pp. 265–288.
27. T Wakamatu, Y Sato, Y Saito. On sodium chloride action in the gelation process of low density lipoprotein (LDL) from hen egg yolk. *Journal of Food Science* 48:507–516, 1983.
28. NF Haard. Product composition and the quality of frozen foods. In: MC Erickson, YC Hung, Eds., *Quality in Frozen Food*. New York: Chapman & Hall, 1997, pp. 275–295.
29. LL Chou, LS Lai, SC Yang. Effect of antifreezing agent, pasteurization and frozen storage on the rheological and functional properties of freeze-thawed egg yolk. *Taiwanese Journal of Agricultural Chemistry and Food Science* 39 (2):135–143, 2001.
30. LL Chou, LS Lai, SC Yang. Effect of antifreezing agent, pasteurization and frozen storage on the rheological and functional properties of freeze-thawed whole egg. *Taiwanese Journal of Agricultural Chemistry and Food Science* 39 (6):437–446, 2001.
31. V Telis, TG Kieckbusch. Viscoelasticity of frozen/thawed egg yolk. *Journal of Food Science* 62: 548–550, 1997.
32. C Kennedy. Developments in freezing. In: P Zeuthen, L Bogh-Sorensen, Eds., *Food Preservation Techniques*. Cambridge: Woodhead publishing Limited, 2003, pp. 228–240.
33. N Hedges. Maintaing the quality of frozen fish. In: H Bremner, Ed., *Safety and Quality Issues in Fish Processing*. Cambridge: Woodhead publishing Limited, 2002, pp. 379–406.
34. SR Payne, D Sandford, A Harris, OA Young. Effects of antifreeze proteins on chilled and frozen meat. *Meat Science* 37 (3):429–438, 1994.
35. SR Payne, OA Young. Effects of pre-slaughter administration of antifreeze proteins on frozen meat quality. *Meat Science* 41 (2):147–155, 1995.
36. Y Sato, T Aoki. Influence of various salts on gelation of low-density lipoprotein (egg yolk) during its freezing and thawing. *Agricultural and Biological Chemistry Journal* 39:29–35, 1975.
37. V Kamat, G Graham, M Barrar, M Stubbs. Freeze–thaw gelation of hen's egg yolk low-density lipoprotein. *Journal of the Science of Food and Agriculture* 27:913–928, 1976.
38. S Jaax, D Travnicek. The effects of pasteurization, selected additives and freezing rate on the gelation of frozen-defrosted egg yolk. *Poultry Science* 46:1013–1022, 1968.
39. V Telis, TG Kieckbusch. Viscoelasticity of frozen/thawed egg yolk as affected by salts, sucrose and glycerol. *Journal of Food Science* 63:20–24, 1998.
40. A Ibarz, J Segales. Influence of freezing on rheology of salted egg yolk. *Alimentaria* 266:69–76, 1995.
41. CY Huang, SC Yang. Combined effect of sodium chloride or sucrose addition and pH on the texture and stability of refrigerated and freeze thawed egg white gels. *Journal of the Chinese Society of Animal Science* 25 (1):117–128, 1996.

42. A Lopez, CR Fellers, WD Powrie. Enzyme inhibition of gelation on frozen egg yolk. *Journal of Milk Food Technology* 18 (3):77–80, 1954.
43. RE Fenney, JR MacDonnell, H Fraenkel-Conrat. Effects of crotoxin (lecithinase A) on egg yolk and yolk constituents. *Archives of Biochemistry and Biophysics* 48:130–140, 1954.
44. M Pehanich. Technology opens egg-citing market opportunities. *Food Processing, USA* 63 (6):52–53, 2002.
45. MP Penfield, AM Campbell. *Experimental Food Science*. 3rd ed. New York: Academic Press, 1990, pp. 153–155.
46. YL Xiong. Protein denaturation and functionality losses. In: MC Erickson, YC Hung, Eds., *Quality in Frozen Food*. New York: Chapman & Hall, 1997, pp. 111–140.
47. M Wooton, NT Hong, HLP Thi. A study on the denaturation of egg white proteins during freezing using differential scanning calorimetry. *Journal of Food Science* 46:1337–1338, 1981.
48. Y Nonami, M Akasawa. Variance of quality of frozen whole egg mass packaged in unit tin container. *Journal of Japanese Society of Food Science and Technology* 42 (10):808–814, 1995.
49. JD Efstathiou. Frozen concentrated liquid whole egg and method of making same. US Patent: US6660321B2, 2003.
50. Anonymous. Frozen egg pellets. Egg products: a concentration method developed in the USA. *Process* 1089: 39, 1993.
51. Anonymous. *National Nutrient Database for Standard Reference, Release 16–1.* Agricultural Research Service, Department of Agriculture, Washington, DC, USA, 2004.
52. R Shafi, OJ Cotterill, ML Nichols. Microbial flora of commercially pasteurized egg products. *Poultry Science* 49: 578–585, 1970.
53. HH Palmer, K Ijichi, H Roff. Partial thermal reversal of gelation in thawed egg yolk products. *Journal of Food Science* 35:403–406, 1970.
54. U Bindrich, H Rohenkohl, U Mueller, I Zuerner. Frozen egg products: Advantages, production and properties. *Lebensmitteltechnik* 28 (9):50–52, 54–55, 1996.
55. MM Manderfeld, JD Efstathiou, AH Voecks. Simulated egg patty. US Patent: US5620735, 1997.
56. LJ Schell, CJ Schell. Method of preparing frozen egg butter sauces. US Patent: US6565910B1, 2003.
57. GE Feiser, OJ Cotterill. Composition of serum from cooked–frozen–thawed–reheated scrambled egg at various pH levels. *Journal of Food Science* 47:1333–1337, 1982.
58. GE Feiser, OJ Cotterill. Composition of serum and sensory evaluation of cooked–frozen–thawed scrambled egg at various salt levels. *Journal of Food Science* 48:794–797, 1982.
59. PW Gossett, RC Baker. Prevention of the green-gray discoloration in cooked liquid whole eggs. *Journal of Food Science* 46:328–331, 1981.
60. SW O'Brien, RC Baker, LF Hood, M Liboff. Water-holding capacity and textural acceptability of precooked, frozen whole egg omelets. *Journal of Food Science* 47:412–417, 1982.
61. HS Chang, LH Ho. Effect of freezing temperature and additives on the appearance and texture of heat-coagulated seasoned whole egg. *Journal of the Chinese Society of Animal Science* 27 (2):283–294, 1998.
62. CT Wang, ZS Lin, YH Chen. The effect of pickling and preserving methods on the qualities of frozen salted egg yolk. *Journal of the Chinese Society of Animal Science* 28 (2):237–247, 1999.

Part IV

Monitoring and Measuring Techniques for Quality and Safety

24 Physical Measurements

Parameswarakumar Mallikarjunan
Virginia Polytechnic Institute and State University, Virginia, USA

CONTENTS

I. INTRODUCTION

The frozen food industry is one of the largest segments in the developed countries and major world food markets. The frozen dinner and entrée category continues to be the largest within the frozen food market with more than \$6 billion in annual supermarket sales in the United States alone. The frozen meat/seafood and the frozen novelties categories experienced the largest growth from 2000 to 2001 in the United States, with sales of the meat/seafood category increasing by 13% and the frozen novelties category increasing by 10.5% [1].

Frozen foods fill important niches in the overall market because they provide convenient prepared or semiprepared products that can be made available all year at consistent cost in markets where fresh or refrigerated products are not available, and can be transported to greater distances with minimal damage and degradation to product quality [2]. Frozen foods have generally been perceived by the consumers as having better quality than foods processed and preserved by other methods. Total retail sales of frozen foods in the United States reached more than \$27 billion in 2001, which is 6.1% higher than 2000, up over \$1 billion from 1999 and \$2 billion from 1998 [1]. However, growth in sales of frozen foods has been slowed down due to the deficiencies in frozen foods as compared with fresh, shelf-stable, and refrigerated alternatives [3]. The consumers expect the quality of the frozen food to be close to that of a fresh product. Thus, the quality of the frozen foods is the most important in deciding its growth and success in the competitive global marketplace.

II. QUALITY ATTRIBUTES IMPORTANT TO FROZEN FOODS

Changes occurring in food quality during freezing, frozen storage, and thawing play a major role in the success of the products to be marketed as frozen foods. The quality changes in the frozen food are mainly due to size and distribution of ice crystals in the food and is affected directly by the rate of freezing, storage temperature, and temperature fluctuations during storage. In addition, the product quality is also affected by the rate of thawing and length of frozen storage time. Thus, it is imperative to determine the quality of frozen foods in all the stages, from the fresh product, frozen product to a thawed product. Furthermore, for many frozen food products, they are consumed after further processing, like cooking, the quality of the food should also be determined after the post-processing. The following quality parameters are studied extensively and considered critical: color, texture, juiciness, and flavor.

III. COLOR

The first and foremost quality by which consumer makes a decision even before purchasing a food product is the visual appearance. In addition to influencing the purchase intent, the color of the food also affects the taste perception of the food. Color can be defined as the energy distribution of the light reflected by or transmitted through a particular food product. Depending on how the light reacts with the food, the food can be classified as opaque, translucent, or transparent. Among these reactions, the reflected light determines the color and appearance of the material. The reflected light is affected by the light source, the angle of viewing, and the characteristics of the background; in turn, all these factors affect the color perception. Often, the color is measured using standard CIE (The Commission International de l'Eclariage) scales. In the food industry, the derived color scales such as Hunter L, a, and b or CIE L^*, a^*, and b^* are commonly used. These parameters represents the lightness (L), and the degree of redness or greenness (a), and the degree of yellowness or blueness (b) of the product been measured. A value of 0 or 100 for L represents black or white, respectively. These instrumental color measurements correspond to the visual assessment of food color and the human perception of the color as hue, saturation, and value can be calculated from L^*, a^*, and b^* values. The color parameters for Hunter L, a, and b are square roots using CIE XYZ, whereas CIE $L^*a^*b^*$ is calculated using cube roots of XYZ, where X, Y, and Z are the tristimulus values from the reflectance of a standard observer.

The hue describes the visual sensation of the color. The hue represents the appearance of a given area in comparison to one or proportions of two or more of the perceived standard colors of red, yellow, green, and blue. The hue is measured in angle (degrees or radians) in the color wheel (Figure 24.1).

Chroma, the purity or saturation, describes the intensity of a fundamental color for the amount of white light that is mixed with it. In other words, converting the coordinate system of a^* and b^* to polar coordinates of r and θ gives chroma and hue, respectively. Thus, the chroma and hue angle can be calculated as

$$\text{Chroma or saturation} = [a^{*2} + b^{*2}]^{0.5} \tag{24.1}$$

$$\text{Hue angle} = \tan^{-1}\left(\frac{b^*}{a^*}\right) \tag{24.2}$$

The value or lightness is an indication of overall light reflectance of that color. Thus, the color parameter L^* represents the value or color lightness.

In many instances, the differences in color parameters between two stages provide more meaningful interpretation of the processes than absolute color values. The color differences between two

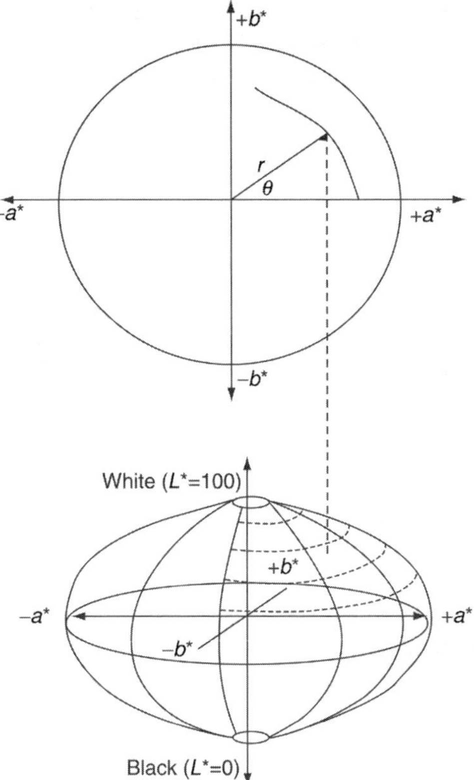

FIGURE 24.1 Color wheel and solid representing color measurement systems.

stages (e.g., between raw and frozen food in frozen, during storage, or thawed stages) can be calculated as

$$\text{Hue angle difference} = \tan^{-1}\left(\frac{b}{a}\right) - \tan^{-1}\left(\frac{b_0}{a_0}\right) \quad (24.3)$$

$$\text{Saturation difference} = [(a - a_0)^2 + (b - b_0)^2]^{0.5} \quad (24.4)$$

$$\text{Brightness difference} = \text{abs}(L - L_0) \quad (24.5)$$

$$\text{Total color difference} = [(L - L_0)^2 + (a - a_0)^2 + (b - b_0)^2]^{0.5} \quad (24.6)$$

where L_0, a_0, and b_0 are the color parameters of the initial stage (or stage 1) and L, a, and b are the corresponding color parameters at other processing stages.

Using either a colorimeter, spectrophotometer, or digital image processing, color can be measured. To duplicate the responses of the human eye, the colorimeter uses a set of three filters (red, green, and blue) with transmission curves similar to the standard X, Y, and Z curves. The light reflected from the object through each filter is recorded and the tristimulus values are obtained. All tristimulus colorimeters available depend on this principle with individual refinements in photocell response, sensitivity, stability, and reproducibility. Nowadays, the meters also provide the color values in other color spaces. Tristimulus colorimeters usually are small in size and portable. They come with special attachments, wide range of aperture openings, and customized for specific applications. The colorimeters are available from Agtron, Inc. (Sparks, NV), BYK-Gardner (Silver Spring, MD), HF Scientific, Inc. (Ft. Myers, FL), Hunter Lab (Reston VA), Minolta Corp. (Ramsey, NJ), and X-Rite, Inc. (Grandville, MI) in the United States.

Recently, use of tristimulus colorimeters to measure the color of frozen foods has increased. Mostly, the meters have been used to measure the surface color of the food products. Lanari et al. [4] measured the surface color of thawed frozen ground beef samples using a colorimeter (Minolta chroma meter). They found repeated freezing and thawing decreased the redness (a^*) and color saturation of semimembranous muscle samples. Andersen and Skibsted [5] and Akamittath et al. [6] measured the surface color of frozen meat to study the effect of salt on lipid oxidation. Battacharya et al. [7] also used a colorimeter (Hunter Lab color difference meter) to measure the surface color of frozen ground beef patties and found that an increase in frozen storage time resulted in a loss of color with products tending to become grayish. Redmond et al. [8] used the color difference meter to measure the color in freeze-chilled lasagna and used a ratio of L^*/b^* as a measure of surface browning during freeze-chilling and subsequent reheating. Forni et al. [9] measured the color changes in peas due to freezing and frozen storage. The color was measured on 100 g of the pea samples using a Hunter Color Difference meter. Along with texture and flavor, the color of the peas was found acceptable ($P < 0.05$) both after freezing and after cold storage. Cano and Martin [10] measured the color of frozen kiwi slices using a Hunter Lab Model D25-9 colorimeter and 50 g of pureed sample. Based on the color parameters L, b, and chroma, the freezing preservation of kiwi fruit slices produced a kiwi fruit product with similar color characteristics and appearance of fresh fruit ($P < 0.05$).

Spectrophotometers measure the ratio of the light reflected or transmitted from a food product to that from a known reference standard. Using multiple sensors sensitive to a particular wavelength, the tristimulus values are calculated mathematically by applying the energy curves of the light source and that of the standard observer. Hwang et al. [11] used a spectrophotometer (Hunter Lab) for measuring the color of frozen precooked beef loin slices. Before the measurement samples were thawed for 6 min and finish cooked for 3 min using a microwave oven. Based on redness (a^*), the exclusion of oxygen in the packages by vacuum and N_2/CO_2 produced little adverse effect on product color. Brewer and Harbers [12] measured the color of frozen ground pork to study the effects of frozen storage on lipid oxidation. Before color measurement using a spectrophotometer (Hunter Lab), all the samples were thawed at 4°C for 48 h. They used a^* for redness and difference in reflectance between 630 and 580 nm for oxymyoglobin content. Ground pork lost its redness and oxymyoglobin during frozen storage and they concluded that the packaging that excluded both oxygen and light were most effective at preventing the loss of red color. Similar results were obtained for frozen ground beef patties over a 12-week storage at −20°C using the same methodology [13]. Chen and Trout [14] measured color (using Perkin–Elmer) of restructured beef steaks to study the effect of various binders during frozen storage on color stability. Cano et al. [15] measured the color of frozen banana slices by determining the reflectance at 485 nm using a spectrophotometer. The color measurement at 485 nm was correlated with oxidation of phenylenediamine. They concluded that the effect of freezing on enzymatic activities of banana slices depended on banana maturity level at the time of processing. Zhang et al. [16] used a spectrophotometer to evaluate the chlorophyll content in areca fruits subjected to different process conditions before subjecting the fruits to a quick freezing in a liquid nitrogen bath.

Both tristimulus colorimeters and spectrophotometers provide product color with reasonable accuracy and reproducibility. The choice of which meter to use will depend on the food material and type of application. Tristimulus colorimeters are primarily used in production settings to monitor product quality during each processing step. Spectrophotometers can provide high accuracy and they are most suited for laboratory applications. The recent development of portable spectrophotometers enables their use in production settings as well. Regardless of the choice of the meter, it is important that the sample preparation and presentation to the instrument are followed to obtain reproducible measurements. Light scattering and sample uniformity can be altered by grinding, mixing, and blending procedures. It is advisable to turn the sample 90° from the first reading and an average of the two readings can be used.

Based on a point-by-point measurement on the surface of the object, the tristimulus colorimeter is more appropriate to determine uniformly distributed color schemes; spectrophotometers are most applicable to the measurement of a liquid where the amount of light transmitted can be determined. However, when there is a broader color change within the sample over a large area, color machine vision will allow for a large area of sampling, which in most instances means the entire sample can be analyzed. A color machine vision system can be inexpensively developed through the use of a standard lighting system, a digital camera, and a computer with photo editing software such as Adobe Photoshop®. Tokuşoğlu et al. [17] unitized a computer vision system to create color profiles for seafood products stored under refrigerated conditions. This system was reported to be successful in separating the contributing color elements that governed the color of food products of varying ages. Similarly, the color machine vision systems has been used to measure color and marbling in beef cuts and products having larger surface area [18].

IV. TEXTURE

One of the most important qualities of frozen food which has a major influence on its sales is its texture. Food texture depends mainly on the structural constituents. Thus the amount and size of ice crystals in the frozen food influences its texture during frozen storage and especially after thawing. Textural properties are related to the deformation, disintegration, and the flow of food under force. The properties can be measured objectively and can be correlated to sensory textural attributes.

Szczesniak [19] gave a system of classification of textural characteristics on fundamental rheological principles. Textural characteristics can be classified into five primary parameters of hardness, cohesiveness, viscosity, elasticity, and adhesiveness, and into three secondary parameters of brittleness, chewiness, and gumminess. These properties referred to the manner in which the food behaved in the mouth.

Food materials generally are viscoelastic in nature, that is, exhibiting properties of both the solid and liquid. The majority of the frozen foods exhibit more solid characteristics, and so the texture is measured using applied compression, shear and tensile forces (e.g., Warner Bratzler shear test, puncture test, and textural profile analysis). It is best to measure the texture of the food material close to the stage as it is consumed, so that it can be compared with the sensory measurements or consumer acceptance of the product. Thus, in foods such as baked goods [20–22], frozen yogurt, and ice creams [23], the texture can be measured in its frozen stage. However, the determination of texture of many other frozen foods, in general, are conducted after thawing; in some cases, such as meat, after postprocessing (e.g., cooking). The evaluation methods depend on the nature of the product and the sensory property that needs to be compared such as mouth feel, chewiness, or tenderness.

The tenderness is a measure of texture for foods like meat, fruits, and vegetables. It has been measured as shear, bite, penetration, tensile, and compressive forces. Some objective methods of evaluating the tenderness include shear methods using Warner Bratzler shear device or Kramer shear press, compression methods (puncture test, texture profile analysis, and stress relaxation tests) and tensile methods.

In measuring texture, force must be recorded accurately and because rapid force fluctuation and slope changes occur, a high-frequency response capability for the data acquisition is required. In addition, accurate records of the probe position are needed to interpret the data properly. A reproducible deformation rate is critical. To achieve a faster data processing and reproducible deformation, use of a universal testing machine for conducting the experiments is necessary. The universal testing machines are available in the market from Instron Corp. (Canton, MA), TA-Instruments, Inc. (New Castle, DE), Testing Machines, Inc. (Amityville, NY); and Texture

Technologies Corp. (Scarsdale, NY) in the United States. The measurement conditions such as cross-head speed, compression ratio, number of loading cycles vary widely in the literature and no uniform test procedures are available for this property for any given food product. The cross-head speeds ranging from 50 to 200 mm/min are very common and many times a load cell of 50 kg or less is ideal for testing food products.

A. Shear Test

One of the widely used texture measurements is Warner Bratzler Shear test. In this test, the sample is sheared using a Warner–Bratzler (WB) shear blade. The shear blade has a opening in the shape of an inverted triangle. The sample is placed inside the hole and the shear blade is moved in an upward direction during testing (Figure 24.2). Detailed discussion on the deformation pattern during testing and interpretations of the testing is widely available in the literature [24].

The following parameters are obtained during a WB shear test:

(1) Maximum or initial yield force representing a tensile rupture signifying cohesiveness
(2) Cross-sectional area of the sample at rupture, which indicates compression required to initiate rupture and can be correlated with firmness
(3) Slope of the force–time curve that provides an index of firmness
(4) Force per unit length of the blade edge cutting the sample at the peak force.

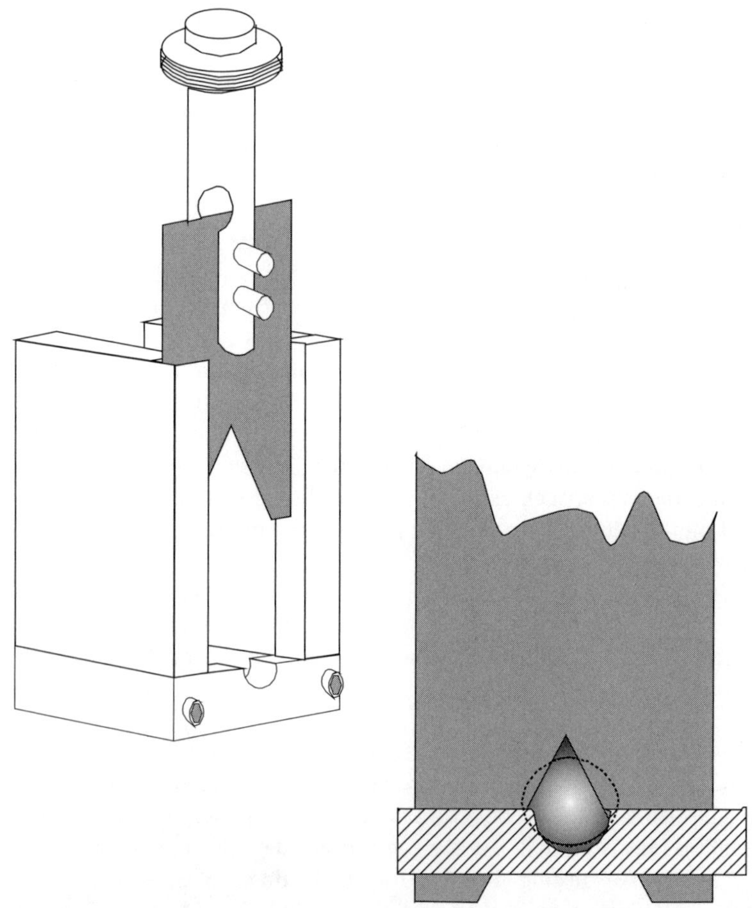

FIGURE 24.2 Warner Bratzler shear.

Modifications have been suggested for texture measurements using the Warner–Bratzler blade such as using a straight-edged blade instead of V-notched blade for frozen food products that are wider or broader in shape. For example, par-fried frozen nuggets or frozen fish fillet can be tested using the straight blade to get better results than using a WB shear blade or a Kramer-shear cell.

Another shear test that is widely used is Kramer-shear test. In Kramer-shear test, the sample is placed in a Kramer-shear cell (Figure 24.3). Kramer-shear cell contains a sample holder and a blade assembly. Kramer-shear blade assembly contains 10 blades each 3.175-mm thick. The sample holder is a metal box ($67 \times 67 \times 63$ mm^3) having 10 slots in its bottom. The sample is filled in the sample holder to the desired level (normally to 30% by volume). The sample holder has a matching slotted cover with 10 slots. The cover is placed on the top of the sample holder. The whole assembly is placed in the base of the Universal testing machine. The blade assembly is lowered so that the bottom of the set of blades is just touching the slotted cover. After setting the universal testing machine to desired test speed, the blades will be lowered until the product gets extruded through the bottom slots. A typical test speed is around 225 mm/min.

Similarly, the following parameters are obtained during a Kramer-shear test:

(1) The initial yield force and yield distance
(2) Peak force, peak force distance
(3) Slope and work (area under the force–time curve) required to shear the sample.

Brewer et al. [25] used WB shear apparatus to determine the effects of different types of blanching on frozen green beans. The texture measurements were obtained on cooked samples. Texture was softest in beans blanched using boiling water and steam blanched beans. Microwave blanching for 3 min (at full power) resulted in a similar texture to steam blanched beans. Forni et al. [9] used

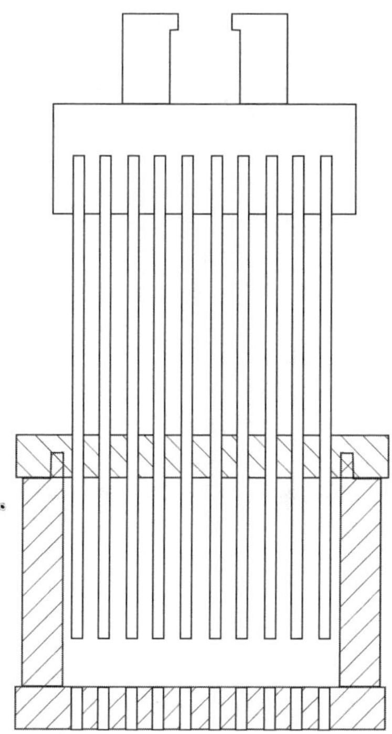

FIGURE 24.3 Cross-sectional view of Kramer-shear compression cell.

Kramer-shear test to evaluate the firmness of frozen peas. An increase in firmness of peas as observed after frozen storage at -20°C. Bhattacharya et al. [7] used Kramer-shear test on cooked beef patties. They found an increase in shear strength in ground beef patties stored at -12 to -35°C. Lamkey et al. [26] used Kramer shear to measure the texture of restructured beef steaks. Adding phosphates along with salt increased the tenderness of beef steak frozen at -30°C. The texture of freeze-chilled lasagna is also measured using a Kramer-shear cell [8].

B. Puncture Test

The puncture test, a compression method, is widely used in texture measurement of fruits and vegetables. The puncture test is popular due to its simplicity as only the force required to push a puncture probe into or through the sample is to be determined. Standard penetration techniques with various shapes of probes have been used. Depending on the size and shape of the probe, the puncture test may only give data related to the sample compressibility and the force necessary to breakdown the sample. In addition to using a puncture probe, the sample can be placed on a die, and so the probe can penetrate through the sample completely.

The following three parameters are obtained during a puncture test:

(1) Maximum or initial yield force that can be correlated with firmness
(2) Sample area at rupture, which indicates energy required to initiate rupture
(3) Slope of the force–deformation curve that provides modulus of elasticity.

Cano et al. [15] studied the effects of thermal treatment before freezing on firmness of banana slices. To evaluate firmness, samples were tested using a Bellevue pressure tester with 5/6 in. plunger. Freeze and thawing without thermal treatment produced an excessively soft product and a prethermal treatment (blanching) resulted in a firm acceptable texture. Further they concluded that, in addition to thermal treatment, selection of banana maturity level for freezing must compromise between the sensorial quality and processing parameters. Hwang et al. [11] used the measurements of puncture force to penetrate a meat piece using a simulated tooth attachment. The puncture measurements were obtained on thawed (using a microwave oven) samples. The frozen precooked beef samples stored at different modified atmosphere conditions did not show any significant differences in textural parameters. Bolin and Huxsoll [27] studied the effects of partial drying before freezing on the texture of cut pears. The peak force required to penetrate a side of thawed pear sample using an 8-mm Magness–Taylor probe was correlated with the texture. Concentrating the pears to 50% weight reduction using osmotic drying improved the texture of frozen–thawed pears 1.5 times better than that obtained using hot-air drying. Similarly, puncture test measurement was used to study the textural changes in frozen strawberries and chicken escallops [28] and areca fruits [16]. In the case of strawberries, fruits with high initial firmness resulted in a better quality in a cryomechanical freezer than fruits with low initial firmness. Quick freezing of areca fruits pretreated with calcium chloride produced a better quality product [16].

As the instrumental texture of frozen food products correlate with sensory texture in an agreeable way, many different test conditions and test cells have been used even for the same food product and it is very difficult to compare the results from one study to another. The deformation and stress rates need further investigation and standard test procedures need to be developed.

C. Textural Profile Analysis

Cyclic loading and unloading to a set deformation during uniaxial compression has been widely used by many investigators to determine several textural parameters. Using texture profile analysis, six different texture properties can be estimated: hardness, brittleness, cohesivenss, elasticity, adhesiveness, chewiness, and gumminess. Figure 24.4 illustrates a typical texture profile curve obtained using Instron Universal testing machine.

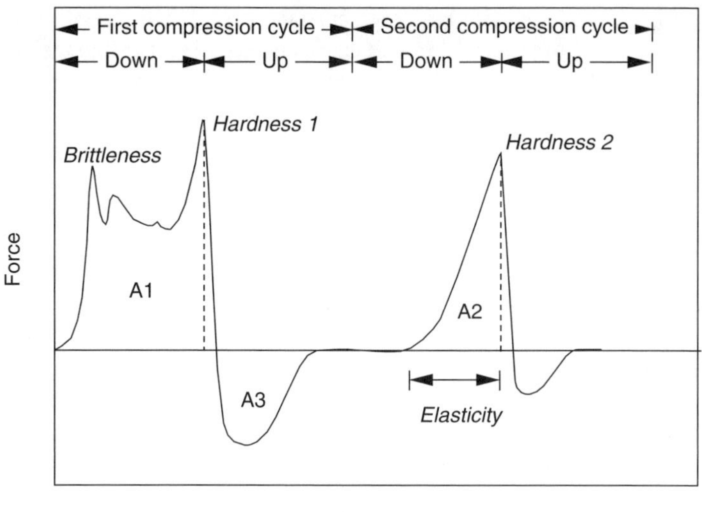

FIGURE 24.4 Typical force–time curve during a texture profile analysis.

Hardness is measured from the profile as the height of the first peak. Cohesiveness is measured as the ratio of the area under the second peak and the area under the first peak. Springiness or elasticity is defined as the height that the food recovers during the time that elapses between the end of the first bite and the start of the second bite. Adhesiveness is measured as the area, A3, of the negative peak beneath the baseline of the profile and represents work necessary to pull the probe from the sample. Brittleness is characterized by the multipeak shape of the first bite, and is measured as the height of the first significant break in the peak. Gumminess is expressed as the product of hardness and cohesiveness. Chewiness is expressed as the product of hardness, cohesiveness, and elasticity.

Jahncke et al. [29] determined the effects of frozen storage on texture of minced cod fillet using texture profile analysis. Before analysis, the samples were thawed overnight at 2°C. They used a modified texture profile analysis using a cross-head speed of 12.7 cm/min and compressed the samples to 65% of their original height. In general, with an increase in frozen storage time, hardness values decreased. They attributed the reduction in hardness to the action of proteases during frozen storage. Dias et al. [30] used texture profile analysis to study the effect of frozen storage on black and silver scabbardfish quality. After thawing and cooking, the minced samples (25-mm diameter and 40-mm height) at 100°C for 2 h, texture measurements were carried out using an Instron equipped with 10-kg load cell at 10 mm/min cross-head speed. They found a significant increase in load at maximum displacement (hardness) with frozen storage and found an inverse correlation with the decrease in salt-soluble proteins. The textural quality of par-baked frozen bread was measured using texture profile analysis and found that the bread texture decreased with storage time and regardless of the variety of bread tested, the maximum shelf-life was found to be between 12 and 20 weeks [22].

V. JUICINESS

Juiciness is another important quality attribute for food products, especially for frozen food products. In frozen foods, the state of water in the product and the size of ice crystals affect the juiciness of the product. Juiciness can be defined as the amount of juices released during consumption. This affects the texture, flavor, and overall acceptance of the product. The product should not be dry or exudative. The frozen products with larger ice crystals, upon thawing, will experience a higher amount of drip loss (i.e., thawing losses) and will cause an excessive drip and result in a less appealing product. The other quality loss that affects product juiciness is cooking loss, that is, the amount of juices released during cooking.

In addition to thawing and cooking losses, the ability to hold the water by the food product plays a crucial role in determining the product juiciness. The capacity of the food product is often called as water-holding capacity (WHC). By definition, WHC is the ability of the food product to hold all or part of its own and added water [31]. A wide variety of methods for measurement of WHC are reported in the literature. The methods used for measuring WHC of meat system have been given by Honikel [31] and Trout [32]. These procedures can be easily adapted to other food systems as well.

There are three basic methods for measuring WHC: methods applying no external force, methods applying external mechanical force, and methods applying thermal force. Most commonly used methods for measuring WHC involve applying external forces such as filter paper press method and centrifugation method. The press method involves pressing the sample between filter papers and measuring the water released. The method is rapid and inexpensive. The sample can be pressed using a hand operated screw press, a high-pressure laboratory press (e.g., Carver press) or a Instron universal testing machine. The latter two pieces of equipment allow better control of the applied pressure, and, consequently, give more reproducible results. The amount of water released from the sample can be measured either directly weighing the filter papers or indirectly measuring the area of filter papers wetter relative to the areas of the pressed sample. Jahncke et al. [29] used Instron universal testing machine for obtaining expressible moisture from frozen cod. The information on amount of load and time were not reported. The length of frozen storage on frozen cod at $-20°C$ for 24 h and stored at $-14°C$ increased the press juice from cod. After 3 weeks of frozen storage, the changes in press juice values were not significantly different ($P > 0.05$). Kenawi [33] measured the press juice of beef during frozen storage. He also found an increase in press juice with storage time in frozen beef. The conditions used to freeze the samples and the temperature during storage were not reported.

Precise and reproducible results can be obtained with centrifugal method. The samples, varying in size from 1 to 20 g, are centrifuged at centrifugal forces between 5000 and $40,000 \times g$. The amount of water released is determined either directly by weighing the amount of water released or indirectly by weighing the sample after centrifugation. Greer and Murray [34] used centrifugal method to evaluate the WHC of pork during frozen storage. The samples were thawed at 1.7°C for 2 days before estimating WHC. They found a decrease in the expressible juice after thawing. Even though, after freezing and thawing, an increase in drip loss was observed, the released juices were reabsorbed by the meat and resulted in a less expressible juice than fresh pork. The drip loss of freeze-chilled lasagna was measured through centrifugal force of $223 \times g$ for 10 min [8]. Storing the lasagna frozen for 9 or more months led to higher drip loss than storing for 3 months for unheated samples. However, for heated samples, a reverse phenomenon was observed.

The methods using no external forces include measurement of drip losses and thawing losses [34] from the product. Chevalier et al. [35] obtained the drip losses to evaluate the effect of pressure shift freezing and air-blast freezing on the quality of turbot. In measuring the WHC by applying thermal forces, it is measured in terms of cooking loss [7,26,36,37]. For most of the frozen foods consumed after thawing, they will be cooked before consumption. Thus the WHC after heating is of interest. Depending on the nature of the ice crystals, and denaturation of proteins and cellular structure in the product, the cooking loss of the product will vary.

VI. FLAVOR

The flavor or aroma of a food product is a complex sensation affected by moisture content and composition of volatile organic compounds. The flavor or aroma of the product is perceived by the nose as well as from the released juices during mastication. The flavor or aroma characteristics of frozen food products change primarily during frozen storage due to continuing chemical reactions such as lipid oxidation and enzymatic reactions. These reactions continue in the frozen food products even during frozen storage but at slower rates thus limiting the storage life of the frozen foods.

The flavor compounds can be evaluated using chromatographic techniques (both gas chromatography and liquid chromatography) either alone or in combination with mass spectroscopic techniques. Sometimes, the gas chromatographic equipment could be fitted with an olfactory port so that trained panelists can smell the volatiles coming at different retention times and identify the aroma to specific known aroma patterns. The alternate method to chromatographic technique is the use of electronic nose or chemosensory systems [38]. Electronic nose sensor systems use a set of sensors that change their property (conductivity, oxidation or reduction potential, or oscillation frequency), which can be monitored and converted to smell prints. Many of these systems rely on statistical procedures like principal component analysis or canonical discriminant analysis to differentiate flavor or aroma characteristics. The use of electronic nose systems in the frozen food industry is still evolving. Additional information about chromatographic techniques or electronic nose systems can be found elsewhere.

VII. CONCLUSIONS

Quality is an important factor in determining the success of the frozen foods in the market. Thus, methods to assess these quality attributes should be able to lead the food industry to identifying process conditions that will affect the end product acceptance. With recent improvement in technologies, hand-held color meters become affordable for the food industry. Different types of universal testing machines under $15,000 are available that can do specific tests on food materials, making food testing as a part of the R & D to quality control in the industry. The methods described in this chapter will provide the basic information for developing methods specific to a food product. Developments in measurement techniques will continue to shape the future improvements in frozen foods which will meet consumer's needs.

NOMENCLATURE

a	color redness (redness–greenness) in the color space
b	color yellowness (yellowness–blueness) in the color space
L	lightness or color value in the color space
X	tristimulus color in the CIE color space under a standard observer
Y	tristimulus color in the CIE color space under a standard observer
Z	tristimulus color in the CIE color space under a standard observer

Superscript

*	CIE color space

Subscript

0	initial value of color coordinates

REFERENCES

1. Anon. *Frozen Food Trends*. McLean: American Frozen Food Institute, 2005.
2. PO Persson, G Londahl. Freezing technology. Chapter 2. In: Mallet CP, Ed., *Frozen Food Technology*. London: Blackie Academic and Professional, 1993, p 20–58.

3. J Dagnoli, JL Erikson. The looming battle for the center of the plate. *Advertising Age* 60 (49):810, 1989.
4. MC Lanari, RG Cassens, DM Schaefer, KK Scheller. Dietary vitamin E enhances color and display life of frozen beef from Holstein steers. *Journal of Food Science* 58:701–704, 1993.
5. HJ Andersen, LH Skibsted. Oxidative stability of frozen pork patties; effect of light and added salt. *Journal of Food Science* 56:1182–1184, 1991.
6. JG Akamittath, CJ Brekke, EG Schanus. Lipid oxidation and color stability in restructured meat systems during frozen storage. *Journal of Food Science* 55:1513–1517, 1990.
7. M Bhattacharya, MA Hanna, RW Mandigo. Effect of frozen storage conditions on yields, shear strength and color of ground beef patties. *Journal of Food Science* 53:696–700, 1988.
8. GA Redmond, TR Gormley, F Butler. Effect of short and long term storage with MAP on the quality of freeze-chilled lasagna. *Lebensmittel-Wissenschaft und-Technologie* 38 (1):81–87, 2005.
9. E Forni, G Crivelli, A Polesello, M Ghezzi. Changes in peas due to freezing and storage. *Journal of Food Processing and Preservation* 15:379–389, 1991.
10. MP Cano, MA Martin. Pigment composition and color of frozen and canned kiwi fruit slices. *Journal of Agriculture and Food Chemistry* 40:2141–2146, 1992.
11. SY Hwang, JA Bowers, DH Kropf. Flavor, texture, color, and hexanal and TBA values of frozen cooked beef packaged in modified atmosphere. *Journal of Food Science* 55:26–29, 1990.
12. MS Brewer, CAZ Harbers. Effect of packaging on color and physical characteristics of ground pork in long-term frozen storage. *Journal of Food Science* 56:363–366, 1991.
13. MS Brewer, SY Wu. Display, packaging, and meat block location effects on color and lipid oxidation of frozen lean ground beef. *Journal of Food Science* 58:1219–1223, 1993.
14. CM Chen, GR Trout. Color and its stability in restructured beef steaks during frozen storage: effects of various binders. *Journal of Food Science* 56:1461–1464, 1475, 1991.
15. P Cano, MA Martin, C Fuster. Freezing of banana slices: influence of maturity level and thermal treatment prior to freezing. *Journal of Food Science* 55:1070–1072, 1990.
16. M Zhang, ZH Duan, JF Zhang, J Peng. Effects of freezing conditions on quality of areca fruits. *Journal of Food Engineering* 61:393–397, 2004.
17. Ö Tokuşoğlu, MO Balaban. Correlation of odor and color profiles of oysters (*Crassostrea virginica*) with electronic nose and color machine vision. *Journal of Shellfish Research* 23 (1):143–148, 2004.
18. J Subbiah, N Ray, GA Kranzler, ST Acton. Computer vision segmentation of the *Longissimus dorsi* for beef quality grading. *Transactions of the ASAE* 47 (4):1261–1268, 2004.
19. AS Szczesniak. The meaning of textural characteristics. *Journal of Texture Studies* 19:51–59, 1988.
20. E Varriano-Martson, KH Hsu, J Mahdi. Rheological and structural changes in frozen dough. *Baker's Digest* 54 (1):32, 1980.
21. SP Cauvain. Improving the control of staling in frozen bakery products. *Trends in Food Science and Technology* 9 (2):56–61, 1998.
22. IR Vulicevic, ESM Abdel-Aal, GS Mittal, X Lu. Quality and storage life of par-baked frozen breads. *Lebensmittel-Wissenschaft und-Technologie* 37:205–213, 2004.
23. HD Goff, B Freslon, ME Sahagian, TD Hauber, AP Stone, DW Stanley. Structural development in ice cream — dynamic rheological measurements. *Journal of Texture Studies* 26:517–536, 1995.
24. PW Voicey. Engineering assessment and critique of instruments used for meat tenderness evaluation. *Journal of Texture Studies* 7:11–48, 1976.
25. MS Brewer, BP Klein, BK Rastogi, AK Perry. Microwave blanching effects on chemical, sensory and color characteristics of frozen green beans. *Journal of Food Quality* 17:245–259, 1994.
26. JW Lamkey, RW Mandigo, CR Calkins. Effect of salt and phosphate on the texture and color stability of restructured beef steaks. *Journal of Food Science* 51:873–875, 911, 1986.
27. HR Bolin, CC Huxsoll. Partial drying of cut pears to improved freeze/thaw texture. *Journal of Food Science* 58:357–360, 1993.
28. ME Agnelli, RH Mascheroni. Quality evaluation of foodstuffs frozen in a cryomechanical freezer. *Journal of Food Engineering* 52:257–263, 2002.
29. M Jahncke, RC Baker, JM Regenstein. Frozen storage of unwashed cod (*Gadus morhua*) frame mince with and without kidney tissue. *Journal of Food Science* 57:575–580, 1992.
30. J Dias, ML Nunes, R Mendes. Effect of frozen storage on the chemical and physical properties of black and silver scabbard fish. *Journal of Science of Food and Agriculture* 66:327–335, 1994.

31. H Honikel. Critical evaluation of methods detecting water holding capacity in meat. In: A Romita, C Valin, AA Taylor, Eds., *Accelerated Processing of Meat*. London, UK: Elsevier Applied Science Publishers, 1987.
32. GR Trout. Techniques for measuring water-binding capacity in muscle foods — a review of methodology. *Meat Science* 23:235–252, 1988.
33. MA Kenawi. Evaluation of some packaging materials and treatments on some properties of beef during frozen storage. *Food Chemistry* 51:69–74, 1994.
34. GG Greer, AC Murray. Freezing effects on quality, bacteriology and retail-case life of pork. *Journal of Food Science* 56:891–894, 912, 1991.
35. D Chevalier, A Sequeira-Munoz, A Le Bail, BK Simpson, M Ghoul. Effect of freezing conditions and storage on ice crystal and drip volume in turbot (*Scophthalmus maximus*): evaluation of pressure shift freezing vs. air-blast freezing. *Innovative Food Science and Emerging Technologies* 1:193–201, 2001.
36. BW Berry. Low fat level effects on sensory, shear, cooking, and chemical properties of ground beef patties. *Journal of Food Science* 57:537–540, 1992.
37. MS Brewer, FK McKeith, K Britt. Fat, soy and carrageenan effects of sensory and physical characteristics of ground beef patties. *Journal of Food Science* 57:1051–1052, 1055, 1992.
38. PN Bartlett, JM Elliott, JW Gardner. Electronic noses and their application in the food industry. *Food Technology* 51 (12):44–48, 1997.

25 Chemical Measurements

Marilyn C. Erickson
Department of Food Science and Technology, University of Georgia,
Griffin, Georgia, USA

CONTENTS

I. INTRODUCTION

Chemicals are the building blocks of foods and hence are responsible for the physical and sensory properties of those foods. During freezing and frozen storage, chemical reactions occur; while they are slower than those occurring at higher temperatures, they often have undesirable consequences. To quantify those changes, analyses have been developed that target either substrates or products of the reactions. This chapter provides an overview of those chemical measurements that assess degradation of vitamins, lipids, proteins, carbohydrates, and pigments. This chapter presents a brief description of the methodology as well as a number of examples of studies where those measurements have been incorporated. The use of chemical measurements to differentiate fresh from frozen product is also presented.

II. EVALUATION OF PREFREEZING TREATMENTS

A. BLANCHING

Analyses to measure the application or effectiveness of prefreezing treatments are critical for maximizing quality retention in frozen products. In vegetable processing, blanching has long been used to slow quality deterioration caused by enzyme activities. Residual activity of those enzymes can therefore be used as an indicator of the adequacy of blanching. As an illustration of this relationship, blanching time was inversely related to residual peroxidase activity in frozen broccoli florets [1]. Surprisingly, only a few studies have demonstrated correlation between residual peroxidase activity and undesirable quality in frozen vegetables [2,3]; therefore, several other enzymes have been targeted as indices to blanching adequacy including lipoxygenase and cystine lyase [4–6]. Targeting either of these enzymes that are known to generate off-odors and off-flavors has resulted in significantly shorter blanch times and improved nutritional and sensory quality of frozen vegetables. The standard method for measuring lipoxygenase activity involves incubation of the sample with linoleic acid and measurement of conjugated dienes or hydroperoxides spectrophotometrically. Coupled reactions with carotenoid bleaching, however, have also been used [5]. At first glance, such reactions suggest that bleaching of endogenous carotenoids in frozen vegetables could also be used to monitor adequacy of blanch treatments. In a recent study, however, rates of pigment decomposition varied in three cultivars of red peppers, suggesting the existence of different inherent stabilities [7].

B. FRESHNESS

Application of freezing is designed to extend product shelf-life, however, final product quality is also based on raw material quality. For example, in the case of fish, initial quality can vary, from live to spoiled, prior to freezing. Hence, several freshness indicators have been adopted for evaluation of frozen product quality. These indicators include the high-performance liquid chromatography (HPLC) or biosensor analysis of K_1 (a ratio based on the changes in the level of catabolites of adenosine triphosphate (ATP) occurring in the muscle after death) and putrescine (a biogenic amine generated via microbial metabolic processes) [8]. Putrescine, at a reject level of 3 ppm, confirmed sensory analysis decisions on frozen Penaeid shrimp of the level of decomposition that had occurred during holding prior to freezing [9]. On the basis of the observation that ATP breakdown and generation of inosine monophosphate (IMP) occurred during frozen storage, caution should be exercised with the use of nucleotide degradation as a measure of prefreezing decomposition [10]. Hypoxanthine has also been shown to increase in scallop adductor muscles during frozen storage when the product was frozen immediately after processing [11].

C. IRRADIATION

Owing to the potential for pathogen contamination and growth in products prior to freezing and the ability of these pathogens to survive freezing and frozen storage, irradiation has been proposed as an effective technology to decontaminate frozen foods. As regulations in many countries require that irradiated food be labeled as such, reliable scientific tests that can detect whether a food has been irradiated or not have been developed. Both thermoluminescence, which measures the light emitted from the inorganic components of a sample as it is rapidly heated, and electron spin resonance spectroscopy, which measures free radicals, depend on the presence of solid matrices (bone, traces of silicate minerals as surface contamination, etc.) [12,13]. Other potential markers for irradiation include long-chain hydrocarbons, 2-alkylcyclobutanones, *o*-tyrosine, radiolytic products of DNA, and radiolytic H_2 and CO gases [14–22]. For example, based on the limit of detection for radiolytic products of 16:2, 17:1, and 17:2, an irradiation dose of 0.25 kGy could be distinguished in irradiated frozen meat and poultry [21]. Similarly, radiolytic H_2 and CO have proven more useful as irradiation markers in frozen foods compared with unfrozen food due to lower diffusion losses during storage. Using those gases as a probe, irradiated frozen shrimp could be distinguished from nonirradiated shrimp for 3 months after 1.1–8.8 kGy irradiation. Irradiated frozen cod slices and oyster could also be distinguished for at least two months at the dose ranges of 1.4–5.5 and 1.2–6.0 kGy, respectively [18]. In the case of the DNA comet assay (a single-cell gel electrophoresis assay measuring DNA damage), however, several restrictions limit its usefulness as an irradiation marker. The DNA comet assay cannot be used with processed meats (e.g., cooked, roasted) as DNA is damaged by this treatment. Also, extensive DNA damage occurs with repeated freeze–thawing. Attempts to circumvent the influence of these other damaging factors have been made with the calculation of a relative damage index (RDI). The RDI has enabled discrimination of treatment levels within each treatment but cannot differentiate samples between treatments [22].

III. METHODS TO ASSESS DIFFERENTIATION OF FRESH FROM FROZEN

Although freezing is an excellent method to extend the storage life of foods, consumers perceive frozen product as inferior to fresh product. As a result of this, lower prices are commonly assigned to frozen foods. To prevent fraudulent sale of frozen products as fresh, methods are needed to differentiate these product forms. Hence, several approaches that target changes in chemical constituents within the product have been advocated.

A. Measurement of Enzyme Activity

On the basis that freezing and thawing lead to cellular disruption and release of bound enzymes into the cellular fluid, one approach for distinguishing fresh from frozen–thawed food includes measuring the activities of several mitochondrial or lysosomal enzymes. Examples of mitochondrial enzymes include β-hydroxyacyl-CoA-dehydrogenase or HADH [23,24], L-malate-NADP-oxidoreductase [25], aspartate aminotransferase [25,26], fumarase, glutamate, or lactate dehydrogenases [27], cytochrome oxidase [28,29], and L-malate dehydrogenase [30]. Lysosomal enzymes that have been investigated include α-glucosidase [24,31–33], β-*N*-acetylglucosaminidase [24,31–33], acid phosphatase [31], β-galactosidase [27], and β-glucuronidase [27]. One disadvantage of a mitochondrial enzyme serving as an indicator of freezing is in the existence of corresponding isoenzymes in the cytoplasm. Therefore electrophoresis or similar separation procedures have to be applied, making it difficult to obtain quantitative results. Lysosomal enzymes, however, can be readily distinguished from cytoplasmic and bacterial enzymes by their activity in the acid pH range. In any event, before using either mitochondrial or lysosomal enzymes as markers, it is important to verify that other stresses imposed on the product do not facilitate perturbations in the enzyme activities. For example, while HADH activity was significantly higher in frozen–thawed frog legs than in unfrozen legs, HADH activity was not affected by the storage time in crushed ice up to 6 days [23]. Assays based on α-glucosidase activity, on the contrary, require that freshness measurements be conducted to avoid confusing a frozen–thawed fish from a fish in an advanced stage of spoilage [24]. Applicability of a marker enzyme may also vary with the product. Although mitochondrial aspartate aminotransferase was not found in the pressed juice of unfrozen bovine muscle, its presence in unfrozen porcine muscle implies mitochondrial damage by other stresses [26]. In some cases, however, enzyme activities may provide additional information beyond differentiation of fresh and frozen product. In the case of the mitochondrial membrane-bound cytochrome oxidase, amplification in the freezing-induced activation occurred with an ice storage period prior to freezing [29]. Multiple freeze–thaw cycles could also be distinguished in rainbow trout through enzyme activities of α-glucosidase and β-*N*-acetylglucosaminidase [32].

B. Measurement of Volatile Composition

Another group of analyses that have been used to differentiate fresh from frozen product has focused on the volatile composition of the products. Using headspace gas chromatography (GC) and electronic nose instruments, three types of commercial orange juice samples (pasteurized not from concentrate, frozen concentrate, and single strength juice reconstituted from concentrate) were evaluated. Initially 25 volatile juice constituents were screened in the GC data but through backward stepwise discriminant analysis, 11 volatiles afforded consistent separation of the orange juice into three distinct groups. In contrast, discriminant analysis of data generated with an electronic nose containing 11 sensors could not distinguish the frozen concentrated orange juice from the other two samples [34]. Ideally, methods to differentiate fresh from frozen should be rapid, nondestructive, accurate, and adaptable to online monitoring in processing plants or portable hand-held instruments in the field. Spectroscopic techniques may be used to measure physical characteristics of frozen foods (see Chapter 24) and also be used to measure changes in one or more chemicals that occur with freezing and frozen storage. For example, near-infrared diffuse reflectance spectroscopy has been used to classify samples in frozen or unfrozen beef using 400–2500 nm spectra on centrifuged meat juice [35]. Similarly, Fourier transform infrared spectral data between 1600 and 800 nm was correlated to dimethylamine (DMA) content in minced red hake and hence successfully distinguished fresh from frozen 90% of the time [36].

IV. METHODS TO ASSESS NUTRITIONAL DEGRADATION DURING FREEZING OR FROZEN STORAGE

Freezing has proven to be a suitable procedure to prolong the shelf-life of many food products. Numerous studies, however, continue to assess whether modifications have occurred in the levels of endogenous components upon freezing or storage. This section reviews some of the recent studies that have examined the nutritional quality of frozen products as it relates to vitamins, phenolic acids, and glucosinolates, whereas changes associated with lipid, protein, carbohydrates, and pigments will be addressed in subsequent sections of this chapter.

A. Ascorbic Acid and Organic Acid Measurements

One of the major nutrients of interest in frozen produce is ascorbic acid. Following extraction, quantification of ascorbic acid is commonly accomplished through HPLC and either ultraviolet (UV), spectrofluorometric, or electrochemical detection [37–39]. Using these analytical methods, ascorbic acid has been used as a marker for the evaluation of different freezing methods and prefreezing operations. For example, freezing in carbon dioxide roughly halved the ascorbic acid degradation rate in Brussel sprouts compared with conventional freezing [40]. Castro et al. [41] demonstrated that keeping the stem intact on strawberries minimized the losses of ascorbic acid during freezing.

In characterizing the responses of different products to freezing and frozen storage, degradation of ascorbic acid is commonly used as an index of the relative stability of that product. To illustrate this point, no losses of ascorbic acid were observed in frozen strawberries after 2 months of frozen storage (−20°C) [42], whereas losses did occur in frozen papayas after 12 months of frozen storage (−18°C) [43] and in frozen, fresh squeezed, unpasteurized, polyethylene-bottled orange juice following a 24-month storage (−23°C) [39]. As ascorbic acid loss is encountered for many produce items, the stability of this vitamin is also commonly used to judge the suitability of varieties and cultivars for freezing. In a study comparing two types of parsley (Hamburg cv. Berlińska and leafy type cv. Paramount), the Hamburg type was considered the better raw material for freezing, but rate of degradation was not the determining factor in that selection. More specifically, ascorbic acid losses following 9 months of frozen storage (−20°C) were similar for the two types of parsley

(35.8–37.6%); however, the Hamburg parsley had a significantly higher content initially [44]. In contrast to this study, significant differences in rates of ascorbic acid degradation were observed in a study comparing four raspberry cultivars. After 365 days of storage (−20°C), "Rubi," "Zeva," "Heritage," and "Autumn Bliss" cultivars suffered ascorbic acid losses of 49, 47, 34, and 56%, respectively [38]. At the same time, significant differences also occurred in degradation of ellagic acid with decreases of 19,16, 21, and 14%. In comparison, losses of ellagic acid, in the range 30–40%, were observed over 9 months of frozen storage (−20°C) for the raspberry cultivars, "Ottawa" and "Muskoka," respectively [45]. Measurement of this acid as well as other organic acids is commonly accomplished with HPLC and UV detection [46] and is of interest for their potential antioxidant, antimicrobial, and flavor-enhancing properties. For these reasons, freezing effects on the organic acid content of low-moisture Mozzarella cheese has also been examined. In this case, organic acid results demonstrated no effect of freezing on Mozzarella samples ripened after frozen storage compared with those samples ripened before freezing [47].

B. Glucosinolate Measurement

Another group of bioactive compounds whose levels have been monitored in foods during frozen storage are the glucosinolates. These secondary plant metabolites and their breakdown products are important aroma and flavor compounds in *Brassica* vegetables. As such, their analysis requires that they be extracted prior to separation by HPLC. In a study evaluating the glucosinolate levels of the principal and secondary influorescences of broccoli, freezing was shown to be the best preservation process [48]. For example, no significant differences were noted in 4-methylsulfinylbutyl glucosinolate content between the fresh harvested material and frozen material for the principal influorescence. Storage at 4°C, however, reduced 4-methylsulfinylbutyl glucosinolate content by 31%, whereas at room temperature (20°C), the reduction was 82%.

C. Folate Measurement

Folates may also potentially be lost during freezing and frozen storage, and thus chemical measurements may be applied for purposes of nutritional evaluation. Using the *Lactobacillus casei* microbiological assay, folate losses have been observed in beef liver during the first 30 days of storage (−20°C) [49]. When HPLC was used for folate quantification, improved packaging was held responsible for the absence of folate losses in beef liver and strawberries after 6 months of frozen storage (−18°C) [50].

V. METHODS TO ASSESS LIPID DEGRADATION DURING FREEZING OR FROZEN STORAGE

Two major mechanisms contribute to the degradation of lipids during frozen storage. They are lipid hydrolysis and lipid oxidation. To assess the contribution of these activities during frozen storage, a wide range of analyses have been undertaken that encompass measurements on their substrates, catalysts, inhibitors, and products. A list of these markers and a short description of the general approaches taken to measure these constituents are provided in Tables 25.1–Table 25.4. Inherent in any of these procedures, however, is attention to sampling and preparation of the sample for analysis. Hence, sample heterogeneity must be considered as it influences the size and number of samples drawn. Generally, to minimize heterogeneity, materials may be ground or mixed prior to removal of the sample. As the process to achieve maximum sample homogeneity may vary, Lichon and James [76] compared 12 methods of homogenization and found that cryogenic homogenization methods (cryogenic milling; top-drive macerator/dry-ice grind) effectively pulverized the sample to a small particle size facilitating milligram subsample masses to be used for analysis. In cases where the sample is too large to analyze in its entirety, however, the location

TABLE 25.1
Description of Methodologies Used to Measure Lipid Substrates Involved in Degradative Reactions in Frozen Foods

Constituent	Description of Analysis
Polyunsaturated fatty acids (PUFAs)	Fatty acids in a lipid extract are esterified either with sulfuric acid in methanol [37], sodium borohydride in methanol, or potassium hydroxide in methanol; once esterified, the fatty acid methyl esters (FAMEs) are injected into a gas chromatograph for separation, and detection is through a flame ionization detector
Phospholipid/triacylglycerol	Lipids are commonly extracted with a chloroform:methanol mixture; the extract is applied to a Sep-Pak cartridge, and nonpolar lipids (i.e., triacylglycerols) are eluted with chloroform whereas polar lipids (i.e., phospholipids) are eluted with methanol; subsequent to hydrolysis, triacylglycerols may be quantified by measuring the levels of glycerol spectrophotometrically, whereas phospholipids are quantified by measuring the levels of phosphate spectrophotometrically; alternatively, either fraction may be subjected to high-pressure liquid chromatography (HPLC) and evaporative scattering detection, or esterified and subjected to gas chromatography (GC) for quantification of their fatty acid methyl esters [37]
Free fatty acids (FFAs)[a]	From a lipid extract, the carboxylic groups of FFA are neutralized with the addition of sodium hydroxide and a change in the color of the metacresol indicator signals the end point [51]; correction must be made for the contribution of the acidic phospholipids to the percent FFA value obtained by alkalimetric titration [52,53]
	FFA may be separated from triacylglycerols and phospholipids using a thin-layer chromatography plate; the spot corresponding to FFA may then be scraped off, subjected to esterification conditions, and the fatty acid methyl esters subjected to gas chromatography (GC) using an internal standard and quantified [37]
	Cupric acetate in pyridine is added to a lipid extract suspended in benzene where it complexes with free fatty acids; the upper layer of the two-phase system is read at 715 nm [54]

[a]FFAs are products of lipolysis reactions, however, they also serve as a substrate for lipid oxidative reactions.

TABLE 25.2
Description of Methodologies Used to Measure Inhibitors Involved in Degradative Reactions of Lipids in Frozen Foods

Constituent	Description of Analysis
Tocopherol	The frozen food material is thawed and then saponified under heat with KOH; tocopherol is extracted with organic solvents from the saponified mixture; the extract sample is dried under nitrogen prior to its reconstitution in the mobile-phase solvent; both normal-phase and reverse-phase high-pressure liquid chromatography (HPLC) may be used for separation (caution: reverse phase does not separate β- and γ-tocopherols); detection is mainly by fluorescence or electrochemical detection [55]
Glutathione	The sample is homogenized with sulfosalicyclic acid; after centrifugation, an aliquot of the supernatant may either be subjected to HPLC and electrochemical detection or is added to a coupled enzyme reaction system for spectrophotometric monitoring at 412 nm [56]
Ascorbic acid	An acidified extract is prepared from the sample and subjected to ion-pairing or reverse-phase HPLC and electrochemical detection [37,56,57]
Glutathione peroxidase	An extract from the sample is reacted with reduced glutathione and the color development is followed spectrophotometrically at 412 nm [58] Glutathione peroxidase activity may also be based on a coupled enzyme system; with glutathione reductase, the oxidation of NADPH (monitored at 340 nm) over time is monitored [59]
Superoxide dismutase (SOD)	With phosphate buffer (pH 7.5), EDTA, xanthine, cytochrome *c*, xanthine oxidase, and an aliquot of a sample homogenate extract, SOD activity is determined by monitoring the inhibition in the reduction of cytochrome *c* between reaction mixtures with and without the sample extract at 550 nm [60]

TABLE 25.3
Description of Methodologies Used to Measure Catalysts Involved in Degradation of Lipids in Frozen Foods.

Constituent	Description of Analysis
Heme proteins/metmyoglobin	A supernatant fraction is obtained via centrifugation of a phosphate-buffered (pH 6.5) homogenized sample; total heme pigments are determined on this supernatant either through direct spectrophotometric measurement at 525 nm [61] or indirectly, following exposure to sodium dithionite, at 432 and 410 nm; heme content in the direct method is calculated using the molar extinction coefficient of myoglobin whereas in the indirect method, the difference in the absorbance at the two wavelengths is compared with a standard curve constructed using hemoglobin [62]
Non-heme iron	The sample is homogenized prior to exposing it to acid hydrolysis conditions at elevated temperatures; the supernatant from this acidified sample is exposed to a solution of bathophenanthroline and thioglycolic acid and then held for a brief period of time; the absorbance of this mixture is read at 540 nm and compared with an iron standard curve [61]
Catalase	A phosphate-buffered (pH 7.0) homogenate is prepared; the supernatant recovered after centrifugation is mixed with hydrogen peroxide; catalase activity is based on the spectrophotometric decrease in hydrogen peroxide at 240 nm [59]
Lipoxygenase	The supernatant recovered after centrifugation of a homogenate is incubated with linoleic acid; the increase in conjugated dienes or hydroperoxides corresponds to the level of lipoxygenase activity [5]
Peroxidase	Oxidation of guaiacol by an extract of the sample is monitored for a designated period at 420 nm [6]
Lipase/phospholipase	A phosphate-buffered (pH 7.5) homogenate is prepared from the sample; after centrifugation, an aliquot of the enzyme extract is incubated with 4-methylumbelliferyl oleate as substrate; the reaction medium is incubated at 37°C and fluorescence periodically monitored at excitation wavelength of 355 nm and emission wavelength of 460 nm [63]

TABLE 25.4
Description of Methodologies Used to Measure Products Generated From Degradative Reactions of Lipids in Frozen Foods

Constituent	Description of Analysis
Hydroperoxides	Both the titrimetric method (peroxide value) and the spectrophotometric method rely on the ability of the iodide ion to reduce hydroperoxides under anaerobic conditions; determination of liberated iodine is either determined by titration with thiosulfate or the absorbance (A_{360}) of the triiodide ion is used [64] Another spectrophotometric method for hydroperoxide is based on the ability of lipid hydroperoxides to oxidize ferrous ions; following addition of xylenol orange to the system, the absorbance is measured at 560 nm that corresponds to the complex formed between xylenol orange and ferric ions [65] Near-infrared (NIR) spectroscopy is applied to lipid extracts and the absorbance at 2076 nm is attributed to the hydroxyl group of hydroperoxides [66]
Conjugated dienes (CD)	The spectrophotometric absorbance of a lipid extract dissolved in isooctane is read at 232 nm; CD levels are determined by applying the extinction coefficient of CD to the absorbance value [37,67–68]; sensitivity of this procedure is limited as the CD absorbance appears as a rather imprecise shoulder on the strong absorption peak of the nonperoxidized fatty acid itself; by using second-derivative spectroscopy [69,70], or ultraviolet difference spectroscopy with tandem cuvettes [71], sensitivity can be increased
Thiobarbituric acid-reactive substances (TBARS)	Following distillation or aqueous acid extraction, malondialdehyde in the sample is reacted with thiobarbituric acid and the absorbance is recorded at 532 nm; addition of a chelator and an antioxidant, such as propyl gallate, to the samples before homogenization is recommended to retard further lipid oxidation [55]
Volatiles	Volatiles may be isolated from thawed frozen foods using three main procedures: purge and trap, solid-phase microextraction (SPME), or direct sampling of headspace; the isolated volatiles are then separated on packed or capillary columns using gas chromatography (GC); packed columns offer faster run times when only a few volatiles are of concern, whereas capillary columns generate complex volatile profiles [56,72,73] A thawed sample may be placed inside a chamber of an electronic nose (gas-sensor array system); either at room temperature or after a short heating time, the vapors are exposed to the sensors; output signals are generated based on the change in resistance as vapors react with the surface of the sensor; the data are subjected to chemometric and artificial neural network software to generate patterns that may be associated with the extent of lipid oxidation; no published studies have been conducted to characterize the oxidative stability of frozen foods but a correlation between sensor response and oxidative stability of ice-stored herring has been reported [74]
Cholesterol oxidation products	Lipids are extracted, typically with chloroform:methanol, and then saponified; after isolation of the nonsaponifiable fraction, cholesterol oxidation products may be purified using Sep-Pak cartridges and a series of solvent mixtures with increasing polarity; the isolated fraction is finally silanized before subjecting the derivatized products to GC and flame ionization detection [75]
Fluorescent pigments	Aqueous and organic fluorescent pigments may be determined on a chloroform:methanol lipid extract; diluted samples of the aqueous and organic layers are taken and fluorescence may be measured either with a fluorometer (excitation and emission filters selected for wavelengths in the range 320–390 and 420–500 nm, respectively) [37]) or a spectrofluorometer (excitation, 367 nm; emission, 420 nm) [57]

of the subsample often influences the outcome of the chemical measurement. In comparing caudal, ventral, and dorsal sections of frozen mackerel, Icekson et al. [77] observed that thiobarbituric acid values were higher in the caudal area than the two other areas. Moreover, when dark and light portions of frozen hake muscle were compared, a lower lipolytic activity but higher oxidative activity was observed in the dark muscle compared with light muscle [78].

Lipid degradative measurements have been incorporated into experimental studies for a variety of reasons. The studies listed in Table 25.5 may be divided into those that aim to examine the effects of raw material, diet, processing, and additives on lipid degradation during freezing and frozen storage. Typically, only one or two measurements are conducted to measure lipid degradation; however, in some studies a wide assortment of chemical measurements have been made to assist in characterizing the different pathways and stages that occurred prior to examination. For example, Erickson [67] determined that thiobarbituric acid-reactive substances (TBARS), headspace volatiles, and degradation of tocopherol could be used to differentiate oxidative stabilities of frozen bass samples during the early phases of lipid oxidation, whereas conjugated dienes, organic fluorescent pigments, headspace volatiles, and degradation of tocopherol differentiated the bass during later stages of storage. Examination of frozen tilapia samples, however, did not find tocopherol losses to be a useful measure of differentiating different strains [68].

Chemical measurements have also proven useful in defining the potential for storage temperatures to affect lipid oxidation of frozen meat. For example, Hansen et al. [87] used electron spin resonance to measure the mobility of the nitroxyl spin probes TEMPO and TEMPOL in fat and lean pork meat to gage mobility of natural constituents. In that study, the mobility of TEMPO in fat increased for temperatures above $-60°C$ and the mobility of TEMPOL in lean meat increased for temperatures above $-40°C$.

Ultimately, any beneficial or negative aspects lipid degradation are judged by sensory evaluation. As constraints often exist for implementation of sensory tests, numerous studies have examined the correlation between chemical and sensory responses in an attempt to use the chemical measurements as a predictor of sensory perception. Using frozen-stored channel catfish, for example, "oxidized oil flavor" was highly correlated to total volatile aldehydes but not individual volatiles [72]. Bak et al. [88] also found that more than one volatile was necessary to predict the score of "rancid taste" and "rancid odor" of shrimp meat. Peroxide values and free fatty acid levels were shown to be the best parameters to describe increases in "train oil taste," "metal taste," and "bitter taste" in frozen Atlantic salmon [89]. Confirmation of this association was later demonstrated upon addition of unsaturated fatty acids to fresh minced salmon [90].

Chemical measurements to measure lipid degradation often involve multiple steps and expensive instrumentation in the laboratory, however, other analyses that could be conducted online or in the field are also being developed. For example, a portable low-resolution gas-phase Fourier transform infrared (FT-IR) analyzer was applied to the analysis of volatile compounds of thawed strawberries [91]. As no two compounds have identical IR-spectra, FT-IR is a highly characteristic measurement. Odor sensors, on the contrary, involve a more simplified approach to measure headspace volatiles. In these portable instruments, one or two metal oxide semiconductor sensor elements are used in conjunction with an internal micro air pump [92]. When one sensor element is present, the odor intensity is displayed as a numeric value. If a second element is present, information about the odor category may also be displayed. Successful application of an odor sensor has been demonstrated in an iced fish storage study [93] and thus the potential exists for this to be a valid measurement in frozen storage studies.

VI. METHODS TO ASSESS PROTEIN DEGRADATION DURING FREEZING OR FROZEN STORAGE

Another major component in frozen foods that is subject to degradation is protein. To measure this denaturation, a number of assays have been examined and these are described in Table 25.6. Some

TABLE 25.5
Selected Studies Using Measurements of Lipid Degradation

Objective of Study	Measurements of Lipid Degradation																	
	Polyunsaturated Fatty Acids	Free Fatty Acids	Phospholipids/Triacylglyerols	Tocopherol	Ascorbic Acid	Glutathione	Glutathione Peroxidase	Superoxide Dismutase	Heme Proteins/Metmyoglobin	Non-heme Iron	Catalase	Lipase/Phospholipase	Hydroperoxides/CD	TBARS (Malonaldehyde)	Cholesterol Oxidation Products	Fluorescent Pigments	Volatiles	References
Comparison of lipid deterioration in cod and haddock during frozen storage		✓											✓	✓		✓		[79]
Lipid characterization in the scallop's adductor muscle during frozen storage	✓	✓	✓															[80]
Characterization of antioxidant profiles for channel catfish during frozen storage	✓		✓	✓	✓	✓											✓	[37]
Effect of freezing on the activity of catalase in apple flesh tissue											✓							[81]
Characterization of lipids and lipolytic activities of pig muscle during frozen storage	✓	✓	✓									✓	✓	✓				[63]
Lipolysis and lipid oxidation in frozen minced mackerel in relation to presence of gelatin		✓											✓	✓				[82]
Effect of dietary α-tocopherol supplementation on cholesterol oxidation in frozen vacuum packaged, cooked beef steaks														✓	✓			[83]

(Table continued)

TABLE 25.5 *Continued*

Objective of Study	Measurements of Lipid Degradation																	
	Polyunsaturated Fatty Acids	Free Fatty Acids	Phospholipids/Triacylglyerols	Tocopherol	Ascorbic Acid	Glutathione	Glutathione Peroxidase	Superoxide Dismutase	Heme Proteins/Metmyoglobin	Non-heme Iron	Catalase	Lipase/Phospholipase	Hydroperoxides/CD	TBARS (Malonaldehyde)	Cholesterol Oxidation Products	Fluorescent Pigments	Volatiles	References
Influence of dietary fat source, and α-tocopherol, and ascorbic supplementation on lipid oxidation in frozen dark chicken meat														✓	✓			[84]
Effect of dietary vitamin E on the oxidative stability of frozen turkey breast meat														✓			✓	[85]
Influence of prefreezing storage on lipid oxidation in fillets of herring during frozen storage				✓	✓		✓						✓			✓		[57]
Effect of washing mackerel fillets on subsequent lipid oxidation during frozen storage									✓				✓	✓				[62]
Lipid stability of frozen horse mackerel with brine pretreatment	✓												✓	✓		✓		[86]
Effect of multiple freeze–thaw cycles on lipid oxidation of catfish muscle									✓	✓				✓				[61]
Influence of NaCl on antioxidant enzyme activity and lipid oxidation in frozen ground pork							✓	✓			✓		✓	✓				[59]
Effect of commercial plant extract on the stability of horse mackerel	✓						✓							✓		✓		[58]

TABLE 25.6
Description of Methodologies Used to Measure Protein Degradation in Frozen Foods.

Assay or Assay Target	Description of Analysis
Amino acids	Samples are acid hydrolyzed at elevated temperatures (110°C), dried, derivatized with phenylisothiocyanate, and then subjected to HPLC [94]
High-resolution nuclear magnetic resonance (NMR) spectroscopy	Water or salt extracts containing D_2O are subjected to the following conditions: proton NMR spectra are run on a Fourier transform spectrometer at either 300 or 400 mHz, data points are collected with a 30° pulse, a 15 ppm spectral width and a repetition time of 3.4 s; when a large water signal is encountered, a presaturation sequence is employed, with low decoupling for 3.4 s and data collection without decoupling for 1.6 s; between 128 and 512 scans are collected as required [95]
Dimethylamine	A trichloroacetic acid extract of fish muscle is reacted with a copper ammonium reagent and carbon disulfide at 40–50°C to produce a yellow-colored copper dimethyl-dithiocarbamate; the solution is acidified with acetic acid and shaken vigorously to facilitate extraction of the colored complex into a benzene layer; after drying with sodium sulfate, absorbance of the benzene layer is measured at 440 nm [96] The majority of procedures for GC involve extraction of volatile amines from samples with trichloroacetic acid or perchloric acid, followed by neutralization of extracts; dimethylamine is then collected either by steam distillation into hydrochloric or sulfuric acid or by sampling the headspace of the neutralized extract; the isolated fraction is then injected into the GC system; alternatively, the perchloric acid extract is followed by alkalinization with 65% KOH; another extraction with toluene is performed prior to injection into the GC system [97]
Formaldehyde	A trichloroacetic acid or perchloric acid extract of the sample is reacted at 60°C with Nash color reagent containing acetylacetone with an excess of ammonium acetate to produce diacetyldihydrolutidine. The solution is measured spectrophotometrically at 415 nm [98]
Protein sulfhydryls	The sample is homogenized and an aliquot of the homogenate dissolved in an SDS-buffered solution; Ellman's reagent, 5,5′-dithiobis-2-nitrobenzoic acid (DTNB) and urea are then added and the absorbance is read at 412 nm; total and surface SH contents may be calculated using a molar extinction coefficient of 11,400/*M* cm [99]; by omitting the SDS detergent, DTNB would react only with exposed sulfhydryls instead of all sulfhydryls contained in proteins
Protein carbonyls	An aliquot of a sample suspension is taken and mixed with 2,4-dinitrophenylhydrazine (DNPH); after a defined period of time, trichloroacetic acid is added to the mixture to precipitate the protein; the protein pellet is washed (i.e., ethanol:ethyl acetate, 1:1) to remove unreacted DNPH before dissolving in 6.0 *M* guanidine hydrochloride; absorbance of the protein solution is measured at 370 nm and the protein carbonyl content calculated using a molar extinction coefficient of 22,400/*M* cm [99]
Protein hydrophobicity	To a sample extract containing soluble or suspended proteins, fluorescence probes are added; to measure aliphatic hydrophobicity, *cis*-parinaric acid is added; to measure aromatic hydrophobicity, 8-anilino-1-naphthalene sulfonic acid is added; the relative fluorescence intensity is plotted against protein concentration and the slope of the regression line is defined as the surface hydrophobicity

(Table continued)

TABLE 25.6 *Continued*

Assay or Assay Target	Description of Analysis
Protein solubility [100]	
NaCl/KCl (disrupts electrostatic bonds)	A solution of NaCl or KCl is added to a sample homogenate; the mixture is held for a defined period before it is centrifuged; the supernatant is collected for protein determination; (on occasion, the sample homogenate is centrifuged before addition of the salt solution; the supernatant from this centrifugation step is designated the water-soluble sarcoplasmic protein fraction while the precipitate is exposed to the salt solution; after centrifugation, the supernatant from this treatment regime is termed the salt-soluble protein fraction)
SDS (disrupts noncovalent bonds)	The insoluble material from the salt extraction is treated with 2% sodium dodecyl sulfate (SDS); the supernatant recovered after centrifugation is collected for protein determination
SDS plus β-MeOH (disrupts noncovalent and disulfide bonds)	The precipitate collected from the SDS-treated sample is exposed to 2% SDS and 5% β-mercaptoethanol (β-MeOH)
Size-exclusion chromatography	The salt-soluble protein fraction of a sample is applied to a gel column; using a flow rate of 0.5 ml/min and a buffered eluent, the protein fractions are detected with a UV detector; the molecular weights of the peaks are estimated by comparing the mobilities of the fractions to known proteins
Electrophoresis	Protein fractions are solubilized in SDS and mercaptoethanol and then applied to polyacrylamide gels; following application of a electric current, the proteins are fixed, then visualized primarily with Coomassie Brilliant blue; the mobility of each band is compared with the mobility of protein standards; quantitative assessment is achieved by scanning the gels on an image analyzer and relating the optical density to those of standard proteins
Raman spectroscopy	The basis for Raman spectroscopic analysis is the inelastic scattering of photons resulting from vibrational transitions of functional groups of a molecule; both the frequency and intensity of molecular vibrations are sensitive to chemical changes and the microenvironment of functional groups, and these parameters would therefore have influences upon the vibrational spectrum
	Raman spectroscopy may be applied to aqueous solutions, nonaqueous liquids, or solid systems; interference from water molecules is minimal as the water molecules exhibit weak Raman scattering, in contrast to the strong signals of water in infrared spectra; raman spectra are recorded following excitation using a laser; assignment of bands to specific vibrational modes of amino acid side chains or the polypeptide backbone is based on published data [101,102]

ATPase	The salt-soluble extract is diluted to give a protein concentration of approximately 2–5 mg/ml; in addition to a buffer (pH 7.0–7.4), the assay medium contains either $CaCl_2$ for Ca^{2+}-ATPase, $MgCl_2$ for Mg^{2+}-ATPase, or $MgCl_2$ and EGTA for Mg^{2+}-EGTA-ATPase; the assay is started with the addition of ATP and terminated with trichloroacetic acid (TCA); the liberated inorganic phosphate is measured to determine the ATPase activity [99,103]
Proteolysis/proteolytic activity	To measure the level of proteolysis that has already occurred, pressed juice from the sample is prepared and TCA is added; following precipitation of the proteins, the level of nonprotein nitrogen in the supernatant is measured using either a standard protein assay (i.e. Lowry procedure) or by measuring the absorbance of the filtrate at 280 nm [104–105] To measure the potential proteolytic activity, extracts are prepared and incubated with either nonspecific protein substrates (i.e., casein or hemoglobin) or substrates designed to measure specific proteases (i.e., glycyl-phenyl-alanine-2-naphthylamide for cathepsin C) [104,106,107] Myofibril fragmentation index (MFI) is a useful indicator of the extent of proteolysis indicating both rupture of the I-band and breakage of intermyofibril linkages; samples are homogenized, filtered to remove connective tissue, protein concentrations taken on the suspension, suspensions diluted to a final protein concentration of 0.5 mg/ml, and then absorbance values (540 nm) multiplied by 150 to give index values for myofibrillar fragmentation [108]
Fourier transform near-infrared (FTNIR) spectroscopy	Slices of 0.9 mm thickness are cut from the sample and warmed to room temperature (10 min) prior to running the spectra; measurements are made in the transmission mode, at a resolution of 8 cm^{-1}, using silica windows, in the spectral range of 1000–1876 nm (the high-absorbance water peak centered at 1927 nm limits sample thickness [109]
Competitive-enzyme-linked immunosorbent assay (ELISA)	Polyclonal antibodies to purified proteins are prepared; in the assay, purified protein is also fixed to sample wells and blocked with bovine serum albumin to prevent nonspecific binding; the antibody and sample extract are added to the well and incubated for a period of time before a tracer antibody (i.e., horseradish peroxidase-IgG) is added to the system for colorimetric measurement of levels of immunological binding (as the concentration of nondenatured sample protein is increased, the probability of binding between antibody and fixed protein is decreased, leading to lower levels of tracer and hence reduced color development); the levels of nondenatured protein in the sample is quantified by reference to a standard curve [110]
Differential scanning calorimetry (DSC)	Powdered, homogenized, or minced sample is placed in a hermetically sealed polymer-coated aluminum pan of a DSC machine while the reference pan is either empty or contains an equal weight of deionized, distilled water; samples are scanned at a designated heating rate (i.e., 10°C/min) over a defined range of temperatures (i.e., 10–100°C); onset melting temperatures (T_m) are determined by constructing a tangent to the leading edge of a transition curve and determining the temperature at the point of intersection with the baseline; the T_m of an endothermic peak indicates the beginning of denaturation of proteins during heating; thus, the smaller the T_m, the lower the thermal stability of the protein [111–113]

of the assays focus on measuring specific constituents that may be present (i.e., formaldehyde, dimethylamine, amino acids, etc.), whereas other measurements are used to understand the interactions between proteins that contribute to their degradation (i.e., protein solubility, protein sulfhydryls, protein hydrophobicity, etc.). Examples of studies incorporating these different type measurements are given in Table 25.7 and may be divided into the following categories: fundamental studies, raw material comparison, process evaluation, additive evaluation, and predictive evaluation of chemical measurements to functional properties or sensory responses.

Although the newer sophisticated analytical measurements (high-resolution nuclear magnetic resonance (NMR) spectroscopy, Raman spectroscopy, Fourier transform near-infrared (FTNIR) spectroscopy) have been incorporated into studies primarily for the fundamental characterization of protein denaturation, traditional assays to measure formaldehyde content, protein solubility, and myofibrillar fragmentation are more commonly employed in applied studies. Although these latter measurements have been used for several decades, improvements in the methodology continue to be implemented. For example, Hopkins et al. [108] recommended that homogenization speeds of 15,000 rpm be applied during myofibrillar fragmentation assays. Moreover, in measurements of formaldehyde, Bechmann [125] demonstrated that the amount of free plus reversibly bound formaldehyde could be predicted by a linear model that was based on levels of free formaldehyde, and these values were similar to those obtained through distillation.

VII. METHODS TO ASSESS CARBOHYDRATE AND PIGMENT DEGRADATION DURING FREEZING OR FROZEN STORAGE

Although the majority of studies examining the effects of freezing and frozen storage target lipid and protein constituents, carbohydrate moieties within frozen foods may also be modified. Hence, assays to measure cell wall polysaccharide composition have shown losses in total sugars of frozen muskmelons [126]. In contrast, little change was noted in pectic substances for frozen persimmon fruit and astringency reduction was attributed to tannin insolubilization [127].

Chemical measurements for pigment degradation are another area of investigation in frozen shelf-life studies. Examples where chlorophyll measurement has been undertaken include studies on apple fruit [128], asparagus spears [129], and stir-fried pea pods [130]. Similarly, carotenoid (astaxanthin, canthaxanthin) assays have been conducted in frozen salmon [131] and rainbow trout [132], whereas heme pigment (myoglobin, oxymyoglobin, metmyoglobin) assays have been conducted in frozen beef [133], ground pork [134], and bluefin tuna [135]. The development of white spots in shrimp, however, required that the constituents responsible for the quality defect first be characterized and identified before they could routinely be monitored. Consequently, IR and Raman spectroscopy determined that the white spots were crystals of calcite and vaterite, two forms of calcium carbonate, in a matrix of chitin [136].

VIII. CHEMICAL MEASUREMENTS TO MONITOR CHEMICAL AND MICROBIAL ADDITIVES/CONTAMINANTS IN FROZEN FOODS

Studies to measure chemical migration from food contact materials to frozen foods has been limited despite test conditions being specified for frozen foods in European directives [137]. In particular, migration of adhesives and substances used in inks are relevant targets. To detect benzophenone (a photoinitiator for UV-cured ink) in frozen retail foods (Cornish pies, breaded fish sticks, potato waffles, cheese, and onion sticks), the samples were extracted with solvent, subjected to size exclusion chromatographic clean-up, and then analyzed for the targeted substance by gas chromatography–mass spectrometry [138]. Similar type analyses were applied to detect model ink components (chlorodecane, butyl benzoate, dimethyl phthalate, benzophenone, and benzybutyl phthalate) in potato chips and hamburgers stored for 1 year at $-20°C$.

TABLE 25.7
Selected Studies Incorporating Protein Degradation Measurements

Objective of Study	Measurements of Protein Degradation																References
	Amino Acids	NMR Spectroscopy	Dimethylamine	Formaldehyde	Protein Sulfhydryls	Protein Carbonyls	Protein Surface Hydrophobicity	Protein Solubility	Size-Exclusion Chromatography	Electrophoresis	Raman Spectroscopy	Myosin ATPase	Proteolysis/Calpain Activity	FTNIR Spectroscopy	ELISA	DSC	
Freeze denaturation of carp myofibrils compared with thermal denaturation								✓				✓					[114]
Characteristics of the salt-soluble fraction of frozen hake fillets					✓		✓		✓	✓		✓					[115]
Characterization of proteins during frozen storage of minced cod								✓	✓	✓							[116]
Response of cod myosin to frozen storage or modification with formaldehyde							✓	✓		✓	✓						[101]
Response of cod proteins to a nonenzymic free-radical-generating system during frozen storage					✓	✓		✓		✓							[117]
Characterization of proteins during frozen storage of hake			✓								✓						[102]
Characterization of proteins during frozen storage of minced red hake			✓											✓			[36]
Characterization of water-soluble metabolites in cod and haddock subjected to frozen storage		✓		✓													[95]
Myosin denaturation during frozen storage monitored by ELISA										✓					✓		[110]
Modification in proteolysis of cheese subjected to frozen storage													✓				[107]
Depolymerization and aggregation of glutenin in bread dough during frozen storage										✓							[118]
Off-flavor production in frozen strawberries													✓				[119]

(*Table continued*)

TABLE 25.7 *Continued*

Objective of Study	Measurements of Protein Degradation																
	Amino Acids	NMR Spectroscopy	Dimethylamine	Formaldehyde	Protein Sulfhydryls	Protein Carbonyls	Protein Surface Hydrophobicity	Protein Solubility	Size-Exclusion Chromatography	Electrophoresis	Raman Spectroscopy	Myosin ATPase	Proteolysis/Calpain Activity	FTNIR Spectroscopy	ELISA	DSC	References
Comparison of protein denaturation in frozen cod and haddock	✓		✓	✓			✓	✓									[94]
Variation of protein denaturation during frozen storage with lamb genotype										✓		✓				✓	[120]
Calpain activity in thawed rigor muscle and association with toughness													✓				[106]
Response of cod muscle proteins to different freeze–thaw cycles					✓		✓	✓		✓		✓					[61]
Modification in type of aggregation as affected by frozen storage temperature				✓				✓		✓							[121]
Protein quality of sardine fillets as affected by slow and quick defrosting	✓				✓			✓									[122]
Effect of pressure shift freezing and air-blast freezing on protein denaturation of frozen turbot								✓		✓							[123]
Protein denaturation in frozen-stored minced blue whiting muscle as affected by cryostabilizers				✓				✓								✓	[109]
Effect of additives on protein denaturation in frozen hake muscle								✓		✓							[100]
Effect of cryoprotectants on protein denaturation in frozen rainbow trout fillets					✓		✓	✓		✓						✓	[113]
Chemical stability of antioxidant-washed beef heart surimi					✓	✓	✓					✓					[99]
Relationship of foaming capacity to protein degradation of fish minces during frozen storage			✓	✓				✓									[124]

In frozen dough, the viability of yeast additives is critical for proper functioning. To assess this viability, various measurements have been applied to assess gas production and leachate composition of frozen yeast [139]. In the case of gassing power, previously frozen yeast is suspended in a media conducive for fermentation. The CO_2 is collected over a period of time in a gas dispersion tube containing NaOH and quantified by back-titrating with HCl. In the case of leachate composition, compressed frozen–thawed yeast is suspended in water, shaken for a period of time, and nitrogen and total reducing substances determined on the supernatant.

IX. CONCLUSIONS

A wide range of chemical measurements are available to measure constituents of frozen foods. Some analyses seek to measure the loss of a component, whereas others target the products of a chemical reaction. Selection of a chemical measurement is dependent on the objective of a study. In applied studies, one or more standard measurements (i.e., TBARS, dimethylamine) may be applied, whereas analyses required for fundamental studies often involve more sophisticated instrumentation (i.e., Raman spectroscopy, high-resolution NMR spectroscopy). Ultimately, the merit of a measurement should be based on its relationship to the sensory response of the product.

REFERENCES

1. MA Murcia, B Lopez-Ayerra, M Martinez-Tome, AM Vera, F Garcia-Carmona. Evolution of ascorbic acid and peroxidase during industrial processing of broccoli. *Journal of the Science of Food and Agriculture* 80:1882–1886, 2000.
2. MA Joslyn. Enzyme activity in frozen vegetable tissue. *Advances in Enzymology and Related Subjects of Biochemistry* 9:613–652, 1949.
3. E Chen, HY Peng, BH Chen. Studies on enzyme selection as blanching index of frozen green beans and carrots. *Food Chemistry* 57:497–503, 1996.
4. DC Williams, MH Lim, AO Chen, RM Pangborn, JR Whitaker. Blanching of vegetables for freezing — which indicator enzyme to choose. *Food Technology* 6:130–140, 1986.
5. JR Whitaker. *Principles of Enzymology for the Food Sciences*. 2nd ed. New York: Marcel Dekker, 1994.
6. DM Barrett, EL Garcia, GF Russell, E Ramirez, A Shirazi. Blanch time and cultivar effects on quality of frozen and stored corn and broccoli. *Journal of Food Science* 65:534–540, 2000.
7. H Morais, P Rodrigues, C Ramos, V Almeida, E Forgács, T Cserháti, JS Oliveira. Effect of blanching and frozen storage on the stability of β-carotene and capxanthin in red pepper (*Capsicum annuum*) fruit. *Food Science and Technology International* 200:55–59, 2002.
8. BM Poli, G Zampacavallo, G Parisi, A Poli, M Mascini. Biosensors applied to biochemical fish quality indicators in refrigerated and frozen sea bass reared in aerated or hyperoxic conditions. *Aquaculture International* 8:335–348, 2000.
9. RA Benner Jr, WF Staruszkiewicz, PL Rogers, WS Otwell. Evaluation of putrescine, cadaverine, and indole as chemical indicators of decomposition in Penaeid shrimp. *Journal of Food Science* 68:2178–2185, 2003.
10. NR Jones. Interconversions of flavorous nucleotide catabolites in chilled and frozen fish. *Progress in Refrigeration Science and Technology* IV (5):917–922, 1963.
11. NV de Mattio, ME Paredi, M Crupkin. Postmortem changes in the adductor muscle of scallop (*Chlamys tehuelchus*) in chilled and frozen storage. *Journal of Aquatic Food Product Technology* 10:49–59, 2001.
12. TF Moriarty, JM Oduko, NM Spyrou. Thermoluminescence in irradiated foodstuffs. *Nature* 332:22, 1988.
13. NJF Dodd, JS Lea, AJ Swallow. Electron spin resonance detection of irradiated food. *Nature* 334:387, 1988.

14. M Furuta, T Dohmaru, T Katayama, H Toratani, A Takeda. Detection of irradiated frozen meat and poultry using carbon monoxide gas as a probe. *Journal of Agricultural and Food Chemistry* 40:1099–1100, 1992.
15. MH Stevenson. Identification of irradiated foods. *Food Technology* 48:141–144, 1994.
16. W Meier, H Hediger, A Artho. Determination of *o*-tyrosine in shrimps, fish, mussels, frog legs, and egg-white. In: CH McMurray, EM Stewart, R Gray, J Pearce, Eds., *Detection Methods for Irradiated Foods: Current Status*. Cambridge: Royal Society of Chemistry, 1996, pp. 303–309.
17. WW Nawar, Z Zhu, H Wan, E DeGroot, Y Chen, T Aciukewicz. Progress in the detection of irradiated foods by measurement of lipid-derived volatiles. In: CH McMurray, EM Stewart, R Gray, J Pearce, Eds., *Detection Methods for Irradiated Foods: Current Status*. Cambridge: Royal Society of Chemistry, 1996, pp. 241–248.
18. M Furuta, T Dohmaru, T Katayama, H Toratani, A Takeda. Detection of irradiated frozen deboned seafood with the level of radiolytic H_2 and CO gases as a probe. *Journal of Agricultural and Food Chemistry* 45:3928–3931, 1997.
19. CHS Hitchcock. Hydrogen as a marker for irradiated food. *Food Science and Technology Today* 12:112–114, 1998.
20. CHS Hitchcock. Determination of hydrogen as a marker in irradiated frozen food. *Journal of the Science of Food and Agriculture* 80:131–136, 2000.
21. L Merino, H Cerda. Control of imported irradiated frozen meat and poultry using the hydrocarbon method and the DNA comet assay. *European Food Research and Technology* 211:298–300, 2000.
22. J-H Park, C-K Hyun, S-K Jeong, M-A Yi, S-T Ji, H-K Shin. Use of the single cell gel electrophoresis assay (Comet assay) as a technique for monitoring low-temperature treated and irradiated muscle tissues. *International Journal of Food Science and Technology* 35:555–561, 2000.
23. A Pavlov, GD Garcia de Fernando, JA Ordoñez, L Hoz. β-Hydroxyacyl-CoA-dehydrogenase (HADH) activity of unfrozen and frozen-thawed frog (*Rana esculenta*) legs. *Journal of the Science of Food and Agriculture* 64:141–143, 1994.
24. G Duflos, B Le Fur, V Mulak, P Becel, P Malle. Comparison of methods of differentiating between fresh and frozen–thawed fish or fillets. *Journal of the Science of Food and Agriculture* 82:1341–1345, 2002.
25. SK Chhatbar, NK Velankar. A biochemical test for the distinction of fresh fish from frozen and thawed fish. *Fish Technology* 14:131–133, 1977.
26. P Vandekerckhove, D Demeyer, H Henderic. Evaluation of a method to differentiate between nonfrozen and frozen and thawed meat. *Journal of Food Science* 37:636–637, 1972.
27. H Rehbein. Development of an enzymatic method to differentiate fresh and sea frozen and thawed fish fillets. *Zeitschrift fur Lebensmittel-Untersuchung und-Forschung* 169:263–265, 1979.
28. C Barbagli, GS Crescenzy. Influence of freezing and thawing on the release of cytochrome oxidase from chicken's liver and from beef and trout muscle. *Journal of Food Science* 46:491–496, 1981.
29. H Godiksen, F Jessen. Cytochrome oxidase as an indicator of ice storage and frozen storage. *Journal of Agricultural and Food Chemistry* 49:4488–4493, 2001.
30. V Salfi, F. Fucetola, V Verticelli, P Arata. Optimized procedures of biochemical analysis for the differentiation between fresh and frozen thawed fish products; test of mitochondrial malate dehydrogenase. *Industrie Alimentari* 25:634–636, 1986.
31. K Nilsson, B Ekstrand. The effect of storage on ice and various freezing treatments on enzyme leakage in muscle tissue of rainbow trout (*Oncorhynchus mykiss*). *Zeitschrift fur Lebensmittel-Untersuchung und-Forschung* 197:3–7, 1993.
32. K Nilsson, B Ekstrand. Sensory and chemically measured effects of different freeze treatments on the quality of farmed rainbow trout. *Journal of Food Quality* 18:177–191, 1995.
33. H Rehbein, G Kress, W Schreiber. An enzymic method for differentiating thawed and fresh fish fillets. *Journal of the Science of Food and Agriculture* 29:1076–1082, 1978.
34. PE Shaw, RL Rouseff, KL Goodner, R Bazemore, HE Nordby, WW Widmer. Comparison of headspace GC and electronic sensor techniques for classification of processed orange juices. *Lebensmittel-Wissenschaft und-Technologie* 33:331–334, 2000.
35. K Thyholt, T Isaksson. Differentiation of frozen and unfrozen beef using near-infrared spectroscopy. *Journal of the Science of Food and Agriculture* 73:525–532, 1997.

36. J Pink, M Naczk, D Pink. Evaluation of the quality of frozen minced red hake: use of Fourier transform infrared spectroscopy. *Journal of Agricultural and Food Chemistry* 46:3667–3672, 1998.
37. MC Erickson. Compositional parameters and their relationship to oxidative stability of channel catfish. *Journal of Agricultural and Food Chemistry* 41:1213–1318, 1993.
38. B de Ancos, EM González, MP Cano. Ellagic acid, vitamin C, and total phenolic contents and radical scavenging capacity affected by freezing and frozen storage in raspberry fruit. *Journal of Agricultural and Food Chemistry* 48:4565–4570, 2000.
39. HS Lee, GA Coates. Vitamin C in frozen, fresh squeezed, unpasteurized, polyethylene-bottled orange juice: a storage study. *Food Chemistry* 65:165–168, 1999.
40. J Klimczak, Z Irzyniec. Rates of vitamin C degradation in Brussel sprouts during freezing by different methods. *Chlodnictwo* 36:40–42, 2001.
41. I Castro, O Gonçalves, JA Teixeira, AA Vicente. Comparative study of *Selva* and *Camarosa* strawberries for the commercial market. *Journal of Food Science* 67:2132–2137, 2002.
42. J Suutarinen, K Honkapää, R-L Heiniö, K Autio, A Mustranta, S Karppinen, T Kiutamo, H Liukkonen-Lilja, M Mokkila. Effects of calcium chloride-based prefreezing treatments on the quality factors of strawberry jams. *Journal of Food Science* 67:884–894, 2002.
43. E Camacho-Salas, C Diez-Marques, MM Camara-Hurtado. Effect of freezing on the vitamin C content of papayas. *Alimentacion Equipos y Tecnologia* 15:63–69, 1996.
44. Z Lisiewska, W Kmiecik. Effect of freezing and storage on quality factors in Hamburg and leafy parsley. *Food Chemistry* 60:633–637, 1997.
45. SH Häkkinen, SO Kärenlampi, HM Mykkänen, IM Heinonen, AR Törrönen. Ellagic acid content in berries: Influence of domestic processing and storage. *European Food Research and Technology* 212:75–80, 2000.
46. MJ González-Castro, MJ Oruña-Concha, J López-Hernández, J Simal-Lozano. Effects of freezing on the organic acid content of frozen green beans and Padrón peppers. *Zeitschrift fur Lebensmittel-Untersuchung und-Forschung A* 204:365–368, 1997.
47. AN Califano, AE Bevilacqua. Freezing low moisture Mozzarella cheese: changes in organic acid content. *Food Chemistry* 64:193–198, 1999.
48. AS Rodrigues, EAS Rosa. Effect of post-harvest treatments on the level of glucosinolates in broccoli. *Journal of the Science of Food and Agriculture* 79:1028–1032, 1999.
49. FM Aramouni, JS Godber. Folate losses in beef liver due to cooking and frozen storage. *Journal of Food Quality* 14:357–365, 1991.
50. LT Vahteristo, KE Lehikoinen, V Ollilainen, PE Koivistoinen, P Varo. Oven-baking and frozen storage affect folate vitamin retention. *Lebensmittel-Wissenschaft und-Technologie* 31:329–333, 1998.
51. PJ Ke, AD Woyewoda. Titrimetric method for determination of free fatty acids in tissues and lipids with ternary solvents and *m*-cresol purple indicator. *Analytical Chimica Acta* 99:387–391, 1978.
52. AJ de Koning, S Milkovitch, TH Mol. The origin of free fatty acids formed in frozen cape hake mince (*Merluccius capensis*, Castelnau) during cold storage at −18°C. *Journal of the Science of Food and Agriculture* 39:79–84, 1987.
53. S Zhou, RG Ackman. Interference of polar lipids with the alkalimetric determination of free fatty acid in fish lipids. *Journal of the American Oil Chemists Society* 73:1019–1023, 1996.
54. RR Lowry, IJ Tinsely. Rapid colorimetric determination of free fatty acids. *Journal of the American Oil Chemists Society* 53:470–472, 1976.
55. MC Erickson. Chemical measurements of frozen foods. In: MC Erickson, Y-C Hung, Eds., *Quality in Frozen Foods*. New York: Chapman & Hall, 1997, pp. 340–356.
56. RG Brannan, MC Erickson. Quantification of antioxidants in channel catfish during frozen storage. *Journal of Agricultural and Food Chemistry* 44:1361–1366, 1996.
57. I Undeland, H Lingnert. Lipid oxidation in fillets of herring (*Clupea harengus*) during frozen storage: influence of prefreezing storage. *Journal of Agricultural and Food Chemistry* 47:2075–2081, 1999.
58. S Aubourg, A Lugasi, J Hóvári, C Piñeiro, V Lebovics, I Jakóczi. Damage inhibition during frozen storage of horse mackerel (*Trachurus trachurus*) fillets by a previous plant extract treatment. *Journal of Food Science* 69:FCT136–141, 2004.
59. SK Lee, L Mei, EA Decker. Influence of sodium chloride on antioxidant enzyme activity and lipid oxidation in frozen ground pork. *Meat Science* 46:349–355, 1997.

60. GE Lester, DM Hodges, RD Meyer, KD Munro. Pre-extraction preparation (fresh, frozen, freeze-dried, or acetone powdered) and long-term storage of fruit and vegetable tissues: effects on antioxidant enzyme activity. *Journal of Agricultural and Food Chemistry* 52:2167–2173, 2004.
61. S Benjakul, F Bauer. Biochemical and physicochemical changes in catfish (*Silurus glanis* Linne) muscle as influenced by different freeze–thaw cycles. *Food Chemistry* 72:207–217, 2001.
62. MP Richards, SD Kelleher, HO Hultin. Effect of washing with or without antioxidants on quality retention of mackerel fillets during refrigerated and frozen storage. *Journal of Agricultural and Food Chemistry* 46:4363–4371, 1998.
63. P Hernández, J Navarro, F Toldrá. Effect of frozen storage on lipids and lipolytic activities in the *longissimus dorsi* muscle of the pig. *Zeitschrift fur Lebensmittel-Untersuchung und-Forschung A* 208:110–115, 1999.
64. E Løvaas. A sensitive spectrophotometric method for lipid hydroperoxide determination. *Journal of the American Oil Chemists Society* 69:777–783, 1992.
65. NC Shanta, EA Decker. Rapid, sensitive, iron-based spectrophotometric methods for determination of peroxide values of food lipids. *Journal of AOAC International* 77:421–424, 1994.
66. J-H Hong, S Yamaoka-Koseki, K Yasumoto. Analysis of peroxide values in edible oils by near-infrared spectroscopy. *Nippon Shokuhin Kogyo Gakkaishi* 41:277–280, 1994.
67. MC Erickson. Ability of chemical measurements to differentiate oxidative stabilities of frozen minced muscle tissue from farm-raised striped bass and hybrid striped bass. *Food Chemistry* 48:381–385, 1993.
68. MC Erickson, ST Thed. Comparison of chemical measurements to differentiate oxidative stability of frozen minced tilapia fish muscle. *International Journal of Food Science and Technology* 29:585–591, 1994.
69. FP Corongiu, A Milia. An improved and simple method for determining diene conjugation in polyunsaturated fatty acids. *Chemico-Biological Interactions* 44:289–297, 1983.
70. FP Corongiu, S Banni, MA Dessi. Conjugated dienes detected in tissue lipid extracts by second derivative spectrophotometry. *Free Radicals Biology and Medicine* 7:183–186, 1989.
71. RCRM Vossen, MCE van Dam-Mieras, G Hornstra, RFA Zwaal. Continuous monitoring of lipid peroxidation by measuring conjugated diene formation in an aqueous liposome suspension. *Lipids* 28:857–861, 1993.
72. RG Brannan, MC Erickson. Sensory assessment of frozen stored channel catfish in relation to lipid oxidation. *Journal of Aquatic Food Product Technology* 5:67–80, 1996.
73. C Milo, W Grosch. Changes in the odorants of boiled salmon and cod as affected by the storage of the raw material. *Journal of Agricultural and Food Chemistry* 44:2366–2371, 1996.
74. J-E Haugen, I Undeland. Lipid oxidation in herring fillets (*Clupea harengus*) during ice storage measured by a commercial hybrid gas-sensor array system. *Journal of Agricultural and Food Chemistry* 51:752–759, 2003.
75. A Grau, R Codony, S Grimpa, MD Baucells, F Guardiola. Cholesterol oxidation in frozen dark chicken meat: influence of dietary fat source, and α-tocopherol and ascorbic acid supplementation. *Meat Science* 57:197–208, 2001.
76. MJ Lichon, KW James. Homogenization methods for analysis of foodstuffs. *Journal of the Association of Official Analytical Chemists* 73:820–825, 1990.
77. I Icekson, V Drabkin, S Aizendorf, A Gelman. Lipid oxidation levels in different parts of the mackerel, *Scomber scombrus. Journal of Aquatic Food Product Technology* 7:17–29, 1998.
78. SP Auborg, M Rey-Mansilla, CG Sotelo. Differential lipid damage in various muscle zones of frozen hake (*Merluccius merluccius). Zeitschrift fur Lebensmittel-Untersuchung und-Forschung A* 208:189–193, 1999.
79. SP Aubourg, I Medina. Influence of storage time and temperature on lipid deterioration during cod (*Gadus morhua*) and haddock (*Melanogrammus aeglefinus*) frozen storage. *Journal of the Science of Food and Agriculture* 79:1943–1948, 1999.
80. B Young, T Ohshima, C Koizumi. Changes in molecular species compositions of glycerophospholipids in the adductor muscle of the giant ezo scallop *Patinopecten yessoensis* during frozen storage. *Journal of Food Lipids* 6:131–147, 1999.
81. Y Gong, PMA Toivonen, PA Wiersma, C Lu. Effect of freezing on the activity of catalase in apple flesh tissue. *Journal of Agricultural and Food Chemistry* 48:5537–5542, 2000.

82. NC Brake, OR Fennema. Lipolysis and lipid oxidation in frozen minced mackerel as related to *Tg′*, molecular diffusion, and presence of gelatin. *Journal of Food Science* 64:25–32, 1999.
83. K Galvin, A-M Lynch, JP Kerry, PA Morrissey, DJ Buckley. Effect of dietary vitamin E supplement on cholesterol oxidation in vacuum-packaged cooked beef steaks. *Meat Science* 55:7–11, 2000.
84. A Grau, F Guardiola, S Grimpa, AC Barroeta, R Codony. Oxidative stability of dark chicken meat through frozen storage: influence of dietary fat and α-tocopherol and ascorbic acid supplementation. *Poultry Science* 80:1630–1642, 2001.
85. BW Sheldon, PA Curtis, PL Dawson, PR Ferket. Effect of dietary vitamin E on the oxidative stability, flavor, color, and volatile profiles of refrigerated and frozen turkey breast. *Poultry Science* 76:634–641, 1997.
86. SP Aubourg, M Ugliana. Effect of brine pre-treatment on lipid stability of frozen horse mackerel (*Trachurus trachurus*). *European Food Research and Technology* 215:91–95, 2002.
87. E Hansen, L Lauridsen, LH Skibsted, RK Moawad, ML Andersen. Oxidative stability of frozen pork patties: effect of fluctuating temperature on lipid oxidation. *Meat Science* 68:185–191, 2004.
88. LS Bak, L Jacobsen, SS Jørgensen. Characterisation of qualitative changes in frozen, unpeeled coldwater shrimp (*Pandalus borealis*) by static headspace gas chromatography and multivariate data analysis. *Zeitschrift fur Lebensmittel-Untersuchung und-Forschung A* 208:10–16, 1999.
89. HHF Refsgaard, PB Brockhoff, B Jensen. Sensory and chemical changes in farmed Atlantic salmon (*Salmo salar*) during frozen storage. *Journal of Agricultural and Food Chemistry* 46:3473–3479, 1998.
90. HHF Refsgaard, PMB Brockhoff, B Jensen. Free polyunsaturated fatty acids cause taste deterioration of salmon during frozen storage. *Journal of Agricultural and Food Chemistry* 48:3280–3285, 2000.
91. M Hakala, M Ahro, J Kauppinen, H Kallio. Determination of strawberry volatiles with low resolution gas phase FT-IR analyser. *European Food Research and Technology* 212:505–510, 2001.
92. J Kita. Attempts at simplified measurement of odors in Japan using odor sensors, 2004. http://www.env.go.jp/en/lar/odo_measure/02_3_5.pdf.
93. A Gelman, V Crabkin, L Glatman. A rapid non-destructive method for fish quality control by determination of smell intensity. *Journal of the Science of Food and Agriculture* 83:580–585, 2003.
94. F Badii, NK Howell. A comparison of biochemical changes in cod (*Gadus morhua*) and haddock (*Melanogrammus aeglefinus*) fillets during frozen storage. *Journal of the Science of Food and Agriculture* 82:87–97, 2001.
95. N Howell, Y Shavila, M Grootveld, S Williams. High-resolution NMR and magnetic resonance imaging (MRI) studies on fresh and frozen cod (*Gadus morhua*) and haddock (*Melanogrammus aeglefinus*). *Journal of the Science of Food and Agriculture* 72:49–56, 1996.
96. CH Castell, B Smith, WJ Dyer. Simultaneous measurements of trimethylamine and dimethylamine in fish, and their use for estimating quality of frozen stored gadoid fillets. *Journal of the Fisheries Research Board of Canada* 31:383–389, 1974.
97. MT Veciana-Nogues, MS Albala-Hurtado, M Izquierdo-Pulido, MC Vidal-Carou. Validation of a gas chromatographic method for volatile amine determination in fish samples. *Food Chemistry* 57:569–573, 1996.
98. T Nash. The colorimetric estimation of formaldehyde by means of the antzsch reaction. *Biochemical Journal* 55:416–421, 1953.
99. B Wang, YL Xiong, S Srinivasan. Chemical stability of antioxidant-washed beef heart surimi during frozen storage. *Journal of Food Science* 62:939–945,991, 1997.
100. A Huidobro, C Alvarez, M Tejada. Hake muscle altered by frozen storage as affected by added ingredients. *Journal of Food Science* 63:638–643, 1998.
101. M Careche, ECY Li-Chan. Structural changes in cod myosin after modification with formaldehyde or frozen storage. *Journal of Food Science* 62:717–723, 1997.
102. AM Herrero, P Carmona, M Careche. Raman spectroscopic study of structural changes in hake (*Merluccius merluccius* L.) muscle proteins during frozen storage. *Journal of Agricultural and Food Chemistry* 52:2147–2153, 2004.
103. S Benjakul, F Bauer. Physicochemical and enzymatic changes of cod muscle proteins subjected to different freeze-thaw cycles. *Journal of the Science of Food and Agriculture* 80:1143–1150, 2000.
104. MC Erickson, DT Gordon, AF Anglemier. Proteolytic activity in the sarcoplasmic fluids of parasitized Pacific whiting (*Merluccius productus*) and unparasitized true cod (*Gadus macrocephalus*). *Journal of Food Science* 48:1315–1319, 1983.

105. OH Lowry, NJ Rosebrough, AL Farr, RJ Randall. Protein measurement with the folin phenol reagent. *Journal of Biological Chemistry* 193:265–275, 1951.
106. E Dransfield. Calpains from thaw rigor muscle. *Meat Science* 43:311–320, 1996.
107. L Tejada, E Sánchez, R Gómez, M Vioque, J Fernández-Salguero. Effect of freezing and frozen storage on chemical and microbiological characteristics in sheep milk cheese. *Journal of Food Science* 67:126–129, 2002.
108. DL Hopkins, PJ Littlefield, JM Thompson. A research note on factors affecting the determination of myofibrillar fragmentation. *Meat Science* 56:19–22, 2000.
109. J Pink, M Naczk, D Pink. Evaluation of the quality of frozen minced red hake: Use of Fourier transform near-infrared spectroscopy. *Journal of Agricultural and Food Chemistry* 47:4280–4284, 1999.
110. J-W Lee, J-H Park, S-B Kim, C-J Kim, C-K Hyun, H-K Shin. Application of competitive indirect enzyme-linked immunosorbent assay (Ci-ELISA) for monitoring the degree of frozen denaturation of bovine myosin. *International Journal of Food Science and Technology* 33:401–410, 1998.
111. S Srinivasan, YL Xiong, SP Blanchard. Effects of freezing and thawing methods and storage time on thermal properties of freshwater prawns (*Macrobrachium rosenbergii*). *Journal of the Science of Food and Agriculture* 75:37–44, 1997.
112. JJ Herrera, L Pastoriza, G Sampedro. A DSC study on the effects of various maltodextrins and sucrose on protein changes in frozen-stored minced blue whiting muscle. *Journal of the Science of Food and Agriculture* 81:377–384, 2001.
113. S Jittinandana, PB Kenney, SD Slider. Cryoprotection affects physiochemical attributes of rainbow trout fillets. *Journal of Food Science* 68:1208–1214, 2003.
114. Y Azuma, K Konno. Freeze denaturation of carp myofibrils compared with thermal denaturation. *Fisheries Science* 64:287–290, 1998.
115. ML del Mazo, P Torrejon, M Careche, M Tejada. Characteristics of the salt-soluble fraction of hake (*Merluccius merluccius*) fillets stored at −20 and −30°C. *Journal of Agricultural and Food Chemistry* 47:1372–1377, 1999.
116. M Tejada, M Careche, P Torrejón, ML del Mazo, MT Solas, ML García, C Barba. Protein extracts and aggregates forming in minced cod (*Gadus morhua*) during frozen storage. *Journal of Agricultural and Food Chemistry* 44:3308–3314, 1996.
117. S Srinivasan, HO Hultin. Chemical, physical, and functional properties of cod proteins modified by a nonenzymic free-radical-generating system. *Journal of Agricultural and Food Chemistry* 45:310–320, 1997.
118. PD Ribotta, AE León, MC Añón. Effect of freezing and frozen storage of doughs on bread quality. *Journal of Agricultural and Food Chemistry* 49:913–918, 2001.
119. H Deng, Y Ueda, K Chachin, H Yamanaka. Off-flavor production in frozen strawberries. *Postharvest Biology and Technology* 9:31–39, 1996.
120. MA Ojeda, JR Wagner, M Crupkin. Biochemical properties of myofibrils from frozen *longissimus dorsi* muscle of three lamb genotypes. *Lebensmittel-Wissenschaft und-Technologie* 34:390–397, 2001.
121. M Careche, ML del Mazo, P Torrejón, M Tejada. Importance of frozen storage temperature in the type of aggregation of myofibrillar proteins in cod (*Gadus morhua*) fillets. *Journal of Agricultural and Food Chemistry* 46:1539–1546, 1998.
122. MT García-Arias, E Alvarez-Pontes, MC García-Fernández, FJ Sánchez-Muniz. Freezing/defrosting/frying of sardine fillets; influence of slow and quick defrosting on protein quality. *Journal of the Science of Food and Agriculture* 83:602–608, 2003.
123. D Chevalier, A Sequeira-Munoz, AL Bail, BK Simpson, M Ghoul. Effect of pressure shift freezing, air-blast freezing and storage on some biochemical and physical properties of turbot (*Scophthalmus maximus*). *Lebensmittel-Wissenschaft und-Technologie* 33:570–577, 2000.
124. A Huidobro, M Tejada. Foaming capacity of fish minces during frozen storage. *Journal of the Science of Food and Agriculture* 60:263–270, 1992.
125. IE Bechmann. Comparison of the formaldehyde content found in boiled and raw mince of frozen saithe using different analytical methods. *Lebensmittel-Wissenschaft und-Technologie* 31:449–453, 1998.

126. V Simandjuntak, DM Barrett, RE Wrolstad. Cultivar and frozen storage effects on muskmelon (*Cucumis melo*) colour, texture and cell wall polysaccharide composition. *Journal of the Science of Food and Agriculture* 71:291–296, 1996.
127. S Taira, M Ono, M Otsuki. Effects of freezing rate on astringency reduction in persimmon during and after thawing. *Postharvest Biology and Technology* 14:317–324, 1998.
128. CF Forney, MA Jordan, KUKG Nicholas, JR DeEll. Volatile emissions and chlorophyll fluorescence as indicators of freezing injury in apple fruit. *HortScience* 35:1283–1287, 2000.
129. U Kidmose, K Kaack. Changes in texture and nutritional quality of green asparagus spears (*Asparagus officinalis* L.) during microwave blanching and cryogenic freezing. *Acta Agriculturae Scandinavica Section B, Soil and Plant Science* 49:110–116, 1999.
130. J-Y Liao, A Shau-Mei Ou. Studies on the colour retention of frozen stir-fried pea pods. *Taiwanese Journal of Agricultural Chemistry and Food Science* 38:248–254, 2000.
131. EM Sheehan, TP O'Connor, PJA Sheehy, DJ Buckley, R FitzGerald. Stability of astaxanthin and canthaxanthin in raw and smoked Atlantic salmon (*Salmo salar*) during frozen storage. *Food Chemistry* 63:313–317, 1998.
132. C Jensen, E Birk, A Jokumsen, LH Skibsted, G Bertelsen. Effect of dietary levels of fat, α-tocopherol and astaxanthin on colour and lipid oxidation during storage of frozen rainbow trout (*Oncorhynchus mykiss*) and during chill storage of smoked trout. *Zeitschrift fur Lebensmittel-Untersuchung und-Forschung* A 207:189–196, 1998.
133. M Ben Abdallah, JA Marchello, HA Ahmad. Effect of freezing and microbial growth on myoglobin derivatives of beef. *Journal of Agricultural and Food Chemistry* 47:4093–4099, 1999.
134. GK Sprouls, MS Brewer. Tocopherol effects on frozen ground pork color. *Journal of Food Quality* 20:1–15, 1997.
135. C-J Chow, Y Ochiai, S Watabe. Effect of frozen temperature on autoxidation and aggregation of bluefin tuna myoglobin in solution. *Journal of Food Biochemistry* 28, 123–134, 2004.
136. A Mikkelson, SB Engelsen, HCB Hansen, O Larsen, L Skibsted. Calcium carbonate crystallization in the α-chitin matrix of the shell of pink shrimp, *Pandalus borealis*, during frozen storage. *Journal of Crystal Growth* 177:125–134, 1997.
137. Commission of the European Communities. Commission Directive 93/8/EEC, amending Council Directive 82/711/EEC, laying down the basic rules for testing migration of constituents of plastics materials and articles intended to come into contact with foodstuffs. *Official Journal of the European Communities* L90:22–25, 1993.
138. SM Johns, SM Jickells, WA Read, L Castle. Studies on functional barriers to migration. 3. Migration of benzophenone and model ink components from carbonboard to food during frozen storage and microwave heating. *Packaging Technology and Science* 13:99–105, 2000.
139. PD Ribotta, AE León, MC Añón. Effects of yeast freezing in frozen dough. *Cereal Chemistry* 80:454–458, 2003.

26 Sensory Analysis of Frozen Foods

Edgar Chambers IV, Sherry McGraw, and Kathleen Smiley
Sensory Analysis Center, Kansas State University, Manhattan, Kansas, USA

CONTENTS

Sensory analysis uses human perception for the measurement of food characteristics and the effects of those attributes on food acceptance. As almost all food is intended to be eaten by people, sensory studies are critical to understanding the characteristics of food and food acceptance. The United States Department of Agriculture (USDA) stated that "quantifications of sensory attributes of consumer goods are the basic elements needed for processing and delivering the value-added quality attributes — odor, flavor, texture — to consumers" [1]. Peryam [2] stated that a "pressing need" exists for sensory information on products. As early as the 1930s, researchers determined that humans were important for evaluating such practical problems as the oxidized flavor in stored milk [3]. Galvin and Waldrop [4] stated that information on sensory properties is essential to make the new food possibilities of the future a reality at the table.

Research surveys continue to suggest that "taste" or sensory quality is the single most important criterion people use for food selection. Clearly, the way a product looks, smells, tastes, and feels in the mouth has a tremendous impact on which foods are consumed. The food industry has considerable capability for modifying and storing food products to increase nutritional quality, reduce costs, and provide other benefits to consumers, but sensory quality may be compromised in some of these situations.

The ultimate aim in much sensory testing is to evaluate consumer expectations of given products and whether specific products have met those expectations. Sensory testing also can point the way to product improvements that can result in greater consumer satisfaction with a product. When ingredient substitutions or changes in manufacturing process have to be made, sensory testing can determine whether these changes have had a perceptible impact on product quality or acceptability. Evaluations of the products available in a category in the marketplace use sensory methods to determine whether opportunities exist for introduction of a new product, as well as how that new product might fare with respect to consumer acceptance, compared with existing products.

When considering the objectives of maintaining product quality and consumer acceptability about various product sensory attributes (e.g., appearance, flavor, aroma, texture) of frozen food items, several things must be considered. These include:

- The type of product (e.g., fruit or vegetable variety/cultivar/degree of ripeness, type of meat, previous handling/processing, such as cooking, blanching, glazing, etc.)
- The initial quality or condition of the item prior to freezing
- The type of packaging
- The temperature and other conditions under which the item was initially frozen
- The temperature at which the item was maintained in transport, storage, and by the consumer after purchase
- The length of storage when frozen (including any temperature fluctuations occurring during storage at the retailer or in the consumer's home freezer)
- Practical shelf-life of the product under specified conditions.

Adverse changes in any of these factors can have a considerable impact on the sensory qualities of the finished product, and thus an equally important effect on acceptability to consumers and subsequent product success in the marketplace. For this reason, sensory monitoring of frozen foods is essential at virtually all stages of production, storage, and preparation, to assess and correct any problems that may arise. Modern consumers are especially demanding of value, quality, and convenience in food items, and competition is such that few appreciable deviations from acceptable product quality may be tolerated without brand switching or loss of market share. A primary objective of sensory analysis is to detect such adverse changes, or the likelihood of such changes occurring, at an early enough stage that the problem can be corrected before market share is impacted. Properly targeted sensory analysis may help in identifying the origin of specific problems, which can facilitate their resolution.

As a simple example, ice cream can be packaged in a number of ways — each of which has advantages and disadvantages for the manufacturer (cost, availability of the packaging material, required alterations in manufacturing process), the consumer (appearance of the product, freshness after purchase), and the degree to which product freshness, taste, and acceptability can be maintained over time. However, regardless of the package, if the ice cream is not properly stabilized so that conditions in the freezer at the retail outlet allow the ice cream to partially melt and refreeze (freeze–thaw cycles) before the consumer purchases it, the product may be unacceptable regardless of the type of packaging or the care taken in manufacturing [5]. Proper sensory testing early in the development cycle would have noted that problem and steps could have been taken to overcome it.

Industrial research on the effects of variations in each parameter for maintaining product quality is ongoing and extensive, primarily as a result of the magnitude of the potential economic impact. Much progress has been made in identifying specific problem areas and devising practical solutions. As the time consumers can or are willing to devote to meal preparation shrinks, convenience and consistent quality (both sensory and nutritional) become relatively more important as selling points for food products, which lead consumers to purchase frozen foods.

The increasing level of health consciousness among consumers also mitigates in favor of increased consumption of high-quality frozen food items, as freezing often maintains both the nutritional and sensory qualities of many foods better than any other method of preservation. In addition, microbiological safety is less of an issue with frozen foods than with fresh or canned items. Another advantage is that entire meals, consisting of multiple types of food items, can be frozen and subsequently prepared as one item, greatly decreasing preparation time and increasing convenience. Prolonged possible storage times for frozen foods — up to several years for many products [5] — means that, after purchase, the product can be used at the consumer's convenience, and the time frame for usage is not limited by the rapid deterioration that occurs with fresh meat or produce.

To fulfill these attractive possibilities for frozen foods, however, product quality must be maintained rigorously, and proper sensory evaluation is essential to this end. Sensory tests applicable to frozen foods include those involving product, processing, and packaging effects (PPP); time and temperature tolerance (TTT) effects; and assessment of practical shelf-life (PSL) for each product (also termed the use-by date) [6]. The first two (TTT and PPP) relate to product marketability, whereas shelf-life determinations (PSL) have a more direct impact on the retailer, as a too-short shelf-life means the product must be moved rapidly, or must be discarded if the indicated shelf-life is exceeded. Modern consumers are well aware of the issue of shelf-life for various products, and often will choose the food item with the longest remaining shelf-life, lessening the likelihood of sale for products approaching the end of their manufacturer-indicated shelf-life, even though actual product shelf-life may be much longer than the use-by date indicated on the package.

Consumer behavior (e.g., purchasing, handling/storage, preparation, consumption) relative to frozen food items must also be taken into account when devising appropriate sensory analysis methods, for the results to be useful to manufacturers, producers, and retailers. Thus, sensory analysis considers factors related to practical shelf-life, product appearance when purchased as well as after preparation by the consumer (with all the variability that may ensue), and product taste, aroma, and texture after preparation, compared with the same variables for other options (e.g., fresh, canned, dehydrated).

I. FACTORS INFLUENCING THE QUALITY OF FROZEN FOODS

Regardless of the type of product (fruit, meat, fish, vegetable, or other prepared foods), the initial quality of the raw materials figures prominently in the quality of the finished product [5]. Determining the initial sensory qualities of each product is essential, because, with few exceptions, frozen storage will not improve the quality of food. Thus, an evaluation of the sensory characteristics of the fresh product, either before freezing (e.g., meat and vegetables) or before frozen storage begins (e.g., ice cream and frozen dough) is important for understanding the maximum sensory quality that might be expected in a given product.

A. Freezing Method and Rate

The method by which a particular product is frozen also has an important impact on overall sensory quality [7,8]. The freezing method must be appropriate for the particular product, must consistently yield a high-quality product, and must be economically feasible, as well, with respect to the price that can be commanded for the finished product and the energy costs inherent in maintaining the product under specified optimal conditions.

When a product is frozen, the size of the resulting ice crystals has a significant effect on the overall integrity of the product, particularly the texture. At one time, conventional wisdom held that the rate of freezing was the main factor determining the size of ice crystals; more rapid freezing rates and lower storage temperatures were thought to result in formation of smaller ice crystals, which led to better product quality [9]. Subsequent research has shown, however, that, in some cases, more rapid freezing and very low storage temperatures produce a more rapid decline in product quality than slower freezing rates and somewhat higher storage temperatures. A careful balance must be struck between ice crystal formation and the possibility of reactivation of microbes or enzymes in the product that may cause unacceptable deterioration over time at higher storage temperatures [10].

The size of ice crystals in a frozen product can change during storage, as can the distribution of solutes and proteins that results when available water in the product is converted to ice during freezing. In addition, compounds not normally closely associated with one another may be brought into apposition as the remaining amount of fluid water shrinks during freezing. This may lead to chemical or physical reactions that adversely affect sensory attributes of the food item. Crystal size may grow during frozen storage, and water may migrate to the outer surface of the product and refreeze, causing dehydration and freezer burn. These processes can alter the appearance and texture of the product, particularly meat products, in a negative manner [11]. As consumers often purchase meat products primarily on appearance, such changes can be very detrimental to consumer acceptability [12,13].

Freezing rate becomes important when considering products having large volumes, such as entire sides of meat or a large pallet of closely packed product. Even when "quick-freezing" is employed, several hours may be required for the innermost areas of the product to reach the ambient freezer temperature for bulky items or large packages of product, compared with the rapid rate at which the outer areas of the item reach this desired temperature. During that time, changes in sensory quality may occur in the interior of the food item, which remains at slightly higher temperatures for a longer period of time [9].

B. Storage Conditions

Storage conditions (e.g., total storage time, stacking arrangement of product containers within the freezer unit, type of freezer unit, base storage temperature, temperature fluctuations/freeze–thaw cycles, permeability and integrity of packaging materials, packaging atmosphere, light exposure/retail display conditions) also are critical for maintaining frozen food quality [5,9,14–16]. Manufacturing plants may strive to control these factors precisely at all points in the process, but transport or retail facilities may fall far short of optimal conditions for a given product. Conditions of storage once the product is in the hands of consumers are highly variable and cannot be controlled to any appreciable extent. Thus, sensory testing using realistic fluctuations (sometimes including abuse) in storage conditions become critical.

Potential interactions between packaging materials used for storage and the food product contained in that package are the subject of many sensory studies. Permeation of flavors from the package to the product itself, resulting in off-flavors, or from the product into or out of the package, resulting in quality degradation of the product, are concerns to most frozen food manufacturers [9].

Depending on the interactions among the factors detailed above, specific molecular species (lipids, proteins, carbohydrates) in the product may oxidize (causing rancidity or off-flavors) or be subject to enzymatic degradation or structural alterations that affect taste or textural properties of the product in a manner that is objectionable to the consumer. The longer the storage time, the greater the likelihood of these types of reactions. This eventuality forms the basis for the concept of shelf-life for frozen foods. Contrary to previous assumptions, very low temperatures do not always ensure better product quality for some products, as deleterious chemical reactions may still proceed at very low temperatures over prolonged storage times, particularly when oxygen or catalysts such as metal ions are present. Lipids are especially prone to this kind of degradation over time and may, indeed, undergo autoxidation over time, as well. Alterations such as these are responsible for many off-flavors in foods [5,9].

C. Temperature Abuse

Temperature abuse is a term that describes conditions of fluctuating temperature such that there are periods during which the product may be subjected to temperatures higher than the range at which the product was designed to be kept optimally [9,17,18]. Partial thawing may occur, resulting in some loss of sensory quality and shelf-life with each episode of increased temperature. In general, loss of sensory quality will occur more rapidly than safety issues will arise in those cases. Unfavorable conditions can occur during transport of frozen foods in refrigerated trucks or in cases of freezer malfunction at retail stores. Such conditions also apply to frozen foods kept in commercial display cases, where frequent opening and closing of the display freezer case by shoppers causes temperature fluctuations that can adversely affect product quality over a period of time. Conditions qualifying as temperature abuse of frozen items may also occur in the consumer's home freezer or during transport by the consumer from the retail store to the consumer's home. In an ideal situation, such conditions would be minimized, but in the real-world this is generally not possible. Therefore, products should be tested under these conditions to ascertain how much sensory quality might be lost in these circumstances. Sensory studies, especially those that look at the abuse that can result from freeze–thaw cycles, are commonly conducted on many frozen food items. Sensory aspects that are studied can range from visual aspects, such as ice crystal formation and the shifting of ingredients (e.g., the shift of toppings on frozen pizza) to textural or flavor degradation over time. Because these variations are known to occur, developers sometimes need to consider formulations that may not be optimal, but that may have better long-term storage abuse stability.

II. SHELF-LIFE

A product's shelf-life, for the purposes of sensory analysis, is that length of time for which the product (in a specific type of packaging and under specified conditions of storage) will remain the same as the "fresh product" or will be acceptable to consumers [5,16,19]. Sensory shelf-life studies should be conducted under varied sets of conditions to gain a more complete understanding of the possible behavior of the product in retail stores, as well as any problems that may arise as a consequence of temperature variations or packaging issues. The manufacturer determines the end-point for shelf-life according to the company's philosophy (i.e., whether the product is to be offered for sale until consumers can detect an unacceptable change in sensory quality or to the point that some sensory characteristics are different in the product). In some countries, however, shelf-life criteria are determined by regulatory authorities and retailers are forbidden by law to sell product that has passed that date.

Depending on the criteria used for evaluation of each product, practical shelf-life can vary widely. Some products kept at very low temperatures in high-quality, oxygen-impermeable packaging may keep effectively for a number of years without appreciable loss of sensory quality [5]. Shelf-life is highly dependent on storage conditions and any temperature abuse that may have occurred during transport or retail storage. For some products, an algorithm can be constructed that will enable the manufacturer to estimate product shelf-life at a given storage temperature. In general, the higher the storage temperature, the shorter the shelf-life for frozen foods and the greater the likelihood of unacceptable sensory changes over time. Product subjected to intermittent periods of storage at different temperatures loses a certain amount of shelf-life for each interval spent at each temperature, regardless of the sequence in which the product experienced each temperature. For example, a product kept at −34.4°C (−30°F) for 2 weeks and then at −23.3°C (−10°F) for another 2 weeks will have the same remaining shelf-life as an identical product kept at −23.3°C (−10°F) for 2 weeks and then at −34.4°C (−30°F) for a further 2 weeks. Sensory changes usually are cumulative, regardless of sequence [5].

Shelf-life testing can be done, for example, by having several freezers set up at different temperatures (−28.9, −17.8, and −6.7°C; or −20, 0, and 20°F). This range of storage temperatures will provide a fuller understanding of product behavior. Measurement or comparison of these products using either descriptive sensory testing or consumer acceptance can help the manufacturer better understand the effect of temperature on the shelf-life of a particular product.

Because sensory shelf-life testing is a lengthy and involved process, it is important to have a general understanding of how ingredients or processes might affect storage. For example, if companies wish to change suppliers or ingredients for cost savings, by the time actual shelf-life testing of the products using the new ingredients has taken place, other aspects of the product or production may have changed as well. Of primary concern is that researchers remember that new ingredients and processes are not cost-saving if the product does not have an adequate shelf-life during storage and consumer acceptability is decreased.

Abuse testing, accelerated shelf-life testing, or freeze–thaw testing can be used to speed-up the process [20], although results do not correlate specifically with shelf-life. However, if the protocol is properly designed and the current product and the new product are treated in the same way, recommendations can be made about whether the new product appears to perform in the same way as the current product.

III. PRODUCT-SPECIFIC ISSUES

A. Fruits and Vegetables

Fruits and vegetables are inherently much less robust to freezing than are meats, because freezing-induced changes to structural molecules in fruits and vegetables can result in unacceptable and irreversible changes in major sensory attributes such as texture and appearance that severely limit consumer acceptance [5,21]. Color change during frozen storage, as a result of degradation of pigment molecules over time, is particularly important because consumers often will reject foods that do not look like they were packed fresh. Growers have managed to produce certain varieties/cultivars of fruits and vegetables that are better able to resist such changes and maintain their texture, appearance, and color during freezing. Manipulation of cultivar choice can also obviate problems that occur as a result of variations in stage of ripening within a single crop (causing color and textural variations in the final product), damage caused by harvesting equipment or postharvest handling or shipping, or adverse changes in appearance once the item is in the store (reducing consumer acceptance and the likelihood of purchase) [22,23]. Of course, the initial sensory quality of the cultivar must be assessed to ensure that it has appropriate sensory characteristics and good consumer acceptance to begin with. Packaging methods (e.g., freezing fruit in highly concentrated sugar solutions) can also alleviate problems with appearance, to some extent, but this introduces another sensory variable (i.e., sweetness) that must be considered, in relation to the potential end use of the frozen product. For vegetables, blanching the raw product can prolong shelf-life by inactivating enzymes that produce degradation over time. Blanching must be carefully controlled, however, to ensure that the blanching process does not have a major negative impact on initial sensory quality [24].

B. Meat, Poultry, and Fish

The ability of meat, poultry, and fish to retain their pleasing sensory qualities when frozen or during prolonged frozen storage can be affected by procedures undertaken prior to slaughter (cattle, swine, poultry) or immediately after fish are caught and brought onboard the ship or harvested from commercial fish farms. The type of feed used for cattle, swine, poultry, and fish impacts sensory qualities of the meat, as do the specific age, sex, and breed of animal used for production [9,25]. Pre- and postslaughter handling (e.g., conditioning of meat, glazing of seafood), freezing, and storage conditions for meat, poultry, and fish can significantly affect the finished product for good or ill. Much fish that is marketed frozen already has been kept for some period of time on ice on the fishing boat before arriving at the processing facility. Initial flavor of those products can be suspect and must be evaluated to determine whether they are appropriate raw material for freezing [5].

Although conventional wisdom long maintained that meat or poultry products should not be thawed and refrozen before sale or consumer use, current research using sensory analysis has shown that thawing and refreezing, some products up to five times, do not adversely affect sensory qualities evaluated after cooking, as long as microbiological safety is maintained and the product is not kept at too high a temperature for prolonged periods of time before refreezing [5,9]. Whether the product is raw or has been previously cooked is also an important variable. Cooked meat that has been frozen has a greater tendency to develop unacceptable off-flavors (i.e., warmed-over flavor, secondary to lipid oxidation) when reheated than does previously frozen raw meat [5,11]. Manufacturers who produce frozen dinners routinely use previously frozen meat products to prepare these items, and previously frozen meat is also offered for sale after thawing with little perceptible impact on sensory qualities after cooking [5].

C. Breads and Pastries

The market for both frozen dough and frozen bakery products has been increasing as the methodology for assuring that sensory quality can be preserved after freezing has improved. Yeast varieties that are resistant to freezing are available for use in raw dough, and microwave reheating can preserve the sensory qualities of baked products [26]. These facts have facilitated sale of frozen bread and pastry products.

With prebaked goods that are frozen, the thawed/reheated product must retain as much as possible the sensory quality of product that has never been frozen. Thus, ingredients, such as fillings and frostings, must be carefully chosen for their stability with respect to appearance, taste, and consistency when frozen and subsequently thawed [19].

The primary sensory issues with frozen dough are based on the subsequent performance of the dough once it is thawed and baked. Frozen doughs that are intended as bake-to-rise in the oven have become increasingly popular with consumers. Such products range from pizza crust to cinnamon rolls to cookies. The sensory qualities expected depend on the product, but often include: light, fluffy, crisp, yeasty, and browned (appearance and flavor). Some of these characteristics may be difficult to achieve in bake-to-rise doughs, especially yeast-raised products using only a single leavening system, but these qualities are essential to consumer acceptance and ultimate product success.

D. Ice Cream

Initial ingredient quality, pasteurization procedures, blending, flavor, freezing rate, homogenization, and packaging are especially important issues for ice cream. Several markets exist, including an ultrapremium market for high-fat, high-quality, and usually rather expensive ice creams; the average quality products that other consumers may buy on a routine basis; and the low-end products that value-conscious shoppers may choose. Each type of consumer has different expectations for the product they are buying. High-end products will be expected to have smaller ice crystals, better flavor, higher-quality ingredients, better mouthfeel, and more protective packaging materials than low-end products, so that the higher price can be justified to consumers. Ice cream is not an essential food item and consumers may be willing to pay a premium price for such a luxury product if it lives up to their expectations based on price. Quality control is, thus, a major factor in ice cream production, as are storage temperatures and packaging integrity, as these two factors can make or break perceived product quality in the retail environment. Products are differentiated in the marketplace on the basis of taste and perceived/expected quality; if these standards are not met, market share can plummet quickly as a result of the high level of competition among available products [5].

E. Frozen Prepared Dinners

Frozen prepared dinners involve somewhat different issues than frozen fruits, vegetables, or meat products. Because these preparations usually involve more than one type of food item in the same package, and each item may have differing freezing requirements or susceptibility to sensory changes during freezing or storage, guaranteeing the sensory quality of these products is more complicated. Issues include those involving the texture or possible separation of sauces during freezing; the alteration of sensory properties of meats in sauce; and the acceptability with respect to appearance, texture, and flavor of complex dishes after freezing and subsequent preparation by the consumer [27–29].

In addition, with frozen meals that use a combination of single items (e.g., a meal with meat loaf, green peas, mashed potatoes, and cake), issues such as consumer-perceived appropriateness are important. The types of foods that are paired together are important to many consumers [27,28,31]. Foods such as meat loaf and waffles may both be acceptable products, but when paired they could produce low overall acceptance even though, from a nutritional standpoint, this combination might be complimentary.

IV. METHODS OF SENSORY ANALYSIS

When initiating sensory analysis of a food item, several questions must be answered before embarking on these procedures. Answers to these questions, primary among them is "what information is needed?" will help to choose the proper testing method; the proper type and number of panelists required; appropriate methods of sample preparation, presentation, and coding; appropriate methods of analysis; as well as focused, relevant, and appropriate interpretation of data acquired in these tests. Many specific and well-validated methods are available to assess sensory quality in food items, and entire books on sensory methods exist [31–34]. Some are more applicable than others to the unique problems encountered with frozen foods.

A. Specific Types of Sensory Tests

Evaluations of sensory quality fall into three main categories: descriptive, discriminative, and affective.

- Descriptive tests are used to categorize the sensory characteristics of a sample or to compare the occurrence and intensity of sensory attributes among product samples. These types of tests require panelists who are highly trained to describe and measure detailed sensory properties, so that even minor flavor or aroma attributes can be recognized and evaluated. In these tests, a set of reference standards (various attributes and various intensities of each attribute) often are used for comparison with the attributes found in the test sample, and the presence and intensity of each attribute is assessed by each panelist or by a consensus reached among panelists.

- Discriminative tests are used to determine whether two or more products differ in any way in terms of their sensory attributes. Panelists are screened to make sure they can discriminate among products and are oriented or trained in particular methods. This is done to be sure that panelists are able to detect differences in samples reliably and reproducibly. Sample preparation and order of presentation are important issues in these types of tests, and rigorous protocols must be followed to ensure that data are accurate, unbiased, and useful.
- Affective tests (hedonic or preference tests) are used to evaluate liking or preference for products. Panelists used for these tests should be untrained, although they generally should have prior experience with the type of product being tested. Products with different amounts of an ingredient of different ages may be tested and the panelists asked to decide which product they like the most or which they prefer.

B. Types of Questions to be Answered

The objectives of a particular sensory evaluation must be determined previously, because that objective helps to determine the types of data that are needed and thus the specific test choices that can be made (Table 26.1). Possible objectives include:

- Determining the key sensory qualities and their intensities for one product under a specified or differing set of conditions requires descriptive sensory analysis. For example, the following parameters may be evaluated in this way:
  - Prefreezing handling or other raw-material variables
  - Freezing rate
  - Type of freezing apparatus or method
  - Storage temperature
  - Storage time
  - Packaging
  - Freeze–thaw cycles
  - Consumer preparation issues
- A comparison of several products to determine the particular sensory properties that are similar or different among the products requires descriptive sensory analysis.

TABLE 26.1
Types of Sensory Tests, Objectives, and Type of Panelists Required

Class	Question	Type	Participants
Affective	How well are products liked or which products are preferred?	Hedonic	Screened for product use, untrained
Discriminative	Are products different in any way?	Analytic	Screened for sensory acuity, oriented to test method, sometimes trained
Descriptive	How do products differ in specific sensory characteristics?	Analytic	Screened for sensory acuity and motivation, trained or highly trained

- Determining whether two products are significantly different from each other overall (requires difference tests) or for a specific attribute requires either difference or descriptive analysis.
- Determining the degree of difference between products to help determine whether small differences or large differences are noted overall or for individual attributes requires degree of difference tests or descriptive analysis.
- Determining which of a particular group of products is liked the most requires hedonic tests. For example:
  - How much do you like this product (asked for each of several products packaged in different ways and stored for 2 months).
  - Which do you prefer (asked for a pair of products in which different stabilizers have been used to increase shelf-life).

C. Type of Panelist Needed

Panels consisting of either ordinary consumers of the product or panelists trained in sensory analytical methods can be used, depending upon the objective of the evaluation and the expected subtlety of the discriminative abilities required. Some idea of the extent of difference that is to be documented generally needs to be known in advance for proper panelist selection.

Economic considerations may influence the type of panelists who are used for commercial sensory analysis. Highly trained panelists will necessarily be more expensive to train and to use than consumer panels in the short term. However, the number of highly trained individuals required will be fewer than the number of consumers needed to provide usable data, because of the decreased variability in results from trained panelists. However, consumers will not have the extensive experience and vocabulary needed in some types of sensory analyses, and thus one cannot avoid using trained panelists for such testing.

A key question in deciding what types of tests and thus what type of panelists are needed is the question of what results are meaningful to the company or researcher. If the company knows that its product (e.g., frozen vegetables) will change over time because of natural variation or other uncontrollable issues and is only interested in making sure that the product is still acceptable to consumers, then the use of affective consumer testing might be the most appropriate procedure. If the company is concerned that some changes might be acceptable, but other changes in attributes (such as icy texture, rancidity, or off-flavors) should not occur (e.g., a frozen meat product that might vary somewhat in overall flavor intensity, but should not become rancid), then descriptive analysis to examine key attributes may be most appropriate. If the company makes a homogeneous product that should look, taste, and feel the same regardless of the storage time (e.g., cake), then difference testing for overall differences could be the most appropriate test to use.

D. Samples to be Used for Evaluation

Samples to be used in sensory analysis must be carefully chosen and prepared to ensure reproducibility of results and the relevance of the information gathered to the objective of the analysis. Conditions of preparation, handling, and presentation to panelists should be carefully documented to enable results to be reproduced in subsequent tests, if necessary. In addition, samples to be used as references or controls must be chosen with an eye to the primary objective of the analysis and subsequent statistical analyses and data interpretation. In all cases, where it is possible, retention samples should be kept from each test to go back to in case of surprising or unusual results.

E. Elimination of Bias

Bias usually is inherent in all situations where judgments are made, and this is no less true in the area of sensory analysis. All potential biases should be accounted for and, if possible, eliminated or minimized so that true differences in sensory properties can be appreciated when data are evaluated. The presence of a high degree of bias will effectively invalidate the results and the usefulness of the data acquired under those conditions. Randomization of sample presentation and assuring that all samples are presented to each panelist in all possible (or different) sequences are standard methods to alleviate possible bias during sensory evaluation. Individual biases may be detected by interviews and questionnaires given to panelists prior to sensory testing. If biases in consumer tests are expected from age, gender, or socioeconomic stratum, those can be minimized by careful selection of consumers to ensure a random distribution of various demographic or socioeconomic characteristics in the group.

F. Statistical Methods

Sensory information gained from descriptive analyses or comparative trials using difference tests or affective testing methods can be evaluated in a number of ways [35,36], both statistically and graphically. Standard histograms can be used to display the intensities of various sensory attributes of one product or to compare the intensities in several samples. Line graphs can be used to follow changes in a sensory attribute over time. Computer programs are available to produce spatial plots of sensory attributes in two dimensions using principal component analyses (PCA), so that interrelationships among various attributes can be examined visually. Factor analysis is another graphic method that correlates attributes or groups of attributes, to identify which factors change in similar ways for different products/samples, based on similar sensory attributes. Cluster analysis [37] looks at samples/products that have similarities sufficient to show that they are related in terms of sensory variables and distinct from other clusters of products/samples having different sensory spectra. For example, Lotong et al. [38] showed that frozen concentrated orange juices, regardless of brand, generally were more similar to each other than were orange juices that had not been previously frozen.

Statistical analyses appropriate for the kinds of data acquired from sensory analysis include parametric analysis such as analysis of variance (where data are assumed to be normally distributed), nonparametric statistics (in the case of a binomial or an unknown type of distribution), and univariate or multivariate analyses (depending on whether one variable is considered alone or many variables are considered together), and one-, two-, or multiway analyses of variance, with or without an interaction term. Paired comparison tests (e.g., binomial distribution) also are useful when evaluating preference data. Linear regression analysis and correlation coefficients can be helpful with sensory data as well [39].

G. Tests for Shelf-Life

Because the shelf-life of a product can, in some cases, be measured in years, the testing procedures can take quite a bit of time, particularly because they must be done for every new or changed product manufactured [40]. Traditional triangle difference tests are a good place to begin. A triangle test is done by presenting three samples to a tester who understands how to use the triangle test ballot (20–30 testers can be sufficient, although more often are needed). Each product sample is labeled with a three-digit code number (i.e., the tester is blind as to product identity). Two of the three samples are the same, but one is different. The tester is asked to choose which of the samples is different. The number of evaluations resulting in a correct answer is counted and a determination is made as to whether there is an actual difference among the samples. Specific differences (e.g., sweetness and crispness) are not identified through triangle testing; only the fact that a difference has been found is relevant here.

If it is important to understand what the difference is or how the product is changing with increasing storage time, descriptive analysis often is used to provide that information. Sensory attributes are defined by a group (5–10 participants) of highly trained panelists. Panelists evaluate products for their attributes (e.g., sweetness and grittiness) and measure the intensity of each attribute in each product tested. Scores are compared among products at each time point and from one time point to another. When changes in attribute intensities are discovered, companies may decide that shelf-life has been reached, or may use consumer testing to determine whether the product has reached the end of its shelf-life.

When a difference is found among products or samples of the same product, either through triangle testing or descriptive testing, the acceptability of the product at that point might then be determined. This can be done in several ways. In some companies a "quality panel" might judge products as being excellent to poor on some scale. A problem with this type of testing is that internal testers often have a different perception of quality than untrained consumers. A better way to determine acceptability is through the use of consumer testing. Consumers who consume the type of product being testing are recruited. Recommendations for numbers vary, but approximately 100 usually are sufficient. Consumers are asked to rate the overall product quality of each of the samples. Often the scale is a 9-point hedonic scale ranging from like extremely to dislike extremely, although for some products (such as spinach) that may never be liked extremely, the scale might go from high to poor quality. Of course, conducting consumer testing on every product as it progresses through shelf-life testing could be very costly, based on product needs, storage space needed, time, and so on, and so attempts to correlate consumer and descriptive data usually are done so that descriptive sensory tests can be used in future studies either in place of consumer tests or to determine when final consumer tests might be needed.

H. Examples

Examples of some basic results of frozen food sensory testing, including descriptive testing, consumer acceptability tests, and shelf-life determinations, are given to help in understanding these basic applications. These studies, all using the same product (biscuits), are simple simulations of different objectives that a single company might have and ways in which they could be approached. The simulations are intentionally simple and are not intended to convey the complexity of the data that could be collected in some studies.

1. Example 1 — Product Development of Frozen Raw Dough Biscuit

The objective for this test was a comparison of the characteristics of a newly developed frozen raw dough biscuit product with that of several other types of biscuits. The company would like to know how the frozen dough biscuit is similar to or different from competitive products in terms of several key visual, texture, and flavor attributes. The product is not intended to match a particular product, but the company hopes that it is more similar to fresh or prebaked biscuits than to canned biscuits. Depending on the results, the company will determine whether changes are needed to the formula.

Figure 26.1, which provides descriptive data for biscuits, represents some key sensory properties for each of the four biscuit types. These properties are assessed by trained descriptive sensory panelists. The radar chart shown demonstrates similarities and differences in specific attributes. These attributes are not ones that consumers would necessarily be able to describe, and thus the need for a trained panel. The chart shows clearly that the frozen raw dough biscuit is more similar to the fresh biscuit or the prebaked biscuit. It is quite different (has lower scores) from the canned biscuit for several key attributes such as firmness, denseness, and leavening flavor. On the basis of this information, the company's goal of making a product similar to fresh or prebaked biscuits has been achieved. Profiles such as these are useful to a product developer for determining differences between their product and a competing product. Such information provides

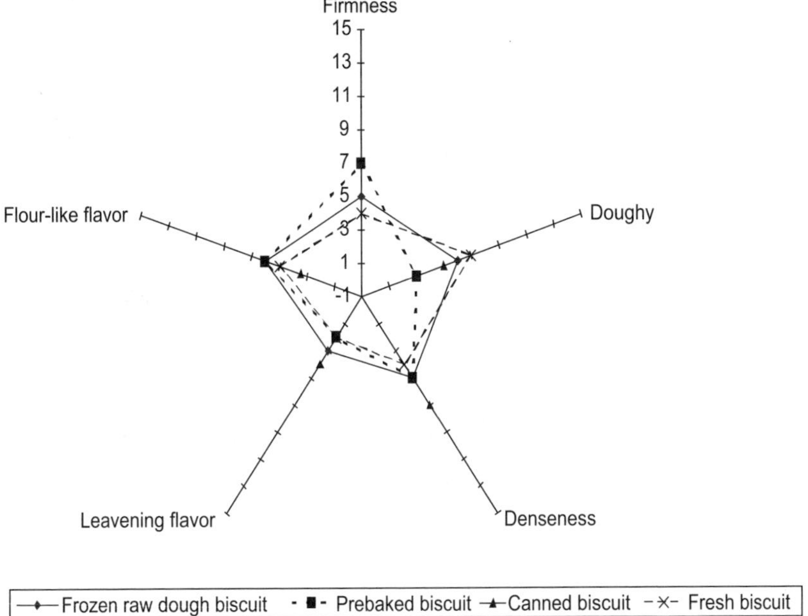

FIGURE 26.1 Graph of descriptive sensory data for several types of biscuits.

guidance as to what ingredients may need to be increased, decreased, or deleted. These data can be paired with consumer data to achieve a greater understanding of what is driving consumer liking of this type of product.

2. Example 2 — Consumer Acceptance of Biscuit Products

The objective for this test was a comparison of the acceptability of a frozen raw dough biscuit product to that of several other types of biscuits. The company would like to know how the frozen dough biscuit fares in a consumer acceptance test. The company is not interested in the specific attributes of the product, but rather is concerned with how much consumers will like the finished product. Depending on the results, the company will determine whether the product should be marketed or needs further development.

Figure 26.2 represents consumer acceptability ratings for the different types of biscuit dough. Consumers were asked to rate each type using a hedonic scale (range: 9 = like extremely to 1 = dislike extremely). The biscuit made from frozen raw dough was liked almost as well as the "fresh" homemade biscuits and more than biscuits that were purchased already baked or purchased and baked from canned biscuit dough. The company now has data to show that the product is liked by consumers and exceeds the liking of several other forms of biscuits available on the market. This information is helpful to the company in deciding to launch the product and as information to potential grocery chains that will need to determine if shelf space should be made available.

3. Example 3 — Shelf-Life of Frozen Biscuit Dough

The objective of this test was to determine how some key sensory attributes associated with storage will change over time. The company would like to know how long the biscuits can be kept in a home self-defrosting freezer set to the recommended temperature. The company is hoping for 6 months of shelf-life, but expects the actual shelf-life to be lower. If shelf-life is too low, a reformulation will be necessary to improve this parameter.

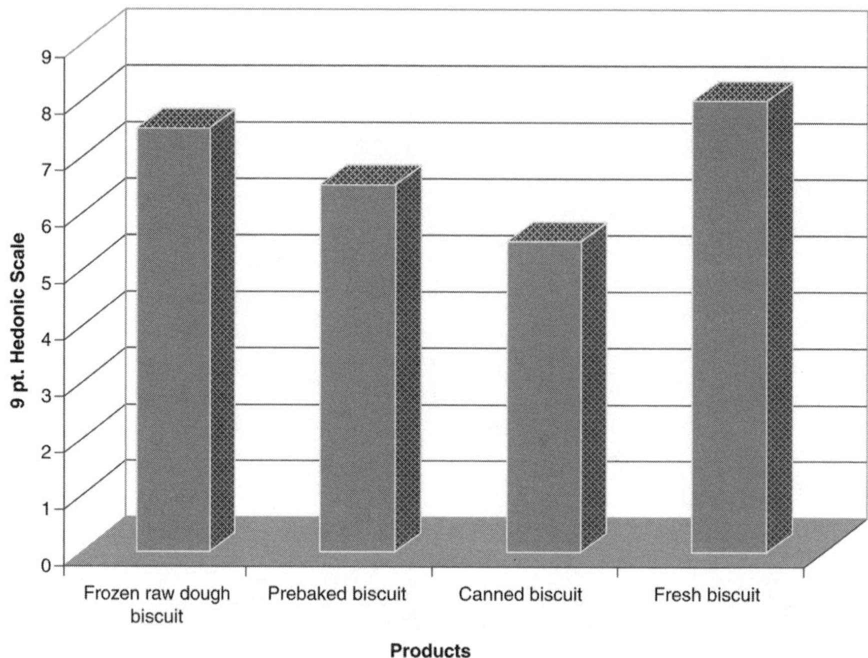

FIGURE 26.2 Graph of consumer acceptability data for several different types of biscuits.

Figure 26.3 represents scores for key sensory attributes over 6 months of time. As shown in the graph, key attributes remain virtually unchanged through 2 months of age, but begin to change by 4 months of age and continue to change through 6 months. Moistness declines and denseness, brownness, and degree of stale/rancidity increase. Only the leavening flavor remains virtually

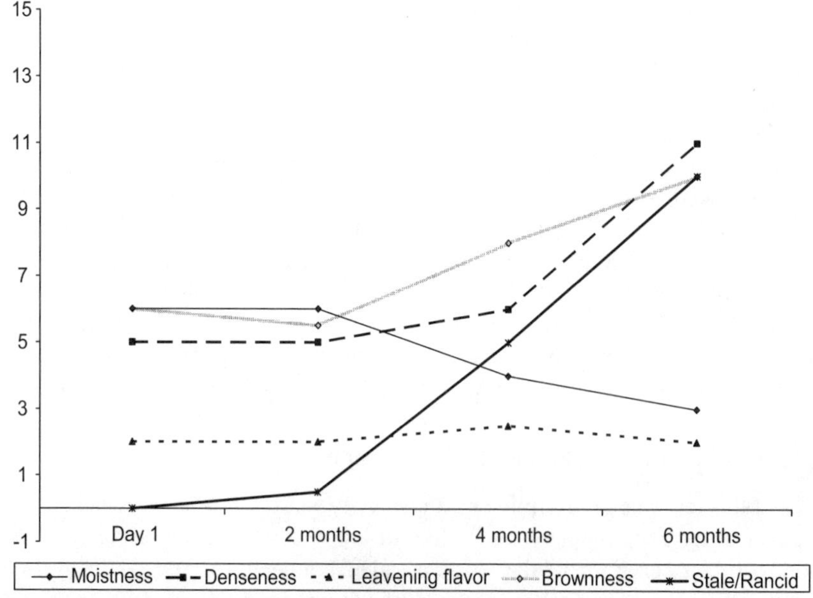

FIGURE 26.3 Graph of shelf-life data for several types of biscuit dough.

unchanged over time. For the company, their hopes of a 6-month shelf-life were dashed, and the reality that the product may last little longer than 2 months at normal home frozen storage means that they have to make a decision: should they accept the short shelf-life and recommend that people use the product relatively quickly or should they revamp the product to improve the shelf-life, but possibly change the initial quality of the product. The key in this case is not a scientific decision, but a business decision that requires the input of product scientists, sensory experts, marketing and sales personnel, and upper management.

V. CONCLUSIONS

Results gained from sensory analysis can provide researchers and management with the information needed to make decisions about a variety of issues, including product quality, production and storage procedures, and shelf-life. Ideally, sensory studies will provide the company, especially R&D and marketing, with information that is consumer-relevant and relates to preferences and purchasing behavior. Such information can be used to focus efforts at product improvement, changes in manufacturing, and product promotion that leads to increased consumer acceptance and market share for the product.

REFERENCES

1. Anonymous. *Research Agenda for the 1990s*. College Station, TX: United States Department of Agriculture and the Texas Agricultural Experiment Station, 1990.
2. DR Peryam. Sensory evaluation — the early days. *Food Technology* 44 (1):86, 1990.
3. GM Trout, PF Sharp. *The Reliability of Flavor Judgement, with Special Reference to the Oxidized Flavor of Milk*. Ithaca, NY: Cornell University Press, 1936.
4. JR Galvin, HL Waldrop, Jr. The future of sensory evaluation in the food industry. *Food Technology* 44 (1):95, 1990.
5. CJ Kennedy, Ed., *Managing Frozen Foods*. Boca Raton, FL: CRC Press, 2000.
6. Anonymous. Expert Panel on Food Safety and Nutrition. Open shelf-life dating of food. A scientific status summary by the Institute of Food Technologists' Expert Panel on Food Safety and Nutrition. *Food Technology* 35 (2):89–96, 1981.
7. M Jul. The intricacies of the freezer chain. *Refrigeration Science and Technology* 1:61–88, 1982.
8. M Syn. Power freezing. *Asia-Pacific Food Industry* 15 (10):32–33, 2003.
9. CP Mallett, ed., *Frozen Food Technology*. London, UK: Blackie Academic and Professional, 1994.
10. CA White, LP Hall, PJ Slade. *The Effect of Wide Fluctuations in Temperature on Food Poisoning Organisms and the Natural Microbial Flora in Frozen Beef and Chicken Substrates*. III. Campden Food Preservation Research Association 298, 1982, 53 pp.
11. E Obuz, ME Dikeman. Effects of cooking beef muscles from frozen or thawed states on cooking traits and palatability. *Meat Science* 65 (3):993–997, 2003.
12. LE Jeremiah. The effects of frozen storage and thawing on the retail acceptability of ham steaks and bacon slices. *Journal of Food Quality* 5 (1):43–58, 1982a.
13. LE Jeremiah. The effects of frozen storage on the retail acceptability of pork loin chops and shoulder roasts. *Journal of Food Quality* 5 (1):73–88, 1982b.
14. HY Gokalp, HW Ockerman, RF Plimpton. Effect of packaging methods on the sensory characteristics of frozen and stored cow beef. *Journal of Food Science* 44 (1):146–150, 1979.
15. TK Gevindan. Packagings for frozen prawn products. I. Corrugated fibre-board master cartons. *Seafood Export Journal* 14 (8):21–24, 1982.
16. BH Chiang, HW Norton, DB Anderson. The effect of hot-processing, seasoning and vacuum packaging on the storage stability of frozen pork patties. *Journal of Food Processing and Preservation* 5 (3):161–168, 1981.
17. RL McBride, KC Richardson. The time–temperature tolerance of frozen foods: sensory methods of assessment. *Journal of Food Technology* 14 (1):57–87, 1979.

18. JER Frijters, SCC Baumer-Stoffer. Comparison of storage time-temperature effects on sensory and hedonic attributes of frozen and deep-frozen chickens. *British Journal of Poultry Science* 19 (2): 225–232, 1978.
19. AB Childers, TJ Kayfus. Determining the shelf life of frozen pizza. *Journal of Food Quality* 5 (1):7–16, 1982.
20. RC Martins, CLM Silva. Computational design of accelerated life testing applied to frozen green beans. *Journal of Food Engineering* 64 (4):455–464, 2004.
21. N Rodrigue, M Guillet, J Fortin, JF Martin. Comparing information obtained from ranking and descriptive tests of four sweet corn products. *Food Quality and Preference* 11 (1/2):47–54, 2000.
22. JF Gallander, RG Hill, Jr. Effect of variety and harvest date on the quality of frozen strawberries. *Ohio Agricultural Research and Development Center* 271:43–48, 1982.
23. WA Sistrunk, RC Wang, JR Morris. Effect of combining mechanically harvested green and ripe puree and sliced fruit, processing methodology and frozen storage on quality of strawberries. *Journal of Food Science* 48 (8):1609–1612, 1983.
24. RH Lane, MD Boschung, M Abdel-Ghany. Sensory comparison of prepared frozen vegetables processed by microwave and conventional methods of blanching. *Journal of Consumer Studies and Home Economics* 8 (1):83–93, 1984.
25. DF Wood, DA Froelich. Sensory evaluation of grain-fed versus milk-fed veal. *Proceedings of the European Meeting of Meat Research Workers 27*, Vol. II (E8):531–533, 1981.
26. EW Davis. Shelf-life studies on frozen doughs. *Baker's Digest* 55 (3):12–13, 1981.
27. R Ahlström, JC Baird, I Jonsson. School children's preferences for food combinations. *Food Quality and Preference* 23 (3):155–166, 1990.
28. M Turner, R Collison. Consumer acceptance of meals and meal components. *Food Quality and Preference* 1 (1):21–24, 1988.
29. Anonymous. The US frozen ready meals market. Mintel International Group Ltd., 2001.
30. D Marshall, R Bell. Meal construction: exploring the relationship between eating occasion and location. *Food Quality and Preference* 14 (1):53–64, 2003.
31. E Chambers IV, MB Wolf, Eds., *Sensory Testing Methods*, 2nd ed. West Conshohocken, PA: American Society for Testing and Materials, 1996.
32. HT Lawless, H Heymann. *Sensory Evaluation of Foods*. New York, NY: Chapman & Hall Publishing, 1998.
33. M Meilgaard, GV Civille, BT Carr. *Sensory Evaluation Techniques*, 3rd Ed. Boca Raton, FL: CRC Press, 1999.
34. H Stone, J Sidel. *Sensory Evaluation Practices*. Burlington, MA: Academic Press and Elsevier, 2004.
35. GB Dijksterhuis. *Multivariate Data Analysis in Sensory and Consumer Science*. Malden, MA: Food and Nutrition Press, Blackwell Publishing, 1997.
36. RC Hootman. *Descriptive Analysis Testing for Sensory Evaluation*. West Conshohocken, PA: American Society for Testing and Materials, 1992.
37. DR Godwin, RE Bargmann, JJ Powers. Use of cluster analysis to evaluate sensory-objective relations of processed green beans. *Journal of Food Science* 43 (4):1229–1234, 1978.
38. V Lotong, E Chambers IV, DH Chambers. Categorization of commercial orange juices based on flavor characteristics. *Journal of Food Science* 68 (2):722–725, 2003.
39. MC Gacula, Jr., J Singh. Statistical methods. In: *Food and Consumer Research*. Malden, MA: Food and Nutrition Press, Blackwell Publishing, 1984.
40. DH Lyon, MA Francombe, TA Hasdell, K Lawson, Eds., *Guidelines for Sensory Analysis in Food Product Development and Quality Control*. New York, NY: Chapman & Hall Publishing, 1992.

27 Foodborne Illnesses and Detection of Pathogenic Microorganisms

Amalia Scannell
Department of Food Science, National University of Ireland,
University College Dublin, Ireland

CONTENTS

I. INTRODUCTION

In spite of increasingly rigorous food hygiene and food safety regulations, *foodborne illness* continues to increase costing the global economy billions of dollars annually [1]. Pathogenic microbes are responsible for approximately 30% of all outbreaks, which make understanding the epidemiology of microbial foodborne illness and the methodology of pathogen detection vital to its control. In 2000, approximately 84% of all microbiological tests performed by the U.S. Food Industry were done to enumerate aerobic colony counts (ACC), yeast and molds (MYC), and coliforms and *Escherichia coli*, leaving only 16% of tests assessing pathogenic microbes [2]. This chapter reviews recent epidemiological data on foodborne illness worldwide and discusses the contribution of frozen foods to illness outbreaks and the factors that affect microbial survival in frozen foods. Finally, standard microbiological methods to enumerate ACC, MYC, Enterobacteriaceae, and food pathogens *Salmonella, Listeria monocytogenes*, and *Staphylococcus aureus* are described.

II. FOODBORNE ILLNESS

Foodborne illness, of infectious or toxic nature, can be loosely defined as any symptoms induced by the direct result of the ingestion of food or water. Biological causes of foodborne disease include infections and intoxications caused by bacteria [3–5]. The most commonly occurring bacterial agents include *Salmonella* sp., *Campylobacter jejuni*, and pathogenic strains of *E. coli*, *Clostridia* sp., *S. aureus*, and *L. monocytogenes*, among others. These agents are listed in Table 27.1 together with the major food categories with which each agent is typically associated. Parasites, including *Cryptospiridium parvum* [6], *Cyclospora cayatenensis* [7], and *Giardia duodenalis* [8], are also implicated in some outbreaks, typically involving animal products eaten raw or insufficiently cooked and plants. Calciviruses, including rotavirus [9], astrovirus [10,11], sapporo-like virus [12,13], and norovirus [14], are also known to be transmissible through food and have a characteristically low infectious dose of infectious units between 10^0 and 10^2 [15]. Because analytical methods are constantly improving [16,17], calciviruses are more commonly associated with outbreaks of foodborne disease and are now believed to be the leading cause of foodborne disease worldwide [18]. Classical gastrointestinal symptoms, such as abdominal pain and diarrhoea, shown in Table 27.2; fever and vomiting are also common as are feelings of nausea, chills, and dehydration [19]. Onset of symptoms may be as rapid as 40 min, in the case of some toxins, or may take up to 7 weeks, in the case of listeriosis [20]. In addition to gastrointestinal symptoms, some foodborne diseases are characterized by mild to life threatening neurological complications including numbness, tingling, temperature sensation reversal, weakness and vertigo, double vision, incoherent speech, difficulty in breathing, and heart and respiratory failure. Typically, neurological symptoms are characteristic of a bacterial intoxication, for example, botulinum toxin produced by *Clostridium botulinium*, and seafood-related dinoflagellate toxins, for example, saxitoxin, gonyautoxin, brevitoxin, dinophysistoxin, domoic acid, and ciguatoxin [21].

The duration, progress, and severity of illness will depend on a number of factors including the amount of the etiological agent ingested, the particular agent implicated, and the medical condition of the patient. In general, groups most vulnerable to foodborne pathogens include the very young — children less than 5 years require a lower infective dose by weight; the elderly — adults greater than 50 to 60 years may have failing immune system or be weakened by chronic ailments; pregnant women — due to altered immunity; and the immunocompromised, either through infection or medical treatment, for example, transplant recipients, cancer patients, and so on. An example of this vulnerability can be clearly seen in persons with acquired immunodeficiency syndrome (AIDS); when compared with otherwise healthy individuals, persons with AIDS are 100 times more likely to suffer from salmnellosis and 35 times more likely to be infected with *C. jejuni.* In 1996, Gerba et al. [22] reported that an estimated 20% of the U.S. population falls into vulnerable groups, and with the aging population dynamics and increases in the number of

TABLE 27.1
Selected Foodborne Pathogens and Associated Foods

	Food Types								
Pathogen	**Poultry**	**Meat and Meat Products**	**Fish and Seafood**	**Dairy**	**Egg**	**Fruit and Vegetables**	**Grain and Flour**	**Herbs, Spices, Cocoa, and Tea**	**Oil Based Condiments**
Bacillus cereus	n	n	n	n	n	YY	YY	YY	n
C. jejuni	YY	YY	n	n	YY	n	n	n	n
C. botulinum	n	y	YY	n	n	YY	YY	n	n
C. perfringens	YY	YY	YY	YY	n	YY	YY	YY	n
Enterohemorrhagic *E. coli*	n	YY	n	n	n	YY	n	n	n
Enteroinvasive *E. coli*	n	YY	n	n	n	YY	n	n	n
Enteropathogenic *E. coli*	YY	YY	YY	YY	n	YY	n	YY	n
Enterotoxigenic *E. coli*	n	YY	n	n	n	n	n	n	n
L. monocytogenes	YY	YY	n	YY	n	n	n	n	n
Salmonella sp.	YY	YY	YY	YY	YY	n	YY	YY	YY
Shigella sp.	n	n	YY	n	n	n	YY	n	n
S. aureus	YY	YY	YY	YY	n	YY	YY	n	YY
Vibrio sp.	n	n	YY	n	n	n	n	n	n
Yersinia enterocolitica	YY	YY	n	YY	n	n	n	n	n
Fungal/mycotoxins	n	n	n	n	n	n	YY	YY	n
Parasites	n	YY	YY	n	n	YY	n	n	n
Toxins	n	n	YY	n	n	YY	n	n	n
Virus	Q	Q	Q	Q	Q	Q	Q	n	Q

Note: YY, very common occurrance; Y, occurance but not common; n, no occurance; Q, possible.

Source: Adapted from M Satin. *Food Sources of Disease. Food Alert! The Ultimate Sourcebook for Food Safety*. New York: Checkmark Books, 1999, pp. 41–96. With permission.

TABLE 27.2
Aetiological Agents of Foodborne Illness and Characteristic Symptoms

Pathogen	Incubation Time	Description	Symptoms
Bacillus cereus	30 min to 5 h	Exoenterotoxin	Nausea; vomiting; occasionally diarrhoea; toxin preformed in food
Emetic	8 to 16 h	Enterotoxin	Nausea; abdominal pain; watery diarrhea; toxin formed in small intestine
Diarrhoeal			
C. jejuni	2 to 7 days	Infection (~400 cells)	Diarrhoea (bloody); fever; severe abdominal pains; vomiting; malaise
C. botulinum	2 h to 8 days	Neurotoxin, A, B, E, F	Gastrointestinal symptoms; vertigo; double vision; muscular weakness; dilated pupils; respiratory paralysis; fatalities occur
C. perfringens	8 to 22 h	Toxico-infection	Diarrhoea; abdominal pain
Pathogenic *E. coli*, for example, enterohemorrhagic, O157:H7	5 to 48 h	Infection: toxin induced bleeding	Severe abdominal pain; diarrhoea (may be bloody); nausea; vomiting; fever; chills; headache; muscular pain; bloody urine (hemorrhagic *E. coli*)
L. monocytogenes	3 to 21 days (or longer)	Infection	Low grade fever; flu-like symptoms, sepsis; still births; menengitis; fatalities
Salmonella sp.	6 to 72 h	Infection	Abdominal pain; diarrhea; fever; chills; nausea; vomiting; malaise
Shigella sp.	27 to 72 h	Infection: toxin induced bleeding	Abdominal pain; diarrhea (bloody/mucoid); fever
S. aureus	1 to 8 h	Exoenterotoxins A, B, C, D, E, F	Nausea; vomiting; empty retching; watery diarrhoea
Vibrio sp.	1 to 3 days	Infection/enterotoxin	Watery diarrhea (profuse for cholera); fever; chills; vomiting; headache; abdominal pain; dehydration; fainting; sunken eyes; reduced skin turgor
Yersinia enterocolitica	24 to 36 h	Infection	Severe abdominal pain (lower right-hand quadrant); fever; headache; sore throat
Calcivirus	16 h to 5 days	Infection	Abdominal pain; vomiting (often projectile); diarrhea; fever; chills; malaise

Source: Adapted from GJ Arnod, BA Munce. Investigation of foodborne disease outbreaks. In: AD Hocking, G Arnold, I Jenson, K Newton, P Sutherland, Eds., *Foodborne Microorganisms of Public Health Significance*, 5th ed. North Sydney, Australia: AIFST (NSW Branch) Food Microbiology Group, 1997, pp. 31–70. With permission.

immunocompromised individuals, this number will continue to grow [22]. In addition to an increased incidence of foodborne disease in the immunocompromised population, complications due to foodborne illness, and even fatalities, are also more common within this group [23]. Other factors that increase risk of foodborne disease include altered gut flora due to antibiotics, poor hygiene, and concomitant infections and alteration of stomach acid due to antacids or excess fluid intake and intake of fatty foods, which may protect pathogens from the action of stomach acids [24].

A. Sequelae of Foodborne Diseases

Even in otherwise healthy individuals, many foodborne illnesses may cause long-term side effects or sequelae [25]. A number of chronic sequelae may result from foodborne infection, including ankylosing spondylitis, arthropathies, renal disease, cardiac and neurologic disorders, and nutritional and other malabsorptive disorders [26]. Reactive arthritis (RA) is an inflammatory condition that occurs in the joints due to presence of bacterial antigens and is thought to occur in ~2% of all cases of food poisoning. Reiter's syndrome (RS) is characterized by symptoms including urethritis, conjunctivitis and psoriasis, septicemia, and meningitis, in addition to joint inflammation. Guillain–Barré syndrome (GBS), which may be described as the destruction of mylien sheath of nerve fibers, results in symptoms ranging from numbness and tingling in the extremities to paralysis and respiratory distress. Hemolytic uremic syndrome (HUS) is a complication, which affects approximately 15% of children suffering from hemorrhagic colitis and results in hemolytic anemia, kidney failure, and sometimes death. Some examples of these complications are given in Table 27.3.

B. Factors Causing Foodborne Disease

For a food to cause an outbreak of foodborne disease, it must first become contaminated with a microorganism and then be placed in such an environment that multiplication and toxin production occurs or be ingested in sufficient quantities to allow the pathogen to establish in the gastrointestinal tract to instigate symptoms of the disease. The World Health Organization [27] has compiled a list of extrinsic factors that most commonly lead to outbreaks of foodborne illness, although in many cases more than one factor was implicated per outbreak. The factors which affect the growth of pathogen in food typically involve poor temperature control and include:

- Preparation too far in advance
- Food left at room temperature
- Improper cooling
- Improper warm holding
- Improper thawing

Pathogen survival is primarily facilitated by inadequate cooking of the foods implicated in the outbreaks or inadequate reheating. Contamination of cooked or ready-to-eat (RTE) foods is a particularly dangerous phenomenon as any control steps to eliminate the pathogens have already been taken at this stage. Sources of contamination are commonly attributed to a contaminated raw ingredient or processed food — particularly processed meats and pies or prepared meals; cross-contamination of cooked food, for example, by utensils or surfaces used in preparation of raw ingredients; and infected food handlers. In addition to inadequate control of the thermal profile (cooking and cooling) during food production, incidence of foodborne illness is also affected by ambient temperatures. Recently, evidence has been reported, indicating an increased incidence in reported cases of salmonellosis and increased climactic temperature. OZFoodNet reported an increase of *Salmonella* notifications with decreasing latitude or increasing yearly temperature for Eastern

TABLE 27.3
Sequelae of Foodborne Diseases

Pathogen	Sequelae									
	TTP	HUS	RS	GBS	RA	Endocarditis	Meningitis	Septacemia	Other	Death
Bacillus cereus						✓				✓ Infant
C. jejuni		✓		✓	✓		✓	✓	✓Reactive colitis Acute chloecystitis	
Pathogenic *E. coli*	✓	✓					✓			
L. monocytogenes									✓ Chrones disease	
Salmonella sp.			✓		✓			✓	✓ Septic arthritis	
Shigella sp.		✓	✓							
Vibrio sp.								✓		
Yersinia sp.					✓				✓Apendictomy Grave's disease	

Note: TTP, thrombocytopenic purpura.

Source: Adapted from Anonymous. In: Bad Bug Book — Foodborne Pathogenic Microorganisms and Natural Toxins Handbook. U.S. Food and Drug Administration, Center for Food Safety and Applied Nutrition, 5600 Fihers Lane, Rockville, MD, 20857-0001, U.S.A M Stain. *Food Sources of Disease. Food Alert! The Ultimate Sourcebook for Food Safety*. New York: Checkmark Books, 1999, pp. 41–96. With permission.

Australia in 2000 [28], whereas more recently, an average linear association between temperature and number of reported cases of salmonellosis was reported throughout ten European countries [29].

C. Trends

In a report estimating the incidence of food-related illnesses in 1999, Mead et al. [30] estimated that 67% is attributable to viruses, 30% to bacteria, and 3% to parasites. Of the bacterial agents, *Campylobacter* sp. (14%) and nontyphoidal *Salmonella* sp. (9.7%) were the most common cause of foodborne illness, whereas noroviruses (66.6%) were largely responsible for the viral outbreaks. In agreement with these figures, norovirus has recently been identified as the leading cause of nonbacterial gastroenteritis in Europe between 1995 and 2000 [31]. In industrialized countries, the total incidence of food-related illness has been estimated to range from 7 to 10% of the population of countries investigated [32]. In general terms, incidence of foodborne illness has been on the decline in recent years. In England and Wales, there was a 53% decrease in total incidence between 1992 and 2000 [33]. With the exception of infection by norovirus and *Campylobacte*r sp., which increased by 125.5 and 45%, respectively, most other etiological agents showed a declining incidence of infection. Similarly, in the United States, preliminary reports from the U.S. Centers for Disease Control account considerable decrease in the incidences of disease caused by *Campylobacter* sp., *E.coli* O157:H7, *Salmonella* sp., and *Yersinia* sp. for the period 1996–2002 [34], whereas reductions in the incidence of listeriosis ceased to decline after 2001. Although the implementation of control measures by both the food industry and government agencies has had a positve impact on the incidence of foodborne disease, there is no room for complacency as the most recent figures issued by FoodNet indicate that for some pathogens, particularly several *Salmonella* serotypes, the incidence is still above the national health objective [34].

D. Economic Burden of Foodborne Illness

To the individual suffering from a mild, uncomplicated bout of foodborne illness, the net result may be a few days of discomfort at home, with the resulting loss of pay that may be incurred and, in the event that medical assistance was sought, medical expenses making up the main cost to that individual. In population terms, estimating the economic burden of foodborne illness is a more complicated task taking into consideration both direct and indirect costs. These would include medical costs (including burial expenses in the case of fatalities), public service costs incurred to investigate the outbreaks, losses to industry due to lost productivity and costs to the food or catering industry incurred through loss of product, correction of cause of the problem leading to the outbreak, and litigation costs. The Economic Research Service (ERS) of the U.S. Department of Agriculture (USDA) has recently estimated the cost of five leading aetiological agents of foodborne disease (*Campylobacter* sp., nontyphoidal *Salmonella* sp., *L. monocytogenes, E. coli* O157:H7, and STEC *E.coli*) to be in the region of $6.9 billion [35]. In a recent report, the Organization for Economic Co-operation and Development (OECD) [36] reported total annual costs of foodborne illness for New Zealand, Australia, Finland, Sweden, Germany, U.K., U.S.A., and Canada (Table 27.4), which range from $25 million to $8.4 billion. Although this is an enormous cost, when one considers that most sporadic incidences of foodborne disease go unreported, that the cause of over 60% of outbreaks that are reported are never identified, and that very often in cost of illness reports the health burden of complications such as arthritis, Guillain–Barré syndrome, and so on are not included, the actual cost is probably well in excess of these estimates. However, despite their limitations, cost estimates give a very clear indication of the impact of foodborne disease on the global economy and the consequent requirement of stringent foodsafety standards in both the food production and food service industries.

TABLE 27.4
Estimates of Total Economic Costs of Foodbrone Illness in Selected Countries

Country	Annual Cost in US Dollars (USD)
Australia	1.99 billion
Canada	977 million
Germany	50 million
New Zealand	25 million
Sweden	123 million
The Netherlands	414 million
United Kingdom	1,137.5 million
United States of America	8.4 billion

Source: Complied from data published in Anonymous. Economic costs of foodborne disease in OECD Countries. In: Foodborne Diseases in OECD Countries: Present State and Economic Costs. Paris, France: OECD Publications, 2003, pp. 61–82; WE van den Brandhof, GA deWitm, YTHP van Duynhoven. Costs of gastroenteritis in the Netherlands. *Epidemiology and Infection* 132 (2):211–221, 2004. With permission.

III. MICROBIOLOGICAL CONSIDERATIONS OF FROZEN FOOD

A. Safety of Frozen Food

Freezing as a means of preservation has been utilized since approximately 1000 BC in the ice cellars of China [37]. However, it was not until the late 1920s that developments in freezing technology were such that large-scale commercial production of frozen food was possible. Since the 1930s, there has been a steady increase in the types of frozen foods available ranging from ice cream and desserts to fish and meat products, vegetables and potato products and more recently including ready meals, snack foods, microwavable products, and ethnic foods. Because frozen foods tend to be stored at temperatures below the minimum temperature required for microbial growth (-2 to -3°C for most spoilage microorganisms and approximately -8°C for molds) once properly stored, frozen products can be kept for months, and in some cases years, without a noticeable increase in the microbial load. In addition, as reducing the temperature of the frozen food will result in a decrease in a_w, the potential microflora will be limited to those microbes which can grow at low temperatures and low a_w [38]. The lack of opportunity of microorganisms to proliferate in frozen food has contributed to the general perception that frozen foods are safe foods. Indeed, frozen foods have a very good food safety record with few associated outbreaks of foodborne illness when compared with other food categories. In most cases where outbreaks have occurred, contaminated raw materials or postprocess contamination, as opposed to freezing *per se*, have been implicated. Of those outbreaks where frozen food has been implicated ice cream has been the main vehicle involved, comprising 0.6% of total outbreaks in the United States from 1988–1997 [39,40]. In 22 U.S. outbreaks between 1966 and 1976, home-made ice cream using nonpasteurized egg resulted in salmonellosis [41]. In more recent ice cream-related outbreaks between 1985 and 1993, the United States had 14 cases of salmonellosis attributable to the same cause [42].

However, as indicated in Table 27.5, *E. coli, L. monocytogenes*, and *G. lambia* have been implicated. Viruses have also been found associated with frozen food vehicles, for example, frozen raspberries contaminated with norovirus [43] and Hepatitis A [44]. Strawberries have also been associated with outbreaks of Hepatitis A [45,46], whereas ice was implicated in norovirus

TABLE 27.5
Incidence of Foodborne Illness Associated with Ice Cream from 1988 to 1997

Period	Aetiology	Number of Outbreaks (Ice Cream)	Number of Cases (Ice Cream)	Number of Outbreaks (Total)	Number of Cases (Total)
1988–1992	*Salmonella* sp.	9	470	2,423	77,373
	E. coli	6			
	L. monocytogenes	1			
	Chemical	2			
	Unknown	2			
	Subtotal	20			
1993–1997	*Salmonella* sp.	11	1,194	2,751	86,058
	Heavy metal	1			
	Unknown	2			
	G. lamblia	1			
	Subtotal	15			
1988–1997	All agents	35	1664	5,174	163,431

Source: Complied from CDC Surveillance surveys — NH Bean, JS Goulding, C Lao, FJ Angulo. Surveillance for foodborne-disease outbreaks — United States, 1988–1992. *Centers for Disease Control (CDC); Morbidity and Mortality Weekly Report* 45 (SS–5):1–55, 1996; SJ Olsen, LC MacKinon, JS Goulding, NH Bean, L Slutsker. Surveillance for foodborne-disease outbreaks United States, 1993–1997. *CDC Morbidity and Mortality Weekly Report* 49 (SS–1):1–67, 2000. With permission.

outbreaks in 1990 and 2000 [47,48]. As most frozen food products, for example, ready meals, pizzas, and fish products, are heated or cooked either directly from frozen or after thawing, provided that the heat treatment provided is adequate, most indigeneous or contaminating bacteria should be inactivated. One such outbreak occurred in 1997 when contaminated frozen beef patties were consumed undercooked leading to 15 cases of *E. coli* O157:H7 [49].

B. Freezing and Microorganisms

In the same way that freezing extends the shelf life and nutritional quality of food, freezing technology is also used as the main method to preserve the viability of microorganisms, both for commercial and institutional culture collections used in research and for cultures of importance to biotechnology and food industries. Starter cultures are normally preserved by freezing, which results in a heavy and bulky product or by freeze drying, producing a light but costly product [50,51]. To protect microorganisms from injury and maintain cell viability, cryoprotectant compounds are often used during the freezing process [52]. These would typically include polyols, polysaccharides, disaccharides, amino acid, proteins, vitamins, and salts. Some examples of cryoprotectants include bovine albumin, dextran, dimethylsulfoxide, gelatine, glycerol, lactose, maltodextrins, sucrose, trehalose and xanthane gum [53,54]. Sodium ascorbate, sodium glutamate and betain have recently been found to act as cryoprotectants for dairy starter culture strains [55]. Microencapsulation in polysaccharide matrices, for example, calcium alginate and κ-carrageenan, prior to exposure to low temperatures, has been used to protect probiotic bacteria *Lactobacillus acidophilus* and *Bifidobacterium* sp., which are sensitive to freeze drying processes or frozen storage [56,57].

It stands to reason that if freezing can be used as a means of long-term storage of bacteria, then bacteria intrinsic to frozen food may also survive for a considerable length of time during product storage. Indeed, some food components and ingredients that make up the complex matrices of frozen food products (e.g., fats, sugars, proteins, tripolyphosphates, etc.) may also serve as cryoprotectants aiding the survival of contaminating bacteria in the frozen product. One possible outcome of this survival is evident in the ice cream-related incidences of salmonellosis mentioned earlier. Salmonellae present in the unpasteurized egg used in the ice cream production remained viable in the product and, on consumption of the product, led to many cases of infection.

From a food safety standpoint, pathogenic microorganisms are of most concern and it may be expected that pathogens associated with fresh ingredients (Table 27.2) may also end up in the frozen product. Additionally, in many food production processes, there may be a possibility of in-line or postprocessing contamination with pathogenic bacteria that go on to contaminate the final product. These situations may be addressed by (1) ensuring that all raw materials used in the production of frozen food are of good microbiological quality and within the limits set in product specifications; (2) incorporation of one or more steps to eliminate pathogenic microorganisms in the process (e.g., pasteurization of ice cream mix prior to freezing); and (3) development and maintenance of a valid Hazard Analysis and Critical Control Point (HACCP) plan for the production process.

In terms of food quality, there is a vast range of microorganisms which can cause spoilage of frozen food. Some of the more important ones include aerobic bacteria, associated with frozen meat, poultry, fish, and eggs (e.g., *Pseudomonas, Alcaligenes, Moraxella, Alteromonas*, and *Flavobacterium*); coryneform bacteria, associated with vegetables and cured meat products (e.g., *Corynebacterium, Kurthia, Arthrobacter*, and *Brochothrix*); Enterobacteriaceae, associated with most proteinaceous foods (e.g., *Escherichia, Citrobacter, Klebsiella, Erwinia*, and *Proteus*); and lactic acid bacteria, associated with milk, meat, and vegetables [58,59].

Frozen foods may be consumed in the frozen state (ice cream, frozen yoghurt) or after reheating or cooking (burger patties,vegetables, pizza, and ready meals); consequently, the final product may be raw, blanched, or fully cooked. Naturally, the microbial load and the microflora will depend on the extent of prefreezing treatments applied to the final product.

C. Surviving the Freezing Process

1. Freeze Injury

An extreme change in the environment of a microorganism, such as occurs during freezing, will cause the microbial cells to be stressed as a result of the impact that the temperature decrease has on metabolic processes. This stress will ultimately lead to lethal or sublethal injury, which is evident from a cessation in growth or an increased lag time with subsequent reduced growth rate [60]. In exponentially growing cells exposed to a rapid decrease in temperature (usually above 0°C), cold-shock response, characterized by the repression of heat-shock proteins and the production of cold-shock proteins, and cold acclimatization proteins, may be induced to aid microbial survival [61]. In rapid freezing methods, for example, individual quick frozen products, temperatures are brought below freezing very quickly to initiate a metabolic response. However, other cooling regimes (e.g., refrigerated hold prior to freezing) used in the food industry could potentially induce either a cold shock or cold adaptation response. Different cold-shock treatments applied before freezing might alter the survival of pathogens or spoilage microorganisms and result in an less safe product or a reduction in post-thawing shelf stability [62]. To illustrate this, *Lactococcus lactis* (an important starter culture in the dairy industry) became more cryotolerant when subjected to mild cold shock and demonstrated resistance to both freezing and lyophilization as a result [63]. The survival of *E. coli* O157:H7 was shown to be enhanced by cold-shock treatment in whole egg, milk, or sausage but not in ground beef or pork [64], while cold adaptation, achieved by gradually reducing the temperature to 4°C and holding for 3 h, also increased the resistance of *E. coli* strains to freeze injury [65].

During the freezing processes itself the food system is progressively cooled so that for periods of time, dependent on the freezing method employed, microorganisms present in the food will be exposed to suboptimal growth temperatures. Such exposure will increase lag phase and induce physiological and structural changes to the cell, which mainly target the cell membrane [62].

A number of factors that cause cell injury and loss of cell viability during freezing and thawing have been proposed [66]. These include:

- Mechanical damage to the cell envelope due to both intracellular and extracellular ice formation
- Electrolyte imbalance caused by dehydration and increasing ice formation causing extracellular and intracellular fluids to become more concentrated
- Denaturation of proteins due to loss of surface water and abnormal interprotein bonding
- Cell shrinkage below a critical minimum level causing rupture
- Ice crystal growth

Microbial injury induced by freezing and thawing may be expressed as an increase in nutritional fastidiousness, increase in lag phase, and a decrease in growth rate and generation time, as well as increased sensitivity to surface active agents and selective media. Leakage of cellular material, for example, enzymes, proteins, peptides, amino acids, ribonucleic acid (RNA), and deoxyribonucleic acid (DNA), is also known to occur [67]. By successive cycling of a food product between the frozen and thawed states, some researchers have found a decrease in the number of recoverable *E. coli*, with loss of viability and injury increasing with each freeze–thaw cycle [68–70]. Such findings have led to the belief that with appropriate understanding of both the systems and the microflora, freezing has the potential to be used to reliably reduce microbial populations as well as to preserve food [37], a belief which has recently been demonstrated in the case of *L. monocytogenes* reduction in media and food systems [71,72].

2. Microbial Survival

In general, a number of factors will determine how a microbial population responds to freezing. The composition of the microflora is important because some microorganisms resist freezing and frozen storage better than others. As a rule of thumb, it may be considered that freezing resistance decreases in the following order: spores > gram-positive bacteria > gram-negative bacteria > ameba, protozoa, and nematodes. This broadly coincides with Mazur's freeze–thaw stress categories a to d [73].

The phase of growth and nutritional status of a microbial population will also impact on its ability to resist and survive freezing conditions. Generally, populations in stationary phase are more resistant than actively growing cells. Cells that have restricted nitrogen accumlulate higher levels of carbohydrates, which confer a degree of cryoprotection during freezing. Survival of microorganisms in freezing conditions is also decreased in the presence of salt or in acidic conditions [70,74–76].

The most rapid cooling rate of food, between 100 and 200°C min^{-1}, is reached in small particles, for example, peas or at the food surface. Generally, however, rates of <100°C min^{-1}, and very often <10°C min^{-1} are used [38]. Optimal cooling rates for survival of *E. coli* have been given in the 10°C min^{-1} range. At cooling rates above this optimum, ice crystal formation decreases survival, whereas at rates below it, extracellular concentration of solutes causes freeze injury [73]. Loss of microbial viability during frozen storage occurs most rapidly in the −2 to −5°C range. As storage temperatures decrease, there is a corresponding decrease in the microbial death rate until a plateau temperature (≤ − 60°C) is reached, at which point cell viability remains relatively constant. In general, rapid freezing rates followed by a low-storage temperature will give optimal microbial survival [77]. Thawing conditions have not been reported to affect the survival of microbes in frozen foods exposed to slow cooling rates; however, survival is reduced in rapidly frozen foods by thawing slowly at refrigeration temperatures, probably due to ice crystal growth [78]. There is not much published information to date describing the impact of thawing on the microbiological quality of food. In the present situation, there is no evidence that a practical significance exists between growth of spoilage organisms on frozen–thawed food when compared with the fresh product. On one hand, if the structure of a product become damaged through ice crystal formation, it is plausible that the microbial growth on the thawed product will be more rapid and the shelf-life reduced; on the other hand, thawed fish that had been frozen at −20°C has a longer shelf life than fresh fish due to inactivation of some of the principal spoilage microorganisms [79]. Although some research has been done to develop predictive models to describe microbial population dynamics in frozen foods [80], more research is needed to understand fully the microbial kinetics of thawed food.

3. Pathogen Survival in Frozen Food Systems

Even though freezing at specified time–temperature conditions is a recommended method to control animal parasites in the United States and Europe, serving to inactivate trematodes and nematodes [38,81], other pathogens are not so readily destroyed. In experimental trials, *L. monocytogenes* has been shown to survive on a variety of frozen food products including chicken breast, spinach, cod, mozzarella, and beef burger for 10 months [82], ice cream for 3 months [83], shrimps for 3 months [84], and frankfurters for 1 month [85]. Repeated freeze–thaw cycles may serve to eliminate *Listeria* but the process also serves to impair the sensory and quality attributes severely [67]. *E. coli* O157:H7 has also shown the ability to survive frozen storage. Survival rates after 10 days at −18°C were estimated at between 30 and 45% in ground beef, depending on the strain, whereas in acid foods, for example, frozen yogurt (pH ~6.1) and apple juice (pH ~3.5), the survival was reduced to approximately 10 and 2%, respectively [86]. In vegetable broth (pH ~5.2) stored for 7 days at −20°C, a survival rate of 4% was observed [87]. Experiments on *Salmonella* survival show no significant reduction in viability on beef trimmings at −18 or −35°C over a 9-month period or on ice cream stored at −20°C for 4 months [88,89]. *C. jejuni* strains were also reported to survive standard storage conditions on beef trimmings without sublethal injury [88]. However, a 5 log reduction of surface inoculated *Salmonnella* was seen after 10 months at −20°C on sterilized chicken patties [90].

D. Microbiological Spoilage of Frozen Foods

Spoilage of frozen foods attributed to molds and yeast occurs because of their ability to grow at low temperatures and low a_w. The most common spoilage reported is mold spoilage of frozen meat, for example, black spot — *Cladiosporium herbarum, C. cladiosporioides, Aureobasidium pullulans,* and *Penicillium hirsutum*; white spot — *Chrysosporium pannorum*; whiskers — *Thamnidium elegans, Mucor racemosus*; and blue green mold — *Penicillium* sp. Meat stored for long periods of time at −2 to −4°C may develop a microflora predominated by yeasts, particularly *Cryptococcus* sp. [91]. Yeasts and molds have also been shown to contaminate frozen yoghurt [92] but no growth was observed throughout the 60 weeks of frozen storage so that their contribution to spoilage is of no practical significance. In general, freezing over a period of time will reduce the population of indigenous microflora and result in an improvement in the overall microbiological quality [93–95]. However, freezing of food with a high initial load of bacteria will have a reduced postthaw shelf life when compared with products produced using good hygienic practices.

IV. MICROBIOLOGICAL EXAMINATION OF FROZEN FOODS

A. Microbiological Sampling Plans and Guidelines

In practical terms, laboratory microbiological analysis of frozen foods is identical to procedures used for the equivalent fresh products; the main differences occur in the methods used to take the samples and the requirement of a resuscitation step to allow stressed or injured cells recover.

Pathogen detection is typically facilitated by the use of selective techniques, be it the use of culture-specific conditions, for example, atmosphere (aerobic, micro-aerophilic, and anerobic) or incubation temperature, or by the addition of selective agents, for example, pH, bile salts, tellurite, antibiotic cocktails, and so on to the media. Selective agents, while inhibiting background flora, usually display a degree of toxicity to the target organisms as well. This is particularly true of freeze-injured cells. Use of a nonselective enrichment or pre-enrichment step may be used for injured cells or for pathogens that may be present at levels below the sensitivity of the detection method to grow a detectable population from a very small initial level. Results from enrichment in broth (unless using the most probable number (MPN) technique) is qualitative and should only be used where presence or absence criteria are required. Ressucitation methods for samples requiring enumeration may be done by placing diluted sample on nonselective agar, for example, plate count agar or tryptic soya agar, and allowing the samples to rest at room temperature for 2–3 h and overlaying with a selective medium before incubation [96]. This method has been used to recover coliform bacteria from milk in which a layer of double strength violet red bile agar (VRBA) was used as the selective agent. Another method used involved incubation of the sample on a membrane filter on nonselective media followed by transfer to a selective media appropriate to the test being carried out [97]. In traditional presence or absence tests for pathogens, for example, *Salmonella* sp., a second selective enrichment step may also be performed prior to isolation on agar, to further promote the target bacteria while repressing competitive background microflora present in the sample.

Pathogens considered likely to be associated with a frozen food product are exactly the same as those associated with its unfrozen counterpart, so that the decision as to which analyses should be done for a given product will largely be determined by the nature of the product and its raw ingredients, by the product specifications between manufacturer and suppliers, and so on, and by legislation. Microbiological guidelines for groups of food products are available from the World Health Organisation (WHO), USDA, and the International Commission for Microbiological Specifications for Foods (ICMSF), among others. ICMSF guidelines prepared in 1986 [98], while valuable, do not take new pathogens into account and, in an effort to deal with these emergent pathogens, a new "Regulation for the Microbiological Criteria for Foodstuffs and Food Production" is currently being developed by the European Commission, in line with the redrafting of the EU Hygiene Regulations. Table 27.6 lists the methods by which pathogens should be detected under the impending legislation.

TABLE 27.6
List of Microbiological Methods as Recommended in EU Commission Draft 9 of EU Regulation for Microbiological Criteria for Foodstuffs and Food Production

Microorganism	Method Recommended
L. monocytogenes	EN/ISO 11290 — 1, 2
Salmonella sp.	EN/ISO 6569
S. aureus (coagulase +)	EN/ISO 6888 — 1, 2
Enterobactericeae	Draft ISO 21528 — 1
E. coli	EN/ISO 7251 or EN/ISO 16644 — 1, 2, 3
Aerobic plate count	EN/ISO 4833

Source: Food Safety Authority of Ireland (Personal Communication).

Because bacterial contaminants are not uniformly distributed throughout a product or a batch of product, it is imperative that a statistically sound method of sampling is used. Most of the internationally accepted methods make use of attribute sampling plans, as adopted by the ICMSF; the attributes being either the presence or absence of a pathogen, presence of that pathogen at levels greater than a specified level, or its presence in greater than a specified number of samples. Two class plans are used when establishing the presence or absence of a pathogen, in which case, n is the number of sample units tested and c is the maximum number of samples above specified limit. The higher the value of n at a given c, the more stringent the criteria. For example, in revision 9 of EU Microbiological Criteria, the suggested plan for *L. monocytogenes* in RTE food is $n = 5$ and $c = 0$, whereas RTE foods for infants and special medical purposes is more stringent at $n = 10$ and $c = 0$. Three class plans make use of three categories of results: those which are acceptable (counts $\leq m$), those which are marginally acceptable (counts $>m$ but $\leq M$), and those which must be rejected (counts $> M$), where m is good manufacturing practice (GMP) limit, M the safety or quality limit, and c the maximum number of samples allowed within the range between m and M, that is, maximum permissible samples of marginal quality [98]. An example of a three-class plan from the EU Microbiological Criteria is the ACC of minced meat, where $n = 5$, $c = 2$, $m = 5 \times 10^5$ cfu/g, and $M = 5 \times 10^6$ cfu/g. In this case, five samples must be tested. If more than two samples have ACCs between 5×10^5 and 5×10^6 cfu/g or if one or more samples have ACCs greater than 5×10^6 cfu/g, then the lot must be rejected. Conversly, if all ACCs are less than 5×10^5 cfu/g, or if two samples have ACCs between 5×10^5 and 5×10^6 cfu/g, whereas the rest are below 5×10^5 cfu/g, then the lot can be accepted.

B. Practicalities of Microbiological Analysis of Frozen Food

Once one has accessed the legislation and consulted the guidelines for a food product that requires testing, one should be armed with the necessary information (associated pathogens and tests) to set up analysis for a frozen food product. In the next section, the steps involved in the microbial analysis of a product as well as some of the more important pathogen detection methods will be described. Methods required by the EU are listed in Table 27.6, however, other standard methods are also commonly used, for example, American Public Health Association (APHA) [99], Association of Analytical Communities (AOAC) [100] and the U.S. Food and Drug Administration's (FDA) Bacteriological Analytical Manual (BAM), and so on. This section will give an example of some analysis methods for the enumeration or detection of a number of microbial groups, which may be of use to the food industry, but for compliance purposes, it is recommended that the user consult relevant local legislation for specific tests.

1. Sample Preparation

When sampling frozen foods, it should ideally be maintained in the frozen state to avoid variations in microbial counts due to variations of thawing methods used and the possibility of microbial growth prior to analysis. Where tempering of the product is required to facilitate sampling, AOAC/BAM recommend that temperatures should be kept between 2 and 5°C and for no longer than 18 h. Rapid thawing using temperatures less than 45°C for less than 15 min may also be used where appropriate.

Sampling a solid block from frozen can be a bit tricky as the frozen material must be broken into small pieces for homogenization. This may be done using sterilized drills or saws. It is vital to ensure that sampling is done under strict aseptic conditions so as not to contaminate the sample with environmental flora. The size of the sample required may vary depending on the tests to be performed, for example, 25 g for *Salmonella*, 50 g for Coliforms and ACC, and so on, so it is important to adhere to the recommendations of the method in use. Once a quantity of sample has been cut, it should be placed aseptically into sterile prechilled containers and kept in the frozen state until analyzed.

In the laboratory, a homogenate of the sample must be made. This serves to distribute bacteria evenly throughout the homogenate and aids the thawing process. The most commonly recommended diluent for enumeration is Butterfield's phosphate — buffered dilution water as it gives satisfactory recovery of injured cells — however, buffered peptone water (0.1% w/v) or sterile peptone saline solution, for example, maximum recovery diluent is also routinely used for enumeration purposes. Most official methods recommend the use of a high-speed blender — one with an autoclavable blender jar would be advantageous — to prepare the mix. A sterile blender jar should be tared on a weighing scales and 50 g (or appropriate sample size) accurately weighed into it. To this, 450 ml of the sterile diluent should be added, to give a 10^{-1} dilution, and the contents should be mixed at high speed for 2 min. Use of a stomacher for 2–5 min to blend the product has also been suggested as the use of sterile bags precludes the need to resterilize equipment between samples [101] and, although it is not suitable for all applications, use of the stomacher is now routinely used for food analyses in many experimental laboratories. To continue the serial dilution, 10 ml of the original homogenate is aseptically pipetted into 90 ml of sterile buffered diluent to give a 10^{-2} dilution; this bottle is shaken and the process is continued until the desired end dilution is reached. For ACC, an end dilution of 10^{-6} should usually suffice.

2. Determination of Microbiological Quality

a. Aerobic Colony Count

The ACC method, otherwise known as the aerobic plate count or standard plate count method, is the most common method used to determine the level of microbial contamination in a product. There are two methods that are generally in use, the spread plate method and the pour plate method, but the ACC using either technique is comparable [102]. Using the spread plate method, autoclaved plate count agar is aspetically poured into sterile petri dishes and allowed to dry. An inoculum of 0.1 ml, taken from each dilution to be tested, is placed on the surface of the agar (typically each dilution would be plated in duplicate) and is aseptically spread along the surface of the plate using a sterile glass rod. The pour plate technique is recommended by AOAC (sec. 966.23) [98] and APHA [103]. In this case, a larger inoculum (1.0 ml) is taken from each dilution (prepared as described earlier), which is placed into duplicate empty sterile petri dishs. Molten plate count agar, sterilized and cooled to 45°C, is poured into the petri dish and the sample is incorporated into the media by alternate gentle rotation and back and forth movement of the covered petri dish on a level surface. Controls, 12–15 ml agar without additions, agar + 1 ml dilution water, and agar + 1 ml pipette water should also be included. Once the agar is solidified, it is inverted and incubated at 35°C for 48 h. Colonies should be counted after 48 h in good light and with the aid of magnification where possible. Plates with colony counts between 25 and 250 should be recorded together with its corresponding dilution. Normally, counts in excess of 250 should be reported as too numerous to count (TNTC), however, in circumstances where the lowest dilution (e.g., 10^{-6}) gives counts greater than 250 or the highest dilution (e.g., 10^{-1}) gives counts less than 25, then results should be reported as estimates (EACC). Plates with no colony forming units (cfu) should be reported as ACC <1 and plates with spreading colonies should be reported as such.

Use of mechanized spiral plate system to reduce the workload in the determination of ACC has also been approved by the AOAC (sec. 977.27) [98] for food and cosmetics. Spiral plating has the advantage of requiring fewer agar plates per sample, reducing costs and preparation time, but because the stylus gets blocked with particulate matter, it is best suited to liquid samples.

It is important to be aware that while ACC is a routinely used measure of production hygiene, in frozen foods results must be interpreted with caution since, without access to time–temperature records for a product, low ACC may merely indicate that a product has been in frozen storage for a long time, causing a decline in the original population or that uncontrolled thawing has occurred leading to cell death.

b. *Total Psychrophilic Aerobes*

Mesophilic microorganisms are not likely to make up the entire microflora of frozen foods, and as such, it may be important to determine the contribution of psychrotrophic or psychrophilic microbes to contamination. The ACC method discussed earlier will not give any indication as to the extent of psychrotrophilic contamination because the incubation conditions are geared to favor mesophiles. In this case, the same method as earlier is used but the incubation temperature used, as recommended by APHA, is 5 to 7°C for 7 to 10 days, or in some cases, up to 28 days [99]. In cases where microflora is heat-sensitive, use of the spread plate technique may be advantageous.

c. *Yeasts and molds*

Because of their broad tolerances to pH, temperature, and their nutritional versatility, yeasts and molds as a group may be found to contaminate many classes of frozen food and may result in a decrease in shelf life during frozen storage at high temperature (−8°C) [104] or after thawing in refrigerated storage. Two methods are used for the detection of yeasts and molds, the direct plating method used for molds on foods which can be transferred with a forceps, for example, coffee beans, or dilution plating which is more generally applicable to foods and which includes the contribution of yeasts [105].

To enumerate yeasts and molds using dilution plating, 25 and 50 g samples are taken as described eariler and weighed into a sterile stomacher bag. Sufficient 0.1% peptone water (225 or 450 ml for 25 and 50 g, respectively) is added to give a 10^{-1} dilution and the homogenate is produced by stomaching for 2 to 5 min, depending on the nature of the sample. Serial dilutions to 10^{-6} should be prepared in 0.1% peptone water as described earlier.

A number of commercially available agar media may be used, including malt extract agar, potato dextrose agar, and dichloran rose bengal chloramphenicol (DRBC) agar. It DRBC be the media of choice, 100 mg/l chloramphenicol (achieved by dissolving 0.1 g chloramphenicol in 40 ml distilled water added to 960 ml agar) should be added prior to autoclaving. Sample dilutions should be analyzed by spread plate technique in triplicate and the plates incubated for 5 to 7 days at 25°C. The presence of molds can make counting the plates difficult, so typically, plates containing between 10 and 150 cfu should be counted to determine the *mold and yeast count* (MYC).

3. Indicator Bacteria: Enterobacteriaceae and *E. coli*

An indicator organism can be defined as a "microorganism or group of microorganisms that are indicative that a food has been exposed to conditions that pose an increased risk that the food has been contaminated with a pathogen or held in conditions conducive to pathogen growth" [106]. As such, indicator bacteria may be used as a means to assess and control the sanitary quality of food or to validate the effectiveness of microbiological control measures in a process designed to inhibit or eliminate a pathogen. To achieve this function, a microorganism must fulfil a number of criteria. Among other requirements, the organism must have a history of always being present in the food when the target pathogen is isolated and, conversely, it must be absent from the food when the target is not present; it must be inactivated by all conditions that would inactivate the target; its growth rate should be the same or greater than the target; it should be easily detected in the food environment and measurable in a short time, usually less than product holding time at each test point; and finally, it should not pose a health risk to analyst if handled properly [107,108]. Even if an indicator fulfils these requirement, it is important to note that the presence of an indicator is not conclusive evidence that a pathogen is present or is its absence proof that a target pathogen is not present. The genus Enterobactericeae may be used as indicators when their presence in a product is considered important and where the organisms share the same sources and routes of contamination as pathogenic species of the same family [109]. Impending EU criteria recommend the detection or enumeration of *E. coli* in vegetable products, shellfish, minced meat and meat preparations, and cheese. Enterobacteriaceae, as opposed to the traditional coliforms, is mooted as a sanitary indicator of animal carcasses, egg products, pasteurized milk, milk and whey powder, and ice cream. The switch from fecal coliform to the more defined Enterobacteriaceae group, which ferment glucose to produce acid and gas, stems from the fact that it includes coliforms, nonlactose-fermenting pathogens *Salmonella* and *Shigella*, enterotoxigenic *E. coli*, as well as heat-resistant *Klebsiella* and *Citrobacter*.

The pour plate method used for the enumeration of Enterobacteriaceae and *E. coli* is similar to that described in the FDA's BAM by Feng et al. [110]. Sample dilutions are prepared as described earlier and 1 ml of each dilution is inoculated in duplicate. Pour a layer (8 to 10 ml) of tryptic soy agar (TSA), which has been sterilized and cooled to 45°C into each plate and swirl gently to mix the sample and allow to solidify. After a lapse of approximately 2 h, a second layer of agar, this time melted and cooled violet red bile glucose (VRBGA) is overlaid onto the TSA. Once this layer has solidified, the plates should be incubated at 44°C for 18 h for *E. coli* or 32°C for 24 h for total Enterobacteriaceae. Count all 1 to 2 mm purple-red colonies with a purple zone of precipitated bile acids, on plates with between 25 and 250 colonies. Confirmatory tests should be done on at least ten representative colonies. These tests include gas production, gram stain, and IMViC, or in the case of *E. coli* API120E, biochemical analyses. By incorporating 4-methylumbelliferyl β-D-glucuronide (MUG) into the the agar, β-glucuronidase activity of *E. coli* can be detected under ultraviolet light (365 nm) as the colonies emit a blue fluorescence. MPN methods are also used in the detection of *E. coli* and use of the lauryl sulfate tryptose–MUG (LST–MUG) test for frozen foods has been adopted as official final action by the AOAC [110].

4. Pathogen Detection

a. *Salmonella sp.*

The main resevoir for *Salmonella* is the intestinal tract of animals, consequently food of animal origin or produced from ingredients of animal organ, particularly pigs and poultry, may become contaminated with this pathogen. The microbiological guidelines for *Salmonella* is usually given as absence in 25 g and under the new EU Microbiological Criteria, all meat, milk, and egg products must adhere to this guideline.

The method used for *Salmonella* detection as directed by the EU Microbiological Criteria is ISO 6579:2002 but other standard methods are also in use (APHA, BAM/AOAC). *Salmonella* isolation is a multistage process and takes 4 days to report a negative result. A test portion is sampled and weighed as described earlier. This is aseptically added to 225 ml of sterile pre-enrichment broth, for example, buffered peptone water [99] or lactose broth (BAM) and incubated at 35°C for 16 to 20 h to allow resuscitation of injured cells.

The next step is a selective enrichment step which promotes the growth of *Salmonella* sp. while repressing the background flora. From the pre-enriched sample bottle, 10 ml is aseptically transferred to 100 ml selenite/cystine broth, which is incubated at 35°C for 24 h. Additionally, 0.1 ml of the culture is aseptically transfered to 10 ml magnesium chloride/malachite green medium (Rappaport–Vassiliadis [RV] medium and tetrathionate [TT] broth are used in the case of BAM protocol) and incubated at 42°C for 24 h.

After incubation is complete, the next stage involves streaking a loopful of culture from each selective enrichment treatment on to a number of selective agars, for example, hectoen enteric (HE), bismuth sulfite (BS) agar, and xylose lysine desoxycholate (XLD) agar. The plates are incubated at 35°C for 24 h. After incubation, five characteristic colonies (blue-green to blue with or without black centers — HE; Pink, with or without black centers — XLD; brown, gray, or black colonies — BS) are streaked on nutrient agar and incubated at 35°C for 18 to 24 h. Colonies are then removed for biochemical and serological analysis.

b. *Staphylococcus aureus*

Staphylococcal food poisoning occurs as a result of heat stable enterotoxin production in the food product. Enterotoxin is not normally produced before the population reaches 10^6 cfu/g at which point, unless it contains a high level of salt, the food is almost invaribly spoilt. However, because *S. aureus* is commonly isolated from the nasal cavity, skin, and infected lesions of man, levels of between 10^2 and 10^3 cfu/g may be used as an indication of poor handling or sanitation practices. In most microbiological specifications, a GMP maximum level (*m*) of $<$10 cfu/g and a safety maximum level (*M*) of 100 cfu/g is recommended for most food products. Coagulase-positive staphylococci are generally regarded as *S. aureus* [111]. The enumeration procedure as specified by EN/ISO 6888 parts 1 and 2 and BAM/AOAC is a spread plate method. Samples and dilutions, are prepared as described earlier. Three Baird parker plates, which have been surface dried are inoculated with 1 ml for each dilution and the sample is evenly spread over the surface of the plate with a sterile glass rod. Plates are inverted and incubated at 37°C for 48 h. Only black, shiny convex colonies with a narrow margin of white precipitate and a zone of clearing are counted as presumptive *S. aureus* and are tested for gram-stain reaction and coagulase activity, or, if preferred, a rapid latex agglutination test (AUREUS TEST, Trisum Corp., Tiawan) may be used [112].

c. *Listeria monocytogenes*

L. monocytogenes is present ubiquitously in food processing environments and is a frequent contaminant of RTE foods. EU Microbiological Criteria will propose that *L. monocytogenes* be absent in 25 g for RTE products after manufacturing and less than 100 cfu/g within the product shelf-life, for the general population. Standard methodology is available from EN/ISO 11290 parts 1 and 2, BAM, AOAC, and APHA. Samples weighing 25 g are pre-enriched in buffered *Listeria* enrichment broth at 30°C for 4 h, after which time selective agents (10 mg/l acriflavin, 40 mg/l sodium naladixate, and an optional 50 mg/l cyclohexamide or 25 mg/l natamycin) are added and pre-enrichment is continued of a further 44 h. Enrichment culture is streaked at 24 and 48 h on Oxford agar to isolate *Listeria* sp. Some rapid method kits are available for specific food products [113]. If *Listeria* sp. are present, enumeration of a reserve sample should be performed by spread plating sample dilutions on Oxford agar to give presumptive *Listeria* counts. Characteristic black presumptive *L. monocytogens* colonies should be streak on trypticase soy agar with 0.6% yeast extract (TSAye) and tested for tumbling motility at $<$30°C — using either microscopic (+) or motility test medium (umbrella-like growth pattern), catalase reaction (+), gram reaction (+), carbohydrate fermentation, and hemolysis on sheep blood agar (+). A number of rapid kits are available, which can confirm *Listeria* colonies from Oxford media as *L. monocytogenes*, for example, VIDAS (bioMeriux, Basingstoke, Hampshire, U.K.), Probelia (Biocontrol, Seattle, WA), and BAX (Qulaicon, Inc., Wilmington DE) *L. monocytogenes* test kits.

5. Rapid Methods

To assure the safety of consumers, it is vital that microbiological methods deliver rapid and reliable results on the presence of pathogens in the global food supply to the food industry and food safety regulatory bodies. As can be seen from the preceeding section, although traditional culture methods requiring growth, isolation, and confirmation steps are indeed reliable and remain the "gold standard" for the detection of pathogens, they are time consuming and often laborious. Since the

TABLE 27.7
Commercially Available Miniaturized Biochemical Kits and Automated Systems

System	Manufacturer	Format	Target
API	bioMerieux	Biochemical (miniaturized kit)	Enterobacteriaceae, *Listeria, Staphylococcus, Campylobacter*, anaerobes, nonfermenters
Vitek	bioMerieux	Automated	Enterobacteriaceae, gram-negative microbes, gram-positive microbes
Gene-Trak	Neogen	Nucleic acid (PCR)	*Listeria, Salmonella*
Assurance	BioControl	Antibody (ELISA)	*Salmonella, Listeria, E. coli* O157:H7
VIP	BioControl	(Ab. ppt)	*Listeria, E. coli* O157:H7
Listeria-Tek	Organon	(ELISA)	*Listeria*
Salmonella-Tek	Teknika	(ELISA)	*Salmonella*
TECRA	TECRA	(ELISA)	*E. coli* O157:H7, *Salmonella, Staphylococcus enterotoxin*
VIDAS	bioMerieux	(ELFA)	*Listeria, Salmonella*
UNIQUE	TECRA	Capture EIA	*Salmonella*
1-2 Test	BioControl	Diffusion	*Salmonella*
Aureus Test	Trisum	Latex agglutination	*S. aureus*

Source: Adapted from P Feng. In: *Bacteriological Analytical Manual Online*, 8th ed. Revision A. U.S. Food and Drug Administration, Center for Food Safety and Applied Nutrition, 5600 Fishers Lane, Rockville, MD, 20857-0001, U.S.A, 2004. With permission.

mid-1960s, the attention of food microbiologists turned to the development of more convenient and rapid methods. Initially, the focus was on the development of miniaturized diagnostic kits, followed by immunological kits, genetic probes and development of qualitative polymerase-chain reaction (PCR) — and now quantitative real-time PCR methods [114]. Currently, the emphasis is on the development of microarrays, biochips, and biosensors [2]. Rapid methods are usually used in sample screening, and while negative results stand, a positive result must be validated by the appropriate standard method [115]. Commercial rapid methods are usually target-specific and return results in a matter of hours and sometimes minutes, which is a great advantage to the food industry. However, because most rapid methods lack sensitivity for direct testing, an enrichment step is often required. Although there is an increasingly large number of commercial kits on the market, only a limited number are officially validated for use in food testing [115]. Table 27.7 and Table 27.8 summarize some commercially available kits.

TABLE 27.8
Commercially Rapid Methods and Speciality Substrate Methods

System	Manufacturer	Format	Target
Isogrid	QA Labs	HGMF	Coliforms, *E. coli, Salmonella*
Redigel	RCR Scientific	Media	Bacteria
Petrifilm	3M	Media film	*E. coli*, Coliforms
Colilert	IDexx	MPN–MUG	*E. coli*, Coliforms
LST–MUG	Difco and GIBCO	MPN media	*E. coli*, Coliforms
Coli complete	BioControl	Mug-Xgal	*E. coli*, Coliforms

Source: Adapted from P. Feng. Rapid methods for detecting foodborne pathogens. In: *Backteriological Analytical Manual Online*. 8th ed. Revision A. U.S. Food and Drug Administration, Center for Food Safety and Applied Nutrition, 5600 Fishers Lane, Rockville, MD, 20857-0001, U.S.A., 2004. With permission.

V. CONCLUSIONS

Frozen foods have enjoyed an excellent safety record, being associated with only a few outbreaks of foodborne illness to date. In general, rapid product freezing and storage at stable low temperatures ($<-18^\circ$C) will give the best quality frozen foods. However, such conditions are also optimal for pathogen survival. Conscientious use of GMPs and HACCP in the production of frozen foods and rigourous monitoring of pathogens, in compliance with food hygiene and safety regulations, will ensure that the record of frozen foods on the market will be maintained for decades to come.

REFERENCES

1. J Schlundt. New directions in foodborne disease prevention. *International Journal of Food Microbiology* 78 (1–2):3–18, 2002.
2. DYC Fung. Rapid methods and automation in microbiology. *Comprehensive Reviews in Food Science and Food Safety* 1:3–21, 2002.
3. H Riemann, FL Bryan. *Foodborne Infections and Intoxications*, 2nd ed. New York, USA: Academic Press, 1979.
4. WM Waites, JP Arbuthhnott. *Food Borne Illness — A Lancet Review*. London, UK: Edward Arnold, 1991.
5. P Gélinas. *Handbook of Foodborne Microbial Pathogens*. Quebec, Canada: Quebec: Polyscience Publications Inc., 1997.
6. S Goh, R Reacher, DP Casemore, NQ Verlander, R Chalmers, M Knowles, J Williams, K Osborn, S Richards. Sporadic Cryptosporidiosis, North Cumbria, England, 1996–2000. *Emerging Infectious Diseases* 10 (6):1007–1015, 2004.
7. C Bern, B Hernandez, M Beatriz Lopez, MJ Arrowood, MA de Mejia, AM de Merida, AW Hightower, L Venczel, BL Herwaldt, RE Klein. Epidemiologic studies of *Cyclospora cayetanensis* in Guatemala. *Emerging Infectious Diseases* 5 (6):766–774, 1999.
8. BW Furness, MJ Beach, JM Roberts. Giardiasis surveillance — United States, 1992–1997. *CDC Morbididity and Mortality Weekly Report* 49 (SS07):1–13, 2000.
9. S Arista, GM Giammanco, S de Grazia, MC Migliore, V Martella, A Cascio. Molecular characterization of the genotype G9 human rotavirus strains recovered in Palermo, Italy, during the winter of 1999–2000. *Epidemiology and Infection* 132 (2):343–349, 2004.
10. P Le Cann, S Ranarijaona, S Monpoeho, F Le Guyader, V Ferre. Quantification of human astroviruses in sewage using real-time RT-PCR. Research in Microbiology 155 (1):11–15, 2004.
11. MD Koci, LA Moser, LA Kelley, D Larsen, CC Brown, SS Cherry. Astrovirus induces diarrhea in the absence of inflammation and cell death. *Journal of Virology* 77 (21):11798–11808, 2003.
12. MAS de Wit, MPG Koopmans, YTHP van Duynhoven. Risk factors for Norovirus, Sapporo-like virus, and group A rotavirus gastroenteritis. *Emerging Infectious Diseases* 9 (12):1563–1570, 2003.
13. S Robinson, IN Clarke, IB Vipond, EO Caul, PR Lambden. Epidemiology of human Sapporo-like caliciviruses in the South West of England: molecular characterisation of a genetically distinct isolate. *Journal of Medical Virology* 67 (2):282–288, 2002.
14. CI Gallimore, J Green, AF Richards, H Cotterill, A Curry, DWG Brown, JJ Gray. Methods for the detection and characterisation of noroviruses associated with outbreaks of gastroenteritis: outbreaks occurring in the North-West of England during two norovirus seasons. *Journal of Medical Virology* 73 (2):280–288, 2004.
15. LA Jaykus. Enteric virusesas 'emerging' agents of foodborne disease. *Irish Journal of Agriculture and Food Research* 39 (2):245–255, 2000.
16. C Beuret. Simultaneous detection of enteric viruses by multiplex real-time RT–PCR. *Journal of Virological Methods* 115 (1):1–8, 2004.
17. CD Kirkwood, RF Bishop. Molecular detection of human calicivirus in young children hospitalized with acute gastroenteritis in Melbourne, Australia, during 1999. *Journal of Clinical Microbiology* 39 (7):2722–2724, 2001.

18. AI Sair, DH d'Sousa, LA Jaykus. Human enteric viuses as causes of foodborne disease. *Comprehensive Reviews in Food Science and Food Safety* 1:74–75, 2002.
19. GJ Arnold, BA Munce. Investigation of foodborne disease outbreaks. In: AD Hocking, G Arnold, I Jenson, K Newton, P Sutherland Eds., *Foodborne Microorganisms of Public Health Significance*. 5th ed. AIFST (NSW Branch) Food Microbiology Group: North Sydney Australia, 1997, pp 31–70.
20. BC Hobbs, D Roberts. Bacterial and other microbial agents of food poisoning and foodborne infection. In: *Food Poisoning and Food Hygiene*, 6th ed. Kent, UK: Edward Arnold, 1993, pp 26–50.
21. Anonymous. In: Bad Bug Book — Foodborne Pathogenic Microorganisms and Natural Toxins Handbook. U.S. Food and Drug Admisintration, Center for Food Safety and Applied Nutrition, 5600 Fishers Lane, Rockville, MD, 20857–0001, U.S.A.
22. CP Gerba, JB Rose, CN Haas. Sensitive populations: Who is at greatest risk? *International Journal of Food Microbiology* 30 (1–2):113 – 123, 1996.
23. Anonymous. Food Safety for Persons with AIDS. Consumer Publications. United States Department of Agriculture Food Safety and Inspection Service, Washington DC, USA, 1996.
24. Council for Agricultural Science and Technology. *Foodborne pathogens: risks and consequences*. Iowa, U.S.A.: Ames,1994, pp. 1–87.
25. M Satin. *Food Sources of Disease. Food Alert! The Ultimate Sourcebook for Food Safety*. New York: Checkmark Books, 1999, pp. 41–96.
26. JA Lindsay. Chronic sequelae of foodborne disease. *Emerging Infectious Disease* 3 (4):443–452, 1997.
27. Anonymous. The Role of Food Safety in health and development. Report of a Joint FAO/WHO Expert Committee on food safety. WHO Technical Report 705. World Health Organisation, Geneva, 1984.
28. Anonymous. Enhancing foodborne disease surveillance across Australia in 2001: the OzFoodNet Working Group. *Communicable Disease Intelligence* 26 (3):375–406, 2002.
29. RS Kovats, SJ Edwards, S Hajat, BG Armstrong, KL Ebi, B Menne, and Collaborating Group. The effect of temperature on food poisoning: a time-series analysis of salmonellosis in ten European Countries. *Epidemiology and Infection* 132 (4):443–453, 2004.
30. PS Mead, L Slutsker, V Dietz, LF McCaig, JS Bresee, C Shapiro, PM Griffen, R Tauxe. Food related illness and death in the United States. *Emerging Infectious Disease* 5 (5):607–625, 1999.
31. BA Lopman, MH Reacher, Y van Duijnhoven, FX Hannon, D Brown, M Koopmans. Viral gastroenteritis outbreaks in Europe, 1995–2000. *Emerging Infectious Disease* 9 (1):90–96, 2003.
32. Anonymous. *Foodborne Disease: A Focus for Health Education.* World Health Organisation, Geneva, Swizerland, 2000, pp. 1–33.
33. GK Adak, SM Long, SJ O'Brien. Trends in indigenous foodborne disease and deaths, England and Wales: 1992 to 2000. *Gut* 51 (6):832–841, 2002.
34. D Vugia, A Cronquist, J Hadler, D Blake, K Smith, D Morse, J Cieslak, T Jones, D Goldman, F Angulo, DV Griffen, R Tauxe, K Kretsinger. Preliminary FoodNet data on the incidence of infection with pathogens transmitted commonly through food — selected sites, United States, 2003. *Morbidity and Mortality Weekly Reports* 53 (16):338–343, 2004.
35. Anonymous. Economics of foodborne disease 2000. Available at Economic Research Service, 1800 M Street NW, Washington DC, 20036–5831, USA, 2004.
36. Anonymous. Economic costs of foodborne disease in OECD Countries, In: *Foodborne Diseases in OECD Countries: Present State and Economic Costs*. Paris, France: OECD Publications, 2003, pp. 61–82.
37. DL Archer Freezing: an underutilized technology. *International Journal of Food Microbiology* 90 (4):127–138, 2004.
38. BM Lund. Freezing. In: BM Lund, TC Baird Parker, GW Gould, Eds. *The Microbiological Safety and Quality of Food*. Vol. 1. Maryland, USA: Aspen Publishers, 2000, pp. 122–145.
39. NH Bean, JS Goulding, C Lao, F J Angulo. Surveillance for foodborne-disease outbreaks — United States, 1988–1992. *Centers for Disease Control (CDC) Morbidity and Mortality Weekly Report* 45 (SS–5):1–55, 1996.
40. SJ Olsen, LC MacKinon, JS Goulding, NH Bean, L Slutsker. Surveillance for foodborne-disease outbreaks United States, 1993–1997. *CDC Morbidity and Mortality Weekly Report* 49 (SS–1):1–67, 2000.

41. RA Gunn, G Markakis. Salmonellosis associated with homemade ice-cream — outbreak report and summary of outbreaks in United-States in 1966 to 1976. *Journal of the American Medical Association* 240 (17):1885–1886, 1978.
42. P Buckner, D Ferguson, F Anzalone, D Anzalone, J Taylor, WG Hlady, RS Hopkins. Outbreak of *Salmonella enteritidis* associated with homemade ice cream — Florida, 1993. *CDC Morbidity and Mortality Weekly Report* 43 (36):669–671, 1994.
43. A Ponk, L Mannula, CH von Bonsdurff, O Lyytilkainene. Outbreak of calcivirus gastroenteritis associated with eating frozen raspberries. *Epidemiology and Infection* 123 (3):469–474, 1999.
44. TMS Reid, HG Robinson. Frozen raspberries and Hepatitis A. *Epidemiology and Infection* 98 (1):109–112, 1987.
45. Anonymous. Hepatitis A associated with the consumption of frozen strawberries — Michigan, March 1997. *CDC Morbidity and Mortality Weekly Report* 46 (13):288–289, 1997.
46. YJF Hutin, V Pool, EH Cramer, OV Nainan, J Weth, I T Williams, ST Goldstein, KF Gensheimer, BP Bell, CN Shapiro, MJ Alter, HS Margolis. A multistate, foodborne outbreak of Hepatitis A. *New England Journal of Medicine* 340 (8):596–602, 1999.
47. RO Cannon, JR Poliner, RB Hirschhorn, DC Rodeheaver, PR Silverman, EA Brown, GH Talbot, SE Stine, SS Monroe, DT Dennis, RI Glass. A multistate outbreak of Norwalk virus gastroenteritis associated with consumption of ice. *Journal of Infectious Disease* 164 (5):860–863, 1991.
48. D Boccia, AE. Tozzi, B Cotter, C Rizzo, T Russo, G Buttinelli, A Caprioli, ML Marziano, FM Ruggeri. Water borne outbreak of Norwalk like virus gastroenteritis at a tourist resort, Italy. *Emerging Infectious Disease* 8 (6):563–568, 2002.
49. Anonymous. *E coli* O157:H7 infections associated with eating a nationally distributes commercial brand of ground patties and burgers. *CDC Morbidity and Mortality Weekly Reports* 46 (33):288 – 289, 1997.
50. CP Champagne, N Gardner, E Brochu, Y Beaulieu. The freeze-drying of lactic acid bacteria. A review. *Canadian Institute of Food Science and Technology Journal* 24 (3–4):118–128, 1991.
51. BCS To, MR Etzel. Survival of *Brevibacterium linens* (ATCC 9174) after spraydrying, freeze drying or freezing. *Journal of Food Science*, 62 (1):167–170, 189, 1997.
52. Z Hubalek. *Cryopreservation of Microorganisms at Ultra-low Temperatures*. Prague: Academia, 1996.
53. CP Champagne, F Mondou, Y Raymond, D Roy. Effect of polymers on the stability of freezedried lactic acid bacteria. *Food Research International* 29 (5–6):555–562, 1996.
54. JM Panoff, B Thammavongs, M Guéguen. Cryoprotectants lead to phenotypic adaptation to freeze-thaw tress in *Lactobacillus delbruekii* ssp. Bulgaricus CIP 101027T. *Cryobiology* 40 (2):264–269. 2000.
55. F Fonseca, C Béal, F Mihoub, M Marin, G Corrieu. Improvement of cryopreservation of *Lactobacillus delbrueckii* subsp. *bulgaricus* CFL1 with additives displaying different cryoprotective effects. *International Dairy Journal* 13 (5):917–926, 2003.
56. NP Shah, RR Ravula. Microencapsulation of probiotic bacteria and their survival in frozen fermented dairy desserts. *The Australian Journal of Dairy Technology* 55 (3):139–144, 2000.
57. JH Tsen, HH Chen, AE King. Survival of freeze-dried *Lactobacillus acidophilus* immobilised in κ-carrageenan gel. *Journal of General Applied Microbiology* 48 (6):237–241, 2002.
58. T Mayes, G Telling. Product safety from factory to consumer. In: CP Mallett, Ed., *Frozen Food Technology*. London, UK: Blackie Academic and Professional, 1996, pp. 93–121.
59. CP Champagne, RR Laing, D Roy, AA Mafu. Psychrotrophs in dairy products: their effects and their control. *Critical Reviews in Food Science and Nutrition* 34 (1):1–30, 2004.
60. B Ray. Impact of bacterial injury and repair in food microbiology: its past, present and future. *Journal of Food Protection* 49 (8):651–655, 1986.
61. ED Berry, PM Foegeding. Cold temperature adaptation and growth of microorganisms. *Journal of Food Protection*, 60 (12):1583–1594, 1997.
62. N Beals. Adaptation of microorganisms to cold temperatures, weak acid preservatives, low pH, and osmotic stress: a review *Comprehensive Reviews in Food Science and Food Safety* 3:1–20, 2004.
63. JR Broadbent, C Lin. Effect of coldshock treatment on the resistance of *Lactococcus lactis* to freezing and lyophilization. *Cryobiology* 39 (1):88–102, 1999.

64. J Bollman, A Ismond, G Blank. Survival of *Escherichia coli* O157:H7 in frozen foods: impact of the cold shock response. *International Journal of Food Microbiology* 64 (10):127–138, 2001.
65. F Mihoub, MY Mistou, A Guillot, JY Leveau, A Boubetra, F Billaux. Cold adaptation of *Escherichia coli*: microbiological and proteomic approaches. *International Journal of Food Microbiology* 89 (2):171–184, 2003.
66. O Geiges. Microbial processes in frozen food. *Advances in Space Research* 18 (12):109–118, 1996.
67. SE El-Kest, EH Marth. Freezing of *Listeria monocytogenes* and other microorganisms: a review. *Journal of Food Protection* 55 (8):639–648, 1992.
68. JR Sage, SC Ingham. Potential use of a hydrophobic grid membrane filter -SD-39 agar method. *Journal of Food Protection* 61 (4):490–494, 1998.
69. JR Sage, SC Ingham. Evaluating the survival of *Escherichia coli* O157:H7 after freezing and thawing in ground beef patties. *Journal of Food Protection* 61 (9):1181–1183, 1998.
70. SA Yammamoto, LJ Harris. The effects of freezing on the survival of *Escherichia coli* O157:H7 in apple juice. *International Journal of Food Microbiology* 67 (1):89–96, 2001.
71. HK Cressy, AR Jerrett, CM Osborne, PJ Bremer. A novel method for the reduction of numbers of *Listeria monocytogenes* cells by freezing in combination with an essential oil in bacteriological media. *Journal of Food Protection* 66 (3):390–395, 2003.
72. L Picart, E Dumay, JP Guiraud, JC Cheftel. Microbial inactivation by pressure-shift freezing: effects on smoked salmon mince inoculated with *Pseudomonas fluorsecens, Micrococcus leuteus* and *Listeria innocua. Lebensmittel-Wissenschaft und Technologie* 37 (1):227–238, 2004.
73. P Mazur. Physical and chemical basis of injury in single-celled microorganisms subjected to freezing and thawing. In: HT Meryman, Ed. *Cryobiology*. London and New York: Academic Press, 1966, pp. 213–315.
74. EL Glovlev. Bacterial cold shock response at the level of DNA transcription, translation and chromosome dynamics. *Microbiology* 72 (1):1–7, 2003.
75. PH Calcott, RA McLeod. Survival of *Escherichia coli* from freeze thaw damage: influence of nutritional status and growth reat. *Canadian Journal of Microbiology* 20 (5):683–689, 1974.
76. TE Minor, EH Marth, Loss of viability by *Staphylococcus aureus* in acidified media 2: inactivation by acids in combination with sodium chloride, freezing and heat. *Journal of Milk and Food Technology* 35 (9):548–555, 1972.
77. F Fonseca, C Béal, G Corrieu. Operating conditions that affect the resistance of lactic acid bacteria to freezing and frozen storage. *Cryobiology* 43 (1):189–198, 2001.
78. R Davies, A Obafemi. Response of microorganismsto freeze–thaw stress. In: RK Robinson, Ed. *Microbiology of Frozen Foods*. London and New York: Elsevier Applied Science Publishers Ltd., 1985, pp. 83–108.
79. L Borg-Sorensen. Maintaining safety in the cold chain. In: CJ Kennedy, Ed., *Managing frozen foods*. Cambridge, England: Woodhead Publishing Ltd., 2002, 5–26.
80. GP Archer, CJ Kennedy, A J Wilson, Position Paper: towards predictive microbiology in frozen food systems — a framework for understanding in microbial population dynamics in frozen structures and in freeze–thaw cycles. *Journal of Food Science and Technology* 30 (6):711–723, 1995.
81. M Abdussalem, FK Käferstein, KE Mott. Food safety measures for the control of foodborne trematode infections. *Food Control* 6 (2):71–79, 1995.
82. M Gianfranceschi, P Aureli. Freezing and frozen storage on the survival of *Listeria monocytogenes* in different foods. *Italian Journal of Food Science* 4 (2):303–309, 1996.
83. JP Dean, EA Zottola. Use of nisin in ice cream and the effect on the survival of *Listeria monocytogenes. Journal of Food Protection* 59 (5):476– 480, 1996.
84. G Jeyasekaran, I Karunasagar. Effect of chilling and freezing on the survival of *Listeria monocytogenes* in shrimps. *Journal of Food Science and Technology* 39 (2):191–193, 2002.
85. ACS Porto, JE Call, JB Luchansky. Effect of reheating on the viability of a five-strain mixture of *Listeria monocytogenes* in vacuum-sealed packages of frankfurters following refrigerated or frozen storage. *Journal of Food Protection* 67 (1):71–76, 2004.
86. D Gradkowska, MW Griffiths. Cryotolerance of *Escherichia coli* O157:H7 in laboratory media and food. *Journal of Food Science* 66 (8):1169–1173, 2001.
87. M Raccach, T Bamiro, J Clince, G Combs. A Giercznski, R Karam, Frozen storage of *Escherichia coli* O157:H7 in vegetable broth. *Food Control* 13 (3):381–385, 2002.

88. GA Dykes, SM Moorhead. Survival of three *Salmonella* serotypes on beef trimmings during simulated commercial freezing and frozen storage. *Journal of Food Safety* 21 (2):87–96, 2001.
89. TA Nassib, MZ El-Din, WM El-Sharoud. Viability of *Salmonella enterica* subsp. *enterica* during the preparation of Egyptian soft cheeses and ice cream. *International Journal of Dairy Technology* 56 (1):30–34, 2003.
90. KS Yoon, TP Oscar. Survival of *Salmonella typhimurium* on sterile ground chicken patties after washing with salt and phosphates and during refrigerated and frozen storage. *Journal of Food Science* 67 (2):772–775, 2002.
91. PD Lowry, CO Gill. Mould growth on meat at freezing temperatures. *International Journal of Refrigeration* 7 (2):133–136, 1984.
92. MC Lopez, LM Medina, MG Cordoba, R Jordano. Fungal contamination of yoghurt ice-cream. *Microbiology Ailments Nutrition* 16 (1):107–112, 1998.
93. EA Moghazy, MOA El-Saaarawy. Quality attributes of beef burger as affected by using propolis and frozen storage. *Egyptian Journal of Agricultural Research* 79 (4); 1149–1151, 2001.
94. L Tejada, E Sánchez, MV Gómez, J Fernández-Salguero. Effect of freezing and frozen storage on chemicals and microbiological characteristics in sheep milk cheese. *Journal of Food Science* 67 (1):126–129, 2002.
95. P Balkir, GF Öztürk. Effect of curd freezing on the physicochemicaland microbiological characteristics of Crottin de Chavignol type lactic goats cheese. *Milschwissenschaft* 58 (11/12):615–619, 2003.
96. BM Mackey, Injured bacteria. In: BM Lund, TC Baird Parker, GW Gould, Eds. *The Microbiological Safety and Quality of Food.* Vol. 1. Maryland, USA: Aspen Publishers, 2000, pp 315–341.
97. R Holbrook, JM Anderson, AC Baird-Parker. Modified direct plate method for counting *Escherichia coli* in food. *Food Technology Australia* 32 (1):78–83, 1980.
98. Anonymous. Microorganisms in Foods 2 — Sampling for Microbiological Analysis Principles and Applications, 2nd ed. International Commission on Microbial Specifications for Food (ICMSF), Canada: University of Toronto Press, 1986.
99. Anonymous. Compendium of methods for the microbiological examination of foods, 4th Ed. FP Downes, K Ito, Eds. Washington DC, USA: American Publlic Health Association (A.P.H.A.) Inc., 2001.
100. Anonymous. Official Methods of Analysis, 17th ed. W Horowiz, Ed. Maryland, USA: Association of Analytical Communities (AOAC) International, 2003.
101. CA White, LP Hall. Laboratory examination of frozen foods. In: RK Robinson, Ed. *Microbiology of Frozen Foods*. London and New York: Elsevier Applied Science Publishers Ltd., 1985, pp. 251–284.
102. YO Thomas, WJ Lulves, AA Kraft. A convenient surface plate method for bacteriological examination of poultry. *Journal of Food Science* 46 (6):1951–1952, 1981.
103. LJ Maturin, JT Peeler. Aerobic plate count. In: *Bacteriological Analytical Manual Online.* 8th ed. Revision A. U.S. Food and Drug Admisintration, Center for Food Safety and Applied Nutrition, 5600 Fishers Lane, Rockville. MD, 20857-0001, U.S.A., 2004.
104. WJ Scott. Water relations of food spoilage microorganisms. *Advances in Food Research* 7 (1):83–127, 1957.
105. V Tournas, ME Stack, HA Kock, R Bandler. Yeasts, moulds and mycotoxins. In: *Bacteriological Analytical Manual Online.* 8th ed. Revision A. U.S. Food and Drug Admisintration, Center for Food Safety and Applied Nutrition, 5600 Fishers Lane, Rockville,. MD, 20857-0001, U.S.A., 2004.
106. FF Busta, TV Suslow, ME Parish, LR Beuchat, JN Farber, EH Garrett, LJ Harris. The use of indicators and surrogate microorganisms for the evaluation of pathogens in fresh and fresh-cut produce. *Comprehensive Reviews in Food Science and Food Safety* 2 (Supplement):179–185, 2003.
107. JM Jay. Indicators of food microbial quality and safety. In: *Modern Food Microbiology*, 6th ed. Maryland, USA: Aspen, 2000, pp. 387–406.
108. RL Buchanan. Acquisition of microbiological data to enhance food safety. *Journal of Food Protection* 63 (6):832–838, 2000.
109. LJ Cox, N Keller, M van Schothorst. The use and misuse of quantitative determinations of Enterobacteriaceae in food microbiology. *Journal of Applied Bacteriology* 65 (Symposium Supplement): S237–S249, 1988.

110. P Feng, JD Weagent, MA Grant. Enumeration of *Escherichia coli* and the coliform bacteria. In: *Bacteriological Analytical Manual Online.* 8th ed. Revision A. U.S. Food and Drug Admisintration, Center for Food Safety and Applied Nutrition, 5600 Fishers Lane, Rockville. MD, 20857-0001, U.S.A., 2004.
111. SHW Notermans, GC Mead. Microbiological contamination of food: analytical aspects. In: K van der Heijden, M Younes, L Fishbein, S Miller, Eds. *International Food Safety Handbook*. New York: Marcel Dekker, Inc., 1999, pp 549–566.
112. RW Bennett, GA Lancette. *Staphylococcus aureus.* In: *Bacteriological Analytical Manual Online. Bacteriological Analytical Manual Online.* 8th ed. Revision A. U.S. Food and Drug Admisintration, Center for Food Safety and Applied Nutrition, 5600 Fishers Lane, Rockville, MD, 20857-0001, U.S.A., 2004.
113. AD Hitchins. Detection and enumeration of *Listeria monocytogenes* in foods. In: *Bacteriological Analytical Manual Online*, 8th ed. Revision A. U.S. Food and Drug Admisintration, Center for Food Safety and Applied Nutrition, 5600 Fishers Lane, Rockville, MD, 20857-0001, U.S.A., 2004.
114. NP Rijpens, LMF Herman. Molecular methods for identification and detection of bacterial food pathogens. *Journal of AOAC International* 85 (4):984–993, 2002.
115. P Feng. Rapid methods for detecting foodborne pathogens. In: *Bacteriological Analytical Manual Online*, 8th ed. Revision A. U.S. Food and Drug Admisintration, Center for Food Safety and Applied Nutrition, 5600 Fishers Lane, Rockville, MD, 20857–0001, U.S.A, 2004.

28 Shelf-Life Prediction of Frozen Foods

Brian McKenna
University College Dublin, Ireland

CONTENTS

I. INTRODUCTION

It is difficult to produce a common method for the prediction of the shelf-life of frozen foods. Fresh or chilled foods normally have a single dominant deterioration mechanism (e.g., microbial spoilage). So, it is relatively easy to model the temperature changes in the product and to superimpose microbial growth and decay models on these temperatures, the integration of which over time will result in a good approximation of when the microbial load will exceed a safe limit and so define the safe shelf-life [1]. Lest one thinks that the foregoing sentence solves the problem for fresh and chilled products, let me quickly add that a deficiency in kinetic data on microbial growth and decay for spoilage organisms at the temperatures involved and their interactions with food composition make this a far from easy task.

For frozen foods, such an approach becomes an impossible task because of the multitude of spoilage mechanisms involved. There is a presumption that freezing stops most deterioration mechanisms. Although this may have some validity in the solid glassy state (see later) reached at very low freezing temperatures, normal frozen food storage temperatures (-18 to -20°C) are significantly higher than the glass transition temperature and will consequently contain some unfrozen water. Blond and Le Meste [2] present a table of typical glass transition temperatures for many foods. These range from -31°C for some juices down to -85°C for beef muscle.

Many publications summarize the spoilage mechanisms prevalent in frozen foods. These include enzymatic deterioration, cell damage and protein and starch interactions, nonenzymatic browning, water migration (both during freezing and storage), water recrystallization and change in crystalline form, solute crystallization, oxidative deterioration (e.g., lipid oxidation in fatty meats and color changes in fish and meat), protein denaturation (which may alter water-binding capacity), and lastly, microbial changes. This last deterioration mechanism is not of major significance because most frozen foods are stored at temperatures below the lower limits of microbial growth (approximately −10°C). However, with temperature fluctuations during storage and distribution, these may become significant.

II. FROZEN FOODS: WHY IT IS DIFFICULT TO PREDICT SHELF-LIFE

A. Unfrozen Water and Glass Transition

The process by which food freezes is now well understood. As heat is removed from the food, ice crystals will start to form once the temperature falls a little below its nominal freezing point which is normally in the range of −1 to −2°C (subcooling). The subsequent release of heat of crystallization will bring the temperature back to the nominal freezing point. However, unlike the freezing of a pure solvent, the effective concentrating effect of ice crystal formation on the liquid phase results in a progressive reduction in the nominal freezing temperature. As a result, temperature continues to decrease and the concentration of the remaining unfrozen liquid rises. Both effects contribute to a significant increase in viscosity of the unfrozen liquid. Also, depending on the rate of freezing, water migration may occur due to osmotic effects.

When the viscosity of the unfrozen liquid surrounding the ice crystals becomes very high (10^{11} to 10^{12} Pa s), solidification or vitrification may quickly occur and the remaining concentrated solution becomes a glass [2]. The temperature at which this occurs is known as the glass transition temperature. No further freezing of water occurs below this temperature. However, for many foods the glass transition temperature is considerably below the temperatures encountered in food freezing and storage, and results in small pockets of unfrozen water within the foods. So, deterioration of the food is not totally inhibited. It should be noted that for many foods the glass transition temperature is independent of the initial moisture content of the food. The reader is referred to Ref. [2] for a treatment on the methods of measuring the glass transition temperature using differential scanning calorimetry and for a discussion of the ambiguities that can arise in the interpretation of the results of such measurements.

Storage temperatures can fluctuate significantly over the complete cold chain and many temperature surveys have been published. Typically, −23°C has been the average for the manufacturer's cold store with a maximum target of no more than −18°C during fluctuations. This may rise to −18°C during distribution to either the wholesaler or the retailer. A further rise in the mean is quoted at retail level. Although −18°C is the target, the norm is closer to −15°C. These fluctuations become greater still when the consumer enters the chain. Transport to the home can be at anything up to 40°C for an hour in the back of a car in a hot climate and surface thawing may begin. The domestic freezer will probably be close to that in the retail outlet (but can have large variations) and will probably be accompanied by a less than ideal refreezing of the surface layer of a slightly thawed product. In addition, temperature changes during the defrost cycles at both retail and domestic levels can have a significant impact. Although the above figures are means extracted from a wide range of publications over the years, it is admitted that fluctuations of 3–5°C may occur in any one part of the chain. This in itself is a serious quality determinant as there is significant evidence that the shelf-life of frozen foods stored at fluctuating temperatures can be much shorter than that of foods stored at the constant mean temperature of the fluctuations. It must, of course, be noted that even at these low but fluctuating temperatures, the product is considerably above its glass transition temperature and, consequently, is not inert and deterioration mechanisms can continue. In particular, it should be noted that vitrification of any supersaturated liquid phase has not taken place.

B. Deterioration Mechanisms

Generally, microbial deterioration is not a problem with frozen foods. Unfortunately for the processor and consumer, this makes the prediction of shelf-life difficult for frozen foods because most of the available model systems are based on microbial prediction of deterioration [1]. In nonfrozen foods, the widespread availability of kinetic data on pathogenic organisms and the relatively sparse kinetic data on nonpathogenic spoilage organisms are the main source of difficulty when prediction spoilage rates in foods. In addition, the normal presence of a mixture of microorganisms may cause difficulties as there is little data on the synergic influences of the mixed flora on kinetic data. Although most bacteria do not grow below $-10^{\circ}C$, other deterioration mechanisms such as enzymatic spoilage may limit the storage life even at temperatures below this. Indeed, there are instances where the enzyme activity may be related to bacteria rather than the product itself and many of these (e.g., lipase and protease) may not be inactivated by the preparatory blanching process.

However, the effective absence of microbial growth at frozen food storage temperatures does not mean that spoilage is absent in such products. Other forms of spoilage, not as well characterized kinetically, will limit the potential storage period. Many of these spoilage mechanisms are influenced partially but not completely by the presence of the unfrozen water pockets within the food. A brief summary of these spoilage mechanisms is presented over the following paragraphs.

Enzymatic spoilage becomes the dominant spoilage mechanisms for frozen foods without microbial spoilage. Unblanched food products will normally encounter spoilage problems. There is a general agreement that blanching before freezing will reduce the problem. These problems obviously vary between products but flavor changes in fruit and vegetables are common. Some products such as meat and poultry may experience cell membrane damage during the blanching process. The freezing process may itself also cause cell damage that limits shelf-life and affects product quality. Drip loss and texture change are the major results. However, blanching to inactivate the enzymes is not universally successful. The thermal inactivation of some enzymes has been reported as reversible and the enzymes can recover their activity under certain conditions. It is reported that the reactivation of enzyme activity after inactivation by heat is one of the properties of lipoxygenase and peroxidase. Although there is evidence of such reactivation in model systems, there is insufficient research on reactivation in stored frozen foods to be definitive [3].

Lipid oxidation is another reaction that will severely limit the shelf-life of a frozen product. This is particularly true for meats (including poultry) and seafood. Even vacuum packaging will not eliminate this problem, as reaction with the molecular oxygen is often the major form of deterioration. Fatty meats and fish, in particular, suffer from this adverse reaction during long-term frozen storage. In addition, enzymatically promoted hydrolysis of the lipids can lead to fatty acid formation and rancidity with the consequent development of an unacceptable flavor. Indeed, pork, having a greater proportion of more reactive unsaturated fatty acids in its fat, will experience a greater degree of such change. This may also occur to a much more limited extent in frozen vegetables.

A less serious but nonetheless undesirable reaction is oxidative color change in a frozen product. The oxidation of myoglobin to meta-myoglobin, common is deterioration of fresh meat and some fish, can also be found after prolonged frozen storage. Although being more inhibitory than dangerous to the consumer, it will definitely lead to "detectable change" by consumer panels, a shelf-life determining factor outlined below. As a consequence, it is common for frozen products in the "prepared consumer foods" category to have antioxidants added to their formulation to inhibit these effects. Fruit products, in particular, can benefit from the addition of ascorbic acid to their formulation because fruit blanching, if used at all, is generally to a much milder heat treatment level than that applied to vegetables so as to avoid thermal degradation. The frozen product may well have a higher level of this desirable vitamin that is present in the corresponding fresh product due to addition of a high level (as an antioxidant) during the frozen product preparation.

Another of the less serious deterioration reactions is protein denaturation which can lead to loss of some properties such as water-binding capacity and protein solubility. The factors promoting such change can obviously vary between products but can sometimes be attributed to the development of high solute concentrations (with associated changes in pH) in the unfrozen phase due to a form of freeze concentration. Solute migration may also result, depending on the rate of freezing, before becoming inhibited due to solidification.

Physical changes and damage to the product structure during freezing and storage is a field large enough to merit a complete chapter in a book such as this. Water migration both within and from the product may occur. Drying may occur during freezing due to vapor pressure differences, whereas sublimation of the ice may occur during long-term frozen storage. In addition to weight and value loss, the color of some products (e.g., meats) may become unattractive to the consumer. This may be due either to desiccation of the meat surface with the consequent development of gray areas (attributed to light scattering effects without ice crystals) or to the darker color of myoglobin compared with oxy-myoglobin. The dehydration effect is commonly termed freezer burn in the frozen food industry — a misnomer in an effect caused by evaporation and sublimation. Surface coating of semiprepared meat and fish products will reduce the effect of this quality loss and may even add value to the product.

The freezing process itself may also cause considerable damage. The cell damage may come about from the physical rupture or crushing of cells by both the size and location of ice crystals. Ice crystal size will initially be determined by the rate of freezing. As a "rule of thumb," slow freezing results in a low rate of nucleation and the production of a small number of large ice crystals, whereas rapid freezing will cause the reverse effect, namely a high rate of nucleation leading to the formation of a large number of very small ice crystals, both processes producing approximately the same ice mass but a different distribution. However, even a rapid freezing process with production of crystals of small size may be reversed during storage as crystals undergo size changes, largely driven by thermodynamic influences. Small crystals have a much larger surface area per unit mass than have large crystals. So, the surface energy of small crystals is significantly higher than that of their larger counterparts. Thermodynamically, the optimum crystal size is one that has the minimum surface energy per unit mass or volume. This driving force causes water migration during storage and the formation of larger ice crystals at the expense of smaller ones, and thus the overall ice crystal mass remaining the same. Ice crystals will also change their shape and crystalline form as they migrate toward the optimum size and shape (a sphere). Variation in temperature during storage and temperature cycling can also enhance the development of larger and larger crystals within the system.

One might ask whether ice crystal size influences the stable shelf-life of the product in any way. In theory it should have little or no effect, but the reality is markedly different. The first obvious effect of ice crystal size is on texture of the product, in particular on the texture of a product that is consumed in the frozen state (e.g., ice cream). A coarse or gritty texture is the normal result. The consumer may often experience this without subjecting the product to long-term storage. Partial melting during transport from the retailer to the domestic freezer cabinet will be followed by a slow refreezing under domestic conditions with the immediate production of coarse large ice crystals instead of the smaller ones from the rapid industrial freezing process. The cell structure of fruits and vegetables (and even meats) may be damaged by the crystals. Large crystals in small cells can also cause damage to the cell walls. However, ice crystal location is also important. A slow freezing process can allow sufficient time for water migration due to osmotic forces from the inner region of a cell to the freeze-concentrated intercell region. This can result in cell desiccation, cell wall disruption, loss of turgor and, ultimately, crushing of the dried cell by the large intercell ice mass. Not only is texture affected but there may also be a significant and sometimes unnoticed drip loss from the product during thawing and cooking before domestic use. Meat products, thawed in the kitchen or microwave oven before cooking, will show a visible water or drip loss. Vegetables, normally put into the cooking water in their frozen state, will have this effect masked. However, significant loss of nutrients may occur through the unseen loss of liquid from the product.

Yet another quality deterioration mechanism results from the crystallization of solutes in the unfrozen pockets within a nominally frozen product. Freeze-concentration effects will concentrate the solutes in these areas and in addition to promoting the solvent (normally water) migration mentioned earlier. These pockets of solution become supersaturated long before the glass transition temperature has been reached. So, solute crystals may be produced in addition to ice crystals. These may have different size, crystal morphology, rates of dissolution, and latent heat requirements to those of the ice crystals which are forming the bulk of the frozen phase. This can give rise to differences in sensory perception of the product (e.g., lactose formation in ice cream). Other effects may be to the product appearance due to ice crystals on the surface.

Other more minor deterioration mechanisms are protein–starch interactions and color changes due to nonenzymatic browning. Labuza and Fu [4] list a range of common deterioration mechanisms for specific foods (Table 28.1).

The foregoing paragraphs illustrate that deterioration in product quality and safety is a complex combination of many changes, unlike in chilled foods where microbiological growth and the consequent effects are the normal product life determinants.

TABLE 28.1
Deterioration Mechanisms for Frozen Foods

Food	Deterioration Process
Frozen meats, poultry, and seafood	Rancidity Toughening (protein denaturation) Discoloration Desiccation (freezer burn)
Frozen fruits and vegetables	Loss of nutrients (vitamins) Loss of texture (temperature abuse) Loss of flavor (lipoxygenase, peroxidase) Loss of tissue moisture (forming package ice) Discoloration
Frozen concentrated juices	Loss of nutrients (vitamins) Loss of flavor Loss of cloudiness Discoloration Yeast growth (upon temperature abuse)
Frozen dairy products (ice cream, yogurt, etc.)	Iciness (recrystallization of ice crystals) Sandiness (lactose crystallization) Loss of flavor Disruption of emulsion system
Frozen convenience foods	Rancidity in meat portions Weeping and curdling of sauces Loss of flavor Discoloration Package ice
Frozen bakery product (raw dough, bread, croissants)	Burst can (upon temperature abuse) (dough) Loss of fermentation capability (dough) Staling (becoming leathery) Loss of fresh aroma

Source: TP Labuza, B Fu. In: YC Hong, Ed., *Frozen Food Quality*. Denver: CRC Press, 1997, pp. 377–415.

The aforementioned deterioration mechanisms may combine to limit the shelf of specific food products. Sikorski [5] published on the deterioration in the organoleptic quality of meat, poultry, and fish products caused by changes in the proteins and fats in the product. Deterioration may be caused by changes in proteins that result in the product exhibiting a loss of extractability of the microfibrillar fraction and loss of functional properties such as water-holding capacity, ease of emulsification of the fats, and the ability to form a gel. Long-term storage may also result in a hardening of the product due to crosslinking of the fibrillar proteins.

Vulicevic et al. [6] have shown that long-term storage of some frozen par-baked bread may change increased firmness, moisture, and flavor, all resulting in an overall product deterioration.

III. SHELF-LIFE DETERMINATION

Most food engineers and technologists like to model shelf-life based on the kinetics of deterioration. As most of the foregoing mechanisms follow either zero-order or first-order kinetics, the mathematical task of shelf-life modeling should be a simple exercise. However, given the multiplicity of deterioration mechanisms present, it is not surprising that kinetic data limitations make the exercise not only difficult but also in many cases impossible. Even when reaction rate constants are available, they have frequently been determined at temperatures well removed from those of frozen food storage conditions. Additionally, many foods may undergo more than one deterioration reaction and the combined effects of these would need to be assessed. So, many laboratory-based procedures have been introduced in an attempt to rectify the situation.

A. Time Temperature Tolerance (TTT)

The first of these were time–temperature–tolerance (TTT) experiments, commonly introduced by the USDA laboratories in the 1960s [7]. The underlying rationale for TTT experiments is that for every food there is a relationship between the storage temperature and the time taken to undergo a certain amount of quality deterioration. Such changes during storage at different temperatures are cumulative and irreversible. Since quality changes are normally smaller at lower temperatures, storage temperature is obviously a dominant quality and shelf-life determinant. However, it is generally agreed that the most detrimental factor influencing frozen food quality is fluctuation in storage temperature because this will significantly reduce the shelf-life of the product.

B. Practical Storage Life

A more commonly used descriptor was later introduced named the practical storage life (PSL). This is defined as the period of storage during which the frozen food retains its quality characteristics and is suitable for consumption [8]. Table 28.2 is reproduced from the IIR publication and demonstrates both the effect of temperature and food type. Such a table has obvious deficiencies. First, the values have been determined for a restricted range of foods. Second, fluctuating storage temperatures can cause problems.

C. High-Quality Life (HQL)

One of the most common shelf-life determinants used in the food industry is the high-quality life parameter. In reality, this is a time–temperature–tolerance variable but differs from the others in that sensory quality is used in its determination. As deterioration during freezing is not usually based on a single set of reactions, it is normally defined as the time elapsed between freezing and the time when a statistically significant difference ($P < 0.1$) can be detected by sensory evaluation. A simpler exercise may be the determination of the elapsed time at which 70% of a trained taste panel can identify a noticeable difference between the frozen food in question and a control when using a triangular test. The control would normally have been stored at $-35^{\circ}C$.

TABLE 28.2
PSL in Months for Selected Food Products

Product	Storage Temperature		
	12°C	−18°C	−24°C
Fruits			
Peaches, apricots, cherries	4	18	>24
Raspberries, strawberries (raw)	5	24	>24
Raspberries, strawberries (in sugar)	3	24	>24
Concentrated fruit juices	—	24	>24
Vegetables			
Asparagus	3	12	>24
Beans (green)	4	15	>24
Broccoli	—	15	24
Brussels sprouts	6	15	>24
Carrots	4	12	>24
Cauliflower	—	12	24
Corn on the cob	4	15	18
Mushrooms	2	8	>24
Peas	6	24	>24
Peppers (red and green)	—	6	12
French fried potatoes	9	24	>24
Spinach	4	18	>24
Onions	—	10	15
Leeks	—	18	—
Meat and meat products			
Beef, ground/minced	6	10	15
Beef steaks	8	18	24
Veal steaks	6	12	15
Lamb steaks	12	18	24
Pork (steaks, cuts, chops)	6	10	15
Bacon (sliced, vacuum packed)	12	12	12
Chicken (whole or cuts)	9	18	>24
Turkey (whole)	8	15	>24
Seafood			
Fatty fish (lazed)	3	5	>9
Lean fish	4	9	>12
Shrimps (cooked/peeled)	—	2	5
Eggs			
Whole egg	—	12	>24
Milk products			
Butter (lactic, unsalted, pH 4.7)	15	18	20
Butter (lactic, salted, pH 4.7)	8	12	14
Cream	—	12	15
Ice cream	1	6	24
Bakery and confectionery			
Cakes (cheese, sponge, chocolate, fruit)	—	15	24
Breads	—	3	—
Raw dough	—	12	18

Source: IIR. *Recommendations for the Processing and Handling of Frozen Foods*. International Institute of Refrigeration, Paris, 1986.

When different storage conditions are used during the life of the product (Table 2 from [9]), the HQL needs to be integrated over the different temperatures. For acceptable quality, it is essential that

$$\sum_{\theta}\left(\frac{t_\theta}{\mathrm{HQL}_\theta}\right) < 1 \tag{28.1}$$

where t_θ is the storage time at a temperature θ and HQL_θ the high-quality life at the same temperature θ. The values of HQL_θ can be read from the chart or, alternatively, the experimental curves from which the chart was derived can be expressed in the form

$$\mathrm{HQL}_\theta = \mathrm{HQL}_{\mathrm{ref}}\ \mathrm{e}^{((\theta_{\mathrm{ref}}-\theta)/D)} \tag{28.2}$$

where D is analogous to the decimal reduction time in bacterial killing. It is found from two points on the semi-log plot of HQL versus θ. In fact D can be calculated as

$$D = \frac{\theta_{\mathrm{ref}} - \theta}{\ln(\mathrm{HQL}/\mathrm{HQL}_{\mathrm{ref}})} \tag{28.3}$$

where $\mathrm{HQL}_{\mathrm{ref}}$ is the high-quality life at a reference temperature θ_{ref}. A typical plot from which D is derived is shown in Figure 28.1 [10].

D. Accelerated Measurement and the Q_{10} Approach

The above type of plot can also be used for the so-called Q_{10} approach. This estimates the effect of temperature on the accelerated deterioration of shelf-life. In its simplest form, it can be expressed as the ratio of the rate of deterioration at a temperature of $\theta + 10°\mathrm{C}$ to that at a temperature of θ. Alternatively, it can be expressed as

$$Q_{10} = \frac{\text{Shelf-life at } \theta}{\text{Shelf-life at } \theta + 10^{\circ}\mathrm{C}} \tag{28.4}$$

An immediate advantage of a knowledge of Q_{10} is the ability to conduct accelerated experimental shelf-life trials at elevated temperatures and then extrapolate the results to normal storage

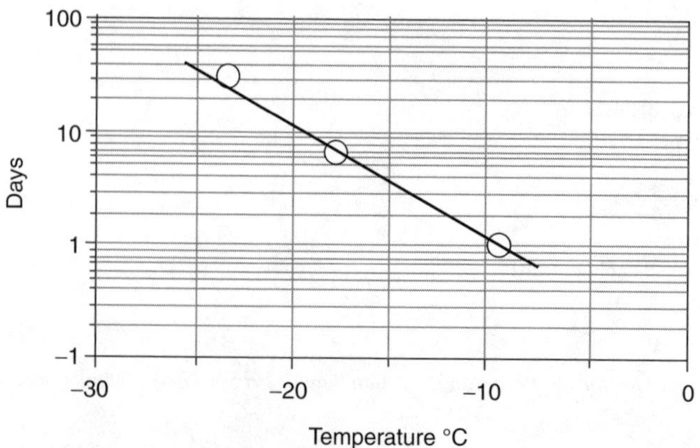

FIGURE 28.1 Plot of shelf-life versus temperature for a typical food.

conditions. Such tests are widely used in the food industry. However, exact values of Q_{10} are difficult to find for many foods and approximate values are frequently used.

In addition, temperature cycling experiments are common in assessing the shelf-life. Either Q_{10} values or reaction rate constants are required to complete the calculation. The method has the advantage of being fast, a characteristic that often outweighs its reduced accuracy over conventional storage testing at normal storage temperatures.

IV. METHODS USED FOR SPECIFIC FOODS

Many researchers have conducted detailed experimental and simulation experiments to predict the shelf-life of various frozen vegetables. It is not intended to provide a comprehensive review of these methods but rather to highlight a few of the more recent methods. In particular, Martins et al. [11] have modeled the deterioration kinetics of green beans. Together with related work, this has allowed estimation of shelf-life even when the product is stored in the variable temperatures of domestic freezer cabinets. They have also shown that temperature fluctuations, inside a refrigerator, influence the accuracy of the kinetic estimates, and if the temperature spectrum is used to derive kinetic estimates, it is possible to apply accurately accelerated methodologies to frozen vegetables.

Reid et al. [12] have developed a rapid assessment method for shelf-life at elevated temperatures. This can be combined with mobility temperature data to produce a plot of expected shelf-life as a function of temperature. They have validated the method using both literature data and their own experimental data and have reduced the experimental period to as low as 60 days.

Frozen breads and doughs are another area in which food-specific trials have been developed. The difficulty here is in determining the shelf-life limiting factor. Often it is not organoleptic but a physical deterioration (through water migration and the location of ice crystals) that may cause such damage as crust flaking and disintegration.

In conclusion, one can state that very significant research efforts have been applied to shelf-life determination but as yet, there is no single, universally accepted method available to the food industry. As is so often the case in calculations related to changes in foods, there are adequate mathematical procedures but all suffer from a deficiency in data. Were rate constants for the common deterioration reactions available for a wide range of frozen foods, there is no doubt that kinetic equations (even as simple as first- or second-order) would predominate the determination of shelf-life. However, sparse data coupled to the multiplicity of deterioration mechanisms make such modeling an aspiration for the future.

V. CONCLUSIONS

Although very significant research efforts have been applied to shelf-life determination, there is as yet no single, universally accepted method available to the food industry. As is so often the case in calculations related to changes in foods, there are adequate mathematical procedures but all suffer from a deficiency in data. Were rate constants for the common deterioration reactions available for a wide range of frozen foods, there is no doubt that kinetic equations (even as simple as first- or second-order) would predominate the determination of shelf-life. This will probably become the preferred method of shelf-life prediction in the fullness of time. However, sparse data coupled to the multiplicity of deterioration mechanisms make such modeling an aspiration for the future. Until then, the food industry will have to rely on less than satisfactory methods such as PSL and HQL determination. These will continue to give good but somewhat inexact predictions. They do, however, have the advantage of being able to handle fluctuating temperatures and will therefore continue to be used by frozen food manufacturers. Pending the accumulation of adequate kinetic data, manufacturers seeking more exact shelf-life predictions will have to rely on experimental methods. In particular, accelerated testing at fluctuating temperatures will be used. In addition, as processors become more proficient at handling complex mathematical relationships (or as software becomes more adept at masking the complexities from the user), modeling using finite element and finite difference methods will become more common in predicting temperatures and changes of state. To this will slowly be added kinetic modeling. In summary, the future of prediction of shelf-life is bright but a lot remains to be done before achieving that goal.

NOMENCLATURE

D	analogue to decimal reduction time (temperature change required for a 10-fold change
HQL	high-quality life
PSL	practical storage life
Q_{10}	ratio of shelf-life at a temperature θ to that at a temperature 10°C higher
t	storage time
θ	temperature

Subscripts

θ	value at a temperature θ
ref	value at the reference temperature

REFERENCES

1. BM McKenna. Prévoir la durée de conservation des produits réfrigérés à traitement minimum *Revue Générale du Froid* 11:36–41, 2000.
2. G Blond, M Le Meste. Principles of frozen storage. In: YH Hui, P Cornillon, Eds., *Handbook of Frozen Foods*. Marcel Dekker, 2004.
3. SK Bahçeci, A Serpen, V Gökmen, J Acar. Study of lipoxygenase and peroxidase as indicator enzymes in green beans: change of enzyme activity, ascorbic acid and chlorophylls during frozen storage. *Journal of Food Engineering* 66 (2):187–192, 2004.
4. TP Labuza, B Fu. Shelf life of frozen foods. Shelf life testing: procedures and prediction methods. In: YC Hong, Ed., *Frozen Food Quality*. Denver: CRC Press, 1997, pp. 377–415.
5. ZE Sikorski. Protein changes in muscle foods due to freezing and frozen storage. *International Journal of Refrigeration* 1 (3):173–180, 1978.
6. IR Vulicevic, ESM Abdel-Aal, GS Mittal, X Lu. Quality and storage life of par-baked frozen breads. *Lebensmicttel-Wissenshaft und-Technologie* 37 (2):205–213, 2004.
7. WB Van Arsdel, MJ Kopley, RL Olsson. *Quality and Stability of Frozen Foods — Time Temperature Tolerance and its Significance*. New York: Wiley, 1971.
8. Anonymous. *Recommendations for the Processing and Handling of Frozen Foods*. Paris: International Institute of Refrigeration, 1986.
9. M Jul. *The Quality of Frozen Foods*. London: Academic Press, 1984.
10. TP Labuza. *Open Shelf Life Dating of Foods*. Westport, CT: Food and Nutrition Press, 1982.
11. RC Martins, IC Lopes, CLM Silva. Accelerated life testing of frozen green beans (*Phaseolus vulgaris* L.) quality loss kinetics: colour and starch. *Journal of Food Engineering* 67 (3):339–346, 2005.
12. DS Reid, K Kotte, P Kilmartin, M Young. A new method for accelerated shelf-life prediction for frozen foods. *Journal of the Science of Food and Agriculture* 83 (10):1018–1021, 2003.

Part V

Packaging of Frozen Foods

29 Introduction to Frozen Food Packaging

John M. Krochta
University of California, Davis

CONTENTS

I. INTRODUCTION

The first four sections of this book provide information essential to the production, storage, transportation, and marketing of safe, quality frozen foods. However, the time and resources devoted to selection of the highest-quality raw materials, accurate application of freezing fundamentals, and use of the most advanced processing and handling facilities are wasted if appropriate packaging systems are not used for frozen foods.

Although freezing is one of the most satisfactory methods of preserving the quality of foods, the conditions of frozen storage are such that frozen foods can lose quality over time [1–5]. Knowledge of the changes that a specific frozen food can undergo is necessary, because selection of the appropriate packaging material and package type options can minimize quality loss [3,6].

Shelf-life of a packaged food is dependent on the nature of the food, the package, and the environment surrounding the packaged food [7–10]. The most common problem for frozen foods is moisture loss through sublimation ("freezer burn"). Sublimation can occur because of the temperature difference (and thus water activity difference) between a frozen food and the colder (and thus frost-accumulating) heat-exchange surfaces of the frozen storage facility. Furthermore, a temperature gradient will also exist within a packaged frozen food, with the resulting formation of ice inside the package ("package ice"). In addition to aggravating the sublimation process, temperature fluctuations in frozen storage increase ice crystal size in the frozen food, due to repeated thawing and refreezing of small amounts of water. Both food desiccation and ice crystal growth produce undesirable food appearance and texture changes. Thus, the ideal packaging material and package design will provide an effective barrier against moisture loss from the food to the environment surrounding the package, minimize moisture movement within the package, and minimize exposure of the food to temperature fluctuations.

Additional quality changes can occur because the small amount of unfrozen water in frozen foods provides an environment in which enzymatic and nonenzymatic oxidation of lipids, colors, flavors, and vitamins can occur, along with both enzymatic and nonenzymatic browning. Desiccation of the food can increase the rates of these changes. The potential loss in quality due to these chemical changes will depend on the particular food product involved. Thus, some frozen foods will have significantly longer storage life when the packaging material and package design provide for removal of oxygen at time of packaging and protection against the incursion of oxygen from the surrounding environment into the package.

Although it is impossible to preserve the quality of frozen foods for an indefinite time, proper packaging can sufficiently delay the rate of quality loss that an acceptable storage life is obtained. Thus, the last section of this book is devoted to selection of packaging materials and package types for frozen foods that provide the final link in fulfilling consumer needs and expectations.

II. FUNCTIONS OF PACKAGING FOR FROZEN FOODS

Generally, packaging is considered to provide four main functions: containment, protection, communication, and convenience (Table 29.1) [8,11,12]. Another function often added is machinability or production efficiency, involving the ability of the package to perform well in rapid filling, closing, and handling operations [7,13]. Additional functions have gained increased importance, including having minimal impact on the environment and maintaining food safety for the consumer [13,14]. Selection of packaging that is appropriate for specific frozen foods must take into consideration all these functions. Thus, packaging materials and package types will be discussed in the context of these package functions for frozen foods.

A discussion of the functions of packaging must also consider the package level, which describes the proximity of the package to the food and the use of the package. The *primary package* is in direct contact with the food product (e.g., plastic-coated paperboard carton or plastic pouch containing frozen food) and usually provides the main protection against the environment. Primary packages are also referred to as retail packages or consumer units, because they provide important communication and convenience in retail sale and consumer use [15,16]. The *secondary package* is the next layer of packaging and generally serves to provide additional protection for the food, usually against physical damage. The secondary packaging can serve as part of the retail package, by working with the primary package (e.g., a paperboard carton that contains a pouch or lidded tray of frozen food) and by unitizing two or more primary packages (e.g., a paperboard carton that unitizes two plastic pouches). Secondary packaging is also sometimes defined as the *distribution, shipping*, or *transport packaging* (e.g., a corrugated box) for a number of primary packages [17]. *Tertiary packaging* and *quarternary packaging* are generally used in the distribution of the packaged food product and not seen by the consumer (e.g., stretch-wrapped pallet of boxes and large metal shipping containers, respectively) [11]. Tertiary and quarternary levels of packaging are also referred to as *logistical packaging, distribution packaging*, or *shipping containers*, because they are used to contain and protect the product during storage, transport, and distribution but have no marketing or consumer use [15,18–20].

A. Containment

The earliest food packaging used by humans served only to contain the food during collection, transportation, and storage. Natural objects such as shells, gourds, leaves, hollowed logs, and animal skins were used. Packaging for frozen foods goes far beyond the basic containment function to provide other functions discussed in this chapter. However, to provide these other functions, frozen food packaging must maintain containment by surviving the abrading, cutting, and puncturing potential of hard, sharp frozen foods. Specifically, frozen food packaging must effectively serve its containment function by maintaining its integrity through the severe environments involved in machine filling (either before or after food freezing), sealing, freezing (unless the food was prefrozen), storage, transportation, thawing, and often cooking. In fact, we increasingly expect frozen

TABLE 29.1
Functions of Food Packaging

Containment (requires packaging integrity)
Protection (against moisture, oxygen, aroma, etc., migration)
Communication (regulatory and marketing aspects)
Convenience (easy opening, dispensing, reclosing, etc.)
Production efficiency (efficient forming filling, closing, handling)
Minimal environmental impact
Maintenance of food safety

food packaging to survive storage temperatures as low as −40°C (−40°F) and then rapid heating to food temperatures up to ∼120°C (250°F) in a microwave oven and air temperatures as high as 230°C (450°F) in a convection oven, all without significant change in integrity or appearance. Different packaging materials will have varying durability in these environments. Besides packaging durability, seal durability is essential to maintain containment.

B. Protection

The needs for protection depend on the food product, but generally include prevention of biological contamination (from microorganisms, insects, rodents), oxidation (of lipids, flavors, colors, vitamins, etc.), moisture change (which affects microbial growth, oxidation rates, food texture, and food appearance), aroma loss or gain, and physical damage (abrasion, fracture, and crushing). In providing protection, packaging for frozen foods maintains the food safety and quality achieved by the freezing process. Biological contamination and microbial growth are generally not problems for frozen foods held in appropriate frozen storage. However, frozen foods are vulnerable to chemical and physical changes due to interaction with the environment.

Modern packaging materials for frozen foods have increasingly provided protection from the environments during transportation, storage, and marketing that cause quality loss (Table 29.2). Frozen foods are quite vulnerable to dehydration (freezer burn) caused by sublimation due to temperature fluctuations in frozen storage. Thus, moisture loss is the biggest problem in the storage of frozen foods. Packaging with a minimum headspace that provides a good moisture barrier prevents moisture loss from the food that ends up as ice in the package and frost on freezer coils.

Because some liquid water remains in frozen foods due to freezing-point depression caused by solute concentration, oxidation of lipids, flavors, colors, and vitamins is a problem for many frozen foods. Thus, packaging that minimizes initial headspace oxygen (by application of vacuum, nitrogen flushing, and oxygen-absorbing active-packaging concepts) and provides a barrier to environmental oxygen can extend shelf-life considerably for many foods. Light can initiate and catalyze oxidative reactions, and so packaging may be called upon to act as a light barrier. Furthermore, even at frozen conditions, foods can lose aromas into and through the packaging material that reduces product quality.

In addition to composition change due to moisture loss, oxidation, and aroma loss, frozen foods are also vulnerable to physical change due to compression, shock, and vibration that can fragment and erode frozen products. Furthermore, temperature fluctuations resulting in thawing and refreezing can cause increase in ice crystal size, with increased damage to cell walls and resulting texture loss. Finally, all foods are vulnerable to tampering. Thus, packaging that provides protection from tampering or obvious evidence of tampering can protect consumers from food contamination.

1. Frozen Food Packaging Materials

Materials used for packaging fall into the general categories of glass, metals, paper, plastics, and combinations of these materials [21–25]. The materials most commonly used for frozen food

TABLE 29.2
Causes of Quality Loss in Frozen Food

Causes of Quality Loss in Frozen Food
Moisture loss (freezer burn)
Oxidation of lipids, flavors, colors, and vitamins
Aroma loss or gain
Light (catalyzes oxidative reactions)
Crushing and fracturing
Tampering

TABLE 29.3
Properties of Materials Used for Food Packaging

Material	Properties
Papberboard and molded pulp	Good structural properties; poor barrier properties
Plastics	Excellent moisture, oxygen, and aroma barriers[a] Semirigid containers have good structural properties
Aluminum	Total moisture, oxygen, and aroma barrier Semirigid containers have good structural properties

[a]Depending on plastic material.

packaging include paperboard and molded paper pulp, several different plastics, and aluminum [3,6] (Table 29.3). Because of their different properties, these materials are often used in combination. Although paperboard and molded pulp provide little barrier to moisture, oxygen, or aromas, they provide good protection against physical damage for frozen foods. Various plastic materials are good moisture, oxygen, and aroma barriers, depending on their polarity and hydrophilicity (Table 29.4). When plastics are used to make thin flexible pouches, they provide no protection from physical damage. However, film-coated paperboard or thicker semirigid or rigid plastic containers provide more protection. Aluminum is a total barrier when it produced without pinholes or cracks. A thin aluminum layer can be used in combination with a plastic layer or plastic and paper or paperboard layers to provide a total barrier.

Steel and glass are not used for frozen food packaging, because of their high cost, weight, and the potential for confusion with shelf-stable products that are packaged in steel or glass containers. In addition, glass containers are vulnerable to cracking due to physical and thermal shock.

TABLE 29.4
Properties of Plastic Materials Used for Food Packaging

Material	Properties
LDPE[a], HDPE[b] and PP[c]	Excellent moisture barriers; poor oxygen barriers
PVDC[d]/PVC copolymer	Excellent moisture, oxygen, & aroma barrier
EVOH[e] copolymer & Polyamide (Nylon)	Excellent oxygen & aroma barrier; poor moisture barriesr
PET[f]	Good moisture, oxygen, and aroma barrier Good heat resistance
CPET[g]	Good moisture, oxygen, and aroma barrier Crystallized PET has improved heat resistance

[a]Low-density polyethylene.

[b]High-density polyethylene.

[c]Polypropylene.

[d]Polyvinylidene chloride/polyvinylchloride.

[e]Ethylene vinyl alcohol.

[f]Polyethylene terephthalate.

[g]Crystallized polyethylene terephthalate.

TABLE 29.5
Important Developments in Frozen Food Packaging [26–28]

Decade	Developments
1960s	LPDE-coated paperboard cartons Spiral-wound composite juice cans with tear-off aluminum ends Flexible pouches Boil-in/microwave-in-bags
1970s–1980s	Stand-up and resealable pouches Microwavable polymers Microwave susceptors for browning and crisping
1990s and 2000s	Microwavable PP and PP-coated paperboard trays Dual-ovenable PET and PET-coated paperboard trays Microwavable-defrosting HDPE frozen juice cans Irradiation and high-pressure processing of plastic packaging Intelligent (communicative and responsive) packaging

2. Frozen Food Package Types

Several types of packaging are used for frozen foods [3,6]. Table 29.5 lists developments in food packaging important for frozen foods over the past half-century [26–28].

a. Boxes/Cartons

The earliest frozen food packages comprised of paperboard boxes/cartons coated with wax. The wax-coated cartons were often supplemented with wax-coated paper or coated cellophane liners and over wraps. These were eventually replaced with low-density polyethylene (LDPE)-coated paperboard boxes. The carton gives protection against physical damage, and the LDPE coating provides water resistance against wet products and protects frozen product from moisture loss (Figure 29.1).

LDPE has a melting point of ~220°F (104°C), which is ideal for heat sealing of LDPE-coated paperboard containers. However, to provide the convenience of microwaving the product in the

FIGURE 29.1 LDPE-coated paperboard carton for frozen food. (Courtesy of Bea Slizewski, Birds Eye Foods, Inc., Rochester, NY.)

package, the LDPE must be replaced with more heat-resistant polypropylene (PP), which has a melt point of ~340°F (171°C). Coating the paperboard with polyethylene terephthalate (PET) (melt point ~500°F (260°C)) provides heat resistance adequate for the higher temperatures of a convection oven, making the container dual-ovenable [29].

The regular shape of a box lends itself to plate freezing of the product after filling into the box. However, the product can also be individually quick-frozen (IQF) in a blast freezer or immersion-frozen with liquid nitrogen or carbon dioxide before filling into a box/carton. Ice cream is partially frozen in a scraped-surface heat exchanger before filling into cylindrical LDPE-coated paperboard cartons. The filled cartons are then placed into cold storage to complete the freezing process.

b. Bags/Pouches

Many frozen products (e.g., frozen vegetables) are now packaged in flexible bags or pouches made of LDPE or HDPE, which give less protection from physical damage than boxes but provide excellent protection from moisture loss to the frozen-storage environment (Figure 29.2). Usually, the food product is IQF before being filled into the pouch, which allows for easy dispensing of all or a portion of the product. Pigment commonly added to the film protects the frozen food contents from light that can initiate and catalyze oxidation. The pigmented, printed film is supplied from a roll to a filling device, which continuously forms side and bottom seams, fills the frozen food, and then seals the top seam. Coating the LDPE or HDPE with PET or polyvinylidene chloride/polyvinyl chloride (PVDC/PVC) copolymer gives the product additional protection from oxygen. Alternatively, sandwiching moisture-sensitive ethylene-vinyl alcohol (EVOH) copolymer or polyamide between layers of PE provides oxygen-barrier protection. Pouches can also be made with bottom gussets and zipper resealing features, which allow them to be "stand-up pouches" with more visibility for marketing and greater convenience for consumers (Figure 29.3).

c. Heat-in-bag Pouches

If a pouch is made from HDPE or PP, it has the heat resistance to allow preparation of the product in the pouch by placing in boiling water or a microwave oven. The product is often made in a sauce, which adds value to the product and increases heat transfer during freezing and subsequent thawing and warming in boiling water or a microwave oven. The pouch is usually contained in a carton to

FIGURE 29.2 Flexible pouch used for IQF frozen food. (Courtesy of Bea Slizewski, Birds Eye Foods, Inc., Rochester, NY.)

FIGURE 29.3 Stand-up pouch used for frozen food. (Courtesy of Steve Ross, Brakebush Brothers, Inc., Westfield, WI.)

maintain the pouch in a sanitary condition (Figure 29.4). Adding a PET or PVDC/PVC copolymer coating or an EVOH copolymer or polyamide layer increases the pouch barrier to oxygen.

d. Lidded Trays and Pans

The first frozen precooked dinner, commercialized in the 1940s, was a frozen pot pie. It was packaged in an aluminum tray with an aluminum lid. In the mid-1950s, complete frozen dinners were made available in multicompartment aluminum trays. Molded aluminum foil trays and pans have become commonly used in packaging of frozen fruit and meat pies, as well as whole frozen dinners (Figure 29.5). These products have the advantage of being heatable directly in the aluminum tray or

FIGURE 29.4 Paperboard carton containing heat-in-bag pouch. (Courtesy of Denise Bosch, reprinted with permission of General Mills, Inc., Minneapolis, MN.)

FIGURE 29.5 Aluminum pans and trays used for frozen foods. (Courtesy of Stephen McEwen Confoil Pty. Ltd., Bayswater, Australia.)

pan in a convection oven. With proper care, foods in aluminum containers can also be heated in microwave ovens. All lids must be removed, the container must be centered in the oven away from walls, and the microwave oven must be less than 25 year old.

Microwavable-molded PP trays, PP-coated paperboard, or PP-coated molded-pulp trays are also commonly used for frozen food packaging. These trays can be used with heat-sealed PP-film or PP-coated paperboard lids, as well as oriented polystyrene (OPS) or PP snap-on lids.

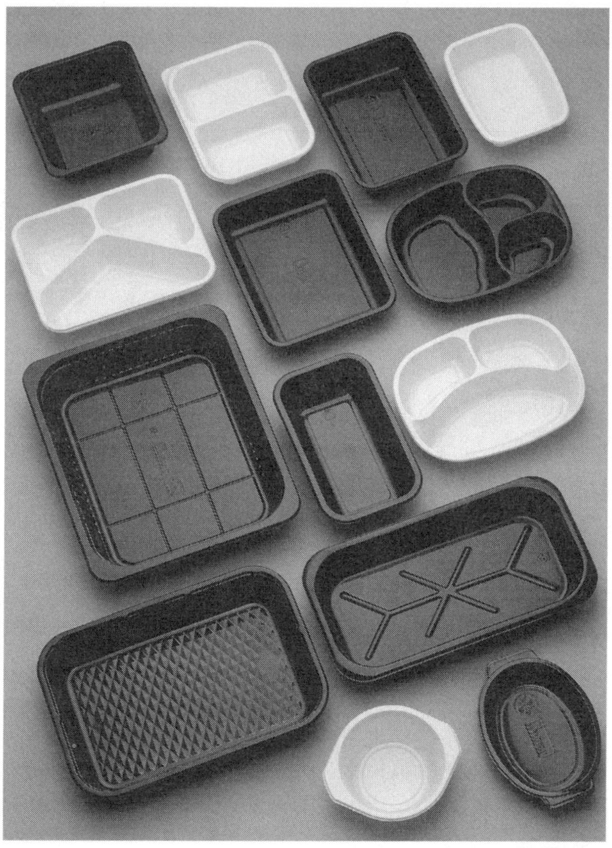

FIGURE 29.6 PET trays used for frozen food. (Courtesy of Linda Braha, Coextruded Plastic Technologies, Inc., Edgerton, WI.)

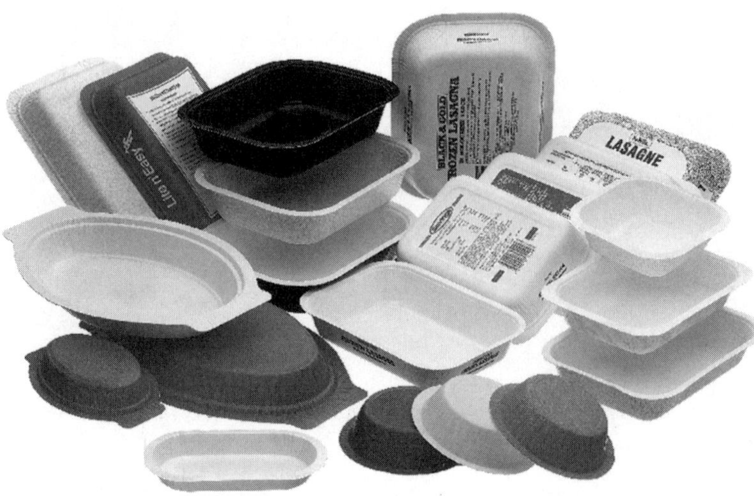

FIGURE 29.7 PET-coated pressed-paperboard trays used for frozen foods. (Courtesy of Stephen McEwen Confoil Pty. Ltd., Bayswater, Australia.)

Crystallized PET (CPET) trays (Figure 29.6), as well as PET-coated paperboard and PET-coated molded-pulp trays (Figure 29.7) have the advantage of being dual-ovenable [29,30]. These trays generally have heat-sealed PET film or PET-coated paperboard lids. Use of PET-coated paperboard trays allows incorporation of aluminum-flake susceptors that absorb to allow browning and crisping and reflect energy for even heating of food products during microwave oven heating [31] (Figure 29.8).

FIGURE 29.8 MICRORITE® — PET-coated paper and paperboard trays, pouches, sleeves, and disks with microwave susceptors used for crisping, browning, and even heating of frozen food in a microwave oven. (Courtesy of Dan Keefe, Graphic Packaging International, Marietta, GA.)

FIGURE 29.9 HDPE can for frozen juices allows microwave thawing. (Courtesy of James Callahan, Welch Foods, Inc., Concord, MA. Image and Package Design © Welch's. All Rights Reserved.)

e. Composite and Plastic Cans

Most frozen juices are packaged in composite containers that consist of an LDPE-coated paperboard body and metal ends, usually provided with an easy-open pull-ring or pull-tab. The body can also incorporate an aluminum foil layer to increase barrier properties. Some frozen juices are packaged in HDPE cans, which have the advantage of microwave thawing and easy recyclability (Figure 29.9). As with ice cream, juices are partially frozen in a scraped surface heat exchanger before filling into the package and then finish-frozen in frozen storage.

C. Communication

After consideration of the containment and protection functions that a package must serve for a particular frozen food product, the manufacturer must consider the information the package must convey. The information provided to consumers on a package fulfills legal requirements on labeling, as well as having marketing objectives.

1. Laws and Regulations on Food Labeling

National food packaging laws in the United States are contained in a number of statutes, including the Food, Drug and Cosmetic Act (FDCA), the Fair Packaging and Labeling Act (FPLA), the Nutrition Labeling Education Act (NLEA), and the Food Allergen Labeling and Consumer Protection Act (FALCPA) [32–34]. Many other laws in the United States affect food packaging, including federal laws on patents, trademarks, and copyrights, and state laws on weights and measures [33]. Other countries also have packaging laws that are generally contained in various statutes. For European countries, the European Union (EU) has had an important impact on developing regulations that are uniform across all the countries in the EU [35–37].

In the United States, several government agencies have the responsibility for developing and enforcing regulations based on the FDCA, FPLA, NLEA, and FALCPA. The Food and Drug Administration (FDA) has authority for labeling of most foods, except that the U.S. Department of Agriculture (USDA) has authority for labeling of foods containing >2% cooked meat or poultry or >3% raw meat or poultry [37,38]. Both FDA and USDA consider foods to be misbranded if the labeling is false, misleading, or incorrect in any manner, or if the package itself is misleading because of the way it is made or filled [39]. The Federal Trade Commission (FTC), which regulates product advertising, can also become involved in cases involving deceptive packaging and labeling.

FDA and USDA food labeling regulations are contained in Title 21 and Title 9 of the *Codes of Federal Regulations* (CFR), respectively. In most respects, the FDA and USDA have the same labeling requirements for food products under their jurisdiction. The U.S. Treasury Department

Bureau of Alcohol, Tobacco and Firearms has authority for labeling of all alcoholic beverages, with regulations found in Title 27 of the CFR. Information on where to find the major food packaging regulations in Title 27 and Title 9, as well as how to locate, read, and interpret CFR regulations, can be found in Hanlon et al. [33]. As the CFR is constantly being updated, it is important to refer to the most recent version. In addition, the *Federal Register* publishes proposed new federal regulations on a daily basis, allowing a specific amount of time for comments and suggestions. The final regulation is then published in the *Federal Register* and then eventually in the *Codes of Federal Regulations.*

In the United States, food labels are required to provide information on the product identity, product manufacturer, ingredients in descending order of amount, net content, specific nutrient contents, and country of origin. In addition, the FALCPA requires easy-to-understand labeling of allergen ingredients on food packaging, including declaration of allergens present in flavoring, coloring, or incidental additives [34]. The principle display panel (PDP), on the front of the package oriented toward the consumer, must include the information on the identity of the product and net quantity of package content [40]. The PDP may include the other required information, but the identity and contact information of the manufacturer, list of ingredients, nutritional facts, and country of origin are usually provided on a separate information panel [40]. The nutritional facts must include information on specific nutritional components, using a format designated by the FDA or USDA [41]. If nutrient contents or health claims are made on the label, they must use specific approved wording that has been defined by FDA or USDA. If fruit- or vegetable-juice-containing beverages are less than 100% juice, they must be named using specific defining terms such as "beverage," "cocktail," or "drink" and the percentage of juice must be declared. Blends of juices must be described by descending order of quantity or relative quantity. If a juice is made from a concentrate, it must be identified as "from concentrate" or "reconstituted" [38]. If a product is processed using irradiation, contains nonpasteurized juice, or contains certain ingredients specified by FDA and USDA (e.g., saccharin), a declaration must be included on the cover [38].

Labels regulated by the USDA have additional information requirements related to food safety. These include an official inspection legend and handling instructions to ensure safety and quality (e.g., "Keep Frozen"). Frozen product labels also generally include information on proper preparation for consumption.

2. Marketing Objectives of Packaging

Package graphics and colors are intended to communicate product quality and, thus, sell the product. Most product-selection decisions are made when consumers are in the store looking at products on the shelf. Thus, packaging provides the main advertising of a product [42]. Package shape also influences purchases, communicating such information as brand identity, product protection, and product amount.

In addition to enticing consumers, packages include bar codes that aid in supply chain tracking, rapid sales checkout, inventory management, automatic reordering, and sales analysis [43,44]. Other package codes allow determination of food production location and date critical to tracking of a product. Various open dating systems inform the consumer about the shelf-life of the food product.

D. Convenience

Providing convenience (sometimes referred to as utility-of-use or functionality) to consumers has become a more important function of packaging. Increasingly, consumers are attracted to convenience features in packaging and are willing to pay for them. Basic convenience issues involve ease of opening, dispensing, and resealing. Additional convenience features can include ease of preparation, serving, and clean up. Many frozen food packages provide a considerable amount of convenience to consumers. Examples include:

- Easy-to-open pouches of IQF ready-to-cook-and-eat whole-meal mixtures of meat and poultry, vegetables, and pasta that allow easy dispensing of the desired amount and then easy resealing of the remaining mixture.
- Boil-in-bag or microwave-in-bag frozen products (e.g., vegetables in cheese sauce) that allow for easy preparation and simple clean up.
- Aluminum tray containers with sections for frozen dinners (i.e., "TV dinner" trays) in which the food can be thawed and heated using a convection oven and then served.
- Polypropylene (PP) tray or PP-coated paperboard tray containers for frozen entrees or dinners in which the food can be heated using a microwave oven and then served.
- Polyethylene terephthalate (PET) tray or PET-coated paperboard tray containers for frozen entrees or dinners in which the food can be heated using either a microwave or convection oven and then served.
- Easy-open pull-tab composite cans (paperboard-laminated body with metal ends) for frozen juices.
- Ease-open pull-tab high-density polyethylene (HDPE) cans for frozen juices that can be thawed in a microwave oven.

E. Production Efficiency

Packages can be preformed at a package production facility separate from the food production facility. Glass and metal containers and some plastic containers are produced in this manner. In this case, the preformed package is then transported to the food production facility, unloaded, conveyed to a filler, filled, and then sealed.

Increasingly, packages are inline formed at the food production facility, where the packages are formed, filled, and sealed in close sequence using high-speed packaging machinery [45]. Much frozen food packaging is handled in this manner. For example, LDPE-coated paperboard pieces (blanks) that are flat, precut, preprinted, and possibly prescored are automatically erected, folded, and sealed into carton form, with one end or a top lid open for filling. After automatic filling, the final seal is made and the filled cartons are conveyed for loading into secondary and possibly tertiary containers for storage and transportation. For production of pouches and bags, preprinted flexible film is automatically unwound, with the proper length folded lengthwise over a form for heat-sealing of a side seal, with simultaneous heat-sealing of a bottom seal to form a pouch. This is followed immediately by filling and then heat-sealing of the top seal, with simultaneous formation of the bottom seal of the next pouch and cutting of the form-filled-sealed pouch from the continuous run of film.

For either preformed or inline formed packages, the package forming, filling, and sealing operations must be automated and very fast and stoppages must be infrequent. Thus, the packaging material and design must lend themselves to rapid handling by automated machines to achieve high production efficiency.

F. Minimal Environmental Impact

1. Life Cycle Analysis

Accessing raw materials (e.g., wood) used in the production of packaging materials, manufacturing the packaging materials (e.g., paperboard), converting packaging materials to containers (e.g., cartons), transporting the packaging as part of the packaged product, and dealing with packaging waste all have impact on the environment [46,47]. Life cycle assessment (LCA) takes into account all the resources consumed in the creation, use, and disposal of a package [48,49]. The goal is to identify areas of environmental concern so that the packaging having least impact can be selected. However, comparing the environmental impacts of competing packaging is complex. Often, there is no clearly superior packaging choice [50].

2. Reduction, Reuse, Recycling, and Recovery

The main approach to minimizing environmental impact has been to reduce the amount of packaging waste that ends up in landfills and as litter. Packaging occupies approximately one-third of municipal solid waste (MSW) volume, with paper and plastic taking up the greatest volume [51,52].

The approaches to minimizing package waste constitute the 4 R's of packaging: source reduction, package reuse, package recycling, and energy recovery [47]. The impact of packaging on the environment is often assessed by the extent to which the packaging achieves one or more of the four R's.

a. Source Reduction

There are strong economic incentives and technical challenges involved in reduction of the amount of packaging required to protect a food. Advances in packaging material properties and in package design have resulted in reduction of the volume and weight of packaging material used for each package type. Development of semirigid and rigid plastic containers, plastic/paperboard/plastic/foil/plastic laminate cartons, and flexible pouches as packaging alternatives has contributed greatly to source reduction [53]. Other examples of source reduction include smaller-diameter can ends and shrink-wrap to replace corrugated board boxes used for transport packaging [54].

For frozen foods, increased strength of paperboard, improved barrier of plastic films, and development of heat-resistant plastic containers have decreased the amount and cost of packaging. Many frozen products are now packaged in lightweight bags and pouches made from thin plastic films that contribute very little to package waste volume.

Besides improvements in packaging material properties and the introduction of new packaging types, other approaches can reduce the amount of packaging per unit weight of product. These include providing the option of a larger amount of product per package (which decreases amount of packaging per unit food weight) as well as concentrated product with reduced volume and thus reduced packaging.

b. Package Reuse

Reusable plastic and paperboard containers are seen as impractical, because they are absorbent and impossible to clean and maintain shape. The hermetic sealing requirement for most metal containers makes reuse of metal containers impossible. Thus, the only retail food packages ever reused are glass beverage bottles. However, with larger regional food manufacturers, returnable glass bottles are generally no longer economically feasible, especially because reusable bottles must have thicker, heavier walls to endure the increased handling. In addition, reintroducing returnable bottles would require education of consumers, who resist returning bottles [54]. However, in closed-loop distribution systems, returnable or reusable transport tote and bulk containers are finding greater application.

No frozen food packaging is returned and reused for frozen food. However, to the degree the consumer can reuse a frozen food package (e.g., aluminum or plastic tray), the package has greater value and reduces consumption of other materials.

c. Package Recycling

After source reduction, package recycling has had the biggest impact on reducing packaging waste. Specifically, the single-material semirigid and rigid packages used for frozen foods such as aluminum trays/pans, HDPE cans and tubs, and PET trays/bowls are recyclable. However, most municipalities exclude multimaterial containers and flexible packaging from their recycling programs. Thus, the plastic-coated paperboard and plastic film pouches used for much frozen food are not presently generally recyclable. However, an increasing number of municipalities are capable of recycling mutlilayer paperboard/plastic and paperboard/plastic/aluminum containers.

Three possible types of recycling technology are possible for packaging: mechanical, chemical, and biological [48,55]. The most common type of recycling is mechanical recycling, which involves processing of recycled package materials by physical means that include cleaning, shredding, grinding, separating, and reforming. These steps result in new metal and glass containers that are acceptable for direct-contact use with foods. However, they generally do not ensure removal of all possible contaminants from paper and plastic materials. Recycled paper material is more often used for secondary food packaging such as paperboard cartons, as well as for nonfood uses. Mechanically recycled plastic materials are more often used for nonfood uses, such as carpeting. clothing, and plastic wood. However, FDA has also approved use of mechanically recycled plastic when it is coextruded with a virgin layer of the plastic that is the food-contact surface. Manufacturers of packaging and other products from recycled material can obtain certification that allows use of a label symbol indicating percentage of recycled content [54].

Chemical recycling involves depolymerization of plastic polymers, followed by repolymerization to the polymer. Several processes have been developed for chemical recycling of PET, which allow removal of all possible contaminants [47,52,56]. Generally, chemical recycling is still more costly than producing packaging from virgin plastic materials. However, as the cost and availability of fossil resources become problems, chemical recycling becomes a more attractive alternative.

Biological recycling involves use of renewable and biodegradable polymers for food packaging. A critical challenge for biodegradable packaging is achieving controlled lifetime. Biodegradable packaging must be stable and functions properly at the conditions of use, so as not to compromise the quality and safety of the food, and then biodegrade efficiently on exposure to the appropriate microorganisms and environment [57]. Food packaging based on cornstarch has been developed for confections, whose low water activity will not support microbial growth [58]. Polylactide (PLA) produced by fermentation of cornstarch-derived sugars has been formed into containers for packaging of refrigerated deli products such as cheese, desserts, fruits, and vegetables [59]. Because such biodegradable packaging materials will not biodegrade in frozen storage, they may well suited for frozen foods. A 2003 survey conducted in Europe found that consumers favor food packaged in renewable materials, even if it is more expensive [59].

Widespread use of biodegradable polymers will require reductions in production costs, easy sorting from nonbiodegradable recyclable polymers, establishment of dedicated composting facilities, and increase in fossil resource costs. However, it seems inevitable that sustainable approaches to the production of packaging materials will be necessary.

d. Energy Recovery

The paper and plastic packaging materials generally used for packaging of frozen foods have energy content that can be captured by incineration to produce electricity or steam. Energy recovery, sometimes called "thermal recycling," is an attractive alternative for mixed plastic and mixed plastic and paper wastes that cannot be easily recycled [48]. Waste incineration with energy recovery is more common in Europe and Asia compared to the United States.

3. Laws and Regulations on Packaging Waste and Recycling

In the United States, regulations governing packaging waste disposal originate from state and local government legislation [32]. Various state laws have involved a range of approaches dealing with plastic waste, including required recycling rates, mandated recycle content, advanced disposal fees, and bans on plastic and composite packaging [54,60]. Most of these approaches have been abandoned, because they were judged impractical or unconstitutional. The approaches eventually deemed most practical involve recycling, including required deposit or refund fees to encourage recycling of beverage containers, curbside recycling, and sorting of MSW. Although there is no national solid-waste-reduction program in the United States, the Environmental Protection Agency (EPA) set a 25% MSW recycling goal that was met on schedule in 1995 and set a 35% MSW recycling goal for 2005 [54].

In the United States, the FDA reviews applications for food-contact applications of recycled paper and plastic materials on a case-by-case basis. The review considers the cleanliness of the recycled materials, the ability of the recycling process to remove possible contaminants, and the proposed food-contact application [56]. The result of the review is an advisory opinion, with a positive opinion taking the form of a "Letter of No Objection." FDA has issued such letters on use of mechanically recycled plastics for several specific short-term food-contact applications, including HDPE grocery bags, PS egg cartons, HDPE and PP crates for transporting fresh fruits and vegetables, and PET pint and quart baskets for fresh fruits and vegetables. All these applications include the expectation that the food is cleaned before use or that the food is protected by a barrier (e.g., egg shell) [56]. In addition to these applications, it has been shown that a mechanical recycling process can produce recycled PET acceptable to FDA for unlimited food contact time [47]. FDA has also approved the use of a methanolysis process for chemically recycling PET for food-contact use [54].

Canada has also largely left packaging waste legislation to provincial and local governments. However, the Canadian Council of Ministers of the Environment (CCME) sets goals for reduction of packaging waste and established a National Task Force on Packaging which set up a National Packaging Protocol (NAPP) [54]. The NAPP is based on a set of Guiding Principles for Packaging Stewardship, which Environment Canada is translating to model procedures to guide provincial source reduction, reuse, and recycling programs.

Europe has taken a more aggressive approach to reduce packaging waste, and Asian countries are developing policies similar to those instituted in Europe [61]. The European Union (EU) has adopted a Packaging and Packaging Waste Directive that requires 50–65% recovery and 25–45% recycling rates for packaging, along with other environmental criteria [54,61]. Germany has more stringent requirements for recycling, along with mandated levels of returnable and reusable beverage containers. Germany also has a Green Dot system that identifies packages that can be collected and sorted for return to the originating company for recovery and recycling. To obtain the green dot, the company must pay a charge based on the type and weight of packaging material, to cover the cost of the program [36].

G. Food Safety

As discussed earlier, the conditions of frozen storage prevent biological contamination and microbial growth. However, selection of packaging materials and package types for frozen foods must consider the possibility of migration of potentially harmful components from the packaging material to the food during storage or preparation. The method of packaging must also ensure that contaminants are not introduced into the food. In addition, the possibility of food tampering means that the tamper-resistance, tamper-evidence, and traceability of the package must be taken into account. All these food safety issues can be addressed with an appropriate Hazard Analysis Critical Control Point (HACCP) plan in place.

1. Food–Package Interaction

To varying degrees, all materials used for food packaging have been found to interact with food [62,63]. Thus, the paper, plastic, and metal used in frozen food packaging each have potential for interacting with food.

a. *Packaging Component Migration*

Migration occurs when a component of a packaging material transfers to a food product. Possible migrating substances of concern for frozen food packaging include plastic monomers and plasticizers [64], as well as paper coating and adhesive components [65].

Migration is not likely to be a problem for frozen foods in storage. However, it has been a concern because of the high temperatures reached in heating of frozen foods in microwave or convection ovens, including microwaving of frozen foods when susceptors are incorporated to produce crispness and browning [66,67]. The high temperatures attained in these instances increase migration rates and can even break down polymers into additional migrating compounds. Thus, proper selection of materials is critical when they are intended to serve as containers during the heating of frozen foods.

b. Laws and Regulations on Packaging Components as Indirect Food Additives

Food packaging is subject to rigorous laws and regulations to ensure food safety [32,33,36]. In the United States, the FDA has primary responsibility for ensuring that food packaging does not contaminate or adulterate food. If a component of a food packaging material migrates into a food, regulatory approval of the migrant as an indirect food additive must be obtained from FDA through a food additive petition. This process requires an estimate of the amount of the substance that will enter the diet and demonstration that the amount is safe. However, the substance may be exempted from the FDA food additive regulations if it has received prior sanction for its intended use by the FDA or USDA before the Food Additive Amendment of 1958 or if it is "generally recognized as safe" (GRAS) by qualified experts [32]. An exemption from FDA regulations can also be obtained if it can be shown that an "insignificant" amount of the substance transfers to the food. Extraction studies that simulate the intended use with a food must be performed on the packaging material. Because of the complexity of foods, solvents that simulate the intended food are generally used. More detailed information on migration, including U.S. and E.C. regulations, tests, food simulants and simulant alternatives, is available [62,68].

Recycled plastic packaging materials are a potential food safety concern, because they may have been exposed to hazardous compounds that they absorbed. The FDA requires that all food-contact surfaces be suitably pure for their intended use. Furthermore, recycled packaging must adhere to food additive regulations. As mentioned earlier, the packaging industry practice for using recycled plastic is to seek a "Letter of No Objection" from FDA, based on proof that any potential contaminants would produce less than the "threshold" dietary level of 0.5 ppb [32].

In Europe, the EU has worked to develop broadly applicable legislation and has published Practice Guide NI, which provides guidance on materials that come in contact with foods [36].

2. Food-Package Tampering

The possibility of food tampering involving addition of a contaminant has been an issue since the well-publicized adulteration of Tylenol in 1982 that led to the death of seven people [69].

a. Laws and Regulations on Tamper-Evident Packaging

The Tylenol adulteration incident quickly led to FDA regulations requiring tamper-evident (TE) packaging for all over-the-counter (OTC) medications. The TE packaging must include at least one barrier to opening the package that will provide clear indication when the package has been opened [33]. The TE component may be on the primary or secondary package, and a printed warning on the package must explain the TE feature to consumers.

The FDA regulations for TE packaging of OTC drugs do not extend to food packaging. However, given the foreseeable possibility of food product tampering, manufacturers are seen by attorneys specializing in packaging and product liability as responsible for protecting consumers [69]. Therefore, an increasing number of food packages now incorporate TE features. Heightened concerns about terrorist contamination of the food supply suggest increased use of TE packaging for foods in the future.

b. Tamper-Evident Packaging for Frozen Foods

Several TE concepts are used in food packaging, including shrink-wraps or bands, breakaway closures, and inner seals that are quite evident when disturbed. Tests have been developed for assessing the difficulty in restoring a violated package to near-original condition [69]. TE concepts particularly useful for frozen foods [33]:

- Tabs or flaps on sealed cartons that are clearly torn when the carton is opened.
- Shrink-film wraps that are obvious when damaged or removed.
- Distinctive patterns and seals on film pouches that are clearly distorted when stretched or torn.

Studies have shown that consumers prefer packages that have shelf-visible TE features and that they are willing to pay more than for a competing product that does not have the TE feature [69].

3. GMPs and HACCP Plans

Good Manufacturing Practices (GMPs) define sanitary practices in food processing facilities necessary to ensure the safety of foods. HACCP is a seven-step preventive approach to identifying potential hazards in the production of foods that establishes critical points in the manufacturing process to be monitored for possible corrective action [70,71].

a. Laws and Regulations on GMPs and HACCP Plans

All aspects of food processing, including the packaging materials and food packaging step, must conform to GMP regulations established by the FDA and USDA. The FDA requires HACCP plans for seafood and juice processing, and USDA requires HACCP plans for meat and poultry processing. However, in 1985, the National Academy of Sciences (NAS) made a strong recommendation that the HACCP approach be adopted by all regulatory agencies and be required for all food processors [70]. In addition, implementation of HACCP plans for nonregulated is being strongly driven by buyer demand.

b. HACCP for Frozen Food Packaging

HACCP plans must be specific for each food product. However, several general packaging-related questions should be asked in conducting a hazard analysis as part of HACCP plan development of all foods, including frozen foods. These include [70]:

- Could the packaging materials introduce any unsafe indirect additives to the food?
- Could the method of filling and sealing the food into the packaging introduce any food contaminants?
- Does the packaging have sufficient integrity to prevent contamination of the food during storage and transportation?
- Does the package design incorporate appropriate tamper-evident features?
- Is the package clearly labeled with "Keep Frozen"?
- Does the package label include accurate instructions for safe handling and preparation of the food?
- Is every package properly labeled and coded for easy identification and tracking?

III. TRENDS IN FROZEN FOOD PACKAGING

Certain trends in frozen food packaging reflect increased understanding of factors that improve frozen food quality. These advances can provide information useful in improving the protection function of food packaging. Other trends reflect social and cultural changes that place more expectations on packaging, involving the communication, convenience, and minimal environmental impact functions of packaging.

A. Flexible Packaging

Flexible packaging is an attractive alternative to more traditional rigid and semirigid containers because of several advantages, including packaging material source reduction, convenience for consumers, and visual and handling appeal [72]. Flexible packaging can be based on a single layer of plastic material or a combination of materials that can include several different plastics, paper, and aluminum. The significant package weight reduction when using flexible packaging has to be balanced against the fact that flexible packaging is not currently recyclable.

Flexible packaging has had an important role in the development of several frozen food packaging concepts [72–74]. These include resealable pouches of IQF foods, stand-up pouches for frozen foods, and modified atmosphere packaging of frozen foods that will later be thawed for sale as frozen-fresh. Stand-up pouches has replaced many applications of lay-down pillow pouches, because their greater visibility and new dispensing and resealing approaches are attractive to consumers.

B. Microwavable Trays, Cartons, and Bowls

Semirigid or rigid plastic trays and bowls and paperboard-based cartons have been developed as alternatives to aluminum for frozen entrees and meals. PP trays and bowls and PP-coated paperboard are convenient because of their microwavability and usefulness as serving containers. Frozen food products in microwavable bowls are convenient items that can be made available in a dispensing machine with nearby microwave oven. Use of PET coating on paperboard allows for incorporation of susceptors that develop higher localized temperatures that produce greater food crisping and browning.

C. Nonthermal Food Processing Techniques in Combination with Freezing

Many frozen foods are blanched or cooked before freezing. Such heating reduces microbial counts on the food and makes the food safer and more easily reheated for serving. Several nonthermal technologies are available that can be applied to raw, blanched, or cooked foods prior to freezing, with resulting improvement of food safety [75–77].

Ionizing radiation has been approved as a treatment before freezing for several foods, including uncooked poultry, meat, and ground meat [78]. Levels of allowed radiation can destroy vegetative food spoilage and pathogenic microorganisms, but are not sufficient to sterilize foods. The irradiation process is performed on prepackaged foods to prevent recontamination. Irradiation has been found to affect the properties of plastic and paper packaging materials [79,80]. The result can be a modification of the mechanical and barrier properties of the plastic material, as well as the strength of heat seals. Thus, it is necessary to be aware of these possible changes and select packaging materials that are compatible with the irradiation process.

High-pressure processing (HPP) has also been found effective for inactivation of vegetative microorganisms with little effect on food quality [81,82]. Thus far, HHP has been applied only to foods before refrigeration, but it could be used to reduce microorganisms in foods before freezing. Like radiation, HPP processing is generally performed on prepackaged foods. The effects of HPP on properties of the packaging materials have generally been found to be small [80,83].

Several other nonthermal processing techniques are being developed that could potentially be used to increase the quality and safety of frozen foods [84,85]. In each case, the packaging involved must be evaluated to ensure food safety and quality.

D. Modified Atmosphere Packaging

The quality and shelf-life of many foods have been improved due to packaging that maintains an atmosphere in the package headspace that is different from air [86–93]. The modified atmosphere, which often excludes oxygen to prevent oxidative rancidity and includes carbon dioxide for antimicrobial effect, compliments refrigeration to retard chemical and microbiological deterioration of the food. Modified atmosphere packaging (MAP) can be used to complement freezing for foods that are quite sensitive to oxidation, such as fish, and for foods that are thawed and marketed as frozen-fresh. Frozen bread packaged in a nitrogen and carbon dioxide atmosphere has a longer shelf-life after thawing, allowing wider distribution and sale as frozen-fresh.

E. Active Packaging

Active packaging has been defined as performing some desired role other than providing an inert (passive) barrier to external conditions [94,95]. Active packaging concepts thus enhance the performance of the package, by changing the condition of the packaged food to improve quality and shelf-life [91]. With consumer interest in ever higher quality and safety in foods, active packaging is a field of high interest and development [96–103].

1. Protective Active Packaging

Most active packaging concepts enhance the protective function of food packaging [91]. Protective active packaging approaches include oxygen-scavenger sachets, labels, closure liners, and films that complement the oxygen-barrier property of the package [104–106]. Such concepts could increase the protection of oxygen-sensitive frozen foods.

2. Convenience Active Packaging

A number of active packaging concepts enhance the convenience of packaged frozen foods. Packaging that is stable to the conditions of a microwave or conventional oven (dual-ovenable) can serve as a convenient container for preparation, service, and consumption of frozen foods. Incorporation of susceptors in microwavable packaging allows crisping and browning of the food.

F. Intelligent Packaging

Intelligent (or smart) packaging can be divided into two types [107,108]. "Simple" intelligent packaging contains components that sense the environment and communicate information important to proper handling of the food product. "Interactive" or "responsive" intelligent packaging has additional capability allowing response to environmental change and, thus, prevention of food deterioration [109].

Several intelligent packaging concepts involve sensors that provide information related to food quality [91,107,108,110]. One category includes temperature sensors that indicate whether a frozen food package has been exposed to temperatures above a critical limit. Time–temperature indicators (TTIs) are also available which provide time-integrated information about the temperature history of the product [111]. TTIs are often self-adhesive color-changing labels that respond gradually and irreversibly to the cumulative exposure of the product to temperature (Figure 29.10). TTIs can be matched to the specific shelf-life characteristics of each product. Such indicators allow more accurate assessment of the remaining product shelf-life [112]. However, the Arrhenius-type temperature behavior of TTIs does not take into account the concentration effect, ice-crystal growth, and glass transition phenomena of frozen foods [111].

Another category of intelligent packaging includes components that range from bar codes to radio frequency transmitters that allow accurate tracking of product for improved supply chain management and rapid traceability [44].

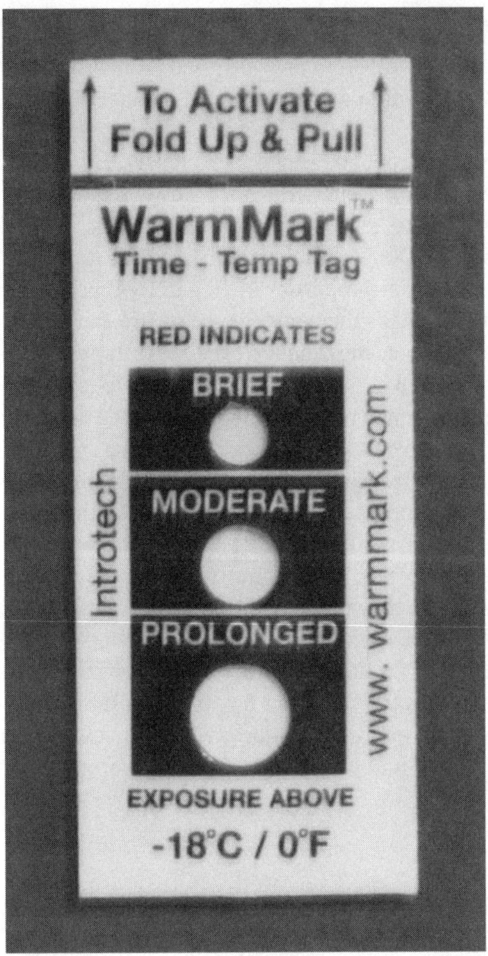

FIGURE 29.10 Time–temperature indicator for indicating food quality. (WarmMark indicator provided courtesy of Peter van Buren, Introtech, The Netherlands.)

Intelligent packaging has been proposed for a future smart kitchen [113]. The cooking appliance system would read a bar code that includes information on optimum cooking conditions and appropriately adjust the oven. The system could also read a TTI to alert the consumer to spoiled food.

G. Consumer-Friendly Packaging

Packaging innovation aimed at increasing convenience has become more important than decreasing package costs [114]. Consumers want packaging that provides a high level of food safety and food security, has an easy-to-read label that helps to guide food choices, is easy-to-open and reseal, provides an easy-to-prepare meal, and uses a minimum amount of material that is recyclable. Package design, including labeling, must also respond to cultural and demographic differences and changes. Concerns about obesity are affecting regulatory considerations impacting food labeling, including reevaluation of serving or portion sizes and calorie labeling, as well as coding systems, symbols, and nutritional categories to help consumers identify products. Packaging must enable electronic control of global distribution, rapid traceability for improved food safety and food security, electronic purchasing for future "smart shopping," and the electronic control of the future "smart kitchen" [113,115–118]. Improvements in packaging materials, design, and intelligence will be necessary to achieve these goals [75,119].

IV. CONCLUSIONS

Packaging for frozen foods has evolved to provide good protection at minimum cost. Paperboard is commonly used to provide protection against physical damage. LDPE is the most common plastic used in frozen food packaging, providing good protection against moisture loss as a coating on paperboard or as a flexible film pouch. Heat-resistant HDPE and PP film pouches are used for boil-in-bag or microwave-in-bag pouches. PP can be used to form trays or coat paperboard trays that are microwavable. Even more heat-resistant PET can be used to form trays or coat paperboard trays that are dual-ovenable. PVDC/PVC coating or EVOH or polyamide sandwiched by PE or PP are used to increase oxygen-barrier properties of frozen food packaging.

Frozen food packaging enhances the convenience of frozen foods. Packages have evolved that are easy-to-open, allow easy dispensing, and are easily resealable. Some frozen food packages also act as containers to heat and even serve the food.

Frozen food packaging has achieved good source reduction and recyclability. These trends will continue with additional improvements in packaging material properties and package design. Future recyclability of frozen food bags/pouches and plastic-coated paperboard will additionally minimize the environmental impact of frozen food packaging. Frozen food packaging also seems to lend itself to the use of biodegradable polymers, because premature biodegradability in frozen storage is not possible.

Frozen food packaging does an excellent job of preserving the quality and safety of food. A concern is potential migration of package material components into the food during heating in a microwave or convection oven. Vigilance concerning this issue will certainly continue.

Several trends, including developments in flexible packaging, microwavable containers, non-thermal food processing, MAP, and active and intelligent packaging, have influenced the nature of food packaging. These trends will certainly continue to improve the effectiveness of packaging for frozen foods.

REFERENCES

1. SSH Rizvi. Requirements for foods packaged in polymeric films. *Critical Reviews in Food Science and Nutrition* 14 (2):111–134, 1981.
2. A Hirsch. *Flexible Food Packaging — Questions and Answers*. New York: Van Nostrand Reinhold, 1991, pp. 130–133.
3. FA Paine, HY Paine. *A Handbook of Food Packaging*. New York: Van Nostrand Reinhold, 1992, pp. 248–264.
4. GL Robertson. *Food Packaging — Principles and Practice*. New York: Marcel Dekker, 1993, pp. 288–297.
5. MC Erickson, Y-C Hung, eds., *Quality in Frozen Food*. New York: Chapman & Hall, 1997, p. 484.
6. VW Balasubramaniam, MS Chinnan. Role of packaging in quality preservation of frozen foods. In: MC Erickson, Y-C Hung, eds., *Quality in Frozen Food*. New York: Chapman & Hall, 1997, pp. 296–309.
7. FA Paine, HY Paine. *A Handbook of Food Packaging*. New York: Van Nostrand Reinhold, 1992, pp. 1–32.
8. KL Yam, JS Paik, CC Lai. Packaging, Part I: General considerations. In: YH Hui, ed., *Encyclopedia of Food Science and Technology*. New York: John Wiley & Sons, 1992, pp. 1971–1975.
9. GL Robertson. *Food Packaging — Principles and Practice*. New York: Marcel Dekker, 1993, pp. 338–380.
10. KS Marsh. Shelf life. In: AL Brody, KS Marsh, eds., *The Wiley Encyclopedia of Packaging Technology*. New York: John Wiley & Sons, 1997, pp. 831–835.
11. GL Robertson. *Food Packaging — Principles and Practice*. New York: Marcel Dekker, 1993, pp. 1–8.
12. KS Marsh. Looking at packaging in a new way to reduce food losses. *Food Technology* 55 (2):48–52, 2001.

13. MD Steven, JH Hotchkiss. Package functions. In: AL Brody, KS Marsh, eds., *Encyclopedia of Agricultural, Food, and Biological Engineering*. New York: Marcel Dekker, 2003, pp. 716–719.
14. R Coles. Introduction. In: R Coles, D McDowell, M Kirwan, eds., *Food Packaging Technology*. Boca Raton, FL: CRC Press, 2003, pp. 1–31.
15. PJ Fellows. *Food Processing Technology — Principles and Practice*. Boca Raton, FL: CRC Press, 2000, pp. 462–510.
16. MS Chinnan, DS Cha. Primary packaging. In: DR Heldman, ed., *Encyclopedia of Agricultural, Food, and Biological Engineering*. New York: Marcel Dekker, 2003, pp. 781–784.
17. RA Bourque. Secondary packaging. In: DR Heldman, ed., *Encyclopedia of Agricultural, Food, and Biological Engineering*. New York: Marcel Dekker, 2003, pp. 873–879.
18. D Twede. Logistical/distribution packaging. In: AL Brody, KS Marsh, eds., *The Wiley Encyclopedia of Packaging Technology*. New York: John Wiley & Sons, 1997, pp. 572–579.
19. W Soroka. *Fundamentals of Packaging Technology*. Naperville, IL: Institute of Packaging Professionals, 2002, pp. 407–436.
20. D Twede, B Harte. Logistical packaging for food marketing systems. In: R Coles, D McDowell, M Kirwan, eds., *Food Packaging Technology*. Boca Raton, FL: CRC Press, 2003, pp. 95–119.
21. T Kadoya, ed., *Food Packaging*. New York: Academic Press, 1990, p. 424.
22. FA Paine, HY Paine. *A Handbook of Food Packaging*. New York: Van Nostrand Reinhold, 1992, p. 497.
23. KL Yam, JS Paik, CC Lai. Packaging, Part III: Materials. In: YH Hui, ed., *Encyclopedia of Food Science and Technology*. New York: John Wiley & Sons, 1992, pp. 1978–1982.
24. GL Robertson. *Food Packaging — Principles and Practice*. New York: Marcel Dekker, 1993, p. 676.
25. JF Hanlon, RJ Kelsey, HE Forcinio. *Handbook of Package Engineering*. Lancaster, PA: Technomic Publishing Co., 1998, p. 698.
26. S Sacharow, AL Brody. *Packaging: An Introduction*. Duluth, MN: Harcourt Brace Jovanovich Publications, Inc., 1987, pp. 35–77.
27. TW Downes. Food packaging in the IFT era: five decades of unprecedented growth and change. *Food Technology* 43 (9):228–229, 232–236, 238–240, 1989.
28. AL Brody, *Personal communication.* 2004.
29. GF Huss. Microwavable packaging and dual-ovenable materials. In: AL Brody, KS Marsh, eds., *The Wiley Encyclopedia of Packaging Technology*. New York: John Wiley & Sons, 1997, pp. 642–646.
30. K Ono. Frozen food and oven-proof trays. In: T Kadoya, ed., *Food Packaging*. New York: Academic Press, 1990, pp. 253–267.
31. SAE Lefeuvre, MBM Audhuy-Peaudecerf. Microwavability of packaging foods. In: M Mathlouthi, ed., *Food Packaging and Preservation*. New York: Chapman & Hall, 1994, pp. 62–87.
32. RA Simmons. Laws and regulations, United States. In: AL Brody, KS Marsh, eds., *The Wiley Encyclopedia of Packaging Technology*. New York: John Wiley & Sons, 1997, pp. 552–558.
33. JF Hanlon, RJ Kelsey, HE Forcinio. *Handbook of Package Engineering*. Lancaster, PA: Technomic Publishing Co., 1998, pp. 607–654.
34. MM Cramer. The time has come for clear food allergen labeling. *Food Safety Magazine*, 18–22, 2004.
35. FA Paine, HY Paine. *A Handbook of Food Packaging*. New York: Van Nostrand Reinhold, 1992, pp. 477–490.
36. RM White, PA Tice. Laws and regulations, Europe. In: AL Brody, KS Marsh, eds., *The Wiley Encyclopedia of Packaging Technology*. New York: John Wiley & Sons, 1997, pp. 541–552.
37. JR Blanchfield, ed., *Food Labeling*. Boca Raton, FL: CRC Press, 2000, p. 320.
38. J Storlie, AL Brody. Mandatory food package labeling in the United States. In: AL Brody, JB Lord, eds., *New Food Products for a Changing Marketplace*. Lancaster, PA: Technomic Publishing, 2000, pp. 409–438.
39. EF Greenberg. *Guide to Packaging Law — A Primer for Packaging Professionals*. Herndon, VA: Institute of Packaging Professionals, 1996, p. 115.
40. YH Hui. Packaging, Part II: Labeling. In: YH Hui, ed., *Encyclopedia of Food Science and Technology*. New York: John Wiley & Sons, 1992, pp. 1975–1978.
41. J Storlie, K Hare. Nutrition labeling. In: AL Brody, KS Marsh, eds., *The Wiley Encyclopedia of Packaging Technology*. New York: John Wiley & Sons, 1997, pp. 674–681.

42. Anonymous. Markaging: your in-store salesman. *Food Processing*, (August): 63–66, 2004.
43. RD LaMoreaux. *Barcodes and Other Automatic Identification Systems*. Leatherhead, UK: Pira International, 1995, p. 343.
44. H Barthel. Code, Bar. In: AL Brody, KS Marsh, eds., *The Wiley Encyclopedia of Packaging Technology*. New York: John Wiley & Sons, 1997, pp. 225–228.
45. FA Paine, HY Paine. *A Handbook of Food Packaging*. New York: Van Nostrand Reinhold, 1992, pp. 97–166.
46. SEM Selke. *Packaging and the Environment — Alternatives, Trends and Solutions*. Lancaster, PA: Technomic Publishing Co., Inc., 1994, p. 179.
47. SEM Selke. *Understanding Plastics Packaging Technology*. Cincinnati, OH: Hanser/Gardner Publications Inc., 1997, pp. 173–198.
48. D Brown. Plastic packaging of food products: the environmental dimension. *Trends in Food Science and Technology* 4 (9):294–300, 1993.
49. WF Franklin, TK Boguski, P Fry. Life-cycle assessment. In: AL Brody, KS Marsh, eds., *The Wiley Encyclopedia of Packaging Technology*. New York: John Wiley & Sons, 1997, pp. 563–569.
50. DT Allen, N Bakshani. Environmental impact of paper and plastic grocery sacks. *Chemical Engineering Education* 26 (2):82–86, 1992.
51. RJ Rowatt. The plastics waste problem. *Chem Tech* 23 (1):56–60, 1993.
52. JK Borchardt. Recycling. In: AL Brody, KS Marsh, eds., *The Wiley Encyclopedia of Packaging Technology*. New York: John Wiley & Sons, 1997, pp. 799–805.
53. T McCormack. Plastics packaging and the environment. In: GA Giles, DR Bain, eds., *Materials and Development of Plastics Packaging for the Consumer Market*. Boca Raton, FL: CRC Press, 2000, pp. 152–176.
54. JF Hanlon, RJ Kelsey, HE Forcinio. *Handbook of Package Engineering*. Lancaster, PA: Technomic Publishing Co., 1998, pp. 655–676.
55. IS Dent. Recycling and reuse of plastics packaging for the consumer market. In: GA Giles, DR Bain, eds., *Materials and Development of Plastics Packaging for the Consumer Market*. Boca Raton, FL: Sheffield Academic Press, 2000, pp. 177–202.
56. HR Thorsheim, DJ Armstrong. Recycled plastics for food packaging. *ChemTech* August:55–58, 1993.
57. DL Kaplan, JM Mayer, D Ball, J McCassie, AL Allen, P Stenhouse. Fundamentals of biodegradable polymers. In: C Ching, D Kaplan, E Thomas, eds., *Biodegradable Polymers and Packaging*. Lancaster, PA: Technomic Publishing Co., Inc., 1993, pp. 1–42.
58. Anonymous. Packaging plastic from plants. *Innova* May/June:4, 2004.
59. Anonymous. Advanced materials inspire new packaging options. *Dairy Foods* 105 (9):30–31, 2004.
60. M Raymond. Environmental regulations, North America. In: AL Brody, KS Marsh, eds., *The Wiley Encyclopedia of Packaging Technology*. New York: John Wiley & Sons, 1997, pp. 351–355.
61. M Raymond. Environmental regulations, international. In: AL Brody, KS Marsh, eds., *The Wiley Encyclopedia of Packaging Technology*. New York: John Wiley & Sons, 1997, pp. 348–351.
62. GL Robertson. *Food Packaging — Principles and Practice*. New York: Marcel Dekker, 1993, pp. 622–662.
63. LL Katan. Introduction. In: LL Katan, ed., *Migration from Food Contact Materials*. London, UK: Blackie Academic & Press, 1996, pp. 1–10.
64. K Figge. Plastics. In: LL Katan, ed., *Migration from Food Contact Materials*. London, UK: Blackie Academic & Professional, 1996, pp. 77–108.
65. L Soderhjelm, T Sipilainen-Malm. Paper and board. In: LL Katan, ed., *Migration from Food Contact Materials*. London, UK: Blackie Academic & Press, 1996, pp. 159–180.
66. SJ Risch, K Heikkila, R Williams. Analysis of volatiles produced in foods and packages. In: SJ Risch, JH Hotchkiss, eds., *Food and Packaging Interactions II*. Washington, DC: American Chemical Society, 1991, pp. 1–10.
67. LL Katan, AD Schwope, H Ishiwata. Real life and other special situations. In: LL Katan, ed., *Migration from Food Contact Materials*. London, UK: Blackie Academic & Professional, 1996, pp. 251–275.
68. AL Baner, R Franz, O Piringer. Alternative fatty food simulants for polymer migration testing. In: M Mathlouthi, ed., *Food Packaging and Preservation*. New York: Chapman & Hall, 1994, pp. 23–47.

69. JL Rosette. Tamper-evident packaging. In: AL Brody, KS Marsh, eds., *The Wiley Encyclopedia of Packaging Technology.* New York: John Wiley & Sons, 1997, pp. 879–882.
70. B Blakistone, D Bernard. HACCP. In: AL Brody, KS Marsh, eds., *The Wiley Encyclopedia of Packaging Technology.* New York: John Wiley & Sons, 1997, pp. 485–489.
71. JL Bricher. Lights, Camera, HACCP. *Food Safety Magazine*, 38–50, 2004.
72. TH Shellhammer. Flexible Packaging. In: DR Heldman, ed., *Encyclopedia of Agricultural, Food, and Biological Engineering.* New York: Marcel Dekker, 2003, pp. 333–336.
73. MJ Greely. Standup flexible pouches. In: AL Brody, KS Marsh, eds., *The Wiley Encyclopedia of Packaging Technology.* New York: John Wiley & Sons, 1997, pp. 852–856.
74. J Dixon. Development of flexible plastics packaging. In: GA Giles, DR Bain, eds., *Materials and Development of Plastics Packaging for the Consumer Market.* Boca Raton, FL: CRC Press, 2000, pp. 79–104.
75. RR Goddard. Packaging — A view of the future. *Packaging Technology and Science* 8:119–126, 1995.
76. S Bolado-Rodriguez, MM CGongora-Nieto, U Pothakamury, GV Barbosa-Canovas, BG Swanson. A review of nonthermal technologies. In: JE Lozano, et al., eds., *Trends in Food Engineering.* Lancaster, PA: Technomic Publishing Co. Inc., 2000, pp. 227–265.
77. GW Gould. Emerging technologies in food preservation and processing in the last 40 years. In: GV Barbosa-Canovas, GW Gould, eds., *Innovations in Food Processing.* Lancaster, PA: Technomic Publishing Co., Inc., 2000, pp. 1–11.
78. DW Thayer. Ionizing irradiation, treatment of food. In: DR Heldman, ed., *Encyclopedia of Agricultural, Food, and Biological Engineering.* New York: Marcel Dekker, 2003, pp. 536–539.
79. Z El Makhzoumi. Effect of irradiation of polymeric packaging material on the formation of volatile compounds. In: M Mathlouthi, ed., *Food Packaging and Preservation.* New York: Chapman & Hall, 1994, pp. 88–99.
80. BF Ozen, JD Floros. Effects of emerging food processing techniques on the packaging materials. *Trends in Food Science and Technology* 12:60–67, 2001.
81. D Knorr. Process aspects of high-pressure treatment of food systems. In: GV Barbosa-Canovas, GW Gould, eds., *Innovations in Food Processing.* Lancaster, PA: Technomic Publishing Co., Inc., 2000, pp. 13–30.
82. VW Balasubramaniam. High pressure food preservation. In: DR Heldman, ed., *Encyclopedia of Agricultural, Food, and Biological Engineering.* New York: Marcel Dekker, 2003, pp. 490–496.
83. Y Lanmbet, G Demazeau, JM Bouvier, S Laborde-Croubit, M Cabannes. Packaging for high-pressure treatments in the food industry. *Packaging Technology and Science* 13:63–71, 2000.
84. GV Barbosa-Canovas, GW Gould, eds., *Innovations in Food Processing. Food Preservation Technology Series.* Lancaster, PA: Technomic Publishing Co., Inc., 2000, p. 260.
85. JE Lozano, C Anon, E Parada-Arias, GV Barbosa-Canovas, eds., *Trends in Food Engineering. Food Preservation Technology.* Lancaster, PA: Technomic Publishing Co. Inc., 2000, p. 347.
86. B Ooraikul, ME Stiles, eds., *Modified Atmosphere Packaging of Food.* Belmont, CA: Thompson International Publishing, 1991.
87. JP Smith, B Simpson, H Ramaswamy. Packaging, Part IV: Modified atmosphere packaging — principles and applications. In: YH Hui, ed., *Encyclopedia of Food Science and Technology.* New York: John Wiley & Sons, 1992, pp. 1982–1992.
88. D Zagory. Modified atmosphere packaging. In: AL Brody, KS Marsh, eds., *The Wiley Encyclopedia of Packaging Technology.* New York: John Wiley & Sons, 1997, pp. 651–656.
89. R Perdue. Vacuum packaging. In: AL Brody, KS Marsh, eds., *The Wiley Encyclopedia of Packaging Technology.* New York: John Wiley & Sons, 1997, pp. 949–955.
90. BA Blakistone, ed., *Principles and Applications of Modified Atmosphere Packaging of Foods.* 2nd ed., Gaithersburg, MD: Aspen Publishers, 1999.
91. R Ahvenainen. Active and intelligent packaging. In: R Ahvenainen, ed., *Novel Food Packaging Techniques.* Boca Raton, FL: CRC Press, 2003, pp. 5–21.
92. AL Brody. Modified atmosphere packaging. In: DR Heldman, ed., *Encyclopedia of Agricultural, Food, and Biological Engineering.* New York: Marcel Dekker, 2003, pp. 666–670.
93. M Mullan, D McDowell. Modified atmosphere packaging. In: R Coles, D McDowell, M Kirwan, eds., *Food Packaging Technology.* Boca Raton, FL: CRC Press, 2003, pp. 303–339.

94. ML Rooney, ed., *Active Food Packaging.* New York: Blackie Academic & Professional, 1995.
95. ML Rooney. Active packaging. In: AL Brody, KS Marsh, eds., *The Wiley Encyclopedia of Packaging Technology.* New York: John Wiley & Sons, 1997, pp. 2–8.
96. TP Labuza, WM Breene. Applications of "active packaging" for improvement of shelf-life and nutritional quality of fresh and extended shelf-life foods. *Journal of Food Processing and Preservation* 13:1–69, 1989.
97. TP Labuza. An introduction to active packaging for foods. *Food Technology* 50 (4):68–71, 1996.
98. L Vermeiren, F Devlieghere, M van Beest, N de Kruijf, J Debevere. Developments in the Active Packaging of Foods. *Trends in Food Science and Technology* 10 (3):77–86, 1999.
99. ML Rooney. Plastics in active packaging. In: GA Giles, DR Bain, eds., *Materials and Development of Plastics Packaging for the Consumer Market.* Boca Raton, FL: CRC Press 2000.
100. AL Brody, ER Strupinsky, LR Kline. *Active Packaging For Food Applications.* Lancaster, PA: Technomic Publishing Co., 2001, p. 218.
101. R Ahvenainen, ed., *Novel Food Packaging Techniques.* Boca Raton, FL: CRC Press, 2003, p. 590.
102. BPF Day. Active packaging. In: R Coles, D McDowell, M Kirwan, eds., *Food Packaging Technology.* Boca Raton, FL: CRC Press, 2003, pp. 282–302.
103. N de Kruijf, MD van Beest. *Active Packaging. Encyclopedia of Agricultural, Food, and Biological Engineering.* New York: Marcel Dekker, 2003, pp. 5–9.
104. Y Harima. Frozen food and oven-proof trays. In: T Kadoya, ed., *Food Packaging.* New York: Academic Press, 1990, pp. 229–252.
105. RC Idol. Oxygen scavengers. In: AL Brody, KS Marsh, eds., *The Wiley Encyclopedia of Packaging Technology.* New York: John Wiley & Sons, 1997, pp. 687–692.
106. L Vermeiren, L Heirlings, F Devlieghere, J Debevere. Oxygen, ethylene and other scavengers. In: R Ahvenainen, ed., *Novel Food Packaging Techniques.* Boca Raton, FL: CRC Press, 2003, pp. 22–49.
107. NDR Goddard, RMJ Kemp, R Lane. An overview of smart technology. *Packaging Technology and Science* 10:129–143, 1997.
108. ET Rodrigues, JH Han. Intelligent packaging. In: DR Heldman, ed., *Encyclopedia of Agricultural, Food, and Biological Engineering.* New York: Marcel Dekker, 2003, pp. 528–535.
109. M Karel. Tasks of food technology in the 21st century. *Food Technology* 54 (6):56–64, 2000.
110. N de Kruijf, M van Beest, R Rijk, T Sipilainen-Malm. Active and intelligent packaging: applications and regulatory aspects. *Food Additives and Contaminants* 19 (Suppl.):144–162, 2002.
111. PS Taoukis, TP Labuza. Time–temperature indicators (TTIs). In: R Ahvenainen, ed., *Novel Food Packaging Techniques.* Boca Raton, FL: CRC Press, 2003, pp. 103–126.
112. DL Johnson. Indicating devices. In: AL Brody, KS Marsh, eds., *The Wiley Encyclopedia of Packaging Technology.* New York: John Wiley & Sons, 1997, pp. 498–503.
113. KL Yam. Intelligent packaging for the future smart kitchen. *Packaging Technology and Science* 13:83–85, 2000.
114. MA Ferrante. Packaging for the next millennium. *Food Engineering International*, 29–34, 1997.
115. PJ Louis. Review paper — food packaging in the next millennium. *Packaging Technology and Science* 12:1–7, 1999.
116. AL Brody. Smart packaging becomes Intellipac^{TM}. *Food Technology* 54 (6):104–107, 2000.
117. K Sonneveld. What drives (food) packaging innovation? *Packaging Technology and Science* 13:29–35, 2000.
118. SD Nightingale. New technologies for food traceability: package and product Markers. *Food Safety Magazine*, 22–23, 58–59, 2004.
119. P Reynolds. Technology to take us forward. *Packaging World*, 70–76, 2002.

30 Plastic Packaging of Frozen Foods

Kwang Ho Lee
Korea Food and Drug Administration, Seoul, Korea

CONTENTS

I. INTRODUCTION

Freezing of foods is used to prevent the growth and activity of microorganisms in food, to retard chemical reactions, and to prevent the action of enzymes at around $-18°C$ [1,2]. However, to maintain frozen foods in perfect condition, packaging should provide the following protections [2]:

1. To avoid dehydration caused by moisture vapor evasion through the wall or seals of the package. This moisture loss dehydrates surface areas of the frozen food and causes desiccation such as freezer burn. The dehydrated surface layer can be very thin, but may affect the appearance and ultimate quality of the product.
2. To limit oxidation promoted by enzymes not eliminated by blanching if air penetrates the package.
3. To inhibit oxidation particularly in food with a high fat content, which can be accelerated by light as heat can induce increased enzyme activity and chemical and bacterial deterioration.

4. To avoid flavor or volatile loss and the absorption of airborne odors which are unlikely to occur at the same time as prepackaged foods remain frozen. Special care is necessary with precooked foods where evaporation of the volatile content could cause flavor loss.
5. To control physical damage caused by compression during storage and transport. Special care should be paid in handling cases containing packs of frozen products. Further damage may occur to the bottom layers of package if the outer containers are dropped onto a hard surface.

Therefore, frozen food packaging materials must withstand low temperatures and sometimes high temperatures, such as microwave and boil-in-bag products. It should be nontoxic and convey no odors or flavors to the food. It should also provide a barrier to the transmission of water vapor or oxygen and be water-resistant, and it should be able to be handled with machine. Package should have the ability for graphic decoration and be tamper-resistant. Furthermore, the package should not disintegrate when it becomes moist on thawing, and in the retail cabinet package should be free of defects such as deterioration or collapsing.

In this chapter plastic packaging materials for frozen foods are discussed in terms of physical and chemical properties, package types, and commercial frozen foods packaged with the plastic.

II. TYPES OF PLASTIC MATERIALS FOR FROZEN FOODS

For frozen foods, various plastic packaging materials, such as polyethylene, ethylene vinyl acetate copolymer, polypropylene, polyvinyl chloride, polyvinylidene chloride, polyethylene terephthalate, polystyrene, and nylon are commercially used now and their chemical structures are shown in Figure 30.1.

A. Polyethylene

Polyethylene (PE) produced by coordination and radical polymerizations are mainly referred to as low-density and high-density polyethylenes, respectively [3].

1. Low-Density Polyethylene

Low-density polyethylene (LDPE) was introduced commercially after World War II. LDPE is made from ethylene at very high pressure (about 170 MPa) and temperature above 150°C and under the control of free radical initiators. At these conditions, free radicals attack the double bond and add to the monomer, leaving a free radical which repeats this addition action on more monomer molecules.

After polymerization is complete, the pressure is reduced to atmospheric, residual raw materials and solvents are recovered from the reactor, meanwhile, the polymer is isolated as solid particles. These are then extruded through a die that cuts the extruded strands into pellets for shipping and later processing.

During this procedure, many side chain branches also can be formed. These side chains hinder crystallization and reduce key properties such as stiffness and impact toughness, however they improve clarity and reduce density, which save the area cost of films made.

Packaging film made from LDPE for frozen foods, with a density of about 910 kg/m^3 is typically soft, flexible, and readily stretched. It has good clearness and provides a good barrier to moisture but a poor barrier to oxygen. It gives no off-odors or flavors to foods and is readily heat-sealed to itself. These desirable features, with its very low cost per unit area, have made LDPE one of the most widely used plastic packaging materials. It also shows excellent cold resistance, withstanding extreme low temperature of -70°C [4].

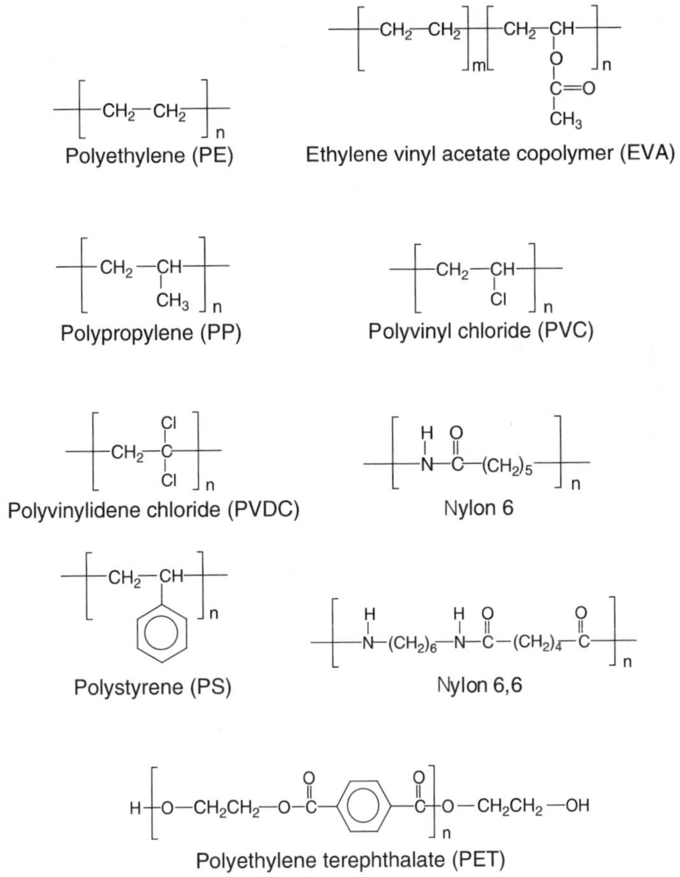

Polyethylene (PE)

Ethylene vinyl acetate copolymer (EVA)

Polypropylene (PP)

Polyvinyl chloride (PVC)

Polyvinylidene chloride (PVDC)

Nylon 6

Polystyrene (PS)

Nylon 6,6

Polyethylene terephthalate (PET)

FIGURE 30.1 Chemical structure of various polymers.

2. High-Density Polyethylene

High-density polyethylene (HDPE) originally polymerized at much lower pressures if a catalyst used to initiate polymerization of ethylene dissolved or slurried in a hydrocarbon medium. Later, gas-phase processes were also developed. HDPE has a slightly higher density of about 940 kg/m^3 than LDPE, with very little long-chain branching and a greater level of crystallinity. As a result, it is stronger in tension, stiffer, harder, and more gas-impermeable than LDPE; however, it has reduced clarity and impact resistance resulting from its greater crystallinity. Strength, perhaps its most important property, is a function of molecular weight.

HDPE is used for packaging films and for applications such as bottles, jars, and vials because of the ease of converting HDPE to blow- or injection-molded containers where it is needed for greater strength, stiffness, and lower clarity.

B. Ethylene Vinyl Acetate Copolymer

Ethylene vinyl acetate (EVA) copolymer is copolymerized vinyl acetate with ethylene, and the resulting plastic resins are widely used as adhesives in coextrusion or to make films that have all the desirable properties of LDPE but are much tougher. Generally up to 8% vinyl acetate content copolymerized EVA is used for frozen foods where toughness is required [5]. Large blocks of ice are packaged in EVA film because it can successfully resist puncturing by the sharp corners of the block and hold the heavy weight [6].

C. Polypropylene

When propylene molecules react to form long polymer chains at about 1.5 MPa in a hydrocarbon solvent and without a catalyst, or in a gas-phase process, the CH_3 side groups usually follow a regular pattern, in which the polymer molecules are lined up head-to-tail, nearly parallel, and packed together in a crystalline structure with a high degree of regularity called "isotactic" or "syndiotactic." If a large number of the molecules do not follow this regular array, the polymer called "atactic" is soft and sticky and is useful only as an adhesive (Figure 30.2). As a heat-seal layer in multiple structures, polypropylene (PP) used to be copolymerized with 1–5% ethylene to give a wider melting range [7].

PP, with a density of 900 kg/m^3, is the lightest resin of all used for packaging. Oriented PP film is clearer than LDPE or HDPE, stiffer and tougher than LDPE, and has lower permeability to moisture and gases than either, and with its higher melting point it is better suited to elevated temperature packaging applications. This combination of properties, including a stiff feel, resembles those of coated cellophane much more closely than does any PE film. For high gas barrier, it can be coated with polymers such as polyvinylidene chloride (PVDC) for oxygen-sensitive products [8].

PP film is also used in some packaging applications, such as a heat-seal layer for retort pouch or boil-in-bag. Like HDPE, PP is stiff enough to be used in rigid containers where its superior clarity gives it an edge over HDPE [6].

However, the major factor, which makes PP one of the most widely used clear plastic for packaging, is temperature. It is not strong enough to resist deformation at the temperatures used to sterilize foods in retort processing or the high temperatures used to radiant oven. Unless it is copolymerized with a maximum ethylene content of 20%, it tends to be brittle at low temperatures for frozen foods [4].

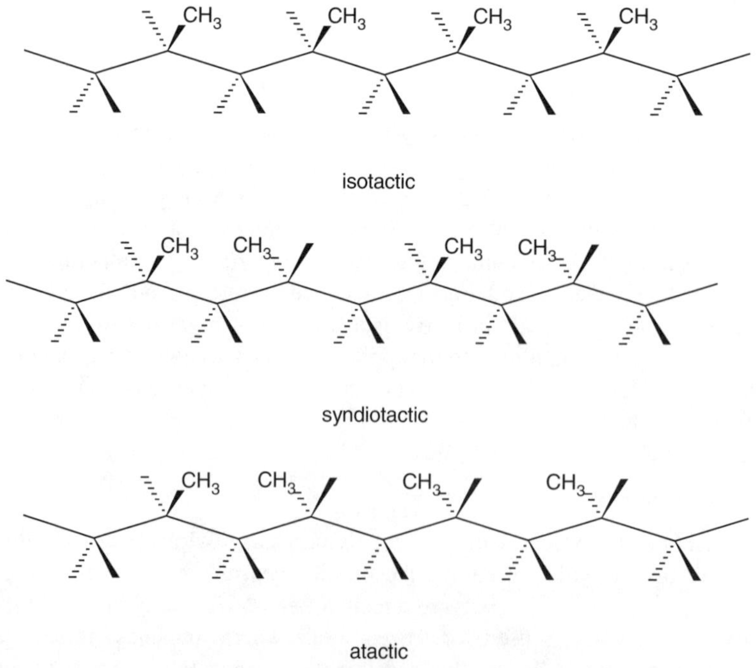

CH3 CH3 CH3 CH3

isotactic

CH3 CH3 CH3 CH3

syndiotactic

CH3 CH3 CH3 CH3

atactic

FIGURE 30.2 Different structures of PP.

D. Polyvinyl Chloride

Polyvinyl chloride (PVC), the so-called "vinyl," was introduced commercially in the late 1920s. It is synthesized by the low-pressure free radical polymerization of vinyl chloride at temperatures in the range of 38–71°C. Vinyl chloride monomer (VCM) contains a double bond that can be broken to allow head-to-tail polymerization to produce the polymer. However, various types of polymerization processes can be used to make polymers for specific applications [6].

PVC is a naturally brittle material and requires the addition of large amounts of other chemical compounds called "plasticizers" to make it useful as packaging film. Plasticized PVC packaging materials are tough and clear, provide a moderate barrier to oxygen and moisture, and can be processed to produce films, such as PP and LDPE, with good shrink properties.

However, concerns about the dangers involved in municipal incinerators have focused particularly on chlorine-containing plastics such as PVC and PVDC because it is supposed that the formation of dioxin, a complex chlorinated toxic molecule, is due to the inflow of these chlorinated plastics into the incinerator [9]. Regardless of the truth of this statement, it nevertheless casts another shadow over PVC and contributes to its growing unpopularity all over the world. It will probably be gradually replaced in some food packaging applications by lower-cost films such as PE, PP, or polyethylene terephthalate (PET) that can match its functional characteristics.

E. Polyvinylidene Chloride

PVDC is made by copolymerizing vinylidene chloride (VDC) with other comonomers such as VCM. Although the product is normally referred to as PVDC, it is always used in the copolymer form in packaging applications [7]. This material finds wide use in packaging because PVDC films are clear, soft, and high barrier materials with excellent cling characteristics. PVDC coatings can be readily applied to plastics when dissolved in solvents or dispersed as emulsions.

These coatings have the lowest permeability to oxygen at high humidity compared with any other polymers used in large-volume food-packaging applications. Ethylene vinyl alcohol (EVOH) can only compete with PVDC for high oxygen barrier; however, its oxygen barrier property is very poor at high humidity due to the swelling of the polar polymer with the moisture molecules.

F. Polyethylene Terephthalate

PET is one of polyester polymers made by the condensation polymerization which are formed by ester bonding and generating a small molecule, like water, from two different reactants, leaving bonding sites being able for the two reactants to join together into long chains.

PET is produced under the catalytic melt polymerization of ethylene glycol and either dimethyl terephthalate (DMT) or terephthalic acid (TPA) [7]. It is important to note that all these monomers have two reacting groups such as —OH on the glycol, —COOH on the TPA, or —$COOCH_3$ on DMT. This is necessary because the formation of a long chain depends on H_2O or CH_3OH being split out from the both end functional groups of reacting molecules.

The clarity and the mechanical properties of PET improve dramatically when it is biaxially oriented. This is done by stretching the film in both the longitudinal and transverse directions. Tenter frame process is usually used for this purpose, although tubular process equipped with biaxial orientation features can also be used. Unoriented PET, with its inferior properties, is hardly ever used in packaging.

PET is a commercially very important food packaging material because at elevated temperatures, it has excellent mechanical properties with inertness to food for reheated frozen foods. Its excellent high-temperature properties led its early use to boil-in-bag packaging and packaging for readymade meals where products are warmed up for consumption without removing them from their package. The latter feature makes it one of the few plastics approved by Food and Drug Administration (FDA) for contact with food at high temperatures.

Especially oriented PET has very excellent strength and toughness, and possesses better oxygen barrier property, especially for fatty foods, and better CO_2 barrier property than any of the common polyolefins such as PE and PP.

A copolymer of PET with cyclohexane dimethanol, called PETG, is tough but not as clear in the unoriented state. Despite its high cost, it is used in some thermoforming applications.

A crystallized form of PET, called CPET, is frequently used for dual-ovenable frozen precooked dinner and entrée plates that must withstand radiant oven temperatures and microwave without deformation [6,10].

G. Polystyrene

Polystyrene (PS) is made by bulk or suspension polymerization of styrene via the double bond in the ethylene group attached to the benzene ring. Polymerization is produced at low pressure and temperature in the range of 120–200°C. PS is an amorphous, crystal clear, hard, brittle, low-strength material with a relatively low melting point of 90°C and poor impact strength. However, it is readily thermoformed and injection molded but PS films have poor moisture and gas barrier properties [6].

Copolymerization with synthetic rubber, such as polybutadiene or styrene butadiene rubber, up to 10% by total volume improves its impact resistance, and such PS is called high-impact polystyrene (HIPS) [4]. HIPS is widely used for deep-draw food packaging such as egg trays, ice cream containers, and drinking cups.

PS can be foamed by adding foaming agents such as hexane to the reaction mixture during the suspension polymerization step. Its foamed form with 10 : 60 ratio is called expanded polystyrene (EPS), which has a very low density but is still a rigid material that is widely used for trays for egg, meat, poultry, and other products. It also has poor ability in conducting, providing insulation against high and low temperature for frozen foods, and can act as heat shock absorbent [4]. The drawback of EPS for the food packaging is its total lack of gas barrier, requiring packagers to overwraps with barrier films when this property is required.

H. Nylon

Nylon belongs to the class of polyamide made by the condensation polymerization of an organic acid and an amine. "Nylon" was the brand name for the most common of these polymers and is now widely used as a general name for them all [6].

In packaging application, nylon 6 and nylon 6.6 are important. Nylon 6 is made by polymerizing a single monomer called caprolactam, which has both the acid group and the amino group on the same molecule. Nylon 6.6 is made by reacting hexamethylene diamine and adipic acid to form an organic salt by eliminating water and form the long chain polymer. Many other nylons made from acids and amines with different structures are used for nonpackaging applications.

Nylon 6 is more common than Nylon 6.6 in food packaging, because it has a lower softening point and wider melting range and is thus easier to heat-seal and coextrude with other thermoplastic resins. Although its optical and mechanical properties are somewhat inferior to nylon 6.6, both uniaxial and biaxial orientations are used to enhance the barrier and mechanical properties of nylon 6 [11].

Nylon 6 is a clear film with pretty good gas and aroma barrier but has poor moisture barrier properties; it also has superior strength and outstanding tear and puncture resistance at low temperature. Furthermore, it maintains its mechanical properties well at elevated temperatures [7,11].

Table 30.1 lists the properties, such as water vapor transmission rate (WVTR), O_2 permeability, and service temperature, of the above-mentioned major plastic packaging materials for frozen foods [6,12].

TABLE 30.1
Properties of Major Packaging Materials for Frozen Foods (Data Based on 25 μm Film Thickness)

	HDPE	LDPE	EVA	PP	PET	PVC	PVDC	PS	Nylon
Density (kg/m^3)	945–967	910–925	930	900	1400	1220–1360	1600–1700	1050	1140
Yield (m^2/kg)	41.2	42.6	41.9	44	28.4	28	24	38	35
Tensile strength (GPa)	0.02–0.04	0.01–0.03	0.01–0.02	0.14–0.20	0.17–0.23	0.03–0.06	0.06–0.11	0.06–0.08	0.17–0.26
Tensile modulus [1% secant (GPa)]	0.86	0.14–0.28	0.06–0.14	2.41	4.83	2.41–4.14	0.35–1.03	2.76–3.28	1.72–2.07
Elongation at break (%)	200–600	200–600	500–800	50–275	70–130	100–400	50–100	2–30	70–120
Tear strength (Graves) (kg/m)	—	1800–8900	1800–8900	17900–26800	17900–35700	1800–5400	40	5400–17900	8900–14300
Tear strength (Elmendorf)(kg/m)	8000–14000	4000–8000	2000–8000	13000	800–4000	16000–28000	400–4000	100–600	600–1200
WVTR at 38°C and 90%RH [g/(m^2 day)]	6	16–31	31–47	6	16–23	31–465	0.8–5	109–155	155
O_2 permeability at 25°C and 0% RH [ml/(m^2 day atm)]	1550–3100	7750	10850–13950	1550–2480	50–90	465–9300	2–16	3100–5430	15–30
Haze (%)	3	5–10	2–10	3	2	1–2	1–5	1	1.5
Light transmission (%)	—	65	55–75	80	88	90	90	92	88
Heat-seal temperature range (°C)	140–150	120–180	70–180	90–150	140–180	140–180	120–150	120–180	120–180
Service temperature (°C)	−40–120	−60–80	−70–70	−40–120	−70–150	−30–70	−20–135	−60–80	−70–200

III. TYPES OF PLASTIC PACKAGE

For frozen food plastic packages, bags, containers, trays, and stretch wrap types are commonly used (Figure 30.3). Bag materials vary from unsupported PE and PP films to laminates of PE, PP, PET, or nylon. Coating the films such as PP or PET with PVDC increases the barrier properties for fatty foods. In boil-in-bags the products are packaged in bags in which they are intended to be cooked before opening. Foods, which produce a strong flavor during cooking, can be cooked very conveniently.

Examples of various materials which can be used for boil-in-bag products are HDPE and PP, which can give a reasonable shelf-life, although the packaging materials are fairly permeable to gases. Laminated materials are used to give a longer shelf-life. More expensive materials are PP/PE, PET/PE, PET/nylon/PP, and a laminated PP/PE or PET/PE coated with PVDC. For containers, HIPS materials are used for frozen desserts and EPS trays, which look whiter and cleaner with stretch wraps such as PVC or PE being used as alternative for paper mold product.

Thus, there is a wide range of package types available for frozen foods. However, the choice of package types has to be made carefully, bearing in mind the cost and storage performance, and the nature of the frozen product.

Tray and stretch wrap for fish

Tray and stretch wrap for shrimp

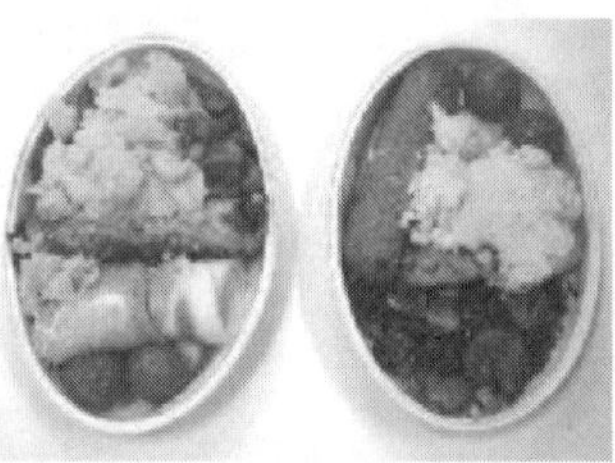

Containers for ready-made meals

Container for icecream

Bag for fruits

Bag for meat

FIGURE 30.3 Various types of plastic package for frozen foods.

IV. FROZEN FOODS PACKAGED WITH PLASTICS

A. Frozen Meat

The spoilage of meat is reduced when the temperature falls below freezing at around −8°C and bacteria and molds stop developing although some still grow at around −10°C. The physical and chemical changes in meat take place more slowly as the temperature falls down; however, they are not completely stopped even when stored at −30°C [1].

The fat in frozen foods will eventually turn to be rancid and, if exposed to light, the red pigment of myoglobin in the lean tissue will be fade. Irreversible dehydration will also occur at the surface of the meat unless it is packaged in air-tight, vapor-resistant material, for example, plastic bags or trays in conjunction with stretch wraps. Special equipment is available for this operation and the film used must combine good water vapor resistance with an oxygen barrier. To maintain the quality in frozen meat stored over long periods, a low temperature is essential. This must be a minimum of −18°C, which is the normal running temperature of the domestic freezer, but temperatures of −25°C or below, which is used for commercial cold store, is better.

Freezer burn is due to the dehydration of the surface of unpackaged or badly packaged frozen meat. Freezer burn becomes progressively worse when badly wrapped frozen meat is stored for a long time and grayish-white marks appear on the lean surfaces of the meat. To protect frozen meat from rancid and freezer burn, PE bags and PVC or PVDC films are used with EPS trays [7].

B. Frozen Poultry

With the development of the skintight PVDC film package, prepared poultry are inserted into bags and transferred to a rotary vacuumizing machine which packages the product with clip closures and bag neck trimming. When shrunk, the bags form a second skin around the exact contours of the birds, which are then either frozen in brine or in blast freezers, according to the preference of the particular processor.

The tight vacuumized and shrunk bags protect the birds in the brine bath and prevent freezer burn during prolonged storage. Bags are available in a variety of films and colors. Water absorption is negligible and the function of oxygen barrier is sufficient to prevent fats from becoming rancid. Materials used for packaging include PVDC film and a range of laminates with PE. Considerable developments have taken place in recent years with the introduction of, for example, frozen uncooked and precooked poultry portions [1,6].

C. Frozen Fish

Most fish begin to freeze at about −1°C and multiplication of putrefactive bacteria is stopped at −9°C. Although bacterial spoilage is suspended, not all bacteria are destroyed. Protection of frozen fish is needed against evaporation in cold storage caused by the transfer of moisture. This is now usually taken care of by glazing based on dipping the frozen fish in water to ice coat the surface or else by sealing the fish in a water and water-vapor-resistant wrapper. Thus the packed weight of the product is maintained, visible-surface dehydration, such as freezer burn, is avoided, and rancidity is retarded.

For packaging frozen fish, PVDC is used in vacuum-packaging some fish such as whole frozen salmon. This system provides a better alternative to glazing process. It eliminates moisture loss on initial freezing, drip loss on thawing, weight loss on glaze, and reduces labor and time needed for traditional glazing. The lightly vacuumized package enables the salmon to retain its fresh characteristics throughout the entire distribution system [6].

Packaging material, such as PET film, is also used for fish cakes in pillow pack style on horizontal form-fill-seal machines. The film is reverse printed on the treated side and laminated with PE. This laminate possesses barrier properties and is puncture-resistant over a wide range of temperatures, giving protection during transportation.

D. Frozen Fruits and Vegetables

Many fruits suffer substantial damage on freezing. Osmotic changes occur as a result of ice formation that destroys the cell membrane. Generally fruits do not require blanching before freezing and can be packaged in sugar or syrup of pureed before freezing. Vegetables, however, need to be blanched before freezing to ensure enzyme inactivation, which would otherwise result in objectionable flavors and loss of nutritional value and color. Most vegetables benefit from quick freezing, which gives a crisper final product.

The majority of frozen fruits and vegetables are packaged in plastics films, such as deep freeze grades of PE, made into pillow-type packs on vertical form-fill-seal machines. Vegetables are packaged commonly with LDPE as moisture barriers. Larger quantities, such as one pound or more, are packaged in EVA bags that are strong enough to carry the weight, offer the necessary moisture barrier, have good heat-seal properties, and remain pliable at freezer temperatures. As the contents of these larger packages are not generally consumed all at once, a reclosable feature on the package is often used. Some soft fruits, such as raspberries, are packaged in lidded plastic containers [1,6].

E. Other Frozen Products

One effective way to inhibit mold growth and greatly extend shelf-life for baked goods is to freeze the product. Frozen bread is packaged in LDPE bags. For frozen desserts such as ice cream, frozen sorbets, mousses, and puddings, thermoformed HIPS containers are often used [6,9].

V. FUTURE TRENDS

Frozen foods depend on the low temperature at which they are kept after being rapidly frozen to preserve them in the best quality condition and their packaging is required to prevent such as dehydration and oxidation of fats, which is often promoted by light, flavor, and aroma loss or gain and physical damage during handling and transport. For this purpose, a variety of primary packaging are currently employed including plain, coated, metallized, laminated plastic films, lidded plastic trays and thermoforms often contained in paperboard sleeves or cartons.

However, to extend keeping times and better quality of frozen food at -25 to -30°C, it requires the capability of withstanding these significantly lower temperatures without embrittlement. In case of merchandizing transparent packaging, antifogging bags are required, which can give an advantage to chilled foods on display, whereas frozen foods will spoil the transparency due to deposits of frost, because of the below-zero temperatures.

The development of ready meals for microwave is not so easy as might be supposed because different meal components require different times for heating unless their processing has taken account of this. The use of susceptors to provide "browning" and "crisping" of pizza or fish stick cases has received much attention and is a field that has still to be developed further.

Much emphasis will also be placed on environmental protections. Considerations should be given during the design and development of plastic packaging for frozen foods for the best packaging material to meet not only the protective needs of the food, but also environmental protections in terms of material resources, energy conservation, and most importantly, recyclability, which will result in minimizing the growing landfill problem [2,9,13,14].

VI. CONCLUSIONS

Plastic packaging materials, such as PE, PP, PET, PS, and nylon, are widely used for frozen foods. These packaging materials should have proper barrier properties such as moisture vapor, oxygen, and flavor or volatile compounds and impact strength and puncture resistant to be used at low temperature, while keeping acceptable food quality of frozen products during processing, storage, and handling of the foods.

Frozen meat, poultry, fish, fruits and vegetables, and desserts such as ice cream are some of frozen products commonly packaged with plastics. In the future, plastic packaging materials will be developed to meet not only functional protective needs of foods, but also material resources, energy conservations, and especially recyclability to handle waste management for environmental protection.

REFERENCES

1. GL Robertson. *Food Packaging Principles and Practice*. New York: Marcel Dekker, 1993, pp. 323–325.
2. FA Paine, HY Pain. *A Handbook of Food Packaging*. 2nd ed. UK: Blackie Academic Professional, 1992, pp. 248–264.
3. JS Kong, SY Han. Polyethylene for food packaging. *Polymer Science and Technology* 12 (2):183–196, 2001.
4. T Kadoya. *Food Packaging*. New York: Academic Press, 1990, pp. 131–137.
5. HUNTSMAN. EVA Copolymer Film Product Chart. 2004.
6. WA Jenkins. *Packaging Food with Plastics*. Lancaster: Technomic, 1991, pp. 35–63, 134–135, 241, 270–284.
7. MH Pack, DS Lee, KH Lee. *Food Packaging Science*. Seoul: Hyungseol, 2002, pp. 84–107.
8. KB Kang. Introduction to market and manufacturing of polyolefin film-focusing on PP film. *Polymer Science and Technology* 14 (2):154–162, 2003.
9. KR Osborn, WA Jenkins. *Plastic Films*. Lancaster: Technomic, 1992, pp. 217–220.
10. YC Kim, CG Park. Application for package material of polyester film. *Polymer Science and Technology* 12 (2):197–209, 2001.
11. IS Cho, HY Woo. Manufacturing and application of Nylon film. *Polymer Science and Technology* 12 (2):223–232, 2001.
12. YW Kim, SW Kim. Technology and application of high performance films. *Polymer Science and Technology* 14 (2):163–173, 2003.
13. G Bureau, JL Multon. *Food Packaging Technology*. New York: VCH, 1996, pp. 221–224.
14. BJ Kelsey. *Packaging in Today's Society*. Lancaster: Technomic, 1989, pp. 101–107.

31 Paper and Card Packaging of Frozen Foods

David Tanner
Food Science Australia, Sydney, Australia

Nevin Amos
ZESPRI International, Mt Maunganui, New Zealand

CONTENTS

I. INTRODUCTION

Paper is the general term used for a wide range of matted or felted webs of vegetable fiber (mostly wood) that have been formed on a screen from a water suspension [1]. This thin, layered network of fibers is adhered together by hydrogen bonds. The ability of the fibers to bond together and to exhibit a random layered structure are two key requirements of paper [2]. Cellulosic fibers exhibit one of the fundamental properties required for successful papermaking, the ability to bond without an adhesive [2]. Other properties of cellulosic fibers that make them an ideal raw material for paper are shown in Table 31.1.

Although there is no firm demarcation between paper and card, ISO standards do state that paper with a base weight (or grammage) $>250\ g/m^2$ should be termed paperboard, board, or card. It must be noted that there are exceptions to the above guide, dependent on country and use of the material.

Paper and card are commonly used for packaging frozen foods. Paper may also be used as a surface coating that provides a smooth surface for high-quality printing. Card or paperboard is used to produce both folding and rigid cartons, often not in direct contact with the food product. The board often consists of plies made from different materials. A widely used board, white-lined chipboard, has a white surface on one side made from a bleached virgin pulp, with the bulk being composed of "chip," which is usually gray and made from a high proportion of recycled

TABLE 31.1
Properties of Cellulosic Fibers [21]

Hydrophilic
High-tensile strength
Inherent bonding ability
Suppleness
Water insoluble
Chemically stable
Relatively colorless (white)
Ability to absorb modifying additives
Wide range of dimensions (length and diameter)

paper [3]. Solid or corrugated paperboard is also used in the production of secondary packaging, that is, outer cases, transport cartons, and so on. Solid board consists mainly of board made from recycled waste with an outer ply of Kraft serving as an outer skin. Corrugated board consists of three or more layers of paperboard that are laminated together. Heavy-duty corrugated board consisting of a double- or triple-wall structure is occasionally used in the frozen food industry for pallet-stabilizing corner posts or other load-bearing applications. Different grades of corrugated boards are shown in Figure 31.1.

The range of configurations for packaging of frozen food products depends on the purpose (e.g., transport or display) the packaging is required to deliver. Figure 31.2 shows some examples of paper and card packaging used in the containment and presentation of frozen foods. Figure 31.3 shows the loading of frozen food products packaged in paper-based packaging materials into a refrigerated shipping container, whereas Figure 31.4 shows commonly used paper-based cartons for bulk meat packaging.

II. DESIGN

The choice of packaging material and package design for a particular frozen food will depend on a range of factors:

- Cost
- Thermal properties of the food and packaging material
- Mechanical properties required to maintain integrity
- The legal and regulatory requirements
- Recyclability
- Product stability when in contact with the material
- Appearance

The cost of packaging can have a major influence on the decision made by a food producer. If the packaging is only required for containment, and print quality is of little importance, a low-cost paper finish may be adequate. However, if the use for the package involves retail display and is to be visually appealing, a higher quality finish (and likely higher cost material) will be required.

It is generally accepted that paper and card packaging is low cost in comparison to other available materials. Before such a statement can be quantified, however, the cost of raw material purchase, packaging conversion, the influence of the packaging on rates of cooling, and product stability as well as the costs of recycling or disposal must be taken into account.

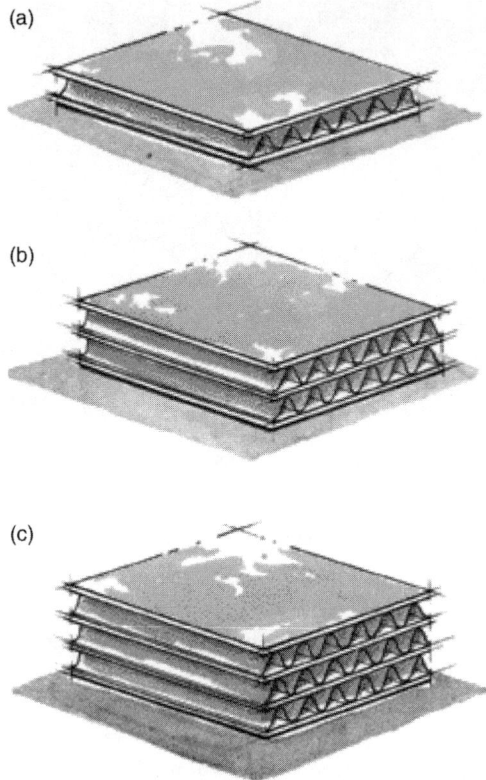

FIGURE 31.1 Three different grades of corrugated paperboard: (a) single-wall, (b) double-wall and (c) triple-wall corrugated. (From Anonymous. What is corrugated? Fibre Box Association. Rolling Meadows, Illinois, USA. With permission.)

III. PHYSICAL PROPERTIES

A. Barrier Properties

Paper and paperboard are poor barriers to gas (including water vapor). To improve the barrier properties and physical characteristics, these materials can be laminated, with plastic (e.g., polyethylene) or can be coated (e.g., with wax or clays). The choice of materials that can be used as barriers is wide and varied, with economic considerations often dictating selection.

Waxed papers have proven to be a reliable alternative in applications requiring direct contact with food as a barrier against penetration of liquids and vapors, as well as heat sealability, lamination, and even printing [1]. A wide assortment of refrigerated and frozen foods have been

FIGURE 31.2 Typical paper-based packaging used for retail packaging of frozen foods.

FIGURE 31.3 Loading paper-based packaged frozen food into a refrigerated shipping container.

packaged in this manner since the 1920s [1]. There are two different waxing processes generally available:

1. Wet waxing — where the wax coating is applied to the surface of the paper or card sheet. This is desirable for heat sealing and lamination and is essential for vapor barrier development.
2. Dry waxing — where the wax is absorbed into the sheet and often does not look or feel waxy. Such treated papers allow free transmission of water vapor and gases.

The formulations commonly used offer a number of functional properties for the carton:

- Protection against loss or gain of moisture
- High initial gloss and good gloss retention
- Scuff and abrasion resistance

FIGURE 31.4 Frozen beef packed into paper-board packaging.

- Grease resistance
- Adequate slip properties to aid high-speed handling on packaging equipment

B. Moisture Sorption Properties

Paper-based packaging can readily adsorb or desorb moisture when introduced into a storage environment, depending on the initial moisture content of the material and the relative humidity of the storage atmosphere. However, by introducing paper-based packaging materials to an environment with high relative humidity, strength properties will be reduced due to moisture uptake [4–6].

Hedlin [7] undertook experiments on the moisture adsorption of wood products at subfreezing temperatures (−12°C). These showed that the product readily adsorbed moisture when the relative humidity increased in the local environment. Figure 31.5 shows the increase in moisture content versus time of Douglas Fir wood shavings in air at −12°C and 30% relative humidity that is moved to air at −12°C and 50% relative humidity. The slope of the curve over the first 24 h suggests that this period has the greatest rate of moisture change.

Paper-based materials in refrigerated facilities can therefore act as both moisture sources and sinks, however the form of the moisture sorption isotherm is dependent on the material. A range of different isotherm models have been proposed for various classes of materials [8]. A form commonly used for modeling paper-based materials is the Guggenheim–Anderson–de Boer (GAB) isotherm [9–11], which can be expressed as follows [12,13]:

$$X = \frac{X_m C K a_w}{(1 - K a_w)(1 - K a_w + C K a_w)} \tag{31.1}$$

The moisture sorption isotherm for a given material is dependent on temperature and on whether adsorption or desorption is occurring [14,15]. Thus, the parameters of Equation (31.1) (X_m, K and C) are both temperature- and process-dependent.

Paper-based products, as with many food products, display hysteresis, as the moisture content is dependent on the moisture content history [15]. This means that paper-based materials adsorbing moisture will have a lower moisture content at the same temperature and relative humidity than paper desorbing moisture (Figure 31.6).

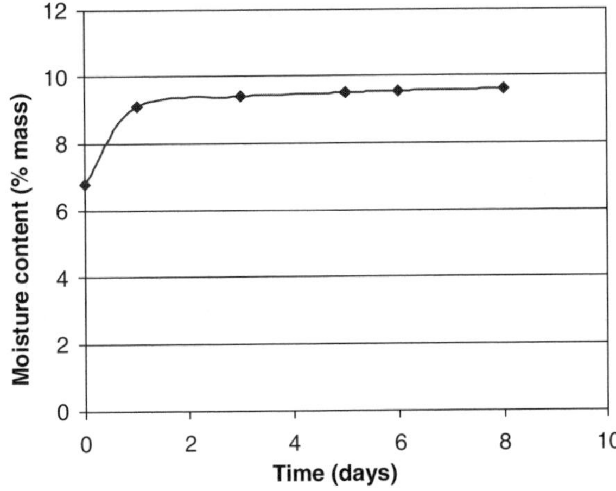

FIGURE 31.5 Rate of change in moisture content of Douglas Fir wood shavings initially in air at −12°C and 30% relative humidity and moved to air at −12°C and 50% relative humidity. (From CP Hedlin. *Forest Products Journal* 17 (12):43–48, 1966. With permission.)

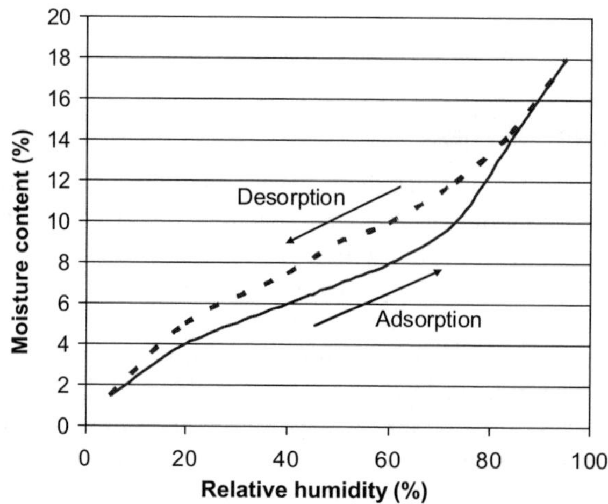

FIGURE 31.6 General moisture sorption isotherm for paper showing adsorption (solid line) and the desorption (dotted line).

There are many reports in the literature detailing isotherm data for a wide range of materials, including packaging materials under various conditions [14–18]. However, there is very little data for packaging materials at temperatures below 0°C. Research is underway in laboratories around the world to collect this information.

C. Insulating Properties

For many food products, the packaging system is required to provide some degree of thermal insulation. This may be in order to offer some protection during periods when the food is outside the freezer or to buffer against temporal variability within the frozen food supply chain. The importance of this relates to the impact of temperature variability on product quality (e.g., ice cream) as discussed in other chapters.

Heat transfer can take place by conduction, convection, and radiation and is affected by the thickness of the material, its thermal properties, its porosity, and its reflectivity. These properties are considered to collectively determine the insulating properties for a packaging system.

Thermal conductivity is a measure of the resistance to heat transfer by conduction through a material. Thermal conductivity depends on many factors, including the kind of substance (metal, solid liquid), composition (impurities, mixtures), structure and structural orientation, temperature, and pressure [19]. Some experimental data are available for some paper-based and wooden packaging materials (Table 31.2).

The specific heat capacity of a material (units of $J\ kg^{-1}\ K^{-1}$) is defined as the amount of heat necessary to raise the temperature of a unit mass of the material by a unit degree. Table 31.3 presents data for the specific heat capacity of paper-based and wooden materials.

The density of a material (units of $kg\ m^{-3}$) is defined as the ratio between the mass of the material and its volume. Table 31.4 presents data for the density of paper-based and wooden materials.

In paper and card packaging, insulation properties can be achieved in three ways:

1. By incorporating other thermally insulating packaging materials (such as polystyrene foam) into the packaging system (generally between the food and the outer paper or card packaging).

TABLE 31.2
Experimental Values of Thermal Conductivity for Some Packaging Materials

Packaging Material	Thermal Conductivity ($W\ m^{-1}\ K^{-1}$)
Corrugated cardboard	0.065 [19]
Solid cardboard	0.07 [31]
Wood—pine, white	0.11 [32]
Hardwoods	0.15 [33]
Natural kraft paper	0.079 [34]

2. By incorporating air into the paperboard matrix through use of corrugated structures, thereby reducing the effective thermal conductivity of the materials.
3. By incorporating reflective materials on the outer surface of the package to aid in reducing the influence of radiative heat transfer, thereby slowing the thawing rate of foods.

D. Mechanical Properties

Mechanical strength of papers and paperboards are largely dependent on the environment under which they are used. A range of mechanical tests can be used on paper-based packaging materials to quantify their strength properties, including tensile strength, compression, impact, tear, and bursting.

"Wet" strength of paper and paperboard is important. The adsorption of moisture causes the paper fibers to swell, forcing them apart, increasing the length, width, and thickness of the sheet [20]. Such an increase in moisture content reduces the ability of paper to withstand compressive and tensile forces [20–22].

The strength consideration for frozen foods takes on a further dimension as the packaging must be able to withstand freeze–thaw cycles without losing shape, tearing, warping, or absorbing too much moisture [23]. The packaging must be able to withstand about −40°C if the product is packaged prior to air blast freezing, and much colder temperatures if cryogenic freezing is employed. When the frozen product is intended to be heated in the package, strength and performance at high temperature must be maintained. The package must be able to withstand up to approximately 200°C for products baked or cooked in oven.

TABLE 31.3
Values for Specific Heat Capacity of Some Packaging Materials

Packaging Material	Specific Heat ($J\ kg^{-1}\ K^{-1}$)
Corrugated board	1700 [35]
Molded pulp "Friday" tray	1340 [36]
Paper	1300 [37]
Softwoods	1630 [37]
Solid cardboard	1260 [38]

TABLE 31.4
Experimental Values of Density for a Range of Horticultural and Packaging Products

Packaging Material	Density (kg m^{-3})
626[a] B — Corrugated board	250 [39]
626[a] C — Corrugated board	195 [39]
6226[b] B — Corrugated board	250 [39]
Molded pulp "Friday" tray	260 [39]
Solid cardboard	802 [32]
Unbleached Liner — Grade 1	706
Unbleached Liner — Grade 2	695
Unbleached Liner — Grade 3	688
Unbleached Liner — Grade 6	659
Wood — pine, white	430 [32]

[a] 290/160/290 g m^{-2} paper grades.

[b] 290/160/160/290 g m^{-2} paper grades.

E. Other Functional Properties

With the advent of the microwave oven came the demand for suitable packaging for use within the oven itself that would allow foods to crisp and brown [24]. To meet this need, susceptor technology was introduced. This usually consists of a thin layer of polyester film lightly vacuum metalized with aluminum, then laminated metal side down to a paper or card substrate. When placed in the microwave oven, a proportion of the radiation is absorbed by the aluminum layer, generating a heating effect that is regulated by the quantity of metal used in the composite. Using this system, a maximum temperature of 240°C can be attained [24] and, under such conditions, food in direct, or very close contact, can be made to crisp and brown.

IV. PRODUCT STABILITY

Product stability is important for maintaining the quality of frozen foods, therefore the influence of paper and card packaging on this will be further outlined here.

An essential requirement of the package is that it does not interact with the product negatively. As previously discussed, the package must have sufficient barrier properties to protect goods from microbial, chemical, and physical contamination, thus minimizing product quality loss. In addition, the packaging must also meet requirements such as maintenance of strength and performance at the range of storage temperatures encountered during each phase of the supply chain, and must be chemically stable to avoid tainting, or scalping of flavor or odor from the product.

When choosing a packaging material for containment, migration of components from the packaging to the product and scalping of flavor and aroma by the packaging material are important considerations. Tice [25] has written a comprehensive review covering migration and legislation associated with paper and board (and regenerated cellulose films) intended for contact with foodstuffs.

Paper and card packaging materials are generally not considered as major contributors to migration (in comparison with polymeric materials), although phthalates and benzophenone have been reported as migratory components [26,27]. Printing inks are applied to the outer surface of food packaging and with few exceptions — they are not intended to make direct contact with the food. If the paper or paperboard acts as a barrier to migration, there should be no transfer of ink components to the food. Woods [28] has shown, however, that benzophenone, a photoinitiator for UV-cured inks, can permeate through printed paperboard during room temperature storage and microwave reheating. Johns et al. [29] studied the migration of this substance from printed paper and board into frozen foods and found that even at $-20^{\circ}C$, migration to foods occurred. They stated that transfer could be considerable over long storage periods (e.g., 1 year). They also stated, however, that if the content of low-molecular-weight volatiles is controlled in the inks used to print food contact materials, then migratory levels could be kept low in low-temperature conditions.

V. RECYCLABILITY

Paper reuse or recycling is good for the environment. It minimizes landfill requirements and reduces the need for imported virgin pulp in some countries. Although recycling often makes economic and environmental sense, waste paper cannot be used in all paper grades, nor can it be used indefinitely. Three criteria must be considered:

- Strength — every time a fiber is recycled it loses some of its strength and the fiber length decreases. After being reused about six times the fibers become too short for papermaking.
- Quality — some grades make little or no use of recycled fiber because they need certain qualities provided only by new pulp.
- Utility — it is not possible to recover all paper.

There are over 60 recognized grades of waste paper in Europe, categorized into five main groups by the Confederation of European Paper Industries (CEPI) and the Bureau of International Recycling (BIR) [30]:

- Ordinary grades: These papers tend to contain a substantial amount of short fibers.
- Medium grades: This category contains articles such as unsold newspapers, sorted office paper, and so on.
- High grades: predominantly white papers made from virgin fibers.
- Kraft grades: Generally come from brown unbleached packaging materials such as paper sacks and corrugated cases.
- Special grades: This a hotchpotch of papers which tend to be uneconomic to sort and so are used in the middle layers of packaging papers and boards.

Recycling programs are highly mature in many countries. In America, it is stated that over 74% of all corrugated packaging is recovered for recycling. In Europe, the EU Packaging Directive from 1994 has been important in promoting increased recycling and recovery rates of packaging materials, as well as waste reduction related to packaging.

VI. MODELING

It is common practice to freeze meat or fish products in their transport packaging because this is usually the only packaging applied to the product that has the necessary rigidity, as the packaging only has to get the product to the freezer; thereafter the solid frozen product itself holds the shape.

Temperature fluctuations during storage can have a major influence on product quality, particularly for diary foods (e.g., ice cream) and for higher value fish. Cardboard packaging can play an important role in buffering the food from many of the air temperature fluctuations experienced in commercial freezing facilities.

Design of packaging involves incorporation of a number of interacting factors such as mass transfer to and from the packaging, heat transfer and possibly mass transfer to and from the food and fluctuating air temperatures and humidity surrounding the packaging. Steady-state designs can be accomplished using simple calculations, however, these are generally limited to simple geometries, heat transfer only, and constant boundary properties. Mathematical models, generally solved by computer, have become of increasing importance over the last two decades for assisting in the design of packaging systems. These enable more comprehensive studies of packaging performance without the need to undertake expensive and time-consuming experimental studies. A number of models are now commercially available for use in packaging design.

As an example, Figure 31.7 shows scenario testing for thermal efficiency during freezing of various package designs. The example is for freezing of fish in a blast freezer operating at $-30^{\circ}C$ with an air velocity of $1\ m\ s^{-1}$ across the packages. The output demonstrates several key points to consider when designing cardboard packaging systems:

- Instead of one large carton, consider multiple smaller cartons.
- Ensure that cartons are frozen when in the smallest logistics unit. If possible, freeze individual cartons. Rows of cartons will freeze slower than individual cartons, but will perform better than slabs of cartons. For freezing slabs or rows of cartons, the height of the carton is important. The smaller the height of the package, the faster the rate of freezing will be.

VII. CONCLUSIONS

The choice of packaging material can play an important role in preserving the quality of frozen foods. Paper-based materials are commonly used in the frozen food industry. These materials

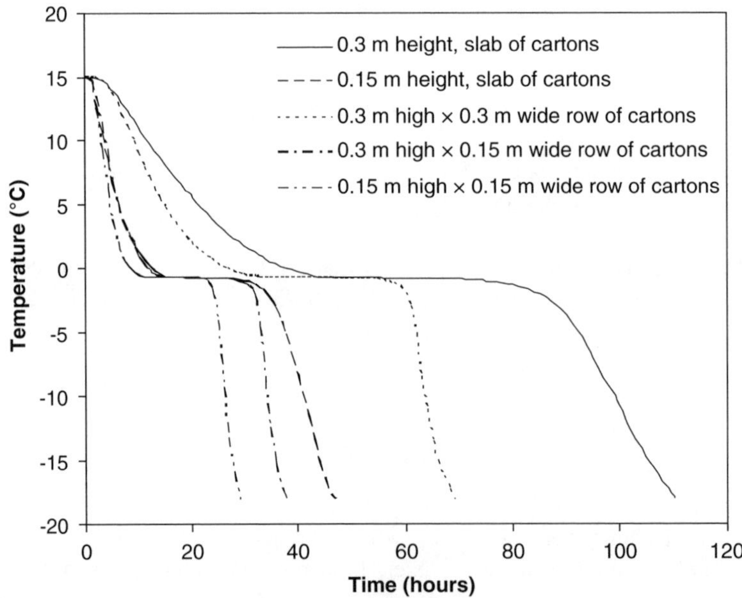

FIGURE 31.7 Estimated center temperatures for various configurations of cartons of fish during freezing at $-30^{\circ}C$ in a blast freezer with an air velocity of 1 m/s.

are able to be used in a number of different ways: either as paper, solid wall card (also known as paperboard), or corrugated paperboard. The designs of paper-based packages need to take into account factors including cost, thermal performance, ability to maintain integrity, legal and or regulatory requirements (increasingly), recyclability, and product stability.

The barrier properties of paper-based materials are generally poor, but can be improved with the addition of waxes or other coatings. These materials can also readily adsorb or desorb moisture when transferred between different temperature and humidity environments. Adsorption or desorption impacts negatively on strength properties of paper-based materials and must be considered when selecting packaging for foods to be stored in frozen conditions. The configuration of the paper-based material (e.g., solid or corrugated) will determine the insulating properties of the package. This important factor needs to be considered at the design stage. In addition, defined functionality, such as the incorporation of properties for microwave heating, need to be considered prior to packaging development.

Increasingly, stability of the product within the package and the recyclability of the package are important considerations. Understanding the views of manufacturers, retailers, and consumers on these factors can assist in packaging designs that better provide a paper-based package with appropriate functionality.

Finally, mathematical modeling as a tool for assessing some of the important physical attributes of packaging, prior to trial-and-error experimentation, is finding increasing favor. This approach can be cost-effective as it can narrow the design spectrum down, by discarding packaging systems that fail to meet the user's demands.

NOMENCLATURE

a_w	water activity
C	adsorption constant
K	adsorption constant
X	moisture content (kg/kg)
X_m	moisture content when each sorption site contains one water molecule (monolayer) (kg/kg)

REFERENCES

1. AL Brody, KS Marsh. *The Wiley Encyclopedia of Packaging Technology*. 2nd ed. New York: John Wiley & Sons, 1997.
2. JE Kline. *Paper and Paperboard: Manufacturing and Converting Fundamentals*. 2nd ed. San Francisco: Miller Freeman Publications, 1991.
3. M George. Selecting packaging for frozen foods. In: C Kennedy, Ed., *Managing Frozen Foods*. Cambridge: Woodhead Publishing, 2000, pp. 195–211.
4. WA Wink. The effect of relative humidity and temperature on paper properties. *TAPPI Journal* 44:171–180, 1961.
5. JA Marcondes. Cushioning properties of corrugated fiberboard and the effects of moisture content. *Transactions of the ASAE* 35:1949–1953, 1992.
6. JA Marcondes. Effect of load history on the performance of corrugated fiberboard boxes. *Packaging Technology and Science* 5:179–187, 1992.
7. CP Hedlin. Sorption isotherms of twelve woods at subfreezing temperatures. *Forest Products Journal* 17 (12):43–48, 1966.
8. J Chirife, HA Iglesias. Equations for fitting water sorption isotherms of foods. Part 1. A review. *Journal of Food Technology* 13:159–174, 1978.
9. EA Guggenheim. *Application of Statistical Mechanics*. Oxford: Clarendon Press, 1966.
10. RB Anderson. Modifications of the Brunauer, Emmett and Teller equation. *Journal of the American Chemical Society* 68:686–691, 1946.

11. JH de Boer. *The Dynamical Character of Adsorption*. Oxford: Clarendon Press, 1953.
12. H Bizot. Using the 'G.A.B.' model to construct sorption isotherms. In: R Jowitt, F Escher, B Hallström, HFTh Meffert, WEL Spiess, G Vos, Eds., *Physical Properties of Foods*. London: Applied Science Publishers, 1983, pp. 43–54.
13. C van den Berg. Description of water activity of foods for engineering purposes by means of the G.A.B. model or sorption. In: BM McKenna Ed., *Engineering and Food: Vol. 1. Engineering Sciences in the Food Industry*. London: Elsevier Applied Science Publishers, 1984, pp. 311–321.
14. RT Skogman, CE Scheie. The effect of temperature on the moisture adsorption of Kraft paper. *TAPPI Journal* 52 (3):489–490, 1969.
15. DG Eagleton, JA Marcondes. Moisture–sorption isotherms for paper-based components of transport packaging for fresh produce. *TAPPI Journal* 77 (7):75–81, 1994.
16. LM Pidgeon, O Maass. The adsorption of water by wood. *Journal of the American Chemical Society* 52:1053–1069, 1930.
17. R Mauritz, F Solar, A Pfitzner. Sorptionsverhalten wohnraumumschließender Materialien. Teil 3: Sorptionsverhalten anderer wohnraumumschließender Materialien im Vergleich zu Holz. (Sorption characteristics of indoor materials. Part 3: Sorption behavior of other indoor materials in comparison to wood). *Holzforschung und Holzverwertung* 42 (1):6–12, 1990.
18. G Sørensen, J Hoffmann. Moisture sorption in moulded fiber trays and effect on static compression strength. *Packaging Technology and Science* 16 (4):159–169, 2003.
19. Anon. *ASHRAE Handbook — Fundamentals*. American Society of Refrigeration and Air Conditioning Engineers Publishers, Atlanta, GA, 1993.
20. K Kawanishi. Estimation of the compression strength of corrugated fiberboard boxes and its application to box design using a personal computer. *Packaging Technology and Science* 2:29–39, 1989.
21. GA Smook. *Handbook for Pulp and Paper Technologists*. 2nd ed. Vancouver: Angus Wilde Publications, 1992.
22. P Harrison, M Croucher. Packaging of frozen foods. In: CP Mallett, Ed., *Frozen Food Technology*. London: Blackie, 1993.
23. LR Hancock, JC Hare. *Blending style and strength — tips for food-package printing on unbleached paperboard*. Flexographic Technical Association, 2004.
24. P Harrison. The role of packaging in achieving microwave browning and crisping. *Packaging Technology and Science* 2 (1):5–10, 1989.
25. P Tice. Paper and board, and regenerated cellulose films intended for contact with foodstuffs. In: R Ashby, I Cooper, S Harvey, P Tice, Eds., *Food Packaging Migration and Legislation*. London: Pira International, 1997, pp. 97–122.
26. B Aurela, H Kulmala, L Soderhjelm. Phthalates in paper and board packaging and their migration into Tenax and sugar. *Food Additives and Contaminants* 16 (12):571–577, 1999.
27. WAC Anderson, L Castle. Benzophenone in cartonboard packaging materials and the factors that influence its migration into food. *Food Additives and Contaminants* 20 (6):607–618, 2003.
28. KD Woods. Food–package interaction safety. In: SJ Risch, JH Hopkiss, Eds., *Food and Packaging Interactions II*. Washington, DC: American Chemical Society, 1991, pp. 111–117.
29. SM Johns, SM Jickells, WA Read, L Castle. Studies on functional barriers to migration. 3. Migration of benzophenone and model ink components from cartonboard to food during frozen storage and microwave heating. *Packaging Technology and Science* 13:99–104, 2000.
30. Anon. *Recovery and recycling of paper and board*. Confederation of Paper Industries, Swindon, Wiltshire, UK.
31. RH Perry. *Chemical Engineers Handbook*. 5th ed. New York: McGraw-Hill, 1973.
32. Anon. Wood Handbook No. 72. Forest Products Laboratory, U.S. Department of Agriculture, 1955.
33. Anon. *Thermal properties of building structures*. The Chartered Institute of Building Services, Vol. A3, 1980, pp. A3-31–A3-42.
34. TR Robertson, FB Thompson, AC Cleland. Measuring thermal resistance of corrugated made simple. *Packaging Technology and Engineering* 7 (8):48–50, 1998.
35. ND Amos. Mathematical modelling of heat transfer and water vapor transport in apple coolstores. Ph.D. Dissertation, Massey University, Palmerston North, New Zealand, 1995.
36. I Merts. Mathematical modelling of modified atmosphere packaging systems for apples. Ph.D. Dissertation, Massey University, Palmerston North, New Zealand, 1996.

37. Anon. *ASHRAE Handbook — Fundamentals*. American Society of Refrigeration and Air Conditioning Engineers Publishers, Atlanta, GA, 1997.
38. RL Earle. *Unit Operations in Food Processing*, 2nd ed. Oxford, England: Pergamon Press, 1983.
39. DJ Tanner. Mathematical modelling for the design of horticultural packaging. Ph.D. Dissertation, Massey University, Palmerston North, New Zealand, 1998.

32 Packaging of Frozen Foods with Other Materials

Gerrit Hasselmann and André Wötzel
Fraunhofer Institute Material Flow und Logistics, Dortmund, Germany

CONTENTS

I. INTRODUCTION

Packaging products made of aluminum and plastic are relatively modern and their development coincided with the appearance of frozen food. Because of their ability to protect the temperature and "aroma" of the goods, they are a better solution than packaging made of paper or cardboard.

In this chapter, the most common types of aluminum and plastic packaging are presented. The forms of the packaging will be studied as well as the specific characteristics of the different materials and their use for the packaging of frozen food will be shown by examples. Furthermore, the different waste disposal methods are shown using the example of German practice. The German waste disposal system is organized by "Das Duale System Deutschland" and is one of the world's best waste disposal systems.

Certainly, the user is interested in the production costs for the different packaging. Unfortunately, such information is not available because there is quite a variety of parameters, which influence the production costs, for example, the different forms and bonded systems, the number of items, and the world price for crude oil.

II. FOIL PACKAGING

In the first section of this chapter, the different possibilities are described to make packaging from plastic foil. Plastics are organically, highly molecular materials, which are mainly produced synthetically. They are produced as polymers (therefore also called polymer materials) from monomers by polymerization, polycondensation, or polyaddition. Monomers are substances built of carbon (C), hydrogen (H), oxygen (O) as well as nitrogen (N), chlorine (Cl), sulfur (S), and fluorine (F). The materials react differently depending on the type of polymer: linear polymers are thermo plastics; linked polymers are a thermosetting plastics, and more or less wide meshed linked polymers are elastic plastics, which are also called elastomers [1].

The advantage of foil packaging is their light weight and their flexibility. In addition to this, the foil is elastic and resistant to chemical and mechanical strains, to name just the most outstanding characteristics. Further advantages will be studied later on. There are two basic technologies for the production of foil packaging: the pouch packaging and the foil-wrapped cup.

A. Pouch Packaging

1. Form

There are many forms of pouch packaging and there is no guideline considering their form. This solely depends on the requirements and needs of the producer. In this section, some forms are discussed.

Some types are the so-called stand-up pouch, the flat bag, the side gusseted bag, the three-side sealed pouch, the block-bottom bags, or the shrink bag. Flat bags are divided into the so-called flat or tubular bags depending on whether the bag is made of flat or tubular foils. Furthermore, bags are distinguished by their sealing. One example shown in Figure 32.1 is the three sides sealed pouch, where the filling side is sealed after the bag has been filled [2]. This kind of bag is heat sealed.

FIGURE 32.1 Stand-up pouch for frozen food. (From Anonymous. Newsletter and company information, Wipf AG, Industriestrasse 29, CH-8604 Volketswil, 2005. With permission.)

The block-bottom bag is a stand-up pouch because of the form of its bottom, which may be rectangular or round.

2. Materials

The earlier mentioned packaging may consist of a mixture of foil materials. Plastic foils are rather seldom used in one layer. Another mixture may be foils and paper or different foil plastics like polyethylene (PE-LD, PE-LLD, PE-HD) or polyamide.

A compound consists of two or more different materials which are combined by means of a chosen laminating technique. The material compound has the following advantages:

- *Combination of characteristics*: In material compound, different characteristics are combined.
- *Division of functions*: The potentials of the different materials are used optimally.
- *Cost optimization*: By distributing the different functions to the individual components, the single components can be produced much more easily and economically.

To maintain the quality of the packaged goods, the steam and gas permeability of the foil packaging is of great importance. Table 32.1 shows some key values for the steam and gas permeability of plastic foils at 20°C and a thickness of 0.1 mm [3]. In addition, the specific density *"d"* of the plastic is shown. It is clearly shown in Table 32.1 that the steam and oxygen permeability of the material increases in line with its specific density. According to literature the temperature resistance improves in line with a higher density.

Another important parameter of packaging for frozen food is the heat transfer coefficient. The heat transfer coefficient (also called heat insulation) describes the heat insulation factor of a material. This value indicates the heat transfer through a stationary surface A of 1 m^2 when the difference between the inner and outer temperature is 1 K. The value thus is a "quality factor" describing the heat insulation of components and is measured in "Watt per meter square multiplied by Kelvin." The smaller the value, the better the heat insulation of the component, for example, the value for plastics is about 3.49 $W/(m^2K)$ [4]. In addition to this coefficient, Table 32.2 shows mechanical characteristics of three different plastics. The puncture resistance of the foils indicates the resistance against mechanical impacts and the average value for plastics is about 2.1 N/mm^2 [5].

TABLE 32.1
Key Values for Plastics

Plastic	Gas Permeability According to DIN 53380 $cm^3/(m^2$ days kPa) (24 h)			Steam Permeability According to DIN 53122 0–85% Relative Humidity $g/(m^2$ day)(24 h)
	O_2	N_2	CO_2	
Polyethylene-LD, $d = 920$ kg/m^3	21	4	60	1.1
Polyethylene-HD, $d = 960$ kg/m^3	5	1	17	0.4
Polypropylene, $d = 1060$ kg/m^3	6	1.4	25	0.5
Polyamide 6, $d = 1140$ kg/m^3	0.2	0.06	1.5	15.0

Source: Anonymous. Wirtschaftsfaktor Aluminium, GDA Gesamtverband der Aluminiumindustrie, Postfach 10 54 63, 40045 Düsseldorf, Deutschland, 2005. With permission.

TABLE 32.2
Mechanical Characteristics of Some Plastics

	E-Module (GPA)	Yield Stress (MPA)	Breakage Stress (MPA)	Breaking Strain (%)
Polyethylene-LD	0.1–0.3	6.9–14	10–17	400–700
Polypropylene	1.0–1.6	23	24–38	200–600
Polyamide 6	3.8	—	800	25

Source: Otto E Ahlhaus. *Verpackung mit Kunststoffen*, 1st ed. Carl Hauser Verlag München, Wien, 1997, pp. 14–203. With permission.

Table 32.3 describes characteristics two examples of plastics, which are often used for the production of foil and their use for frozen food. Products out of the series "Cryovac" by the company sealed air GmbH are here used as an example for different laminated foils. They are used for the packaging of frozen food because of their good shrinking and sealing behavior and their good oxygen resistance. Table 32.4 lists the key values for some Cryovac foils [6].

3. Characteristics

Pouch packaging is an ideal packaging, and owing to their form, they require only a minimum of storage space when empty. When they are filled, they can be arranged geometrically, for example, in stacks or covering boxes. Pouch packaging can be filled manually, semiautomatically, or fully automatically. Furthermore, the pouch packaging optimally protects the products against mechanical strain, which may occur during transport and transhipment.

TABLE 32.3
Characteristics of the Plastics PA and PE

Plastic	Description
Polyamide	In packaging, PA is used for the production of foils. Because of its resistance to low temperatures (up to −40°C), this material is often used for packaging for frozen food. As it is almost impermeable for oxygen and odor, PA foil is also used for vacuum packaging. Amorphous polyamides produce a crystal clear foil, whereas foils made of undyed semicrystalline polyamides are turbid Plasticizers are added to some PA foils so that they cannot be used as food packaging. This may also have a negative effect on the absorption and transfer of water
Polyethylene	PE belongs to the group of polyolefins, (semicrystalline thermoplastics). It is differentiated between polyethylene-LD and polyethylene-HD, according to the density PE-LD and PE-HD are turbid (nearly crystal clear, only when they are processed into thin foils) and steam resistant. The lowest temperature where PE can be used is about −50 to 60°C for PE-LD. Because of its higher density, PE-HD is resistant to temperatures of about +90°C PE foils are highly steam resistant. A disadvantage, however, is their permeability to gas and odor. Because of its high density, the resistance of PE-HD with regard to oxygen, carbondioxide, steam, and odor is better than that of PE-LD Other than for foils (PE-foils, laminated foils, and shrink foils), PE is used for the production of bottles, bottle crates, barrels, jerry cans, tins, cups, and so on

Note: PA, polyamide; PE, polyethylene.

TABLE 32.4
Key Values for the Cryovac Foils

Foil Data	T9225	T9230	T9235	T9240	T9250	T9260
			Dimensions			
Thickness (μ)	65	75	90	100	125	150
Grammage (g/m^2)	62.7	72.4	86.85	96.5	120.6	144.8
			Characteristics			
Deformation temperature (°C)			For underfoil 70–90			
Sealing temperature (°C)			130–170			
Heat treatment (°C, max)/h	95/1	95/1	95/1	95/1	95/1	95/1
			Physical characteristics			
E-Module at 23°C (kg/cm^2 L/T)	4650/4600	4650/4600	4650/4600	4650/4600	4650/4600	4650/4600
Elongation at break (% L/T)	600/590	600/590	600/590	600/590	600/590	600/590
Tearing strength (kg/cm^2 L/T)	400/350	400/350	400/350	400/350	400/350	400/350
Seal strength (kg/25 mm)	4	4	4	4	4	4
			Permeability			
O_2 at 23°C, 0% RH (cm^3/m^2, 24 h, bar)	76	70	61	51	45	35
H_2 O-D at 38°C, 90% RH (g/24 h, m^2)	≤10	≤10	≤10	≤10	≤10	≤10

Source: Anonymous. Product information, Capro Großhandel und Handelsagentur, Verwaltung Osnabrück, Am Natruper Holz 15 49076 Osnabrück, Deutschland, 2005. With permission.

Another benefit is the high-temperature resistance. The products are well protected at temperatures from about −40 to +90°C. In addition to this, pouch packaging is very hygienic. Hygiene means that the companies take measures to avoid negative impacts on the product. According to the German Food Hygiene Regulation, for example, "... negative effects may be a nauseating or other impairment of the proper hygienic condition of food by micro-organisms, contamination, climatic influences, smell, temperature, gas, steam, smoke, aerosols, varmints, human and animal excrements as well as wastes, waste water, detergents, disinfectants, pesticides or unsuitable treatment and processing" [7]. Also according to the German Food Hygiene Regulation, foods considered as perishables are those which "from the micro-biological point of view are easily perishable and should be transported only at certain temperatures or under certain conditions" [7]. All foil packaging are turbid to transparent, whereas monofoils are light-sensitive.

4. Use

The appearance of a sales packaging, especially for frozen food, has a great influence on the customer's purchasing behavior. Owing to its various designs, the pouch is an ideal display packaging because the foil can easily be printed.

Because the foil pouches are "close," they can be used for liquid, pulverized, pasteurized, or small products, for example, frozen instant noodle dishes, fruit, vegetables, or fish.

5. Waste Disposal

In Germany, the disposal of packaging is regulated by the EC regulation 94/62/EG of 20th December 1994 on packaging and packaging waste. The aim of the regulation for the avoidance and recycling of packaging waste is "... to avoid or reduce the impact of packaging wastes on the environment. Packaging wastes have to be avoided; the reuse and recycling of packaging has to be preferred to the mere disposition of packaging wastes. Up to 30th June 2001, 65% of the complete packaging wastes should be reused and 35% should be recycled. This regulation applies to all packaging distributed within the scope of the Packaging Law, no matter if they arise in industry, trade, administration or the service sector, in households, or elsewhere and independent of the materials they are made" [8].

In 1994, the Packaging Law required the food industry for the first time to take back and recycle used packaging (since then the communities were solely responsible for the waste disposal). By holding manufacturers and distributors also responsible for the disposal of packaging wastes, this regulation set the basis for a reduction of packaging.

Because of this "liability" in Germany, a nationwide collection and disposal system, the "Duale System Deutschland AG" ("Der Grüne Punkt"), was founded on behalf of the economy. Since 1994, more than 36 million tons of sales packaging has been collected and recycled by this system [9].

Foil packaging comprises only a small volume and the plastics can be easily recycled and "reproduced" several times. To recycle a material, the different plastics have to be separated either before or after the collection. This can hardly be done in private households, since the user does not know the different material combinations. For a direct recycling, the foil packagings have to be uncontaminated. So-called "clean" packaging wastes from industrial production generally can always be recycled.

If it is impossible to reduce the degree of contamination by an economically acceptable treatment, the waste can be deposited or treated thermally, that is, incinerated for energy production. Foil plastics have a high fuel value and are thus preferred in power plants as extra loading.

A separate collection has considerable advantages because disposal costs as well as the quantity of unutilizable waste can be reduced.

B. Wrapped Cups

1. Form

Wrapped cups are plastic packaging consisting of a rigid cup sealed with a foil. The packaging manufacturers decide, independently or according to customer requirements, on the form and design of the cup because there are no binding standards or regulations.

This kind of packaging is available on the market in many different forms and designs (Figure 32.2) [10]. The cup also called tray can be oval, round, or rectangular, of different height and with different edges. Furthermore, the bottom may be flat or separated into compartments.

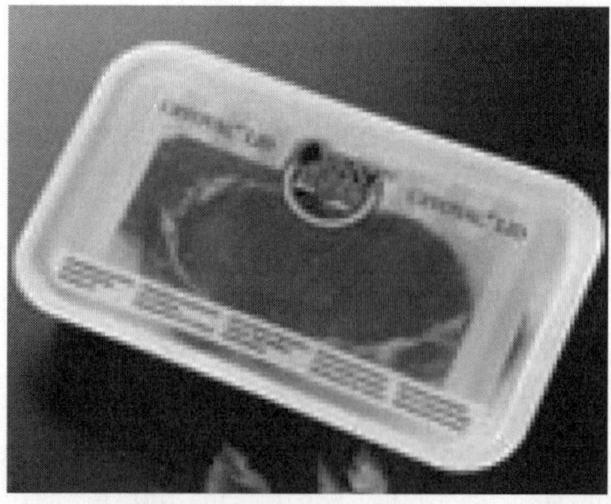

FIGURE 32.2 Wrapped tray with sealed foil. (From Anonymous. "Frostige Zeiten? Tiefkühlverpackungen können mehr" Verpackungs-Rundschau (magazine) Issue 9, 1999, pp. 272–279. With permission.)

When they have been filled, the trays are sealed with a foil, which may consist of different plastics. Both components are then laminated. This protects the frozen goods against humidity and germs better than a stretch foil. The foil cover may be transparent or printed.

2. Materials

The materials used for a wrapped tray have to be studied, which includes the material of the tray as well as the material of the foil.

The trays mainly consist of polymers, aluminum, or pressed ground wood. Ground wood is made "mechanically by grinding (on a grinder) barked softwood by adding water. According to the used measure, this results in white or brown mechanical pulp or chemical pulp [11].

Information on the insulating performance, the mechanical and thermal load of the ground wood tray is hardly available, because the range of materials and designs changes continuously and is highly complex. The foil consists of plastics like polyethylene, polypropylene, or polyester and may be separated or laminated. Monofoils are thermally less resistant than laminated foils. A laminated foil may consist of polyamide polyester, polypropylene, polyamide, or other combinations. Table 32.5 shows some variations of laminated foils and their specifications.

3. Characteristics

Wrapped cups consisting of tray and foil can be used at deep freeze as well as higher temperature. They are nearly impermeable for gas and water and protect the goods against mechanical strain during transport and storage because the bottom structures of the trays are dimensionally stable.

The foil is highly flexible but also highly impermeable for germs and other substances like dust. Owing to these characteristics, this kind of packaging meets the requirements, for example, of the German hygiene regulation.

TABLE 32.5
Structure and Characteristics of Laminated Foils

Material Combination	Special Packaging Features	Use for Packaged Goods
Polypropylene/copolymer sat	Easy to weld, Decoratable surface	Packaging for bread
Polyamide 6/ionomer	Easy to seal	Packaging foil for meat and sausages
Polypropylene–copolymer/ polypropylene + regenerat/ polypropylene– statistical copolymer	Resistant to low temperatures; low raw material costs	Packaging for frozen food
Polyamide 6/ethylene with vinylalcohol/adhesive agent/polyethylene-LD	Highly insulating; impermeable for aroma; good sealing performance	Vacuum packaging
Polypropylen/adhesive agent/ethylene with vinylalcohol/adhesive agent/polypropylene	Impermeable for gas, steam, and aroma; sterilizable	Convenient food
Polyethylene-LD/adhesive agent/polyamide 6/adhesive agent/polyethylene-LD	Highly insulating; humidity protection	Packagings for meat, sausages
Polyethylene-LD/adhesive agent/polyamide	Inpermeable for gas, water, and aroma	Packaging for meat, cheese, fish, and convenient food

4. Use

The combination of tray and foil is often used as a packaging for frozen goods. Prints can be made on the foil and also the tray so that information about the contents (product, manufacturer, batch, weight, date of packaging, date of use, etc.) can be displayed. In addition to this, an eye-catching print can improve the marketability of a product.

This packaging can be used for hygienically sensitive products, for example, all kinds of frozen meat and fish but also fresh fruit and vegetables. The packaging can be only used for goods, which contain no or only a small proportion of liquid before the freezing.

5. Waste Disposal

Wrapped trays are disposed off according to the national packaging laws. Because wrapped trays consist of at least two different materials, they have to be collected separately to allow for the reuse of used "packaging wastes." The foil may be reused or disposed.

It is important to know from which material the trays are made up of. Trays made of ground wood or plastic can be either deposited or incinerated for power production. The disposal of aluminum cups is described previously.

III. ALUMINUM PACKAGING

In this section, aluminum packaging for frozen food are described. After oxygen and silicon, aluminum is the third frequent element on Earth. Despite large deposits, the metal has been known and used only since the 19th century. Processing companies had to face two main problems. Aluminum only occurs in a mixed very stable form, and for its separation, a high energy consumption is necessary [12].

Today, this material is used for a wide range of packaging. Pure aluminum, for example, is used as a wrapping foil for chocolate, as cover for yoghourt cups, spray tins, or tins for pet food. Furthermore, aluminum combined with other materials is used for packaging, for example, cardboard boxes (Tetra Pak) for beverages, vacuum packaging for coffee, and bags for instant soups.

During the period from 1980 to 1995, the share of aluminum packaging in the United States, Europe, and Japan increased from 17.8 to 20.6% [13–15]. This development shows the important role that aluminum packaging plays in the packaging industry.

A. Form

Aluminum is used for a large variety of different packaging, mainly in the form of relatively thin foil of 20–200 μm in thickness, which retains its stability by its form. There are "wrinkle walled" and "smooth walled" forms: the wrinkle walled form is a packaging made of wrinkled aluminum foil and the smooth walled from smooth aluminum foil. For most aluminum packaging, the easily processable metal is formed into a cup, which takes up the goods. This form, also called tray, is covered either by a cardboard, a foil, or an aluminum lid. Figure 32.3 illustrates some examples [10].

B. Materials

Aluminum packaging is either mono-material or composite systems. Composites consist of different single materials, which are placed in layers and have specific functions within the packaging system. Generally, aluminum packaging for frozen food are composites where aluminum foils may be laminated with paper as well as with plastic foils, with 75–80% paper accounting for the largest share and is used to improve the stability of packaging.

Often, polyethylene foils are used. This plastic layer should improve the natural impermeability of the aluminum because this metal always has nonmetallic characteristics. The nonmetal causes pores and holes within the material, which reduces the density of the packaging. The plastic foil layer builds a dense surface. It may also be decorated by printing.

FIGURE 32.3 Aluminum cup for convenient food. (From Anonymous. "Frostige Zeiten? Tiefkühlverpackungen können mehr" Verpackungs-Rundschau (magazine) Issue 9, 1999, pp. 272–279. With permission.)

The steam permeability of the aluminum composite is less than 0.05 g/m^2 per day, better than that of plastic. Because of the different mixtures, no values are available for the gas permeability.The temperature resistance depends on whether the material is monoform or a composite. In monoform, it can be used for deep freeze products at temperatures between −70 and +600°C. This is impossible for a composite, which covers temperatures from about −40 to +80°C. Aluminum has a relatively low tensile strength of 40–160 N/mm^2 and a high ductility, especially for pure aluminum. The tensile strength values for composites as well as the puncture resistance, however, may vary considerably.

C. Characteristics

Aluminum packaging is also used because of their low specific density of 2700 kg/m^3, which makes them more permeable for air and water than plastic as indicated in Table 32.1. Aluminum is also used for highly corrosion-resistant packaging. Its corrosion resistance results from an oxide layer, which covers the metal when in contact with air or liquids. When this layer is damaged, oxidation leads to an automatic repair. The more oxidizing the acids, the more resistant is the layer, so that aluminum is resistant even against concentrated saltpetre acid. It is not resistant, however, against substances which damage its oxidation layer, for example, concentrated alkali, which dissolves the aluminum oxide [11].

Aluminum is an ideal packaging material, as it is tasteless and odorless and thus physiologically harmless. Aluminum is photoresistant, that is, the packaging protects the goods against light and UV rays. In addition to this, aluminum is air- and water-resistant and protects the goods against extreme temperatures, that is, −70°C up to +600°C for pure aluminum and −40°C up to +80°C for composites. Accordingly, packaged goods can be frozen and then be baked, steamed, or cooked in their packaging. Furthermore, aluminum protects goods against oil and fats and can be further decorated.

Another advantage for aluminum packaging is its low specific density, resulting in sufficient stability and a low packaging weight. Because of its "lightweight," it can be easily processed and treated so that the packaging can be formed according to the product.

D. Use

With the existence of the oxidation layer, aluminum packaging can protect food against salts and acids so that products like salted fish can be packed without problems. Foods, like baked goods, ice cream, and so on, are protected against light, especially UV rays using aluminum packaging.

As aluminum can be easily formed into mono or composite systems, the packaging can be adapted to meet the specific requirements of frozen foods. This feature is of great importance for a large variety of meat products.

E. Waste Disposal

It was recognized long ago that because of its value, aluminum is too precious to be deposited. For this reason, the raw material has to be recovered from the waste. Today, about 30% of the aluminum is gained by recycling. Only 5% of the production energy is necessary to melt it down.

Unprinted and uncoated aluminum packagings are directly melted down. Another proven method is the "pyrolysis," where packaging remains like glue are gasified in a smouldering process (about 500°C). Solid pyrolysis remains (pyrolysis coke) are then separated from the basic material with sieves.

First of all, the single components of composite packaging have to be separated. At first, this packaging is pressed into bales in a recycling center. The materials are separated with so-called pulpers, that is, water-filled containers where the packaging waste is stirred with rotors at a temperature of about +40°C. The rotors tear the materials apart so that water can intrude and separate the different materials. The paper is sucked off at the bottom and reused, for example, for sanitary paper or paper bags.

The plastic polyethylene is used for energy production because of its high fuel value. Bauxite can also be retrieved from the aluminum remains and is used in cement production. Furthermore, secondary aluminum can be used for new products, for example, in the automotive industry. These examples show that recycled aluminum is highly economical and ecological.

IV. COSTS OF DIFFERENT TYPES OF PACKAGING

It is very difficult to give exact information about the different packagings made of plastic, foil, and aluminum because of the variety of factors that influence the production of the packaging materials and the packaging itself. The recycling process is another important cost factor.

Factors which have an impact on the production costs are batch size, purchasing price for raw material, energy costs during production, different mixtures for laminated foils, labor costs, and so on.

The purchasing costs for raw materials strongly depend on the market and may vary considerably. The raw material for all kinds of plastic and foils is crude oil. In recent years, prices for raw materials fluctuate continuously according to the political situation. Exact information about laminated packaging is also not available due to the different mixtures.

The disposal is also influenced by different factors, for example, country-specific waste collection, energy costs for reuse/recycling, and so on. In Germany, for example, waste is already separated in private households into residual waste, glass, plastic, and so on. As this specific waste separation is not common in all countries, the energy costs differ considerably.

V. CONCLUSIONS

In this chapter, packaging of aluminum and plastic foils are studied. Packaging, which is used for frozen goods, has to be resistant to temperatures of up to −40 °C. This concerns their mechanical characteristics (tearing resistance, stretch resistance, and puncture resistance) as well as their impermeability with regard to gas and steam.

Single foils of different plastics like polyethylene, polypropylene, or polyamide are unsuitable as packaging for frozen goods. However, a combination of selected single foils may well be used. The same applies to aluminum. Foils of pure aluminum are often combined with plastic foil and paper to benefit from the qualities of the single foils. Aluminum may also be used as single material, for example, as tray, if the packaging should be heated together with the product.

In this chapter, two sections contain a description of packaging of foil and aluminum and some of their possible forms as well as a comparison of material key values. Another section deals with the characteristic features of the packaging. Then follow two sections about the typical use of the packaging and their disposal afterwards. Users certainly would prefer to obtain information about the costs of the different types of packaging. Detailed figures, however, are not available, as the costs are influenced by a variety of parameters and fluctuate considerably.

The different packaging of foil and aluminum, which are described in this chapter, are typical examples. It has to be said, however, that this list is not exhaustive, because new variants are continuously developed throughout the world.

REFERENCES

1. W Weitz, K-H Grote. *Dubbel, Taschenbuch für den Maschinenbau*, 19th ed. Springer-Verlag, 1997, pp. E66–E70.
2. Anonymous. Newsletter and company information, Wipf AG, Industriestrasse 29, CH-8604 Volketswil, 2005.
3. Otto E Ahlhaus. *Verpackung mit Kunststoffen*, 1st ed. Carl Hauser Verlag München, Wien, 1997, pp. 14–203.
4. Anonymous. *Wärmedurchgangskoeffizient Wikipedia, The Free Encyclopedia*, Wikimedia Foundation Inc., Petersburg, U.S.A., 2004.
5. Anonymous. Product information, Capro Großhandel und Handelsagentur, Verwaltung Osnabrück, Am Natruper Holz 15 49076 Osnabrück, Deutschland, 2005.
6. Anonymous. Product information, Sealed Air GmbH, D-22844 Norderstedt, Deutschland, 2001.
7. Anonymous. Lebensmittelhygiene-Verordnung, Bundesministerium für Verbraucherschutz, Ernährung und Landwirtschaft, Wilhelmstr. 54, 10117 Berlin, Deutschland, 1997.
8. Anonymous. Verpackungsverordnung, Bundesministerium für Umwelt, Naturschutz und Reaktorsicherheit, Alexanderplatz 6, D-10178 Berlin, Deutschland, 1998.
9. Anonymous. Verpackungsverordnung, Niedersächsischer Bildungsserver, Medien- und Computer-Centrum des NiLS, Richthofenstraße 29, 31137 Hildesheim, Deutschland, 2003.
10. Anonymous. "Frostige Zeiten? Tiefkühlverpackungen können mehr" Verpackungs-Rundschau (magazine) Issue 9, 1999, pp. 272–279.
11. Anonymous. Schneidersöhne Unternehmensgruppe, Encyclopedia, Wolfgang Walenski, 2002.
12. H-J Bargel, G Schulz. Werkstoffkunde, 7th ed. Springer-Verlag, 2000, pp. 272–279.
13. Anonymous. Wirtschaftsfaktor Aluminium, GDA Gesamtverband der Aluminiumindustrie, Postfach 10 54 63, 40045 Düsseldorf, Deutschland, 2005.
14. Anonymous. Newsletter from the company, Aluminium-Verband Schweiz, Hallenstrasse 15, CH-8024 Zürich, 2005.
15. Anonymous. *Aluminum in Packaging*, 2nd ed. European Aluminum Association, Brussels, 1997.

33 Packaging Machinery

Rajeshwar S Matche
Central Food Technological Research Institute, Mysore, Karnataka, India

CONTENTS

I. INTRODUCTION

In the production of packaged goods, varying degrees of mechanization are required, depending on the type of manufacturing and the volume and diversity of the product line. Many businesses start out in a small way, with hand assembly of the product into its container, and as the volume increases, they find it necessary to add mechanical equipment. This may take the form of a conveyor belt or holding fixtures, or it may be a sophisticated assembling machine, according to the needs of the situation. There should be a good compatibility of packaging materials with the equipment for coordination of various machines and combining them into the finished units.

One of the most important factors in the packaging system is the nature of the product that must be handled which in turn depends on its physical properties and chemical characteristics. There may be a need to look into the possibility of a change in the product to suit the requirements of a higher-speed operation, or the tighter limitations on the physical properties of the product for a more trouble-free packaging system. It is highly important to know whether the packaging machine is used for test marketing operation as a possible option to upgrade in size and speed or integrated into a full-scale production line at a later date. It is also necessary to know the anticipated pack size or product variations, and to plan for any changes that are likely to occur in the requirements of the marketplace. Some of the factors that can influence such decision are space limitations, available utilities, safety requirements, sanitation problems, dust or fumes, and the level of skill of the operators who are available to run the equipment, including operation and maintenance. The packaging materials need to be accurately specified before the equipment for handling them can be selected. It would also be wise to try to anticipate the future changes when formulating these specifications [1].

There are a large number of packaging machines used for frozen food processing. It may not be possible to discuss all of them in detail in this chapter. Some of them are discussed here on the basis of their importance and applications in the food industry.

II. BAG MAKING

Bag is one of the oldest forms of packaging and the most popular one. It performs all basic functions of packaging such as containment, protection, and communication at the lowest cost. A flexible

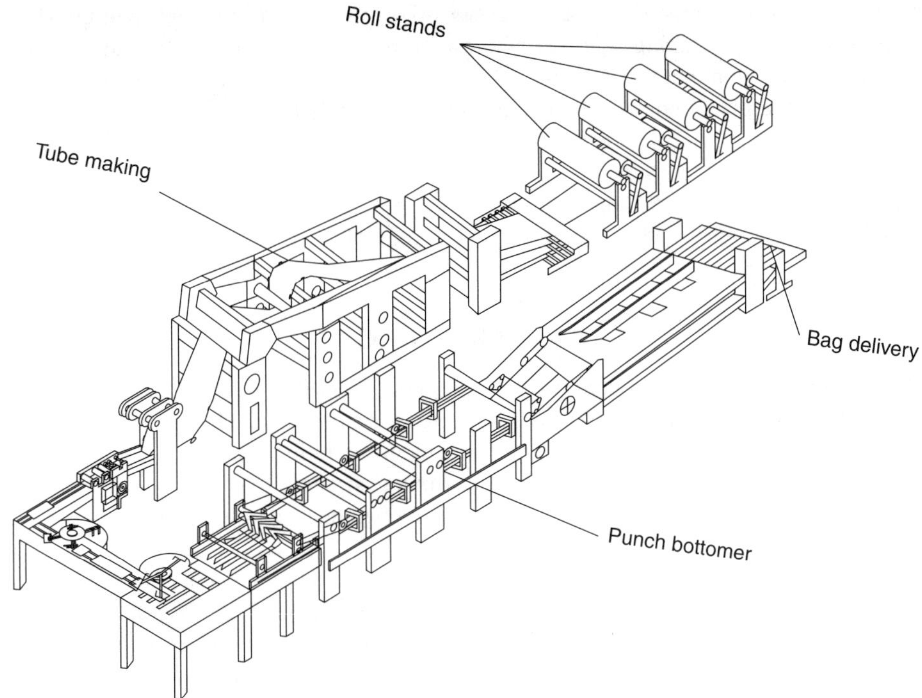

FIGURE 33.1 Online bag-making machine.

container which is open at one end is called as a bag. Bag holding 50 lb or more is called a sack. Bag-making machines have two parts: tube forming and bottoming equipment. Rolls of film or paper are formed into the tube. In case of multiwall bags, spout or adhesive are applied between the plies to hold them together if paper is used. The material is then formed into the tube. The tube can be with gussets or without gussets. The tube is converted into bags by closing one or both the ends of the tube. In case of both the ends closed, one of the bottoms provided with an opening or valve through which the bag is filled by insertion of the spout. The valve is closed by the pressure of the bag content. Inline tube forming and tube bottoming (Figure 33.1) give one-to-one operation of tubing and bottoming as increasing the speed up to 250 bags per min [2].

III. BOTTLING

Bottling includes different machines connected together by a synchronized drive arrangement for maximum output. Bottling lines should handle not only glass bottles but also variety of plastics with different shape. Differently shaped containers need different ways of handling and container filling.

The various steps involved in bottling process are: (1) bottle unpacking, (2) bottle cleaning, (3) bottle filling, (4) bottle closing, (5) bottle labeling, and (6) collating and packing for transport [2].

IV. CANNING

Canning may be defined as the packaging of perishable foods in hermetically sealed containers that are to be stored at ambient temperatures for extended times (months or years). The objective is to produce a commercially sterile food product. Commercially sterile does not mean that the food is free of microorganisms, but rather that the food does not contain viable organisms that might be a public health risk or might multiply under normal storage conditions leading to spoilage. The food product may be made commercially sterile either prior to or after filling and sealing. Three conditions must be met for canning safe and wholesome food: (1) sufficient heat must be applied to the food to render it commercially sterile; (2) the container must prevent recontamination of the product; and (3) the filled and sealed container must be handled in a manner which prevents loss of integrity [9].

Canning of processed foods may be divided into eight unit or basic operations: (1) handling and storage of empty cans, (2) cleaning empty cans, (3) product preparation, (4) filling, (5) closing, (6) processing, (7) cooling, and (8) handling and storage of filled cans [10].

V. CARTONING

A cartoning operation is to erect carton from a flat condition, fill with product, and close it. The machinery varies from simple hand-fed machines to automatic coupled with packing the cartons directly into cases for dispatch. There are three main operations in the process. Erecting the container, filling the container, and closing. Cartoning systems may be required to carry out other operations such as handling paper liners, embossing codes, and inserting leaflets [3].

The semiautomatic or fully automatic machines are differentiated based on the filling or loading of the carton. If loading is done directly into the cartons, even though an operator inserts it into the infeed conveyor by hand, the system is classified as automatic. If the rest of the operation is automatic but the load is inserted directly into the carton by hand, the system is described as semiautomatic. Most systems requiring high speeds use continuous-motion machines; lower speed systems usually use intermittent-motion machines. The latter can also be of advantage where the nature of the product demands a stationery carton at the moment of filling [3].

A. Cartons for Liquid Products

The packaging requirements about barrier properties and hygiene are stringent when packing liquid food products. Seal integrity and seal area contamination are very important. There are two types of machines: (1) form a reel, form the package, and fill it in a continuous operation, and (2) a premanufactured blank [3].

In the Tetra-Pak system, the containers are continuously formed from a roll of packaging material, aseptically filled, sealed, and formed into bricks. The packaging material is a multilayered sheet comprised of printed cardboard, aluminum foil, and plastic layers. An outer polyethylene layer protects the ink layer and enables the flaps to be sealed in the formation of the final brick form. Next, there is a bleached cardboard layer for the graphics and printed material, which provides mechanical rigidity. A laminated polyethylene layer follows that binds the next aluminum layer to the cardboard. The very thin aluminum foil layer provides the required barrier to gases and light. This is followed by two inner polyethylene layers that provide a liquid barrier. The outside of the package is imprinted with a design that will be exactly in line with the package dimensions. The packaging material is pretreated to facilitate the forming process.

All food contact surfaces of both filling and packaging must be sterilized. Two methods are used by the Tetra-Pak to sterilize the food contact machinery surfaces: sterilization with hot (i.e., 360°C maximum) air by spraying the surfaces with 35% hydrogen peroxide, followed by drying with sterile air. The sterilization time for either procedure is 30 min. The packaging material is continuously sterilized during production by the application of hydrogen peroxide and drying. In one method, a thin film of hydrogen peroxide solution (15–35%) is applied to the inner surface of the packaging material by passing it over a roll wetted with the sterilant. Excess fluid is squeezed off by a roller and the wetted surface is dried under a stream of hot sterile air. In another method, the packaging material is fed through a deep bath filled with a 30–40% hydrogen peroxide solution maintained at a minimum of 70°C [4].

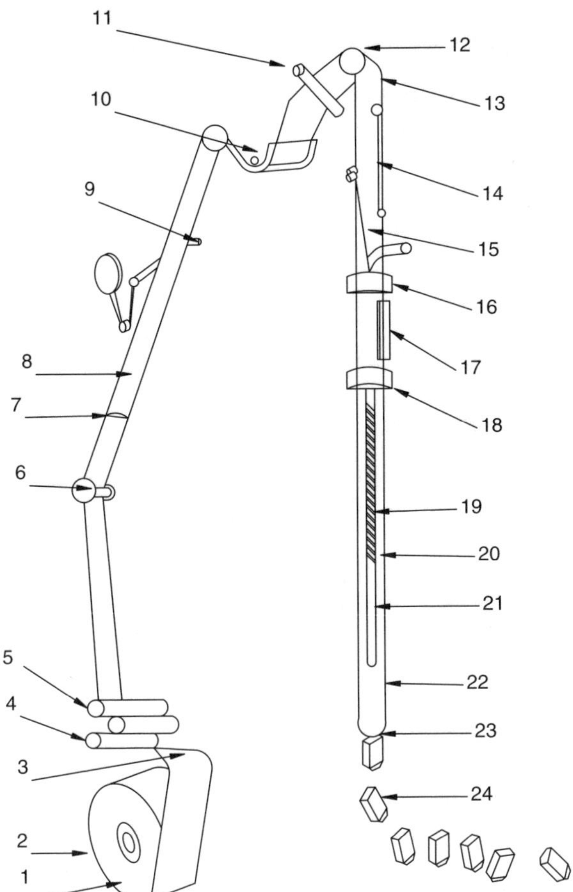

FIGURE 33.2 Tetra Brik aseptic packaging unit. 1. Roll, 2. Photocell, 3. Bending rolls, 4. Brake, 5. Printer, 6. Bending roller, 7. Splicer, 8. Paper guide, 9. Application of plastic strip, 10. Hydrogen peroxide bath, 11. Rotating cylinders, 12. Cover for collection of hot air, 13. Overhead bending roll, 14. Lower tube support, 15. Stainless steel filling tube, 16. Upper forming ring, 17. Longitudinal sealer, 18. Forming collar, 19. Heating tube, 20. Level controller, 21. Float valve, 22. Fill tube. 23. Final sealer, 24. Final folder unit.

Figure 33.2 shows the Tetra-Pak machine, using the principle of forming from a reel, the different stages being combined in one machine. This machine permits carton-filling under fully aseptic conditions. Developments in this system have led to the Tetra Brik package which is oblong, and not a tetrahedron in shape. The Purepak which starts from carton blanks, heat sealed along one side [3].

B. Cartons for Solid Products

Cartoning systems can load the product through the end of the preformed or erected carton sidewise (Figure 33.3) or fill the product vertically (Figure 33.4 and Figure 33.5). These may be performed as continuous processes. The precut board may be supplied to the systems which will erect, fill and close either a prelined carton (Figure 33.6) or the one that makes its own liner on the machine (Figure 33.7). Both these types of lined carton can be used for vac/gas and vacuum packaging [3].

There are several factors to be considered when selecting a cartoning system: (1) variations in the product; (2) postpackaging handling of the product, protective barriers requirement;

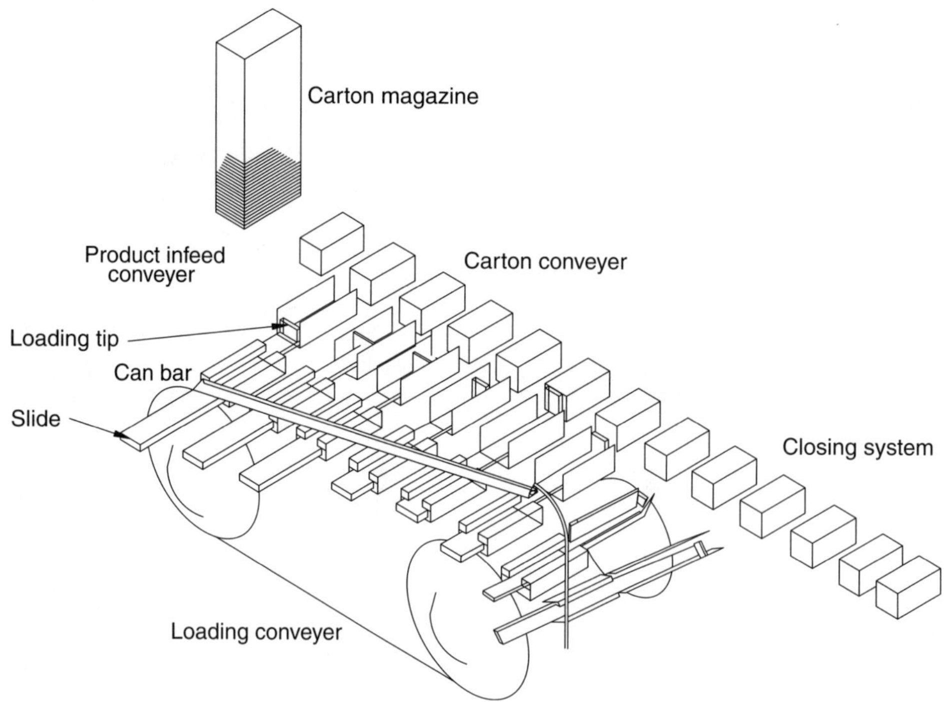

FIGURE 33.3 Fully automatic horizontal carton-making machine.

(3) the product protection requirement in terms of barrier to moisture, oxygen, odors, and so on; (4) product handleability; (5) the number of package sizes required; (6) the frequency of size changes; (7) the product type or number within the size variation anticipated; (8) production rate requirement at regular production and at peak demand in future; (9) the availability of labor for the machine lines; (10) the space available at regular work and at peak hours; (11) the packaging

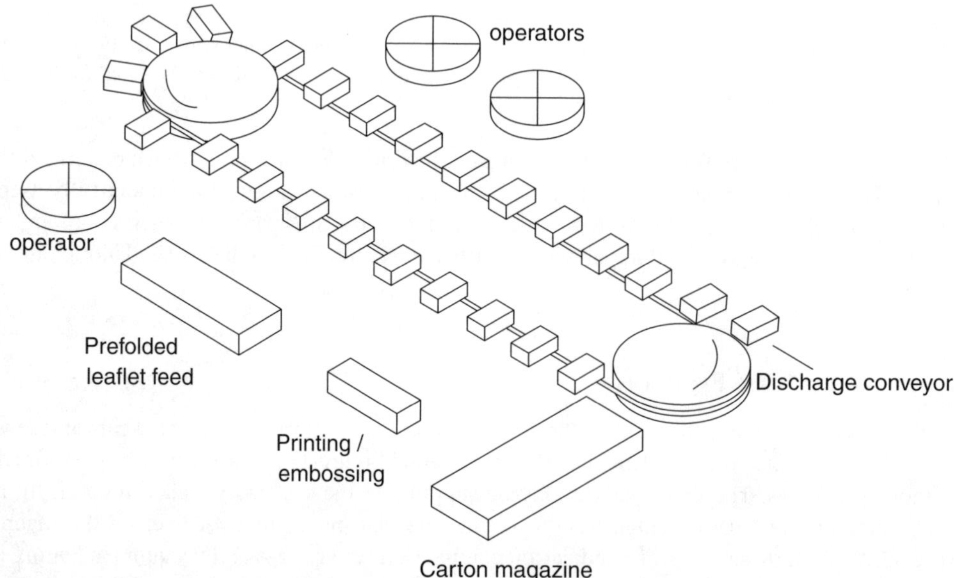

FIGURE 33.4 Semiautomatic vertical carton-making machine.

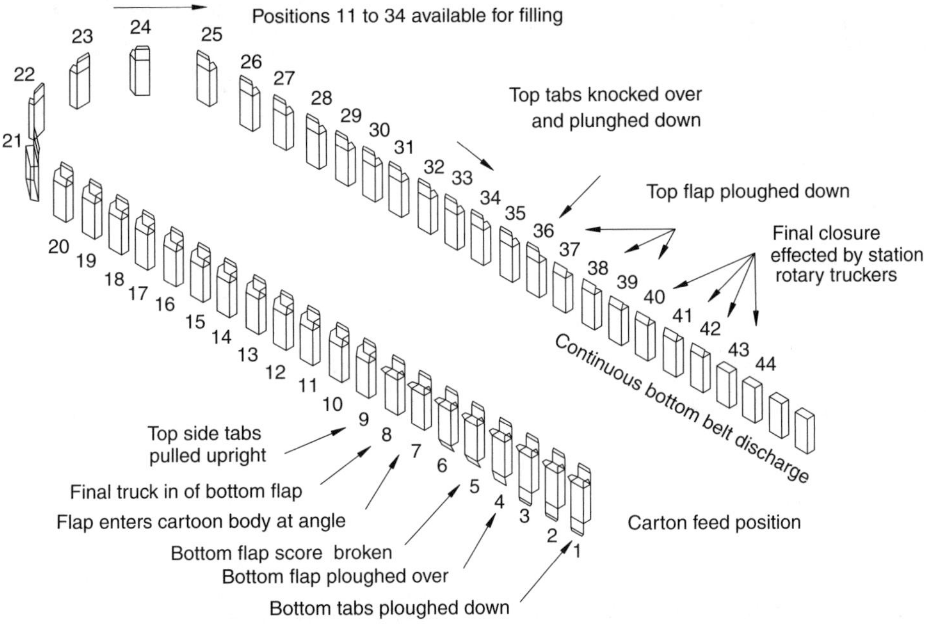

FIGURE 33.5 Constant-motion cartoning machines — vertical.

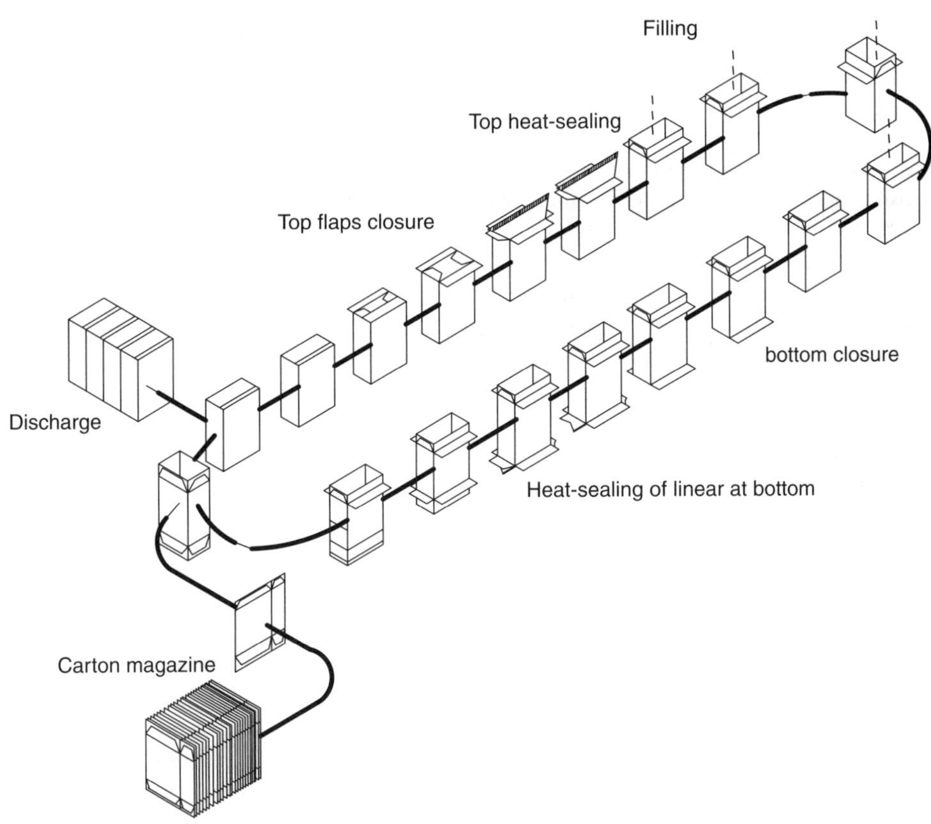

FIGURE 33.6 Carton-making machine for prelined cartons.

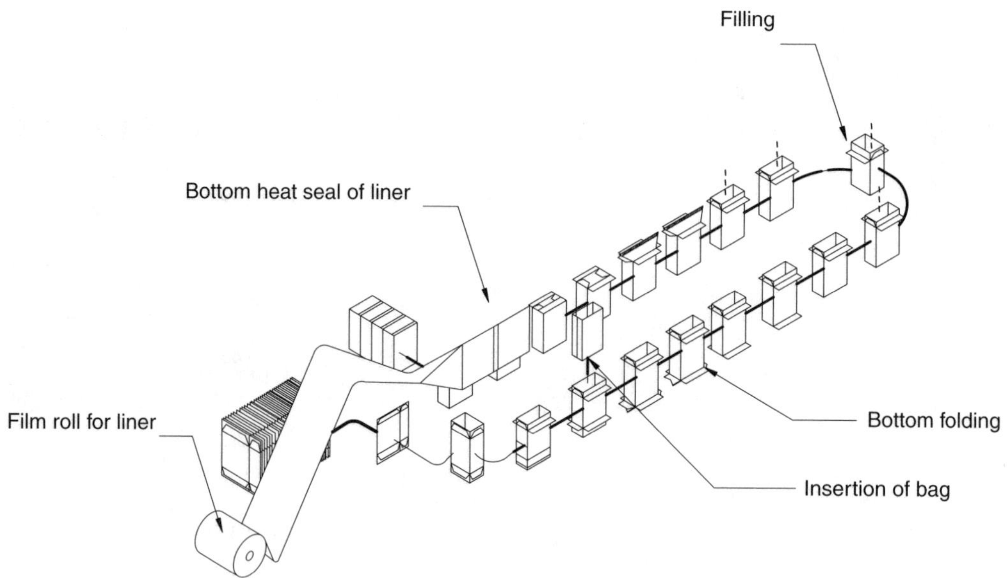

FIGURE 33.7 Carton-making machine for lining cartons with bags made inline.

material requirements; (12) the protection the carton must provide to its contents; (13) the property requirement of the board for the ease of functioning of the carton system from the feeding, erecting, and closing standpoints; and (14) linking up with any other existing equipment, such as case packers or filling heads [3].

1. Semiautomatic Vertical Cartoning Machine

In most cases the semiautomatic vertical cartoning machine is arranged as shown in Figure 33.4. The tubular carton is fed from a horizontal magazine, expanded, and transferred into the carton conveyor. The bottom of the carton is closed by tucking or gluing and then conveyed past one or more operators who manually place the product in the carton. The machine then closes the top of the carton by tucking or gluing. The semiautomatic cartoning machine can be equipped with a number of attachments such as a leaflet feed mechanism, code impresser, or printing mechanism for lot numbers, expiration dates, or prices. Many packages require a leaflet or coupon to be placed inside the carton with the product. This may be accomplished by an operator stationed along the carton conveyor who can place a prefolded leaflet into the carton next to the product.

An automatic mechanism can be installed to feed a prefolded leaflet from a magazine and partially insert it into the carton. When the product is loaded by the operator, the leaflet is pushed down to the bottom of the carton [5].

2. Semiautomatic Horizontal Cartoning Machine

The semiautomatic horizontal cartoning machine is similar to the vertical cartoning machine except that the carton is carried through the machine lying on its back panel. "Hand set-up" end-load style carton sealing machine is used in short production runs or for limited test market promotions or for very low production rate requirement. There are intermittent motions, sealing machine cycled by an electrically controlled single revolution clutch. Carton set-up and product insertion is the responsibility of the operator before insertion into the machine. The carton must be placed in position by hand with the trailing minor flaps folded in and forward. When this operation is complete, the operator removes his hands from the carton and depresses the dual cycle buttons. Speeds of

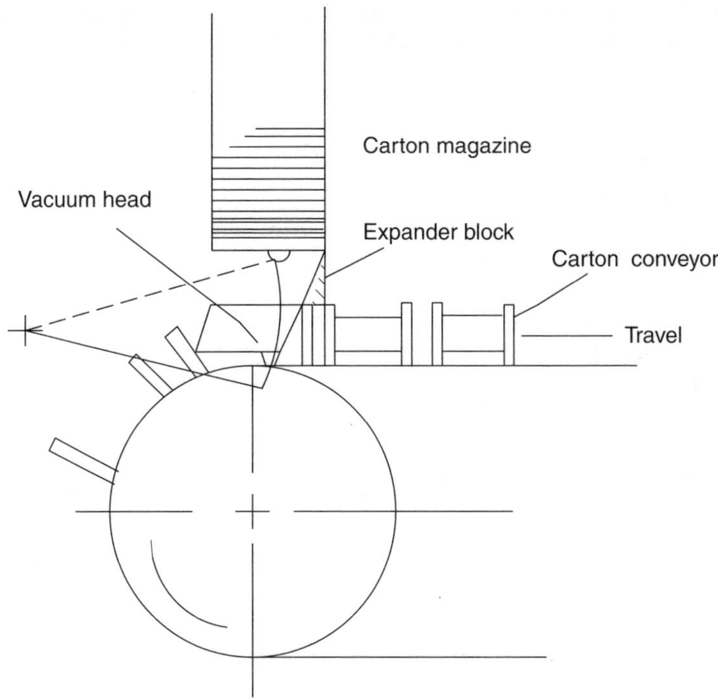

FIGURE 33.8 Carton feed for cartons.

the machine are totally dependent on the carton size, product type, and operator dexterity. Average attainable machine speeds range from 15 to 25 cartons/min. Hot melt glue pots for carton sealing [5].

3. Automatic Vertical Cartoning Machine

These machines are similar to semiautomatic vertical cartoning machines except in these machines the loading is done automatically.

4. Automatic Horizontal Cartoning Machine

These machines are similar to semiautomatic horizontal machines except, in the automatic machines loading is done automatically.

In automatic machines operating speed can be 50 to 60 packages per minute. These types of machines have less flexibility in size range. It consists of a product infeed conveyor, carton feed, carton conveyor, loading mechanism, and closing system (Figure 33.3).

a. Carton Feeds

The basic carton feed for tubular cartons consists of a magazine to hold the supply of unexpanded cartons, a vacuum head, and a transfer system to place the open carton into a conveyor. A simple carton feed is illustrated in Figure 33.8. The suction head, moving downward, pulls the carton from the ledges, and expansion begins when the carton contacts the beveled expander block. As the vacuum head reaches the bottom of its stroke, it straddles or goes between the carton conveyor chains, and as the vacuum is released, the carton is transferred into the conveyor lugs which are moving around the chain sprocket. Expansion is completed as the lugs level out. The flaps on the loading side of the carton are usually guided outward, and so they mesh with the mouth of the bucket, carrying the product to create a funnel for loading the carton [5].

b. Carton Closing

1. *Tucking*: The tucking operation is accomplished over a number of positions on the machine; side (dust flap) folding, prebreak of the tuck flap score line, possible slitting of the dust flaps to assure clearance if lock slits are used, alignment with the carton, first and usually second tucking, followed by a final contact to seat the lock slits. Usually these operations are done with the tucking parts on a bar carried by parallel cranks with each tucking position on a separate pitch of the machine. Tucking by means of belts is also done successfully, eliminating the need for cranks, although the trailing side flaps must still be folded over by a separate mechanism prior to tucking. The final tuck for locking the lock slit is accomplished by a separate rotary mechanism [6].

2. *Gluing*: A glue end carton may be closed by single or double gluing. When single gluing, the two outer end flaps are glued together; when double gluing, the inner end flap glued to the side flaps and then the outer end flap is glued to the inner. Single gluing results in a slight crack between the glued end flaps and the folded side flap, which may be acceptable for a bottle or other solid object. Cartons for food products or facial tissue are usually double glued to protect the contents [6].

VI. FILLING MACHINERY

A. Filling of Liquids

Filling machines for liquids can be divided into five basic types, namely: (1) vacuum filling, (2) constant volume, (3) gravity filling, (4) pressure filling, and (5) pressure–volume filling.

1. Vacuum Filling

Filling by vacuum is the cleanest and most economical way to handle many products. In spite of the care which is taken in making bottles and cleaning them, there is always a percentage of defective bottles — one with holes, chips, and cracks. These are not easily detected in the prehandling of bottles before filling, but vacuum filling machines automatically avoid such bottles. Moreover, with vacuum filling there is no drip or other losses. There is little loss of product and it is unnecessary to wash or wipe the bottles before labeling [7].

Vacuum fillers are of three types: rotary, tray, and automatic feed. On a rotary machine every bottle is handled individually. It is centered under a filling stem, raised, and then filled as it travels around the machine, independent of all other bottles. On a tray-type machine, bottles are placed abreast in trays and rolled on conveyors under the filling head which may consist of one to eight feeding stems. The automatic feed type will operate by means of a lever that discharges the group of filled bottles and moves the empty bottles into position under these stems [7].

The vacuum system requires a supply tank which is below the level of the bottles to be filled, filling stem which consists of a tube through which the product flows into the container and a tube connected to the vacuum system. There are dozens of styles of filling tubes with the final design dictated by the product, container, and filling conditions. When the filling stem is inserted into the container, the top of the bottle contacts the seal ring and immediately air is drawn from the container through the vacuum system. This decreases the pressure in the container compared with the atmospheric pressure on the product in the supply tank, and thus forcing the liquid from the supply tank, through the filling tube and into the container. The fill cycle continues until the product level covers the vacuum tube inlet. The product at this level is drawn through the vacuum system. The level at which the liquid is drawn out of the container by the vacuum system has been preset and is known as the "fill height level." The product drawn off through the vacuum system is separated

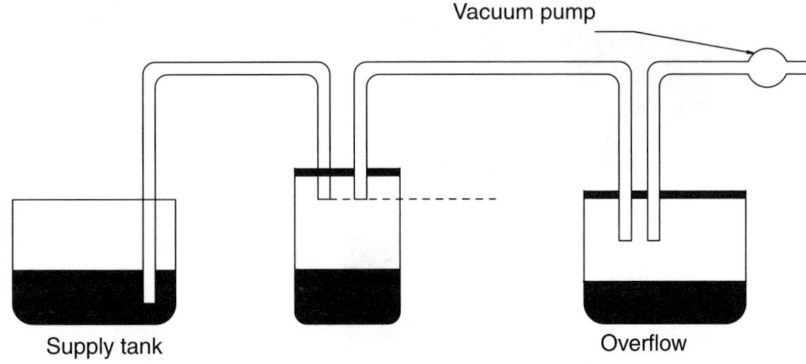

FIGURE 33.9 Vacuum-filling machinery.

from the air being drawn from the other containers (which may not have yet reached their "fill height level") into a separator jar where the product, being heavier than the air, falls to the bottom and is either returned to the supply tank if no foam was produced or, alternatively, pumped into a settling tank (Figure 33.9). The filling stem is then withdrawn and the bottle passes to the closure plant.

2. Constant Volume Filling

Predetermined volume filler measures the liquid into a cylinder, before it is put into the bottles. This type of equipment performs its function by turning on a positive displacement type pump. During the period the pump is operating, a measured amount of product is dispensed. The method is not widely used for nonviscous liquids in glass bottles because it is generally slower than control by filling height, and not necessarily more accurate [8] (Figure 33.10).

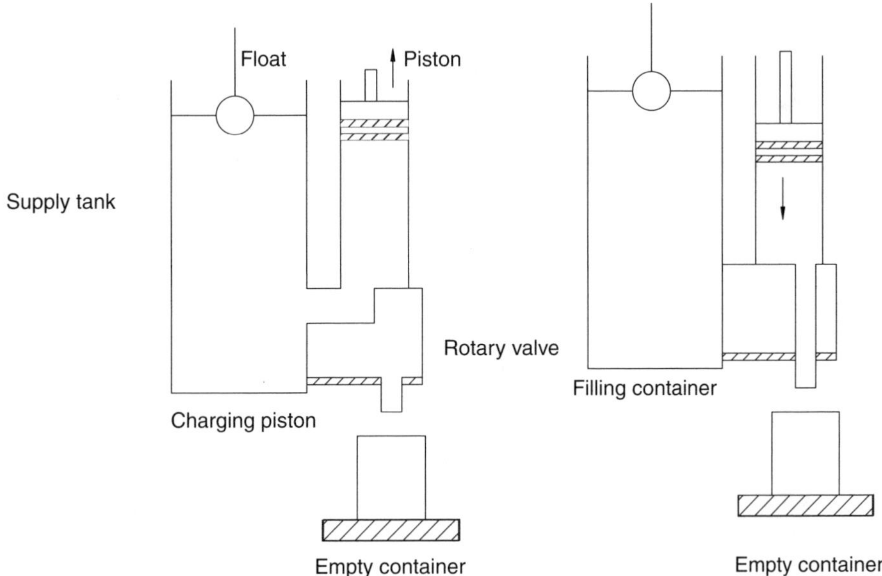

FIGURE 33.10 Constant-volume-filling machinery.

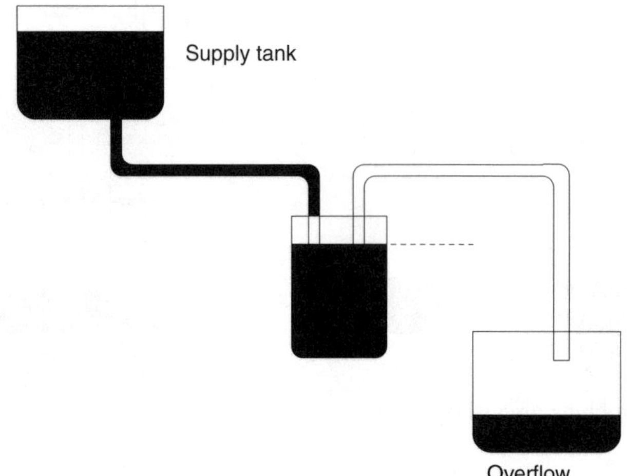

FIGURE 33.11 Gravity-filling machinery.

3. Gravity Filling

There are three types of gravity-filling machines, one of which fills on a controlled time cycle, the second using a measuring chamber, and the third the vacuum-assisted gravity filler. In the first, gravity filling employs the force of gravity to achieve container fill. A supply tank is elevated above the filling tube. The product falls into the container when the filling tube value is opened. The product flow into the container by gravity continues until the "fill height level" is reached, after which the product entering the container is forced up through the overflow tube into an overflow tank where it can be held or returned to the product supply tank (Figure 33.11).

Gravity filling may be employed for filling products which have foaming characteristics. The flow rate of the product into the container is slower and product velocity is minimal. This reduces the agitation which is the cause of foam. If the product is exceptionally foamy, however, gravity filling should not be employed. In the second type of gravity filler, a supply valve opens to admit liquid to a calibrated chamber. When a container is correctly presented at the filling head, the supply valve closes and a delivery valve opens, thus charging the container.

Vacuum-assisted gravity filler is one of the methods; it combines the possibility of high-speed filling, with the simplicity of design essential for really thorough machine cleaning. Modern machines can fill up to 10 pint milk bottles per filling head per minute. In one group of machines, the bottle is lifted to the filling head, and in another the bottle stays at a fixed height and the filling head descends [9].

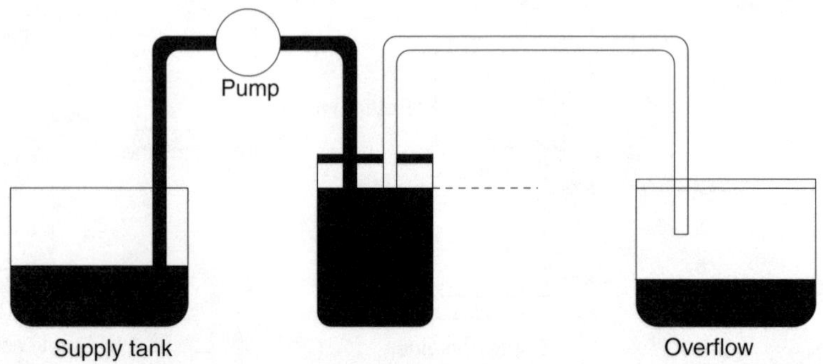

FIGURE 33.12 Pressure-filling machinery.

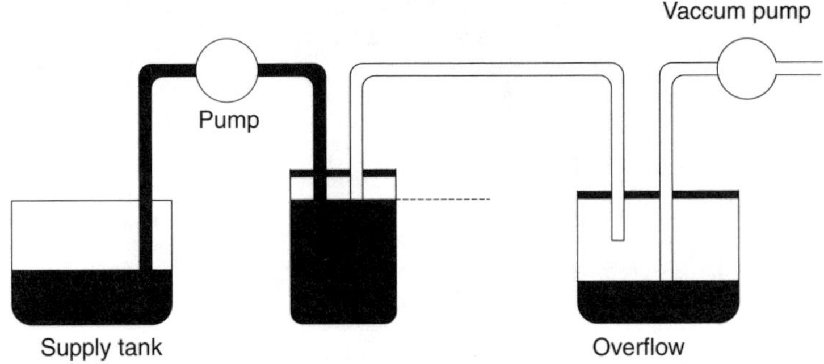

FIGURE 33.13 Pressure–vacuum-filling machinery.

4. Pressure Filling

Pressure filling is used where the liquid is not free flowing and where product agitation is not desirable. This system is similar to the gravity system except that a pump is used to increase the flow rate over that obtained by gravity (Figure 33.12). The best way to minimize foaming is to immerse the end of the filling tube in the liquid during the fill cycle. This minimizes the turbulence of the product during the fill cycle.

5. Pressure–Vacuum Filling

All purpose fillers which can be adapted to employ any combination of pressure, gravity, and vacuum (Figure 33.13). This machine uses the method, described earlier, of controlling the filling heights accurately by lowering the bottle slightly for the final suck-off [7]. Fine adjustments to filling heights can be made but the machine is running, by means of a hand-wheel, which alters the distance through which the bottles are lowered (Figure 33.14 and Figure 33.15).

This combines the vacuum system with the pressure system. It is used when the product flow rate is too slow on a straight vacuum system due to the inability of the container to withstand high vacuum.

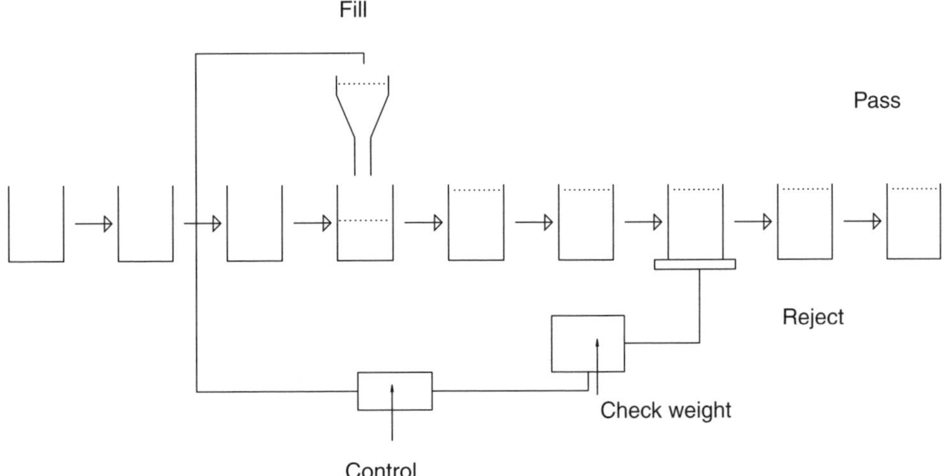

FIGURE 33.14 Automatic filling operation.

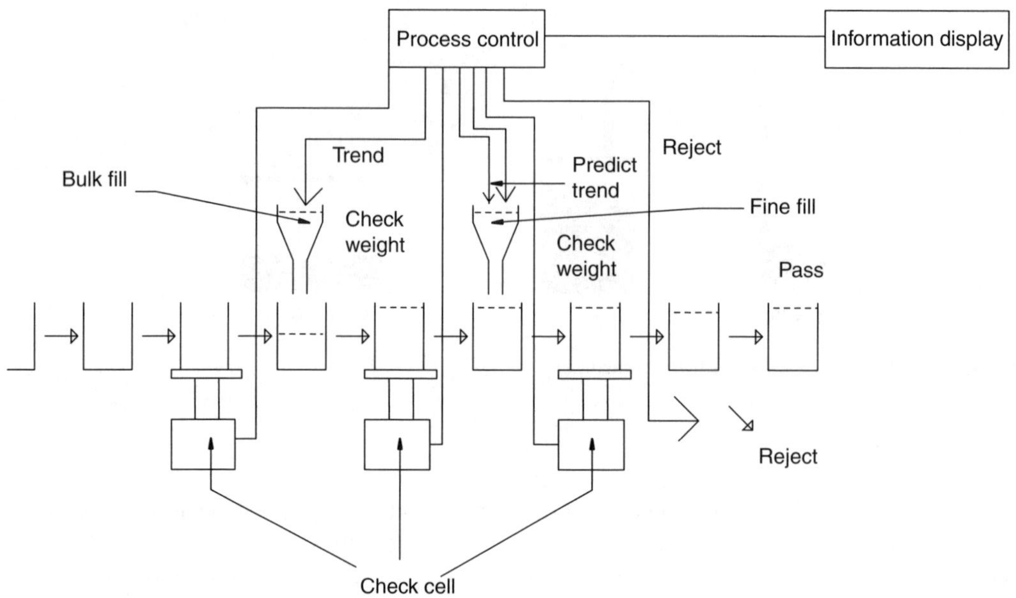

FIGURE 33.15 Fully automated filling control system.

B. Filling of Solids

There are four basic methods of filling solid products into containers. These are volumetric filling, vacuum filling, filling after weighing, and counting. The choice of the method depends on the product to be filled.

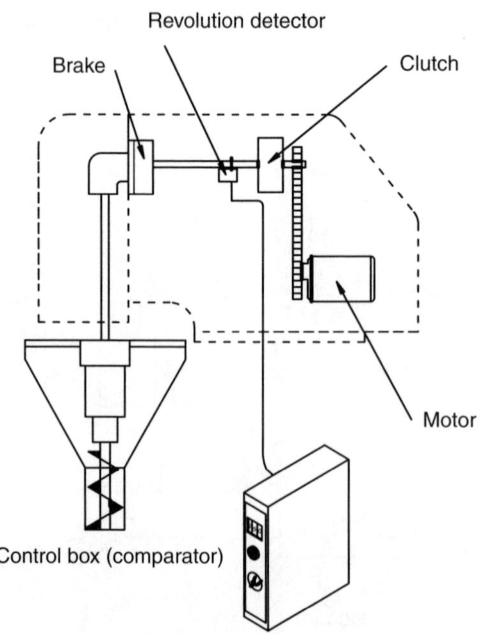

FIGURE 33.16 Auger filler.

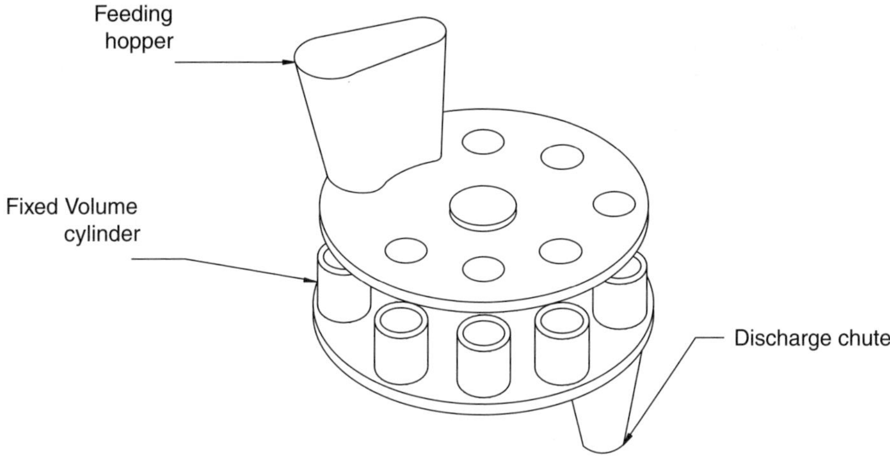

FIGURE 33.17 Volumetric cup filler.

1. Volumetric Filling

The main methods are filling by auger, flask (or cup) fillers, and vacuum filling. In auger fillers, an auger is fitted into a sleeve mounted below a hopper containing the product, which should be granular and not too powdery. The side walls of the hopper are funneled to form the sleeve in which the auger runs, and the product is discharged from the end of the tube into the container. Usually, augers are used for nonfree-flowing products, but free-flowing products may be handled, giving the equipment more versatility. The fall through a product which is free flowing in nature is prevented by a saucer or disc which is attached to the lower end of the auger, when it stops. As the auger spins, the centrifugal force throws the product off the disc. This type of auger incorporates a funnel which catches the product and directs it into the container. Quantity delivered is controlled by the number of turns the auger makes in one cycle (Figure 33.16).

Cup or flask fillers will handle powders and granular materials. This is the simplest and least expensive type of filling equipment for dry products. It incorporates a number of cylindrical-shaped cups, each equipped with a trap door, rotating under a supply hopper. As the cup passes under the hopper, it is filled with the product. The cylindrical cup then rotates to a position, where the cup is scraped or brushed level, giving an even measured cupful. The cup then rotates, to the areas where the discharge spout is located. As the cup passes this area, the trap door opens and the measured product is discharged down the spout into the container. The cup then continues its rotation back around to the filling station and the cycle repeats (Figure 33.17).

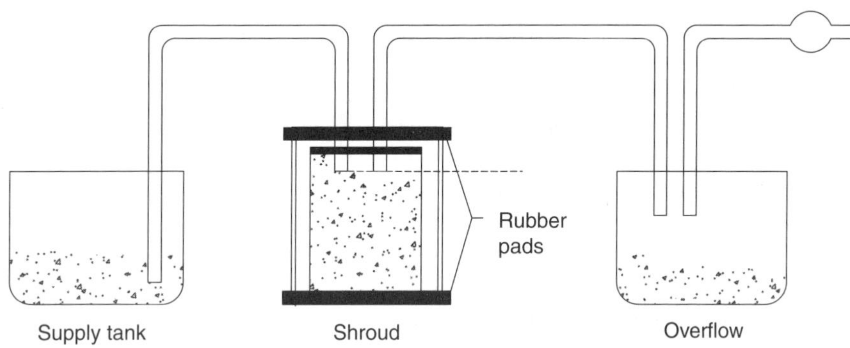

FIGURE 33.18 Vacuum-filling machinery.

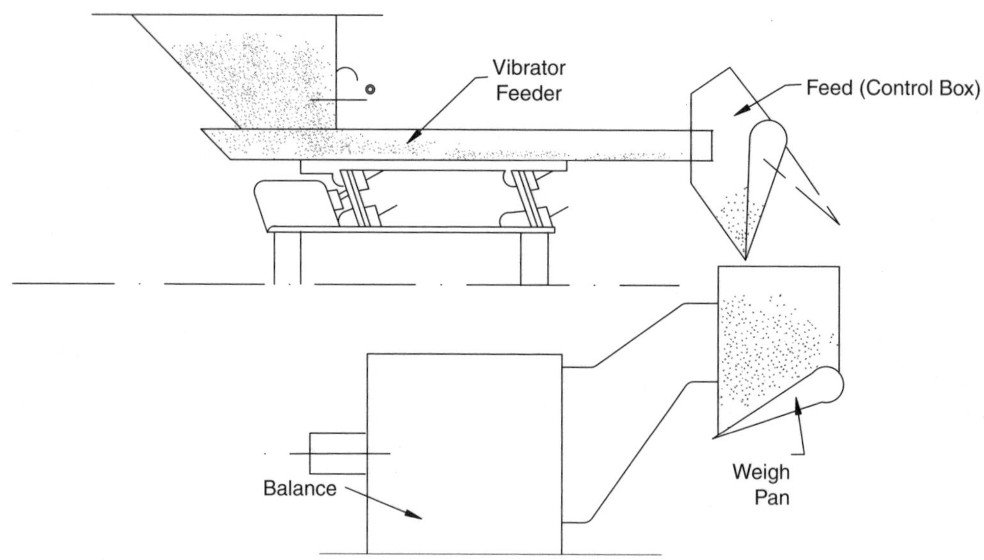

FIGURE 33.19 Filling by weight.

2. Vacuum Filling

Fine powder products can be filled using the vacuum principle, which is basically the same as that used for liquids. A vacuum is drawn on the container and the product flows from the hopper at atmospheric pressure into the container which is less than an atmospheric pressure. If a glass, plastic, or metal container is used which will withstand a vacuum, then no additional problems occur. However, much of this type of product is marketed in a fiber board or paper cylindrical container which is porous and on which no vacuum should be pulled. To overcome this problem, a rigid shroud is placed over the container and the vacuum drawn on it (Figure 33.18). Thin-wall containers that would collapse with normal vacuum are also used in this manner, for example, metal, plastic, fiber, or paper. Wide-body containers with very small necks can readily be filled by vacuum. Vacuum filling is used for fine powders.

3. Filling by Weight

Net weight filling is widely used even though it represents the most expensive method of filling dry products and is obviously the most satisfactory way of meeting the requirements of weights and measures regulations. There are different weighting techniques which use single or double-action scale beams, or heads, which are operated by compressed air and most recently, electronic- and microprocessor-controlled systems [10]. An automatic scale is quite simple in principle. The product is discharged from the end of a vibratory trough into the scale bucket. When the scale comes up to weight, the scale lever opens an electrical contact and the vibration moving the product is not constant in density (Figure 33.19). Therefore, even if this vibration is constant due to the amount of product suspended in the air at the time the electric vibrator is stopped. For this reason, most net weight scales employ a double vibratory trough system with the bulk of the product being placed in the net weight scale bucket for the net weight introduced through the "dribble" trough.

There are other weight-sensing systems such as the electronic weight cell, air balance system, and liquid displacement system which offer greater scale sensitivity and are able to detect very fine changes in weight. The weight-sensing device, however, is not necessarily the whole answer for

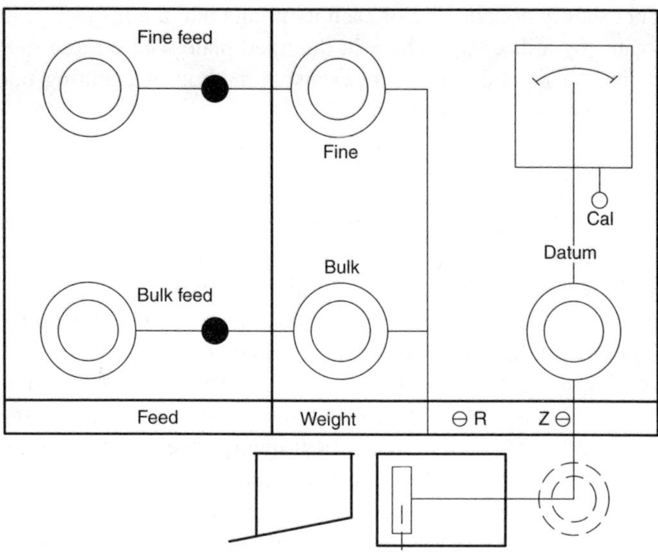

FIGURE 33.20 Filling by weight — bulk and fine feeds.

accurate weight. Net weight scale systems can only be as accurate as the flow control of the product to the weight bucket.

The basic principle is to divide the supply of product into a main feed and fine feed above the weighing head. At the beginning of the weighing cycle, both the bulk and the fine feed are operating to fill the weighing pan until about 80–90% of the required material has been added. When this is reached, the bulk feed stops and the dribble feed continues until the exact balance is reached. At that moment the fine feed ceases and the load is discharged, usually by tipping into the container [10] (Figure 33.20).

4. Counting

Some products require an exact number of discrete pieces in the package. Products like candy are marketed according to the number of units in the package. There are four ways commonly used to count a product.

a. Electric Eye Counting

Counting of the product is done by electric eye. The item is fed by means of a vibratory bowl feeder, belt, or some other means of stringing out the product past an electric eye scanner. As each piece falls past the scanner, it is recorded. Usually, the discharge beyond the electric eye scanner is diverted to the waiting container. On the completion of the count, the gate diverting the product to a particular container moves and directs the product into a second container placed in position to receive the product. This type of counting is used on some hardware items and tablets where versatility and quick change from one product to another is a necessity.

b. Perforated Disc

A disc-type counting mechanism is used for tablets, capsules, and some other items. This is a simple mechanism incorporating a disc manufactured from plastic or wood which is drilled out, allowing room for an individual item to locate as the disc revolves over a fixed plate. The count required is obtained by placing groups of holes of the correct count around the disc and as this group of holes rotates by the product supply hopper, one of each item falls into the respective hole. The pattern of filled holes then rotates by a discharge chute in the fixed plate where these items fall through the chute into the container. This is a simple, inexpensive method of counting but cannot guarantee a count as one or more holes may not fill.

c. Chutes or Channels

The most popular machine for high-speed counting of tablets feeds the product from a hopper down a series of chutes or channels. The diameter of the tablet or pill is normally consistent and, therefore by multiplying this dimension by the number of pills required a measured row of tablets may be controlled and discharged to a chute filling the container. Where a larger count is required, three, four, or five premeasured lengths of tablets in their respective chutes may be released one after the other, forming a long line of tablets giving counts of 100 or more per dump. This method is used to handle items such as headache tablets and other high-volume tablet products.

The latest high-speed tablet-counting equipment incorporates a rotary bottle feeding principle with a number of counting heads mounted on a rotary turntable. The count is performed and automatically released into the container during its filling cycle.

Straightline tablet counters (units where the bottles stop on their conveyor during the fill cycle) normally employ a person who acts as an inspector to ensure that the chutes of the tablet filler are all completely filled before tripping the mechanism releasing the count into each bottle. The bottles are normally indexed automatically, leaving the operator complete time to inspect the tablets looking for such things as "capping" and broken tablets.

d. Orientation Devices

Items such as screws, flat washers, nuts, and bolts are counted on machines which either tumble or vibrate mass of the product in the hopper in a manner which allows some of the items to locate in a track leading out of the hopper. The items caught in the track move down the track either by vibration, gravity, or a combination of both, to the counting station. Here a group is released (in the same manner as tablet-counting machines) by holding the items above a certain point and releasing those below, allowing them to fall into the container.

VII. FORM, FILL, AND SEAL

In general, form and fill machines either take a web of material and form it into a tube which is filled and sealed at intervals, or take a web of material which is folded along its length, sealed at intervals to form a series of open sachets which are then filled and closed [11]. The form, fill, and seal machines can be classified as vertical and horizontal machines.

A. Vertical Form/Fill/Seal Machine

In this type of packging, a machine takes a web (or two webs for duplex application) of packaging material and forms it into a vertical tube. The tube is filled and sealed intermittently to produce a pack sealed at both ends and down the length of one face. This long seal can be made as an overlap seal or as a face-to-face "flip" seal. The web of material is unwound and after passing over various tensioning rollers are formed into a vertical tube by passing over a specially shaped forming shoulder. The tube of material formed encloses the filling tube and at this stage the long seal is made. The filled packs are cut off at the same time as the cross-seals are made [11]. Filling can be done in a number of ways, depending on the characteristics of the product. Fillers include liquid fillers, paste fillers, augers, pocket fillers, vibratory, orifice-type, and gravimetric units as described under bottle fillers [12] (see Section VI).

Some of these machines can produce a gusseted bag by having a "tucking" station just before the cross-seals are made and such packs are especially useful where they have to be subsequently

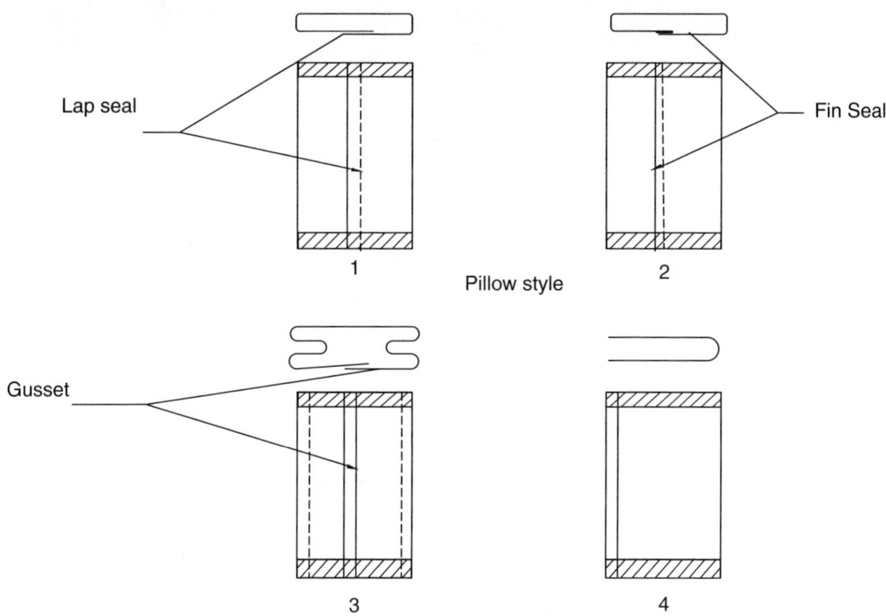

FIGURE 33.21 Package styles on vertical form, fill, and seal machinery.

placed into a carton. By the introduction of a small tube into the filling tube, gas flushing can be carried out with nitrogen or other gas if required [11]. Vertical form/fill/seal machines can make a number of different bag styles. A pillow-style bag with conventional seals on the top and bottom, and a seal fin or lap in the center of the back panel from top to bottom, gusseted bag with tucks on both sides and three- or four-sided seal package (Figure 33.21) [13].

Thermoplastic and "heat-sealable" materials are required for form, fill and seal. Nonheat-sealable materials require a heat-seal layer that provides a seal with the right combination of time, temperature, and pressure. A fin seal can be made of materials with sealing properties on one side only. A lap seal uses slightly less materials, but it requires sealing properties on both sides. The design of the bag-forming collar can be engineered to get the optimum efficiency from metallized materials, heavy paper laminates, and so on, as the wrapping material moves down around the forming tube, the film is overlapped for either the fin or lap seal. The overlapped material moving down (vertically) along the bag-forming tube will be sealed. The packaging material/film advances a predetermined distance that equals the desired bag-length dimension. After the film advance is completed, the bag-sealing and filling completes the remainder of one cycle [13] (Figure 33.22).

The mode of operation of form and fill machines makes certain demands on the flexible packaging material employed. The most important considerations are:

(a) The material must be thin and flexible. In forming a web of material into a tube, the material is turned through acute angles and must withstand this treatment without creasing or breaking. Cellulose film, laminations of cellulose film to itself, and polythene and polythene-coated paper are satisfactory.
(b) It is essential that the material has good slip characteristics. This is especially important where machines have filling tubes. The material is wrapped around this tube and if the friction between it and the tube is too high, the web may break. This trouble is experienced with polythene coatings or laminations, and in these instances an additive can sometimes relieve the problem. Sometimes it may be caused by static electricity.
(c) The packaging material must exhibit high tack properties when hot, that is to say the heat-seal coating or film must be very tacky when molten as, just after the cross-seals

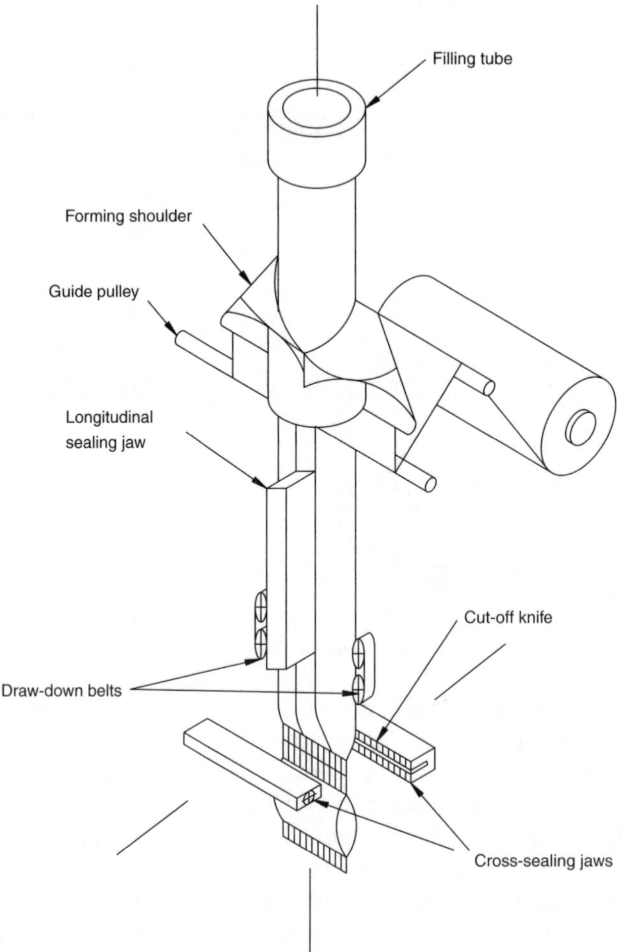

FIGURE 33.22 Vertical form, fill, and seal machine.

have been made, the product is dropped down the filling tube on to it. Sometimes this drop can be over 2 ft and the seal must not break apart even when still not fully cooled and set. A polymer such as polythene is ideal in this respect because it forms a solid weld when heat-sealed and sets quickly.

(d) The material should not be tacky on outside the web or have a coating on the outside which becomes tacky at the sealing temperature needed. This fault can lead to excessive drag on the forming shoulder in the first instance and sticking to the heat seal jaws in the latter.

(e) Unsupported plastic films such as polythene can be handled on some of these machines, but this material requires impulse sealing [11].

B. Horizontal Form/Fill/Seal Machines

Horizontal form/fill/seal equipment can be classified as follows: (1) Pouch form/cut/fill/seal, pouch vertical: inline equipment; and rotary equipment; (2) pouch form/fill/seal/cut, pouch vertical: single-lane, continuous motion; multilane, continuous motion; and single-lane, intermittent motion; (3) pouch form/fill/seal/cut, pouch horizontal: single-lane, intermittent, and continuous motion; and multilane, intermittent, and continuous motion; (4) thermoform/fill/seal equipment: multilane, intermittent, and continuous motion; and (5) horizontal form/fill/seal bag-in-box equipment: single-lane, intermittent motion [12].

The horizontal type takes a web of material and forms this into a horizontal tube. The product is fed into the tube by conveyers. The long seal is made by passing the material past heater blocks and then through cold rollers. The motion is maintained by friction belt drives on the tube of material containing the product. The cross-seals are made by rotary crimping sealers which also cut off the packs. Speeds up to 180 packs per minute can be achieved and the machines will handle such products as biscuits, ice lollies, and chocolate bars. The speed is mainly determined by how quickly the product can be fed (Figure 33.23) [11].

Machines to handle unsupported polythene for textile and bakery goods have been produced which takes two webs of material and produces a flat sachet sealed around four sides. This machine is used for such products as stockings and motor spares. Some models are capable of gas flushing.

(a) Material must be flexible and thin, flexible for ease of forming and thin so that the necessary heat for heat sealing can pass through the wrapper in the very short dwell time allowed. Foil laminations can be run to a limited extent and in some instances two sets of end crimpers are provided to get the heat through at speed.
(b) High tack when hot is essential. With the very small dwell times the heat-sealing medium must melt and then set very quickly. Polythene, PVDC, and similar coatings are ideal although in some instances cold-seal coatings are used to avoid the necessity for heating. Such cold-seal coatings allow the machines to run very fast and will permit the packaging of such articles as chocolate and ice lollies without fear of melting.

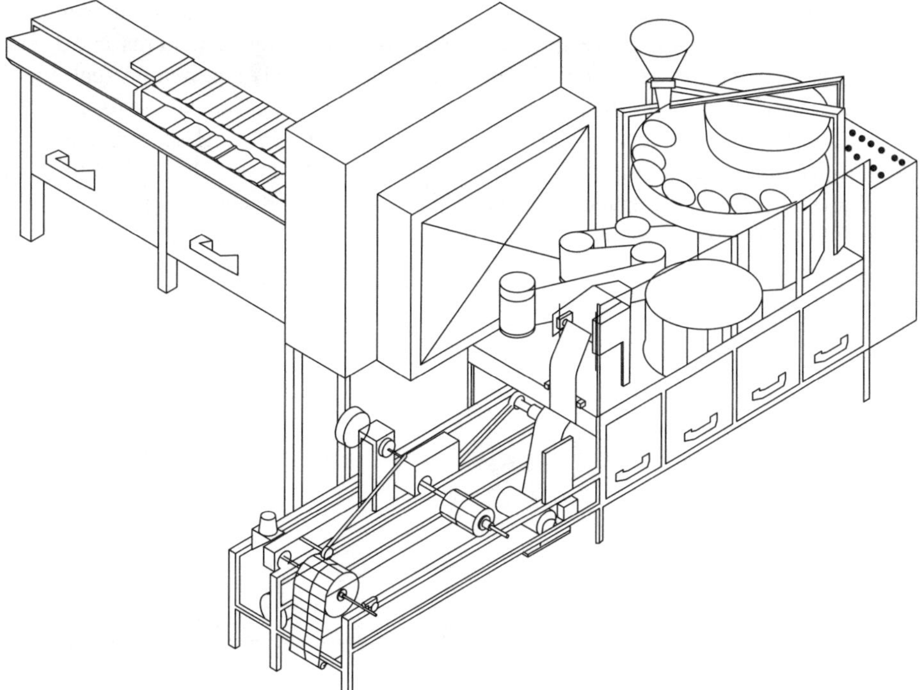

FIGURE 33.23 High-speed horizontal/fill/seal machine.

(c) Slip characteristics are not quite so important with this type of machines. These are filling tubes in some vertical flowpak-type machines.
(d) Coated materials must not become tacky on the outside either when hot or at room conditions, as this will result in sticking to the heaters or a build-up of the coating on the machine [11].

C. Sachet-Making Machines

A sachet may be defined as a rectangular pouch which is sealed on all four edges. There are two broad divisions of machines used for making sachets, which may be loosely referred to as horizontal and vertical.

1. Horizontal Machines

The sachet is usually formed from a single web of material. The web is passed over a triangular-shaped shoulder to produce a double thickness. Heat seals are made in the vertical plane and in some instances along the folded edge as well to produce a series of pockets along the web which are cut into individuals, opened, filled, and finally heat-sealed along the top edge to complete the package (Figure 33.24).

A range of products can be handled on these machines (powders, granules, creams, and liquids). They are principally used for powders and granules when the very short drop is a big advantage in keeping the heat-seal areas free of the product. Speeds vary but can be increased by using a number of filling heads. The speed with one filling head was 60 per min [11].

2. Vertical Machines

The majority of vertical sachet machines use one web of material folded over a triangular-forming shoulder in exactly the same way as the horizontal machine (Figure 33.25).

The material is folded to pass on both sides of a filling tube and sealed on three edges of the sachet (side seals and bottom) by reciprocating platens with crimping patterns and filled. The final top seal is made at the same time as the next set of side seals and bottom seal, the sachet being separated by cutting the horizontal seals along the center line. The side seals and end seals may be made in two separate operations or in one operation using an H-shaped sealing platen. The end-seal platens do not move up and down, for this time the material is pulled through by a series of draw rollers, usually situated between the side-seal and end-seal jaws. The finished sachets are cut off by scissors which are situated below the end-seal platens.

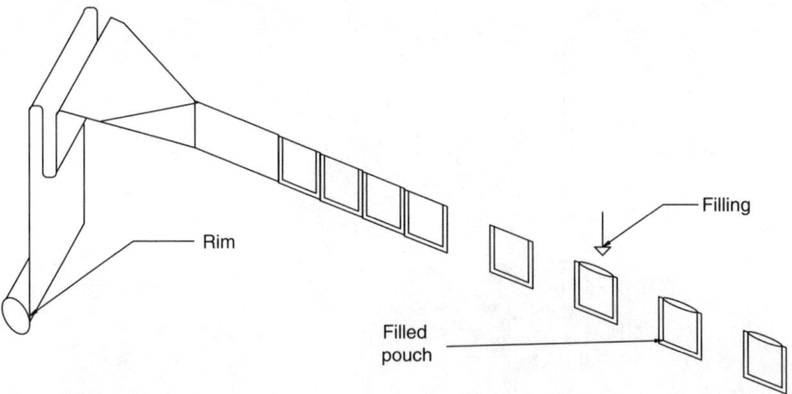

FIGURE 33.24 Horizontal sachet machine.

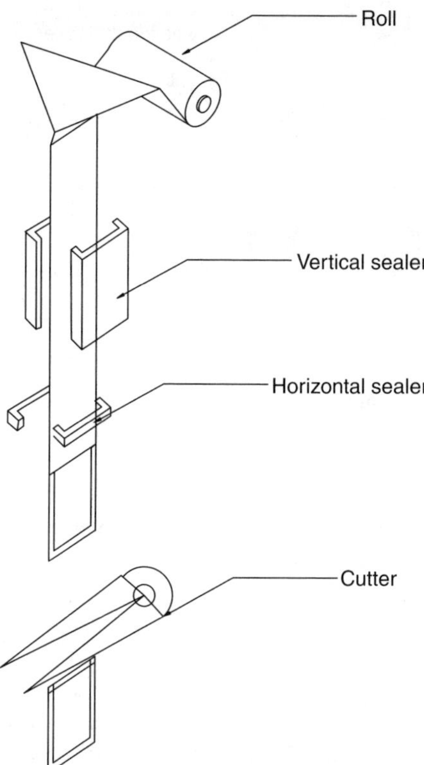

FIGURE 33.25 Vertical sachet machine.

The vertical machine dispenses with the triangular shoulder and in its place introduces a slitting operation to divide the web into two and then turns the slit webs to bring them together face to face. The fact that the packaging material only requires folding means that a very much greater range of laminates and coated materials is available for the sachet type of machine than for other types. As already mentioned, a good degree of rigidity in the packaging material is essential for good running on a horizontal machine and it is also true to say that rigidity is helpful on a vertical machine.

A sachet has an obvious disadvantage compared with a pouch in that the area of packaging material required for a given quantity of product is greater with a sachet. Sachets are used therefore either for very small units or materials not suitable for running on other machines. Aluminum foil laminates are ideally suited for making sachets and are capable of giving a very high degree of protection indeed — as good as a lidded can in some instances.

Other materials suitable for sachet machines include film laminations such as cellulose film/polythene, cellulose acetate/pliofilm. Heat-seal-coated papers, polythene-coated papers, moisture-resistant cellulose film, and coated oriented polypropylene are also suitable, but waxed papers are not suitable because of their lack of tack when hot. Single plastic films also cause difficulties and are not generally used [11].

D. Thermoform/Fill/Seal Machine

In this machine, thermoplastic web is heated and formed. The depressions are indexed forward as a web. The cavity is filled, lid is applied, and cut from the web. In some machines, a pressure-forming die is used for heat-form to form a soft aluminum tray (Figure 33.26). Lid material can be plastic

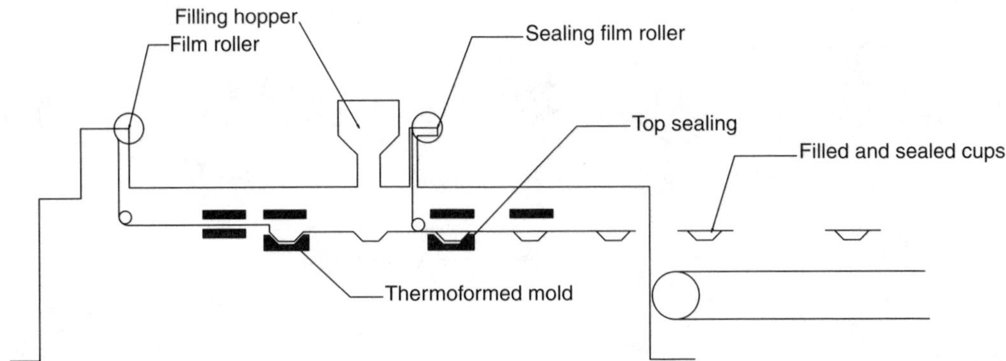

FIGURE 33.26 Thermoforming machine.

film or film laminate. Printed metal foil reverse coated for heat sealing is also used. The products packed are butter, cream, yogurt, jams, and so on [14].

VIII. HEAT SEALERS

Thermoplastics are being widely used as construction material in flexible packaging. When heat is applied, the thermoplastics melt and act like glue in effecting a seal. The fundamental principle of heat sealing is to provide heat at the interfaces, pressure to bring them intimately in contact, and complete a weld, all within an acceptable time period. The only exception to this principle is in radiant heat sealing, which relies on film orientation and surface tension.

During sealing of a thin material, generally it is sufficient to introduce heat from one side of the construction. When using thicker materials or if higher speeds are required with thinner materials, heat may be introduced from both sides.

To get good heat seals, the time, temperature, and pressure of the bar are all important factors. Nowadays we can find a variety of heat sealers which are designed on different principles and at different capacities and which can even seal packages at the rate of 1200 to 1500 per hour. A few of them are listed subsequently.

A. Bar Sealing

It is the most widely used sealing method. Bar sealing is the least expensive technique compared with other methods. It is used both to make and seal pouches. It is also used in most form–fill–seal equipment as well.

A simple hot bar covered with antiadhesive agent like Teflon-coated cloth and silicone rubber on the other face effectively seals the thermoplastic material (Figure 33.27). As it is important to avoid wrinkles in the seals, serrated bars are sometimes used to offset the wrinkles. Serrated bars are also useful to get good mechanical strength as well as to tolerate tiny leaks in the seal. Temperature-measuring and -controlling equipment such as thermostats are essential for this type of heat sealers for good results. Dies have to be accurate so that the die contacts all parts of the work surface. The die cannot be made of soft material which cannot withstand the abrasion of the film. Brass and steel are generally used for this purpose. All the material within the hot bar should be made of the same material to avoid thermal buckling. The use of silicone rubber as resilient surface corrects the variation in the pressure distribution along the length of the bar.

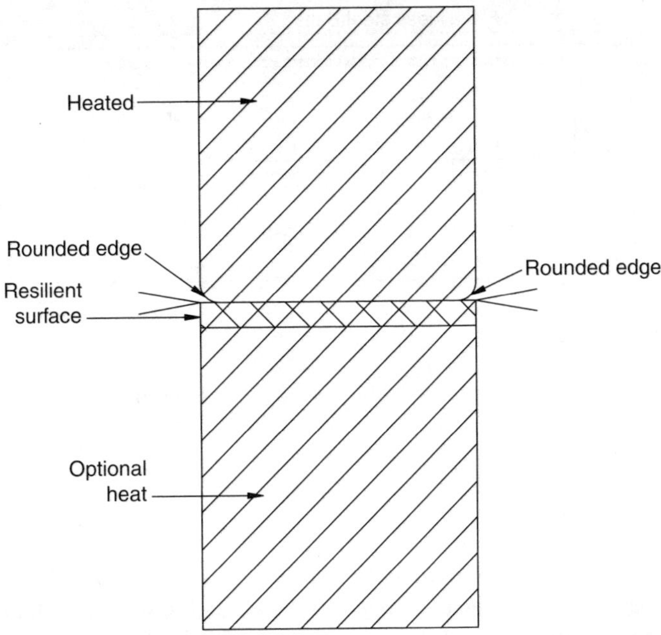

FIGURE 33.27 Bar sealer.

Bar sealers find application in manufacture and closing of laminated pouches, cup lidding, form–fill–seal packaging machines (Figure 33.28).

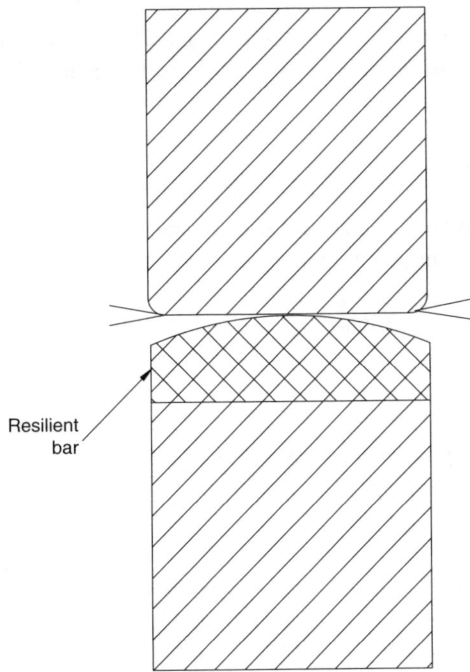

FIGURE 33.28 Rounded resilient bar sealer.

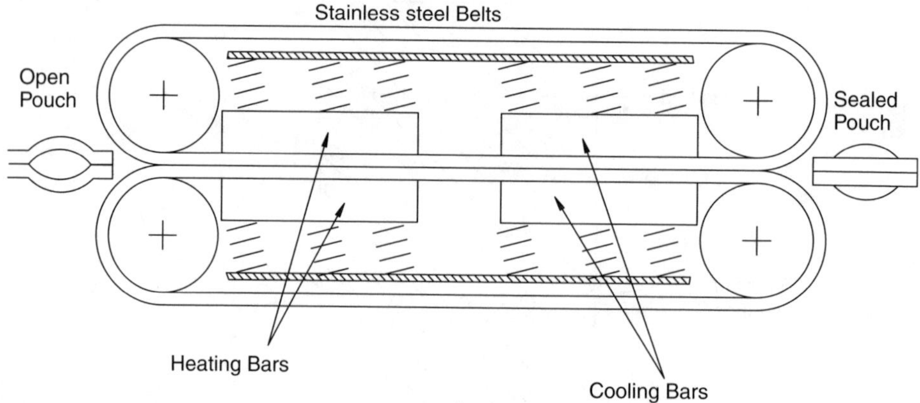

FIGURE 33.29 Band sealer.

B. Band Sealing

In band sealing, a pouch mouth is introduced between two moving bands, which are pressed together by heated bars. The heat passes through the bands and into the pouch material, softening it for sealing. As the pouch continues along between the bands, the bands are next pressed together by chilled bars that withdraw heat from the pouch seal through the bands (Figure 33.29). The bands then progress to release the pouch. Band sealing is normally used to close the mouth of the pouch filled with product. This type of sealing provides a continuous method for sealing. These find applications in sealing-filled pouches including those made from unsupported materials.

C. Impulse Heat Sealing

Impulse sealers have the same general configuration and mechanical construction used for bar sealers; the difference lies in the sealing jaws. Each opposed jaw is generally covered with a resilient surface, such as silicone rubber-coated fiberglass, Teflon-coated fiberglass, and Teflon-coated kapton (Figure 33.30). A pouch mouth is placed between the jaws, and the jaws are closed. An electric current passes through the nichrome ribbon for a brief period of time and is then turned off.

The advantage of impulse sealing over the bar sealing is that the pressure is continued during the cooling cycle also, by which adequate seal strength can be obtained. The dominant drawback of the impulse heat sealers is the high maintenance cost. Hence, impulse sealing should be used only where bar sealing does not do the job properly. Impulse heat sealers find applications in sealing sacky materials, unsupported thermoplastic films, for example, frozen vegetable bags.

D. Ultrasonic Sealing

In this method, the sealing heat is produced by mechanically hammering or rubbing the packaging materials together at a high frequency. The theory that applies to the use of ultrasonic is that any two materials will bond together provided their surfaces contact within the atomic distance of each other (Figure 33.31).

The frequency range normally employed for this purpose ranges between 20 and 100 kcycles per second. This method is quite useful for highly oriented films, where sufficient heat is generated to melt the interface and also useful to materials that are too thick to permit heat transfer through them for sealing. This method finds applications in sealing biaxially oriented films, thick webs, aluminum foil, and rigid container components.

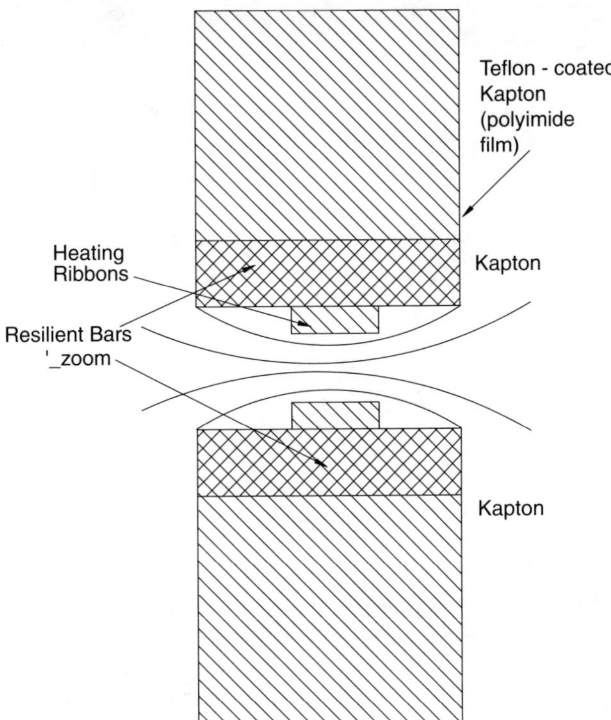

FIGURE 33.30 Impulse sealer.

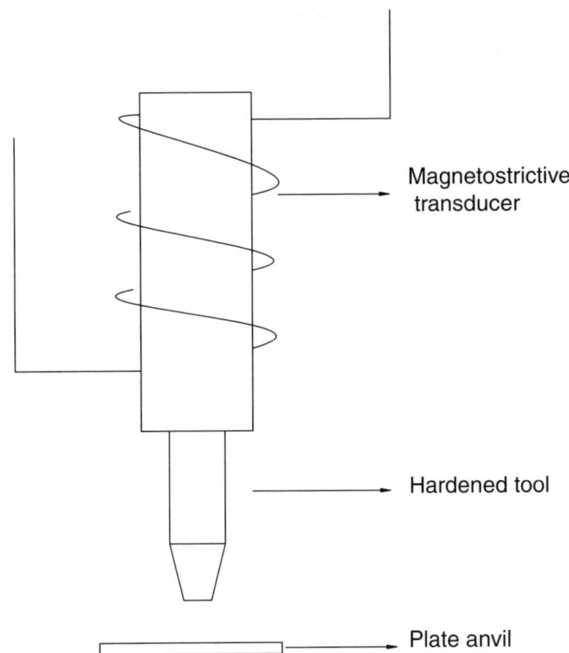

FIGURE 33.31 Ultrasonic sealer.

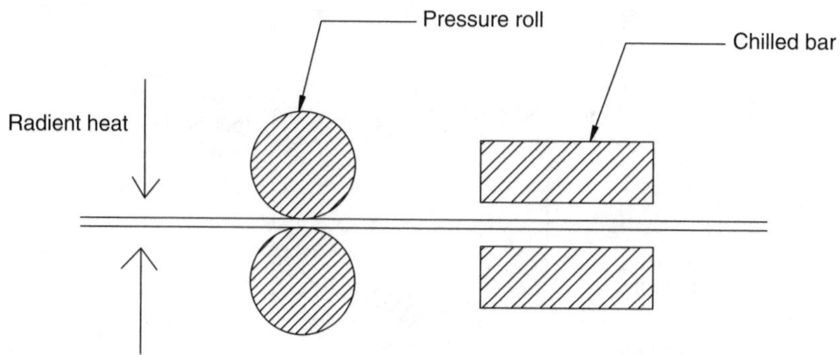

FIGURE 33.32 Radiant sealer.

E. Friction Sealing

Cylindrical objects are sealed together by this method. The two members to be welded are placed into the machine, with the surfaces to be welded in contact with each other. One member is rotated rapidly, whereas the other is held by a friction brake. The heat generated by friction at the interface between the two cylinders melts their surfaces and the viscous interface causes the other cylinder to rotate in unison with the rotary cylinder. This effects the seal.

Thermoplastics with surfaces that become slippery when heated are best suited for friction sealing; those that become dough-like create problems. Ultrasonic welding is actually a method of friction welding. These types of sealers find applications in assembly of round containers and sealing ends of strapping.

F. Radiant Sealing

Heavier packaging materials which are considered either difficult or impossible to seal without coatings can be sealed by radiant heat sealing technique. Here the radiant heat is applied to both the surfaces and then pressed together. The seal must be clamped between chilled bars until it sets (Figure 33.32). Polyester films which are used for packaging constructions and many nonwoven plastics can be sealed with this technique. These find applications in sealing uncoated highly oriented films and nonwovens including polyester, nylon, polyolefins.

G. Contact Sealing

In this method, a heated plate is placed against the surface to be sealed and then both surfaces are pressed together by chilled bars. This is used for strapping as an alternative to the friction sealing and also used for sealing ends of strapping and tubing.

H. Hot Melt Sealing

In this method, a continuous stream of molten thermoplastic material is applied between the surfaces to be sealed. Then they pass through pressure rollers. The hot melt contains sufficient heat to cause it to adhere to the package members to be sealed together. Hot melt sealing is intensively used where applied heat might damage a packaging member or where peelability is desired. It finds applications in paperboard containers, peelable seals, and case packers.

I. Pneumatic Sealing

In pneumatic sealing, the air pressure from the atmosphere pushes a hot, tacky film into a close conformity with an object and seals the film to a substrate on which the object is placed. It finds usage in skin packaging.

J. Dielectric Sealing

A high-frequency electric field generated between two metal bars is used, in principle, for dielectric sealing. This method is normally used to seal PVC material. The members are pressed together between a cold bar, generally made of brass, and a cold flat metal surface. The field is generated between the bar and the lower surface.

K. Magnetic Sealing

Magnetic sealing is used where heavy polyolefin package members are to be sealed. A gasket shaped to fit between the sealing faces of the two package members is made from the same thermoplastic as the package, but has milled into it an iron-containing compound that has a high hysteresis loss. The package portions are pressed together with the gasket between their sealing surfaces and placed in a magnetic field oscillating at a frequency high enough to melt the gasket. This in turn melts the sealing surfaces and causes them to weld together.

L. Solvent Sealing

Solvent sealing is used to join together package configurations that do not lend themselves to heat sealing. It is also used for joining materials that are either not susceptible to heat sealing, or may be damaged by the application of heat.

IX. VACUUM PACKAGING

Vacuum packaging, which alters the environment surrounds the food products, retard rate of biological and biochemical deterioration significantly. Packaging under vacuum retards oxidation and aerobic microbial growth. The vacuum may be displaced by inert gases such as nitrogen in situation in which the package might collapse or the product might be damaged from differential pressure of air on the outside, for example, in packaging coffee, dry milk, hydrogenated vegetable shortening, and so on. The presence of nitrogen acts as a cushion or pillow which prevents the product damage in transportation and in handling.

The vacuum packaging requires very high gas barrier packaging materials. The vacuum is created to reduce the oxygen percentage in package to increase shelf-life. There are four types of vacuum packaging machines: nozzle, chamber, skin, and thermoforming type [15].

A. Nozzle-Type Vacuum Packaging Machine

In nozzle-type vacuum packaging machine, a nozzle connected to a vacuum pump is inserted in the pouch. Air in the pouch is evacuated through the nozzle. Open part of the pouch is sealed by sealer as shown in Figure 33.33. However, for blocks such as meat and meat products, the pouch is clipped with aluminum wire. The two processes of vacuumizing and sealing on most of this type of machines are automatically done, initiated by stepping on a foot switch. The degree of vacuum of bags packaged by this type of machine is lower than other machinery [15]. Nozzle vacuumizing is used for packaging whole fresh and frozen poultry, fresh-cut vegetables, and bulk packaging of fresh meats, poultry, fish, processed meats, nuts, and so on. For whole poultry, manual and high-speed rotary nozzle machines that provide an aluminum clip closure are used. The twisting of

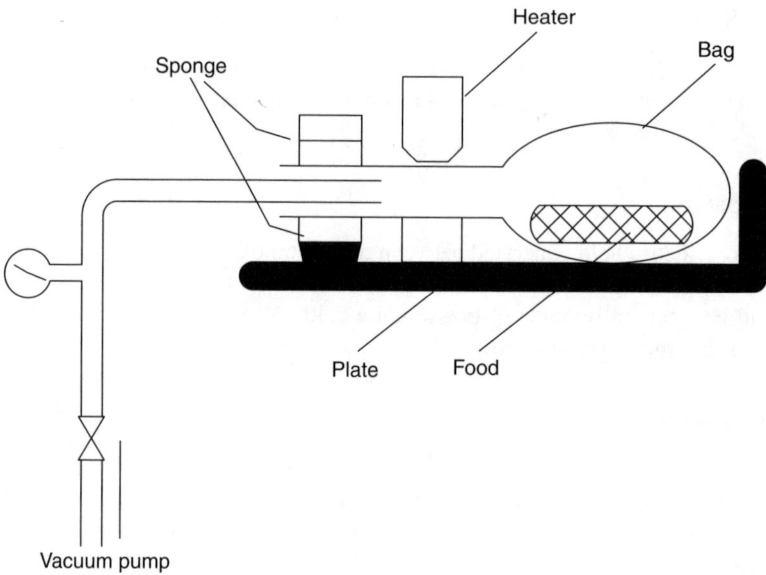

FIGURE 33.33 Nozzle-type vacuum-packaging machine.

the bag neck and clipping provides for good whole-bird shaping and sealing under high-moisture conditions [16].

B. Chamber-Type Vacuum Packaging Machine

The chamber type of vacuum packaging machine is shown in Figure 33.34. In this machine, the filled pouch is kept on the lower jaw of the sealing bar and is clamped. The knobs provided on the front side of the machine are then adjusted to get the desired amount of vacuum and the sealing time. The lid is closed and after the operation, the lid will open automatically.

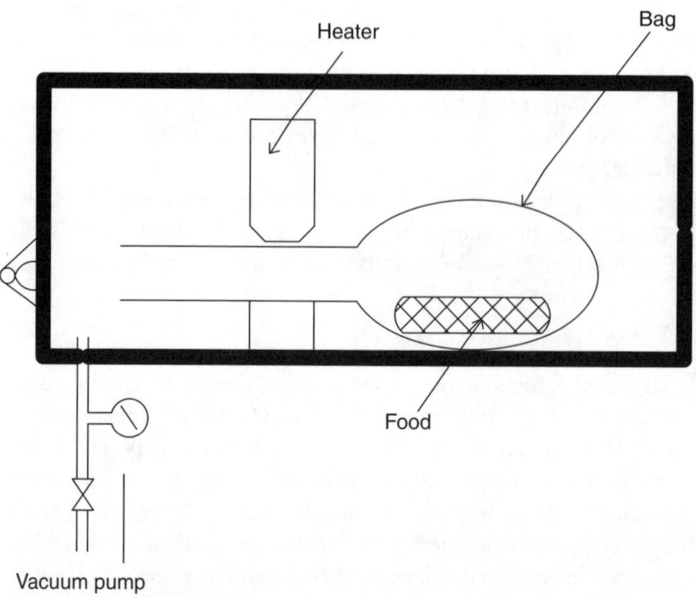

FIGURE 33.34 Chamber-type vacuum-packaging machine.

The latest vacuum packing machines are based on the above type, in which the operating conditions like percentage of vacuum, sealing time, percentage of gas flushing, and so on, can be preprogrammed and can be kept in memory. This provision is very useful to keep the operating conditions for different products and different packaging materials, which are often available in any food industry. The desired program can be retrieved with the help of the menu provided on the front panel of the machine. The machine can be used with and without gas flushing. Vacuum can be achieved in two stages with 90% evacuation in the first stage and 99.8% in the second stage, which is optional. As a rule of thumb, a package should be evacuated in a chamber to an absolute pressure of 25 mm. But to some meat products even low levels may be required to avoid discoloration, which cannot be obtained either with reciprocating piston or with ordinary vane pump, the so-called rotating piston pump is essential. These factors should be kept in mind when selecting the machine for the desired purpose.

C. Skin-Type Vacuum-Packaging Machine

In this vacuum-packaging machine, heated and softened upper film is applied skin-tight over food and a lower film (or tray or cardboard) in the vacuum chamber. The appearance of food vacuum-packaged by this type of machine increases food value and display effect [15]. The top web is fed

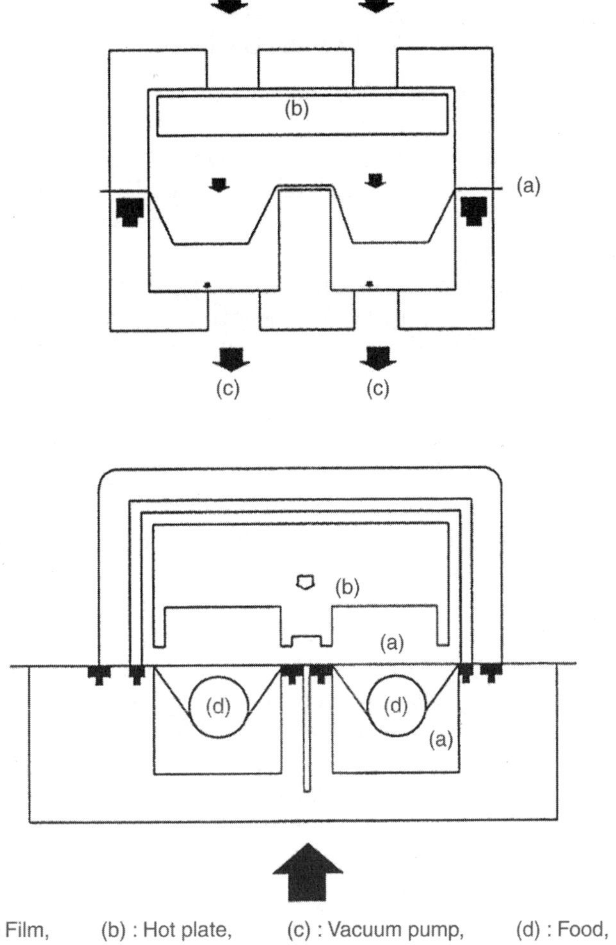

FIGURE 33.35 Thermoforming type vacuum-packaging machine.

down and into the vacuum skin-packaging chamber; it is heated and partially thermoformed to fit over the product. When the chamber closes, the bottom web is pulled against the bottom of the vacuum skin-packaging chamber with a high vacuum. The top web is either suspended over the product by drawing an equal vacuum over both sides of the web and around the product; or is drawn against the heated surfaces of the upper half of the vacuum skin-packaging chamber by a high vacuum and the area around the product between the top and bottom webs is vacuumized [16].

D. Thermoforming-Type Vacuum-Packaging Machine

The method of forming the bag and packaging in the vacuum chamber is shown in Figure 33.35. The lower film is warmed by a hot plate and then deep drawn by vacuumizing through a pump. After food loaded in the bag is covered by the upper film, both films are sealed in the vacuum chamber. The newest machine of this type is controlled by computer, and deep drawing, loading, vacuumizing, sealing, trimming, and so on are set going automatically [15].

The product is manually or automatically loaded into the formed pockets. The top web is then fed down over the bottom web between the two halves of the vacuum-sealing chamber. The two halves of the chamber closed and a high vacuum ($\geq$29 in.Hg) is drawn on both sides of both webs and around the product. When the desired vacuum level is reached, the webs are heat-sealed together [16].

X. CONCLUSIONS

There are number of packaging machines available in the market. Right choice of packaging machinery is very important. Wrong choice may lead to financial losses and poor quality of the package product. Packaging machines are capital-intensive investment. Changing packaging machinery at a later stage may not be an economically feasible proposition. Packaging machines include bottling, bag-making, canning, cartoning, filling machinery, form, fill and seal, heat sealers, vacuum packaging, and so on.

The choice of packaging machinery depends on a number of factors which can be classified into four groups: machine requirement, product characteristics, packaging material requirement, and general requirement. In the machine requirement, the production rate requirement at regular time and peak demand, package sizes, changes in package sizes, product types, and labor requirement of the machine are important criteria to be considered. In product characteristics, variation in product size, postpackaging handling of the product like deep frozen, protection requirement of the product from oxygen, moisture, and so on influences the packaging machinery selection. There may be a need to look into the possibility of a change in the product to suit the requirements of a high-speed operation or the tighter limitations on the physical properties of the product for a more trouble-free packaging system. It is highly important to know whether the packaging machine is used for test marketing operation as a possible option to upgrade in size and speed or integrated into a full-scale production line at a later date. It is also necessary to know the anticipated pack size or product variations, and to plan for any changes that are likely to occur for the requirements of the marketplace. Some of the factors that can influence such decision are space limitations, available utilities, safety requirements, sanitation problems, dust or fumes, and the level of skill of the operators who are available to run the equipment, including operation and maintenance.

The packaging materials need to be accurately specified before the equipment for handling them can be selected. It would also be wise to try to anticipate the future changes when formulating these specifications. Packaging material requirement in terms of consumers demand, sales appeal, and properties of the packaging material to protect the product in terms of oxygen, water vapor or odor barrier, machinability of the packaging material to get required packaging speed also influences the packaging machinery selection. General considerations in terms of linking the new machinery with existing production line or existing packaging machinery for synergy and space utilization should be considered for packaging machinery selection. Understanding all these factors are very important in this very competitive and consumer-driven industry for success.

REFERENCES

1. JL Hanlon. *Handbook of Package Engineering*. New York: McGraw-Hill, 1971, Chapter 20, pp. 1–5.
2. JF Hanlon. *Handbook of Package Engineering*. USA: McGraw-Hill, 1976, Chapter 5, 1 pp.
3. FA Paine, HY Paine. *A Handbook of Food Packaging*. Glasgow: Leonard Hill, 1983, pp. 113–122.
4. J Larousse, B Brown. *Food Canning Technology*. New York: Wiley-VCH, 1997, 589 pp.
5. AL Brody, KS Marsh. *The Wiley Encyclopedia of Packaging Technology*, 2nd ed. New York: John Wiley & Sons, 1997, pp. 580–583.
6. AL Brody, KS Marsh. *The Wiley Encyclopedia of Packaging Technology*, 2nd ed. New York: John Wiley & Sons, 1997, 587 pp.
7. FA Paine, HY Paine. *A Handbook of Food Packaging*. Glasgow: Leonard Hill, 1983, pp. 86–87.
8. BE Moody. *Packaging in Glass*. London: Hutchinson & Co. (Publishers) Limited, 1972, pp. 166–167.
9. BE Moody. *Packaging in Glass*. London: Hutchinson & Co. (Publishers) Limited, 1972, 156 pp.
10. FA Paine, HY Paine. *A Handbook of Food Packaging*. Glasgow: Leonard Hill, 1983, pp. 88–92.
11. FA Paine. *Packaging Materials and Containers*. London: Blackie & Son Limited, 1967, pp. 311–319.
12. AL Brody, KS Marsh. *The Wiley Encyclopedia of Packaging Technology*, 2nd ed. New York: John Wiley & Sons, 1997, 465 pp.
13. AL Brody, KS Marsh. *The Wiley Encyclopedia of Packaging Technology*, 2nd ed. New York: John Wiley & Sons, 1997, pp. 468–469.
14. FA Paine. *Packaging Materials and Containers*. London: Blackie & Son Limited, 1967, pp. 320–321.
15. Takashi Kadoya. *Food Packaging*. California: Academic Press, 1991, pp. 279–280.
16. AL Brody, KS Marsh. *The Wiley Encyclopedia of Packaging Technology*, 2nd ed. New York: John Wiley & Sons, 1997, pp. 949–951.

34 Future Developments in Frozen Food Packaging

Martin George
Campden and Chorleywood Food Research Association,
Chipping Campden, Gloucestershire, UK

CONTENTS

I. INTRODUCTION

Innovations in frozen food packaging have, to date, largely been driven by the needs for product differentiation, reduced product costs, and consumer convenience. Other issues have generally received much less attention, although many end-users of frozen food packaging will regard them as no less important. Examples might be the use of packaging to extend product quality and safety shelf-life, the role of packaging to help assure product integrity, and provide evidence of tampering, or the use of food packages to indicate product abuse and remaining optimal product shelf-life. Packaging also provides the medium to display product information for the consumer and, within the global frozen food supply business, delivers the opportunity to uniquely label and identify individual packages for purposes of traceability and protection from counterfeiting.

Although some attention has been given to the role of packaging for quality retention and shelf-life extension for frozen foods, there are still many unexplored opportunities to improve the interaction between packaging and frozen foods. This contrasts with the chilled foods sector, where package innovation and novel packaging formats are major components of the product line. In particular, the chilled foods sector has hugely benefited from "interactive" packaging formats such as modified atmosphere packaging (MAP). In MAP, the air is evacuated from the package and replaced with an inert mixture of, generally, carbon dioxide or nitrogen. The evacuation of oxygen from the microenvironment surrounding the food helps to retard undesirable biochemical spoilage reactions and can dramatically extend quality shelf-life. The current focus of attention for chilled foods is the application of active and intelligent packaging. These are themselves experiencing a new surge in functionality and application due to advances in polymer science, biotechnology, microelectronics, and information technologies. It is widely recognized that these emerging technologies could raise the standards of integrity, quality, and safety of refrigerated foods in the future. The use of such innovation in packaging should also be a key consideration to meet the future packaging needs of the frozen food manufacturer.

Active packaging encompasses a wide range of technologies such as oxygen-scavenging technology, moisture scavenging, antimicrobial films, and temperature-activated polymers. Additionally, "intelligent packaging" provides a number of package technologies that can "sense and inform" the user on the state and status of the food within the package. Typical examples are monitoring of product history, measurements of product safety or quality over time, evidence of product tampering, and the tracking of a food product throughout its journey from the factory to the consumer to improve "traceability." Intelligent packaging includes temperature indicators, time–temperature indicators, freshness indicators, microbiological indicators, and radio frequency identification (RFID) tags.

This chapter highlights the new and emerging packaging developments that will impact the frozen food sector and focuses on some of the benefits available to the frozen food manufacturer for improved safety, increased quality, extended shelf-life, and enhanced product traceability through the frozen food chain.

II. POTENTIAL FUTURE TECHNOLOGIES FOR FROZEN FOOD PACKAGING

Across all food sectors, it is generally recognized that one of the most important catalysts for packaging advancement has been the development of multilayer plastics possessing high barrier properties, low cost, and ease of fabrication. The advancements are also partially due to the ability to co-extrude plastics into a wide range of formats to complement the needs of developing food-preservation technologies, such as MAP and dual-ovenable packaging technology. The impact on the frozen food sector has been that a number of high-barrier plastic packaging options have become available and the degree of protection offered against, for example, ingress of oxygen or water vapor has never been greater. This contributes to the long quality shelf-lives that frozen foods currently enjoy and is a major reason for consumer acceptance for both chilled and frozen foods in preference to heat-processed alternatives.

The further advancement of frozen food packaging will be undoubtedly drawn upon improvements to current, conventional packaging strategies, but also on the application of other innovative packaging technologies. These include active packaging, intelligent packaging, and relevant advances in the range and application of other conventional packaging formats, such as vacuum packaging, skin packaging, or edible coatings and films.

A. Active Packaging Technologies

Most barrier packaging is considered as "passive" as its major function is to provide a barrier to the ingress of substances likely to degrade the food product, such as oxygen and moisture. However, like many barriers, the integrity of the package can occasionally fail. Active packaging offers a host of new packaging technologies that can actively sense and then neutralize the presence of product degraders. The technology can also be considered as a means of indicating when the barriers presented by the food packaging have broken down. Depending upon the type of device chosen, active packaging technologies can alter the condition of the packed food, notably to extend shelf-life, maintain product safety, and help to preserve sensory qualities over a longer shelf-life period. The term active packaging describes a collection of technologies that have been developed to solve specific problems in food preservation. Traditional concepts where the packaging merely act as a barrier to protect the food are reaching the limit of their ability to maximize product shelf-life and, consequently, innovative packaging concepts, such as active packaging, are being sought. One of the major features of active packaging is their ability to respond to changes that occur within the food during the passage from factory to consumer. It is well known that such changes alter storage lifetimes as a result of a number of actions. These may include physiological changes (e.g., due to respiration of fresh produce), physical changes (e.g., due to desiccation of the product surface), chemical changes (e.g., due to the oxidation of lipids within the food), microbiological changes (e.g., due to the growth of spoilage bacteria), and the ingress of infestation (due to the penetration of insects or pests). It should be recognized that active packaging is not a single technology capable of responding to all these and other effects. Active packaging is a generic term for a collection of technologies and each active packaging element or device can act only on its specific target. It has been suggested [1] that active packaging encompasses 10 broad-spectrum functions, namely:

- Oxygen scavengers
- Antimicrobials
- Carbon dioxide controllers
- Ethylene controllers
- Microwave susceptors
- Moisture control
- Flavor enhancers
- Odor generators
- Oxygen-permeable films
- Oxygen generators

Oxygen scavengers are, currently, the most common forms of active packaging. The removal and control of oxygen within the headspace of a food package has been a target of food technologists for many years. In chilled foods, the use of MAP or controlled atmosphere packaging (CAP) has been highly successful in this area, and significant quality shelf-life extensions have been reported [2]. However, it is recognized that aerobic spoilage can still occur as a result of the residual oxygen present in the headspace, possibly as a result of the small but finite oxygen permeability of the packaging material, small leakages due to poor sealing of the package, air that may be entrained within the food, or inadequate evacuation and gas flushing at the packing stage. The range of applications for oxygen scavengers is growing rapidly, including chilled foods, dried snack foods, soft and alcoholic drinks, bakery products and meat products, with the major benefits being an increase in quality shelf-life. The advantageous consequences of reduced oxygen include protection against growth of mold/yeast and aerobic microorganisms, minimizing effects of lipid oxidation, discoloration and loss of sensory properties, and greater retention of nutritional properties of the packaged food.

Oxygen scavengers or absorbers are normally in the form of small sachets containing easily oxidizable materials. Treated iron powder is the most common oxygen absorber as it meets many requirements for use with food substrates, such as safety, ease of handling, low cost, stability, compactness, and a large oxygen absorbing capacity. Iron-based oxygen absorbers use the principle that iron consumes oxygen during a chemical corrosion reaction according to the reaction:

$$\mathrm{Fe} \longrightarrow \mathrm{Fe^{2+}} + 2\mathrm{e^-}$$

$$\frac{1}{2}\mathrm{O_2} + \mathrm{H_2O} + 2\mathrm{e^-} \longrightarrow 2\mathrm{OH^-}$$

$$\mathrm{Fe^{2+}} + 2(\mathrm{OH})^- \longrightarrow \mathrm{Fe(OH)_2}$$

$$\mathrm{Fe(OH)_2} + \frac{1}{4}\mathrm{O_2} + \frac{1}{2}\mathrm{H_2O} \longrightarrow \mathrm{Fe(OH)_3}$$

A difficulty with iron-based sachets of oxygen absorber is that they are detected by metal detection equipment. In applications where this might pose a problem, oxygen absorbent materials based on organic materials such as ascorbic acid can be used. Oxygen absorbers based on polymeric reactions and enzymic scavengers are also available. The normal method of using these absorbers is to place them as loose sachets within the packaged food although other types are becoming available, including label formats, which can be hidden within the food package, for example, on the reverse of the package label, and not be visible to the consumer. A criticism of these devices is that the inclusion of a sachet boldly stating "do not eat" can dissuade the consumer from purchasing the product. A flat card format for an oxygen scavenger can be used for thin, flat packages or for foods which have a close fitting package. The technology of oxygen scavenging is also being assisted by advances in polymer technology; plastic wraps and bags can be made to incorporate inner layers of oxidizable polymers that trap oxygen in the same way as iron.

Oxygen-scavenging technologies may also have a role to play in the future direction for frozen-food-packaging developments. Flavor and nutritional deterioration resulting from the oxidation of the food's lipid component often limit the quality shelf-life of frozen foods. Lipid oxidation is often influenced by a number of factors, including temperature, pH, water activity, light, and the availability of oxygen. In the case of frozen foods, as temperature is reduced to below the freezing point of the water within the food, reaction rates for lipid oxidation are retarded. However, even in the temperature range of 0 to $-10°C$, it has been noted that the temperature-induced deceleration of oxidation is more than offset by the increased concentration of reactants available to take part in undesirable oxidative reactions as water is removed in the form of ice. The need for limiting the availability of oxygen remains at frozen food temperatures. In addition, the formation of ice crystals will often result in damage to the cell walls and cellular interfaces, thereby increasing the intermixing of cell contents and an increased likelihood of intercellular lipid interactions [3]. The size and distribution of ice crystals, and the subsequent associated disruption caused to cellular structure of the food, will vary with both the rate of temperature reduction during freezing and the stability of low temperatures during postfreezing handling and storage. Freezing as rapidly as possible and maintaining consistent low frozen storage temperatures are essential elements in preventing oxidation in frozen foods. In packaging terms, this can be assisted by minimizing the availability of oxygen within the food package. By selecting a packaging material that provides a high barrier to the ingress of oxygen and water vapor, the opportunities for oxidative deterioration can be significantly reduced. However, as frozen storage periods often extend to several weeks or months, the combined effects of the finite oxygen permeability of the package, small leakages in poorly formed package seals, air enclosed within the food and inadequate evacuation prior to filling will contribute to unavoidable residual oxygen within the headspace. In such cases, active packaging elements such as oxygen scavengers have a role to play in minimizing quality deterioration of the frozen food product.

Moisture controllers cover a wide range of products, all of which act to prevent the build-up of excessive levels of moisture within the package. Water vapor trapped within the food product and water naturally exuded by foods such as meat, fish, fruit, and vegetables can cause undesirably high moisture levels in-pack. In addition, moisture ingress through the packaging material can create an environment that leads to loss of quality and the potential for microbial growth and other hygiene problems. In frozen foods, excess water within the package often manifests as ice that settles on the inner package surfaces or on the surface of the food product to create an unsightly product. Moisture-absorbent pads represent the single most common example of moisture controllers used in food products, although there also exist other moisture controllers that work on different principles, such as humectants. An added advantage of the application of moisture-absorbent pads in frozen foods is their ability to absorb the exudate (drip loss) released from the food product during thawing, making a cleaner, more appealing product for the consumer.

Recently, antimicrobials have experienced much R&D, although commercial examples are relatively scarce. This has been due mostly to the lack of scientifically validated data regarding the efficacy of such products, but also to the questions arising regarding safety of such technologies, particularly those which involve the migration of antimicrobials to the food materials contained within the package. In general, antimicrobials effectively seek out and destroy harmful aerobic bacteria, which could multiply over time on the surface of a range of foods at a rate dependent on the growth conditions, for example, temperature, oxygen, and humidity. Antimicrobial substances can either be held onto the inner layer of the food package, targeting only surface bacteria, or released into the pack to interact more closely with the food. For frozen foods, although the temperature ranges associated with frozen storage is most often well below the minimum growth temperature of most bacteria, the presence of bacterial contamination within the food or package prior to freezing needs to be controlled; otherwise the freezing operation merely preserves the bacterial load alongside the food. In this respect, antimicrobials may have future application to the frozen food area.

Flavor enhancers are among the most controversial areas of active packaging due to concerns regarding their ability to mask product taint. In some applications, off-flavors and off-odors are the only by-products of food product spoilage and their removal does impair the consumers' ability to recognize the point at which a food has become a safety hazard. These devices have been used to mask the impact on food flavor of plastic and other packaging materials by the release of neutralizing substances from the active element. Such devices have been applied to the positive effect of reinforcing and improving the flavors of foods, for example, to the packing of vitamins with activated carbon used to absorb naturally occurring off-odors. Although much interest has been shown in this technology, only relatively little R&D has been undertaken and this situation is likely to remain until the principle of flavor/odor absorption becomes more widely accepted.

Microwave susceptors have seen significant commercial opportunities for many years. The active nature of these packages is that they become functional during the heating of the food and package in the microwave oven and can create greater uniformity of microwave heating and the possibility of browning and crisping the surface of microwave-heated foods. Due to the nature of microwave heating, these active packaging devices are particularly well suited to the frozen foods sector.

Frozen foods are relatively transparent to microwave energy, allowing the microwaves to pass through the food and interact with the susceptor elements of the packaging. In turn, this allows the susceptor to reach high temperatures, up to 250°C, which then acts as a radiant heat source at the surface of the food. As the food starts to thaw, the microwave energy then interacts more strongly with the food material itself and the product benefits from the rapid, volumetric nature of microwave heating, and the surface-heating assistance of the microwave susceptor. Waite [2] gives a review of the commercial opportunities presented by susceptor technologies.

Other active packaging technologies, such as ethylene controllers, odor generators, and gas generators, are less advantageous to the frozen food sector. Ethylene is a by-product of food product respiration during ripening and, although important in chilled food applications where the fruit or vegetable experiences a significant degree of ripening during chilled storage, the low temperatures associated with frozen foods have precluded this technology making any impact in this sector.

B. Intelligent Packaging Technologies

Intelligent packaging can, in general terms, be defined as packaging that can sense either the status of the food inside the package or the state of the environment surrounding the package, and has the ability to convey this information to the user. The range of "intelligent" technologies that have been researched in recent years has been enormous, however only a few developments have made an impact on the commercial food-packaging sector. It is evident that the major applications of intelligent packaging to date have focused on the security and traceability of the food. Examples are tamper evident packaging, packaging designed to protect from theft, and electronic-based tagging devices incorporated into the food package.

Generally, there have been two approaches to the concepts of intelligent packaging, using either the application of electronics and microelectromechanical systems (MEMS), or by the use of novel synthetic materials such as biopolymers.

Among the key applications for intelligent packaging within the frozen foods sector are devices that can measure and indicate parameters that influence the quality and safety of the frozen food. The most common development for this application has been the temperature indicator (TI) or time–temperature indicator (TTI). The usefulness of these devices relies on the fact that the quality and safety of foods are strongly influenced by temperature. Although it is widely recognized within the frozen food industry that temperature control and temperature monitoring are essential elements for the control of hygienic, nutritional, and sensory quality of foods, the ability to maintain an unbroken cold chain from the producer to the consumer is not a straightforward matter. Manual readings of time and temperature histories are time-consuming and require special knowledge and suitable equipment, and the cost of such checks often leads to low frequency of monitoring. The modern demand for robust quality management systems, in particular hazard analysis and critical control point (HACCP), has been an important element in improving the safety and quality of food worldwide, and demands that food product temperatures must be kept at recommended levels and checked at regular intervals, particularly at each critical point in the distribution chain. TIs indicate whether the food product temperature has exceeded a set value, for example, for frozen foods, this may be the point at which the food starts to thaw. TTIs measure both time and temperature and integrate them into a single result, thus indicating the cumulative time–temperature history of the product to which they are attached. For frozen foods, TTIs may be used, for example, for indication of temperature cycling throughout the distribution chain, or to monitor the efficacy of temperature control and integrity throughout distribution. Both TIs and TTIs are simple, inexpensive devices, generally in the form of self-adhesive labels attached to the surface of the packaged food product. Commonly, the indication is a simple, irreversible color change that can be easily recognized by the food handler and the type of device can be selected to suit the temperature range of interest.

The fundamental operating principle of TIs and TTIs is a temperature-dependent process, which, depending on the type of device, can be mechanical, electronic, physical, chemical, biochemical, or electrochemical. Recently, over 100 patents have been filed for TIs and TTIs, although only a few have reached commercialization. The mechanisms usually result in one of the three methods to indicate the temperature or time–temperature history experienced, namely color change, diffusion, and radio-frequency. One commercial type of color change indicator has been based on controlled enzymic hydrolysis of a lipid substrate, which results in a change of pH with temperature history and an associated color change, for example, green to yellow. Another color change indicator is based on polymerization of a diacetylene monomer, which produces a colored polymer whose depth of color is a result of the experienced time–temperature history. Specially adapted barcode readers can read these indicators or, when used on consumer packages, the active element can be placed next to a color reference for easy comparison with a "standard" color. Diffusion-based indicators rely on the effects of temperature to influence the rate of diffusion of a dye along a semiabsorbent surface. The speed of diffusion is temperature-dependent, such that the distance moved by the dye is a measure of the cumulative time–temperature history. Recently, an indicator based on viscous elastic polymers that migrate into a porous light-reflecting matrix was introduced. Radio-frequency systems can use many of the mechanisms of TI and TTI operation, whereas capturing the information electronically and conveying the time–temperature information into an electronic scanner before transferring data into a software program. This information can be stored or transferred onto computing systems that can calculate and display remaining shelf-life, perhaps even changing the expiry date marked on the package accordingly.

Some of the potential advantages in the use of TIs and TTIs are highlighted below [4]:

- With regard to product safety and quality, indicators allow improved cold chain performance without increasing the costs compared with increased use of traditional temperature measurements. The infrastructural changes in today's society involving fewer and larger processing and distribution plants and a need for longer transportation make this even more important.

- In some cases, reading and recording of the indicator response as part of the control system can be automated, which would be useful for future industrial applications.
- Indicators of the time–temperature history of a food product during distribution from production to retailing/catering will provide the retailer/caterer with important feedback, assure food safety and high quality and minimize product and economic losses.
- In the future, product marketing could be based on the indicated remaining shelf-life rather than the time spent at a given temperature only, as with current expiry markings.
- Depending on the construction and function of the indicators, the use of these devices is likely to reduce control program costs throughout the distribution chain.

However, it should be recognized that the application of such indicators relies upon several assumptions regarding their performance, accuracy, and functionality. The major problems associated with TIs and TTIs are highlighted below [4]:

- Surface placement of the indicators for ease of readability means that they react to changes in the surrounding temperature, which are normally more extreme than those actually occurring in the product. The relationship between surface temperature and product temperature varies from product to product, depending on the packaging material, physical properties of the product, headspace, and so on. Hence, adjustment of the indicator results to represent the exact condition of the product is difficult.
- The concept of TIs and TTIs is still not commonly understood throughout the distribution chain and by the consumer. Application of such indicators might be limited in use to master cartons, monitoring the temperature during distribution from producer to retailer/caterer. The use of consumer packs will probably not be common in the very near future.
- The cost of a single TI or TTI can be significant relative to the value of some products when used on consumer packs.
- Potential conflict between the TI and TTI indications and the mandatory expiry dates required in some countries may occur. Until devices are certified as a method to measure remaining shelf-life, controlling authorities and legislation will continue to use expiry date markings. Hence, the use of TIs or TTIs cannot completely eliminate ordinary temperature measurements.
- Standardization of TTIs is difficult. The time–temperature relationship between temperature history and shelf-life is not the same for all food products. Hence, it will be necessary to have a number of different TTIs to suit the wide range of food products.
- The performance of TTIs in terms of precisely mimicking the product quality deterioration as a function of time–temperature response has been shown to be acceptable in a number of cases. However, standards specifying acceptable levels of accuracy have not yet been defined and adopted, making it difficult to compare devices and manufacturers in an objective manner.

Commercial applications for TIs and TTIs have mostly been limited to short-term trials for monitoring the efficacy of the producer-to-retailer end of the cold chain. Several leading UK food retailers have used TTIs for the in-house monitoring of the cold chain, and at least two European food retailers (in France and Spain) have trialed TTIs on consumer chilled food packages. The results have been promising, with positive consumer response and low product returns [5]. The food industry, however, has been reluctant to widely adopt these devices, mostly due to concerns that large variations in the biological materials, processing and packaging processes make it difficult to measure actual food product quality and safety for any food using TTIs. A specification for defining the technical standards demanded by the food industry from TIs and TTIs has been prepared [6]. The advantages of using both TIs and TTIs for monitoring food product flows and as an inexpensive method for frequent checking of temperatures during food handling and distribution are likely to facilitate a much greater interest and widespread adoption in the future. The potential ability of TTIs to illustrate remaining product quality shelf-life is of great interest, but the most likely adoption of this technology will be in the quest to improve the efficiency of control systems between the food manufacturer and the retailer for the handling, storage, and distribution of perishable foods.

C. Vacuum-Packaging Technologies

Vacuum-packaging technologies are well suited to many of the needs of frozen foods. Vacuum packaging removes air from the packaging around the food and hermetically seals the package so that a near-perfect vacuum remains inside. A common variation of the process is vacuum skin packaging, where the entire package forms a skin-tight seal around the food product. Vacuum packaging inhibits the growth of many food spoilage bacteria that would normally have an adverse effect on the shelf-life of the product. The removal of oxygen can also prevent degradation or oxidative processes that limit product shelf-life, for example, oxidative rancidity in fats and oils, or color deterioration in raw meats. An added advantage for frozen foods is that the sealing of the food within a skin-tight package prevents dehydration and evaporative water loss from the surface of the food, and can minimize the effects of "freezer burn" (excessive dehydration loss from the product surface) and post-thaw exudate (drip loss) that often limit the quality shelf-life of frozen foods.

Reduced oxygen packaging (ROP), which provides an environment that contains little or no oxygen, offers particular advantages but also raises many microbiological concerns. Packaging conditions that exclude oxygen favor the growth of microorganisms that thrive in anaerobic environments, for example, *Clostridium botulinum.* Although frozen storage temperatures are well below the minimum growth temperatures for these microorganisms, care still has to be taken in the prefreezing steps, for example, use of high-quality and safe raw materials, assurance of the efficacy of heat treatment steps, and rigorous hygiene during handling. Freezing is an excellent means of long-term food preservation, but it is also an excellent means of preserving microorganisms already present in the food at the outset. Producing and distributing such food products with an HACCP approach will help to assure food safety. ROP is a fundamental requirement of the *sous vide* process, which reduces bacterial load but is not in itself sufficient to make the food shelf-stable. *Sous vide* processing involves five key steps, namely: (1) preparation of the food raw materials, (2) packaging of the product, application of vacuum, and sealing of the package, (3) pasteurization of the product for a specified time–temperature treatment, (4) rapid and monitored cooling of the product to below 3°C or frozen, and (5) reheating the packages to a specified temperature before opening and service. As the applied heat treatment does not achieve commercial sterility, the food requires refrigeration to prevent spoilage and ensure product safety. For this reason, *sous vide* products are frequently flash frozen in liquid nitrogen immediately after heating and held in frozen storage until required for use.

D. Edible Films and Coatings

Edible films and coatings can function to prevent quality losses in foods in various ways, one of the most important being the prevention of migration of moisture or oxygen. This has potentially significant advantages that can be conferred onto frozen foods. Edible coatings (formed directly on foods) and edible films (preformed, and then placed between food components or on foods) have not been adopted to any appreciable extent in frozen foods despite the potential to improve long-term stability and quality retention for many categories of frozen foods. A major reason for this is the uncertainty that the films or coatings can provide a robust, unbroken barrier around the low temperatures associated with frozen storage, without the fear of fracture or brittleness that would compromise the films during the food product's transit through the cold chain.

However, if these technical difficulties can be overcome, the application of edible films and coatings can benefit frozen foods in several ways.

Moisture migration is the major physical change occurring in frozen foods and has major effects on the chemical and biochemical properties. Moisture migration can manifest in several forms, the two most influential in promoting quality deterioration being moisture loss by sublimation and moisture absorption and redistribution. Moisture loss has important economic consequences and is often described as the main shelf-life limiting factor in frozen meats, where desiccation of the product surface can lead to undesirable odors during subsequent cooking. Moisture loss can also lead to the characteristic "freezer burn," a glassy, or sometimes whitish, appearance on the surface of some meat products produced by the presence of small cavities left behind by sublimated ice. Moisture migration inside the food packaging results in unattractive ice formation on the inner surfaces of the food package and moisture loss during thawing manifests as drip loss, where the food loses appearance, juiciness, and texture. The economic value of moisture loss also cannot be overstated—the water that is lost has the same economic value as the product it represents. Moisture absorption and redistribution is particularly important for certain prepared foods such as frozen pizzas, pies, and pastries where the different food components have markedly different water activities and transfer of moisture from areas of high water activity to low water activity readily takes place if allowed.

Edible coatings have been proposed as barriers to minimize moisture loss in frozen foods. An edible coating formed at the surface of a food component can reduce the rate of moisture transfer between the food and the surrounding environment, thus retarding the rate of package ice formation and dehydration of the product surface. Lipid-based films, for example, waxes, oils, and surfactants, form edible coatings that are good moisture barriers, although they often have poor structural properties and are not well suited to the rigors of freezing and frozen storage. However, their application to meat sausage, poultry, and fish showed advantages for improved textural and sensory properties over storage periods of several months by retaining the moisture inherent within the frozen foodstuff [7].

Edible coatings have also been applied to restrict the movement of oxygen within the food and minimize the availability of oxygen at the surface of the food, thus minimizing undesirable reactions influenced by oxidative rancidity. In this application, polysaccharide films and their derivatives (e.g., alginates, pectins, agars, carageenans, cellulose, and starch-based films) have exhibited good oxygen and carbon dioxide barrier properties. However, due to their hydrophilic nature, they are poor moisture barriers. Studies have shown improvements of frozen fish and meat quality. Protein-based films (e.g., casein, whey proteins, gluten, soy proteins, gelatin, and collagen derivatives) are reported to have similar gas barrier properties, but have been much less well studied. Applications reported have included meat and fish, and significant protection against rancidification has been found [7].

Major activity in this area is currently concentrating on multicomponent coatings, which are a blend of polysaccharides, proteins, and lipids formed from emulsions in a one-step process or formed as laminates in a multistep process. Such coatings confer the advantages of each individual film type to the final coating. Edible coatings formed from emulsions are desirable in an industrial setting because they are convenient to use and often more resilient than a single component coatings. Laminate coatings are more complicated to form and may therefore be less desirable.

The limited information available at this time indicates that edible films and coatings have the potential to contribute to improved quality retention in frozen foods. Their notable advantages include reduction of moisture loss, reduced diffusion of oxygen throughout the food, and improved structural properties. Their greatest potential may be within the prepared foods sector, where the multicomponent nature of the food contains components having widely different tolerances to frozen storage temperatures. However, before such coatings can become widely applied across the frozen food sector, much work has to be done in developing and optimizing the coatings to work consistently at frozen temperatures.

III. CONCLUSIONS

The preservation of foods at frozen temperatures has revolutionized the domestic and foodservice food supply systems and provides high-quality products, at relatively low cost and with an enviable record for food safety. Packaging developments have played a significant role in this revolution. The type of package selected for a particular frozen food is dependent upon many factors, including the product type, its niche in the marketplace, the company's packaging philosophy, and the availability of existing equipment. However, there is still much to be learned from the chilled foods sector, where dynamic packaging evolution has been a key factor in the rapid growth and consumer acceptance of chilled foods.

Active packaging technologies such as MAP and its host of "associated" technologies (e.g., oxygen control devices, antimicrobial films, moisture absorbers, and flavor and odor generators) have shown some potential for application in the frozen food area, but are as yet largely untried in this area. Their ability to extend quality shelf-life of chilled foods has been well demonstrated over many years, although there appears to be only limited scientific understanding of the precise mechanisms for retarding spoilage, and much reported work in the chilled foods area has been based upon empirical trials and observations. To transfer such success to the frozen foods area, these technologies must be complemented by an improvement in the quality and integrity of the current range of packaging used with frozen foods; many frozen food packages currently have poor seal strength and readily expose the food to the external environment, or become brittle at low freezer temperatures and easily fracture upon handling.

Intelligent-packaging technologies are closer to becoming adopted by the frozen foods sector, for example, TIs, TTIs, and RFID packaging technologies to monitor the temperature abuse, temperature history, or manufacturing status of the food product throughout the frozen food chain. The possibility of integrating the outputs of intelligent sensors with the remote and automated data capturing capabilities of RFID technologies makes these systems an attractive proposition for monitoring the passage of product through the component parts of the frozen food chain. Perhaps the greatest difficulty to overcome is the need to ensure that the food manufacturing sectors adopt a common technology platform for this purpose, so that each separate product can be monitored at each separate part of the chain, and also that different product lines (perhaps for different customers) can be monitored using a single generic system.

Edible films and coatings have shown some promise for food quality preservation, and their limited application has been most beneficial for multicomponent foods, for example, ready meals, which are a major part of the frozen food sector. However, their application across all food sectors (e.g., chilled meals, frozen meals, and bakery goods) has been low and is likely to remain that way until more robust films can be made, which are easy to apply and are stable over periods of storage. Their application at the low temperatures associated with frozen foods is even more difficult, as the current range of films generally exhibit a brittleness at frozen temperatures, further reducing their benefits.

Vacuum packaging, and reduced oxygen packaging, is an excellent way of preserving frozen food quality and safety. The exclusion of oxygen from the microenvironment surrounding the food has benefits to prevent the growth of aerobic microorganisms and also minimizes the effects of detrimental oxidative biochemical reactions, which would normally set a limit on the achievable quality shelf-life of the food. Vacuum packaging is also an excellent means of minimizing surface moisture loss, both in terms of dehydration during freezing and drip loss during thawing. The technology is also beneficial from a processing perspective, as the skin-tight package on the food readily allows heat transfer from the food to the freezing medium. This means that foods can be packaged prior to the freezing operation, which makes the essential requirement of good hygiene practice a little easier for the food manufacturer.

With all new packaging formats, particularly those which are untried and untested, the food manufacturer will have many concerns. Technological considerations, which have been the main focus of attention in this chapter, are just one aspect of the successful package choice. However, equal importance will need to be placed on other issues, such as legislative requirements, food contact regulations, compatibility with the existing packing formats within the factory, and the costs of the package. But, of course, the ultimate decision on acceptance of any new, innovative and technically advanced frozen food package lies with the consumer.

REFERENCES

1. A Brody. *Active Packaging: Beyond Barriers*. West Chester, Pennsylvania, USA: Packaging Strategies, 2003.
2. N Waite. *Active Packaging*. Leatherhead, UK: Pira International Ltd., 2003.
3. MC Erickson. Lipid oxidation: flavour and nutritional quality deterioration in frozen foods. In: MC Erickson, YC Hung, Eds., *Quality in Frozen Foods*. New York: Chapman & Hall, 1997.
4. Anonymous. *Temperature Indicators and Time–Temperature Integrators*. Third Informatory Note on Refrigeration and Foods. Paris, France: International Institute of Refrigeration, 2004.
5. RM George. Food Industry specifications for the performance evaluation of temperature indicators (TI) and time–temperature indicators (TTI). *International Conference on Active and Intelligent Packaging*, 7–8 September, Chipping Campden, UK: Campden & Chorleywood Food Research Association, 2000.
6. RM George, R Shaw. A food industry specification for defining the technical standards and procedures for the evaluation of temperature indicators and time-temperature indicators. Technical Manual No. 35. Chipping Campden, UK: Campden & Chorleywood Food Research Association, 1992.
7. YM Stuchell, JM Krochta. Edible coatings and films. In: MC Erickson, YC Hung, Eds., *Quality in Frozen Foods*. New York: Chapman & Hall, 1997.

Index

D

E

F

G

H

I

J

L

M

N

O

P

Q

R

U

V

W

Y